MANUAL
OF STEEL
CONSTRUCTION

LOAD &
RESISTANCE
FACTOR
DESIGN

Volume II

Connections

Second Edition

FOREWORD

The American Institute of Steel Construction, founded in 1921, is the non-profit technical specifying and trade organization for the fabricated structural steel industry in the United States. Executive and engineering headquarters of AISC are maintained in Chicago, Illinois.

The Institute is supported by three classes of membership: Active Members totaling 400 companies engaged in the fabrication and erection of structural steel, Associate Members who are allied product manufacturers, and Professional Members who are individuals or firms engaged in the practice of architecture or engineering. Professional members also include architectural and engineering educators. The continuing financial support and active participation of Active Members in the engineering, research, and development activities of the Institute make possible the publishing of this Second Edition of the *Load and Resistance Factor Design Manual of Steel Construction*.

The Institute's objectives are to improve and advance the use of fabricated structural steel through research and engineering studies and to develop the most efficient and economical design of structures. It also conducts programs to improve product quality.

To accomplish these objectives the Institute publishes manuals, textbooks, specifications, and technical booklets. Best known and most widely used are the *Manuals of Steel Construction,* LRFD (Load and Resistance Factor Design) and ASD (Allowable Stress Design), which hold a highly respected position in engineering literature. Outstanding among AISC standards are the *Specifications for Structural Steel Buildings* and the *Code of Standard Practice for Steel Buildings and Bridges.*

The Institute also assists designers, contractors, educators, and others by publishing technical information and timely articles on structural applications through two publications, *Engineering Journal* and *Modern Steel Construction.* In addition, public appreciation of aesthetically designed steel structures is encouraged through its award programs: Prize Bridges, Architectural Awards of Excellence, Steel Bridge Building Competition for Students, and student scholarships.

Due to the expanded nature of the material, the Second Edition of the LRFD Manual has been divided into two complementary volumes. Volume I contains the LRFD Specification and Commentary, tables, and other design information for structural members. Volume II contains all of the information on connections. Like the LRFD Specification upon which they are based, both volumes of this LRFD Manual apply to buildings, not bridges.

The Committee gratefully acknowledges the contributions of Roger L. Brockenbrough, Louis F. Geschwindner, Jr., and Cynthia J. Zahn to this Manual.

By the Committee on Manuals, Textbooks, and Codes,

William A. Thornton, Chairman

Barry L. Barger, Vice Chairman

Horatio Allison	Mark V. Holland	David T. Ricker
Robert O. Disque	William C. Minchin	Abraham J. Rokach
Joseph Dudek	Thomas M. Murray	Ted W. Winneberger
William G. Dyker	Heinz J. Pak	Charles J. Carter, Secretary
Ronald L. Hiatt	Dennis F. Randall	

REFERENCED SPECIFICATIONS, CODES, AND STANDARDS

Part 6 (Volume I) of this LRFD Manual contains the full text of the following:

American Institute of Steel Construction, Inc. (AISC)
Load and Resistance Factor Design Specification for Structural Steel Buildings, December 1, 1993
Specification for Load and Resistance Factor Design of Single-Angle Members, December 1, 1993
Seismic Provisions for Structural Steel Buildings, June 15, 1992
Code of Standard Practice for Steel Buildings and Bridges, June 10, 1992

Research Council on Structural Connections (RCSC)
Load and Resistance Factor Design Specifications for Structural Joints Using ASTM A325 or A490 Bolts, June 8, 1988

Additionally, the following other documents are referenced in Volumes I and II of the LRFD Manual:

American Association of State Highway and Transportation Officials (AASHTO)
AASHTO/AWS D1.5–88

American Concrete Institute (ACI)
ACI 349–90

American Iron and Steel Institute (AISI)
Load and Resistance Factor Design Specification for Cold-Formed Steel Structural Members, 1991

American National Standards Institute (ANSI)
ANSI/ASME B1.1–82   ANSI/ASME B18.2.2–86
ANSI/ASME B18.1–72   ANSI/ASME B18.5–78
ANSI/ASME B18.2.1–81

American Society of Civil Engineers (ASCE)
ASCE 7-88

American Society for Testing and Materials (ASTM)

ASTM A6–91b	ASTM A490–91	ASTM A617–92
ASTM A27–87	ASTM A500–90a	ASTM A618–90a
ASTM A36–91	ASTM A501–89	ASTM A668–85a
ASTM A53–88	ASTM A502–91	ASTM A687–89
ASTM A148–84	ASTM A514–91	ASTM A709–91
ASTM A153–82	ASTM A529–89	ASTM A770–86
ASTM A193–91	ASTM A563–91c	ASTM A852–91
ASTM A194–91	ASTM A570–91	ASTM B695–91
ASTM A208(A239–89)	ASTM A572–91	ASTM C33–90
ASTM A242–91a	ASTM A588–91a	ASTM C330–89
ASTM A307–91	ASTM A606–91a	ASTM E119–88
ASTM A325–91c	ASTM A607–91	ASTM E380–91
ASTM A354–91	ASTM A615–92b	ASTM F436–91
ASTM A449–91a	ASTM A616–92	

American Welding Society (AWS)

AWS A2.4–93	AWS A5.25–91
AWS A5.1–91	AWS A5.28–79
AWS A5.5–81	AWS A5.29–80
AWS A5.17–89	AWS B1.0–77
AWS A5.18–79	AWS D1.1–92
AWS A5.20–79	AWS D1.4–92
AWS A5.23–90	

PART 8

BOLTS, WELDS, AND CONNECTED ELEMENTS

OVERVIEW

Part 8 contains general information, design considerations, examples, and design aids for the design of bolts, anchor rods, other mechanical fasteners, welds, and connected elements in connections. It is based on the provisions of the 1993 LRFD Specification. Supplementary information may also be found in the Commentary on the LRFD Specification.

Following is a detailed overview of the topics addressed.

BOLTED CONSTRUCTION

High-Strength Bolts

LRFD Specification Section A3.3 permits the use of ASTM A325 and A490 high-strength bolts. ASTM A325 bolts are available in diameters from ½-in. to 1½-in. in two types. Type 1 medium-carbon-steel bolts are for general purpose use and use in elevated temperatures; they may be galvanized. Type 3 bolts offer improved atmospheric corrosion resistance and weathering characteristics similar to those of ASTM A242 or A588 steels.

ASTM A490 bolts are available in diameters from ½-in. to 1½-in. in two types. Type 1 bolts are alloy-steel bolts. Type 3 are alloy-steel bolts with improved atmospheric corrosion resistance and weathering characteristics similar to those of ASTM A242 or A588 steels. ASTM A490 bolts should not be galvanized and caution should be exercised if used in highly corrosive environments.

Type 2 (martensite) bolts, popular for many years, have been discontinued. Information on this type can be found in previous editions of the AISC *Manual of Steel Construction*.

When bolts of diameter larger than 1½ in. are required, ASTM A449 bolts are permitted to be used for snug-tightened and fully tensioned bearing-type connections; this material is not recognized in LRFD Specification Section A3.3 for use in slip-critical connections nor for use as bolts in diameters not greater than 1½ in. ASTM A449 bolts may be galvanized.

When an ASTM A449 bolt is used in tension or bearing and is tightened in excess of 50 percent of its minimum specified tensile strength, LRFD Specification Section J3.1 requires that an ASTM F436 washer be installed under the head of the bolt. The nut must be from the approved list in RCSC Specification Section 2(c). Since ASTM A325 nuts and washers for use with high-strength bolts are available only up to 1½-in. diameter, reference should be made to ASTM A563 for nuts and ASTM F436 for washers to select suitable sizes and grades for the intended application.

While ASTM A449 seems to be the equal of ASTM A325, there are two important differences which should be noted. First, ASTM A449 bolts are not produced to the same inspection and quality assurance requirements as ASTM A325 bolts. Second, ASTM A449 bolts are not produced to the same heavy-hex head and nut dimensions.

Alternative Design Bolts

RCSC Specification Section 2d permits the use of other fasteners when they meet the requirements as outlined therein. Figure 8-1 shows a tension-control or "twist-off" bolt which is installed with a special tool which twists off the splined end when the proper

Fig. 8-1. Tension-control or "twist-off" bolt.

Table 8-1.
Compatability of High-Strength Bolts, Nuts, Washers

ASTM Bolt Desig.	Type	Coating	A563 Heavy Hex Nut Grade		F436 Washer Grade
			Recommended	Suitable	Recommended
A325	1	plain	C	C3, D, DH, DH3	1
		galvanized	DH	—	1
	3	plain	C3	DH3	3
A490	1	plain	DH	DH3	1
	3	plain	DH3	—	3
A449	1	plain	A	C, C3, D, DH, DH3	1
		galvanized	DH	D	1

bolt tension is achieved. Tension-control bolts are commonly available to meet the specifications of ASTM A325 and A490.

Compatible Nuts and Washers
The compatibility of ASTM A563 nuts and F436 washers with the aforementioned high-strength bolt specifications is as listed in Table 8-1. Alternatively, appropriate ASTM A194 nuts may be used. RCSC Specification Section 7c gives general requirements for when washers are required for high-strength bolts.

Economical Considerations
Since the material cost per unit of strength of ASTM A490 bolts is comparable with that of ASTM A325 bolts, it might seem more cost effective to reduce the number of bolts in a given connection by specifying ASTM A490 bolts. However, ASTM A490 bolts are more difficult to tighten and raise inventory and quality control issues associated with the use of multiple fastener grades; mixing of ASTM A325 and A490 bolts of the same diameter should be avoided to assure that the ASTM A490 bolts are installed in the proper location. Thus, the net benefit of specifying ASTM A490 bolts may be less than expected; cost ratios should be considered by the designer.

Similarly, cost ratios between grades of alternative design bolts will vary from those of conventional high-strength bolts. Thus, the decision regarding fastener selection will vary accordingly.

Regardless of the bolt type selected, the normal sizes of ¾-in., ⅞-in., and 1-in. diameter are usually preferred. Diameters above one inch are not commonly available, nor are they practical since special tools may be required to achieve fully tensioned installation.

Bearing-type connections should be specified whenever possible. Slip-critical connections with coatings other than clean mill scale incur appreciable extra costs associated with blasting, painting, drying, assembling, reblasting, and abrasion touch-up. If slip-critical connections are required for the proper serviceablity of the structure, care should be taken to avoid requiring the faying surfaces to be masked as this also contributes great expense; coatings which provide a Class A or Class B slip coefficient may be an economical alternative to masking.

Dimensions and Weights
ASTM A325 and A490 bolts, A563 nuts, and F436 washers are given identifying marks as illustrated in Figure 8-2. A detailed description of identifying marks may be found in the RCSC Specification. Dimensions of ASTM A325 and A490 bolts, A563 nuts, and F436 washers are given and illustrated in Table 8-2. Threading dimensions of high-strength bolts are given in Table 8-7. Weights of conventional ASTM A325 and A490 bolts, A563 nuts, and F436 washers are given in Table 8-3. For dimensions and weights of tension-control ASTM A325 and A490 bolts, refer to manufacturers' literature or IFI. For dimensions and weights of ASTM A449 bolts, refer to Table 8-6.

Threads for high-strength bolts may be rolled or cut. Note that thread lengths for high-strength bolts are shorter than those for non-high-strength bolts. This allows the threads to be excluded from the shear plane when the thickness of the connected ply closest to the nut is as shown in Figure 8-3. While the RCSC Specification permits some thread run-out into the shear plane, it is important to provide sufficient thread to avoid jamming the nut into the run-out when tightening the bolt. Inspection controversy will be reduced by recognizing that bolts intentionally have a limited thread length, a manufacturing tolerance, and limited length increments; as with all manufactured items, dimensional tolerances must be considered.

The RCSC Specification recognizes these tolerances in two ways. First, additional washers are permitted to be used under the nut or under the head when circumstances permit. Second, there is no specified bolt "stick-through" requirement since only full-thread engagement of the nut is required; from RCSC Specification Section 2(b), "...The length of bolts shall be such that the end of the bolt will be flush with or outside the face of the nut when properly installed." A requirement for "stick-through", sometimes written in project specifications, increases the risk of jamming the nut on the thread run-out, and thus, of preventing tightening. A "stick-through" requirement will not enhance the performance of the bolt and should not be included in a project specification.

Alternatively, ASTM A325 bolts with length less than or equal to four times the nominal diameter may be ordered as fully threaded with the designation ASTM A325 T. Fully threaded ASTM A325 T bolts are not for use in bearing-type X connections since it would be impossible to exclude the threads from the shear plane. While this supplementary provision exists for ASTM A325 bolts, there is no similar supplementary provision made in ASTM A490 for full-length threading.

The ordered length of ASTM A325 and A490 bolts should be calculated as the grip (see Figure 8-2) plus the thickness of the washer(s) plus the allowance from Table 8-2. A thickness of $5/32$-in. for circular washers and $5/16$-in. for beveled washers should be provided per washer used; refer to the RCSC Specification for washer requirements. This total should be rounded to the next higher one-quarter inch. Note that bolts longer than five inches are generally available only in $1/2$-in. increments, except by special arrangement with the manufacturer or vendor. While longer lengths may be ordered, an 8-in. length is generally the maximum stock length available. Clipped washers are available for use in areas of tight clearance.

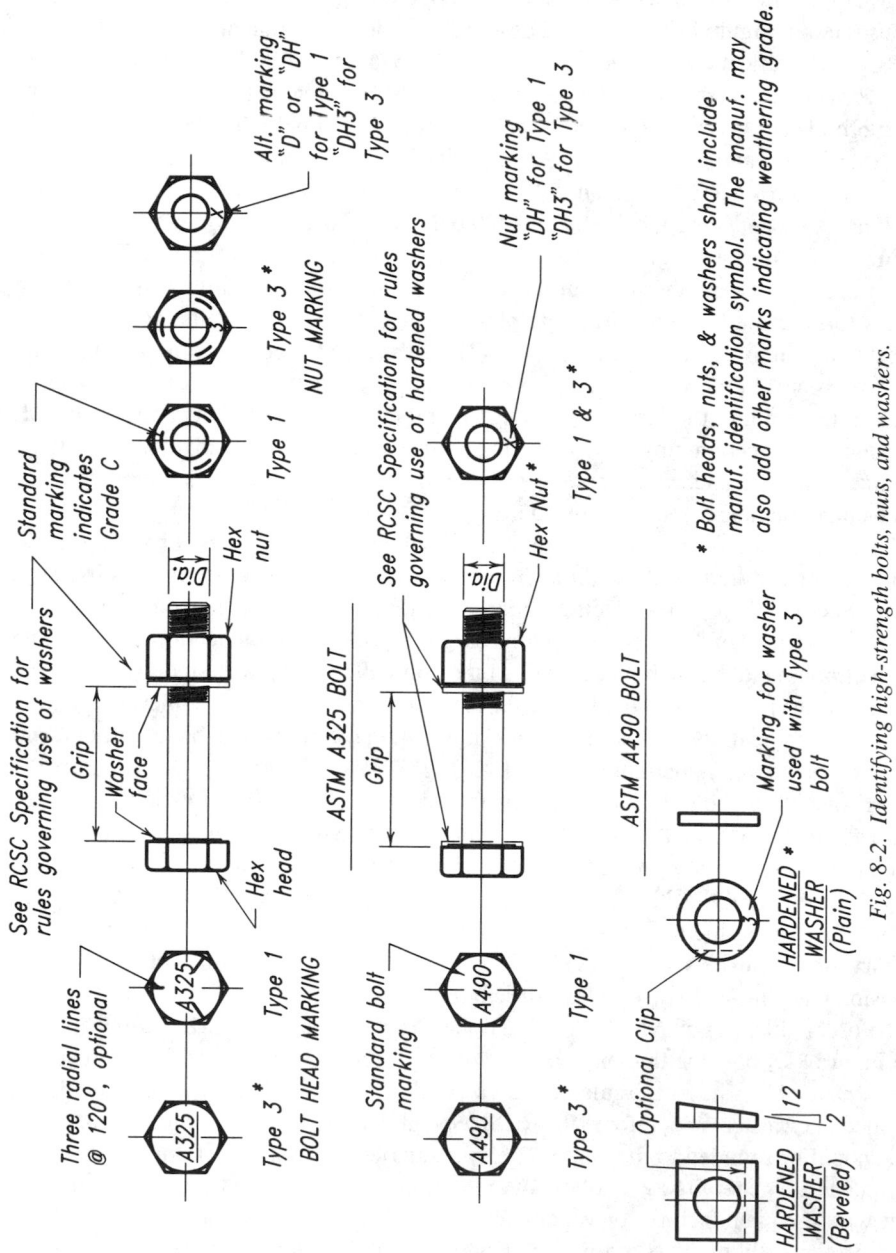

Fig. 8-2. Identifying high-strength bolts, nuts, and washers.

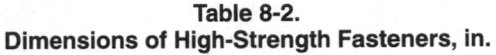

Table 8-2. Dimensions of High-Strength Fasteners, in.

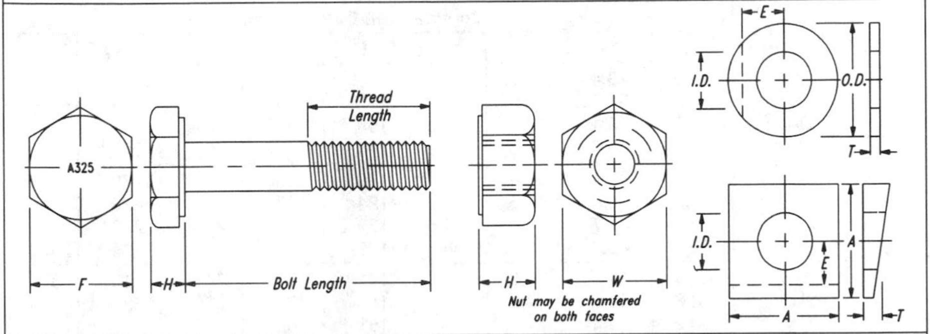

Measurement		Nominal Bolt Diameter, in. ½	⅝	¾	⅞	1	1⅛	1¼	1⅜	1½
A325 and A490 Bolts[a]	Width Across Flats *F*	7/8	1 1/16	1 1/4	1 7/16	1 5/8	1 13/16	2	2 3/16	2 3/8
	Height *H*	5/16	25/64	15/32	35/64	39/64	11/16	25/32	27/32	15/16
	Thread Length	1	1 1/4	1 3/8	1 1/2	1 3/4	2	2	2 1/4	2 1/4
	Bolt Length[f] =Grip + →	11/16	7/8	1	1 1/8	1 1/4	1 1/2	1 5/8	1 3/4	1 7/8
A563 Nuts[b]	Width Across Flats *W*	7/8	1 1/16	1 1/4	1 7/16	1 5/8	1 13/16	2	2 3/16	2 3/8
	Height *H*	31/64	39/64	47/64	55/64	63/64	1 7/64	1 7/32	1 11/32	1 15/32
F436 Circular Washers[c]	Nom. Outside Diameter *OD*	1 1/16	1 5/16	1 15/32	1 3/4	2	2 1/4	2 1/2	2 3/4	3
	Nom. Inside Diameter *ID*	17/32	11/16	13/16	15/16	1 1/8	1 1/4	1 3/8	1 1/2	1 5/8
	Thckns. *T* Min.	0.097	0.122	0.122	0.136	0.136	0.136	0.136	0.136	0.136
	Thckns. *T* Max.	0.177	0.177	0.177	0.177	0.177	0.177	0.177	0.177	0.177
	Min. Edge Distance *E*[d]	7/16	9/16	21/32	25/32	7/8	1	1 3/32	1 7/32	1 5/16
F436 Square or Rect. Washers[c,e]	Min. Side Dimension *A*	1 3/4	1 3/4	1 3/4	1 3/4	1 3/4	2 1/4	2 1/4	2 1/4	2 1/4
	Mean Thckns. *T*	5/16	5/16	5/16	5/16	5/16	5/16	5/16	5/16	5/16
	Taper in Thickness	2:12	2:12	2:12	2:12	2:12	2:12	2:12	2:12	2:12
	Min. Edge Distance *E*[d]	7/16	9/16	21/32	25/32	7/8	1	1 3/32	1 7/32	1 5/16

a Tolerances as specified in ASTM A325 and A490.
b Tolerances as specified in ASTM A563.
c ASTM F436 Washer Tolerances, in.:
Nominal Outside Diameter –1/32; +1/32
Nominal Diameter of Hole –0; +1/32
Flatness: max. deviation from straight-edge placed on cut side shall not exceed 0.010
Concentricity: center of hole to outside diameter (full indicator runout) 0.030
Burr shall not project above immediately adjacent washer surface more than 0.010
d For clipped washers only.
e For use with American standard beams (S) and channels (C).
f Tabular value does not include thickness of washer(s).

Table 8-3.
Weights of High-Strength Fasteners, pounds per 100 count

Bolt Length, in.	Nominal Bolt Diameter, in.								
	1/2	5/8	3/4	7/8	1	1 1/8	1 1/4	1 3/8	1 1/2
1	16.5	29.4	47.0	—	—	—	—	—	—
1 1/4	17.8	31.1	49.6	74.4	104	—	—	—	—
1 1/2	19.2	33.1	52.2	78.0	109	148	197	—	—
1 3/4	20.5	35.3	55.3	81.9	114	154	205	261	333
2	21.9	37.4	58.4	86.1	119	160	212	270	344
2 1/4	23.3	39.8	61.6	90.3	124	167	220	279	355
2 1/2	24.7	41.7	64.7	94.6	130	174	229	290	366
2 3/4	26.1	43.9	67.8	98.8	135	181	237	300	379
3	27.4	46.1	70.9	103	141	188	246	310	391
3 1/4	28.8	48.2	74.0	107	146	195	255	321	403
3 1/2	30.2	50.4	77.1	111	151	202	263	332	416
3 3/4	31.6	52.5	80.2	116	157	209	272	342	428
4	33.0	54.7	83.3	120	162	216	280	353	441
4 1/4	34.3	56.9	86.4	124	168	223	289	363	453
4 1/2	35.7	59.0	89.5	128	173	230	298	374	465
4 3/4	37.1	61.2	92.7	133	179	237	306	384	478
5	38.5	63.3	95.8	137	184	244	315	395	490
5 1/4	39.9	65.5	98.9	141	190	251	324	405	503
5 1/2	41.2	67.7	102	146	196	258	332	416	515
5 3/4	42.6	69.8	105	150	201	265	341	426	527
6	44.0	71.9	108	154	207	272	349	437	540
6 1/4	—	74.1	111	158	212	279	358	447	552
6 1/2	—	76.3	114	163	218	286	367	458	565
6 3/4	—	78.5	118	167	223	293	375	468	577
7	—	80.6	121	171	229	300	384	479	589
7 1/4	—	82.8	124	175	234	307	392	489	602
7 1/2	—	84.9	127	179	240	314	401	500	614
7 3/4	—	87.1	130	183	246	321	410	510	626
8	—	89.2	133	187	251	328	418	521	639
8 1/4	—	—	—	192	257	335	427	531	651
8 1/2	—	—	—	196	262	342	435	542	664
8 3/4	—	—	—	—	—	—	444	552	676
9	—	—	—	—	—	—	453	563	689
Per inch add'tl. add	5.5	8.6	12.4	16.9	22.1	28.0	34.4	42.5	49.7
100, F436 Circular Washers	2.1	3.6	4.8	7.0	9.4	11.3	13.8	16.8	20.0
100, F436 Square Washers	23.1	22.4	21.0	20.2	19.2	34.0	31.6	31.2	32.9

Left axis label: 100, Conventional A325 or A490 Bolts with A563 Nuts

This table conforms to weight standards adopted by the Industrial Fasteners Institute (IFI), updated for washer weights.

Entering and Tightening Clearances

The assembly of high-strength bolted connections requires clearance for entering and tightening the bolts with an impact wrench. The clearance requirements for conventional high-strength bolts are as given in Table 8-4. When high-strength tension-control bolts are specified, the entering and tightening clearances are as specified in Table 8-5.

Snug-Tightened and Fully Tensioned Installation

When subjected to shear only, high-strength bolts may be used in snug-tightened bearing-type, fully tensioned bearing-type, and slip-critical connections. When subjected

Table 8-4.
Entering and Tightening Clearances, in.
Conventional ASTM A325 and A490 Bolts

Aligned Bolts

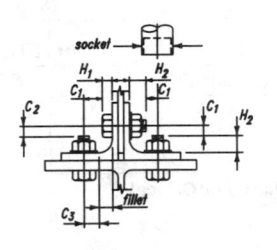

Nominal Bolt Dia., in.	Socket Dia., in.	H_1	H_2	C_1	C_2	C_3 Circular	C_3 Clipped
$5/8$	$1\ 3/4$	$25/64$	$1\ 1/4$	1	$11/16$	$11/16$	$9/16$
$3/4$	$2\ 1/4$	$15/32$	$1\ 3/8$	$1\ 1/4$	$3/4$	$3/4$	$11/16$
$7/8$	$2\ 1/2$	$35/64$	$1\ 1/2$	$1\ 3/8$	$7/8$	$7/8$	$13/16$
1	$2\ 5/8$	$39/64$	$1\ 5/8$	$1\ 7/16$	$15/16$	1	$7/8$
$1\ 1/8$	$2\ 7/8$	$11/16$	$1\ 7/8$	$1\ 9/16$	$1\ 1/16$	$1\ 1/8$	1
$1\ 1/4$	$3\ 1/8$	$25/32$	2	$1\ 11/16$	$1\ 1/8$	$1\ 1/4$	$1\ 1/8$
$1\ 3/8$	$3\ 1/4$	$27/32$	$2\ 1/8$	$1\ 3/4$	$1\ 1/4$	$1\ 3/8$	$1\ 1/4$
$1\ 1/2$	$3\ 1/2$	$15/16$	$2\ 1/4$	$1\ 7/8$	$1\ 5/16$	$1\ 1/2$	$1\ 5/16$

Staggered Bolts

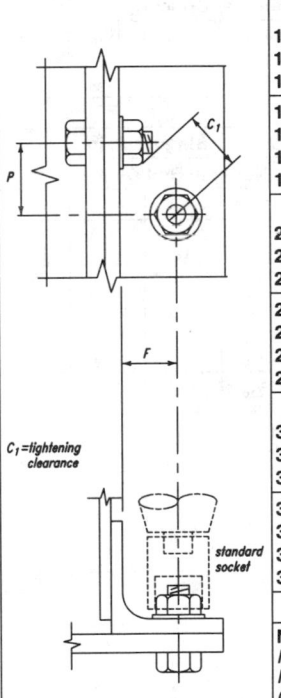

				Stagger P, in.				
				Nominal Bolt Diameter, in.				
F	$5/8$	$3/4$	$7/8$	1	$1\ 1/8$	$1\ 1/4$	$1\ 3/8$	$1\ 1/2$
1	$1\ 5/8$							
$1\ 1/8$	$1\ 1/2$							
$1\ 1/4$	$1\ 1/2$	$1\ 15/16$						
$1\ 3/8$	$1\ 7/16$	$1\ 7/8$	$2\ 3/16$					
$1\ 1/2$	$1\ 1/4$	$1\ 13/16$	$2\ 1/8$	$2\ 5/16$				
$1\ 5/8$	$1\ 1/4$	$1\ 3/4$	$2\ 1/16$	$2\ 5/16$	$2\ 9/16$			
$1\ 3/4$	$1\ 3/16$	$1\ 11/16$	2	$2\ 1/4$	$2\ 9/16$	$2\ 13/16$	3	
$1\ 7/8$	$1\ 1/8$	$1\ 9/16$	$1\ 15/16$	$2\ 3/16$	$2\ 1/2$	$2\ 3/4$	3	$3\ 3/4$
2	1	$1\ 1/2$	$1\ 13/16$	$2\ 1/8$	$2\ 7/16$	$2\ 3/4$	$2\ 15/16$	$3\ 1/4$
$2\ 1/8$	$13/16$	$1\ 3/8$	$1\ 11/16$	2	$2\ 3/8$	$2\ 11/16$	$2\ 15/16$	$3\ 3/16$
$2\ 1/4$		$1\ 1/4$	$1\ 9/16$	$1\ 7/8$	$2\ 1/4$	$2\ 5/8$	$2\ 7/8$	$3\ 3/16$
$2\ 3/8$		$1\ 1/8$	$1\ 1/2$	$1\ 3/4$	$2\ 1/8$	$2\ 1/2$	$2\ 13/16$	$3\ 1/8$
$2\ 1/2$		$7/8$	$1\ 3/8$	$1\ 5/8$	2	$2\ 7/16$	$2\ 3/4$	$3\ 1/16$
$2\ 5/8$			$1\ 13/16$	$1\ 1/2$	$1\ 15/16$	$2\ 5/16$	$2\ 7/8$	3
$2\ 3/4$			$15/16$	$1\ 3/8$	$1\ 7/8$	$2\ 1/8$	$2\ 1/2$	$2\ 7/8$
$2\ 7/8$				$1\ 3/16$	$1\ 3/4$	$2\ 1/16$	$2\ 3/8$	$2\ 13/16$
3				$7/8$	$1\ 5/8$	2	$2\ 1/4$	$2\ 11/16$
$3\ 1/8$					$1\ 1/2$	$1\ 7/8$	$2\ 1/8$	$2\ 1/2$
$3\ 1/4$					$1\ 1/4$	$1\ 3/4$	2	$2\ 3/8$
$3\ 3/8$					$15/16$	$1\ 5/8$	$1\ 15/16$	$2\ 1/4$
$3\ 1/2$						$1\ 3/8$	$1\ 3/4$	$2\ 1/8$
$3\ 5/8$						$1\ 1/16$	$1\ 9/16$	2
$3\ 3/4$							$1\ 5/16$	$1\ 7/8$
$3\ 7/8$								$1\ 11/16$
4								$1\ 3/8$

Notes:
H_1 = height of head, in.
H_2 = maximum shank extension,* in.
C_1 = clearance for tightening, in.
C_2 = clearance for entering, in.
C_3 = clearance for fillet,* in.
P = bolt stagger, in.
F = clearance for tightening staggered bolts, in.
*Based on one standard hardened washer.

Table 8-5.
Entering and Tightening Clearances, in.
Tension-Control ASTM A325 and A490 Bolts

Aligned Bolts

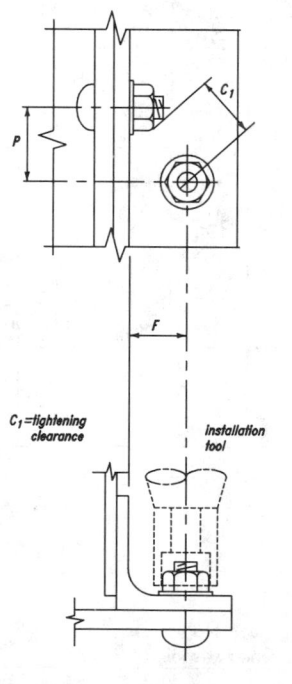

Tools	Nominal Bolt Dia, in.	H_1	H_2	C_1	C_2	C_3 Circular	C_3 Clipped
Large Tools			3 3/8-in. Diameter Critical				
	3/4	1/2	1 3/8	1 7/8	7/8	3/4	—
	7/8	9/16	1 1/2	1 7/8	1	7/8	—
	1	5/8	1 3/4	1 7/8	1 1/8	1	—
			2 1/2-in. Diameter Critical				
	3/4	1/2	1 3/8	1 3/8	7/8	3/4	—
	7/8	9/16	1 1/2	1 3/8	1	7/8	—
	1	5/8	1 3/4	1 3/8	1 1/8	1	—
Small Tools			3-in. Diameter Critical				
	5/8	7/16	1 1/4	1 5/8	13/16	11/16	—
	3/4	1/2	1 3/8	1 5/8	7/8	3/4	—
	7/8	9/16	1 1/2	1 5/8	1	7/8	—
			2 3/16-in. Diameter Critical				
	5/8	7/16	1 1/4	1 1/8	13/16	11/16	—
	3/4	1/2	1 3/8	1 1/8	7/8	3/4	—
	7/8	9/16	1 1/2	1 1/8	1	7/8	—

Staggered Bolts

	Stagger P, in.			
	Nominal Bolt Diameter, in.			
F	5/8	3/4	7/8	1
1 1/4	1 13/16			
1 3/8	1 3/4	2 1/16	2 1/4	2 7/16
1 1/2	1 11/16	2	2 3/16	2 3/8
1 5/8	1 9/16	1 7/8	2 1/16	2 1/4
1 3/4	1 1/2	1 13/16	2	2 3/16
1 7/8	1 7/16	1 3/4	1 7/8	2 1/8
2	1 5/16	1 5/8	1 3/4	2
2 1/8	1 1/4	1 9/16	1 11/16	1 15/16
2 1/4	1 3/16	1 1/2	1 9/16	1 7/8
2 3/8	1 1/8	1 3/8	1 1/2	1 3/4
2 1/2	1	1 5/16	1 3/8	1 11/16
2 5/8		1 3/16	1 5/16	1 9/16
2 3/4		1 1/8	1 3/16	1 1/2
2 7/8			1 1/8	1 3/8
3				1 5/16
3 3/8				1 5/16

C_1 = tightening clearance   installation tool

Notes:
H_1 = height of head, in.
H_2 = maximum shank extension,* in.
C_1 = clearance for tightening, in.
C_2 = clearance of entering, in.
C_3 = clearance for fillet,* in.
P = bolt stagger, in.
F = clearance for tightening staggered bolts, in.

*Based on one standard hardened washer.

to tension or combined shear and tension, high-strength bolts must be used in fully tensioned bearing-type or slip-critical connections.

Bearing-type connections are typically used for shear, moment, and diagonal bracing connections in buildings. Bolts in bearing-type connections are installed in the snug-tightened condition unless required in LRFD Specification Section J1.11 to be fully tensioned. Note that bolts in bearing-type connections required to be fully tensioned must not be confused with fully tensioned bolts in slip-critical connections. Fully tensioned bolts in bearing-type connections have no requirements regarding the slip resistance of the contact surfaces. Thus, painted surfaces in fully tensioned bearing-type connections need not meet the slip resistance requirements of slip-critical connections.

Slip-critical connections are used when slip would be detrimental to the serviceability of the structure; this is essentially fatigue related and is primarily of concern in bridge design. From LRFD Specification Section K4, "The occurrence of full design wind or earthquake loads is too infrequent to warrant consideration in fatigue design." Consequently, slip-critical connections are not normally required or used for wind or seismic loading in buildings.

Slip-critical shear connections are required, however, in applications such as those involving oversized holes, fatigue loading, or in craneway and bridge connections. High-strength bolts in slip-critical connections are always fully tensioned to resist slip on the faying surface(s) of the connection. While faying surfaces in slip-critical connections are not normally painted, painted surfaces in accordance with RCSC Specification Section 3(b) are permitted.

When subjected to tension only or combined shear and tension, high-strength bolts must be used in fully tensioned bearing-type or slip-critical connections. Examples of these applications are hanger connections, extended end-plate FR moment connections, and diagonal bracing connections.

Fully tensioned bolts in bearing-type or slip-critical connections must meet the minimum tensioning requirements for ASTM A325 and A490 bolts as specified in Table 4 of the RCSC Specification. Fully tensioned bolts in either case may be tightened by the same methods. The methods approved by the RCSC are: (1) turn-of-nut method; (2) calibrated wrench method; (3) alternative design bolt method; and, (4) direct tension indicator method. It is important to note that the RCSC prohibits the use of any published relationship between torque and tension.

Inspection of Fully Tensioned High-Strength Bolts

When a joint with fully tensioned high-strength bolts is assembled, the RCSC Specification requires that all joint surfaces, including surfaces adjacent to the bolt head and nut be free of scale, except tight mill scale, and of dirt or other foreign material. Burrs need not be removed unless they prevent solid seating of the connected parts in the snug-tightened condition.

ASTM A6 lists tolerances for straightness and flatness. These tolerances can prevent the faying surfaces from sufficiently contacting in medium- to large-size connections. Section C8 of the Commentary on the RCSC Specification states: "...Even after being fully tightened, some thick parts with uneven surfaces may not be in contact over the entire faying surface. In itself, this is not detrimental to the performance of the joint. As long as the specified bolt tension is present in all bolts of the completed connection, the clamping force equal to the total of the tensions in all bolts will be transferred at the locations that are in contact and be fully effective in resisting slip through friction."

Table 8-6.
Dimensions of Non-High-Strength Bolts and Nuts, in.

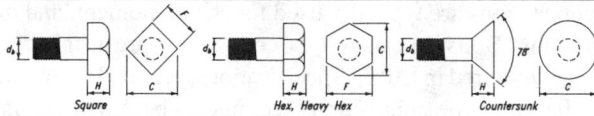

Bolts

Bolt Dia. d_b, in.	Square F, in.	Square C, in.	Square H, in.	Hex F, in.	Hex C, in.	Hex H, in.	Heavy Hex F, in.	Heavy Hex C, in.	Heavy Hex H, in.	Countersunk C, in.	Countersunk H, in.	Min. Thrd. Length L ≤ 6 in.	Min. Thrd. Length L > 6 in.
1/4	3/8	1/2	3/16	7/16	1/2	3/16	—	—	—	1/2	1/8	3/4	1
3/8	9/16	13/16	1/4	9/16	5/8	1/4	—	—	—	11/16	3/16	1	1 1/4
1/2	3/4	1 1/16	5/16	3/4	7/8	3/8	7/8	1	3/8	7/8	1/4	1 1/4	1 1/2
5/8	15/16	1 5/16	7/16	15/16	1 1/16	7/16	1 1/16	1 1/4	7/16	1 1/8	5/16	1 1/2	1 3/4
3/4	1 1/8	1 9/16	1/2	1 1/8	1 5/16	1/2	1 1/4	1 7/16	1/2	1 3/8	3/8	1 3/4	2
7/8	1 5/16	1 7/8	5/8	1 5/16	1 1/2	9/16	1 7/16	1 11/16	9/16	1 9/16	7/16	2	2 1/4
1	1 1/2	2 1/8	11/16	1 1/2	1 3/4	11/16	1 5/8	1 7/8	11/16	1 13/16	1/2	2 1/4	2 1/2
1 1/8	1 11/16	2 3/8	3/4	1 11/16	1 15/16	3/4	1 13/16	2 1/16	3/4	2 1/16	9/16	2 1/2	2 3/4
1 1/4	1 7/8	2 5/8	7/8	1 7/8	2 3/16	7/8	2	2 5/16	7/8	2 1/4	5/8	2 3/4	3
1 3/8	2 1/16	2 15/16	15/16	2 1/16	2 3/8	15/16	2 3/16	2 1/2	15/16	2 1/2	11/16	3	3 1/4
1 1/2	2 1/4	3 3/16	1	2 1/4	2 5/8	1	2 3/8	2 3/4	1	2 11/16	3/4	3 1/4	3 1/2
1 3/4	—	—	—	2 5/8	3	1 3/16	2 3/4	3 3/16	1 3/16	—	—	3 3/4	4
2	—	—	—	3	3 7/16	1 3/8	3 1/8	3 5/8	1 3/8	—	—	4 1/4	4 1/2
2 1/4	—	—	—	3 3/8	3 7/8	1 1/2	3 1/2	4 1/16	1 1/2	—	—	4 3/4	5
2 1/2	—	—	—	3 3/4	4 5/16	1 11/16	3 7/8	4 1/2	1 11/16	—	—	5 1/4	5 1/2
2 3/4	—	—	—	4 1/8	4 3/4	1 13/16	4 1/4	4 15/16	1 13/16	—	—	5 3/4	6
3	—	—	—	4 1/2	5 3/16	2	4 5/8	5 5/16	2	—	—	6	6 1/2
3 1/4	—	—	—	4 7/8	5 5/8	2 3/16	—	—	—	—	—	6	7
3 1/2	—	—	—	5 1/4	6 1/16	2 5/16	—	—	—	—	—	6	7 1/2
3 3/4	—	—	—	5 5/8	6 1/2	2 1/2	—	—	—	—	—	6	8
4	—	—	—	6	6 15/16	2 11/16	—	—	—	—	—	6	8 1/2

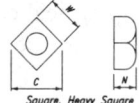
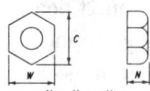

Square, Heavy Square   Hex, Heavy Hex

Nuts

Nut Size, in.	Square W, in.	Square C, in.	Square N, in.	Hex W, in.	Hex C, in.	Hex N, in.	Heavy Square W, in.	Heavy Square C, in.	Heavy Square N, in.	Heavy Hex W, in.	Heavy Hex C, in.	Heavy Hex N, in.
1/4	7/16	5/8	1/4	7/16	1/2	3/16	1/2	11/16	1/4	1/2	9/16	1/4
3/8	5/8	7/8	5/16	9/16	5/8	1/4	11/16	1	3/8	11/16	13/16	3/8
1/2	13/16	1 1/8	7/16	3/4	7/8	3/8	7/8	1 1/4	1/2	7/8	1	1/2
5/8	1	1 7/16	9/16	15/16	1 1/16	7/16	1 1/16	1 1/2	5/8	1 1/16	1 1/4	5/8
3/4	1 1/8	1 9/16	11/16	1 1/8	1 5/16	1/2	1 1/4	1 3/4	3/4	1 1/4	1 7/16	3/4
7/8	1 5/16	1 7/8	3/4	1 5/16	1 1/2	9/16	1 7/16	2 1/16	7/8	1 7/16	1 11/16	7/8
1	1 1/2	2 1/8	7/8	1 1/2	1 3/4	11/16	1 5/8	2 5/16	1	1 5/8	1 7/8	1
1 1/8	1 11/16	2 3/8	1	1 11/16	1 15/16	3/4	1 13/16	2 9/16	1 1/8	1 13/16	2 1/16	1 1/8
1 1/4	1 7/8	2 5/8	1 1/8	1 7/8	2 3/16	7/8	2	2 13/16	1 1/4	2	2 5/16	1 1/4
1 3/8	2 1/16	2 15/16	1 1/4	2 1/16	2 3/8	15/16	2 3/16	3 1/8	1 3/8	2 3/16	2 1/2	1 3/8
1 1/2	2 1/4	3 3/16	1 5/16	2 1/4	2 5/8	1	2 3/8	3 3/8	1 1/2	2 3/8	2 3/4	1 1/2
1 3/4	—	—	—	—	—	—	—	—	—	2 3/4	3 3/16	1 3/4
2	—	—	—	—	—	—	—	—	—	3 3/8	3 5/8	2
2 1/4	—	—	—	—	—	—	—	—	—	3 1/2	4 1/16	2 3/16
2 1/2	—	—	—	—	—	—	—	—	—	3 7/8	4 1/2	2 7/16
2 3/4	—	—	—	—	—	—	—	—	—	4 1/4	4 15/16	2 11/16
3	—	—	—	—	—	—	—	—	—	4 5/8	5 5/16	2 15/16
3 1/4	—	—	—	—	—	—	—	—	—	5	5 3/4	3 3/16
3 1/2	—	—	—	—	—	—	—	—	—	5 3/8	6 3/16	3 7/16
3 3/4	—	—	—	—	—	—	—	—	—	5 3/4	6 5/8	3 11/16
4	—	—	—	—	—	—	—	—	—	6 1/8	7 1/16	3 15/16

Notes:
For high-strength bolt and nut dimensions, refer to Table 8-2.
Square, hex, and heavy hex bolt dimensions, rounded to nearest 1/16-in., are in accordance with ANSI B18.2.1.
Countersunk bolt dimensions, rounded to the nearest 1/16-in., are in accordance with ANSI 18.5.
Minimum thread length = $2d_b$ + 1/4-in. for bolts up to 6-in. long, and $2d_b$ + 1/2-in. for bolts longer than 6-in.

Table 8-7.
Threading Dimensions for High-Strength and Non-High-Strength Bolts, in.

SCREW THREADS
Unified Standard Series–UNC/UNRC and 4UN/4UNR
ANSI B1.1

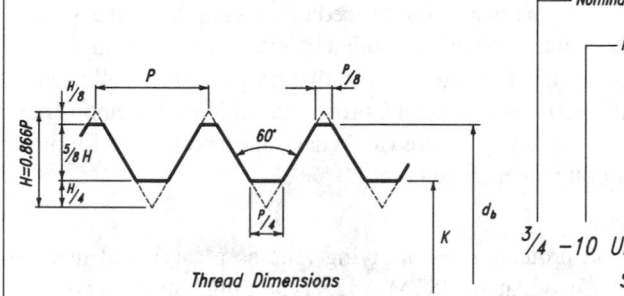

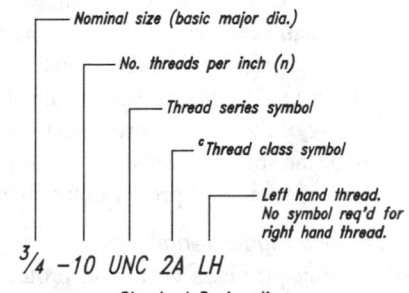

Thread Dimensions

Standard Designations

$\frac{3}{4}$ -10 UNC 2A LH

Diameter		Area			
Bolt Diameter d_b, in.	Min. Root K, in.	Gross Bolt Area, in.^{2}	Min. Root Area, in.^{2}	Net Tensile Area, in.$^{2\,a}$	Threads per inch, n^b
$\frac{1}{4}$	0.189	0.049	0.029	0.032	20
$\frac{3}{8}$	0.298	0.110	0.070	0.078	16
$\frac{1}{2}$	0.406	0.196	0.129	0.142	13
$\frac{5}{8}$	0.514	0.307	0.207	0.226	11
$\frac{3}{4}$	0.627	0.442	0.309	0.334	10
$\frac{7}{8}$	0.739	0.601	0.429	0.462	9
1	0.847	0.785	0.563	0.606	8
$1\frac{1}{8}$	0.950	0.994	0.709	0.763	7
$1\frac{1}{4}$	1.08	1.23	0.908	0.969	7
$1\frac{3}{8}$	1.17	1.49	1.08	1.16	6
$1\frac{1}{2}$	1.30	1.77	1.32	1.41	6
$1\frac{3}{4}$	1.51	2.41	1.78	1.90	5
2	1.73	3.14	2.34	2.50	$4\frac{1}{2}$
$2\frac{1}{4}$	1.98	3.98	3.07	3.25	$4\frac{1}{2}$
$2\frac{1}{2}$	2.19	4.91	3.78	4.00	4
$2\frac{3}{4}$	2.44	5.94	4.69	4.93	4
3	2.69	7.07	5.70	5.97	4
$3\frac{1}{4}$	2.94	8.30	6.80	7.10	4
$3\frac{1}{2}$	3.19	9.62	8.01	8.33	4
$3\frac{3}{4}$	3.44	11.0	9.31	9.66	4
4	3.69	12.6	10.7	11.1	4

Notes:

[a] Net tensile area $= 0.7854 \left(d_b - \dfrac{0.9743}{n} \right)^2$

[b] For diameters listed, thread series is UNC (coarse). For larger diameters, thread series is 4UN.
[c] 2A denotes Class 2A fit applicable to external threads;
2B denotes corresponding Class 2B fit for internal threads.

It should be noted that, even when bolts in bearing-type connections are required to be fully tensioned, high bolt tension is not normally required for proper connection performance. Thus, a significant reduction in inspection costs will be achieved by relying on visual inspection of the bolt head or nut to note the peening marks signifying that the tightening wrench was applied.

From RCSC Specification Commentary Section C9, "It is apparent from the commentary on installation procedures that the inspection procedures giving the best assurance that bolts are properly installed and tensioned is provided by inspector observation of the calibration testing of the bolts using the selected installation procedure followed by monitoring of the work in progress to assure that the procedure which was demonstrated to provide the specified tension is routinely adhered to. When such a program is followed, no further evidence of proper bolt tension is required."

Galvanizing High-Strength Bolts

Galvanizing provides corrosion protection by applying zinc as a sacrificial metal to protect the base metal. As previously stated, ASTM A325 Type 1 high-strength bolts and A449 bolts are permitted to be galvanized; A490 bolts are not permitted to be galvanized.

There are two methods of galvanizing: hot-dip galvanizing and mechanical galvanizing. Hot-dip galvanizing is a process whereby the bolt is dipped in molten zinc and spun in a centrifuge to remove the excess. This process is described in detail in ASTM A153. In contrast, mechanical galvanizing utilizes a combination of powdered zinc, chemicals, and water with the bolts in a spun hopper. As result of collisions between the bolts, zinc, and glass beads, the zinc is cold-welded to the surface of the bolts. This process is described in detail in ASTM B695. For more information, refer to AISC (1993).

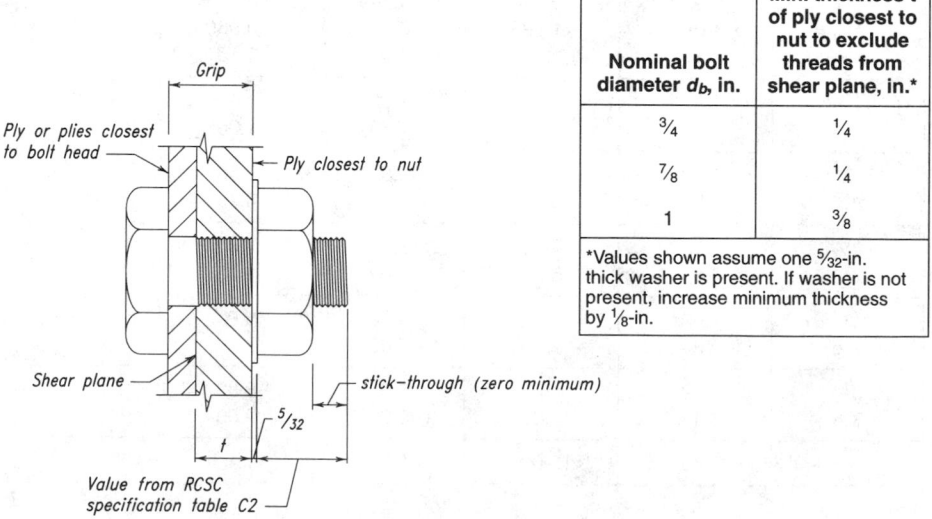

Nominal bolt diameter d_b, in.	Min. thickness t of ply closest to nut to exclude threads from shear plane, in.*
$3/4$	$1/4$
$7/8$	$1/4$
1	$3/8$

*Values shown assume one $5/32$-in. thick washer is present. If washer is not present, increase minimum thickness by $1/8$-in.

Fig. 8-3. Minimum thickness of ply closest to nut to exclude threads from shear plane.

Reuse of High-Strength Bolts

From RCSC Specification Section 8f, ASTM A490 bolts and galvanized ASTM A325 bolts shall not be reused. Other A325 bolts are permitted to be reused if approved by the engineer of record.

A simple rule based on the prevention of excessive plastic deformation of the bolt is that non-galvanized A325 bolts are satisfactory for reuse, regardless of previous use, if the nut can be placed on the threads and run down the full length of the thread by hand (AISC, 1988). Kulak, et al. (1987) recommends that non-galvanized ASTM A325 bolts may be reused once or twice, provided that proper control on the number of reuses can be established; adequate nut rotation capacity will be present as long as there is some lubricant on the bolt. This lubricant can be the original lubrication or oil, grease, or wax, or a lubricant that is added later. For a detailed assessment of the performance of repetitively tightened high-strength bolts, refer to Bowman and Betancourt (1991).

Non-High-Strength Bolts

LRFD Specification Section A3.3 permits the use of ASTM A307 non-high-strength bolts for structural applications not requiring fully tensioned installation, that is, snug-tightened bearing-type connections. ASTM A307 bolts are available with both hex and square heads in diameters from $\frac{1}{4}$-in. to four inches in two grades: Grade A for general applications and Grade B for cast-iron-flanged piping joints. ASTM A563 Grade A nuts are recommended for use with ASTM A307 bolts. Other suitable grades are listed in ASTM A563 Table X1.1.

Dimensions and Weights

Typical non-high-strength bolt head and nut dimensions are given in Table 8-6. Thread lengths listed in this table may be calculated for non-high-strength bolts as $2d_b + \frac{1}{4}$-in. for bolts up to six inches long and $2d_b + \frac{1}{2}$-in. for bolts over six inches long, where d_b is the bolt diameter. Note that these thread lengths are longer than those given previously for high-strength bolts in Table 8-2. Threading dimensions are given in Table 8-7. Weights of non-high-strength bolts are given in Tables 8-8, 8-9, and 8-10.

Entering and Tightening Clearances

As with high-strength bolts, clearance is required for entering and tightening the bolts with an impact wrench. The required clearances are the same as those given for high-strength bolts in Table 8-4.

Design Strength of Bolts

The design strength of bolts is determined in accordance with the provisions of LRFD Specification Section J3. LRFD Specification requirements are based upon the provisions of the RCSC Specification.

For bolts in bearing-type connections subjected to shear only, the limit states of bolt shear strength and bearing strength at bolt holes must be checked. For bolts in bearing-type connections subjected to tension only, the limit state of bolt tensile strength, including the effect of prying action, must be checked. For bolts in bearing-type connections subjected to combined shear and tension, the limit states of bolt tensile strength, including the effects of both the bolt shear stress present and prying action, and bearing strength at bolt holes must be checked.

Table 8-8.
Weights of Non-High-Strength Fasteners, pounds

100 Square Bolts with Hexagonal Nuts

Bolt Length, in.	Nominal Bolt Diameter, in.								
	$\frac{1}{4}$	$\frac{3}{8}$	$\frac{1}{2}$	$\frac{5}{8}$	$\frac{3}{4}$	$\frac{7}{8}$	1	$1\frac{1}{8}$	$1\frac{1}{4}$
1	2.38	6.11	13.0	24.1	38.9	—	—	—	—
$1\frac{1}{4}$	2.71	6.71	14.0	25.8	41.5	—	—	—	—
$1\frac{1}{2}$	3.05	7.47	15.1	27.6	44.0	67.3	95.1	—	—
$1\frac{3}{4}$	3.39	8.23	16.5	29.3	46.5	70.8	99.7	—	—
2	3.73	8.99	17.8	31.4	49.1	74.4	104	143	—
$2\frac{1}{4}$	4.06	9.75	19.1	33.5	52.1	77.9	109	149	—
$2\frac{1}{2}$	4.40	10.5	20.5	35.6	55.1	82.0	114	155	206
$2\frac{3}{4}$	4.74	11.3	21.8	37.7	58.2	86.1	119	161	213
3	5.07	12.0	23.2	39.8	61.2	90.2	124	168	221
$3\frac{1}{4}$	5.41	12.8	24.5	41.9	64.2	94.4	129	174	229
$3\frac{1}{2}$	5.75	13.5	25.9	44.0	67.2	98.5	135	181	237
$3\frac{3}{4}$	6.09	14.3	27.2	46.1	70.2	103	140	188	246
4	6.42	15.1	28.6	48.2	73.3	107	145	195	254
$4\frac{1}{4}$	6.76	15.8	29.9	50.3	76.3	111	151	202	262
$4\frac{1}{2}$	7.10	16.6	31.3	52.3	79.3	115	156	208	271
$4\frac{3}{4}$	7.43	17.3	32.6	54.4	82.3	119	162	215	279
5	7.77	18.1	33.9	56.5	85.3	123	167	222	288
$5\frac{1}{4}$	8.11	18.9	35.3	58.6	88.4	127	172	229	296
$5\frac{1}{2}$	8.44	19.6	36.6	60.7	91.4	131	178	236	304
$5\frac{3}{4}$	8.78	20.4	38.0	62.8	94.4	136	183	242	313
6	9.12	21.1	39.3	64.9	97.4	140	188	249	321
$6\frac{1}{4}$	9.37	21.7	40.4	66.7	100	143	193	255	329
$6\frac{1}{2}$	9.71	22.5	41.8	68.7	103	147	198	262	337
$6\frac{3}{4}$	10.1	23.3	43.1	70.8	106	151	204	269	345
7	10.4	24.0	44.4	72.9	109	156	209	275	354
$7\frac{1}{4}$	10.7	24.8	45.8	75.0	112	160	214	282	362
$7\frac{1}{2}$	11.0	25.5	47.1	77.1	115	164	220	289	371
$7\frac{3}{4}$	11.4	26.3	48.5	79.2	118	168	225	296	379
8	11.7	27.0	49.8	81.3	121	172	231	303	387
$8\frac{1}{2}$	—	28.6	52.5	85.5	127	180	241	316	404
9	—	30.1	55.2	89.7	133	189	252	330	421
$9\frac{1}{2}$	—	31.6	57.9	93.9	139	197	263	343	438
10	—	66.1	60.6	98.1	145	205	274	357	454
$10\frac{1}{2}$	—	34.6	63.3	102	151	213	284	371	471
11	—	36.2	66.0	106	157	221	295	384	488
$11\frac{1}{2}$	—	37.7	68.7	110	163	230	306	398	505
12	—	39.2	71.3	115	170	238	316	411	522
$12\frac{1}{2}$	—	—	74.0	119	176	246	327	425	538
13	—	—	76.7	123	182	254	338	439	556
$13\frac{1}{2}$	—	—	79.4	127	188	263	349	452	572
14	—	—	82.1	131	194	271	359	466	589
$14\frac{1}{2}$	—	—	84.8	135	200	279	370	479	605
15	—	—	87.5	140	206	287	381	493	622
$15\frac{1}{2}$	—	—	90.2	144	212	296	392	507	639
16	—	—	92.9	148	218	304	402	520	656
Per inch add'tl. add	1.3	3.0	5.4	8.4	12.1	16.5	21.4	27.2	33.6

Notes:
For weights of high-strength fasteners, see Table 8-3.
This table conforms to weight standards adopted by the Industrial Fasteners Institute (IFI).
*Square bolt per ANSI B18.2.1, hexagonal nut per ANSI B18.2.2. For other non-high-strength fasteners, refer to Tables 8-9 and 8-10.

Table 8-9.
Weight Adjustments for Combinations of Non-High-Strength Fasteners Other than Tabulated in Table 8-8

Combinations of 100:		Add or Subtr.	Nominal Bolt Diameter, in.								
			$\frac{1}{4}$	$\frac{3}{8}$	$\frac{1}{2}$	$\frac{5}{8}$	$\frac{3}{4}$	$\frac{7}{8}$	1	$1\frac{1}{8}$	$1\frac{1}{4}$
Square Bolts with	Square Nuts	+	0.1	1.0	2.0	3.4	3.5	5.5	8.0	12.2	16.3
	Heavy Square Nuts	+	0.6	2.1	4.1	7.0	11.6	17.2	23.2	32.1	41.2
	Heavy Hex Nuts	+	0.4	1.5	2.8	4.6	7.6	10.7	14.2	18.9	24.3
100, Square Bolts with Hexagonal Nuts*	Square Nuts	+	0.1	0.6	1.1	1.4	0.2	0.5	-0.2	-0.1	-1.7
	Hex Nuts	−	0.0	0.4	0.9	2.0	3.3	5.0	8.2	12.3	18.0
	Heavy Square Nuts	+	0.6	1.7	3.2	5.0	8.3	12.2	15.0	19.8	23.2
	Heavy Hex Nuts	+	0.4	1.1	1.9	2.6	4.3	5.7	6.0	6.6	6.3
100, Hex Bolts	Heavy Square Nuts	+	—	—	4.7	7.3	11.3	16.5	20.7	27.0	33.6
	Heavy Hex Nuts	+	—	—	3.4	4.9	7.3	10.0	11.7	13.8	16.7

Notes:
For weights of high-strength fasteners, see Table 8-3.
This table conforms to weight standards adopted by the Industrial Fasteners Institute (IFI).

*Add or subtract value in this table to or from the value in Table 8-8.

Table 8-10.
Weights of Non-High-Strength Bolts of Diameter Greater Than 1¼-in., pounds

Weight of 100 Each:		Nominal Bolt Diameter, in.											
		$1\frac{3}{8}$	$1\frac{1}{2}$	$1\frac{3}{4}$	2	$2\frac{1}{4}$	$2\frac{1}{2}$	$2\frac{3}{4}$	3	$3\frac{1}{4}$	$3\frac{1}{2}$	$3\frac{3}{4}$	4
Heads of:	Square Bolts	105	130	—	—	—	—	—	—	—	—	—	—
	Hex Bolts	84.0	112	178	259	369	508	680	900	1120	1390	1730	2130
	Heavy Hex Bolts	95.0	124	195	280	397	541	720	950	—	—	—	—
One Linear Inch, Unthreaded Shank		42.0	50.0	68.2	89.0	113	139	168	200	235	272	313	356
One Linear Inch, Threaded Shank		35.0	42.5	57.4	75.5	97.4	120	147	178	210	246	284	325
Square Nuts		94.5	122	—	—	—	—	—	—	—	—	—	—
Heavy Square Nuts		125	161	—	—	—	—	—	—	—	—	—	—
Heavy Hex Nuts		102	131	204	299	419	564	738	950	1190	1530	1910	2180

Notes:
For weights of high-strength fasteners, see Table 8-3.
This table conforms to weight standards adopted by the Industrial Fasteners Institute (IFI).

For bolts in slip-critical connections subjected to shear only, the limit states of slip resistance, bolt shear strength, and bearing strength at bolt holes must be checked. For bolts in slip-critical connections subjected to combined shear and tension, the limit states of slip resistance, including the effect of the tensile force present, bolt shear strength, and bearing strength at bolt holes must be checked.

Bolt Shear Strength

As illustrated in Figure 8-4a, this limit state considers a shear failure of the bolt shank on plane **cdef**. Since there is one shear plane, the bolt is in single shear (S). Additional plies of material may increase the number of shear planes and, therefore, the shear strength of the bolt. This condition, as illustrated in Figure 8-4b, is called double shear (D).

Additionally, high-strength bolts may be specified with the threads included (N) or excluded (X) from the shear plane of the connection. Note that the shear strength of bolts with the threads included is about 25 percent less than that of bolts with the threads excluded. In spite of this, many designers prefer to specify N bolts when possible due to the difficulty in assuring that threads are excluded from the shear plane in the as-built condition. If, however, the threads are to be excluded from the shear plane, care must be taken to specify a bolt of sufficient overall length given the thread length and required bolt length from Table 8-2. Note that additional washers may be required to accomplish this; refer to Figure 8-3.

From LRFD Specification Section J3.6, the design bolt shear strength is ϕR_n, where $\phi = 0.75$ and:

$$R_n = (F_v A_b)n$$

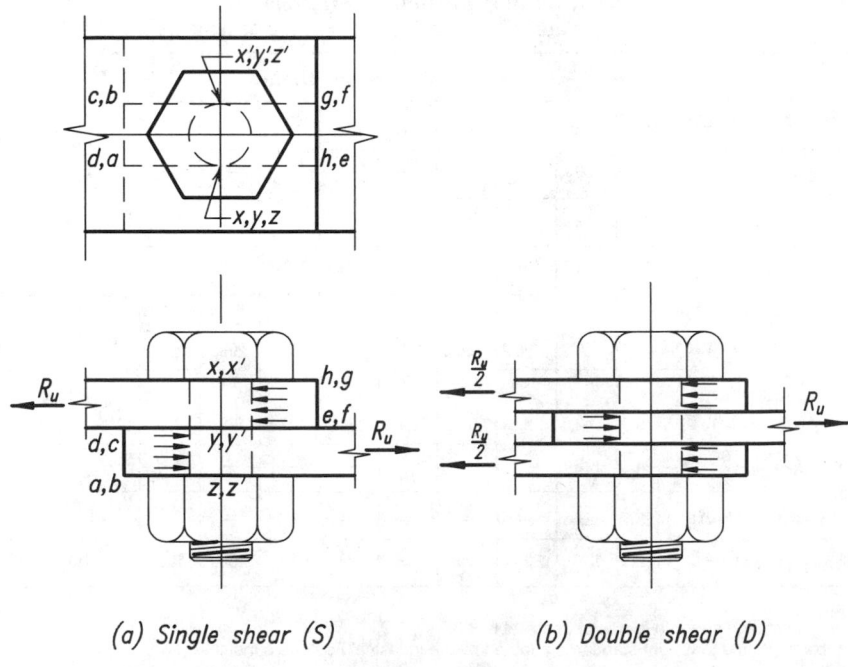

(a) Single shear (S)   *(b) Double shear (D)*

Fig. 8-4. Bolt shear.

In the above equation, n is the number of bolts in the connection, F_v is the nominal shear strength, and A_b is the nominal bolt area. For convenience, the design bolt shear strengths of various bolts are summarized in Table 8-11; design bolt shear strengths of vertical rows of n bolts are summarized in Table 8-12.

Bearing Strength at Bolt Holes
As illustrated in Figure 8-5, this limit state considers both a tear fracture of the connected material and deformation around the bolt holes. Bearing strength is a function of the material being connected, the type of bolt hole, and the spacing and edge distance; it is independent of both the type of bolt and the presence or absence of threads on the bearing area.

From LRFD Specification Section J3.10, when deformation around the bolt holes is a design consideration for standard holes, oversized holes, short-slotted holes, and long-slotted holes parallel to the line of force, the design bearing strength at bolt holes is ϕR_n, where $\phi = 0.75$ and, for two or more bolts in the line of force, when $L_e \geq 1.5d$ and $s \geq 3d$:

$$R_n = (2.4 dt F_u)n$$

For a single bolt in the line of force or when $L_e < 1.5d$ or $s < 3d$:

$$R_n = \left[L_e + \left(s - \frac{d}{2} \right)(n-1) \right](t F_u) \leq (2.4 dt F_u)n$$

In the above equations, n is the number of bolts in the connection, d is the nominal bolt diameter, t is the thickness in bearing, and L_e is the edge distance. If deformation around the bolt hole is not a design consideration, or for long-slotted holes perpendicular to the line of force, refer to LRFD Specification Section J3.10.

For convenience, the design bearing strength at bolt holes is tabulated for the foregoing conditions in Tables 8-13 and 8-14, respectively. Note that these tables may be applied to bolts with countersunk heads, by subtracting one-half the depth of the countersink from the material thickness t. As illustrated in Figure 8-6, this is equivalent to subtracting one-quarter the diameter of the bolt from the material thickness t.

Bolt Tensile Strength
From LRFD Specification Section J3.6, when subjected to tension only, the design bolt tensile strength is ϕR_n,

where

$\phi = 0.75$
$R_n = (F_t A_b)n$

In the above equation, n is the number of bolts in the connection. For convenience, the design bolt tensile strengths of various bolts is summarized in Table 8-15. When subjected to combined shear and tension, the design bolt tensile strength is reduced by a function of the shear stress present in the bolt as specified in LRFD Specification Section J3.7.

LRFD Specification Section J3.6 states that any tension resulting from prying action must be considered in determining the required strength of the bolts. Prying action is a phenomenon (in bolted construction only) whereby the deformation of a fitting under a tensile force increases the tensile force in the bolt. The required strength per bolt is the

Table 8-11.
Design Shear Strength of One Bolt, kips

ASTM Desig.	Thread Cond.	ϕF_v (ksi)	Loading	Nominal Bolt Diameter d, in.							
				$5/8$	$3/4$	$7/8$	1	$1\frac{1}{8}$	$1\frac{1}{4}$	$1\frac{3}{8}$	$1\frac{1}{2}$
				Nominal Bolt Area, in.^{2}							
				0.3068	0.4418	0.6013	0.7854	0.9940	1.227	1.485	1.767
A325	N	36.0	S	11.0	15.9	21.6	28.3	35.8	44.2	53.5	63.6
			D	22.1	31.8	43.3	56.5	71.6	88.4	107	127
	X	45.0	S	13.8	19.9	27.1	35.3	44.7	55.2	66.8	79.5
			D	27.6	39.8	54.1	70.7	89.5	110	134	159
A490	N	45.0	S	13.8	19.9	27.1	35.3	44.7	55.2	66.8	79.5
			D	27.6	39.8	54.1	70.7	89.5	110	134	159
	X	56.3	S	17.3	24.9	33.9	44.2	56.0	69.1	83.6	99.5
			D	34.5	49.7	67.7	88.4	112	138	167	199
A307	—	18.0	S	5.52	7.95	10.8	14.1	17.9	22.1	26.7	31.8
			D	11.0	15.9	21.6	28.3	35.8	44.2	53.5	63.6

N = Threads included in shear plane
X = Threads excluded from shear plane
S = Single shear
D = Double shear

Table 8-12.
Design Shear Strength of n Bolts in Double Shear*

	ASTM A325						ASTM A490					
	N			X			N			X		
n	$3/4$	$7/8$	1	$3/4$	$7/8$	1	$3/4$	$7/8$	1	$3/4$	$7/8$	1
12	382	520	679	477	649	848	477	649	848	596	812	1060
11	350	476	622	437	595	778	437	595	778	547	744	972
10	318	433	565	398	541	707	398	541	707	497	676	884
9	286	390	509	358	487	636	358	487	636	447	609	795
8	254	346	452	318	433	565	318	433	565	398	541	707
7	223	303	396	278	379	495	278	379	495	348	474	619
6	191	260	339	239	325	424	239	325	424	298	406	530
5	159	216	283	199	271	353	199	271	353	249	338	442
4	127	173	226	159	216	283	159	216	283	199	271	353
3	95.4	130	170	119	162	212	119	162	212	149	203	265
2	63.6	86.6	113	79.5	108	141	79.5	108	141	99.4	135	177
1	31.8	43.3	56.5	39.8	54.1	70.7	39.8	54.1	70.7	49.7	67.7	88.4

N = Threads included in shear plane
X = Threads excluded in shear plane
*For design strength of bolts in single shear, divide tabular value by 2.

sum of r_{ut}, the factored force per bolt due to the tensile force, and q_u, the additional tension per bolt resulting from prying action produced by deformation of the connected parts.

While the effect of prying action is considered in the design of the bolts, it is primarily a function of the connected elements; thus, the connected elements must possess adequate flexural strength and it is their stiffness which is the key to satisfactory performance. Refer to "Hanger Connections" in Part 11 for treatment of prying action.

Slip Resistance

In slip-critical connections, the fully tensioned bolt creates resistance to slip through friction on the faying surface between two connected parts. This slip resistance is a function of the slip coefficient μ of the faying surface.

Clean mill scale with no coating is defined as a Class A surface with $\mu = 0.33$. Blast-cleaned surfaces with no coatings are defined as Class B surfaces with $\mu = 0.50$. Hot-dip galvanized and roughened surfaces are defined as Class C surfaces with $\mu = 0.40$.

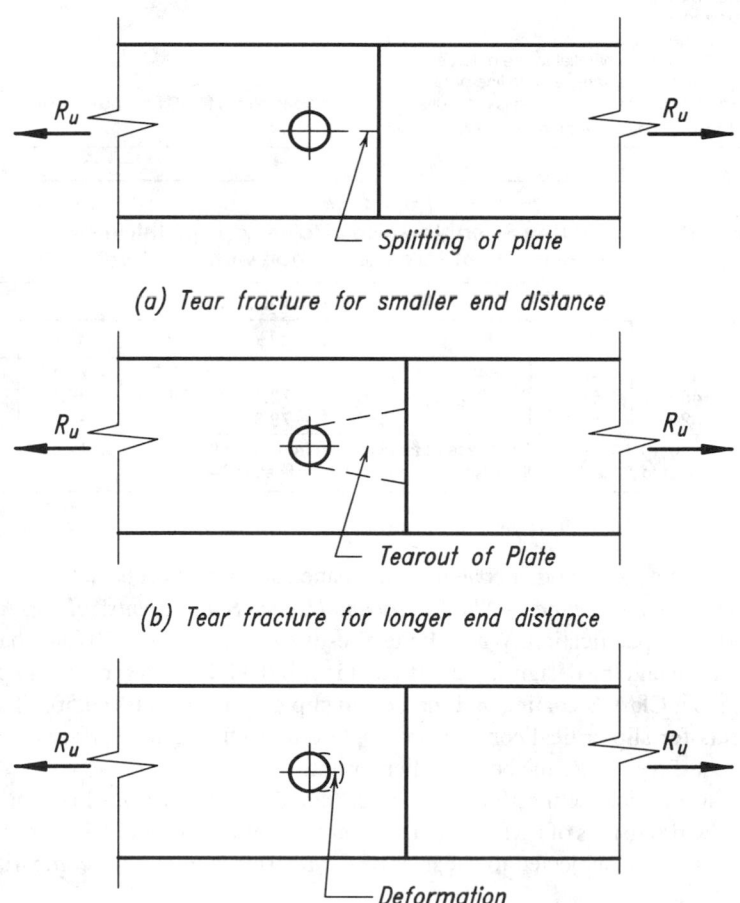

(a) Tear fracture for smaller end distance

(b) Tear fracture for longer end distance

(c) Deformation of material at bolt hole

Fig. 8-5. Bearing strength at bolt holes.

Table 8-13.
Design Bearing Strength at Bolt Holes, kips/in. thickness
Two or more holes in line of force with $L_e \geq 1.5d$
and $s \geq 3d$; hole deformation considered*

Hole Type	F_u, ksi	Nominal Bolt Diameter d, in.							
		$5/8$	$3/4$	$7/8$	1	$1\frac{1}{8}$	$1\frac{1}{4}$	$1\frac{3}{8}$	$1\frac{1}{2}$
		1.5d							
		$15/16$	$1\frac{1}{8}$	$15/16$	$1\frac{1}{2}$	$1^{11}/16$	$1\frac{7}{8}$	$2\frac{1}{16}$	$2\frac{1}{4}$
		3d							
		$1\frac{7}{8}$	$2\frac{1}{4}$	$2\frac{5}{8}$	3	$3\frac{3}{8}$	$3\frac{3}{4}$	$4\frac{1}{8}$	$4\frac{1}{2}$
STD, OVS	58	65.3	78.3	91.4	104	117	131	144	157
SSL, LSLP	65	73.1	87.8	102	117	132	146	161	176
	70	78.8	94.5	110	126	142	158	173	189
LSLT	58	54.4	65.3	76.1	87.0	97.9	109	120	131
	65	60.9	73.1	85.3	97.5	110	122	134	146
	70	65.6	78.8	91.9	105	118	131	144	158

STD = Standard Hole
OVS = Oversized Hole
SSL = Short-Slotted Hole
LSLP = Long-Slotted Hole parallel to line of force
LSLT = Long-Slotted Hole transverse to line of force
*When $s < 3d$, or when hole deformation is not a design consideration, refer to LRFD Specification Section J3.10.
When $L_e < 1.5d$ or for one hole in the line of force, refer to Table 8-14.

Table 8-14.
Design Bearing Strength at Bolt Holes, kips/in. thickness
One hole in line of force or top bolt with $L_e < 1.5d$*

F_u, ksi	Edge Distance L_e, in.							
	1	$1\frac{1}{8}$	$1\frac{1}{4}$	$1\frac{3}{8}$	$1\frac{1}{2}$	$1\frac{5}{8}$	$1\frac{3}{4}$	$1\frac{7}{8}$
58	43.5	48.9	54.4	59.8	65.3	70.7	76.1	81.6
65	48.8	54.8	60.9	67.0	73.1	79.2	85.3	91.4
70	52.5	59.1	65.6	72.2	78.8	85.3	91.9	98.4

*Design strength from Table 8-14 shall not exceed tabular value from Table 8-13. For remaining bolts, when $s - d/2 > 2.4d$, refer to Table 8-13; otherwise refer to LRFD Specification Section J3.10.

Slip coefficients for all other coated blast-cleaned surfaces must be determined by the *Testing Method to Determine the Slip Coefficient Used in Bolted Joints*; refer to Appendix A of the RCSC Specification. When the test results in $0.33 \leq \mu < 0.50$, the coating is a Class A coating and the design slip coefficient is $\mu = 0.33$. If the test results in $\mu \geq 0.50$, the coating is a Class B coating and the design slip coefficient is $\mu = 0.50$. The surface requirements for slip-critical connections apply only to the faying surfaces and do not include the surfaces under the bolt, washer, or nut.

Bolts in slip-critical connections may be designed at either service loads or factored loads with the provisions of LRFD Specification Section J3.8. From LRFD Specification Section J3.8a, when subjected to shear only, the resistance to slip for comparison with service loads is ϕR_n,

where

$$R_n = (F_v A_b)n$$

		Nominal Bolt Diameter d, in.							
		$\frac{5}{8}$	$\frac{3}{4}$	$\frac{7}{8}$	1	$1\frac{1}{8}$	$1\frac{1}{4}$	$1\frac{3}{8}$	$1\frac{1}{2}$
ASTM Desig.	ϕF_t, ksi	Nominal Bolt Area, in.^{2}							
		0.3068	0.4418	0.6013	0.7854	0.9940	1.227	1.485	1.767
A325	67.5	20.7	29.8	40.6	53.0	67.1	82.8	100	119
A490	84.8	26.0	37.4	51.0	66.6	84.2	104	126	150
A307	33.8	10.4	14.9	20.3	26.5	33.5	41.4	50.1	59.6

Table 8-15.
Design Tensile Strength of Bolts, kips

and $\phi = 1.0$ for standard holes, oversized holes, short-slotted holes, and long-slotted holes perpendicular to the direction of the load; $\phi = 0.85$ for long-slotted holes parallel to the direction of the load. In the above equation, n is the number of bolts in the connection. In general, slip is likely to occur at 1.4 to 1.5 times the service loads.

Note that the values of F_v tabulated in LRFD Specification Table J3.6 for bolts in slip-critical connections assume Class A surfaces with $\mu = 0.33$. As stated in LRFD Specification Section J3.8a, it is permissible to increase F_v to the applicable value in the RCSC Specification for other surfaces. When subjected to combined shear and tension, the slip capacity for comparison with service loads must be reduced by the factor:

$$\left(1 - \frac{T}{T_b}\right)$$

as specified in LRFD Specification Section J3.9a, where T is the unfactored force on the connection and T_b is the minimum bolt tension from LRFD Specification Table J3.1.

From LRFD Specification Appendix J3.8a, the design slip resistance for comparison with factored loads is ϕR_{str},

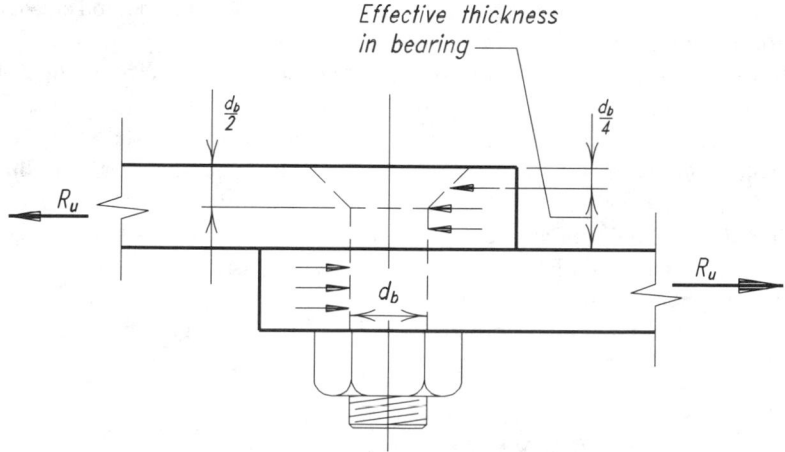

Fig. 8-6. Effective thickness for bearing of countersunk bolts.

where

$$R_{str} = 1.13\mu T_m N_b N_s$$

and ϕ is equal to 1.0 for standard holes, 0.85 for oversized and short-slotted holes, 0.70 for long-slotted holes perpendicular to the direction of the load, and 0.60 for long-slotted holes parallel to the direction of the load. When subjected to combined tension and shear, the design slip resistance for comparison with factored loads must be reduced by the factor:

$$\left(1 - \frac{T_u}{1.13T_m N_b}\right)$$

as specified in LRFD Specification Appendix J3.8b. In the above equations, T_u is the factored force on the connection, T_m is the minimum bolt tension from LRFD Specification Table J3.1, and N_b is the number of bolts in the connection.

For convenience, slip capacities for comparison with service loads and design slip resistances for comparison with factored loads are tabulated in Tables 8-16 and 8-17, respectively.

ECCENTRICALLY LOADED BOLT GROUPS

When the line of action of an applied load does not pass through the center of gravity (CG) of a bolt group, the load is eccentric and results in a moment which must be considered in the design of the connection.

Eccentricity in the Plane of the Faying Surface

Eccentricity in the plane of the faying surface produces additional shear. The bolts must be designed to resist the combined effect of the direct shear from the applied load P_u and the additional shear from the induced moment $P_u e$. Two methods of analysis for this type of eccentricity will be discussed: (1) the instantaneous center of rotation method; and, (2) the elastic method.

Instantaneous Center of Rotation Method

Also known as the ultimate strength method (Crawford, 1968), this method considers the load-deformation relationship of each bolt and, thus, more accurately predicts the ultimate strength of the eccentrically loaded connection. Eccentricity produces both a rotation about the centroid of the bolt group and a translation of one connected element with respect to the other. The combined effect of this rotation and translation is equivalent to a rotation about a point defined as the instantaneous center of rotation (IC) as illustrated in Figure 8-7a. The location of the IC depends on the geometry of the bolt group as well as the direction and point of application of the load. The individual resistance of each bolt is assumed to act on a line perpendicular to a ray passing through the IC and the centroid of that bolt as illustrated in Figure 8-7b.

The load-deformation relationship of one bolt is illustrated in Figure 8-8,

where

$$R = R_{ult}(1 - e^{-10\Delta})^{0.55}$$

In the above equation,

Table 8-16.
Slip-Critical Connections
Design Resistance to Shear at Service Loads,* kips
(Class A faying surface, $\mu = 0.33$)

ASTM Desig.	Hole Type	Loading	\[Nominal Bolt Diameter, in.\] $5/8$	$3/4$	$7/8$	1	$1 1/8$	$1 1/4$	$1 3/8$	$1 1/2$
			\[Nominal Bolt Area, in.^{2}\] 0.3068	0.4418	0.6013	0.7854	0.9940	1.227	1.485	1.767
A325	STD	S	5.22	7.51	10.2	13.4	16.9	20.9	25.2	30.0
		D	10.4	15.0	20.4	26.7	33.8	41.7	50.5	60.1
	OVS SSL	S	4.60	6.63	9.02	11.8	14.9	18.4	22.3	26.5
		D	9.20	13.3	18.0	23.6	29.8	36.8	44.5	53.0
	LSLP	S	3.13	4.51	6.13	8.01	10.1	12.5	15.1	18.0
		D	6.26	9.01	12.3	16.0	20.3	25.0	30.3	36.0
	LSLT	S	3.68	5.30	7.22	9.42	11.9	14.7	17.8	21.2
		D	7.36	10.6	14.4	18.8	23.9	29.5	35.6	42.4
A490	STD	S	6.44	9.28	12.6	16.5	20.9	25.8	31.2	37.1
		D	12.9	18.6	25.3	33.0	41.7	51.5	62.4	74.2
	OVS SSL	S	5.52	7.95	10.8	14.1	17.9	22.1	26.7	31.8
		D	11.0	15.9	21.6	28.3	35.8	44.2	53.5	63.6
	LSLP	S	3.93	5.65	7.70	10.1	12.7	15.7	19.0	22.6
		D	7.85	11.3	15.4	20.1	25.4	31.4	38.0	45.2
	LSLT	S	4.60	6.63	9.02	11.8	14.9	18.4	22.3	26.5
		D	9.20	13.3	18.0	23.6	29.8	36.8	44.5	53.0

STD = Standard Hole
OVS = Oversized Hole
SSL = Short-Slotted Hole
LSLP = Long-Slotted Hole parallel to line of force
LSLT = Long-Slotted Hole transverse to line of force
S = Single Shear
D = Double Shear
*For design slip resistance at factored loads, refer to Table 8-17.

R = shear force in one bolt at a deformation Δ, kips.
R_{ult} = ultimate shear strength of one bolt, kips.
Δ = total deformation of a bolt, including shearing, bearing, and bending deformation, plus local bearing deformation of the plate, in.
e = 2.718..., base of the natural logarithm.

Applying a maximum deformation Δ_{max} to the bolt most remote from the IC, the maximum shear strength of that bolt may be determined. For other bolts, deformations are assumed to vary linearly with distance from the IC, and shear strengths can be obtained from this relationship. The strength of the bolt group is, then, the sum of the

Table 8-17.
Slip-Critical Connections
Design Slip Resistance at Factored Loads, kips
(Class A faying surface, $\mu = 0.33$)

ASTM Desig.	Hole Type	Loading	Nominal Bolt Diameter, in.							
			$\frac{5}{8}$	$\frac{3}{4}$	$\frac{7}{8}$	1	$1\frac{1}{8}$	$1\frac{1}{4}$	$1\frac{3}{8}$	$1\frac{1}{2}$
			Minimum ASTM A325 Bolt Tension, kips							
			19.0	28.0	39.0	51.0	56.0	71.0	85.0	103
A325	STD	S	7.09	10.4	14.5	19.0	20.9	26.5	31.7	38.4
		D	14.2	20.9	29.1	38.0	41.8	53.0	63.4	76.8
	OVS SSL	S	6.02	8.88	12.4	16.2	17.8	22.5	26.9	32.6
		D	12.0	17.8	24.7	32.3	35.5	45.0	53.9	65.3
	LSLP	S	4.25	6.26	8.73	11.4	12.5	15.9	19.0	23.0
		D	8.50	12.5	17.5	22.8	25.1	31.8	38.0	46.1
	LSLT	S	4.96	7.31	10.2	13.3	14.6	18.5	22.2	26.9
		D	9.92	14.6	20.4	26.6	29.2	37.1	44.4	53.8
			Minimum ASTM A490 Bolt Tension, kips							
			24.0	35.0	49.0	64.0	80.0	102	121	148
A490	STD	S	8.95	13.1	18.3	23.9	29.8	38.0	45.1	55.2
		D	17.9	26.1	36.5	47.7	59.7	76.1	90.2	110
	OVS SSL	S	7.61	11.1	15.5	20.3	25.4	32.3	38.4	46.9
		D	15.2	22.2	31.1	40.6	50.7	64.7	76.7	93.8
	LSLP	S	5.37	7.83	11.0	14.3	17.9	22.8	27.1	33.1
		D	10.7	15.7	21.9	28.6	35.8	45.6	54.1	66.2
	LSLT	S	6.26	9.14	12.8	16.7	20.9	26.6	31.6	38.6
		D	12.5	18.3	25.6	33.4	41.8	53.3	63.2	77.3

STD = Standard Hole
OVS = Oversized Hole
SSL = Short-Slotted Hole
LSLP = Long-Slotted Hole parallel to line of force
LSLT = Long-Slotted Hole transverse to line of force
S   = Single Shear
D   = Double Shear

individual strengths of all bolts. If the correct location of the IC has been selected, the three equations of in-plane statics will be satisfied; i.e., $\Sigma F_x = 0$, $\Sigma F_y = 0$, and $\Sigma M = 0$.

Tables 8-18 through 8-25 employ the instantaneous center of rotation method for the bolt patterns and eccentric conditions indicated and inclined loads at 0°, 15°, 30°, 45°, 60°, and 75°. The load-deformation relationship is based on data obtained experimentally for $\frac{3}{4}$-in. diameter ASTM A325 bolts, where $R_{ult} = 74$ kips, and $\Delta_{max} = 0.34$ in. The non-dimensional coefficient C is obtained by dividing the factored eccentric force P_u by R_{ult}.

For any of the bolt group geometries shown, the design strength of the eccentrically loaded bolt group is ϕR_n,

where

$$\phi R_n = C \times \phi r_n$$

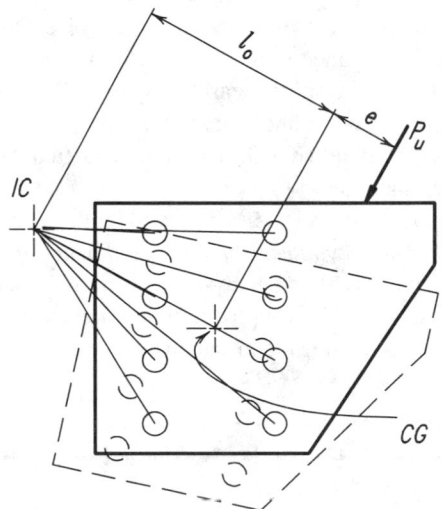

(a) Instantaneous center of rotation (IC)

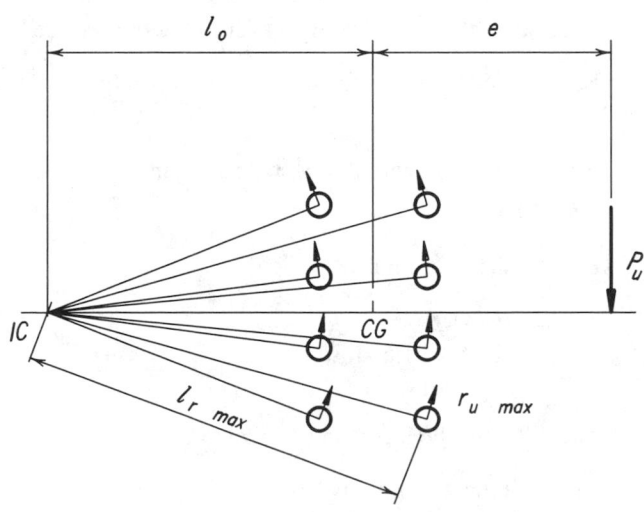

(b) Forces on bolts in group

Fig. 8-7. Instantaneous center of rotation method.

In the above equation, ϕr_n is the least design strength of one bolt determined from the limit states of bolt shear strength, bearing strength at bolt holes, and slip resistance (if the connection is to be slip-critical). The design strength ϕR_n must be greater than or equal to the required strength P_u. Thus, by dividing P_u by ϕr_n, the minimum coefficient C is obtained, and a bolt group can be selected for which the coefficient is of that magnitude or greater.

These tables may be used with any bolt diameter and are conservative when used with ASTM A490 bolts. Linear interpolation within a given table between adjacent values of e_x is permitted. Design strengths determined with these tables provide a factor of safety equivalent to that for bolts in connections less than 50 inches long, subjected to shear produced by a concentric load in either bearing-type or slip-critical connections. Although this procedure is based on connections which may experience slip under load, both load tests and analytical studies (Kulak, 1975) indicate that it may be conservatively extended to slip-critical connections.

A convergence criterion of one percent was employed for the tabulated iterative solutions. *Straight line interpolation between values for loads at different angles may be significantly unconservative.* Therefore, unless a direct analysis is performed, use only the values for the next lower angle for design. For bolt group patterns not treated by these tables, a special ultimate strength analysis is required if the instantaneous center of rotation method is to be used.

Example 8-1

Given: Refer to Figure 8-9. Determine the largest eccentric force P_u for which the design shear strength of the bolts in the connection is adequate using the instantaneous center of rotation method. Use $\frac{7}{8}$-in. diameter A325-N bolts, $\phi r_n = 21.6$ kips/bolt.

A. Assume the load is vertical as illustrated in Figure 8-9 ($\theta = 0°$)

B. Assume the load acts at an angle of 75° with respect to vertical ($\theta = 75°$)

Solution A: From Table 8-20 with $\theta = 0°$, with $s = 3$ in., $e = 16$ in., and $n = 6$:

$C = 3.55$

Design Shear Strength

$\phi R_n = C \times \phi r_n$
     $= 3.55 \times 21.6$ kips/bolt
     $= 76.7$ kips

Thus, P_u must be less than or equal to 76.7 kips.

Comment: Note that this eccentricity has effectively reduced the shear strength of this bolt group by about 70 percent when compared with the concentrically loaded case.

Solution B: From Table 8-20 with $\theta = 75°$, $s = 3$ in., $e = 16$ in., and $n = 6$:

$$C = 7.90$$

Design shear strength

$$\phi R_n = C \times \phi r_n$$
$$= 7.90 \times 21.6 \text{ kips/bolt}$$
$$= 171 \text{ kips}$$

Thus, P_u must be less than or equal to 171 kips.

Comment: In Solution B, the vertical component of the design strength is

$$\phi R_n \sin 75° = (171 \text{ kips})(0.966)$$
$$= 165 \text{ kips}$$

and the horizontal component of the design strength is

$$\phi R_n \cos 75° = (171 \text{ kips})(0.259)$$
$$= 44.3 \text{ kips}$$

Elastic Method
Alternatively, the elastic method may be used to analyze eccentrically loaded bolt groups. It offers a simplified, conservative approach but does not render a consistent factor of safety and, in some cases, provides excessively conservative results. Furthermore, the elastic method ignores both the ductility of the bolt group and the load redistribution which occurs. Refer to Higgins (1971).

In the elastic method, for a force applied parallel to the Y principal axis of the bolt group as illustrated in Figure 8-10, the eccentric force P_u is resolved into a force P_u acting through the center of gravity (CG) of the bolt group and a moment $P_u e$ where e is the eccentricity. Each bolt is then assumed to support an equal share of the concentric force P_u, and a share of the eccentric moment $P_u e$ which is proportional to its distance from the

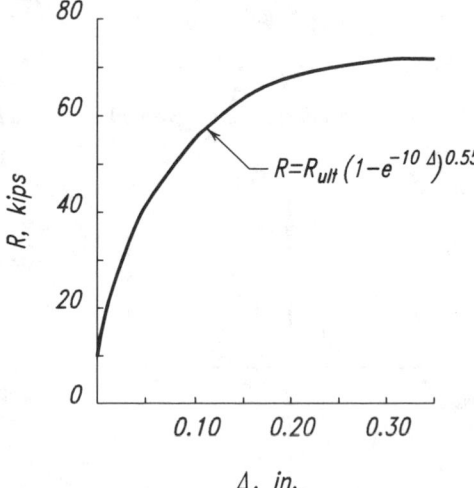

Fig. 8-8. *Load-deformation relationship for bolts.*

CG. The bolt most remote from the CG, then, is the most highly stressed. The resultant vectorial sum of these forces r_u is the required strength for the bolt.

The direct shear force per bolt due to the concentric force P_u is r_1,

where

$$r_1 = \frac{P_u}{n}$$

and n is the number of bolts.

The shear force in each bolt due to the moment $P_u e$ varies with distance from the CG and will be maximum in the bolt which is must remote from the CG. The maximum shear due to the moment $P_u e$ is r_m,

where

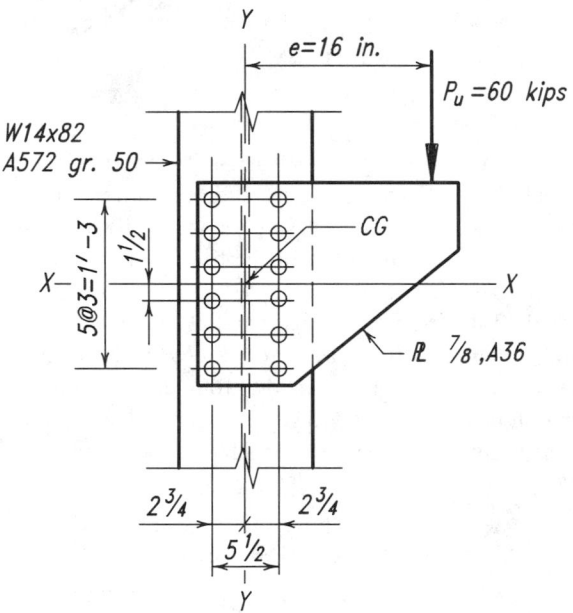

Fig. 8-9. Bolted bracket plate for Examples 8-1 and 8-2.

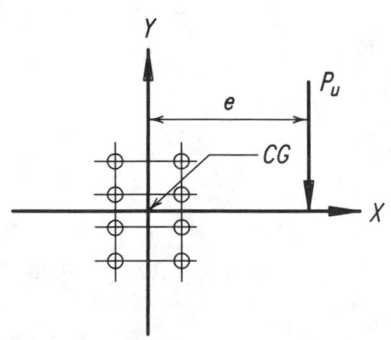

Figure 8-10

$$r_m = \frac{P_u e c}{I_p}$$

In the above equation,

    c = distance from CG to center of bolt most remote from CG, in.
    I_p = polar moment of inertia of the bolt group, in.^{4} per in.^{2} (see any text on statics).

To determine the resultant force on the most highly stressed bolt, r_m must be resolved into vertical component r_2 and horizontal component r_3,

where

$$r_2 = \frac{P_u e c_x}{I_p}$$

$$r_3 = \frac{P_u e c_y}{I_p}$$

In the above equation, c_x and c_y are the horizontal and vertical components of the diagonal distance c. Thus, the resultant factored force is r_u,

where

$$r_u = \sqrt{(r_1 + r_2)^2 + (r_3)^2}$$

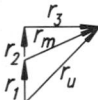

and the bolts must be chosen such that the design strength ϕr_n exceeds the required strength r_u.

For the more general case of an inclined eccentric force, i.e., not parallel to the Y principal axes of the bolt group, the effect of the X-direction component of the direct shear must also be included. Refer to Iwankiw (1987).

Example 8-2

Given: Refer to Example 8-1. Recalculate the largest eccentric force P_u for which the design shear strength of the bolts in the connection is adequate using the elastic method. Compare the result with that of Example 8-1. Use 7/8-in. diameter A325-N bolts, $\phi r_n = 21.6$ kips.

$$I_p = 406 \text{ in.}^4 \text{ per in.}^2$$

Solution: Direct shear force per bolt:

$$r_1 = \frac{P_u}{n}$$

$$= \frac{P_u}{12}$$

Additional shear force on bolt due to eccentricity:

$$r_2 = \frac{P_u e c_x}{I_p}$$

$$= \frac{P_u(16 \text{ in.}) \left(\dfrac{5\frac{1}{2}\text{-in.}}{2}\right)}{406 \text{ in.}^4 \text{ per in.}^2}$$

$$= 0.108 \, P_u$$

$$r_3 = \frac{P_u \, e c_y}{I_p}$$

$$= \frac{P_u(16 \text{ in.}) \, (7\frac{1}{2}\text{-in.})}{406 \text{ in.}^4 \text{ per in.}^2}$$

$$= 0.296 \, P_u$$

Resultant shear force:

$$r_u = \sqrt{(r_1 + r_2)^2 + (r_3)^2}$$

$$= \sqrt{\left(\frac{P_u}{12} + 0.108 P_u\right)^2 + (0.296 P_u)^2}$$

$$= 0.352 \, P_u$$

Since r_u must be less than or equal to ϕr_n,

$$P_u \leq \frac{\phi r_n}{0.352}$$

$$\leq \frac{21.6 \text{ kips}}{0.352}$$

$$\leq 61.3 \text{ kips}$$

This 20 percent reduction in the strength predicted by the instantaneous center of rotation method in Example 8-1a is indicative of the conservatism of the elastic method.

Eccentricity Normal to the Plane of the Faying Surface

Eccentricity normal to the plane of the faying surface produces tension above and compression below the neutral axis of the bracket connection illustrated in Figure 8-11. The eccentric load P_u can be resolved into a concentric force P_u acting at the faying surface of the connection and a moment $P_u e$ normal to the plane of the faying surface where e is the eccentricity. Each bolt is then assumed to support an equal share of the concentric force P_u, and the moment is resisted by tension in the bolts above the neutral axis and compression between the lower part of the bracket and the column flange.

The forces for which the bolts in this connection must be designed must be determined by balancing the tensile forces in the bolts above the neutral axis with the resultant compressive force below the neutral axis. The analysis of such a connection is straightforward and usually begins with one of two assumptions: Case I assumes the neutral axis is not at the center of gravity (CG) while Case II assumes the neutral axis is at the CG.

For a bearing-type connection, the limit state of bolt tension, including the effect of prying action and the shear stress present, must still be checked as specified in LRFD Specification Section J3.7. For a slip-critical connection, the bolts above the neutral axis subject to tension would lose a portion of their clamping force. The overall connection, however, would experience no reduction in total clamping force because the clamping force below the neutral axis is increased by an equivalent amount. Therefore, it would be unnecessary to reduce the strength of this connection for the interaction of tension and shear above the neutral axis. However, the limit state of bolt tension, including the effect of prying action and the shear stress present, must still be checked as specified in LRFD Specification Section J3.9.

Case I—Neutral Axis Not at Center of Gravity
The shear force per bolt due to the concentric force P_u is r_{uv},

where

$$r_{uv} = \frac{P_u}{n}$$

and n is the number of bolts in the connection.

To determine the location of the neutral axis, assume a trial position of the neutral axis at one-sixth of the total bracket depth, measured upward from the bottom. In Figure 8-12a, this is indicated by the line X-X. To provide for reasonable proportions and to recognize that the effective bearing area will depend upon the bracket flange or support flange bending stiffness, the effective width of the compression block W_{eff} should be taken as:

$$W_{eff} = 8t_f \le b_f$$

where

   t_f = lesser of bracket flange and support flange thicknesses, in.
   b_f = bracket flange width, in.

This effective width is valid for bracket flanges made from W or S shapes, welded plates, and angles. Where the bracket flange thickness is not constant, the average flange thickness should be used.

Having assumed the width of the compression block, it is possible to check an assumed location of the neutral axis by checking static equilibrium assuming an elastic stress

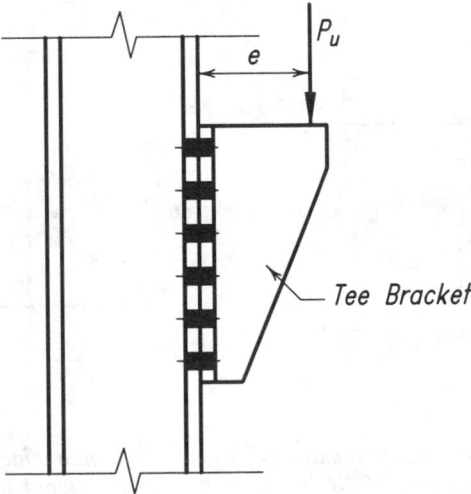

Fig. 8-11. Bolts subjected to eccentricity normal to the plane of the faying surface.

distribution. Equating the moment of the bolt area above the neutral axis with the moment of the compression block area below the neutral axis,

$$\Sigma A_b \times y = W_{eff} \times d \times \frac{d}{2}$$

In the above equation,

    ΣA_b = sum of the areas of all bolts above the neutral axis, in.^{2}
    y = distance from line X-X to CG of the bolt group above neutral axis, in.
    d = depth of compression block, in.

The value of d may then be adjusted until a reasonable equality exists.

Once the neutral axis has been located, the tensile force per bolt r_{ut}, as illustrated in Figure 8-12b may be determined as:

$$r_{ut} = \frac{P_u ec}{I_x} \times A_b$$

where

    c = distance from neutral axis to most remote bolt in group, in.
    I_x = combined moment of inertia of bolt group and compression block about neutral axis, in.^{4}

Bolts above the neutral axis are subjected to the shear force r_{uv}, the tensile force r_{ut}, and the effect of prying action; bolts below the neutral axis are subjected to the shear force r_{uv} only.

Case II—Neutral Axis at Center of Gravity
This method provides a more direct, but also a more conservative result. As for Case I, the shear force per bolt due to the concentric force P_u is r_{uv}, where

$$r_{uv} = \frac{P_u}{n}$$

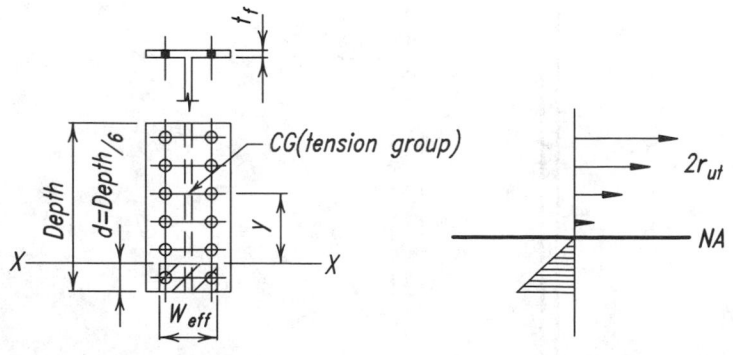

(a) Initial approximation
of location of NA

(b) Force diagram with
final location of NA

Fig. 8-12. Case I—Neutral axis (NA) not at center of gravity (CG).

and n is the number of bolts in the connection.

The neutral axis is assumed to be located at the CG of the bolt group as illustrated in Figure 8-13. The bolts above the neutral axis are in tension and the bolts below the neutral axis are said to be in "compression." To obtain a more accurate result, a plastic stress distribution is assumed; this assumption is justified because this method is still more conservative than Case I. Accordingly, the tensile force r_{ut} in each bolt above the neutral axis due to the moment P_{ue} is:

$$r_{ut} = \frac{P_u e}{n' d_m}$$

where

n' = number of bolts above the neutral axis

d_m = moment arm between resultant tensile force and resultant compressive force, in.

Bolts above the neutral axis are subjected to the shear force r_{uv}, the tensile force r_{ut}, and the effect of prying action; bolts below the neutral axis are subjected to the shear force r_{uv} only.

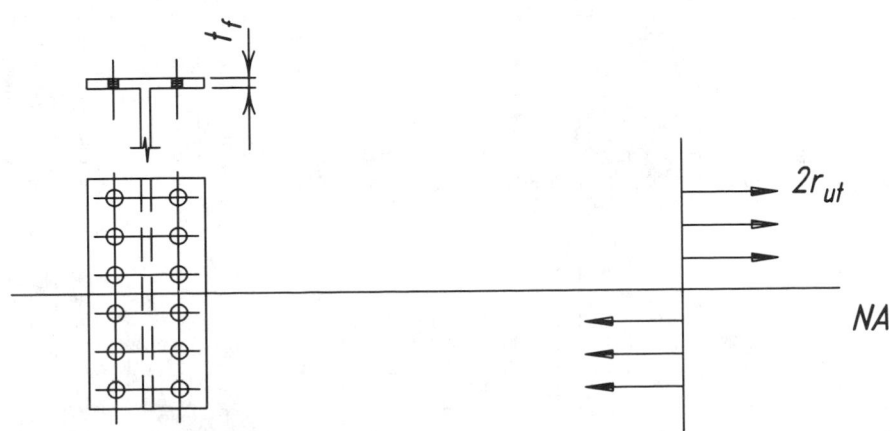

Fig. 8-13. Case II—Neutral axis (NA) at center of gravity (CG).

Table 8-18.
Coefficients C for Eccentrically Loaded Bolt Groups
Angle = 0°

$$C_{req} = \frac{P_u}{\phi r_n} \text{ or } \phi R_n = C \times \phi r_n$$

where

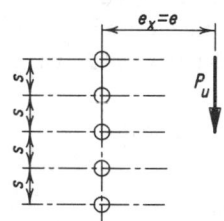

P_u = factored force, kips

ϕr_n = design strength per bolt, kips

ϕR_n = design strength of bolt group, kips

e = eccentricity of P_u with respect to centroid of bolt group, in. (not tabulated, may be determined by geometry.)

e_x = horizontal component of e, in.

s = bolt spacing, in.

C = coefficient tabulated below.

s, in.	e_x, in.	\multicolumn{11}{c}{Number of bolts in one vertical row, n}
		2	3	4	5	6	7	8	9	10	11	12
	2	1.18	2.23	3.32	4.39	5.45	6.48	7.51	8.52	9.53	10.5	11.5
	3	0.88	1.75	2.81	3.90	4.98	6.06	7.12	8.17	9.21	10.2	11.3
	4	0.69	1.40	2.36	3.40	4.47	5.56	6.64	7.72	8.78	9.84	10.9
	5	0.56	1.15	2.01	2.96	3.98	5.05	6.13	7.22	8.30	9.38	10.4
	6	0.48	0.97	1.73	2.59	3.55	4.57	5.63	6.70	7.79	8.87	9.96
	7	0.41	0.83	1.51	2.28	3.17	4.13	5.15	6.20	7.28	8.36	9.44
	8	0.36	0.73	1.34	2.04	2.85	3.75	4.72	5.73	6.78	7.85	8.93
	9	0.32	0.65	1.21	1.83	2.59	3.42	4.34	5.31	6.32	7.36	8.42
	10	0.29	0.59	1.09	1.66	2.36	3.14	4.00	4.92	5.89	6.90	7.94
3	12	0.24	0.49	0.92	1.40	2.00	2.68	3.44	4.27	5.15	6.09	7.06
	14	0.21	0.42	0.79	1.21	1.74	2.33	3.01	3.75	4.55	5.41	6.31
	16	0.18	0.37	0.70	1.06	1.53	2.06	2.67	3.33	4.06	4.85	5.68
	18	0.16	0.33	0.62	0.95	1.37	1.84	2.39	3.00	3.66	4.38	5.15
	20	0.15	0.29	0.56	0.85	1.24	1.67	2.16	2.72	3.33	3.99	4.70
	24	0.12	0.25	0.47	0.71	1.03	1.40	1.82	2.29	2.81	3.37	3.99
	28	0.11	0.21	0.40	0.61	0.89	1.20	1.57	1.97	2.42	2.92	3.45
	32	0.09	0.18	0.35	0.54	0.78	1.05	1.37	1.73	2.13	2.57	3.04
	36	0.08	0.16	0.31	0.48	0.69	0.94	1.22	1.54	1.90	2.29	2.72
	2	1.63	2.71	3.75	4.77	5.77	6.77	7.76	8.75	9.74	10.7	11.7
	3	1.39	2.48	3.56	4.60	5.63	6.65	7.65	8.66	9.66	10.7	11.6
	4	1.18	2.23	3.32	4.39	5.45	6.48	7.51	8.52	9.53	10.5	11.5
	5	1.01	1.98	3.07	4.15	5.23	6.28	7.33	8.36	9.38	10.4	11.4
	6	0.88	1.75	2.81	3.90	4.98	6.06	7.12	8.17	9.21	10.2	11.2
	7	0.77	1.56	2.58	3.64	4.73	5.81	6.89	7.95	9.00	10.1	11.1
	8	0.69	1.40	2.36	3.40	4.47	5.56	6.64	7.72	8.78	9.84	10.9
	9	0.62	1.26	2.17	3.17	4.22	5.30	6.39	7.47	8.55	9.61	10.7
	10	0.56	1.15	2.01	2.96	3.98	5.05	6.13	7.22	8.30	9.38	10.4
6	12	0.48	0.97	1.73	2.59	3.55	4.57	5.63	6.70	7.79	8.87	9.96
	14	0.41	0.83	1.51	2.28	3.17	4.13	5.15	6.20	7.28	8.36	9.44
	16	0.36	0.73	1.34	2.04	2.85	3.75	4.72	5.73	6.78	7.85	8.93
	18	0.32	0.65	1.21	1.83	2.59	3.42	4.34	5.31	6.32	7.36	8.42
	20	0.29	0.59	1.09	1.66	2.36	3.14	4.00	4.92	5.89	6.90	7.94
	24	0.24	0.49	0.92	1.40	2.00	2.68	3.44	4.27	5.15	6.09	7.06
	28	0.21	0.42	0.79	1.21	1.74	2.33	3.01	3.75	4.55	5.41	6.31
	32	0.18	0.37	0.70	1.06	1.53	2.06	2.67	3.33	4.06	4.85	5.68
	36	0.16	0.33	0.62	0.95	1.37	1.84	2.39	3.00	3.66	4.38	5.15

Table 8-18 (cont.).
Coefficients C for Eccentrically Loaded Bolt Groups
Angle = 15°

$$C_{req} = \frac{P_u}{\phi r_n} \text{ or } \phi R_n = C \times \phi r_n$$

where

P_u	= factored force, kips	
ϕr_n	= design strength per bolt, kips	
ϕR_n	= design strength of bolt group, kips	
e	= eccentricity of P_u with respect to centroid of bolt group, in. (not tabulated, may be determined by geometry.)	
e_x	= horizontal component of e, in.	
s	= bolt spacing, in.	
C	= coefficient tabulated below.	

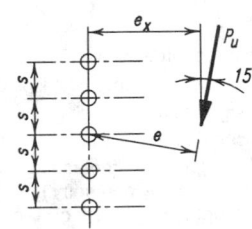

s, in.	e_x, in.	\multicolumn{11}{c}{Number of bolts in one vertical row, n}
		2	3	4	5	6	7	8	9	10	11	12
	2	1.15	2.20	3.28	4.34	5.39	6.42	7.45	8.46	9.47	10.5	11.5
	3	0.86	1.76	2.78	3.85	4.92	5.98	7.03	8.08	9.11	10.1	11.2
	4	0.67	1.42	2.35	3.36	4.41	5.48	6.55	7.61	8.67	9.72	10.8
	5	0.55	1.17	2.00	2.94	3.94	4.98	6.04	7.11	8.18	9.24	10.3
	6	0.47	0.99	1.73	2.58	3.52	4.52	5.55	6.61	7.67	8.74	9.81
	7	0.41	0.86	1.52	2.30	3.16	4.11	5.10	6.13	7.18	8.24	9.30
	8	0.36	0.75	1.35	2.06	2.86	3.74	4.69	5.68	6.70	7.74	8.80
	9	0.32	0.67	1.22	1.86	2.60	3.43	4.32	5.27	6.26	7.28	8.31
	10	0.29	0.61	1.10	1.69	2.38	3.16	4.00	4.90	5.85	6.84	7.85
3	12	0.24	0.51	0.93	1.43	2.03	2.71	3.46	4.28	5.15	6.06	7.01
	14	0.21	0.43	0.81	1.24	1.76	2.37	3.04	3.78	4.57	5.41	6.30
	16	0.19	0.38	0.71	1.09	1.56	2.10	2.70	3.37	4.09	4.87	5.69
	18	0.17	0.34	0.63	0.97	1.39	1.88	2.43	3.04	3.70	4.42	5.18
	20	0.15	0.30	0.57	0.88	1.26	1.70	2.20	2.76	3.37	4.03	4.74
	24	0.12	0.25	0.48	0.73	1.06	1.43	1.86	2.33	2.86	3.43	4.04
	28	0.11	0.22	0.41	0.63	0.91	1.23	1.60	2.02	2.47	2.97	3.51
	32	0.09	0.19	0.36	0.55	0.80	1.08	1.41	1.77	2.18	2.62	3.10
	36	0.08	0.17	0.32	0.49	0.71	0.96	1.26	1.58	1.95	2.34	2.78
	2	1.61	2.69	3.72	4.74	5.74	6.74	7.73	8.73	9.71	10.7	11.7
	3	1.36	2.45	3.52	4.56	5.59	6.60	7.61	8.61	9.61	10.6	11.6
	4	1.15	2.20	3.28	4.34	5.39	6.42	7.45	8.46	9.47	10.5	11.5
	5	0.98	1.96	3.03	4.10	5.16	6.21	7.25	8.28	9.30	10.3	11.3
	6	0.86	1.76	2.78	3.85	4.92	5.98	7.03	8.08	9.11	10.1	11.1
	7	0.75	1.57	2.55	3.60	4.66	5.73	6.80	7.85	8.90	9.94	11.0
	8	0.67	1.42	2.35	3.36	4.41	5.48	6.55	7.61	8.67	9.72	10.8
	9	0.61	1.29	2.16	3.14	4.17	5.23	6.30	7.36	8.43	9.49	10.5
	10	0.55	1.17	2.00	2.94	3.94	4.98	6.04	7.11	8.18	9.24	10.3
6	12	0.47	0.99	1.73	2.58	3.52	4.52	5.55	6.61	7.67	8.74	9.81
	14	0.41	0.86	1.52	2.30	3.16	4.11	5.10	6.13	7.18	8.24	9.30
	16	0.36	0.75	1.35	2.06	2.86	3.74	4.69	5.68	6.70	7.74	8.80
	18	0.32	0.67	1.22	1.86	2.60	3.43	4.32	5.27	6.26	7.28	8.31
	20	0.29	0.61	1.10	1.69	2.38	3.16	4.00	4.90	5.85	6.84	7.85
	24	0.24	0.51	0.93	1.43	2.03	2.71	3.46	4.28	5.15	6.06	7.01
	28	0.21	0.43	0.81	1.24	1.76	2.37	3.04	3.78	4.57	5.41	6.30
	32	0.19	0.38	0.71	1.09	1.56	2.10	2.70	3.37	4.09	4.87	5.69
	36	0.17	0.34	0.63	0.97	1.39	1.88	2.43	3.04	3.70	4.42	5.18

Table 8-18 (cont.).
Coefficients C for Eccentrically Loaded Bolt Groups
Angle = 30°

$$C_{req} = \frac{P_u}{\phi r_n} \text{ or } \phi R_n = C \times \phi r_n$$

where

P_u	= factored force, kips
ϕr_n	= design strength per bolt, kips
ϕR_n	= design strength of bolt group, kips
e	= eccentricity of P_u with respect to centroid of bolt group, in. (not tabulated, may be determined by geometry.)
e_x	= horizontal component of e, in.
s	= bolt spacing, in.
C	= coefficient tabulated below.

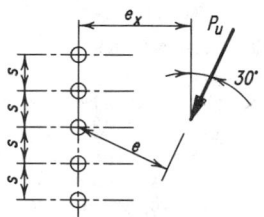

s, in.	e_x, in.	\multicolumn{11}{c}{Number of bolts in one vertical row, n}
		2	3	4	5	6	7	8	9	10	11	12
	2	1.14	2.20	3.25	4.30	5.33	6.36	7.38	8.39	9.40	10.4	11.4
	3	0.86	1.80	2.79	3.83	4.87	5.92	6.96	7.99	9.02	10.0	11.1
	4	0.69	1.50	2.40	3.39	4.41	5.45	6.49	7.53	8.57	9.61	10.6
	5	0.57	1.27	2.08	3.00	3.98	4.99	6.02	7.06	8.11	9.15	10.2
	6	0.49	1.09	1.82	2.68	3.60	4.57	5.58	6.60	7.64	8.68	9.72
	7	0.43	0.95	1.61	2.40	3.27	4.20	5.17	6.17	7.18	8.21	9.25
	8	0.38	0.83	1.44	2.17	2.98	3.86	4.79	5.76	6.75	7.77	8.79
	9	0.34	0.75	1.30	1.98	2.74	3.57	4.46	5.39	6.35	7.34	8.35
	10	0.31	0.67	1.19	1.82	2.52	3.31	4.15	5.05	5.98	6.95	7.93
3	12	0.26	0.56	1.01	1.55	2.17	2.87	3.64	4.46	5.33	6.24	7.17
	14	0.23	0.48	0.87	1.35	1.90	2.53	3.23	3.98	4.78	5.63	6.51
	16	0.20	0.42	0.77	1.20	1.69	2.26	2.89	3.58	4.33	5.11	5.94
	18	0.18	0.38	0.69	1.07	1.52	2.04	2.62	3.25	3.94	4.67	5.45
	20	0.16	0.34	0.62	0.97	1.37	1.85	2.38	2.97	3.61	4.30	5.02
	24	0.14	0.28	0.52	0.81	1.16	1.57	2.02	2.53	3.09	3.69	4.33
	28	0.12	0.24	0.45	0.70	1.00	1.36	1.75	2.20	2.69	3.22	3.79
	32	0.10	0.21	0.40	0.61	0.88	1.19	1.54	1.94	2.38	2.85	3.37
	36	0.09	0.19	0.35	0.55	0.78	1.07	1.38	1.74	2.13	2.56	3.03
	2	1.59	2.66	3.69	4.70	5.71	6.70	7.70	8.69	9.68	10.7	11.7
	3	1.34	2.43	3.48	4.52	5.54	6.55	7.55	8.56	9.55	10.6	11.5
	4	1.14	2.20	3.25	4.30	5.33	6.36	7.38	8.39	9.40	10.4	11.4
	5	0.98	1.99	3.02	4.06	5.11	6.14	7.17	8.20	9.22	10.2	11.2
	6	0.86	1.80	2.79	3.83	4.87	5.92	6.96	7.99	9.02	10.0	11.1
	7	0.77	1.64	2.59	3.60	4.64	5.68	6.73	7.77	8.80	9.83	10.9
	8	0.69	1.50	2.40	3.39	4.41	5.45	6.49	7.53	8.57	9.61	10.6
	9	0.63	1.37	2.23	3.19	4.19	5.22	6.26	7.30	8.34	9.38	10.4
	10	0.57	1.27	2.08	3.00	3.98	4.99	6.02	7.06	8.11	9.15	10.2
6	12	0.49	1.09	1.82	2.68	3.60	4.57	5.58	6.60	7.64	8.68	9.72
	14	0.43	0.95	1.61	2.40	3.27	4.20	5.17	6.17	7.18	8.21	9.25
	16	0.38	0.83	1.44	2.17	2.98	3.86	4.79	5.76	6.75	7.77	8.79
	18	0.34	0.75	1.30	1.98	2.74	3.57	4.46	5.39	6.35	7.34	8.35
	20	0.31	0.67	1.19	1.82	2.52	3.31	4.15	5.05	5.98	6.95	7.93
	24	0.26	0.56	1.01	1.55	2.17	2.87	3.64	4.46	5.33	6.24	7.17
	28	0.23	0.48	0.87	1.35	1.90	2.53	3.23	3.98	4.78	5.63	6.51
	32	0.20	0.42	0.77	1.20	1.69	2.26	2.89	3.58	4.33	5.11	5.94
	36	0.18	0.38	0.69	1.07	1.52	2.04	2.62	3.25	3.94	4.67	5.45

Table 8-18 (cont.).
Coefficients *C* for Eccentrically Loaded Bolt Groups
Angle = 45°

$$C_{req} = \frac{P_u}{\phi r_n} \text{ or } \phi R_n = C \times \phi r_n$$

where

P_u	= factored force, kips
ϕr_n	= design strength per bolt, kips
ϕR_n	= design strength of bolt group, kips
e	= eccentricity of P_u with respect to centroid of bolt group, in. (not tabulated, may be determined by geometry.)
e_x	= horizontal component of e, in.
s	= bolt spacing, in.
C	= coefficient tabulated below.

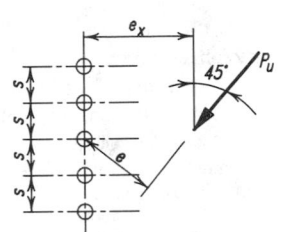

		Number of bolts in one vertical row, *n*										
s, in.	e_x, in.	2	3	4	5	6	7	8	9	10	11	12
	2	1.17	2.23	3.26	4.28	5.29	6.30	7.31	8.32	9.32	10.3	11.3
	3	0.92	1.89	2.87	3.87	4.88	5.90	6.91	7.93	8.94	9.95	11.0
	4	0.75	1.63	2.54	3.50	4.49	5.49	6.51	7.52	8.53	9.55	10.6
	5	0.64	1.42	2.25	3.17	4.13	5.11	6.11	7.11	8.12	9.14	10.2
	6	0.55	1.25	2.01	2.88	3.80	4.76	5.73	6.73	7.73	8.73	9.74
	7	0.49	1.11	1.81	2.63	3.51	4.43	5.38	6.36	7.34	8.34	9.34
	8	0.44	0.99	1.64	2.41	3.25	4.14	5.06	6.01	6.98	7.96	8.96
	9	0.40	0.90	1.49	2.22	3.02	3.87	4.77	5.69	6.64	7.61	8.58
	10	0.36	0.81	1.37	2.06	2.82	3.63	4.50	5.39	6.32	7.27	8.23
3	12	0.31	0.68	1.17	1.79	2.47	3.22	4.02	4.87	5.74	6.65	7.58
	14	0.27	0.59	1.03	1.58	2.20	2.88	3.62	4.41	5.24	6.11	6.99
	16	0.24	0.52	0.91	1.41	1.97	2.60	3.29	4.03	4.81	5.63	6.48
	18	0.21	0.46	0.82	1.27	1.78	2.36	3.00	3.70	4.43	5.21	6.02
	20	0.19	0.41	0.74	1.16	1.62	2.16	2.76	3.41	4.10	4.84	5.61
	24	0.16	0.35	0.63	0.98	1.38	1.85	2.37	2.94	3.56	4.22	4.92
	28	0.14	0.30	0.54	0.85	1.19	1.61	2.08	2.58	3.14	3.73	4.37
	32	0.12	0.26	0.48	0.75	1.05	1.43	1.84	2.30	2.80	3.34	3.92
	36	0.11	0.23	0.43	0.67	0.94	1.28	1.65	2.07	2.53	3.02	3.55
	2	1.57	2.64	3.66	4.67	5.67	6.66	7.66	8.65	9.64	10.6	11.6
	3	1.35	2.43	3.46	4.48	5.49	6.49	7.50	8.49	9.49	10.5	11.5
	4	1.17	2.23	3.26	4.28	5.29	6.30	7.31	8.32	9.32	10.3	11.3
	5	1.03	2.05	3.06	4.07	5.09	6.10	7.12	8.13	9.13	10.1	11.1
	6	0.92	1.89	2.87	3.87	4.88	5.90	6.91	7.93	8.94	9.95	11.0
	7	0.83	1.75	2.70	3.68	4.68	5.69	6.71	7.72	8.74	9.75	10.8
	8	0.75	1.63	2.54	3.50	4.49	5.49	6.51	7.52	8.53	9.55	10.6
	9	0.69	1.52	2.39	3.33	4.30	5.30	6.30	7.31	8.33	9.34	10.4
	10	0.64	1.42	2.25	3.17	4.13	5.11	6.11	7.11	8.12	9.14	10.2
6	12	0.55	1.25	2.01	2.88	3.80	4.76	5.73	6.73	7.73	8.73	9.74
	14	0.49	1.11	1.81	2.63	3.51	4.43	5.38	6.36	7.34	8.34	9.34
	16	0.44	0.99	1.64	2.41	3.25	4.14	5.06	6.01	6.98	7.96	8.96
	18	0.40	0.90	1.49	2.22	3.02	3.87	4.77	5.69	6.64	7.61	8.58
	20	0.36	0.81	1.37	2.06	2.82	3.63	4.50	5.39	6.32	7.27	8.23
	24	0.31	0.68	1.17	1.79	2.47	3.22	4.02	4.87	5.74	6.65	7.58
	28	0.27	0.59	1.03	1.58	2.20	2.88	3.62	4.41	5.24	6.11	6.99
	32	0.24	0.52	0.91	1.41	1.97	2.60	3.29	4.03	4.81	5.63	6.48
	36	0.21	0.46	0.82	1.27	1.78	2.36	3.00	3.70	4.43	5.21	6.02

Table 8-18 (cont.).
Coefficients C for Eccentrically Loaded Bolt Groups
Angle = 60°

$$C_{req} = \frac{P_u}{\phi r_n} \text{ or } \phi R_n = C \times \phi r_n$$

where

P_u = factored force, kips

ϕr_n = design strength per bolt, kips

ϕR_n = design strength of bolt group, kips

e = eccentricity of P_u with respect to centroid of bolt group, in. (not tabulated, may be determined by geometry.)

e_x = horizontal component of e, in.

s = bolt spacing, in.

C = coefficient tabulated below.

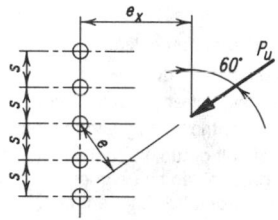

s, in.	e_x, in.	\multicolumn{11}{c}{Number of bolts in one vertical row, n}										
		2	3	4	5	6	7	8	9	10	11	12
3	2	1.27	2.32	3.32	4.31	5.30	6.30	7.29	8.27	9.27	10.3	11.3
	3	1.05	2.05	3.02	4.00	4.98	5.97	6.96	7.94	8.94	9.93	10.9
	4	0.89	1.83	2.77	3.72	4.69	5.66	6.64	7.62	8.61	9.60	10.6
	5	0.77	1.65	2.54	3.47	4.41	5.37	6.34	7.32	8.29	9.28	10.3
	6	0.68	1.49	2.34	3.24	4.16	5.10	6.06	7.02	7.99	8.97	9.95
	7	0.61	1.37	2.17	3.03	3.93	4.85	5.79	6.74	7.71	8.67	9.64
	8	0.56	1.26	2.01	2.83	3.71	4.61	5.54	6.48	7.43	8.39	9.35
	9	0.51	1.16	1.87	2.66	3.51	4.39	5.30	6.23	7.17	8.12	9.07
	10	0.47	1.07	1.74	2.50	3.32	4.19	5.08	5.99	6.92	7.86	8.81
	12	0.40	0.93	1.52	2.22	3.00	3.82	4.67	5.55	6.45	7.37	8.30
	14	0.35	0.81	1.35	2.00	2.73	3.50	4.32	5.16	6.03	6.92	7.83
	16	0.32	0.72	1.21	1.81	2.49	3.23	4.00	4.81	5.65	6.51	7.40
	18	0.29	0.65	1.09	1.66	2.30	2.98	3.72	4.50	5.31	6.14	7.00
	20	0.26	0.58	1.00	1.53	2.12	2.77	3.47	4.21	4.99	5.80	6.63
	24	0.22	0.49	0.85	1.32	1.84	2.41	3.05	3.73	4.45	5.21	5.99
	28	0.19	0.42	0.74	1.15	1.61	2.13	2.71	3.34	4.00	4.70	5.44
	32	0.17	0.37	0.65	1.02	1.43	1.91	2.44	3.02	3.63	4.28	4.97
	36	0.15	0.33	0.59	0.92	1.29	1.72	2.21	2.74	3.31	3.92	4.57
6	2	1.60	2.65	3.65	4.64	5.64	6.63	7.62	8.61	9.60	10.6	11.6
	3	1.42	2.48	3.48	4.48	5.47	6.46	7.45	8.44	9.44	10.4	11.4
	4	1.27	2.32	3.32	4.31	5.30	6.30	7.29	8.27	9.27	10.3	11.3
	5	1.15	2.18	3.17	4.15	5.14	6.13	7.12	8.11	9.10	10.1	11.1
	6	1.05	2.05	3.02	4.00	4.98	5.97	6.96	7.94	8.94	9.93	10.9
	7	0.96	1.93	2.89	3.86	4.83	5.81	6.80	7.78	8.77	9.76	10.8
	8	0.89	1.83	2.77	3.72	4.69	5.66	6.64	7.62	8.61	9.60	10.6
	9	0.83	1.73	2.65	3.59	4.55	5.51	6.49	7.47	8.45	9.43	10.4
	10	0.77	1.65	2.54	3.47	4.41	5.37	6.34	7.32	8.29	9.28	10.3
	12	0.68	1.49	2.34	3.24	4.16	5.10	6.06	7.02	7.99	8.97	9.95
	14	0.61	1.37	2.17	3.03	3.93	4.85	5.79	6.74	7.71	8.67	9.64
	16	0.56	1.26	2.01	2.83	3.71	4.61	5.54	6.48	7.43	8.39	9.35
	18	0.51	1.16	1.87	2.66	3.51	4.39	5.30	6.23	7.17	8.12	9.07
	20	0.47	1.07	1.74	2.50	3.32	4.19	5.08	5.99	6.92	7.86	8.81
	24	0.40	0.93	1.52	2.22	3.00	3.82	4.67	5.55	6.45	7.37	8.30
	28	0.35	0.81	1.35	2.00	2.73	3.50	4.32	5.16	6.03	6.92	7.83
	32	0.32	0.72	1.21	1.81	2.49	3.23	4.00	4.81	5.65	6.51	7.40
	36	0.29	0.65	1.09	1.66	2.30	2.98	3.72	4.50	5.31	6.14	7.00

Table 8-18 (cont.).
Coefficients C for Eccentrically Loaded Bolt Groups
Angle = 75°

$$C_{req} = \frac{P_u}{\phi r_n} \text{ or } \phi R_n = C \times \phi r_n$$

where

P_u = factored force, kips

ϕr_n = design strength per bolt, kips

ϕR_n = design strength of bolt group, kips

e = eccentricity of P_u with respect to centroid of bolt group, in. (not tabulated, may be determined by geometry.)

e_x = horizontal component of e, in.

s = bolt spacing, in.

C = coefficient tabulated below.

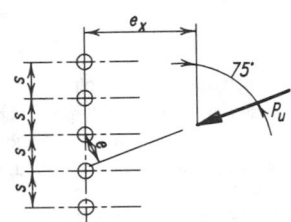

		Number of bolts in one vertical row, n										
s, in.	e_x, in.	2	3	4	5	6	7	8	9	10	11	12
	2	1.49	2.51	3.49	4.46	5.44	6.42	7.40	8.38	9.36	10.3	11.3
	3	1.32	2.33	3.30	4.27	5.24	6.21	7.18	8.15	9.13	10.1	11.1
	4	1.18	2.18	3.14	4.09	5.05	6.01	6.98	7.95	8.92	9.89	10.9
	5	1.07	2.04	2.99	3.93	4.88	5.84	6.79	7.75	8.72	9.68	10.7
	6	0.98	1.92	2.85	3.79	4.73	5.67	6.62	7.57	8.53	9.49	10.5
	7	0.90	1.82	2.73	3.65	4.58	5.52	6.46	7.40	8.36	9.31	10.3
	8	0.84	1.72	2.62	3.52	4.44	5.37	6.30	7.24	8.19	9.14	10.1
	9	0.78	1.63	2.51	3.40	4.31	5.23	6.16	7.09	8.03	8.97	9.92
	10	0.73	1.55	2.41	3.29	4.19	5.10	6.02	6.94	7.88	8.81	9.76
3	12	0.65	1.41	2.23	3.08	3.95	4.84	5.75	6.66	7.59	8.51	9.45
	14	0.58	1.30	2.06	2.88	3.73	4.60	5.50	6.40	7.31	8.23	9.16
	16	0.53	1.20	1.92	2.70	3.52	4.38	5.26	6.15	7.05	7.96	8.88
	18	0.48	1.11	1.78	2.53	3.33	4.17	5.03	5.91	6.80	7.70	8.61
	20	0.44	1.03	1.66	2.38	3.16	3.97	4.82	5.69	6.56	7.45	8.35
	24	0.38	0.89	1.46	2.12	2.85	3.63	4.44	5.27	6.13	6.99	7.87
	28	0.34	0.79	1.29	1.90	2.59	3.33	4.11	4.91	5.73	6.57	7.43
	32	0.30	0.70	1.16	1.73	2.38	3.08	3.81	4.58	5.37	6.19	7.02
	36	0.27	0.62	1.05	1.58	2.19	2.85	3.55	4.28	5.05	5.84	6.65
	2	1.71	2.72	3.70	4.69	5.67	6.66	7.64	8.79	9.78	10.8	11.7
	3	1.60	2.61	3.59	4.57	5.55	6.53	7.52	8.50	9.48	10.5	11.5
	4	1.49	2.51	3.49	4.46	5.44	6.42	7.40	8.38	9.36	10.3	11.3
	5	1.40	2.42	3.39	4.37	5.34	6.31	7.29	8.26	9.24	10.2	11.2
	6	1.32	2.33	3.30	4.27	5.24	6.21	7.18	8.15	9.13	10.1	11.1
	7	1.25	2.25	3.22	4.18	5.14	6.11	7.07	8.05	9.01	10.0	11.0
	8	1.18	2.18	3.14	4.09	5.05	6.01	6.98	7.95	8.92	9.89	10.9
	9	1.13	2.11	3.06	4.01	4.97	5.92	6.88	7.85	8.81	9.78	10.8
	10	1.07	2.04	2.99	3.93	4.88	5.84	6.79	7.75	8.72	9.68	10.7
6	12	0.98	1.92	2.85	3.79	4.73	5.67	6.62	7.57	8.53	9.49	10.5
	14	0.90	1.82	2.73	3.65	4.58	5.52	6.46	7.40	8.36	9.31	10.3
	16	0.84	1.72	2.62	3.52	4.44	5.37	6.30	7.24	8.19	9.14	10.1
	18	0.78	1.63	2.51	3.40	4.31	5.23	6.16	7.09	8.03	8.97	9.92
	20	0.73	1.55	2.41	3.29	4.19	5.10	6.02	6.94	7.88	8.81	9.76
	24	0.65	1.41	2.23	3.08	3.95	4.84	5.75	6.66	7.59	8.51	9.45
	28	0.58	1.30	2.06	2.88	3.73	4.60	5.50	6.40	7.31	8.23	9.16
	32	0.53	1.20	1.92	2.70	3.52	4.38	5.26	6.15	7.05	7.96	8.88
	36	0.48	1.11	1.78	2.53	3.33	4.17	5.03	5.91	6.80	7.70	8.61

Table 8-19.
Coefficients C for Eccentrically Loaded Bolt Groups
Angle = 0°

$$C_{req} = \frac{P_u}{\phi r_n} \text{ or } \phi R_n = C \times \phi r_n$$

where

P_u = factored force, kips

ϕr_n = design strength per bolt, kips

ϕR_n = design strength of bolt group, kips

e = eccentricity of P_u with respect to centroid of bolt group, in. (not tabulated, may be determined by geometry.)

e_x = horizontal component of e, in.

s = bolt spacing, in.

C = coefficient tabulated below.

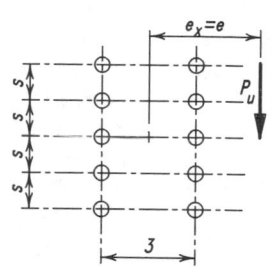

s, in.	e_x, in.	Number of bolts in one vertical row, n											
		1	2	3	4	5	6	7	8	9	10	11	12
	2	0.84	2.54	4.48	6.59	8.72	10.8	12.9	15.0	17.0	19.0	21.0	23.0
	3	0.65	2.03	3.68	5.67	7.77	9.91	12.1	14.2	16.3	18.3	20.4	22.5
	4	0.54	1.67	3.06	4.86	6.84	8.93	11.1	13.2	15.4	17.5	19.6	21.7
	5	0.45	1.42	2.59	4.21	6.01	8.00	10.1	12.2	14.4	16.5	18.7	20.8
	6	0.39	1.22	2.25	3.69	5.32	7.17	9.16	11.2	13.4	15.5	17.7	19.8
	7	0.35	1.08	1.99	3.27	4.74	6.46	8.33	10.3	12.4	14.5	16.7	18.8
	8	0.31	0.96	1.78	2.93	4.27	5.86	7.60	9.50	11.5	13.6	15.7	17.8
	9	0.28	0.86	1.60	2.65	3.87	5.34	6.97	8.75	10.7	12.7	14.7	16.8
	10	0.26	0.78	1.46	2.42	3.53	4.90	6.42	8.10	9.91	11.8	13.8	15.9
3	12	0.22	0.66	1.24	2.06	3.01	4.19	5.51	7.01	8.63	10.4	12.2	14.2
	14	0.19	0.57	1.08	1.78	2.62	3.66	4.82	6.15	7.61	9.19	10.9	12.7
	16	0.17	0.51	0.95	1.57	2.32	3.24	4.27	5.47	6.79	8.23	9.78	11.4
	18	0.15	0.45	0.85	1.41	2.07	2.90	3.83	4.92	6.11	7.43	8.85	10.4
	20	0.14	0.41	0.77	1.27	1.88	2.63	3.48	4.47	5.55	6.76	8.07	9.48
	24	0.12	0.34	0.65	1.07	1.58	2.21	2.93	3.77	4.69	5.72	6.85	8.06
	28	0.10	0.29	0.56	0.92	1.36	1.90	2.53	3.25	4.05	4.95	5.93	7.00
	32	0.09	0.26	0.49	0.80	1.19	1.67	2.22	2.86	3.57	4.36	5.23	6.18
	36	0.08	0.23	0.43	0.72	1.06	1.49	1.98	2.55	3.18	3.90	4.67	5.52
	2	0.84	3.24	5.39	7.47	9.51	11.5	13.5	15.5	17.5	19.5	21.5	23.4
	3	0.65	2.79	4.93	7.08	9.17	11.2	13.3	15.3	17.3	19.3	21.3	23.3
	4	0.54	2.41	4.44	6.60	8.75	10.9	12.4	15.0	17.0	19.1	21.1	23.1
	5	0.45	2.10	3.97	6.11	8.27	10.4	12.5	14.6	16.7	18.7	20.8	22.8
	6	0.39	1.85	3.55	5.62	7.77	9.93	12.1	14.2	16.3	18.4	20.4	22.5
	7	0.35	1.64	3.18	5.17	7.27	9.43	11.6	13.7	15.9	18.0	20.1	22.1
	8	0.31	1.47	2.87	4.75	6.79	8.92	11.1	13.3	15.4	17.5	19.6	21.7
	9	0.28	1.34	2.61	4.39	6.34	8.43	10.6	12.7	14.9	17.1	19.2	21.3
	10	0.26	1.22	2.39	4.06	5.92	7.96	10.1	12.2	14.4	16.6	18.7	20.9
6	12	0.22	1.04	2.04	3.52	5.20	7.10	9.12	11.2	13.4	15.6	17.7	19.9
	14	0.19	0.90	1.77	3.09	4.61	6.36	8.27	10.3	12.4	14.5	16.7	18.9
	16	0.17	0.80	1.57	2.75	4.12	5.74	7.52	9.44	11.7	13.5	15.7	17.8
	18	0.15	0.71	1.41	2.48	3.72	5.21	6.87	8.68	10.6	12.6	14.7	16.8
	20	0.14	0.64	1.28	2.25	3.38	4.77	6.31	8.02	9.85	11.8	13.8	15.9
	24	0.12	0.54	1.07	1.90	2.86	4.06	5.40	6.91	8.55	10.3	12.2	14.1
	28	0.10	0.46	0.93	1.64	2.47	3.52	4.70	6.05	7.52	9.12	10.8	12.6
	32	0.09	0.41	0.81	1.44	2.18	3.11	4.16	5.37	6.69	8.15	9.71	11.4
	36	0.08	0.36	0.73	1.29	1.94	2.78	3.72	4.81	6.02	7.34	8.78	10.3

Table 8-19 (cont.).
Coefficients C for Eccentrically Loaded Bolt Groups
Angle = 15°

$$C_{req} = \frac{P_u}{\phi r_n} \text{ or } \phi R_n = C \times \phi r_n$$

where

P_u	= factored force, kips
ϕr_n	= design strength per bolt, kips
ϕR_n	= design strength of bolt group, kips
e	= eccentricity of P_u with respect to centroid of bolt group, in. (not tabulated, may be determined by geometry.)
e_x	= horizontal component of e, in.
s	= bolt spacing, in.
C	= coefficient tabulated below.

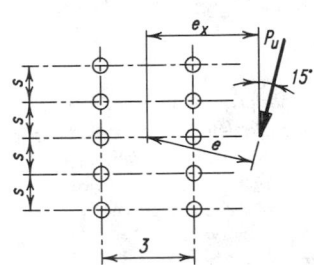

s, in.	e_x, in.	Number of bolts in one vertical row, n											
		1	2	3	4	5	6	7	8	9	10	11	12
3	2	0.87	2.54	4.47	6.54	8.63	10.7	12.8	14.8	16.9	18.9	20.9	22.9
	3	0.68	2.04	3.71	5.63	7.69	9.80	11.9	14.0	16.1	18.2	20.2	22.3
	4	0.55	1.69	3.11	4.85	6.79	8.84	10.9	13.0	15.2	17.3	19.4	21.5
	5	0.47	1.44	2.66	4.21	6.00	7.94	9.98	12.1	14.2	16.3	18.4	20.5
	6	0.41	1.25	2.31	3.70	5.34	7.15	9.09	11.1	13.2	15.3	17.4	19.6
	7	0.36	1.10	2.04	3.29	4.79	6.46	8.30	10.2	12.3	14.3	16.4	18.6
	8	0.32	0.98	1.83	2.96	4.32	5.87	7.60	9.45	11.4	13.4	15.5	17.6
	9	0.29	0.88	1.65	2.68	3.94	5.37	6.99	8.74	10.6	12.6	14.6	16.6
	10	0.27	0.81	1.51	2.45	3.61	4.93	6.45	8.11	9.88	11.8	13.7	15.7
	12	0.23	0.68	1.28	2.09	3.08	4.24	5.58	7.05	8.66	10.4	12.2	14.1
	14	0.20	0.59	1.11	1.82	2.69	3.71	4.90	6.21	7.67	9.23	10.9	12.7
	16	0.17	0.52	0.98	1.61	2.38	3.29	4.36	5.54	6.86	8.29	9.83	11.5
	18	0.16	0.47	0.88	1.44	2.13	2.96	3.92	4.99	6.20	7.51	8.93	10.4
	20	0.14	0.42	0.79	1.31	1.93	2.68	3.56	4.54	5.65	6.85	8.17	9.57
	24	0.12	0.35	0.67	1.10	1.62	2.26	3.00	3.84	4.79	5.82	6.96	8.17
	28	0.10	0.30	0.57	0.94	1.40	1.95	2.60	3.32	4.15	5.05	6.05	7.12
	32	0.09	0.27	0.50	0.83	1.23	1.72	2.28	2.93	3.66	4.46	5.34	6.29
	36	0.08	0.24	0.45	0.74	1.10	1.53	2.04	2.61	3.27	3.98	4.78	5.64
6	2	0.87	3.21	5.35	7.42	9.45	11.5	13.5	15.5	17.4	19.4	21.4	23.4
	3	0.68	2.76	4.88	7.00	9.09	11.1	13.2	15.2	17.2	19.2	21.2	23.2
	4	0.55	2.38	4.40	6.53	8.65	10.7	12.8	14.9	16.9	18.9	20.9	22.9
	5	0.47	2.07	3.96	6.04	8.17	10.3	12.4	14.5	16.5	18.6	20.6	22.6
	6	0.41	1.83	3.56	5.56	7.67	9.80	11.9	14.0	16.1	18.2	20.3	22.3
	7	0.36	1.63	3.22	5.12	7.19	9.30	11.4	13.6	15.7	17.8	19.9	21.9
	8	0.32	1.47	2.92	4.73	6.72	8.81	10.9	13.1	15.2	17.3	19.4	21.5
	9	0.29	1.34	2.66	4.37	6.29	8.33	10.4	12.6	14.7	16.8	18.9	21.0
	10	0.27	1.23	2.45	4.05	5.90	7.88	9.95	12.1	14.2	16.3	18.5	20.6
	12	0.23	1.05	2.09	3.53	5.21	7.06	9.04	11.1	13.2	15.3	17.5	19.6
	14	0.20	0.91	1.83	3.11	4.64	6.35	8.22	10.2	12.2	14.3	16.5	18.6
	16	0.17	0.81	1.62	2.78	4.17	5.75	7.51	9.38	11.4	13.4	15.5	17.6
	18	0.16	0.72	1.45	2.50	3.77	5.24	6.88	8.66	10.5	12.5	14.5	16.6
	20	0.14	0.66	1.32	2.28	3.45	4.80	6.34	8.02	9.82	11.7	13.7	15.7
	24	0.12	0.55	1.11	1.93	2.93	4.10	5.46	6.95	8.57	10.3	12.1	14.0
	28	0.10	0.48	0.96	1.67	2.54	3.57	4.78	6.11	7.58	9.15	10.8	12.6
	32	0.09	0.42	0.84	1.47	2.24	3.16	4.24	5.44	6.77	8.21	9.75	11.4
	36	0.08	0.37	0.75	1.32	2.00	2.83	3.80	4.89	6.10	7.42	8.85	10.4

Table 8-19 (cont.).
Coefficients *C* for Eccentrically Loaded Bolt Groups
Angle = 30°

$$C_{req} = \frac{P_u}{\phi r_n} \text{ or } \phi R_n = C \times \phi r_n$$

where

P_u	= factored force, kips	
ϕr_n	= design strength per bolt, kips	
ϕR_n	= design strength of bolt group, kips	
e	= eccentricity of P_u with respect to centroid of bolt group, in. (not tabulated, may be determined by geometry.)	
e_x	= horizontal component of e, in.	
s	= bolt spacing, in.	
C	= coefficient tabulated below.	

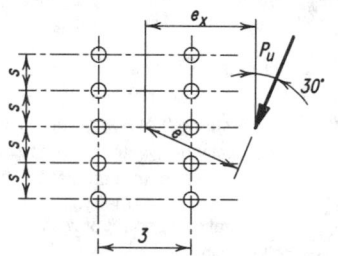

s, in.	e_x, in.	\multicolumn{12}{c}{Number of bolts in one vertical row, n}
		1	2	3	4	5	6	7	8	9	10	11	12
	2	0.97	2.60	4.52	6.54	8.59	10.6	12.9	14.7	16.7	18.8	20.8	22.8
	3	0.75	2.12	3.83	5.71	7.71	9.75	11.8	13.9	15.9	18.0	20.0	22.1
	4	0.62	1.78	3.29	4.99	6.88	8.87	10.9	13.0	15.1	17.1	19.2	21.3
	5	0.52	1.53	2.85	4.39	6.16	8.06	10.0	12.1	14.1	16.2	18.3	20.4
	6	0.45	1.34	2.51	3.89	5.54	7.33	9.23	11.2	13.2	15.3	17.3	19.4
	7	0.40	1.19	2.23	3.48	5.01	6.70	8.51	10.4	12.4	14.4	16.4	18.5
	8	0.36	1.07	2.00	3.15	4.57	6.14	7.86	9.68	11.6	13.6	15.6	17.6
	9	0.32	0.97	1.81	2.87	4.19	5.66	7.28	9.02	10.9	12.8	14.7	16.7
	10	0.30	0.88	1.66	2.64	3.87	5.24	6.77	8.43	10.2	12.0	13.9	15.9
3	12	0.25	0.75	1.41	2.27	3.34	4.54	5.92	7.43	9.04	10.8	12.5	14.4
	14	0.22	0.65	1.23	1.98	2.93	3.99	5.24	6.61	8.09	9.67	11.4	13.1
	16	0.19	0.58	1.08	1.76	2.60	3.56	4.69	5.94	7.30	8.77	10.3	12.0
	18	0.17	0.52	0.97	1.58	2.34	3.21	4.24	5.38	6.64	8.00	9.45	11.0
	20	0.16	0.47	0.88	1.43	2.12	2.92	3.87	4.92	6.08	7.34	8.70	10.1
	24	0.13	0.39	0.74	1.21	1.79	2.48	3.29	4.18	5.19	6.29	7.48	8.75
	28	0.12	0.34	0.64	1.04	1.55	2.14	2.85	3.63	4.52	5.49	6.54	7.68
	32	0.10	0.30	0.56	0.92	1.36	1.89	2.51	3.21	4.00	4.87	5.81	6.83
	36	0.09	0.26	0.50	0.82	1.21	1.69	2.25	2.87	3.59	4.37	5.22	6.15
	2	0.97	3.20	5.31	7.37	9.39	11.4	13.4	15.4	17.4	19.4	21.3	23.3
	3	0.75	2.75	4.86	6.95	9.01	11.1	13.1	15.1	17.1	19.1	21.1	23.1
	4	0.62	2.39	4.42	6.49	8.57	10.6	12.7	14.7	16.8	18.8	20.8	22.8
	5	0.52	2.10	4.02	6.04	8.11	10.2	12.3	14.3	16.4	18.4	20.4	22.5
	6	0.45	1.87	3.67	5.61	7.66	9.73	11.8	13.9	16.0	18.0	20.1	22.1
	7	0.40	1.69	3.36	5.21	7.21	9.27	11.4	13.4	15.5	17.6	19.6	21.7
	8	0.36	1.53	3.08	4.84	6.79	8.82	10.9	13.0	15.1	17.1	19.2	21.3
	9	0.32	1.40	2.84	4.51	6.40	8.39	10.4	12.5	14.6	16.7	18.7	20.8
	10	0.30	1.29	2.63	4.21	6.04	7.98	9.99	12.0	14.1	16.2	18.3	20.4
6	12	0.25	1.12	2.28	3.70	5.39	7.23	9.16	11.2	13.2	15.3	17.3	19.4
	14	0.22	0.98	2.00	3.29	4.86	6.57	8.41	10.3	12.3	14.4	16.4	18.5
	16	0.19	0.87	1.78	2.95	4.40	6.01	7.75	9.60	11.5	13.5	15.5	17.6
	18	0.17	0.79	1.60	2.68	4.02	5.52	7.17	8.93	10.8	12.7	14.7	16.7
	20	0.16	0.71	1.45	2.45	3.70	5.09	6.65	8.33	10.1	12.0	13.9	15.9
	24	0.13	0.60	1.23	2.08	3.17	4.39	5.79	7.32	8.95	10.7	12.5	14.4
	28	0.12	0.52	1.06	1.82	2.77	3.85	5.11	6.49	7.99	9.59	11.3	13.0
	32	0.10	0.46	0.93	1.61	2.45	3.42	4.56	5.82	7.20	8.68	10.3	11.9
	36	0.09	0.41	0.83	1.44	2.20	3.08	4.12	5.27	6.53	7.91	9.37	10.9

Table 8-19 (cont.).
Coefficients C for Eccentrically Loaded Bolt Groups
Angle = 45°

$$C_{req} = \frac{P_u}{\phi r_n} \text{ or } \phi R_n = C \times \phi r_n$$

where

P_u = factored force, kips

ϕr_n = design strength per bolt, kips

ϕR_n = design strength of bolt group, kips

e = eccentricity of P_u with respect to centroid of bolt group, in. (not tabulated, may be determined by geometry.)

e_x = horizontal component of e, in.

s = bolt spacing, in.

C = coefficient tabulated below.

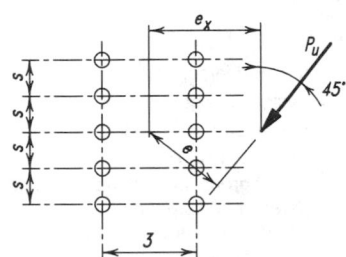

s, in.	e_x, in.	Number of bolts in one vertical row, n											
		1	2	3	4	5	6	7	8	9	10	11	12
	2	1.17	2.79	4.67	6.62	8.61	10.6	12.6	14.6	16.6	18.6	20.6	22.6
	3	0.92	2.32	4.06	5.92	7.86	9.83	11.8	13.9	15.9	17.9	19.9	21.9
	4	0.75	1.99	3.57	5.31	7.16	9.09	11.1	13.1	15.1	17.1	19.1	21.1
	5	0.64	1.74	3.17	4.78	6.53	8.39	10.3	12.3	14.3	16.3	18.3	20.3
	6	0.55	1.54	2.84	4.33	5.98	7.76	9.63	11.6	13.5	15.5	17.5	19.5
	7	0.49	1.38	2.57	3.93	5.49	7.20	9.00	10.9	12.8	14.8	16.8	18.7
	8	0.44	1.25	2.33	3.60	5.06	6.70	8.43	10.3	12.1	14.0	16.0	18.0
	9	0.40	1.14	2.13	3.31	4.69	6.25	7.91	9.67	11.5	13.4	15.3	17.2
	10	0.36	1.05	1.96	3.06	4.36	5.85	7.44	9.14	10.9	12.7	14.6	16.5
3	12	0.31	0.90	1.68	2.65	3.83	5.17	6.63	8.20	9.86	11.6	13.4	15.2
	14	0.27	0.78	1.47	2.33	3.40	4.61	5.95	7.41	8.97	10.6	12.3	14.1
	16	0.24	0.69	1.31	2.08	3.05	4.16	5.38	6.74	8.20	9.75	11.4	13.1
	18	0.21	0.62	1.17	1.88	2.76	3.77	4.91	6.18	7.55	9.00	10.5	12.1
	20	0.19	0.56	1.06	1.71	2.52	3.45	4.51	5.69	6.97	8.34	9.80	11.3
	24	0.16	0.48	0.90	1.45	2.14	2.94	3.87	4.91	6.04	7.26	8.57	9.95
	28	0.14	0.41	0.77	1.26	1.86	2.56	3.38	4.30	5.30	6.41	7.59	8.85
	32	0.12	0.36	0.68	1.11	1.64	2.27	3.00	3.82	4.73	5.73	6.80	7.94
	36	0.11	0.32	0.61	0.99	1.47	2.03	2.70	3.44	4.26	5.17	6.15	7.20
	2	1.17	3.24	5.30	7.32	9.33	11.3	13.3	15.3	17.3	19.3	21.3	23.2
	3	0.92	2.84	4.90	6.93	8.96	11.0	13.0	15.0	17.0	19.0	21.0	23.0
	4	0.75	2.51	4.52	6.53	8.56	10.6	12.6	14.6	16.6	18.6	20.6	22.6
	5	0.64	2.24	4.17	6.15	8.15	10.2	12.2	14.2	16.2	18.3	20.3	22.3
	6	0.55	2.03	3.86	5.78	7.76	9.77	11.8	13.8	15.8	17.9	19.9	21.9
	7	0.49	1.85	3.59	5.45	7.39	9.38	11.4	13.4	15.4	17.5	19.5	21.5
	8	0.44	1.70	3.35	5.13	7.03	9.00	11.0	13.0	15.0	17.1	19.1	21.1
	9	0.40	1.57	3.13	4.85	6.70	8.63	10.6	12.6	14.6	16.7	18.7	20.7
	10	0.36	1.46	2.94	4.58	6.38	8.28	10.2	12.2	14.2	16.3	18.3	20.3
6	12	0.31	1.28	2.60	4.11	5.81	7.64	9.54	11.5	13.5	15.6	17.5	19.5
	14	0.27	1.13	2.32	3.71	5.31	7.06	8.89	10.8	12.7	14.7	16.7	18.7
	16	0.24	1.01	2.09	3.36	4.88	6.55	8.31	10.2	12.0	14.0	15.9	17.9
	18	0.21	0.92	1.90	3.07	4.50	6.09	7.78	9.56	11.4	13.3	15.2	17.2
	20	0.19	0.84	1.73	2.83	4.18	5.69	7.31	9.02	10.8	12.7	14.6	16.5
	24	0.16	0.72	1.47	2.43	3.64	5.00	6.48	8.08	9.76	11.5	13.3	15.2
	28	0.14	0.62	1.28	2.13	3.22	4.45	5.80	7.28	8.86	10.5	12.2	14.0
	32	0.12	0.55	1.13	1.90	2.88	3.99	5.24	6.62	8.09	9.65	11.3	13.0
	36	0.11	0.49	1.01	1.71	2.61	3.62	4.77	6.05	7.43	8.90	10.4	12.0

Table 8-19 (cont.).
Coefficients *C* for Eccentrically Loaded Bolt Groups
Angle = 60°

$$C_{req} = \frac{P_u}{\phi r_n} \text{ or } \phi R_n = C \times \phi r_n$$

where

P_u　= factored force, kips

ϕr_n　= design strength per bolt, kips

ϕR_n　= design strength of bolt group, kips

e　= eccentricity of P_u with respect to centroid of bolt group, in. (not tabulated, may be determined by geometry.)

e_x　= horizontal component of e, in.

s　= bolt spacing, in.

C　= coefficient tabulated below.

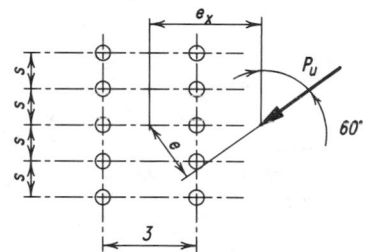

s, in.	e_x, in.	Number of bolts in one vertical row, n											
		1	2	3	4	5	6	7	8	9	10	11	12
	2	1.51	3.17	4.97	6.85	8.77	10.7	12.7	14.6	16.6	18.6	20.6	22.5
	3	1.24	2.76	4.47	6.30	8.19	10.1	12.0	14.0	16.0	17.9	19.9	21.9
	4	1.04	2.43	4.04	5.81	7.65	9.53	11.5	13.4	15.3	17.3	19.3	21.2
	5	0.89	2.16	3.70	5.39	7.17	9.01	10.9	12.8	14.7	16.7	18.6	20.6
	6	0.77	1.95	3.40	5.01	6.73	8.52	10.4	12.3	14.2	16.1	18.0	20.0
	7	0.68	1.77	3.13	4.67	6.33	8.07	9.88	11.7	13.6	15.5	17.4	19.4
	8	0.61	1.62	2.90	4.37	5.96	7.65	9.42	11.2	13.1	15.0	16.9	18.8
	9	0.56	1.49	2.70	4.09	5.62	7.26	8.98	10.8	12.6	14.5	16.3	18.2
	10	0.51	1.38	2.52	3.84	5.31	6.89	8.58	10.3	12.1	14.0	15.8	17.7
3	12	0.43	1.20	2.21	3.40	4.76	6.25	7.85	9.53	11.3	13.0	14.9	16.7
	14	0.38	1.06	1.96	3.05	4.30	5.71	7.23	8.83	10.5	12.2	14.0	15.8
	16	0.34	0.95	1.76	2.75	3.92	5.24	6.68	8.20	9.79	11.5	13.2	14.9
	18	0.30	0.85	1.60	2.51	3.59	4.84	6.19	7.64	9.16	10.8	12.4	14.1
	20	0.27	0.78	1.46	2.30	3.32	4.48	5.76	7.14	8.60	10.1	11.7	13.4
	24	0.23	0.66	1.24	1.97	2.87	3.90	5.04	6.29	7.64	9.06	10.6	12.1
	28	0.20	0.57	1.07	1.72	2.52	3.44	4.47	5.61	6.85	8.17	9.55	11.0
	32	0.18	0.50	0.95	1.52	2.24	3.07	4.01	5.06	6.20	7.41	8.70	10.1
	36	0.16	0.45	0.85	1.37	2.02	2.77	3.63	4.59	5.65	6.77	7.98	9.26
	2	1.51	3.39	5.36	7.33	9.31	11.3	13.3	15.2	17.2	19.2	21.2	23.2
	3	1.24	3.08	5.04	7.01	8.98	11.0	12.9	14.9	16.9	18.9	20.9	22.8
	4	1.04	2.80	4.73	6.69	8.66	10.6	12.6	14.6	16.6	18.6	20.5	22.5
	5	0.89	2.57	4.45	6.39	8.35	10.3	12.3	14.3	16.2	18.2	20.2	22.2
	6	0.77	2.37	4.20	6.11	8.05	10.0	12.0	13.9	15.9	17.9	19.9	21.8
	7	0.68	2.19	3.98	5.85	7.76	9.70	11.7	13.6	15.6	17.6	19.5	21.5
	8	0.61	2.04	3.77	5.61	7.49	9.41	11.6	13.3	15.3	17.2	19.2	21.2
	9	0.56	1.91	3.59	5.38	7.24	9.13	11.1	13.0	15.0	16.9	18.9	20.9
	10	0.51	1.80	3.42	5.17	7.00	8.87	10.8	12.7	14.7	16.6	18.6	20.5
6	12	0.43	1.60	3.11	4.78	6.54	8.37	10.2	12.1	14.1	16.0	18.0	19.9
	14	0.38	1.44	2.85	4.43	6.13	7.91	9.74	11.6	13.5	15.4	17.4	19.3
	16	0.34	1.31	2.63	4.12	5.74	7.48	9.27	11.1	13.0	14.9	16.8	18.7
	18	0.30	1.20	2.43	3.84	5.40	7.08	8.84	10.7	12.5	14.4	16.3	18.2
	20	0.27	1.10	2.26	3.58	5.08	6.71	8.43	10.2	12.0	13.9	15.7	17.6
	24	0.23	0.95	1.97	3.15	4.53	6.06	7.69	9.39	11.2	12.9	14.8	16.6
	28	0.20	0.84	1.73	2.80	4.08	5.52	7.06	8.68	10.4	12.1	13.9	15.7
	32	0.18	0.74	1.54	2.52	3.71	5.05	6.51	8.05	9.66	11.3	13.1	14.8
	36	0.16	0.67	1.39	2.28	3.39	4.65	6.02	7.49	9.03	10.7	12.3	14.0

Table 8-19 (cont.).
Coefficients C for Eccentrically Loaded Bolt Groups
Angle = 75°

$$C_{req} = \frac{P_u}{\phi r_n} \text{ or } \phi R_n = C \times \phi r_n$$

where

P_u	= factored force, kips
ϕr_n	= design strength per bolt, kips
ϕR_n	= design strength of bolt group, kips
e	= eccentricity of P_u with respect to centroid of bolt group, in. (not tabulated, may be determined by geometry.)
e_x	= horizontal component of e, in.
s	= bolt spacing, in.
C	= coefficient tabulated below.

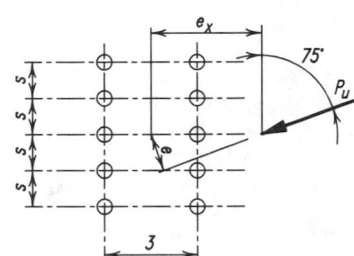

s, in.	e_x, in.	\multicolumn{12}{c}{Number of bolts in one vertical row, n}
		1	2	3	4	5	6	7	8	9	10	11	12
	2	1.84	3.63	5.44	7.29	9.17	11.1	13.0	14.9	16.9	18.8	20.8	22.7
	3	1.71	3.41	5.17	6.97	8.82	10.7	12.6	14.5	16.4	18.4	20.3	22.3
	4	1.57	3.19	4.90	6.67	8.50	10.4	12.2	14.1	16.0	18.0	19.9	21.8
	5	1.44	2.98	4.65	6.39	8.19	10.0	11.9	13.8	15.7	17.6	19.5	21.4
	6	1.31	2.79	4.41	6.12	7.90	9.71	11.6	13.4	15.3	17.2	19.1	21.0
	7	1.20	2.61	4.19	5.88	7.62	9.42	11.3	13.1	15.0	16.9	18.8	20.7
	8	1.10	2.45	3.99	5.65	7.37	9.14	11.0	12.8	14.7	16.5	18.4	20.3
	9	1.01	2.31	3.81	5.43	7.14	8.89	10.7	12.5	14.3	16.2	18.1	20.0
	10	0.93	2.18	3.63	5.23	6.91	8.65	10.4	12.2	14.1	15.9	17.8	19.6
3	12	0.81	1.95	3.33	4.86	6.49	8.19	9.94	11.7	13.5	15.3	17.2	19.0
	14	0.71	1.77	3.06	4.53	6.11	7.76	9.47	11.2	13.0	14.8	16.6	18.4
	16	0.63	1.61	2.83	4.23	5.75	7.36	9.03	10.8	12.5	14.3	16.1	17.9
	18	0.57	1.48	2.63	3.96	5.42	6.98	8.61	10.3	12.0	13.8	15.6	17.4
	20	0.52	1.36	2.45	3.72	5.12	6.63	8.23	9.88	11.6	13.3	15.1	16.9
	24	0.44	1.18	2.15	3.30	4.60	6.02	7.53	9.12	10.8	12.4	14.2	15.9
	28	0.38	1.04	1.91	2.95	4.16	5.49	6.93	8.45	10.0	11.7	13.3	15.0
	32	0.34	0.92	1.71	2.67	3.78	5.04	6.41	7.86	9.37	10.9	12.6	14.2
	36	0.30	0.83	1.55	2.43	3.47	4.65	5.94	7.32	8.78	10.3	11.9	13.5
	2	1.84	3.66	5.55	7.48	9.42	11.4	13.3	15.3	17.6	19.6	21.5	23.5
	3	1.71	3.49	5.36	7.27	9.20	11.2	13.1	15.1	17.0	19.0	21.0	22.9
	4	1.57	3.32	5.18	7.08	9.00	10.9	12.9	14.8	16.8	18.7	20.7	22.7
	5	1.44	3.16	5.01	6.89	8.81	10.7	12.7	14.6	16.6	18.5	20.5	22.4
	6	1.31	3.02	4.84	6.72	8.62	10.5	12.5	14.4	16.3	18.3	20.2	22.2
	7	1.20	2.88	4.69	6.55	8.44	10.4	12.3	14.2	16.1	18.1	20.0	22.0
	8	1.10	2.75	4.54	6.39	8.27	10.2	12.1	14.0	15.9	17.9	19.8	21.8
	9	1.01	2.63	4.40	6.24	8.11	10.0	11.9	13.8	15.7	17.7	19.6	21.5
	10	0.93	2.52	4.27	6.09	7.95	9.83	11.7	13.6	15.6	17.5	19.4	21.3
6	12	0.81	2.32	4.03	5.82	7.66	9.52	11.4	13.3	15.2	17.1	19.0	20.9
	14	0.71	2.15	3.82	5.57	7.38	9.22	11.1	13.0	14.9	16.7	18.7	20.6
	16	0.63	2.00	3.62	5.35	7.13	8.95	10.8	12.7	14.5	16.4	18.3	20.2
	18	0.57	1.87	3.44	5.14	6.90	8.69	10.5	12.4	14.2	16.1	18.0	19.9
	20	0.52	1.75	3.28	4.94	6.67	8.45	10.3	12.1	13.9	15.8	17.7	19.5
	24	0.44	1.55	2.98	4.57	6.24	7.98	9.75	11.6	13.4	15.2	17.1	18.9
	28	0.38	1.40	2.74	4.24	5.85	7.54	9.28	11.1	12.9	14.7	16.5	18.3
	32	0.34	1.27	2.52	3.95	5.49	7.13	8.83	10.6	12.4	14.1	16.0	17.8
	36	0.30	1.16	2.33	3.68	5.16	6.75	8.41	10.1	11.9	13.7	15.4	17.3

Table 8-20.
Coefficients C for Eccentrically Loaded Bolt Groups
Angle = 0°

$$C_{req} = \frac{P_u}{\phi r_n} \text{ or } \phi R_n = C \times \phi r_n$$

where

P_u = factored force, kips

ϕr_n = design strength per bolt, kips

ϕR_n = design strength of bolt group, kips

e = eccentricity of P_u with respect to centroid of bolt group, in. (not tabulated, may be determined by geometry.)

e_x = horizontal component of e, in.

s = bolt spacing, in.

C = coefficient tabulated below.

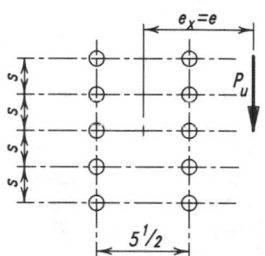

s, in.	e_x, in.	Number of bolts in one vertical row, n											
		1	2	3	4	5	6	7	8	9	10	11	12
	2	1.14	2.75	4.59	6.61	8.69	10.8	12.9	14.9	17.0	19.0	21.0	23.0
	3	0.94	2.32	3.92	5.80	7.82	9.90	12.0	14.1	16.2	18.3	20.4	22.4
	4	0.80	1.99	3.39	5.10	6.98	9.00	11.1	13.2	15.3	17.4	19.6	21.7
	5	0.70	1.74	2.96	4.51	6.24	8.15	10.2	12.3	14.4	16.5	18.6	20.8
	6	0.62	1.54	2.62	4.03	5.60	7.39	9.30	11.3	13.4	15.5	17.7	19.8
	7	0.55	1.38	2.36	3.63	5.07	6.72	8.53	10.5	12.5	14.6	16.7	18.8
	8	0.50	1.25	2.14	3.30	4.61	6.15	7.84	9.67	11.6	13.6	15.7	17.8
	9	0.46	1.14	1.96	3.01	4.22	5.66	7.23	8.97	10.8	12.8	14.8	16.9
	10	0.42	1.04	1.80	2.78	3.89	5.23	6.70	8.34	10.1	12.0	13.9	15.9
3	12	0.37	0.90	1.55	2.39	3.36	4.53	5.82	7.28	8.87	10.6	12.4	14.2
	14	0.32	0.79	1.36	2.10	2.96	3.99	5.13	6.44	7.87	9.42	11.1	12.8
	16	0.29	0.70	1.21	1.87	2.64	3.55	4.58	5.76	7.05	8.47	9.99	11.6
	18	0.26	0.63	1.09	1.68	2.37	3.20	4.14	5.21	6.38	7.68	9.08	10.6
	20	0.24	0.57	0.99	1.53	2.16	2.91	3.77	4.75	5.82	7.02	8.30	9.69
	24	0.20	0.48	0.84	1.29	1.83	2.46	3.19	4.03	4.94	5.97	7.07	8.28
	28	0.18	0.42	0.73	1.11	1.58	2.13	2.77	3.49	4.29	5.19	6.15	7.21
	32	0.16	0.37	0.64	0.98	1.39	1.88	2.44	3.08	3.79	4.58	5.44	6.38
	36	0.14	0.33	0.57	0.88	1.24	1.68	2.18	2.75	3.39	4.10	4.87	5.72
	2	1.14	3.25	5.37	7.45	9.49	11.5	13.5	15.5	17.5	19.5	21.4	23.4
	3	0.94	2.86	4.93	7.05	9.14	11.2	13.2	15.3	17.3	19.3	21.3	23.3
	4	0.80	2.52	4.47	6.59	8.72	10.8	12.9	15.0	17.0	19.0	21.0	23.0
	5	0.70	2.24	4.04	6.12	8.25	10.4	12.5	14.6	16.7	18.7	20.8	22.8
	6	0.62	2.00	3.65	5.66	7.77	9.91	12.1	14.2	16.3	18.4	20.4	22.5
	7	0.55	1.80	3.31	5.23	7.29	9.42	11.6	13.7	15.8	17.9	20.0	22.1
	8	0.50	1.64	3.02	4.84	6.83	8.93	11.1	13.2	15.4	17.5	19.6	21.7
	9	0.46	1.50	2.77	4.49	6.39	8.45	10.6	12.7	14.9	17.0	19.2	21.3
	10	0.42	1.38	2.56	4.18	5.99	7.99	10.1	12.2	14.4	16.5	18.7	20.8
6	12	0.37	1.19	2.21	3.65	5.29	7.16	9.15	11.2	13.4	15.5	17.7	19.8
	14	0.32	1.04	1.95	3.24	4.72	6.44	8.32	10.3	12.4	14.5	16.7	18.8
	16	0.29	0.93	1.74	2.90	4.24	5.83	7.59	9.48	11.5	13.6	15.7	17.8
	18	0.26	0.84	1.57	2.62	3.84	5.31	6.95	8.74	10.7	12.6	14.7	16.8
	20	0.24	0.76	1.43	2.39	3.50	4.87	6.39	8.08	9.89	11.8	13.8	15.9
	24	0.20	0.64	1.21	2.02	2.98	4.16	5.49	6.99	8.61	10.4	12.2	14.1
	28	0.18	0.55	1.05	1.76	2.59	3.63	4.80	6.13	7.59	9.18	10.9	12.7
	32	0.16	0.49	0.93	1.55	2.29	3.21	4.25	5.45	6.77	8.21	9.76	11.4
	36	0.14	0.43	0.83	1.38	2.05	2.88	3.81	4.90	6.09	7.41	8.83	10.4

Table 8-20 (cont.).
Coefficients C for Eccentrically Loaded Bolt Groups
Angle = 15°

$$C_{req} = \frac{P_u}{\phi r_n} \text{ or } \phi R_n = C \times \phi r_n$$

where

P_u = factored force, kips

ϕr_n = design strength per bolt, kips

ϕR_n = design strength of bolt group, kips

e = eccentricity of P_u with respect to centroid
     of bolt group, in. (not tabulated, may be
     determined by geometry.)

e_x = horizontal component of e, in.

s = bolt spacing, in.

C = coefficient tabulated below.

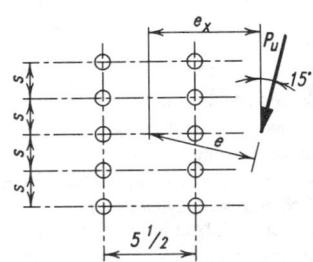

s, in.	e_x, in.	Number of bolts in one vertical row, n											
		1	2	3	4	5	6	7	8	9	10	11	12
3	2	1.18	2.78	4.61	6.59	8.64	10.7	12.8	14.8	16.8	18.9	20.9	22.9
	3	0.97	2.34	3.97	5.80	7.78	9.83	11.9	14.0	16.1	18.1	20.2	22.2
	4	0.83	2.02	3.45	5.11	6.97	8.94	11.0	13.1	15.2	17.3	19.3	21.4
	5	0.72	1.77	3.03	4.54	6.26	8.12	10.1	12.1	14.2	16.3	18.4	20.5
	6	0.64	1.57	2.70	4.06	5.65	7.39	9.27	11.2	13.3	15.4	17.5	19.6
	7	0.57	1.41	2.43	3.66	5.13	6.74	8.52	10.4	12.4	14.4	16.5	18.6
	8	0.52	1.28	2.20	3.34	4.68	6.18	7.86	9.65	11.6	13.5	15.6	17.6
	9	0.48	1.17	2.01	3.06	4.30	5.70	7.27	8.97	10.8	12.7	14.7	16.7
	10	0.44	1.07	1.85	2.82	3.98	5.27	6.76	8.36	10.1	11.9	13.8	15.8
	12	0.38	0.93	1.60	2.44	3.44	4.58	5.90	7.34	8.91	10.6	12.4	14.2
	14	0.33	0.81	1.40	2.15	3.03	4.05	5.22	6.51	7.94	9.47	11.1	12.8
	16	0.30	0.72	1.25	1.91	2.70	3.62	4.68	5.84	7.14	8.54	10.1	11.7
	18	0.27	0.65	1.13	1.72	2.44	3.27	4.23	5.28	6.48	7.77	9.16	10.7
	20	0.25	0.59	1.02	1.57	2.22	2.98	3.86	4.83	5.93	7.11	8.40	9.78
	24	0.21	0.50	0.87	1.33	1.88	2.53	3.27	4.11	5.05	6.07	7.19	8.39
	28	0.18	0.43	0.75	1.15	1.63	2.19	2.84	3.57	4.39	5.29	6.28	7.33
	32	0.16	0.38	0.66	1.01	1.43	1.93	2.50	3.15	3.88	4.68	5.56	6.50
	36	0.14	0.34	0.59	0.90	1.28	1.73	2.24	2.82	3.48	4.19	4.99	5.84
6	2	1.18	3.24	5.34	7.40	9.43	11.5	13.5	15.4	17.4	19.4	21.4	23.4
	3	0.97	2.85	4.90	6.99	9.07	11.1	13.2	15.2	17.2	19.2	21.2	23.1
	4	0.83	2.51	4.45	6.53	8.63	10.7	12.8	14.8	16.87	18.9	20.9	23.0
	5	0.72	2.23	4.05	6.07	8.16	10.3	12.4	14.5	16.5	18.6	20.6	22.6
	6	0.64	2.00	3.68	5.62	7.69	9.80	11.9	14.0	16.1	18.2	20.2	22.3
	7	0.57	1.81	3.36	5.20	7.22	9.31	11.4	13.5	15.7	17.7	19.8	21.9
	8	0.52	1.65	3.08	4.82	6.78	8.83	10.9	13.1	15.2	17.3	19.4	21.4
	9	0.48	1.52	2.83	4.48	6.36	8.37	10.5	12.6	14.7	16.8	18.9	21.0
	10	0.44	1.40	2.62	4.18	5.98	7.93	9.97	12.1	14.2	16.3	18.4	20.6
	12	0.38	1.21	2.27	3.66	5.31	7.13	9.08	11.1	13.2	15.3	17.4	19.6
	14	0.33	1.07	2.00	3.25	4.76	6.44	8.28	10.2	12.3	14.3	16.4	18.6
	16	0.30	0.95	1.79	2.92	4.29	5.85	7.58	9.43	11.4	13.4	15.5	17.6
	18	0.27	0.86	1.62	2.65	3.90	5.34	6.97	8.72	10.6	12.5	14.6	16.6
	20	0.25	0.78	1.47	2.42	3.58	4.91	6.43	8.09	9.87	11.7	13.7	15.7
	24	0.21	0.66	1.25	2.06	3.05	4.21	5.55	7.03	8.64	10.4	12.2	14.1
	28	0.18	0.57	1.08	1.79	2.66	3.68	4.87	6.19	7.65	9.22	10.9	12.6
	32	0.16	0.50	0.95	1.58	2.35	3.26	4.33	5.52	6.84	8.27	9.81	11.4
	36	0.14	0.45	0.85	1.42	2.11	2.93	3.90	4.97	6.18	7.49	8.91	10.4

Table 8-20 (cont.).
Coefficients C for Eccentrically Loaded Bolt Groups
Angle = 30°

$$C_{req} = \frac{P_u}{\phi r_n} \text{ or } \phi R_n = C \times \phi r_n$$

where

P_u = factored force, kips

ϕr_n = design strength per bolt, kips

ϕR_n = design strength of bolt group, kips

e = eccentricity of P_u with respect to centroid of bolt group, in. (not tabulated, may be determined by geometry.)

e_x = horizontal component of e, in.

s = bolt spacing, in.

C = coefficient tabulated below.

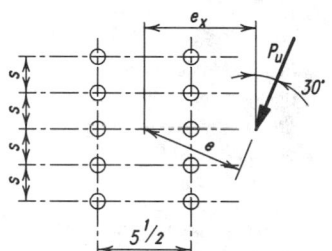

s, in.	e_x, in.	Number of bolts in one vertical row, n												
		1	2	3	4	5	6	7	8	9	10	11	12
3	2	1.30	2.90	4.72	6.66	8.65	10.7	12.7	14.7	16.7	18.7	20.8	22.8
	3	1.08	2.47	4.13	5.94	7.86	9.85	11.9	13.9	16.0	18.0	20.0	22.1
	4	0.92	2.14	3.64	5.30	7.12	9.04	11.0	13.0	15.1	17.1	19.2	21.2
	5	0.80	1.89	3.24	4.76	6.46	8.29	10.2	12.2	14.2	16.3	18.3	20.4
	6	0.71	1.69	2.91	4.29	5.88	7.61	9.45	11.4	13.4	15.4	17.4	19.5
	7	0.64	1.53	2.63	3.90	5.38	7.01	8.76	10.6	12.5	14.5	16.5	18.6
	8	0.58	1.39	2.40	3.57	4.95	6.49	8.14	9.92	11.8	13.7	15.7	17.7
	9	0.53	1.28	2.20	3.29	4.58	6.02	7.59	9.29	11.1	12.9	14.9	16.8
	10	0.49	1.18	2.03	3.04	4.26	5.61	7.09	8.72	10.4	12.2	14.1	16.0
	12	0.42	1.02	1.76	2.65	3.72	4.92	6.25	7.73	9.31	11.0	12.8	14.6
	14	0.37	0.90	1.55	2.34	3.29	4.37	5.58	6.93	8.38	9.93	11.6	13.3
	16	0.33	0.80	1.38	2.09	2.95	3.92	5.03	6.26	7.59	9.03	10.6	12.2
	18	0.30	0.72	1.25	1.89	2.67	3.55	4.57	5.70	6.93	8.27	9.70	11.2
	20	0.27	0.66	1.13	1.73	2.43	3.25	4.19	5.23	6.36	7.62	8.95	10.4
	24	0.23	0.56	0.96	1.46	2.07	2.77	3.57	4.47	5.47	6.56	7.73	8.99
	28	0.20	0.48	0.83	1.27	1.79	2.41	3.11	3.90	4.78	5.75	6.78	7.91
	32	0.18	0.43	0.73	1.12	1.58	2.13	2.76	3.46	4.25	5.11	6.04	7.06
	36	0.16	0.38	0.66	1.00	1.42	1.91	2.47	3.10	3.81	4.59	5.44	6.36
6	2	1.30	3.27	5.33	7.36	9.38	11.4	13.4	15.4	17.4	19.3	21.3	23.3
	3	1.08	2.89	4.91	6.96	9.01	11.0	13.1	15.1	17.1	19.1	21.1	23.0
	4	0.92	2.56	4.50	6.53	8.58	10.6	12.7	14.7	16.8	18.8	20.8	22.8
	5	0.80	2.29	4.13	6.10	8.14	10.2	12.3	14.3	16.4	18.4	20.4	22.5
	6	0.71	2.08	3.80	5.69	7.70	9.75	11.8	13.9	15.9	18.0	20.0	22.1
	7	0.64	1.89	3.51	5.31	7.27	9.30	11.4	13.4	15.5	17.6	19.6	21.7
	8	0.58	1.74	3.25	4.96	6.86	8.86	10.9	13.0	15.0	17.1	19.2	21.3
	9	0.53	1.61	3.02	4.64	6.49	8.44	10.5	12.5	14.6	16.7	18.7	20.8
	10	0.49	1.49	2.81	4.35	6.13	8.04	10.0	12.1	14.1	16.2	18.3	20.4
	12	0.42	1.30	2.47	3.85	5.51	7.31	9.22	11.2	13.2	15.3	17.3	19.4
	14	0.37	1.15	2.19	3.44	4.98	6.67	8.49	10.4	12.4	14.4	16.4	18.5
	16	0.33	1.03	1.96	3.11	4.54	6.12	7.83	9.66	11.6	13.5	15.6	17.6
	18	0.30	0.93	1.78	2.83	4.16	5.63	7.26	9.00	10.8	12.8	14.7	16.7
	20	0.27	0.85	1.62	2.60	3.83	5.21	6.74	8.41	10.2	12.0	13.9	15.9
	24	0.23	0.72	1.38	2.23	3.30	4.51	5.89	7.40	9.02	10.7	12.5	14.4
	28	0.20	0.63	1.20	1.95	2.89	3.96	5.21	6.59	8.07	9.66	11.3	13.1
	32	0.18	0.55	1.06	1.73	2.57	3.53	4.67	5.92	7.28	8.75	10.3	12.0
	36	0.16	0.50	0.95	1.55	2.31	3.18	4.22	5.36	6.61	7.98	9.43	11.0

Table 8-20 (cont.).
Coefficients C for Eccentrically Loaded Bolt Groups
Angle = 45°

$$C_{req} = \frac{P_u}{\phi r_n} \text{ or } \phi R_n = C \times \phi r_n$$

where

P_u = factored force, kips

ϕr_n = design strength per bolt, kips

ϕR_n = design strength of bolt group, kips

e = eccentricity of P_u with respect to centroid of bolt group, in. (not tabulated, may be determined by geometry.)

e_x = horizontal component of e, in.

s = bolt spacing, in.

C = coefficient tabulated below.

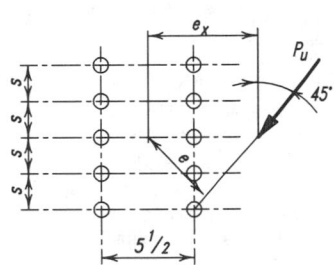

s, in.	e_x, in.	\multicolumn{12}{c}{Number of bolts in one vertical row, n}
		1	2	3	4	5	6	7	8	9	10	11	12
	2	1.53	3.18	4.96	6.84	8.77	10.7	12.7	14.7	16.7	18.7	20.7	22.6
	3	1.30	2.76	4.42	6.22	8.09	10.0	12.0	14.0	15.9	17.9	19.9	21.9
	4	1.11	2.43	3.97	5.67	7.46	9.32	11.2	13.2	15.2	17.2	19.2	21.2
	5	0.98	2.17	3.60	5.19	6.89	8.68	10.6	12.5	14.4	16.4	18.4	20.4
	6	0.87	1.95	3.28	4.77	6.37	8.09	9.90	11.8	13.7	15.6	17.6	19.6
	7	0.78	1.78	3.01	4.40	5.91	7.56	9.31	11.1	13.0	14.9	16.9	18.8
	8	0.71	1.63	2.77	4.07	5.50	7.07	8.76	10.5	12.4	14.2	16.2	18.1
	9	0.65	1.50	2.57	3.78	5.13	6.64	8.26	9.97	11.8	13.6	15.5	17.4
3	10	0.60	1.39	2.39	3.52	4.81	6.25	7.81	9.45	11.2	13.0	14.8	16.7
	12	0.52	1.22	2.08	3.09	4.26	5.58	7.01	8.54	10.2	11.9	13.6	15.4
	14	0.45	1.08	1.85	2.75	3.82	5.02	6.34	7.76	9.28	10.9	12.6	14.3
	16	0.41	0.96	1.65	2.48	3.45	4.55	5.77	7.09	8.53	10.1	11.6	13.3
	18	0.37	0.87	1.50	2.25	3.14	4.16	5.29	6.53	7.87	9.30	10.8	12.4
	20	0.33	0.79	1.37	2.06	2.88	3.82	4.87	6.04	7.30	8.65	10.1	11.6
	24	0.28	0.68	1.16	1.76	2.47	3.28	4.21	5.23	6.35	7.55	8.85	10.2
	28	0.25	0.59	1.01	1.53	2.15	2.87	3.69	4.61	5.61	6.69	7.87	9.11
	32	0.22	0.52	0.89	1.35	1.91	2.55	3.29	4.11	5.01	6.00	7.07	8.20
	36	0.20	0.46	0.80	1.21	1.71	2.29	2.96	3.70	4.53	5.43	6.40	7.44
	2	1.53	3.39	5.36	7.35	9.35	11.3	13.3	15.3	17.3	19.3	21.3	23.2
	3	1.30	3.04	4.99	6.98	8.98	11.0	13.0	15.0	17.0	19.0	21.0	22.9
	4	1.11	2.74	4.64	6.60	8.60	10.6	12.6	14.6	16.6	18.6	20.6	22.6
	5	0.98	2.49	4.31	6.24	8.21	10.2	12.2	14.2	16.3	18.3	20.3	22.3
	6	0.87	2.28	4.02	5.89	7.84	9.82	11.8	13.8	15.9	17.9	19.9	21.9
	7	0.78	2.10	3.76	5.57	7.48	9.44	11.4	13.4	15.5	17.5	19.5	21.5
	8	0.71	1.94	3.53	5.28	7.13	9.07	11.0	13.0	15.1	17.1	19.1	21.1
	9	0.65	1.81	3.32	5.00	6.81	8.71	10.7	12.7	14.7	16.7	18.7	20.7
6	10	0.60	1.69	3.13	4.74	6.50	8.37	10.3	12.3	14.3	16.3	18.3	20.3
	12	0.52	1.50	2.80	4.29	5.94	7.74	9.61	11.5	13.5	15.5	17.5	19.5
	14	0.45	1.34	2.52	3.89	5.45	7.17	8.98	10.9	12.8	14.7	16.7	18.7
	16	0.41	1.21	2.29	3.55	5.02	6.67	8.41	10.2	12.1	14.0	16.0	17.9
	18	0.37	1.10	2.09	3.26	4.65	6.22	7.89	9.65	11.5	13.4	15.3	17.2
	20	0.33	1.01	1.92	3.01	4.33	5.82	7.42	9.11	10.9	12.7	14.6	16.5
	24	0.28	0.86	1.64	2.61	3.79	5.13	6.60	8.17	9.84	11.6	13.4	15.2
	28	0.25	0.75	1.44	2.30	3.36	4.58	5.92	7.38	8.95	10.6	12.3	14.1
	32	0.22	0.67	1.27	2.05	3.02	4.12	5.35	6.72	8.18	9.73	11.4	13.0
	36	0.20	0.60	1.14	1.85	2.73	3.74	4.88	6.15	7.52	8.98	10.5	12.1

Table 8-20 (cont.).
Coefficients C for Eccentrically Loaded Bolt Groups
Angle = 60°

$$C_{req} = \frac{P_u}{\phi r_n} \text{ or } \phi R_n = C \times \phi r_n$$

where

P_u	= factored force, kips
ϕr_n	= design strength per bolt, kips
ϕR_n	= design strength of bolt group, kips
e	= eccentricity of P_u with respect to centroid of bolt group, in. (not tabulated, may be determined by geometry.)
e_x	= horizontal component of e, in.
s	= bolt spacing, in.
C	= coefficient tabulated below.

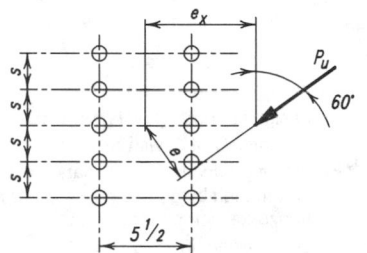

s, in.	e_x, in.	\multicolumn{12}{c}{Number of bolts in one vertical row, n}
		1	2	3	4	5	6	7	8	9	10	11	12
	2	1.78	3.55	5.34	7.17	9.04	10.9	12.9	14.8	16.7	18.7	20.6	22.6
	3	1.62	3.26	4.95	6.71	8.53	10.4	12.3	14.2	16.1	18.1	20.0	22.0
	4	1.45	2.97	4.57	6.27	8.04	9.86	11.7	13.6	15.5	17.5	19.4	21.4
	5	1.31	2.71	4.23	5.86	7.58	9.36	11.2	13.1	15.0	16.9	18.8	20.7
	6	1.18	2.48	3.93	5.50	7.16	8.90	10.7	12.5	14.4	16.3	18.2	20.1
	7	1.07	2.28	3.66	5.18	6.79	8.48	10.2	12.0	13.9	15.7	17.6	19.5
	8	0.98	2.11	3.43	4.88	6.45	8.09	9.80	11.6	13.4	15.2	17.1	19.0
	9	0.90	1.97	3.22	4.61	6.12	7.72	9.39	11.1	12.9	14.7	16.6	18.4
	10	0.83	1.84	3.03	4.37	5.82	7.37	9.00	10.7	12.5	14.2	16.1	17.9
3	12	0.72	1.62	2.70	3.93	5.28	6.73	8.28	9.91	11.6	13.4	15.1	16.9
	14	0.64	1.45	2.43	3.56	4.81	6.19	7.66	9.22	10.9	12.5	14.3	16.0
	16	0.57	1.31	2.21	3.24	4.42	5.71	7.11	8.60	10.2	11.8	13.5	15.2
	18	0.52	1.19	2.02	2.98	4.07	5.29	6.63	8.05	9.55	11.1	12.7	14.4
	20	0.47	1.09	1.85	2.75	3.77	4.93	6.19	7.55	8.98	10.5	12.1	13.7
	24	0.40	0.93	1.59	2.37	3.28	4.32	5.46	6.69	8.01	9.41	10.9	12.4
	28	0.35	0.82	1.39	2.08	2.90	3.83	4.86	5.99	7.21	8.51	9.88	11.3
	32	0.31	0.72	1.24	1.86	2.59	3.43	4.37	5.41	6.54	7.75	9.02	10.4
	36	0.28	0.65	1.11	1.67	2.34	3.11	3.97	4.93	5.98	7.10	8.29	9.55
	2	1.78	3.59	5.48	7.41	9.36	11.3	13.3	15.3	17.2	19.2	21.2	23.2
	3	1.62	3.35	5.20	7.12	9.06	11.0	13.0	15.0	16.9	18.9	20.9	22.9
	4	1.45	3.11	4.93	6.82	8.75	10.7	12.7	14.6	16.6	18.6	20.6	22.5
	5	1.31	2.89	4.66	6.53	8.45	10.4	12.3	14.3	16.3	18.2	20.2	22.2
	6	1.18	2.70	4.42	6.26	8.16	10.1	12.0	14.0	15.9	17.9	19.9	21.9
	7	1.07	2.52	4.19	6.01	7.88	9.79	11.7	13.7	15.6	17.6	19.6	21.5
	8	0.98	2.36	3.99	5.77	7.62	9.51	11.4	13.4	15.3	17.3	19.2	21.2
	9	0.90	2.23	3.81	5.55	7.37	9.24	11.1	13.1	15.0	17.0	18.9	20.9
	10	0.83	2.10	3.64	5.35	7.13	8.98	10.9	12.8	14.7	16.7	18.6	20.6
6	12	0.72	1.89	3.34	4.97	6.70	8.49	10.3	12.2	14.1	16.1	18.0	19.9
	14	0.64	1.71	3.08	4.63	6.29	8.04	9.85	11.7	13.6	15.5	17.4	19.3
	16	0.57	1.57	2.85	4.32	5.92	7.62	9.39	11.2	13.1	15.0	16.9	18.8
	18	0.52	1.44	2.65	4.04	5.58	7.22	8.95	10.7	12.6	14.4	16.3	18.2
	20	0.47	1.33	2.47	3.79	5.26	6.86	8.55	10.3	12.1	13.9	15.8	17.7
	24	0.40	1.16	2.17	3.36	4.71	6.21	7.82	9.50	11.2	13.0	14.8	16.7
	28	0.35	1.02	1.92	3.00	4.26	5.67	7.19	8.80	10.5	12.2	14.0	15.8
	32	0.31	0.91	1.72	2.71	3.88	5.20	6.64	8.17	9.77	11.4	13.1	14.9
	36	0.28	0.82	1.56	2.46	3.55	4.80	6.16	7.61	9.14	10.7	12.4	14.1

Table 8-20 (cont.).
Coefficients C for Eccentrically Loaded Bolt Groups
Angle = 75°

$$C_{req} = \frac{P_u}{\phi r_n} \text{ or } \phi R_n = C \times \phi r_n$$

where

P_u = factored force, kips

ϕr_n = design strength per bolt, kips

ϕR_n = design strength of bolt group, kips

e = eccentricity of P_u with respect to centroid of bolt group, in. (not tabulated, may be determined by geometry.)

e_x = horizontal component of e, in.

s = bolt spacing, in.

C = coefficient tabulated below.

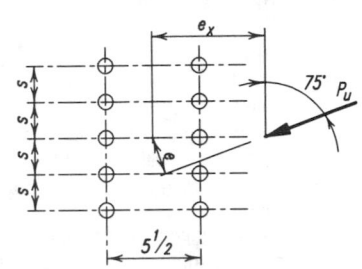

s, in.	e_x, in.	Number of bolts in one vertical row, n											
		1	2	3	4	5	6	7	8	9	10	11	12
3	2	1.92	3.82	5.70	7.57	9.45	11.3	13.2	15.2	17.1	19.0	20.9	22.9
	3	1.87	3.72	5.54	7.36	9.19	11.1	12.9	14.8	16.7	18.6	20.5	22.5
	4	1.82	3.60	5.37	7.14	8.94	10.8	12.6	14.5	16.3	18.2	20.1	22.1
	5	1.75	3.47	5.18	6.92	8.68	10.5	12.3	14.1	16.0	17.9	19.8	21.7
	6	1.68	3.33	5.00	6.69	8.42	10.2	12.0	13.8	15.7	17.5	19.4	21.3
	7	1.60	3.19	4.81	6.47	8.17	9.92	11.7	13.5	15.3	17.2	19.1	20.9
	8	1.52	3.06	4.63	6.26	7.93	9.66	11.4	13.2	15.0	16.9	18.7	20.6
	9	1.45	2.93	4.46	6.05	7.70	9.41	11.2	12.9	14.7	16.5	18.4	20.3
	10	1.38	2.80	4.29	5.85	7.48	9.16	10.9	12.6	14.4	16.2	18.1	19.9
	12	1.25	2.57	3.98	5.48	7.07	8.71	10.4	12.1	13.9	15.7	17.5	19.3
	14	1.13	2.36	3.70	5.15	6.69	8.29	9.96	11.7	13.4	15.2	16.9	18.7
	16	1.03	2.18	3.45	4.85	6.34	7.90	9.53	11.2	12.9	14.7	16.4	18.2
	18	0.95	2.02	3.23	4.57	6.01	7.54	9.13	10.8	12.5	14.2	15.9	17.7
	20	0.87	1.88	3.03	4.32	5.71	7.19	8.75	10.4	12.0	13.7	15.4	17.2
	24	0.75	1.65	2.69	3.87	5.17	6.57	8.05	9.60	11.2	12.9	14.5	16.2
	28	0.66	1.46	2.42	3.50	4.71	6.03	7.44	8.93	10.5	12.1	13.7	15.4
	32	0.59	1.31	2.18	3.19	4.32	5.56	6.90	8.32	9.81	11.4	12.9	14.6
	36	0.53	1.19	1.99	2.92	3.98	5.15	6.42	7.78	9.21	10.7	12.2	13.8
6	2	1.92	3.80	5.69	7.59	9.51	11.5	13.4	15.4	17.6	19.6	21.5	23.5
	3	1.87	3.70	5.55	7.42	9.32	11.2	13.2	15.1	17.1	19.0	21.0	23.0
	4	1.82	3.59	5.40	7.25	9.14	11.1	13.0	14.9	16.9	18.8	20.8	22.7
	5	1.75	3.48	5.26	7.09	8.96	10.9	12.8	14.7	16.6	18.6	20.5	22.5
	6	1.68	3.36	5.11	6.93	8.78	10.7	12.6	14.5	16.4	18.4	20.3	22.2
	7	1.60	3.24	4.97	6.77	8.62	10.5	12.4	14.3	16.2	18.1	20.1	22.0
	8	1.52	3.13	4.84	6.62	8.45	10.3	12.2	14.1	16.0	17.9	19.9	21.8
	9	1.45	3.02	4.71	6.47	8.29	10.2	12.0	13.9	15.8	17.7	19.7	21.6
	10	1.38	2.91	4.58	6.33	8.14	9.98	11.9	13.7	15.6	17.6	19.5	21.4
	12	1.25	2.72	4.34	6.07	7.85	9.67	11.5	13.4	15.3	17.2	19.1	21.0
	14	1.13	2.54	4.13	5.82	7.57	9.38	11.2	13.1	15.0	16.8	18.7	20.6
	16	1.03	2.38	3.92	5.59	7.32	9.10	10.9	12.8	14.6	16.5	18.4	20.3
	18	0.95	2.24	3.74	5.38	7.09	8.85	10.7	12.5	14.3	16.2	18.1	19.9
	20	0.87	2.11	3.57	5.17	6.87	8.61	10.4	12.2	14.0	15.9	17.7	19.6
	24	0.75	1.88	3.27	4.80	6.44	8.15	9.90	11.7	13.5	15.3	17.1	19.0
	28	0.66	1.70	3.00	4.47	6.06	7.72	9.43	11.2	13.0	14.8	16.6	18.4
	32	0.59	1.55	2.77	4.17	5.70	7.31	8.99	10.7	12.5	14.3	16.1	17.9
	36	0.53	1.42	2.57	3.90	5.37	6.93	8.57	10.3	12.0	13.8	15.5	17.3

Table 8-21.
Coefficients C for Eccentrically Loaded Bolt Groups
Angle = 0°

$$C_{req} = \frac{P_u}{\phi r_n} \text{ or } \phi R_n = C \times \phi r_n$$

where

P_u	= factored force, kips
ϕr_n	= design strength per bolt, kips
ϕR_n	= design strength of bolt group, kips
e	= eccentricity of P_u with respect to centroid of bolt group, in. (not tabulated, may be determined by geometry.)
e_x	= horizontal component of e, in.
s	= bolt spacing, in.
C	= coefficient tabulated below.

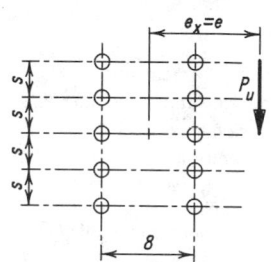

s, in.	e_x, in.	\multicolumn{12}{c}{Number of bolts in one vertical row, n}											
		1	2	3	4	5	6	7	8	9	10	11	12
3	2	1.31	2.91	4.71	6.66	8.69	10.8	12.8	14.9	16.9	18.9	21.0	23.0
	3	1.12	2.54	4.14	5.95	7.90	9.93	12.0	14.1	16.2	18.2	20.3	22.4
	4	0.98	2.24	3.66	5.33	7.15	9.10	11.1	13.2	15.3	17.4	19.5	21.6
	5	0.87	1.99	3.27	4.80	6.48	8.33	10.3	12.3	14.4	16.5	18.6	20.7
	6	0.79	1.80	2.95	4.35	5.90	7.63	9.49	11.5	13.5	15.6	17.7	19.8
	7	0.71	1.63	2.68	3.97	5.40	7.02	8.77	10.7	12.6	14.6	16.7	18.8
	8	0.65	1.49	2.46	3.65	4.97	6.48	8.13	9.91	11.8	13.8	15.8	17.9
	9	0.60	1.38	2.27	3.37	4.59	6.01	7.55	9.24	11.1	13.0	14.9	17.0
	10	0.56	1.28	2.11	3.13	4.27	5.59	7.04	8.64	10.4	12.2	14.1	16.1
	12	0.49	1.11	1.84	2.73	3.73	4.90	6.19	7.63	9.18	10.9	12.6	14.5
	14	0.44	0.99	1.64	2.42	3.31	4.36	5.50	6.80	8.20	9.73	11.4	13.1
	16	0.39	0.89	1.47	2.17	2.98	3.91	4.95	6.13	7.40	8.80	10.3	11.9
	18	0.36	0.80	1.33	1.97	2.70	3.55	4.50	5.57	6.73	8.02	9.39	10.9
	20	0.33	0.73	1.22	1.80	2.47	3.25	4.12	5.10	6.17	7.35	8.62	9.99
	24	0.28	0.63	1.04	1.53	2.10	2.77	3.51	4.35	5.28	6.30	7.39	8.59
	28	0.25	0.55	0.91	1.33	1.83	2.41	3.06	3.79	4.60	5.50	6.46	7.51
	32	0.22	0.48	0.80	1.18	1.62	2.13	2.71	3.36	4.08	4.87	5.73	6.67
	36	0.20	0.43	0.72	1.06	1.45	1.91	2.43	3.01	3.66	4.37	5.15	5.99
6	2	1.31	3.28	5.35	7.42	9.47	11.5	13.5	15.5	17.5	19.5	21.4	23.4
	3	1.12	2.93	4.94	7.03	9.12	11.2	13.2	15.3	17.3	19.3	21.3	23.3
	4	0.98	2.63	4.52	6.59	8.70	10.8	12.9	14.9	17.0	19.0	21.0	23.0
	5	0.87	2.37	4.13	6.15	8.25	10.4	12.5	14.6	16.6	18.69	20.7	22.8
	6	0.79	2.15	3.78	5.72	7.78	9.90	12.0	14.1	16.2	18.3	20.4	22.4
	7	0.71	1.97	3.47	5.32	7.33	9.43	11.6	13.7	15.8	17.9	20.0	22.1
	8	0.65	1.81	3.19	4.95	6.89	8.95	11.1	13.2	15.4	17.5	19.6	21.7
	9	0.60	1.67	2.95	4.62	6.48	8.49	10.6	12.7	14.9	17.0	19.1	21.3
	10	0.56	1.55	2.75	4.33	6.10	8.05	10.1	12.2	14.4	16.5	18.7	20.8
	12	0.49	1.35	2.40	3.82	5.43	7.25	9.21	11.3	13.4	15.5	17.7	19.8
	14	0.44	1.20	2.14	3.41	4.86	6.56	8.40	10.4	12.4	14.5	16.7	18.8
	16	0.39	1.08	1.92	3.07	4.40	5.96	7.69	9.56	11.5	13.6	15.7	17.8
	18	0.36	0.97	1.75	2.79	4.00	5.46	7.06	8.83	10.7	12.7	14.7	16.8
	20	0.33	0.89	1.60	2.56	3.67	5.02	6.52	8.18	9.97	11.9	13.9	15.9
	24	0.28	0.76	1.37	2.18	3.14	4.32	5.62	7.11	8.71	10.4	12.3	14.2
	28	0.25	0.66	1.19	1.90	2.75	3.78	4.93	6.26	7.70	9.27	11.0	12.7
	32	0.22	0.58	1.05	1.68	2.44	3.35	4.38	5.58	6.88	8.31	9.85	11.5
	36	0.20	0.52	0.95	1.51	2.19	3.01	3.94	5.02	6.21	7.52	8.93	10.4

Table 8-21 (cont.).
Coefficients C for Eccentrically Loaded Bolt Groups
Angle = 15°

$$C_{req} = \frac{P_u}{\phi r_n} \text{ or } \phi R_n = C \times \phi r_n$$

where

P_u	= factored force, kips
ϕr_n	= design strength per bolt, kips
ϕR_n	= design strength of bolt group, kips
e	= eccentricity of P_u with respect to centroid of bolt group, in. (not tabulated, may be determined by geometry.)
e_x	= horizontal component of e, in.
s	= bolt spacing, in.
C	= coefficient tabulated below.

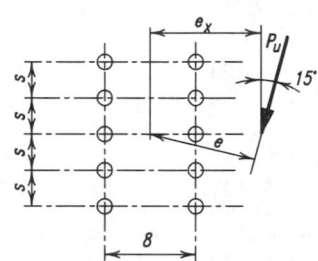

s, in.	e_x, in.	Number of bolts in one vertical row, n											
		1	2	3	4	5	6	7	8	9	10	11	12
3	2	1.35	2.96	4.75	6.67	8.67	10.7	12.7	14.8	16.8	18.8	20.9	22.9
	3	1.16	2.58	4.20	5.98	7.90	9.89	11.9	14.0	16.0	18.1	20.2	22.2
	4	1.02	2.28	3.73	5.37	7.17	9.08	11.1	13.1	15.2	17.3	19.3	21.4
	5	0.90	2.03	3.35	4.85	6.53	8.34	10.3	12.2	14.3	16.3	18.4	20.5
	6	0.81	1.84	3.03	4.40	5.96	7.66	9.48	11.4	13.4	15.4	17.5	19.6
	7	0.74	1.67	2.76	4.02	5.48	7.06	8.79	10.6	12.6	14.5	16.6	18.6
	8	0.68	1.53	2.53	3.70	5.05	6.53	8.17	9.91	11.8	13.7	15.7	17.7
	9	0.63	1.42	2.34	3.43	4.68	6.07	7.61	9.27	11.0	12.9	14.8	16.8
	10	0.58	1.31	2.17	3.19	4.36	5.66	7.12	8.69	10.4	12.2	14.0	16.0
	12	0.51	1.15	1.90	2.79	3.82	4.97	6.28	7.69	9.23	10.9	12.6	14.4
	14	0.45	1.02	1.69	2.48	3.40	4.43	5.61	6.88	8.29	9.79	11.4	13.1
	16	0.41	0.91	1.51	2.23	3.05	3.99	5.05	6.21	7.50	8.88	10.4	11.9
	18	0.37	0.83	1.37	2.02	2.77	3.63	4.60	5.66	6.84	8.11	9.48	11.0
	20	0.34	0.76	1.26	1.85	2.54	3.32	4.21	5.19	6.28	7.45	8.73	10.1
	24	0.29	0.65	1.07	1.58	2.16	2.84	3.60	4.45	5.39	6.40	7.52	8.71
	28	0.25	0.56	0.93	1.37	1.89	2.47	3.14	3.88	4.71	5.61	6.59	7.64
	32	0.23	0.50	0.83	1.22	1.67	2.19	2.78	3.44	4.18	4.98	5.86	6.80
	36	0.20	0.45	0.74	1.09	1.50	1.96	2.49	3.09	3.75	4.47	5.27	6.12
6	2	1.35	3.29	5.33	7.39	9.42	11.4	13.4	15.4	17.4	19.4	21.4	23.4
	3	1.16	2.94	4.93	6.99	9.05	11.1	13.1	15.2	17.2	19.2	21.2	23.2
	4	1.02	2.64	4.52	6.55	8.63	10.7	12.8	14.8	16.9	18.9	20.9	22.9
	5	0.90	2.38	4.15	6.12	8.18	10.3	12.4	14.4	16.5	18.5	20.6	22.6
	6	0.81	2.17	3.82	5.70	7.72	9.80	11.9	14.0	16.1	18.2	20.2	22.3
	7	0.74	1.99	3.52	5.31	7.28	9.33	11.4	13.5	15.6	17.7	19.8	21.9
	8	0.68	1.83	3.25	4.95	6.86	8.87	11.0	13.1	15.2	17.3	19.4	21.5
	9	0.63	1.69	3.02	4.63	6.46	8.43	10.5	12.6	14.7	16.8	18.9	21.0
	10	0.58	1.58	2.81	4.34	6.10	8.00	10.0	12.1	14.2	16.3	18.4	20.5
	12	0.51	1.38	2.47	3.84	5.45	7.23	9.15	11.2	13.2	15.3	17.4	19.6
	14	0.45	1.23	2.20	3.44	4.91	6.56	8.38	10.3	12.3	14.4	16.5	18.6
	16	0.41	1.10	1.98	3.11	4.46	5.99	7.69	9.52	11.5	13.5	15.5	17.6
	18	0.37	1.00	1.80	2.83	4.08	5.49	7.09	8.82	10.7	12.6	14.6	16.6
	20	0.34	0.92	1.65	2.60	3.75	5.06	6.56	8.20	9.96	11.8	13.8	15.7
	24	0.29	0.78	1.41	2.23	3.22	4.36	5.70	7.15	8.74	10.4	12.2	14.1
	28	0.25	0.68	1.23	1.95	2.82	3.83	5.02	6.32	7.76	9.31	11.0	12.7
	32	0.23	0.60	1.09	1.73	2.50	3.41	4.47	5.64	6.96	8.38	9.90	11.5
	36	0.20	0.54	0.97	1.55	2.25	3.07	4.03	5.09	6.30	7.60	9.01	10.5

Table 8-21 (cont.).
Coefficients C for Eccentrically Loaded Bolt Groups
Angle = 30°

$$C_{req} = \frac{P_u}{\phi r_n} \text{ or } \phi R_n = C \times \phi r_n$$

where

P_u　= factored force, kips

ϕr_n　= design strength per bolt, kips

ϕR_n　= design strength of bolt group, kips

e　= eccentricity of P_u with respect to centroid of bolt group, in. (not tabulated, may be determined by geometry.)

e_x　= horizontal component of e, in.

s　= bolt spacing, in.

C　= coefficient tabulated below.

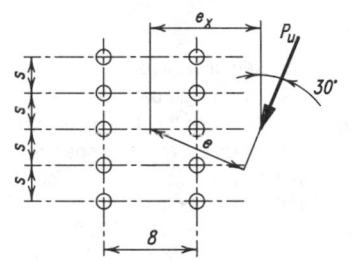

s, in.	e_x, in.	Number of bolts in one vertical row, n											
		1	2	3	4	5	6	7	8	9	10	11	12
	2	1.49	3.12	4.91	6.80	8.75	10.7	12.7	14.7	16.7	18.7	20.8	22.7
	3	1.29	2.74	4.39	6.16	8.04	9.98	12.0	14.0	16.0	18.0	20.0	22.1
	4	1.13	2.43	3.95	5.60	7.37	9.24	11.2	13.2	15.2	17.2	19.2	21.3
	5	1.00	2.18	3.58	5.10	6.77	8.55	10.4	12.4	14.3	16.3	18.4	20.4
	6	0.90	1.98	3.26	4.67	6.23	7.93	9.72	11.6	13.5	15.5	17.5	19.5
	7	0.82	1.81	2.99	4.30	5.76	7.37	9.08	10.9	12.8	14.7	16.7	18.7
	8	0.75	1.67	2.76	3.97	5.35	6.87	8.49	10.2	12.0	13.9	15.9	17.8
	9	0.70	1.55	2.56	3.69	4.98	6.42	7.96	9.62	11.4	13.2	15.1	17.0
	10	0.65	1.44	2.38	3.44	4.66	6.02	7.49	9.07	10.8	12.5	14.4	16.2
3	12	0.57	1.26	2.09	3.03	4.13	5.34	6.66	8.12	9.67	11.3	13.0	14.8
	14	0.50	1.12	1.86	2.71	3.69	4.78	5.99	7.33	8.75	10.3	11.9	13.6
	16	0.45	1.01	1.67	2.44	3.33	4.33	5.44	6.66	7.98	9.39	10.9	12.5
	18	0.41	0.92	1.52	2.22	3.03	3.95	4.97	6.10	7.32	8.64	10.1	11.5
	20	0.38	0.84	1.39	2.03	2.78	3.62	4.57	5.62	6.75	7.98	9.30	10.7
	24	0.32	0.72	1.19	1.74	2.38	3.11	3.93	4.84	5.83	6.92	8.08	9.32
	28	0.28	0.63	1.04	1.52	2.08	2.72	3.44	4.24	5.13	6.09	7.12	8.24
	32	0.25	0.56	0.92	1.35	1.84	2.41	3.06	3.77	4.57	5.43	6.36	7.37
	36	0.23	0.50	0.83	1.21	1.66	2.17	2.75	3.40	4.11	4.89	5.74	6.66
	2	1.49	3.36	5.36	7.37	9.38	11.4	13.4	15.4	17.4	19.3	21.3	23.3
	3	1.29	3.02	4.97	6.99	9.01	11.0	13.1	15.1	17.1	19.1	21.1	23.1
	4	1.13	2.73	4.60	6.58	8.61	10.7	12.7	14.7	16.7	18.8	20.8	22.8
	5	1.00	2.48	4.26	6.18	8.18	10.2	12.3	14.3	16.4	18.4	20.4	22.4
	6	0.90	2.27	3.96	5.80	7.76	9.79	11.8	13.9	15.9	18.0	20.0	22.1
	7	0.82	2.09	3.68	5.44	7.36	9.35	11.4	13.5	15.5	17.6	19.6	21.7
	8	0.75	1.93	3.43	5.11	6.97	8.93	11.0	13.0	15.1	17.1	19.2	21.2
	9	0.70	1.80	3.21	4.81	6.61	8.53	10.5	12.6	14.6	16.7	18.7	20.8
	10	0.65	1.68	3.01	4.53	6.27	8.14	10.1	12.1	14.2	16.2	18.3	20.4
6	12	0.57	1.49	2.67	4.05	5.67	7.43	9.31	11.3	13.3	15.3	17.4	19.4
	14	0.50	1.33	2.39	3.65	5.15	6.81	8.60	10.5	12.4	14.4	16.5	18.5
	16	0.45	1.20	2.16	3.31	4.71	6.27	7.96	9.76	11.7	13.6	15.6	17.6
	18	0.41	1.09	1.97	3.03	4.34	5.79	7.39	9.12	10.9	12.8	14.8	16.8
	20	0.38	1.00	1.81	2.80	4.01	5.37	6.89	8.53	10.3	12.1	14.0	15.9
	24	0.32	0.86	1.55	2.41	3.48	4.68	6.04	7.53	9.14	10.8	12.6	14.5
	28	0.28	0.75	1.35	2.12	3.06	4.13	5.36	6.72	8.19	9.76	11.4	13.2
	32	0.25	0.67	1.20	1.89	2.73	3.69	4.81	6.05	7.40	8.86	10.4	12.0
	36	0.23	0.60	1.08	1.70	2.46	3.34	4.36	5.50	6.74	8.09	9.53	11.1

Table 8-21 (cont.).
Coefficients C for Eccentrically Loaded Bolt Groups
Angle = 45°

$$C_{req} = \frac{P_u}{\phi r_n} \text{ or } \phi R_n = C \times \phi r_n$$

where

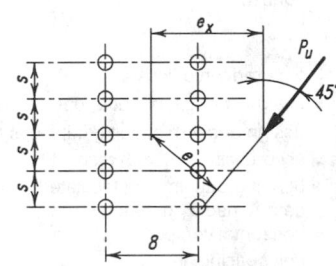

P_u = factored force, kips

ϕr_n = design strength per bolt, kips

ϕR_n = design strength of bolt group, kips

e = eccentricity of P_u with respect to centroid of bolt group, in. (not tabulated, may be determined by geometry.)

e_x = horizontal component of e, in.

s = bolt spacing, in.

C = coefficient tabulated below.

s, in.	e_x, in.	\multicolumn{12}{c}{Number of bolts in one vertical row, n}
		1	2	3	4	5	6	7	8	9	10	11	12
3	2	1.70	3.43	5.22	7.06	8.95	10.9	12.8	14.8	16.8	18.7	20.7	22.7
	3	1.51	3.09	4.76	6.52	8.35	10.2	12.2	14.1	16.1	18.0	20.0	22.0
	4	1.35	2.78	4.34	6.01	7.78	9.60	11.5	13.4	15.3	17.3	19.3	21.3
	5	1.21	2.52	3.97	5.57	7.25	9.01	10.8	12.7	14.6	16.6	18.5	20.5
	6	1.10	2.30	3.67	5.17	6.78	8.47	10.2	12.1	13.9	15.9	17.8	19.8
	7	1.00	2.12	3.40	4.82	6.35	7.97	9.67	11.5	13.3	15.2	17.1	19.0
	8	0.92	1.96	3.17	4.51	5.96	7.51	9.15	10.9	12.7	14.5	16.4	18.3
	9	0.85	1.82	2.96	4.23	5.60	7.08	8.68	10.4	12.1	13.9	15.7	17.6
	10	0.79	1.70	2.78	3.97	5.28	6.70	8.24	9.86	11.5	13.3	15.1	17.0
	12	0.69	1.50	2.46	3.54	4.73	6.04	7.46	8.97	10.6	12.2	14.0	15.7
	14	0.61	1.34	2.21	3.18	4.27	5.48	6.80	8.21	9.70	11.3	12.9	14.6
	16	0.55	1.21	2.00	2.88	3.89	5.01	6.23	7.54	8.95	10.4	12.0	13.6
	18	0.50	1.11	1.82	2.64	3.56	4.60	5.74	6.97	8.30	9.71	11.2	12.7
	20	0.46	1.02	1.67	2.42	3.29	4.25	5.31	6.47	7.73	9.06	10.5	11.9
	24	0.40	0.87	1.43	2.09	2.84	3.68	4.62	5.65	6.77	7.96	9.23	10.6
	28	0.35	0.76	1.26	1.83	2.49	3.24	4.07	5.00	6.00	7.08	8.24	9.47
	32	0.31	0.68	1.12	1.63	2.22	2.89	3.64	4.47	5.38	6.37	7.43	8.56
	36	0.28	0.61	1.00	1.46	2.00	2.60	3.29	4.04	4.87	5.78	6.75	7.79
6	2	1.70	3.52	5.44	7.40	9.37	11.4	13.3	15.3	17.3	19.3	21.3	23.2
	3	1.51	3.23	5.11	7.06	9.03	11.0	13.0	15.0	17.0	19.0	21.0	22.9
	4	1.35	2.96	4.79	6.70	8.67	10.7	12.7	14.6	16.6	18.6	20.6	22.6
	5	1.21	2.72	4.48	6.36	8.30	10.3	12.3	14.3	16.3	18.3	20.3	22.3
	6	1.10	2.51	4.20	6.03	7.94	9.90	11.9	13.9	15.9	17.9	19.9	21.9
	7	1.00	2.33	3.96	5.73	7.60	9.53	11.5	13.5	15.5	17.5	19.5	21.5
	8	0.92	2.18	3.73	5.45	7.27	9.17	11.1	13.1	15.1	17.1	19.1	21.1
	9	0.85	2.04	3.53	5.19	6.96	8.83	10.8	12.7	14.7	16.7	18.7	20.7
	10	0.79	1.92	3.35	4.94	6.67	8.50	10.4	12.4	14.3	16.3	18.3	20.3
	12	0.69	1.71	3.02	4.50	6.13	7.88	9.73	11.6	13.6	15.5	17.5	19.5
	14	0.61	1.55	2.75	4.12	5.65	7.33	9.11	11.0	12.9	14.8	16.8	19.8
	16	0.55	1.41	2.51	3.78	5.22	6.83	8.55	10.3	12.2	14.1	16.0	18.0
	18	0.50	1.29	2.31	3.49	4.85	6.39	8.04	9.77	11.6	13.4	15.3	17.3
	20	0.46	1.19	2.13	3.24	4.53	6.00	7.57	9.25	11.0	12.8	14.7	16.6
	24	0.40	1.03	1.84	2.82	3.99	5.32	6.76	8.32	9.97	11.7	13.5	15.3
	28	0.35	0.90	1.62	2.50	3.56	4.76	6.09	7.53	9.08	10.7	12.4	14.2
	32	0.31	0.80	1.44	2.24	3.20	4.30	5.52	6.86	8.32	9.85	11.5	13.1
	36	0.28	0.72	1.30	2.02	2.90	3.92	5.04	6.30	7.66	9.10	10.6	12.2

Table 8-21 (cont.).
Coefficients *C* for Eccentrically Loaded Bolt Groups
Angle = 60°

$C_{req} = \dfrac{P_u}{\phi r_n}$ or $\phi R_n = C \times \phi r_n$

where

P_u	=	factored force, kips
ϕr_n	=	design strength per bolt, kips
ϕR_n	=	design strength of bolt group, kips
e	=	eccentricity of P_u with respect to centroid of bolt group, in. (not tabulated, may be determined by geometry.)
e_x	=	horizontal component of e, in.
s	=	bolt spacing, in.
C	=	coefficient tabulated below.

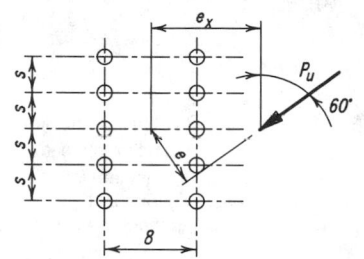

s, in.	e_x, in.	Number of bolts in one vertical row, *n*											
		1	2	3	4	5	6	7	8	9	10	11	12
3	2	1.86	3.71	5.56	7.41	9.28	11.2	13.1	15.0	16.9	18.8	20.8	22.7
	3	1.77	3.52	5.29	7.07	8.88	10.7	12.6	14.5	16.4	18.3	20.2	22.1
	4	1.66	3.31	4.99	6.70	8.45	10.3	12.1	13.9	15.8	17.7	19.6	21.6
	5	1.54	3.10	4.70	6.34	8.04	9.79	11.6	13.4	15.3	17.1	19.0	21.0
	6	1.43	2.90	4.41	6.00	7.64	9.35	11.1	12.9	14.7	16.6	18.5	20.4
	7	1.33	2.71	4.15	5.68	7.27	8.94	10.7	12.4	14.2	16.1	17.9	19.8
	8	1.24	2.54	3.92	5.39	6.94	8.56	10.3	12.0	13.8	15.6	17.4	19.3
	9	1.16	2.38	3.70	5.12	6.63	8.22	9.86	11.6	13.3	15.1	16.9	18.7
	10	1.08	2.24	3.51	4.88	6.34	7.89	9.49	11.2	12.9	14.6	16.4	18.2
	12	0.96	2.00	3.17	4.44	5.82	7.28	8.81	10.4	12.1	13.8	15.5	17.3
	14	0.86	1.81	2.88	4.07	5.36	6.73	8.19	9.72	11.3	13.0	14.7	16.4
	16	0.77	1.64	2.64	3.74	4.95	6.25	7.64	9.11	10.7	12.2	13.9	15.6
	18	0.70	1.51	2.43	3.46	4.59	5.83	7.15	8.56	10.0	11.6	13.2	14.8
	20	0.65	1.39	2.25	3.21	4.28	5.45	6.71	8.06	9.48	11.0	12.5	14.1
	24	0.56	1.20	1.95	2.80	3.76	4.81	5.96	7.19	8.50	9.88	11.3	12.8
	28	0.49	1.06	1.72	2.48	3.34	4.29	5.34	6.47	7.68	8.97	10.3	11.7
	32	0.43	0.94	1.54	2.22	3.00	3.87	4.83	5.87	6.99	8.19	9.46	10.8
	36	0.39	0.85	1.39	2.01	2.72	3.52	4.40	5.36	6.41	7.53	8.71	9.96
6	2	1.86	3.72	5.59	7.50	9.43	11.4	13.3	15.3	17.3	19.2	21.2	23.2
	3	1.77	3.55	5.37	7.25	9.16	11.1	13.0	15.0	17.0	18.9	20.9	22.9
	4	1.66	3.36	5.14	6.98	8.88	10.8	12.7	14.7	16.7	18.6	20.6	22.6
	5	1.54	3.17	4.90	6.72	8.59	10.5	12.4	14.4	16.3	18.3	20.3	22.2
	6	1.43	2.99	4.67	6.46	8.31	10.2	12.1	14.1	16.0	18.0	19.9	21.9
	7	1.33	2.82	4.46	6.21	8.05	9.92	11.8	13.8	15.7	17.7	19.6	21.6
	8	1.24	2.67	4.26	5.98	7.79	9.65	11.5	13.5	15.4	17.3	19.3	21.3
	9	1.16	2.52	4.08	5.76	7.55	9.39	11.3	13.2	15.1	17.0	19.0	20.9
	10	1.08	2.40	3.91	5.56	7.32	9.14	11.0	12.9	14.8	16.7	18.7	20.6
	12	0.96	2.17	3.61	5.20	6.90	8.66	10.5	12.4	14.2	16.1	18.1	20.0
	14	0.86	1.98	3.35	4.87	6.51	8.23	10.0	11.8	13.7	15.6	17.5	19.4
	16	0.77	1.82	3.11	4.57	6.15	7.81	9.56	11.4	13.2	15.1	16.9	18.9
	18	0.70	1.69	2.91	4.30	5.81	7.43	9.13	10.9	12.7	14.5	16.4	18.3
	20	0.65	1.57	2.72	4.05	5.50	7.07	8.73	10.5	12.2	14.1	15.9	17.8
	24	0.56	1.37	2.41	3.61	4.96	6.43	8.00	9.67	11.4	13.2	15.0	16.8
	28	0.49	1.22	2.15	3.25	4.49	5.88	7.38	8.97	10.6	12.3	14.1	15.9
	32	0.43	1.09	1.94	2.94	4.10	5.41	6.83	8.34	9.92	11.6	13.3	15.0
	36	0.39	0.99	1.76	2.69	3.77	5.00	6.35	7.78	9.30	10.9	12.5	14.2

Table 8-21 (cont.).
Coefficients *C* for Eccentrically Loaded Bolt Groups
Angle = 75°

$$C_{req} = \frac{P_u}{\phi r_n} \text{ or } \phi R_n = C \times \phi r_n$$

where

P_u	= factored force, kips
ϕr_n	= design strength per bolt, kips
ϕR_n	= design strength of bolt group, kips
e	= eccentricity of P_u with respect to centroid of bolt group, in. (not tabulated, may be determined by geometry.)
e_x	= horizontal component of e, in.
s	= bolt spacing, in.
C	= coefficient tabulated below.

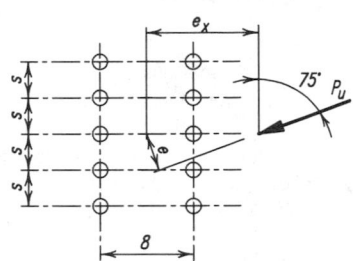

s, in.	e_x, in.	Number of bolts in one vertical row, *n*											
		1	2	3	4	5	6	7	8	9	10	11	12
3	2	1.94	3.87	5.79	7.70	9.61	11.5	13.4	15.3	17.3	19.2	21.1	23.0
	3	1.92	3.82	5.70	7.58	9.45	11.3	13.2	15.1	17.0	18.9	20.8	22.7
	4	1.89	3.75	5.60	7.43	9.26	11.1	12.9	14.8	16.7	18.5	20.4	22.3
	5	1.85	3.67	5.48	7.28	9.07	10.9	12.7	14.5	16.4	18.2	20.1	22.0
	6	1.81	3.59	5.35	7.11	8.87	10.6	12.4	14.2	16.1	17.9	19.8	21.6
	7	1.76	3.50	5.22	6.94	8.67	10.4	12.2	14.0	15.8	17.6	19.4	21.3
	8	1.71	3.40	5.08	6.76	8.46	10.2	11.9	13.7	15.5	17.3	19.1	21.0
	9	1.66	3.30	4.94	6.59	8.26	9.96	11.7	13.4	15.2	17.0	18.8	20.6
	10	1.61	3.20	4.80	6.42	8.06	9.73	11.4	13.2	14.9	16.7	18.5	20.3
	12	1.51	3.01	4.53	6.08	7.67	9.30	11.0	12.7	14.4	16.2	17.9	19.7
	14	1.41	2.82	4.27	5.76	7.31	8.90	10.5	12.2	13.9	15.6	17.4	19.2
	16	1.31	2.65	4.03	5.47	6.96	8.52	10.1	11.8	13.4	15.2	16.9	18.6
	18	1.23	2.48	3.80	5.19	6.64	8.16	9.73	11.3	13.0	14.7	16.4	18.1
	20	1.15	2.34	3.60	4.93	6.34	7.82	9.36	10.9	12.6	14.2	15.9	17.7
	24	1.01	2.08	3.23	4.48	5.80	7.20	8.67	10.2	11.8	13.4	15.0	16.7
	28	0.90	1.87	2.93	4.08	5.33	6.65	8.06	9.52	11.0	12.6	14.2	15.9
	32	0.81	1.69	2.67	3.75	4.91	6.17	7.51	8.91	10.4	11.9	13.5	15.1
	36	0.73	1.54	2.45	3.45	4.55	5.74	7.01	8.36	9.77	11.2	12.8	14.3
6	2	1.94	3.86	5.77	7.68	9.60	11.5	13.5	15.4	17.6	19.6	21.5	23.5
	3	1.92	3.80	5.68	7.55	9.45	11.4	13.3	15.2	17.2	19.1	21.1	23.0
	4	1.89	3.74	5.57	7.42	9.29	11.2	13.1	15.0	16.9	18.9	20.8	22.8
	5	1.85	3.66	5.46	7.29	9.14	11.0	12.9	14.8	16.7	18.7	20.6	22.6
	6	1.81	3.58	5.35	7.15	8.98	10.8	12.7	14.6	16.5	18.5	20.4	22.3
	7	1.76	3.49	5.23	7.01	8.83	10.7	12.5	14.4	16.3	18.3	20.2	22.1
	8	1.71	3.40	5.12	6.88	8.68	10.5	12.4	14.3	16.2	18.1	20.0	21.9
	9	1.66	3.31	5.00	6.74	8.53	10.4	12.2	14.1	16.0	17.9	19.8	21.7
	10	1.61	3.22	4.89	6.61	8.38	10.2	12.0	13.9	15.8	17.7	19.6	21.5
	12	1.51	3.05	4.67	6.36	8.10	9.89	11.7	13.6	15.4	17.3	19.2	21.1
	14	1.41	2.88	4.46	6.12	7.84	9.61	11.4	13.3	15.1	17.0	18.9	20.8
	16	1.31	2.73	4.26	5.89	7.59	9.33	11.1	12.9	14.8	16.6	18.5	20.4
	18	1.23	2.58	4.08	5.68	7.35	9.08	10.8	12.7	14.5	16.3	18.2	20.1
	20	1.15	2.45	3.90	5.47	7.13	8.84	10.6	12.4	14.2	16.0	17.9	19.7
	24	1.01	2.21	3.59	5.10	6.71	8.38	10.1	11.9	13.6	15.5	17.3	19.1
	28	0.90	2.01	3.32	4.77	6.32	7.96	9.65	11.4	13.1	14.9	16.7	18.5
	32	0.81	1.84	3.08	4.47	5.97	7.56	9.21	10.9	12.7	14.4	16.2	18.0
	36	0.73	1.70	2.87	4.19	5.64	7.19	8.80	10.5	12.2	13.9	15.7	17.5

Table 8-22.
Coefficients C for Eccentrically Loaded Bolt Groups
Angle = 0°

$$C_{req} = \frac{P_u}{\phi r_n} \text{ or } \phi R_n = C \times \phi r_n$$

where

P_u = factored force, kips
ϕr_n = design strength per bolt, kips
ϕR_n = design strength of bolt group, kips
e = eccentricity of P_u with respect to centroid of bolt group, in. (not tabulated, may be determined by geometry.)
e_x = horizontal component of e, in.
s = bolt spacing, in.
C = coefficient tabulated below.

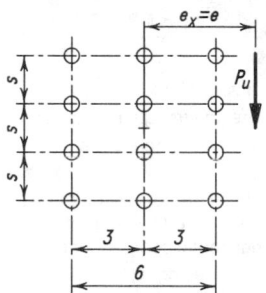

| s, in. | e_x, in. | Number of bolts in one vertical row, n ||||||||||| |
		1	2	3	4	5	6	7	8	9	10	11	12
	2	1.71	4.07	6.81	9.86	13.0	16.1	19.3	22.3	25.4	28.5	31.5	34.5
	3	1.42	3.40	5.79	8.61	11.7	14.8	18.0	21.1	24.3	27.4	30.5	33.6
	4	1.21	2.90	4.97	7.53	10.4	13.4	16.6	19.8	23.0	26.1	29.3	32.5
	5	1.05	2.51	4.34	6.64	9.24	12.1	15.2	18.3	21.5	24.7	27.9	31.1
	6	0.92	2.21	3.85	5.91	8.27	11.0	13.9	16.9	20.0	23.2	26.4	29.7
	7	0.81	1.96	3.44	5.31	7.46	9.95	12.7	15.6	18.6	21.8	25.0	28.2
	8	0.72	1.76	3.11	4.80	6.78	9.09	11.6	14.4	17.3	20.4	23.5	26.7
	9	0.64	1.60	2.83	4.38	6.20	8.34	10.7	13.3	16.1	19.1	22.1	25.2
	10	0.58	1.46	2.59	4.02	5.71	7.70	9.91	12.4	15.0	17.9	20.8	23.8
3	12	0.49	1.24	2.21	3.44	4.91	6.65	8.59	10.8	13.2	15.7	18.5	21.3
	14	0.42	1.08	1.92	3.00	4.30	5.83	7.57	9.53	11.7	14.0	16.5	19.2
	16	0.37	0.95	1.70	2.66	3.82	5.19	6.75	8.51	10.5	12.6	14.9	17.3
	18	0.33	0.85	1.52	2.39	3.43	4.67	6.08	7.68	9.45	11.4	13.5	15.8
	20	0.29	0.77	1.37	2.16	3.11	4.24	5.53	6.99	8.61	10.4	12.3	14.4
	24	0.24	0.64	1.15	1.82	2.62	3.57	4.67	5.92	7.30	8.8	10.5	12.3
	28	0.21	0.55	0.99	1.57	2.26	3.08	4.04	5.12	6.33	7.67	9.13	10.7
	32	0.18	0.49	0.87	1.38	1.98	2.71	3.55	4.51	5.58	6.77	8.06	9.47
	36	0.16	0.43	0.77	1.23	1.77	2.42	3.17	4.03	4.99	6.05	7.21	8.48
	2	1.71	4.85	8.04	11.2	14.2	17.3	20.3	23.2	26.2	29.2	32.2	35.1
	3	1.42	4.24	7.36	10.6	13.7	16.8	19.9	22.9	25.9	28.9	31.9	34.9
	4	1.21	3.72	6.66	9.86	13.1	16.2	19.4	22.4	25.5	28.5	31.6	34.6
	5	1.05	3.29	6.00	9.14	12.4	15.6	18.7	21.9	25.0	28.1	31.1	34.2
	6	0.92	2.93	5.41	8.44	11.6	14.9	18.1	21.2	24.4	27.5	30.6	33.7
	7	0.81	2.63	4.90	7.79	10.9	14.1	17.3	20.6	23.7	26.9	30.0	33.2
	8	0.72	2.38	4.46	7.20	10.2	13.4	16.6	19.8	23.0	26.2	29.4	32.6
	9	0.64	2.17	4.09	6.67	9.54	12.6	15.8	19.1	22.3	25.5	28.7	31.9
	10	0.58	2.00	3.78	6.20	8.94	12.0	15.1	18.3	21.6	24.8	28.0	31.2
6	12	0.49	1.71	3.27	5.41	7.88	10.7	13.7	16.8	20.0	23.3	26.5	29.8
	14	0.42	1.49	2.87	4.78	7.01	9.61	12.4	15.4	18.6	21.8	25.0	28.2
	16	0.37	1.32	2.55	4.28	6.29	8.69	11.3	14.2	17.2	20.3	23.5	26.7
	18	0.33	1.19	2.30	3.86	5.70	7.91	10.4	13.1	15.9	18.9	22.0	25.2
	20	0.29	1.08	2.09	3.51	5.20	7.25	9.54	12.1	14.8	17.7	20.7	23.8
	24	0.24	0.91	1.76	2.97	4.42	6.19	8.19	10.4	12.9	15.5	18.3	21.2
	28	0.21	0.78	1.52	2.57	3.84	5.39	7.14	9.15	11.4	13.7	16.3	19.0
	32	0.18	0.69	1.33	2.27	3.39	4.77	6.33	8.13	10.1	12.3	14.6	17.1
	36	0.16	0.61	1.19	2.03	3.03	4.27	5.67	7.30	9.10	11.1	13.2	15.5

Table 8-22 (cont.).
Coefficients C for Eccentrically Loaded Bolt Groups
Angle = 15°

$$C_{req} = \frac{P_u}{\phi r_n} \text{ or } \phi R_n = C \times \phi r_n$$

where

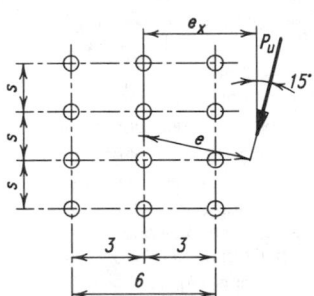

P_u = factored force, kips

ϕr_n = design strength per bolt, kips

ϕR_n = design strength of bolt group, kips

e = eccentricity of P_u with respect to centroid of bolt group, in. (not tabulated, may be determined by geometry.)

e_x = horizontal component of e, in.

s = bolt spacing, in.

C = coefficient tabulated below.

s, in.	e_x, in.	\multicolumn{12}{c}{Number of bolts in one vertical row, n}

s, in.	e_x, in.	1	2	3	4	5	6	7	8	9	10	11	12
	2	1.77	4.10	6.84	9.82	12.9	16.0	19.1	22.2	25.2	28.3	31.3	34.3
	3	1.47	3.45	5.86	8.61	11.6	14.7	17.8	20.9	24.1	27.2	30.3	33.3
	4	1.25	2.95	5.07	7.55	10.4	13.3	16.4	19.5	22.7	25.8	29.0	32.1
	5	1.08	2.57	4.44	6.67	9.26	12.1	15.1	18.1	21.3	24.4	27.6	30.7
	6	0.94	2.26	3.93	5.96	8.33	11.0	13.8	16.8	19.8	23.0	26.1	29.3
	7	0.83	2.01	3.52	5.37	7.55	9.97	12.7	15.5	18.5	21.5	24.7	27.8
	8	0.74	1.81	3.18	4.87	6.88	9.13	11.7	14.4	17.2	20.2	23.2	26.4
	9	0.66	1.64	2.90	4.45	6.31	8.40	10.8	13.3	16.1	18.9	21.9	25.0
	10	0.60	1.50	2.65	4.10	5.81	7.77	9.99	12.4	15.0	17.8	20.7	23.6
3	12	0.50	1.28	2.27	3.52	5.01	6.74	8.71	10.9	13.2	15.8	18.4	21.2
	14	0.43	1.11	1.98	3.08	4.40	5.93	7.69	9.62	11.8	14.1	16.5	19.1
	16	0.38	0.98	1.75	2.73	3.91	5.29	6.87	8.62	10.6	12.7	15.0	17.4
	18	0.34	0.88	1.57	2.45	3.52	4.77	6.20	7.80	9.59	11.5	13.6	15.9
	20	0.30	0.79	1.42	2.22	3.19	4.33	5.65	7.12	8.76	10.5	12.5	14.6
	24	0.25	0.67	1.19	1.87	2.69	3.66	4.78	6.04	7.45	8.99	10.7	12.5
	28	0.22	0.57	1.02	1.61	2.32	3.17	4.14	5.24	6.47	7.82	9.31	10.9
	32	0.19	0.50	0.90	1.42	2.04	2.79	3.65	4.62	5.72	6.92	8.24	9.66
	36	0.17	0.45	0.80	1.26	1.82	2.49	3.26	4.13	5.11	6.20	7.38	8.66
	2	1.77	4.83	7.98	11.1	14.1	17.2	20.2	23.2	26.1	29.1	32.1	35.0
	3	1.47	4.22	7.31	10.5	13.6	16.7	19.7	22.8	25.8	28.8	31.8	34.8
	4	1.25	3.71	6.64	9.77	12.9	16.1	19.2	22.3	25.3	28.3	31.4	34.4
	5	1.08	3.28	6.01	9.06	12.2	15.4	18.5	21.7	24.8	27.8	30.9	33.9
	6	0.94	2.94	5.45	8.38	11.5	14.7	17.8	21.0	24.1	27.2	30.3	33.4
	7	0.83	2.65	4.97	7.75	10.8	13.9	17.1	20.3	23.5	26.6	29.7	32.8
	8	0.74	2.40	4.55	7.17	10.1	13.2	16.4	19.6	22.7	25.9	29.1	32.2
	9	0.66	2.20	4.18	6.66	9.49	12.5	15.6	18.8	22.0	25.2	28.4	31.5
	10	0.60	2.02	3.86	6.20	8.92	11.9	14.9	18.1	21.3	24.5	27.6	30.8
6	12	0.50	1.74	3.34	5.43	7.91	10.6	13.6	16.6	19.8	23.0	26.1	29.3
	14	0.43	1.52	2.94	4.82	7.07	9.60	12.4	15.3	18.4	21.5	24.6	27.3
	16	0.38	1.35	2.62	4.32	6.38	8.71	11.3	14.1	17.0	20.1	23.2	26.3
	18	0.34	1.22	2.36	3.91	5.79	7.95	10.4	13.0	15.8	18.8	21.8	24.9
	20	0.30	1.10	2.14	3.57	5.30	7.31	9.60	12.1	14.8	17.6	20.5	23.5
	24	0.25	0.93	1.81	3.03	4.52	6.26	8.28	10.5	12.9	15.5	18.2	21.1
	28	0.22	0.80	1.56	2.63	3.93	5.47	7.26	9.24	11.4	13.8	16.3	18.9
	32	0.19	0.71	1.37	2.32	3.47	4.85	6.45	8.23	10.2	12.4	14.7	17.1
	36	0.17	0.63	1.23	2.08	3.11	4.35	5.80	7.41	9.23	11.2	13.3	15.6

Table 8-22 (cont.).
Coefficients C for Eccentrically Loaded Bolt Groups
Angle = 30°

$$C_{req} = \frac{P_u}{\phi r_n} \text{ or } \phi R_n = C \times \phi r_n$$

where

P_u	= factored force, kips
ϕr_n	= design strength per bolt, kips
ϕR_n	= design strength of bolt group, kips
e	= eccentricity of P_u with respect to centroid of bolt group, in. (not tabulated, may be determined by geometry.)
e_x	= horizontal component of e, in.
s	= bolt spacing, in.
C	= coefficient tabulated below.

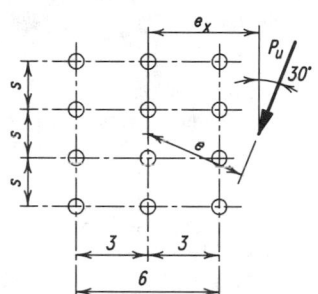

		Number of bolts in one vertical row, n											
s, in.	e_x, in.	1	2	3	4	5	6	7	8	9	10	11	12
	2	1.94	4.26	6.99	9.90	12.9	16.0	19.0	22.0	25.1	28.1	31.1	34.1
	3	1.61	3.63	6.09	8.80	11.7	14.7	17.7	20.8	23.9	27.0	30.0	33.1
	4	1.37	3.15	5.35	7.83	10.6	13.5	16.5	19.5	22.6	25.7	28.7	31.8
	5	1.19	2.77	4.74	7.00	9.54	12.3	15.2	18.2	21.2	24.3	27.4	30.5
	6	1.04	2.45	4.23	6.30	8.67	11.3	14.1	17.0	19.9	23.0	26.0	29.1
	7	0.92	2.19	3.81	5.71	7.92	10.4	13.0	15.8	18.7	21.7	24.7	27.8
	8	0.82	1.98	3.45	5.22	7.27	9.58	12.1	14.8	17.6	20.5	23.4	26.4
	9	0.74	1.80	3.16	4.79	6.71	8.88	11.2	13.8	16.5	19.3	22.2	25.2
	10	0.67	1.65	2.90	4.42	6.22	8.26	10.5	12.9	15.5	18.2	21.1	24.0
3	12	0.56	1.41	2.49	3.82	5.41	7.22	9.23	11.5	13.8	16.4	19.0	21.8
	14	0.48	1.23	2.18	3.36	4.78	6.40	8.22	10.3	12.4	14.8	17.2	19.8
	16	0.42	1.08	1.93	2.99	4.26	5.73	7.40	9.25	11.3	13.4	15.7	18.2
	18	0.38	0.97	1.73	2.69	3.85	5.18	6.71	8.41	10.3	12.3	14.4	16.7
	20	0.34	0.88	1.57	2.44	3.50	4.73	6.14	7.70	9.42	11.3	13.3	15.4
	24	0.28	0.74	1.32	2.06	2.96	4.01	5.22	6.58	8.08	9.72	11.5	13.4
	28	0.24	0.64	1.14	1.78	2.56	3.48	4.54	5.73	7.05	8.51	10.1	11.8
	32	0.21	0.56	1.00	1.57	2.26	3.07	4.01	5.07	6.25	7.55	8.96	10.5
	36	0.19	0.50	0.89	1.40	2.02	2.75	3.59	4.54	5.61	6.78	8.06	9.44
	2	1.94	4.86	7.96	11.0	14.1	17.1	20.1	23.1	26.0	29.0	32.0	35.0
	3	1.61	4.27	7.32	10.4	13.5	16.6	19.6	22.6	25.6	28.6	31.6	34.6
	4	1.37	3.78	6.70	9.75	12.9	15.9	19.0	22.1	25.1	28.1	31.1	34.2
	5	1.19	3.39	6.14	9.10	12.2	15.3	18.4	21.5	24.5	27.6	30.6	33.7
	6	1.04	3.06	5.64	8.48	11.5	14.6	17.7	20.8	23.9	27.0	30.1	33.1
	7	0.92	2.78	5.19	7.91	10.9	13.9	17.0	20.1	23.2	26.3	29.4	32.5
	8	0.82	2.54	4.80	7.38	10.3	13.3	16.3	19.4	22.6	25.7	28.8	31.9
	9	0.74	2.34	4.45	6.90	9.67	12.6	15.7	18.7	21.9	25.0	28.1	31.2
	10	0.67	2.16	4.14	6.46	9.14	12.0	15.0	18.1	21.2	24.3	27.4	30.5
6	12	0.56	1.87	3.61	5.71	8.20	10.9	13.8	16.8	19.8	22.9	26.0	29.1
	14	0.48	1.65	3.20	5.10	7.41	9.95	12.7	15.6	18.5	21.5	24.6	27.7
	16	0.42	1.47	2.86	4.60	6.74	9.12	11.7	14.5	17.3	20.3	23.3	26.4
	18	0.38	1.33	2.58	4.19	6.17	8.39	10.8	13.5	16.2	19.1	22.0	25.0
	20	0.34	1.21	2.35	3.84	5.68	7.75	10.1	12.6	15.2	18.0	20.9	23.8
	24	0.28	1.02	2.00	3.29	4.89	6.71	8.78	11.1	13.5	16.1	18.8	21.6
	28	0.24	0.88	1.73	2.86	4.28	5.90	7.77	9.83	12.1	14.5	17.0	19.6
	32	0.21	0.78	1.52	2.54	3.80	5.25	6.95	8.83	10.9	13.1	15.4	17.9
	36	0.19	0.70	1.36	2.27	3.41	4.73	6.28	8.00	9.88	11.9	14.1	16.4

Table 8-22 (cont.).
Coefficients C for Eccentrically Loaded Bolt Groups
Angle = 45°

$$C_{req} = \frac{P_u}{\phi r_n} \text{ or } \phi R_n = C \times \phi r_n$$

where

P_u = factored force, kips

ϕr_n = design strength per bolt, kips

ϕR_n = design strength of bolt group, kips

e = eccentricity of P_u with respect to centroid of bolt group, in. (not tabulated, may be determined by geometry.)

e_x = horizontal component of e, in.

s = bolt spacing, in.

C = coefficient tabulated below.

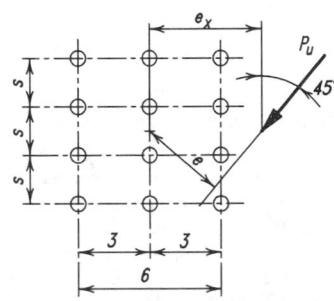

s, in.	e_x, in.	Number of bolts in one vertical row, n											
		1	2	3	4	5	6	7	8	9	10	11	12
	2	2.23	4.67	7.33	10.2	13.1	16.0	19.0	22.0	25.0	28.0	31.0	33.9
	3	1.89	4.06	6.50	9.19	12.0	14.9	17.9	20.9	23.9	26.9	29.9	32.9
	4	1.63	3.57	5.84	8.36	11.1	13.9	16.8	19.7	22.7	25.7	28.7	31.7
	5	1.42	3.17	5.27	7.63	10.2	12.9	15.7	18.6	21.5	24.5	27.5	30.5
	6	1.25	2.84	4.78	6.99	9.40	12.0	14.7	17.6	20.4	23.4	26.3	29.3
	7	1.11	2.57	4.36	6.42	8.70	11.2	13.8	16.6	19.4	22.3	25.2	28.2
	8	0.99	2.33	3.99	5.92	8.09	10.5	13.0	15.7	18.4	21.2	24.1	27.0
	9	0.90	2.13	3.68	5.49	7.54	9.80	12.2	14.8	17.5	20.3	23.1	26.0
	10	0.81	1.96	3.40	5.10	7.05	9.21	11.6	14.0	16.6	19.3	22.1	24.9
3	12	0.68	1.68	2.95	4.46	6.22	8.19	10.4	12.7	15.1	17.7	20.3	23.0
	14	0.59	1.47	2.59	3.95	5.55	7.35	9.34	11.5	13.8	16.2	18.7	21.3
	16	0.52	1.31	2.31	3.54	4.99	6.65	8.49	10.5	12.7	14.9	17.3	19.8
	18	0.46	1.17	2.08	3.20	4.54	6.06	7.77	9.64	11.7	13.8	16.1	18.5
	20	0.41	1.06	1.89	2.92	4.15	5.56	7.15	8.90	10.8	12.8	15.0	17.2
	24	0.35	0.90	1.60	2.48	3.54	4.76	6.15	7.70	9.39	11.2	13.1	15.2
	28	0.30	0.77	1.38	2.15	3.08	4.16	5.39	6.77	8.28	9.91	11.7	13.5
	32	0.26	0.68	1.22	1.90	2.72	3.68	4.79	6.03	7.39	8.87	10.5	12.2
	36	0.23	0.61	1.08	1.69	2.44	3.30	4.30	5.42	6.66	8.02	9.49	11.1
	2	2.23	5.02	8.01	11.0	14.0	17.0	20.0	23.0	25.9	28.9	31.9	34.8
	3	1.89	4.50	7.44	10.4	13.5	16.5	19.5	22.5	25.5	28.4	31.4	34.4
	4	1.63	4.05	6.89	9.86	12.9	15.9	18.9	21.9	24.9	27.9	30.9	33.9
	5	1.42	3.68	6.40	9.30	12.3	15.3	18.3	21.3	24.4	27.4	30.4	33.4
	6	1.25	3.36	5.96	8.78	11.7	14.7	17.7	20.7	23.8	26.8	29.8	32.8
	7	1.11	3.09	5.57	8.29	11.2	14.1	17.1	20.1	23.2	26.2	29.2	32.3
	8	0.99	2.86	5.22	7.84	10.6	13.6	16.5	19.5	22.6	25.6	28.6	31.7
	9	0.90	2.65	4.90	7.43	10.2	13.0	16.0	19.0	22.0	25.0	28.0	31.1
	10	0.81	2.47	4.61	7.04	9.69	12.5	15.4	18.4	21.4	24.4	27.4	30.4
6	12	0.68	2.16	4.11	6.35	8.85	11.6	14.4	17.3	20.2	23.2	26.2	29.2
	14	0.59	1.92	3.69	5.76	8.11	10.7	13.4	16.2	19.1	22.1	25.0	28.0
	16	0.52	1.72	3.34	5.25	7.47	9.94	12.6	15.3	18.1	21.0	23.9	26.9
	18	0.46	1.56	3.04	4.82	6.91	9.26	11.8	14.4	17.2	20.0	22.9	25.8
	20	0.41	1.43	2.79	4.44	6.43	8.66	11.1	13.6	16.3	19.0	21.9	24.7
	24	0.35	1.22	2.38	3.84	5.62	7.64	9.84	12.2	14.7	17.3	20.0	22.8
	28	0.30	1.06	2.08	3.37	4.98	6.81	8.82	11.0	13.4	15.8	18.4	21.1
	32	0.26	0.94	1.84	3.00	4.46	6.12	7.97	10.0	12.2	14.6	17.0	19.5
	36	0.23	0.84	1.65	2.71	4.04	5.56	7.27	9.18	11.2	13.4	15.7	18.1

Table 8-22 (cont.).
Coefficients C for Eccentrically Loaded Bolt Groups
Angle = 60°

$$C_{req} = \frac{P_u}{\phi r_n} \text{ or } \phi R_n = C \times \phi r_n$$

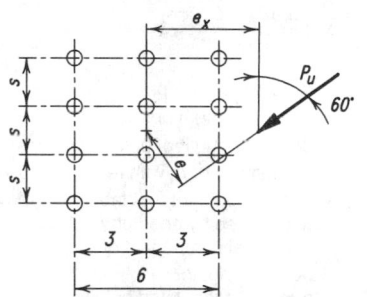

where

P_u	= factored force, kips
ϕr_n	= design strength per bolt, kips
ϕR_n	= design strength of bolt group, kips
e	= eccentricity of P_u with respect to centroid of bolt group, in. (not tabulated, may be determined by geometry.)
e_x	= horizontal component of e, in.
s	= bolt spacing, in.
C	= coefficient tabulated below.

s, in.	e_x, in.	\multicolumn{12}{c}{Number of bolts in one vertical row, n}
		1	2	3	4	5	6	7	8	9	10	11	12
3	2	2.59	5.21	7.88	10.6	13.4	16.3	19.2	22.1	25.0	28.0	30.9	33.9
	3	2.32	4.73	7.27	9.91	12.7	15.5	18.3	21.2	24.1	27.0	30.0	32.9
	4	2.07	4.29	6.69	9.23	11.9	14.6	17.5	20.3	23.2	26.1	29.0	32.0
	5	1.84	3.90	6.18	8.63	11.2	13.9	16.6	19.5	22.3	25.2	28.1	31.0
	6	1.65	3.56	5.73	8.08	10.6	13.2	15.9	18.7	21.5	24.3	27.2	30.1
	7	1.49	3.27	5.32	7.59	10.0	12.6	15.2	17.9	20.7	23.5	26.3	29.2
	8	1.35	3.01	4.95	7.13	9.48	12.0	14.5	17.2	19.9	22.7	25.5	28.4
	9	1.23	2.78	4.63	6.71	8.98	11.4	13.9	16.5	19.2	22.0	24.7	27.6
	10	1.12	2.58	4.34	6.33	8.52	10.9	13.3	15.9	18.5	21.2	24.0	26.8
	12	0.95	2.25	3.84	5.67	7.70	9.91	12.3	14.7	17.3	19.9	22.6	25.3
	14	0.83	1.98	3.43	5.11	7.00	9.08	11.3	13.7	16.1	18.7	21.3	23.9
	16	0.73	1.77	3.09	4.64	6.40	8.36	10.5	12.7	15.1	17.5	20.1	22.6
	18	0.65	1.60	2.81	4.24	5.89	7.73	9.74	11.9	14.2	16.5	19.0	21.5
	20	0.59	1.46	2.57	3.90	5.44	7.19	9.09	11.1	13.3	15.6	17.9	20.4
	24	0.49	1.24	2.20	3.35	4.72	6.27	7.99	9.85	11.9	14.0	16.2	18.5
	28	0.42	1.07	1.91	2.93	4.15	5.55	7.10	8.81	10.7	12.6	14.7	16.8
	32	0.37	0.95	1.69	2.60	3.70	4.97	6.38	7.95	9.65	11.5	13.4	15.4
	36	0.33	0.85	1.51	2.34	3.34	4.49	5.79	7.23	8.81	10.5	12.3	14.2
6	2	2.59	5.32	8.17	11.1	14.0	17.0	19.9	22.9	25.8	28.8	31.8	34.7
	3	2.32	4.94	7.73	10.6	13.5	16.5	19.4	22.4	25.4	28.3	31.3	34.3
	4	2.07	4.57	7.31	10.2	13.1	16.0	19.0	21.9	24.9	27.8	30.8	33.8
	5	1.84	4.25	6.91	9.73	12.6	15.5	18.5	21.4	24.4	27.4	30.3	33.3
	6	1.65	3.95	6.55	9.32	12.2	15.1	18.0	20.9	23.9	26.9	29.8	32.8
	7	1.49	3.69	6.22	8.94	11.8	14.6	17.5	20.5	23.4	26.4	29.3	32.3
	8	1.35	3.46	5.92	8.58	11.4	14.2	17.1	20.0	22.9	25.9	28.8	31.8
	9	1.23	3.25	5.64	8.25	11.0	13.8	16.7	19.6	22.5	25.4	28.4	31.3
	10	1.12	3.06	5.39	7.94	10.6	13.4	16.3	19.1	22.0	24.9	27.9	30.8
	12	0.95	2.73	4.92	7.37	9.97	12.7	15.5	18.3	21.2	24.1	27.0	29.9
	14	0.83	2.46	4.52	6.85	9.36	12.0	14.7	17.5	20.3	23.2	26.1	29.0
	16	0.73	2.23	4.18	6.39	8.80	11.4	14.0	16.8	19.6	22.4	25.3	28.1
	18	0.65	2.04	3.87	5.97	8.28	10.8	13.4	16.1	18.8	21.6	24.4	27.3
	20	0.59	1.88	3.60	5.59	7.81	10.2	12.8	15.4	18.1	20.9	23.7	26.5
	24	0.49	1.63	3.15	4.94	6.99	9.25	11.7	14.2	16.8	19.5	22.2	25.0
	28	0.42	1.43	2.79	4.41	6.31	8.44	10.7	13.1	15.7	18.2	20.9	23.6
	32	0.37	1.27	2.49	3.97	5.74	7.74	9.90	12.2	14.6	17.1	19.7	22.3
	36	0.33	1.15	2.25	3.61	5.26	7.13	9.17	11.4	13.7	16.1	18.6	21.1

Table 8-22 (cont.).
Coefficients *C* for Eccentrically Loaded Bolt Groups
Angle = 75°

$$C_{req} = \frac{P_u}{\phi r_n} \text{ or } \phi R_n = C \times \phi r_n$$

where

P_u	= factored force, kips
ϕr_n	= design strength per bolt, kips
ϕR_n	= design strength of bolt group, kips
e	= eccentricity of P_u with respect to centroid of bolt group, in. (not tabulated, may be determined by geometry.)
e_x	= horizontal component of e, in.
s	= bolt spacing, in.
C	= coefficient tabulated below.

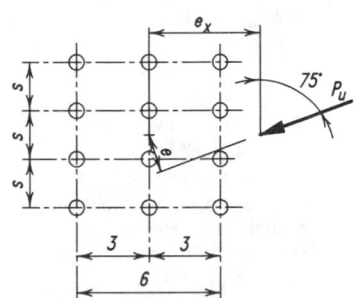

s, in.	e_x, in.	Number of bolts in one vertical row, n											
		1	2	3	4	5	6	7	8	9	10	11	12
	2	2.86	5.68	8.47	11.3	14.1	16.9	19.8	22.6	25.5	28.4	31.3	34.2
	3	2.77	5.49	8.19	10.9	13.7	16.4	19.2	22.1	24.9	27.8	30.7	33.6
	4	2.66	5.27	7.89	10.5	13.2	16.0	18.8	21.6	24.4	27.2	30.1	33.0
	5	2.53	5.04	7.58	10.2	12.8	15.5	18.3	21.0	23.9	26.7	29.5	32.4
	6	2.40	4.81	7.27	9.81	12.4	15.1	17.8	20.6	23.3	26.2	29.0	31.8
	7	2.26	4.57	6.97	9.47	12.0	14.7	17.4	20.1	22.9	25.6	28.4	31.3
	8	2.13	4.35	6.69	9.13	11.7	14.3	16.9	19.6	22.4	25.1	27.9	30.7
	9	2.00	4.13	6.41	8.82	11.3	13.9	16.5	19.2	21.9	24.7	27.4	30.2
	10	1.89	3.93	6.15	8.51	11.0	13.5	16.1	18.8	21.5	24.2	27.0	29.8
3	12	1.67	3.57	5.67	7.95	10.4	12.9	15.4	18.0	20.7	23.4	26.1	28.8
	14	1.49	3.25	5.25	7.44	9.77	12.2	14.7	17.3	19.9	22.6	25.3	28.0
	16	1.34	2.97	4.87	6.98	9.23	11.6	14.1	16.6	19.2	21.8	24.5	27.2
	18	1.21	2.73	4.54	6.56	8.74	11.1	13.5	16.0	18.5	21.1	23.7	26.4
	20	1.10	2.53	4.24	6.18	8.28	10.5	12.9	15.3	17.8	20.4	23.0	25.6
	24	0.93	2.19	3.75	5.52	7.48	9.59	11.8	14.2	16.6	19.1	21.6	24.2
	28	0.80	1.93	3.34	4.97	6.79	8.78	10.9	13.2	15.5	17.9	20.4	22.9
	32	0.71	1.72	3.01	4.51	6.20	8.08	10.1	12.3	14.5	16.8	19.2	21.7
	36	0.63	1.55	2.74	4.12	5.70	7.47	9.40	11.5	13.6	15.9	18.2	20.6
	2	2.86	5.66	8.48	11.3	14.2	17.1	20.1	23.0	26.4	29.3	32.3	35.2
	3	2.77	5.49	8.25	11.1	13.9	16.8	19.7	22.7	25.6	28.5	31.5	34.4
	4	2.66	5.30	8.02	10.8	13.6	16.5	19.4	22.3	25.2	28.2	31.1	34.0
	5	2.53	5.10	7.79	10.6	13.4	16.2	19.1	22.0	24.9	27.8	30.8	33.7
	6	2.40	4.91	7.56	10.3	13.1	15.9	18.8	21.7	24.6	27.5	30.4	33.3
	7	2.26	4.72	7.34	10.1	12.9	15.7	18.5	21.4	24.3	27.2	30.1	33.0
	8	2.13	4.54	7.14	9.83	12.6	15.4	18.3	21.1	24.0	26.9	29.8	32.7
	9	2.00	4.37	6.94	9.61	12.4	15.2	18.0	20.8	23.7	26.6	29.5	32.4
	10	1.89	4.21	6.75	9.40	12.1	14.9	17.7	20.6	23.4	26.3	29.2	32.1
6	12	1.67	3.90	6.39	9.00	11.7	14.4	17.2	20.0	22.9	25.7	28.6	31.5
	14	1.49	3.63	6.06	8.63	11.3	14.0	16.8	19.6	22.4	25.2	28.1	30.9
	16	1.34	3.39	5.75	8.29	10.9	13.6	16.3	19.1	21.9	24.7	27.5	30.4
	18	1.21	3.17	5.47	7.96	10.6	13.2	15.9	18.7	21.4	24.2	27.0	29.9
	20	1.10	2.98	5.22	7.66	10.2	12.9	15.5	18.2	21.0	23.8	26.6	29.4
	24	0.93	2.65	4.76	7.10	9.57	12.2	14.8	17.5	20.2	22.9	25.7	28.5
	28	0.80	2.38	4.37	6.60	8.99	11.5	14.1	16.7	19.4	22.1	24.8	27.6
	32	0.71	2.16	4.03	6.15	8.45	10.9	13.4	16.0	18.7	21.3	24.0	26.8
	36	0.63	1.97	3.73	5.75	7.96	10.3	12.8	15.3	17.9	20.6	23.3	26.0

Table 8-23.
Coefficients C for Eccentrically Loaded Bolt Groups
Angle = 0°

$$C_{req} = \frac{P_u}{\phi r_n} \text{ or } \phi R_n = C \times \phi r_n$$

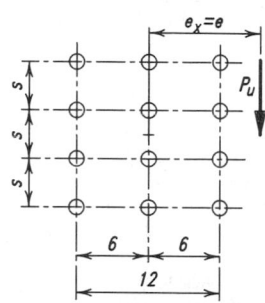

where

P_u	=	factored force, kips
ϕr_n	=	design strength per bolt, kips
ϕR_n	=	design strength of bolt group, kips
e	=	eccentricity of P_u with respect to centroid of bolt group, in. (not tabulated, may be determined by geometry.)
e_x	=	horizontal component of e, in.
s	=	bolt spacing, in.
C	=	coefficient tabulated below.

s, in.	e_x, in.	\multicolumn{12}{c}{Number of bolts in one vertical row, n}
		1	2	3	4	5	6	7	8	9	10	11	12
3	2	2.15	4.55	7.17	10.0	13.0	16.0	19.1	22.2	25.3	28.3	31.4	34.4
	3	1.91	4.06	6.43	9.06	11.9	14.9	17.9	21.0	24.1	27.2	30.3	33.4
	4	1.71	3.65	5.80	8.23	10.9	13.7	16.7	19.8	22.9	26.0	29.1	32.3
	5	1.55	3.31	5.27	7.51	9.97	12.7	15.5	18.5	21.5	24.7	27.8	31.0
	6	1.42	3.02	4.82	6.88	9.16	11.7	14.4	17.3	20.3	23.3	26.4	29.6
	7	1.31	2.77	4.44	6.34	8.46	10.8	13.4	16.1	19.0	22.0	25.1	28.2
	8	1.21	2.56	4.10	5.87	7.85	10.1	12.5	15.1	17.9	20.7	23.7	26.8
	9	1.12	2.38	3.81	5.46	7.31	9.39	11.7	14.1	16.8	19.6	22.5	25.5
	10	1.05	2.21	3.55	5.09	6.84	8.79	10.9	13.3	15.8	18.5	21.3	24.2
	12	0.92	1.94	3.12	4.48	6.03	7.78	9.70	11.8	14.1	16.6	19.1	21.9
	14	0.81	1.72	2.77	3.99	5.38	6.95	8.69	10.6	12.7	14.9	17.3	19.9
	16	0.72	1.53	2.48	3.58	4.84	6.27	7.85	9.60	11.5	13.6	15.8	18.1
	18	0.64	1.38	2.25	3.25	4.40	5.70	7.15	8.75	10.5	12.4	14.4	16.6
	20	0.58	1.26	2.05	2.96	4.02	5.21	6.55	8.03	9.65	11.4	13.3	15.3
	24	0.49	1.06	1.73	2.52	3.42	4.45	5.60	6.88	8.29	9.82	11.5	13.2
	28	0.42	0.92	1.50	2.19	2.97	3.87	4.88	6.00	7.24	8.59	10.1	11.6
	32	0.37	0.81	1.32	1.93	2.63	3.42	4.32	5.32	6.42	7.62	8.93	10.3
	36	0.33	0.72	1.18	1.72	2.35	3.06	3.87	4.77	5.76	6.84	8.02	9.29
6	2	2.15	4.94	7.98	11.1	14.2	17.2	20.2	23.2	26.2	29.2	32.1	35.1
	3	1.91	4.48	7.39	10.5	13.6	16.7	19.8	22.8	25.8	28.9	31.9	34.8
	4	1.71	4.07	6.81	9.86	13.0	16.1	19.3	22.3	25.4	28.5	31.5	34.5
	5	1.55	3.71	6.27	9.22	12.3	15.5	18.6	21.8	24.9	28.0	31.0	34.1
	6	1.42	3.40	5.79	8.61	11.7	14.8	18.0	21.1	24.3	27.4	30.5	33.6
	7	1.31	3.13	5.35	8.05	11.0	14.1	17.3	20.5	23.6	26.8	29.9	33.1
	8	1.21	2.90	4.97	7.53	10.4	13.4	16.6	19.8	23.0	26.1	29.3	32.5
	9	1.12	2.69	4.64	7.07	9.78	12.8	15.9	19.0	22.2	25.4	28.6	31.8
	10	1.05	2.51	4.34	6.64	9.24	12.1	15.2	18.3	21.5	24.7	27.9	31.1
	12	0.92	2.21	3.85	5.91	8.27	11.0	13.9	16.9	20.0	23.2	26.4	29.7
	14	0.81	1.96	3.44	5.31	7.46	9.95	12.7	15.6	18.6	21.8	25.0	28.2
	16	0.72	1.76	3.11	4.80	6.78	9.09	11.6	14.4	17.3	20.4	23.5	26.7
	18	0.64	1.60	2.83	4.38	6.20	8.34	10.7	13.3	16.1	19.1	22.1	25.2
	20	0.58	1.46	2.59	4.02	5.71	7.70	9.91	12.4	15.0	17.9	20.8	23.8
	24	0.49	1.24	2.21	3.44	4.91	6.65	8.59	10.8	13.2	15.7	18.5	21.3
	28	0.42	1.08	1.92	3.00	4.30	5.83	7.57	9.53	11.7	14.0	16.5	19.2
	32	0.37	0.95	1.70	2.66	3.82	5.19	6.75	8.51	10.5	12.6	14.9	17.3
	36	0.33	0.85	1.52	2.39	3.43	4.67	6.08	7.68	9.45	11.4	13.5	15.8

Table 8-23 (cont.).
Coefficients C for Eccentrically Loaded Bolt Groups
Angle = 15°

$$C_{req} = \frac{P_u}{\phi r_n} \text{ or } \phi R_n = C \times \phi r_n$$

where

P_u = factored force, kips

ϕr_n = design strength per bolt, kips

ϕR_n = design strength of bolt group, kips

e = eccentricity of P_u with respect to centroid of bolt group, in. (not tabulated, may be determined by geometry.)

e_x = horizontal component of e, in.

s = bolt spacing, in.

C = coefficient tabulated below.

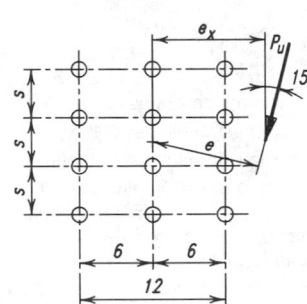

s, in.	e_x, in.	Number of bolts in one vertical row, n											
		1	2	3	4	5	6	7	8	9	10	11	12
3	2	2.22	4.62	7.25	10.1	13.0	16.0	19.0	22.1	25.1	28.2	31.2	34.2
	3	1.97	4.13	6.53	9.13	11.9	14.9	17.9	20.9	24.0	27.1	30.1	33.2
	4	1.77	3.72	5.91	8.31	10.9	13.7	16.7	19.7	22.7	25.8	28.9	32.0
	5	1.61	3.38	5.39	7.60	10.1	12.7	15.5	18.4	21.4	24.5	27.6	30.7
	6	1.47	3.10	4.93	6.98	9.28	11.8	14.4	17.2	20.2	23.2	26.2	29.3
	7	1.35	2.85	4.54	6.45	8.59	10.9	13.5	16.1	19.0	21.9	24.9	27.9
	8	1.25	2.63	4.21	5.98	7.98	10.2	12.6	15.1	17.8	20.7	23.6	26.6
	9	1.16	2.44	3.91	5.57	7.45	9.51	11.8	14.2	16.8	19.5	22.4	25.3
	10	1.08	2.28	3.65	5.21	6.97	8.92	11.1	13.4	15.9	18.5	21.2	24.1
	12	0.94	2.00	3.20	4.59	6.16	7.91	9.84	11.9	14.2	16.6	19.2	21.9
	14	0.83	1.77	2.85	4.09	5.50	7.08	8.84	10.8	12.8	15.0	17.4	19.9
	16	0.74	1.58	2.56	3.68	4.96	6.40	8.00	9.75	11.7	13.7	15.9	18.2
	18	0.66	1.43	2.31	3.34	4.51	5.83	7.30	8.91	10.7	12.6	14.6	16.8
	20	0.60	1.30	2.11	3.05	4.13	5.34	6.70	8.19	9.82	11.6	13.5	15.5
	24	0.50	1.10	1.79	2.59	3.52	4.56	5.74	7.03	8.45	10.0	11.7	13.4
	28	0.43	0.95	1.55	2.25	3.06	3.98	5.01	6.15	7.40	8.77	10.2	11.8
	32	0.38	0.84	1.37	1.99	2.70	3.52	4.43	5.45	6.57	7.79	9.12	10.5
	36	0.34	0.75	1.22	1.78	2.42	3.15	3.98	4.89	5.90	7.01	8.20	9.49
6	2	2.22	4.97	7.97	11.0	14.1	17.1	20.1	23.1	26.1	29.1	32.1	35.0
	3	1.97	4.50	7.40	10.5	13.5	16.6	19.7	22.7	25.7	28.7	31.7	34.7
	4	1.77	4.10	6.84	9.82	12.9	16.0	19.1	22.2	25.2	28.3	31.3	34.3
	5	1.61	3.75	6.32	9.20	12.3	15.4	18.5	21.6	24.7	27.8	30.8	33.9
	6	1.47	3.45	5.86	8.61	11.6	14.7	17.8	20.9	24.1	27.2	30.3	33.3
	7	1.35	3.18	5.44	8.06	11.0	14.0	17.1	20.3	23.4	26.5	29.6	32.7
	8	1.25	2.95	5.07	7.55	10.4	13.3	16.4	19.5	22.7	25.8	29.0	32.1
	9	1.16	2.75	4.73	7.09	9.78	12.7	15.7	18.8	22.0	25.1	28.3	31.4
	10	1.08	2.57	4.44	6.67	9.26	12.1	15.1	18.1	21.3	24.4	27.6	30.7
	12	0.94	2.26	3.93	5.96	8.33	11.0	13.8	16.8	19.8	23.0	26.1	29.3
	14	0.83	2.01	3.52	5.37	7.55	9.97	12.7	15.5	18.5	21.5	24.7	27.8
	16	0.74	1.81	3.18	4.87	6.88	9.13	11.7	14.4	17.2	20.2	23.2	26.4
	18	0.66	1.64	2.90	4.45	6.31	8.40	10.8	13.3	16.1	18.9	21.9	25.0
	20	0.60	1.50	2.65	4.10	5.81	7.77	9.99	12.4	15.0	17.8	20.7	23.6
	24	0.50	1.28	2.27	3.52	5.01	6.74	8.71	10.9	13.2	15.8	18.4	21.2
	28	0.43	1.11	1.98	3.08	4.40	5.93	7.69	9.62	11.8	14.1	16.5	19.1
	32	0.38	0.98	1.75	2.73	3.91	5.29	6.87	8.62	10.6	12.7	15.0	17.4
	36	0.34	0.88	1.57	2.45	3.52	4.77	6.20	7.80	9.59	11.5	13.6	15.9

Table 8-23 (cont.).
Coefficients C for Eccentrically Loaded Bolt Groups
Angle = 30°

$$C_{req} = \frac{P_u}{\phi r_n} \text{ or } \phi R_n = C \times \phi r_n$$

where

P_u = factored force, kips

ϕr_n = design strength per bolt, kips

ϕR_n = design strength of bolt group, kips

e = eccentricity of P_u with respect to centroid of bolt group, in. (not tabulated, may be determined by geometry.)

e_x = horizontal component of e, in.

s = bolt spacing, in.

C = coefficient tabulated below.

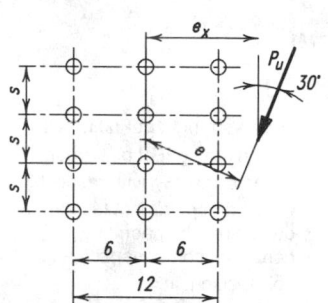

s, in.	e_x, in.	Number of bolts in one vertical row, n											
		1	2	3	4	5	6	7	8	9	10	11	12
3	2	2.40	4.89	7.53	10.3	13.2	16.1	19.1	22.1	25.1	28.1	31.1	34.1
	3	2.15	4.40	6.84	9.45	12.2	15.1	18.0	21.0	24.0	27.0	30.0	33.0
	4	1.94	3.99	6.24	8.69	11.3	14.0	16.9	19.8	22.8	25.8	28.8	31.9
	5	1.76	3.65	5.74	8.02	10.5	13.1	15.8	18.7	21.6	24.6	27.6	30.6
	6	1.61	3.35	5.29	7.42	9.72	12.2	14.8	17.6	20.4	23.4	26.3	29.3
	7	1.49	3.10	4.90	6.89	9.06	11.4	13.9	16.6	19.3	22.2	25.1	28.1
	8	1.37	2.87	4.55	6.42	8.47	10.7	13.1	15.6	18.3	21.1	23.9	26.9
	9	1.28	2.67	4.24	6.00	7.94	10.1	12.4	14.8	17.4	20.0	22.8	25.7
	10	1.19	2.49	3.97	5.63	7.47	9.49	11.7	14.0	16.5	19.1	21.8	24.6
	12	1.04	2.19	3.50	4.98	6.64	8.48	10.5	12.6	14.9	17.3	19.9	22.5
	14	0.92	1.95	3.12	4.46	5.97	7.64	9.46	11.4	13.6	15.8	18.2	20.7
	16	0.82	1.75	2.81	4.03	5.40	6.93	8.61	10.4	12.4	14.5	16.7	19.1
	18	0.74	1.58	2.55	3.66	4.92	6.33	7.89	9.59	11.4	13.4	15.5	17.7
	20	0.67	1.44	2.33	3.35	4.52	5.82	7.27	8.85	10.6	12.4	14.4	16.4
	24	0.56	1.22	1.98	2.86	3.87	5.00	6.26	7.65	9.16	10.8	12.5	14.4
	28	0.48	1.06	1.72	2.49	3.37	4.37	5.48	6.71	8.06	9.51	11.1	12.8
	32	0.42	0.93	1.52	2.20	2.99	3.88	4.87	5.97	7.18	8.49	9.91	11.4
	36	0.38	0.83	1.36	1.97	2.68	3.48	4.38	5.38	6.47	7.66	8.95	10.3
6	2	2.40	5.11	8.05	11.1	14.1	17.1	20.1	23.0	26.0	29.0	32.0	34.9
	3	2.15	4.66	7.51	10.5	13.5	16.5	19.6	22.6	25.6	28.6	31.6	34.6
	4	1.94	4.26	6.99	9.90	12.9	16.0	19.0	22.0	25.1	28.1	31.1	34.1
	5	1.76	3.92	6.52	9.34	12.3	15.3	18.4	21.5	24.5	27.6	30.6	33.6
	6	1.61	3.63	6.09	8.80	11.7	14.7	17.7	20.8	23.9	27.0	30.0	33.1
	7	1.49	3.38	5.70	8.30	11.1	14.1	17.1	20.2	23.2	26.3	29.4	32.5
	8	1.37	3.15	5.35	7.83	10.6	13.5	16.5	19.5	22.6	25.7	28.7	31.8
	9	1.28	2.95	5.03	7.40	10.0	12.9	15.8	18.8	21.9	25.0	28.1	31.2
	10	1.19	2.77	4.74	7.00	9.54	12.3	15.2	18.2	21.2	24.3	27.4	30.5
	12	1.04	2.45	4.23	6.30	8.67	11.3	14.1	17.0	19.9	23.0	26.0	29.1
	14	0.92	2.19	3.81	5.71	7.92	10.4	13.0	15.8	18.7	21.7	24.7	27.8
	16	0.82	1.98	3.45	5.22	7.27	9.58	12.1	14.8	17.6	20.5	23.4	26.4
	18	0.74	1.80	3.16	4.79	6.71	8.88	11.2	13.8	16.5	19.3	22.2	25.2
	20	0.67	1.65	2.90	4.42	6.22	8.26	10.5	12.9	15.5	18.2	21.1	24.0
	24	0.56	1.41	2.49	3.82	5.41	7.22	9.23	11.5	13.8	16.4	19.0	21.8
	28	0.48	1.23	2.18	3.43	4.78	6.40	8.22	10.3	12.4	14.8	17.2	19.8
	32	0.42	1.08	1.93	2.99	4.26	5.73	7.40	9.25	11.3	13.4	15.7	18.2
	36	0.38	0.97	1.73	2.69	3.85	5.18	6.71	8.41	10.3	12.3	14.4	16.7

Table 8-23 (cont.).
Coefficients *C* for Eccentrically Loaded Bolt Groups
Angle = 45°

$$C_{req} = \frac{P_u}{\phi r_n} \text{ or } \phi R_n = C \times \phi r_n$$

where

P_u = factored force, kips

ϕr_n = design strength per bolt, kips

ϕR_n = design strength of bolt group, kips

e = eccentricity of P_u with respect to centroid of bolt group, in. (not tabulated, may be determined by geometry.)

e_x = horizontal component of e, in.

s = bolt spacing, in.

C = coefficient tabulated below.

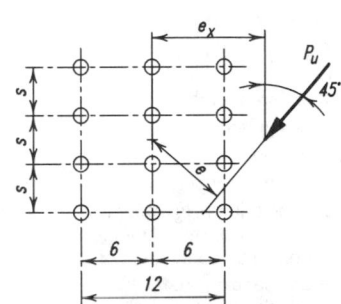

s, in.	e_x, in.	Number of bolts in one vertical row, n											
		1	2	3	4	5	6	7	8	9	10	11	12
3	2	2.64	5.30	8.01	10.8	13.6	16.4	19.3	22.3	25.2	28.1	31.1	34.0
	3	2.43	4.90	7.44	10.1	12.8	15.6	18.4	21.3	24.2	27.2	30.1	33.1
	4	2.23	4.52	6.89	9.38	12.0	14.7	17.5	20.3	23.2	26.1	29.0	32.0
	5	2.05	4.17	6.40	8.75	11.2	13.9	16.6	19.3	22.2	25.0	27.9	30.9
	6	1.89	3.86	5.96	8.20	10.6	13.1	15.7	18.4	21.2	23.99	26.9	29.8
	7	1.75	3.59	5.57	7.70	9.99	12.4	14.9	17.5	20.2	23.0	25.8	28.7
	8	1.63	3.35	5.22	7.25	9.43	11.7	14.2	16.7	19.3	22.1	24.8	27.7
	9	1.52	3.13	4.90	6.83	8.91	11.1	13.5	15.9	18.5	21.2	23.9	26.7
	10	1.42	2.94	4.61	6.45	8.44	10.6	12.8	15.2	17.7	20.3	23.0	25.7
	12	1.25	2.60	4.11	5.78	7.60	9.58	11.7	14.0	16.3	18.8	21.3	23.9
	14	1.11	2.32	3.69	5.21	6.90	8.73	10.7	12.8	15.0	17.4	19.8	22.3
	16	0.99	2.09	3.34	4.74	6.29	8.00	9.85	11.8	13.9	16.1	18.5	20.9
	18	0.90	1.90	3.04	4.33	5.77	7.36	9.10	10.96	12.9	15.0	17.3	19.5
	20	0.81	1.73	2.79	3.98	5.33	6.81	8.44	10.2	12.1	14.1	16.2	18.4
	24	0.68	1.47	2.38	3.42	4.60	5.91	7.35	8.91	10.6	12.4	14.3	16.3
	28	0.59	1.28	2.08	2.99	4.03	5.20	6.49	7.90	9.42	11.1	12.8	14.6
	32	0.52	1.13	1.84	2.65	3.59	4.63	5.80	7.07	8.46	9.95	11.6	13.3
	36	0.46	1.01	1.65	2.38	3.23	4.17	5.23	6.40	7.67	9.04	10.5	12.1
6	2	2.64	5.38	8.22	11.1	14.1	17.0	20.0	22.97	25.9	28.9	31.9	34.8
	3	2.43	5.02	7.78	10.7	13.6	16.6	19.5	22.5	25.5	28.5	31.4	34.4
	4	2.23	4.67	7.33	10.2	13.1	16.0	19.0	22.0	25.0	28.0	31.0	34.0
	5	2.05	4.34	6.90	9.66	12.5	15.5	18.4	21.4	24.4	27.4	30.4	33.4
	6	1.89	4.06	6.50	9.19	12.0	14.9	17.9	20.9	23.9	26.9	29.9	32.9
	7	1.75	3.80	6.16	8.76	11.5	14.4	17.3	20.3	23.3	26.3	29.3	32.3
	8	1.63	3.57	5.84	8.36	11.1	13.9	16.8	19.7	22.7	25.7	28.7	31.7
	9	1.52	3.36	5.54	7.99	10.6	13.4	16.2	19.2	22.1	25.1	28.1	31.1
	10	1.42	3.17	5.27	7.63	10.2	12.9	15.7	18.6	21.5	24.5	27.5	30.5
	12	1.25	2.84	4.78	6.99	9.40	12.0	14.7	17.6	20.4	23.4	26.3	29.3
	14	1.11	2.57	4.36	6.42	8.70	11.2	13.8	16.6	19.4	22.3	25.2	28.2
	16	0.99	2.33	3.99	5.92	8.09	10.5	13.0	15.7	18.4	21.2	24.1	27.1
	18	0.90	2.13	3.68	5.49	7.54	9.80	12.2	14.8	17.5	20.3	23.1	26.0
	20	0.81	1.96	3.40	5.10	7.05	9.21	11.6	14.0	16.6	19.3	22.1	24.9
	24	0.68	1.68	2.95	4.46	6.22	8.19	10.4	12.7	15.1	17.7	20.3	23.0
	28	0.59	1.47	2.59	3.95	5.55	7.35	9.34	11.5	13.8	16.2	18.7	21.3
	32	0.52	1.31	2.31	3.54	4.99	6.65	8.49	10.5	12.7	14.9	17.3	19.8
	36	0.46	1.17	2.08	3.20	4.54	6.06	7.77	9.64	11.7	13.8	16.1	18.5

Table 8-23 (cont.).
Coefficients C for Eccentrically Loaded Bolt Groups
Angle = 60°

$$C_{req} = \frac{P_u}{\phi r_n} \text{ or } \phi R_n = C \times \phi r_n$$

where

P_u = factored force, kips

ϕr_n = design strength per bolt, kips

ϕR_n = design strength of bolt group, kips

e = eccentricity of P_u with respect to centroid of bolt group, in. (not tabulated, may be determined by geometry.)

e_x = horizontal component of e, in.

s = bolt spacing, in.

C = coefficient tabulated below.

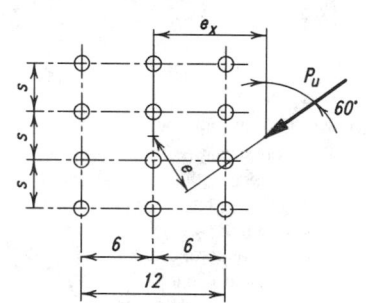

s, in.	e_x, in.	Number of bolts in one vertical row, n											
		1	2	3	4	5	6	7	8	9	10	11	12
3	2	2.83	5.64	8.45	11.3	14.1	16.9	19.8	22.6	25.5	28.4	31.3	34.2
	3	2.72	5.43	8.13	10.8	13.6	16.3	19.1	21.9	24.8	27.6	30.5	33.4
	4	2.59	5.18	7.77	10.4	13.0	15.7	18.5	21.2	24.0	26.8	29.7	32.5
	5	2.46	4.92	7.40	9.92	12.5	15.1	17.8	20.5	23.2	26.0	28.9	31.7
	6	2.32	4.66	7.03	9.46	12.0	14.5	17.1	19.8	22.5	25.2	28.0	30.8
	7	2.19	4.41	6.68	9.02	11.4	13.9	16.5	19.1	21.8	24.5	27.2	30.0
	8	2.07	4.17	6.35	8.61	11.0	13.4	15.9	18.4	21.1	23.7	26.5	29.2
	9	1.95	3.95	6.04	8.22	10.5	12.9	15.3	17.8	20.4	23.0	25.7	28.5
	10	1.84	3.74	5.75	7.86	10.1	12.4	14.8	17.3	19.8	22.4	25.0	27.7
	12	1.65	3.38	5.22	7.19	9.28	11.5	13.8	16.2	18.6	21.1	23.7	26.3
	14	1.49	3.06	4.76	6.61	8.58	10.7	12.9	15.2	17.5	20.0	22.5	25.0
	16	1.35	2.79	4.37	6.09	7.95	9.93	12.0	14.2	16.5	18.9	21.3	23.8
	18	1.23	2.55	4.02	5.64	7.39	9.28	11.3	13.4	15.6	17.9	20.3	22.7
	20	1.12	2.35	3.72	5.24	6.90	8.69	10.6	12.6	14.8	17.0	19.3	21.7
	24	0.95	2.02	3.22	4.57	6.06	7.68	9.43	11.3	13.3	15.4	17.5	19.8
	28	0.83	1.76	2.84	4.04	5.39	6.86	8.47	10.2	12.0	14.0	16.0	18.1
	32	0.73	1.56	2.53	3.61	4.84	6.19	7.66	9.26	11.0	12.8	14.7	16.7
	36	0.65	1.40	2.27	3.26	4.38	5.62	6.98	8.46	10.1	11.7	13.5	15.4
6	2	2.83	5.64	8.47	11.3	14.2	17.1	20.0	23.0	25.9	28.9	31.8	34.8
	3	2.72	5.44	8.19	11.0	13.8	16.7	19.6	22.6	25.5	28.4	31.4	34.3
	4	2.59	5.21	7.88	10.6	13.4	16.3	19.2	22.1	25.0	28.0	30.9	33.9
	5	2.46	4.97	7.57	10.3	13.1	15.9	18.8	21.7	24.6	27.5	30.4	33.4
	6	2.32	4.73	7.27	9.91	12.7	15.5	18.3	21.2	24.1	27.0	30.0	33.0
	7	2.19	4.51	6.97	9.56	12.3	15.0	17.9	20.8	23.7	26.6	29.5	32.4
	8	2.07	4.29	6.69	9.23	11.9	14.6	17.5	20.3	23.2	26.1	29.0	32.0
	9	1.95	4.09	6.43	8.92	11.5	14.3	17.0	19.9	22.8	25.6	28.6	31.5
	10	1.84	3.90	6.18	8.63	11.2	13.9	16.6	19.5	22.3	25.2	28.1	31.0
	12	1.65	3.56	5.73	8.08	10.6	13.2	15.9	18.7	21.5	24.3	27.2	30.1
	14	1.49	3.27	5.32	7.59	10.0	12.6	15.2	17.9	20.7	23.5	26.3	29.2
	16	1.35	3.01	4.95	7.13	9.48	12.0	14.5	17.2	19.9	22.7	25.5	28.4
	18	1.23	2.78	4.63	6.71	8.98	11.4	13.9	16.5	19.2	22.0	24.7	27.6
	20	1.12	2.58	4.34	6.33	8.52	10.9	13.3	15.9	18.5	21.2	24.0	27.0
	24	0.95	2.25	3.84	5.67	7.70	9.91	12.3	14.7	17.3	19.9	22.6	25.3
	28	0.83	1.98	3.43	5.11	7.00	9.08	11.3	13.7	16.1	18.7	21.3	23.9
	32	0.73	1.77	3.09	4.64	6.40	8.36	10.5	12.7	15.1	17.5	20.1	22.6
	36	0.65	1.60	2.81	4.24	5.89	7.73	9.74	11.9	14.2	16.5	19.0	21.5

Table 8-23 (cont.).
Coefficients *C* for Eccentrically Loaded Bolt Groups
Angle = 75°

$$C_{req} = \frac{P_u}{\phi r_n} \text{ or } \phi R_n = C \times \phi r_n$$

where

P_u = factored force, kips

ϕr_n = design strength per bolt, kips

ϕR_n = design strength of bolt group, kips

e = eccentricity of P_u with respect to centroid of bolt group, in. (not tabulated, may be determined by geometry.)

e_x = horizontal component of e, in.

s = bolt spacing, in.

C = coefficient tabulated below.

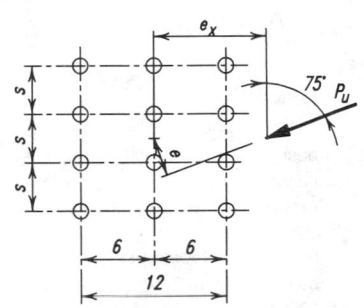

s, in.	e_x, in.	Number of bolts in one vertical row, *n*											
		1	2	3	4	5	6	7	8	9	10	11	12
3	2	2.92	5.83	8.73	11.6	14.5	17.4	20.3	23.1	26.0	28.9	31.8	34.7
	3	2.89	5.77	8.63	11.5	14.3	17.2	20.0	22.8	25.7	28.5	31.4	34.2
	4	2.86	5.70	8.51	11.3	14.1	16.9	19.7	22.5	25.3	28.1	30.9	33.7
	5	2.82	5.61	8.38	11.1	13.9	16.6	19.4	22.1	24.9	27.7	30.5	33.3
	6	2.77	5.51	8.23	10.9	13.6	16.3	19.0	21.8	24.5	27.2	30.0	32.8
	7	2.72	5.40	8.06	10.7	13.4	16.0	18.7	21.4	24.1	26.8	29.6	32.3
	8	2.66	5.29	7.89	10.5	13.1	15.7	18.3	21.0	23.7	26.4	29.1	31.9
	9	2.60	5.16	7.71	10.3	12.8	15.4	18.0	20.6	23.3	26.0	28.7	31.4
	10	2.53	5.04	7.53	10.1	12.6	15.1	17.7	20.3	22.9	25.6	28.3	31.0
	12	2.40	4.78	7.16	9.57	12.0	14.5	17.0	19.6	22.1	24.8	27.4	30.1
	14	2.26	4.52	6.80	9.12	11.5	13.9	16.4	18.9	21.4	24.0	26.6	29.3
	16	2.13	4.27	6.45	8.68	11.0	13.3	15.8	18.2	20.7	23.3	25.9	28.5
	18	2.00	4.03	6.12	8.27	10.5	12.8	15.2	17.6	20.1	22.6	25.1	27.7
	20	1.89	3.81	5.80	7.88	10.1	12.3	14.6	17.0	19.4	21.9	24.4	27.0
	24	1.67	3.41	5.24	7.18	9.22	11.4	13.6	15.9	18.2	20.7	23.1	25.6
	28	1.49	3.06	4.75	6.56	8.49	10.5	12.6	14.9	17.1	19.5	21.9	24.3
	32	1.34	2.77	4.33	6.02	7.84	9.77	11.8	13.9	16.1	18.4	20.7	23.1
	36	1.21	2.52	3.97	5.56	7.27	9.10	11.1	13.1	15.2	17.4	19.7	22.0
6	2	2.92	5.82	8.71	11.6	14.5	17.4	20.3	23.5	26.4	29.3	32.3	35.2
	3	2.89	5.76	8.60	11.4	14.3	17.1	20.0	22.9	25.8	28.7	31.7	34.6
	4	2.86	5.68	8.47	11.3	14.1	16.9	19.8	22.6	25.5	28.4	31.3	34.2
	5	2.82	5.59	8.34	11.1	13.9	16.7	19.5	22.4	25.2	28.1	31.0	33.9
	6	2.77	5.49	8.19	10.9	13.7	16.4	19.2	22.1	24.9	27.8	30.7	33.6
	7	2.72	5.39	8.04	10.7	13.4	16.2	19.0	21.8	24.6	27.5	30.4	33.3
	8	2.66	5.27	7.89	10.5	13.2	16.0	18.8	21.6	24.4	27.2	30.1	33.0
	9	2.60	5.16	7.74	10.4	13.0	15.8	18.5	21.3	24.1	27.0	29.8	32.7
	10	2.53	5.04	7.58	10.2	12.8	15.5	18.3	21.0	23.9	26.7	29.5	32.4
	12	2.40	4.81	7.27	9.81	12.4	15.1	17.8	20.6	23.3	26.2	29.0	31.8
	14	2.26	4.57	6.97	9.47	12.0	14.7	17.4	20.1	22.9	25.6	28.4	31.3
	16	2.13	4.35	6.69	9.13	11.7	14.3	16.9	19.6	22.4	25.1	27.9	30.7
	18	2.00	4.13	6.41	8.82	11.3	13.9	16.5	19.2	21.9	24.7	27.4	30.2
	20	1.89	3.93	6.15	8.51	11.0	13.5	16.1	18.8	21.5	24.2	27.0	29.8
	24	1.67	3.57	5.67	7.95	10.4	12.9	15.4	18.0	20.7	23.4	26.1	28.8
	28	1.49	3.25	5.25	7.44	9.77	12.2	14.7	17.3	19.9	22.6	25.3	28.0
	32	1.34	2.97	4.87	6.98	9.23	11.6	14.1	16.6	19.2	21.8	24.5	27.2
	36	1.21	2.73	4.54	6.56	8.74	11.1	13.5	16.0	18.5	21.1	23.7	26.4

Table 8-24.
Coefficients *C* for Eccentrically Loaded Bolt Groups
Angle = 0°

$$C_{req} = \frac{P_u}{\phi r_n} \text{ or } \phi R_n = C \times \phi r_n$$

where

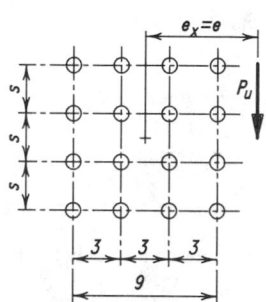

P_u = factored force, kips

ϕr_n = design strength per bolt, kips

ϕR_n = design strength of bolt group, kips

e = eccentricity of P_u with respect to centroid of bolt group, in. (not tabulated, may be determined by geometry.)

e_x = horizontal component of e, in.

s = bolt spacing, in.

C = coefficient tabulated below.

s, in.	e_x, in.	Number of bolts in one vertical row, n											
		1	2	3	4	5	6	7	8	9	10	11	12
3	2	2.60	5.70	9.24	13.2	17.3	21.4	25.6	29.7	33.8	37.9	41.9	45.9
	3	2.23	4.92	8.05	11.7	15.6	19.7	23.9	28.1	32.3	36.5	40.6	44.7
	4	1.94	4.30	7.09	10.4	14.0	18.0	22.1	26.3	30.5	34.7	38.9	43.1
	5	1.69	3.79	6.30	9.29	12.6	16.4	20.3	24.4	28.6	32.9	37.1	41.4
	6	1.49	3.37	5.65	8.37	11.5	14.9	18.7	22.7	26.7	30.9	35.2	39.4
	7	1.32	3.03	5.10	7.59	10.4	13.7	17.2	21.0	24.9	29.0	33.2	37.5
	8	1.18	2.74	4.63	6.92	9.56	12.6	15.9	19.5	23.3	27.3	31.4	35.5
	9	1.07	2.50	4.24	6.35	8.81	11.6	14.7	18.1	21.7	25.6	29.6	33.7
	10	0.98	2.29	3.89	5.86	8.15	10.8	13.7	16.9	20.3	24.1	27.9	31.9
	12	0.83	1.96	3.34	5.06	7.06	9.37	12.0	14.9	17.9	21.3	24.9	28.6
	14	0.73	1.72	2.92	4.44	6.21	8.27	10.6	13.2	16.0	19.1	22.4	25.8
	16	0.65	1.52	2.59	3.95	5.54	7.39	9.48	11.9	14.4	17.2	20.2	23.4
	18	0.58	1.37	2.33	3.55	4.99	6.67	8.57	10.7	13.1	15.7	18.4	21.4
	20	0.53	1.24	2.11	3.23	4.53	6.07	7.81	9.77	11.9	14.3	16.9	19.6
	24	0.44	1.04	1.78	2.72	3.83	5.14	6.62	8.30	10.2	12.2	14.4	16.8
	28	0.38	0.90	1.54	2.35	3.31	4.45	5.73	7.20	8.82	10.6	12.6	14.7
	32	0.34	0.79	1.36	2.07	2.91	3.92	5.05	6.35	7.79	9.38	11.1	13.0
	36	0.30	0.71	1.21	1.85	2.60	3.50	4.51	5.68	6.96	8.39	9.95	11.6
6	2	2.60	6.48	10.7	14.9	18.9	23.0	27.0	31.0	34.9	38.9	42.9	46.8
	3	2.23	5.75	9.79	14.0	18.2	22.3	26.4	30.5	34.5	38.5	42.5	46.5
	4	1.94	5.12	8.91	13.1	17.4	21.6	25.7	29.9	33.9	38.0	42.0	46.1
	5	1.69	4.58	8.10	12.2	16.5	20.7	24.9	29.1	33.2	37.4	41.4	45.5
	6	1.49	4.13	7.37	11.3	15.5	19.7	24.0	28.3	32.5	36.7	40.8	44.9
	7	1.32	3.74	6.74	10.5	14.5	18.8	23.1	27.4	31.6	35.8	40.0	44.1
	8	1.18	3.41	6.20	9.73	13.6	17.8	22.1	26.4	30.6	34.9	39.1	43.3
	9	1.07	3.13	5.73	9.05	12.8	16.9	21.1	25.4	29.7	34.0	38.3	42.5
	10	0.98	2.89	5.31	8.45	12.1	16.0	20.1	24.4	28.7	33.0	37.3	41.5
	12	0.83	2.50	4.63	7.43	10.7	14.3	18.3	22.5	26.7	31.0	35.3	39.6
	14	0.73	2.19	4.09	6.60	9.53	12.9	16.7	20.6	24.7	29.0	33.3	37.6
	16	0.65	1.95	3.65	5.93	8.59	11.7	15.2	19.0	22.9	27.1	31.3	35.5
	18	0.58	1.76	3.29	5.37	7.81	10.8	14.0	17.5	21.3	25.3	29.4	33.6
	20	0.53	1.60	2.99	4.90	7.15	9.85	12.9	16.2	19.8	23.6	27.6	31.7
	24	0.44	1.35	2.53	4.16	6.10	8.44	11.1	14.0	17.3	20.8	24.5	28.3
	28	0.38	1.17	2.19	3.61	5.31	7.37	9.69	12.3	15.2	18.4	21.8	25.3
	32	0.34	1.03	1.93	3.19	4.69	6.53	8.61	11.0	13.6	16.5	19.6	22.9
	36	0.30	0.92	1.72	2.85	4.20	5.85	7.73	9.89	12.3	14.9	17.8	20.8

Table 8-24 (cont.).
Coefficients C for Eccentrically Loaded Bolt Groups
Angle = 15°

$$C_{req} = \frac{P_u}{\phi r_n} \text{ or } \phi R_n = C \times \phi r_n$$

where

P_u	= factored force, kips
ϕr_n	= design strength per bolt, kips
ϕR_n	= design strength of bolt group, kips
e	= eccentricity of P_u with respect to centroid of bolt group, in. (not tabulated, may be determined by geometry.)
e_x	= horizontal component of e, in.
s	= bolt spacing, in.
C	= coefficient tabulated below.

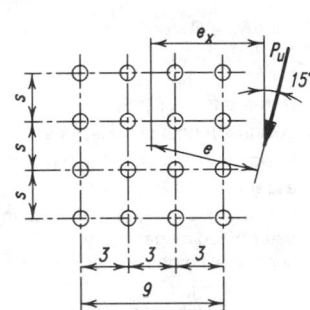

s, in.	e_x, in.	\multicolumn{12}{c}{Number of bolts in one vertical row, n}
		1	2	3	4	5	6	7	8	9	10	11	12
3	2	2.68	5.77	9.31	13.2	17.2	21.3	25.4	29.5	33.6	37.7	41.7	45.7
	3	2.30	5.00	8.17	11.7	15.6	19.6	23.7	27.9	32.0	36.2	40.2	44.3
	4	1.99	4.38	7.22	10.4	14.1	17.9	22.0	26.0	30.2	34.4	38.5	42.7
	5	1.74	3.88	6.43	9.37	12.8	16.4	20.2	24.2	28.3	32.5	36.7	40.9
	6	1.53	3.45	5.77	8.47	11.6	15.0	18.6	22.5	26.6	30.6	34.8	39.0
	7	1.36	3.10	5.21	7.71	10.6	13.7	17.2	20.9	24.8	28.8	32.9	37.1
	8	1.22	2.81	4.74	7.05	9.70	12.7	15.9	19.5	23.2	27.1	31.1	35.2
	9	1.11	2.57	4.34	6.48	8.95	11.8	14.8	18.1	21.7	25.5	29.4	33.4
	10	1.01	2.36	4.00	5.98	8.29	10.9	13.8	17.0	20.4	24.0	27.7	31.6
	12	0.86	2.02	3.44	5.18	7.21	9.52	12.2	15.0	18.1	21.4	24.9	28.5
	14	0.75	1.77	3.01	4.55	6.36	8.43	10.8	13.3	16.1	19.2	22.4	25.8
	16	0.67	1.57	2.68	4.05	5.67	7.54	9.66	12.0	14.6	17.3	20.3	23.5
	18	0.60	1.41	2.40	3.65	5.12	6.81	8.74	10.9	13.3	15.8	18.6	21.5
	20	0.54	1.28	2.18	3.32	4.66	6.21	7.98	9.95	12.1	14.5	17.1	19.8
	24	0.46	1.08	1.84	2.80	3.94	5.26	6.78	8.47	10.4	12.4	14.6	17.0
	28	0.40	0.93	1.59	2.43	3.41	4.56	5.89	7.37	9.02	10.9	12.8	14.9
	32	0.35	0.82	1.40	2.14	3.00	4.03	5.19	6.51	7.98	9.59	11.3	13.2
	36	0.31	0.73	1.25	1.91	2.68	3.60	4.65	5.83	7.15	8.59	10.2	11.9
6	2	2.68	6.48	10.6	14.7	18.8	22.9	26.9	30.9	34.9	38.8	42.8	46.7
	3	2.30	5.75	9.75	13.9	18.1	22.2	26.3	30.3	34.4	38.3	42.4	46.3
	4	1.99	5.13	8.91	13.0	17.2	21.4	25.5	29.7	33.7	37.7	41.8	45.8
	5	1.74	4.61	8.14	12.1	16.3	20.5	24.7	28.9	33.0	37.1	41.2	45.2
	6	1.53	4.17	7.45	11.2	15.3	19.5	23.7	27.9	32.2	36.3	40.4	44.5
	7	1.36	3.79	6.84	10.4	14.5	18.6	22.8	27.0	31.3	35.4	39.6	43.7
	8	1.22	3.46	6.30	9.71	13.6	17.6	21.8	26.0	30.3	34.5	38.7	42.9
	9	1.11	3.19	5.83	9.05	12.8	16.8	20.9	25.1	29.3	33.5	37.8	42.0
	10	1.01	2.94	5.42	8.47	12.1	15.9	20.0	24.1	28.3	32.6	36.8	41.0
	12	0.86	2.55	4.73	7.47	10.7	14.3	18.2	22.2	26.4	30.6	34.8	39.1
	14	0.75	2.24	4.18	6.66	9.62	12.9	16.6	20.5	24.5	28.6	32.9	37.1
	16	0.67	2.00	3.74	6.00	8.71	11.8	15.2	18.9	22.8	26.8	30.9	35.1
	18	0.60	1.80	3.38	5.45	7.94	10.8	14.0	17.5	21.2	25.1	29.1	33.2
	20	0.54	1.64	3.08	4.98	7.28	9.92	13.0	16.2	19.8	23.5	27.4	31.4
	24	0.46	1.39	2.60	4.25	6.23	8.54	11.2	14.1	17.3	20.8	24.4	28.1
	28	0.40	1.20	2.26	3.69	5.43	7.48	9.85	12.5	15.4	18.5	21.8	25.3
	32	0.35	1.06	1.99	3.26	4.81	6.65	8.77	11.1	13.8	16.6	19.7	22.9
	36	0.31	0.94	1.78	2.92	4.31	5.97	7.89	10.0	12.5	15.1	17.9	20.9

Table 8-24 (cont.).
Coefficients C for Eccentrically Loaded Bolt Groups
Angle = 30°

$$C_{req} = \frac{P_u}{\phi r_n} \text{ or } \phi R_n = C \times \phi r_n$$

where

P_u	= factored force, kips	
ϕr_n	= design strength per bolt, kips	
ϕR_n	= design strength of bolt group, kips	
e	= eccentricity of P_u with respect to centroid of bolt group, in. (not tabulated, may be determined by geometry.)	
e_x	= horizontal component of e, in.	
s	= bolt spacing, in.	
C	= coefficient tabulated below.	

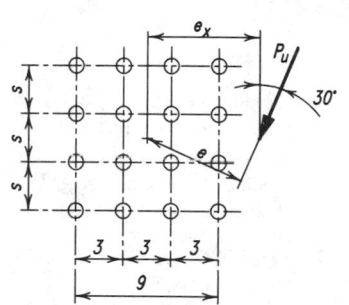

s, in.	e_x, in.	Number of bolts in one vertical row, n											
		1	2	3	4	5	6	7	8	9	10	11	12
	2	2.90	6.06	9.59	13.4	17.3	21.3	25.3	29.4	33.4	37.5	41.4	45.4
	3	2.50	5.31	8.52	12.1	15.8	19.8	23.8	27.8	31.9	35.9	40.0	44.0
	4	2.18	4.70	7.62	10.9	14.4	18.2	22.1	26.1	30.1	34.2	38.3	42.4
	5	1.91	4.18	6.85	9.86	13.2	16.8	20.6	24.5	28.4	32.5	36.6	40.7
	6	1.69	3.75	6.19	8.98	12.1	15.5	19.1	22.9	26.8	30.7	34.8	38.9
	7	1.51	3.38	5.63	8.21	11.2	14.4	17.8	21.4	25.2	29.1	33.1	37.1
	8	1.36	3.07	5.14	7.55	10.3	13.3	16.6	20.0	23.7	27.5	31.4	35.4
	9	1.23	2.81	4.73	6.97	9.54	12.4	15.5	18.8	22.3	26.1	29.9	33.7
	10	1.13	2.59	4.37	6.46	8.88	11.6	14.6	17.7	21.1	24.7	28.3	32.2
3	12	0.96	2.23	3.78	5.62	7.78	10.2	12.9	15.8	18.9	22.2	25.7	29.3
	14	0.84	1.95	3.32	4.96	6.90	9.08	11.6	14.2	17.1	20.2	23.4	26.8
	16	0.74	1.73	2.96	4.43	6.19	8.17	10.4	12.9	15.5	18.4	21.4	24.6
	18	0.67	1.56	2.66	4.00	5.60	7.41	9.46	11.8	14.2	16.9	19.7	22.7
	20	0.61	1.42	2.42	3.65	5.11	6.77	8.67	10.8	13.1	15.5	18.2	21.0
	24	0.51	1.20	2.04	3.09	4.34	5.77	7.41	9.22	11.2	13.4	15.7	18.2
	28	0.44	1.03	1.77	2.68	3.77	5.01	6.46	8.05	9.83	11.8	13.9	16.1
	32	0.39	0.91	1.56	2.36	3.32	4.43	5.71	7.14	8.72	10.5	12.3	14.4
	36	0.35	0.81	1.39	2.11	2.97	3.97	5.12	6.40	7.84	9.41	11.1	13.0
	2	2.90	6.59	10.6	14.7	18.7	22.7	26.7	30.8	34.7	38.7	42.6	46.6
	3	2.50	5.88	9.83	13.9	18.0	22.0	26.1	30.1	34.1	38.2	42.2	46.1
	4	2.18	5.30	9.05	13.0	17.1	21.2	25.4	29.4	33.5	37.5	41.5	45.5
	5	1.91	4.81	8.35	12.3	16.3	20.4	24.5	28.6	32.7	36.8	40.8	44.9
	6	1.69	4.38	7.72	11.4	15.4	19.5	23.6	27.8	31.9	35.9	40.0	44.1
	7	1.51	4.01	7.15	10.7	14.6	18.6	22.7	26.9	31.0	35.1	39.2	43.3
	8	1.36	3.69	6.64	10.0	13.8	17.7	21.8	25.9	30.0	34.2	38.3	42.4
	9	1.23	3.41	6.19	9.41	13.0	16.9	20.9	25.0	29.1	33.3	37.4	41.6
	10	1.13	3.16	5.79	8.85	12.4	16.2	20.1	24.1	28.2	32.4	36.5	40.6
6	12	0.96	2.76	5.09	7.88	11.2	14.7	18.5	22.4	26.4	30.6	34.6	38.8
	14	0.84	2.44	4.54	7.08	10.1	13.4	17.0	20.9	24.7	28.8	32.9	36.9
	16	0.74	2.18	4.08	6.41	9.21	12.3	15.7	19.4	23.2	27.1	31.1	35.1
	18	0.67	1.97	3.70	5.85	8.45	11.4	14.6	18.1	21.8	25.6	29.4	33.4
	20	0.61	1.80	3.38	5.37	7.80	10.5	13.6	16.9	20.4	24.1	27.9	31.8
	24	0.51	1.53	2.87	4.61	6.74	9.16	11.9	14.9	18.1	21.5	25.1	28.8
	28	0.44	1.32	2.49	4.02	5.91	8.07	10.5	13.3	16.3	19.4	22.7	26.2
	32	0.39	1.17	2.20	3.57	5.26	7.20	9.45	11.9	14.6	17.6	20.7	23.9
	36	0.35	1.05	1.97	3.21	4.73	6.49	8.55	10.9	13.4	16.0	18.9	22.0

Table 8-24 (cont.).
Coefficients C for Eccentrically Loaded Bolt Groups
Angle = 45°

$$C_{req} = \frac{P_u}{\phi r_n} \text{ or } \phi R_n = C \times \phi r_n$$

where

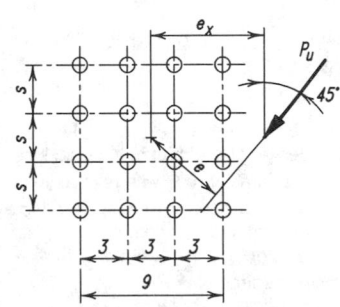

P_u	= factored force, kips
ϕr_n	= design strength per bolt, kips
ϕR_n	= design strength of bolt group, kips
e	= eccentricity of P_u with respect to centroid of bolt group, in. (not tabulated, may be determined by geometry.)
e_x	= horizontal component of e, in.
s	= bolt spacing, in.
C	= coefficient tabulated below.

s, in.	e_x, in.	Number of bolts in one vertical row, n											
		1	2	3	4	5	6	7	8	9	10	11	12
3	2	3.26	6.62	10.2	13.9	17.7	21.6	25.5	29.4	33.4	37.4	41.3	45.3
	3	2.87	5.92	9.19	12.7	16.4	20.2	24.0	27.96	32.0	35.9	39.9	43.9
	4	2.54	5.31	8.36	11.7	15.2	18.9	22.6	26.5	30.5	34.4	38.4	42.4
	5	2.25	4.78	7.63	10.8	14.1	17.6	21.3	25.1	29.0	32.9	36.9	40.8
	6	2.01	4.33	6.99	9.94	13.1	16.5	20.1	23.8	27.5	31.4	35.3	39.3
	7	1.81	3.93	6.42	9.20	12.2	15.5	18.9	22.5	26.2	30.0	33.9	37.7
	8	1.64	3.60	5.92	8.55	11.4	14.6	17.9	21.3	24.9	28.6	32.4	36.3
	9	1.49	3.31	5.49	7.96	10.7	13.7	16.9	20.3	23.8	27.4	31.1	34.9
	10	1.37	3.06	5.10	7.44	10.1	13.0	16.0	19.2	22.7	26.2	29.9	33.6
	12	1.17	2.65	4.46	6.55	8.93	11.6	14.4	17.5	20.7	24.0	27.5	31.1
	14	1.03	2.33	3.95	5.83	8.00	10.5	13.1	15.9	19.0	22.1	25.4	28.8
	16	0.91	2.08	3.54	5.24	7.23	9.47	12.0	14.6	17.5	20.4	23.6	26.8
	18	0.82	1.88	3.20	4.75	6.59	8.66	10.9	13.5	16.1	18.9	21.9	25.0
	20	0.74	1.71	2.92	4.35	6.04	7.96	10.1	12.5	15.0	17.6	20.5	23.5
	24	0.63	1.45	2.48	3.71	5.18	6.84	8.71	10.8	13.0	15.4	18.0	20.7
	28	0.54	1.26	2.15	3.23	4.52	5.99	7.65	9.50	11.5	13.7	16.0	18.5
	32	0.48	1.11	1.90	2.86	4.00	5.31	6.81	8.48	10.3	12.3	14.4	16.7
	36	0.43	0.99	1.69	2.56	3.59	4.77	6.13	7.64	9.30	11.2	13.1	15.2
6	2	3.26	6.89	10.8	14.8	18.7	22.7	26.6	30.6	34.6	38.5	42.5	46.5
	3	2.87	6.28	10.1	14.0	18.0	22.0	26.0	30.0	33.9	37.9	41.9	45.9
	4	2.54	5.74	9.38	13.3	17.3	21.2	25.3	29.2	33.2	37.2	41.2	45.2
	5	2.25	5.27	8.75	12.6	16.5	20.4	24.5	28.5	32.5	36.5	40.5	44.5
	6	2.01	4.85	8.20	11.9	15.7	19.7	23.7	27.7	31.7	35.7	39.7	43.8
	7	1.81	4.49	7.70	11.3	15.1	19.0	22.9	26.9	30.9	34.9	39.0	43.0
	8	1.64	4.16	7.25	10.7	14.4	18.2	22.2	26.1	30.1	34.1	38.2	42.2
	9	1.49	3.87	6.83	10.2	13.8	17.5	21.4	25.4	29.4	33.3	37.4	41.4
	10	1.37	3.62	6.45	9.65	13.1	16.9	20.7	24.6	28.5	32.5	36.6	40.6
	12	1.17	3.19	5.78	8.75	12.0	15.6	19.3	23.2	27.0	31.0	35.0	39.0
	14	1.03	2.84	5.21	7.97	11.1	14.5	18.1	21.8	25.6	29.5	33.4	37.4
	16	0.91	2.56	4.74	7.30	10.3	13.5	16.9	20.6	24.3	28.1	32.0	35.9
	18	0.82	2.33	4.33	6.72	9.48	12.6	15.9	19.4	23.0	26.7	30.6	34.4
	20	0.74	2.13	3.98	6.21	8.83	11.8	15.0	18.4	21.9	25.5	29.2	33.1
	24	0.63	1.82	3.42	5.38	7.74	10.4	13.4	16.5	19.8	23.2	26.8	30.5
	28	0.54	1.59	2.99	4.74	6.87	9.30	12.0	14.9	18.0	21.3	24.7	28.2
	32	0.48	1.41	2.65	4.22	6.17	8.38	10.9	13.6	16.5	19.5	22.8	26.1
	36	0.43	1.26	2.38	3.81	5.59	7.62	9.89	12.5	15.2	18.0	21.1	24.3

Table 8-24 (cont.).
Coefficients C for Eccentrically Loaded Bolt Groups
Angle = 60°

$$C_{req} = \frac{P_u}{\phi r_n} \text{ or } \phi R_n = C \times \phi r_n$$

where

P_u = factored force, kips

ϕr_n = design strength per bolt, kips

ϕR_n = design strength of bolt group, kips

e = eccentricity of P_u with respect to centroid of bolt group, in. (not tabulated, may be determined by geometry.)

e_x = horizontal component of e, in.

s = bolt spacing, in.

C = coefficient tabulated below.

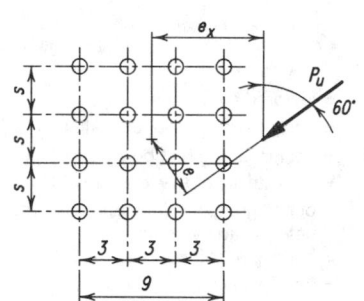

s, in.	e_x, in.	Number of bolts in one vertical row, n											
		1	2	3	4	5	6	7	8	9	10	11	12
3	2	3.63	7.25	10.9	14.6	18.3	22.1	25.9	29.7	33.6	37.5	41.4	45.3
	3	3.38	6.77	10.3	13.8	17.4	21.1	24.8	28.6	32.4	36.3	40.2	44.1
	4	3.10	6.27	9.55	13.0	16.5	20.1	23.7	27.5	31.3	35.1	38.9	42.8
	5	2.84	5.80	8.92	12.2	15.6	19.1	22.7	26.4	30.1	33.9	37.8	41.6
	6	2.60	5.36	8.33	11.5	14.8	18.2	21.7	25.4	29.1	32.8	36.6	40.4
	7	2.38	4.96	7.79	10.9	14.1	17.4	20.9	24.4	28.0	31.8	35.5	39.3
	8	2.19	4.60	7.30	10.2	13.4	16.7	20.0	23.5	27.1	30.7	34.4	38.2
	9	2.02	4.28	6.85	9.68	12.7	15.9	19.2	22.6	26.1	29.7	33.4	37.1
	10	1.87	3.99	6.45	9.17	12.1	15.2	18.4	21.8	25.3	28.8	32.4	36.1
	12	1.62	3.51	5.75	8.27	11.0	13.9	17.0	20.3	23.6	27.0	30.6	34.1
	14	1.43	3.12	5.18	7.50	10.1	12.9	15.8	18.9	22.1	25.4	28.9	32.4
	16	1.27	2.81	4.70	6.85	9.23	11.9	14.7	17.6	20.7	24.0	27.3	30.7
	18	1.15	2.56	4.29	6.28	8.52	11.0	13.7	16.5	19.5	22.6	25.9	29.1
	20	1.04	2.34	3.95	5.80	7.89	10.2	12.8	15.5	18.4	21.4	24.5	27.7
	24	0.88	2.00	3.39	5.01	6.87	8.98	11.3	13.8	16.4	19.2	22.1	25.2
	28	0.76	1.74	2.96	4.39	6.07	7.97	10.1	12.3	14.8	17.4	20.1	23.0
	32	0.67	1.54	2.63	3.91	5.43	7.15	9.06	11.2	13.5	15.9	18.4	21.1
	36	0.60	1.38	2.36	3.52	4.91	6.48	8.22	10.2	12.3	14.5	16.9	19.4
6	2	3.63	7.29	11.1	14.9	18.8	22.7	26.6	30.5	34.5	38.4	42.4	46.3
	3	3.38	6.88	10.6	14.3	18.2	22.1	26.0	29.9	33.9	37.8	41.8	45.7
	4	3.10	6.46	10.0	13.8	17.6	21.5	25.4	29.3	33.3	37.2	41.1	45.1
	5	2.84	6.06	9.55	13.2	17.0	20.9	24.7	28.7	32.6	36.5	40.4	44.4
	6	2.60	5.69	9.09	12.7	16.4	20.3	24.2	28.1	31.9	35.9	39.8	43.8
	7	2.38	5.34	8.66	12.2	15.9	19.7	23.6	27.4	31.3	35.2	39.2	43.1
	8	2.19	5.03	8.27	11.7	15.4	19.1	22.9	26.8	30.7	34.6	38.5	42.4
	9	2.02	4.74	7.90	11.3	14.9	18.6	22.4	26.2	30.1	34.0	37.9	41.8
	10	1.87	4.47	7.55	10.9	14.5	18.1	21.9	25.7	29.5	33.4	37.3	41.2
	12	1.62	4.01	6.93	10.1	13.6	17.2	20.8	24.5	28.3	32.2	36.0	39.9
	14	1.43	3.63	6.38	9.46	12.8	16.2	19.9	23.5	27.3	31.0	34.9	38.7
	16	1.27	3.31	5.91	8.84	12.1	15.4	18.9	22.6	26.3	30.0	33.8	37.6
	18	1.15	3.04	5.49	8.28	11.3	14.6	18.0	21.6	25.2	28.9	32.7	36.5
	20	1.04	2.81	5.12	7.77	10.8	13.9	17.2	20.8	24.3	28.0	31.7	35.4
	24	0.88	2.44	4.49	6.90	9.62	12.6	15.8	19.1	22.6	26.1	29.8	33.4
	28	0.76	2.15	3.99	6.18	8.70	11.5	14.5	17.7	21.1	24.5	28.0	31.6
	32	0.67	1.91	3.58	5.58	7.93	10.6	13.4	16.5	19.7	23.0	26.4	29.9
	36	0.60	1.73	3.24	5.08	7.27	9.76	12.5	15.4	18.4	21.6	24.9	28.3

Table 8-24 (cont.).
Coefficients C for Eccentrically Loaded Bolt Groups
Angle = 75°

$$C_{req} = \frac{P_u}{\phi r_n} \text{ or } \phi R_n = C \times \phi r_n$$

where

P_u = factored force, kips

ϕr_n = design strength per bolt, kips

ϕR_n = design strength of bolt group, kips

e = eccentricity of P_u with respect to centroid of bolt group, in. (not tabulated, may be determined by geometry.)

e_x = horizontal component of e, in.

s = bolt spacing, in.

C = coefficient tabulated below.

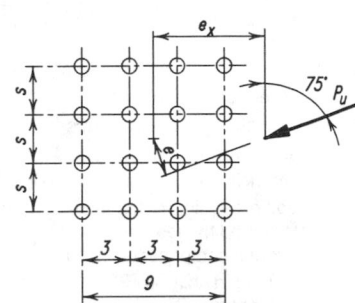

s, in.	e_x, in.	Number of bolts in one vertical row, n											
		1	2	3	4	5	6	7	8	9	10	11	12
3	2	3.86	7.69	11.5	15.3	19.1	22.9	26.7	30.5	34.3	38.2	42.1	45.9
	3	3.79	7.53	11.2	14.9	18.6	22.4	26.1	29.9	33.6	37.5	41.3	45.1
	4	3.70	7.34	11.0	14.6	18.2	21.9	25.5	29.2	33.0	36.7	40.6	44.3
	5	3.59	7.13	10.6	14.2	17.7	21.3	24.9	28.6	32.3	36.1	39.8	43.6
	6	3.47	6.89	10.3	13.8	17.2	20.8	24.4	28.0	31.7	35.4	39.1	42.9
	7	3.34	6.65	9.98	13.4	16.8	20.3	23.8	27.4	31.1	34.8	38.5	42.2
	8	3.20	6.40	9.64	12.9	16.4	19.8	23.3	26.9	30.4	34.1	37.8	41.5
	9	3.07	6.16	9.31	12.6	15.9	19.3	22.8	26.3	29.9	33.5	37.1	40.8
	10	2.94	5.91	8.98	12.2	15.4	18.8	22.2	25.7	29.3	32.9	36.6	40.2
	12	2.68	5.45	8.36	11.5	14.6	17.9	21.3	24.8	28.3	31.8	35.4	39.0
	14	2.45	5.03	7.79	10.7	13.9	17.1	20.4	23.8	27.3	30.8	34.3	37.9
	16	2.24	4.65	7.28	10.1	13.2	16.3	19.6	22.9	26.3	29.8	33.2	36.8
	18	2.06	4.31	6.81	9.55	12.5	15.5	18.8	22.0	25.4	28.8	32.2	35.8
	20	1.90	4.01	6.40	9.03	11.9	14.9	18.0	21.2	24.5	27.9	31.3	34.8
	24	1.63	3.51	5.69	8.13	10.8	13.6	16.6	19.7	22.9	26.2	29.5	32.9
	28	1.43	3.11	5.11	7.36	9.83	12.5	15.4	18.3	21.4	24.6	27.8	31.1
	32	1.27	2.79	4.62	6.71	9.02	11.5	14.3	17.1	20.0	23.2	26.3	29.5
	36	1.14	2.53	4.22	6.15	8.31	10.7	13.3	16.0	18.9	21.8	24.9	28.0
6	2	3.86	7.67	11.5	15.3	19.1	23.0	26.9	30.8	35.2	39.1	43.0	47.0
	3	3.79	7.51	11.2	15.0	18.8	22.6	26.4	30.4	34.3	38.1	42.1	46.0
	4	3.70	7.32	11.0	14.7	18.4	22.2	26.0	29.9	33.8	37.7	41.6	45.5
	5	3.59	7.12	10.7	14.4	18.1	21.9	25.6	29.5	33.3	37.3	41.1	45.0
	6	3.47	6.92	10.4	14.1	17.7	21.5	25.3	29.1	32.9	36.8	40.7	44.6
	7	3.34	6.70	10.2	13.8	17.4	21.1	24.9	28.7	32.6	36.4	40.2	44.1
	8	3.20	6.49	9.92	13.5	17.1	20.8	24.5	28.3	32.1	36.0	39.8	43.7
	9	3.07	6.28	9.66	13.2	16.8	20.5	24.2	28.0	31.8	35.6	39.5	43.3
	10	2.94	6.08	9.42	12.9	16.5	20.2	23.9	27.6	31.4	35.2	39.0	42.9
	12	2.68	5.69	8.95	12.4	15.9	19.5	23.2	26.9	30.7	34.5	38.3	42.1
	14	2.45	5.33	8.51	11.9	15.4	19.0	22.6	26.3	30.1	33.8	37.6	41.4
	16	2.24	4.99	8.10	11.5	14.9	18.5	22.1	25.7	29.4	33.1	36.9	40.7
	18	2.06	4.69	7.72	11.0	14.4	17.9	21.5	25.1	28.8	32.5	36.2	40.0
	20	1.90	4.42	7.36	10.6	13.9	17.4	21.0	24.6	28.2	31.9	35.6	39.3
	24	1.63	3.95	6.74	9.83	13.1	16.5	20.0	23.6	27.1	30.7	34.4	38.1
	28	1.43	3.57	6.21	9.16	12.4	15.7	19.0	22.5	26.1	29.7	33.3	36.9
	32	1.27	3.25	5.74	8.56	11.6	14.9	18.2	21.6	25.1	28.6	32.2	35.9
	36	1.14	2.98	5.33	8.02	11.0	14.1	17.3	20.7	24.1	27.6	31.2	34.8

Table 8-25.
Coefficients *C* for Eccentrically Loaded Bolt Groups
Angle = 0°

$$C_{req} = \frac{P_u}{\phi r_n} \text{ or } \phi R_n = C \times \phi r_n$$

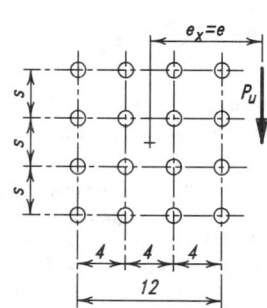

where

P_u = factored force, kips

ϕr_n = design strength per bolt, kips

ϕR_n = design strength of bolt group, kips

e = eccentricity of P_u with respect to centroid of bolt group, in. (not tabulated, may be determined by geometry.)

e_x = horizontal component of e, in.

s = bolt spacing, in.

C = coefficient tabulated below.

| s, in. | e_x, in. | \multicolumn{12}{c}{Number of bolts in one vertical row, n} |||||||||||
		1	2	3	4	5	6	7	8	9	10	11	12
	2	2.82	5.98	9.46	13.3	17.3	21.3	25.5	29.6	33.7	37.7	41.8	45.8
	3	2.50	5.31	8.43	12.0	15.7	19.7	23.8	28.0	32.2	36.3	40.4	44.6
	4	2.23	4.74	7.58	10.8	14.3	18.2	22.2	26.3	30.4	34.6	38.8	43.0
	5	2.01	4.27	6.86	9.82	13.1	16.7	20.5	24.5	28.6	32.8	37.0	41.3
	6	1.81	3.86	6.24	8.96	12.0	15.4	19.0	22.9	26.9	31.0	35.2	39.4
	7	1.64	3.52	5.70	8.22	11.1	14.2	17.6	21.3	25.2	29.2	33.3	37.5
	8	1.49	3.22	5.24	7.57	10.2	13.2	16.4	19.9	23.6	27.5	31.5	35.6
	9	1.36	2.96	4.83	7.01	9.48	12.3	15.3	18.6	22.1	25.9	29.8	33.8
	10	1.25	2.73	4.47	6.51	8.83	11.4	14.3	17.5	20.8	24.4	28.2	32.1
3	12	1.07	2.37	3.89	5.68	7.74	10.1	12.6	15.5	18.5	21.8	25.3	29.0
	14	0.94	2.08	3.42	5.02	6.86	8.95	11.3	13.8	16.6	19.6	22.8	26.2
	16	0.83	1.86	3.05	4.49	6.15	8.04	10.2	12.5	15.0	17.8	20.7	23.9
	18	0.75	1.67	2.75	4.06	5.56	7.29	9.22	11.4	13.7	16.3	19.0	21.9
	20	0.68	1.52	2.50	3.70	5.07	6.65	8.43	10.4	12.6	14.9	17.5	20.2
	24	0.58	1.29	2.12	3.14	4.30	5.66	7.18	8.88	10.8	12.8	15.0	17.4
	28	0.50	1.12	1.84	2.72	3.73	4.92	6.24	7.73	9.37	11.2	13.1	15.2
	32	0.44	0.98	1.62	2.40	3.30	4.34	5.51	6.84	8.29	9.90	11.6	13.5
	36	0.40	0.88	1.45	2.15	2.95	3.89	4.94	6.13	7.43	8.88	10.4	12.1
	2	2.82	6.54	10.6	14.8	18.9	22.9	26.9	30.9	34.9	38.9	42.8	46.8
	3	2.50	5.90	9.81	14.0	18.1	22.3	26.4	30.4	34.5	38.5	42.5	46.5
	4	2.23	5.33	9.01	13.1	17.3	21.5	25.7	29.8	33.9	37.9	42.0	46.0
	5	2.01	4.84	8.27	12.2	16.4	20.6	24.8	29.0	33.2	37.3	41.4	45.5
	6	1.81	4.42	7.60	11.4	15.5	19.7	24.0	28.2	32.4	36.6	40.7	44.8
	7	1.64	4.05	7.02	10.6	14.6	18.8	23.0	27.3	31.5	35.7	39.9	44.1
	8	1.49	3.73	6.51	9.94	13.7	17.8	22.0	26.3	30.6	34.8	39.1	43.3
	9	1.36	3.45	6.06	9.30	13.0	16.9	21.1	25.3	29.6	33.9	38.2	42.4
	10	1.25	3.20	5.66	8.72	12.2	16.1	20.2	24.4	28.6	32.9	37.2	41.5
6	12	1.07	2.80	4.98	7.73	10.9	14.5	18.4	22.5	26.7	30.9	35.2	39.5
	14	0.94	2.47	4.43	6.92	9.81	13.2	16.8	20.7	24.8	29.0	33.2	37.5
	16	0.83	2.21	3.98	6.25	8.90	12.0	15.4	19.1	23.0	27.1	31.3	35.5
	18	0.75	2.00	3.60	5.68	8.13	11.0	14.2	17.7	21.4	25.3	29.4	33.6
	20	0.68	1.82	3.29	5.21	7.47	10.1	13.1	16.4	20.0	23.7	27.7	31.7
	24	0.58	1.55	2.79	4.45	6.40	8.72	11.3	14.3	17.5	20.9	24.5	28.3
	28	0.50	1.34	2.42	3.87	5.59	7.64	9.96	12.6	15.5	18.6	21.9	25.5
	32	0.44	1.18	2.14	3.43	4.95	6.79	8.87	11.2	13.8	16.7	19.7	23.0
	36	0.40	1.06	1.92	3.07	4.44	6.10	7.98	10.1	12.5	15.1	17.9	20.9

Table 8-25 (cont.).
Coefficients C for Eccentrically Loaded Bolt Groups
Angle = 15°

$$C_{req} = \frac{R_u}{\phi r_n} \text{ or } \phi R_n = C \times \phi r_n$$

where

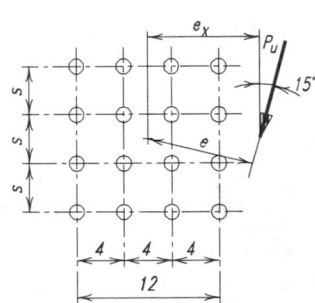

P_u = factored force, kips

ϕr_n = design strength per bolt, kips

ϕR_n = design strength of bolt group, kips

e = eccentricity of P_u with respect to centroid of bolt group, in. (not tabulated, may be determined by geometry.)

e_x = horizontal component of e, in.

s = bolt spacing, in.

C = coefficient tabulated below.

s, in.	e_x, in.	Number of bolts in one vertical row, n											
		1	2	3	4	5	6	7	8	9	10	11	12
3	2	2.91	6.06	9.56	13.3	17.2	21.3	25.3	29.4	33.5	37.5	41.6	45.6
	3	2.57	5.40	8.57	12.0	15.8	19.7	23.7	27.8	31.9	36.1	40.2	44.3
	4	2.30	4.84	7.72	10.9	14.4	18.2	22.1	26.1	30.2	34.3	38.5	42.6
	5	2.06	4.37	6.99	9.93	13.2	16.7	20.5	24.4	28.5	32.6	36.7	40.9
	6	1.86	3.96	6.37	9.09	12.1	15.5	19.0	22.8	26.7	30.8	34.9	39.0
	7	1.69	3.61	5.83	8.36	11.2	14.3	17.7	21.3	25.1	29.0	33.1	37.2
	8	1.53	3.31	5.36	7.72	10.4	13.3	16.5	19.9	23.6	27.4	31.3	35.3
	9	1.40	3.04	4.95	7.15	9.64	12.4	15.4	18.7	22.2	25.8	29.7	33.6
	10	1.29	2.81	4.59	6.65	9.00	11.6	14.5	17.6	20.9	24.4	28.1	31.9
	12	1.11	2.44	4.00	5.82	7.90	10.2	12.8	15.6	18.7	21.9	25.3	28.9
	14	0.97	2.15	3.52	5.15	7.02	9.12	11.5	14.0	16.8	19.8	22.9	26.3
	16	0.86	1.92	3.15	4.61	6.30	8.21	10.3	12.7	15.2	18.0	20.9	24.0
	18	0.78	1.73	2.84	4.17	5.71	7.45	9.41	11.6	13.9	16.5	19.2	22.1
	20	0.71	1.57	2.59	3.80	5.21	6.81	8.61	10.6	12.8	15.2	17.7	20.4
	24	0.60	1.33	2.19	3.23	4.43	5.80	7.36	9.07	11.0	13.0	15.3	17.6
	28	0.52	1.15	1.90	2.80	3.85	5.05	6.41	7.91	9.59	11.4	13.4	15.5
	32	0.46	1.02	1.68	2.48	3.40	4.46	5.67	7.01	8.50	10.1	11.9	13.8
	36	0.41	0.91	1.50	2.22	3.04	4.00	5.08	6.29	7.63	9.09	10.7	12.4
6	2	2.91	6.57	10.6	14.7	18.8	22.8	26.8	30.8	34.8	38.8	42.7	46.7
	3	2.57	5.93	9.81	13.9	18.0	22.1	26.2	30.3	34.3	38.3	42.3	46.3
	4	2.30	5.37	9.04	13.0	17.2	21.3	25.5	29.6	33.6	37.7	41.7	45.8
	5	2.06	4.89	8.33	12.2	16.3	20.5	24.6	28.8	32.9	37.0	41.1	45.1
	6	1.86	4.48	7.70	11.4	15.4	19.5	23.7	27.9	32.1	36.2	40.3	44.4
	7	1.69	4.12	7.13	10.6	14.5	18.6	22.8	27.0	31.2	35.4	39.5	43.7
	8	1.53	3.80	6.62	9.95	13.7	17.7	21.8	26.0	30.2	34.4	38.6	42.8
	9	1.40	3.52	6.17	9.32	12.9	16.8	20.9	25.1	29.3	33.5	37.7	41.9
	10	1.29	3.27	5.77	8.76	12.2	16.0	20.0	24.1	28.3	32.5	36.8	41.0
	12	1.11	2.86	5.09	7.80	11.0	14.5	18.3	22.3	26.4	30.6	34.8	39.0
	14	0.97	2.54	4.53	7.00	9.92	13.2	16.8	20.6	24.6	28.7	32.8	37.1
	16	0.86	2.27	4.08	6.34	9.02	12.0	15.4	19.0	22.9	26.9	30.9	35.1
	18	0.78	2.06	3.70	5.78	8.26	11.1	14.2	17.7	21.3	25.2	29.1	33.2
	20	0.71	1.88	3.38	5.30	7.60	10.2	13.2	16.4	19.9	23.6	27.5	31.4
	24	0.60	1.59	2.88	4.54	6.54	8.84	11.5	14.4	17.5	20.9	24.5	28.2
	28	0.52	1.38	2.50	3.96	5.72	7.77	10.1	12.7	15.6	18.7	22.0	25.4
	32	0.46	1.22	2.21	3.51	5.08	6.92	9.03	11.4	14.0	16.8	19.9	23.1
	36	0.41	1.09	1.98	3.15	4.56	6.23	8.15	10.3	12.7	15.3	18.1	21.1

Table 8-25 (cont.).
Coefficients C for Eccentrically Loaded Bolt Groups
Angle = 30°

$C_{req} = \dfrac{P_u}{\phi r_n}$ or $\phi R_n = C \times \phi r_n$

where

P_u	= factored force, kips
ϕr_n	= design strength per bolt, kips
ϕR_n	= design strength of bolt group, kips
e	= eccentricity of P_u with respect to centroid of bolt group, in. (not tabulated, may be determined by geometry.)
e_x	= horizontal component of e, in.
s	= bolt spacing, in.
C	= coefficient tabulated below.

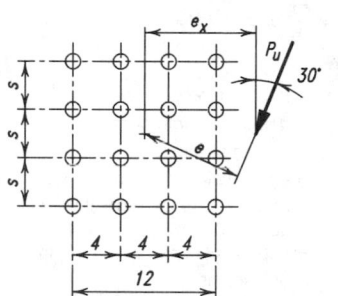

s, in.	e_x, in.	\multicolumn{12}{c}{Number of bolts in one vertical row, n}											
		1	2	3	4	5	6	7	8	9	10	11	12
	2	3.14	6.41	9.91	13.6	17.5	21.4	25.4	29.4	33.4	37.4	41.4	45.4
	3	2.79	5.75	8.95	12.4	16.1	20.0	23.9	27.9	31.9	35.9	40.0	44.0
	4	2.50	5.19	8.16	11.4	14.9	18.5	22.4	26.3	30.3	34.3	38.4	42.4
	5	2.25	4.71	7.45	10.5	13.7	17.2	20.9	24.7	28.6	32.6	36.7	40.7
	6	2.04	4.29	6.83	9.65	12.7	16.0	19.6	23.3	27.1	31.0	35.0	39.0
	7	1.85	3.93	6.28	8.92	11.8	15.0	18.3	21.9	25.6	29.4	33.3	37.3
	8	1.69	3.61	5.80	8.27	11.0	14.0	17.2	20.6	24.2	27.9	31.7	35.6
	9	1.55	3.33	5.38	7.70	10.3	13.1	16.2	19.4	22.9	26.5	30.2	34.0
	10	1.43	3.08	5.00	7.19	9.64	12.3	15.3	18.4	21.7	25.2	28.8	32.5
3	12	1.23	2.68	4.37	6.32	8.52	11.0	13.6	16.5	19.6	22.8	26.2	29.8
	14	1.08	2.36	3.88	5.62	7.61	9.83	12.3	14.9	17.8	20.8	24.0	27.3
	16	0.96	2.11	3.47	5.05	6.86	8.89	11.1	13.6	16.2	19.0	22.0	25.2
	18	0.87	1.91	3.14	4.57	6.24	8.10	10.2	12.4	14.9	17.5	20.3	23.3
	20	0.79	1.74	2.86	4.18	5.71	7.43	9.35	11.5	13.8	16.2	18.9	21.6
	24	0.67	1.48	2.43	3.56	4.88	6.36	8.03	9.87	11.9	14.1	16.4	18.9
	28	0.58	1.28	2.11	3.10	4.25	5.55	7.02	8.65	10.4	12.4	14.5	16.7
	32	0.51	1.13	1.87	2.74	3.76	4.92	6.23	7.69	9.29	11.0	12.9	14.9
	36	0.46	1.01	1.67	2.45	3.37	4.41	5.60	6.91	8.36	9.95	11.7	13.5
	2	3.14	6.75	10.7	14.7	18.7	22.7	26.7	30.7	34.7	38.6	42.6	46.6
	3	2.79	6.12	9.94	13.9	18.0	22.0	26.1	30.1	34.1	38.1	42.1	46.1
	4	2.50	5.58	9.23	13.1	17.2	21.2	25.3	29.4	33.4	37.5	41.5	45.5
	5	2.25	5.13	8.58	12.4	16.3	20.4	24.5	28.6	32.7	36.7	40.8	44.8
	6	2.04	4.73	8.00	11.6	15.5	19.5	23.6	27.7	31.8	35.9	40.0	44.1
	7	1.85	4.38	7.47	10.9	14.7	18.7	22.7	26.8	31.0	35.1	39.2	43.3
	8	1.69	4.06	6.98	10.3	14.0	17.9	21.9	25.9	30.1	34.2	38.3	42.4
	9	1.55	3.78	6.55	9.72	13.3	17.1	21.0	25.1	29.2	33.3	37.4	41.5
	10	1.43	3.53	6.15	9.18	12.6	16.3	20.2	24.2	28.3	32.4	36.5	40.6
6	12	1.23	3.10	5.47	8.25	11.4	14.9	18.6	22.5	26.5	30.6	34.7	38.8
	14	1.08	2.76	4.90	7.46	10.4	13.7	17.2	21.0	24.9	28.8	32.9	37.0
	16	0.96	2.48	4.43	6.79	9.55	12.6	16.0	19.6	23.3	27.2	31.2	35.2
	18	0.87	2.25	4.04	6.22	8.79	11.7	14.9	18.3	21.9	25.7	29.5	33.5
	20	0.79	2.06	3.70	5.72	8.14	10.9	13.9	17.1	20.6	24.2	28.0	31.9
	24	0.67	1.76	3.17	4.93	7.06	9.48	12.2	15.2	18.3	21.7	25.3	28.9
	28	0.58	1.53	2.76	4.32	6.22	8.38	10.8	13.5	16.5	19.6	22.9	26.3
	32	0.51	1.35	2.45	3.84	5.54	7.50	9.73	12.2	14.9	17.8	20.9	24.1
	36	0.46	1.21	2.19	3.46	5.00	6.77	8.82	11.1	13.6	16.3	19.1	22.2

Table 8-25 (cont.).
Coefficients C for Eccentrically Loaded Bolt Groups
Angle = 45°

$$C_{req} = \frac{P_u}{\phi r_n} \text{ or } \phi R_n = C \times \phi r_n$$

where

P_u	=	factored force, kips
ϕr_n	=	design strength per bolt, kips
ϕR_n	=	design strength of bolt group, kips
e	=	eccentricity of P_u with respect to centroid of bolt group, in. (not tabulated, may be determined by geometry.)
e_x	=	horizontal component of e, in.
s	=	bolt spacing, in.
C	=	coefficient tabulated below.

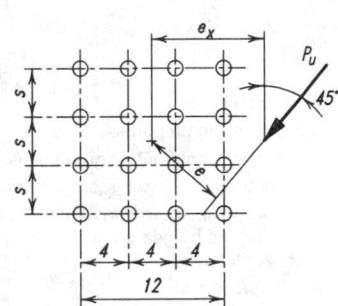

s, in.	e_x, in.	\multicolumn{12}{c}{Number of bolts in one vertical row, n}
		1	2	3	4	5	6	7	8	9	10	11	12
	2	3.46	6.96	10.5	14.2	18.0	21.8	25.7	29.6	33.5	37.4	41.4	45.3
	3	3.15	6.38	9.73	13.2	16.8	20.6	24.4	28.2	32.1	36.1	40.0	44.0
	4	2.87	5.84	8.97	12.3	15.7	19.3	23.1	26.9	30.7	34.6	38.6	42.5
	5	2.61	5.36	8.30	11.4	14.7	18.2	21.8	25.5	29.3	33.2	37.1	41.0
	6	2.39	4.93	7.69	10.7	13.9	17.2	20.7	24.3	28.0	31.8	35.6	39.5
	7	2.19	4.55	7.15	9.98	13.0	16.2	19.6	23.1	26.7	30.4	34.2	38.1
	8	2.01	4.21	6.66	9.34	12.2	15.3	18.6	22.0	25.5	29.2	32.9	36.7
	9	1.86	3.90	6.21	8.76	11.5	14.5	17.7	21.0	24.4	27.9	31.6	35.3
	10	1.72	3.63	5.82	8.24	10.9	13.8	16.8	20.0	23.3	26.8	30.4	34.0
3	12	1.49	3.18	5.14	7.33	9.76	12.4	15.2	18.3	21.4	24.7	28.1	31.6
	14	1.32	2.82	4.59	6.58	8.81	11.3	13.9	16.7	19.7	22.8	26.1	29.5
	16	1.17	2.53	4.14	5.95	8.00	10.3	12.7	15.4	18.2	21.2	24.3	27.5
	18	1.06	2.29	3.76	5.43	7.32	9.44	11.7	14.2	16.9	19.7	22.7	25.7
	20	0.96	2.10	3.44	4.98	6.74	8.71	10.9	13.2	15.7	18.4	21.2	24.2
	24	0.82	1.79	2.94	4.26	5.81	7.53	9.43	11.5	13.8	16.2	18.7	21.4
	28	0.71	1.56	2.56	3.73	5.09	6.61	8.31	10.2	12.2	14.4	16.7	19.2
	32	0.63	1.38	2.26	3.31	4.52	5.89	7.42	9.11	11.0	12.9	15.1	17.3
	36	0.56	1.23	2.03	2.97	4.06	5.30	6.69	8.23	9.91	11.7	13.7	15.8
	2	3.46	7.09	10.9	14.8	18.7	22.7	26.7	30.6	34.6	38.5	42.5	46.5
	3	3.15	6.58	10.3	14.1	18.1	22.0	26.0	30.0	33.9	37.9	41.9	45.9
	4	2.87	6.09	9.65	13.4	17.3	21.3	25.3	29.3	33.3	37.3	41.2	45.2
	5	2.61	5.66	9.07	12.8	16.6	20.6	24.5	28.5	32.5	36.5	40.5	44.5
	6	2.39	5.26	8.54	12.1	15.9	19.8	23.8	27.8	31.8	35.8	39.8	43.8
	7	2.19	4.91	8.07	11.6	15.3	19.1	23.0	27.0	31.0	35.0	39.0	43.0
	8	2.01	4.59	7.63	11.0	14.6	18.4	22.3	26.2	30.2	34.2	38.2	42.2
	9	1.86	4.30	7.23	10.5	14.0	17.7	21.5	25.5	29.4	33.4	37.4	41.4
	10	1.72	4.04	6.85	10.0	13.4	17.1	20.8	24.7	28.6	32.6	36.6	40.6
6	12	1.49	3.59	6.19	9.14	12.4	15.9	19.5	23.3	27.2	31.1	35.1	39.1
	14	1.32	3.22	5.62	8.38	11.4	14.8	18.3	22.0	25.8	29.6	33.5	37.5
	16	1.17	2.91	5.13	7.71	10.6	13.8	17.2	20.8	24.4	28.2	32.1	36.0
	18	1.06	2.66	4.71	7.12	9.87	12.9	16.2	19.6	23.2	26.9	30.7	34.6
	20	0.96	2.44	4.35	6.61	9.22	12.1	15.3	18.6	22.1	25.7	29.4	33.2
	24	0.82	2.10	3.76	5.76	8.11	10.8	13.7	16.7	20.0	23.4	27.0	30.6
	28	0.71	1.83	3.30	5.08	7.22	9.64	12.3	15.2	18.3	21.5	24.9	28.4
	32	0.63	1.63	2.94	4.54	6.50	8.71	11.2	13.9	16.7	19.8	23.0	26.3
	36	0.56	1.46	2.64	4.11	5.90	7.93	10.2	12.7	15.4	18.3	21.3	24.5

Table 8-25 (cont.).
Coefficients C for Eccentrically Loaded Bolt Groups
Angle = 60°

$$C_{req} = \frac{P_u}{\phi r_n} \text{ or } \phi R_n = C \times \phi r_n$$

where

P_u	=	factored force, kips
ϕr_n	=	design strength per bolt, kips
ϕR_n	=	design strength of bolt group, kips
e	=	eccentricity of P_u with respect to centroid of bolt group, in. (not tabulated, may be determined by geometry.)
e_x	=	horizontal component of e, in.
s	=	bolt spacing, in.
C	=	coefficient tabulated below.

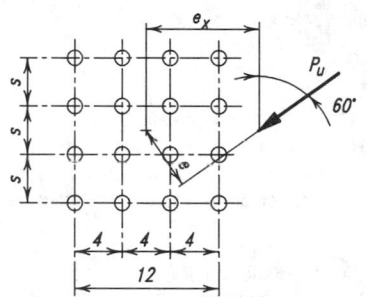

s, in.	e_x, in.	Number of bolts in one vertical row, n											
		1	2	3	4	5	6	7	8	9	10	11	12
	2	3.74	7.46	11.2	14.9	18.6	22.4	26.2	30.0	33.9	37.7	41.6	45.5
	3	3.57	7.12	10.7	14.3	17.9	21.6	25.3	29.0	32.8	36.7	40.5	44.4
	4	3.38	6.75	10.2	13.6	17.1	20.7	24.3	28.0	31.8	35.6	39.4	43.2
	5	3.17	6.36	9.61	12.9	16.4	19.8	23.4	27.0	30.7	34.5	38.2	42.0
	6	2.97	5.99	9.09	12.3	15.6	19.0	22.5	26.1	29.7	33.4	37.1	40.9
	7	2.78	5.63	8.59	11.7	14.9	18.2	21.6	25.1	28.7	32.3	36.0	39.8
	8	2.60	5.29	8.13	11.1	14.2	17.5	20.8	24.3	27.8	31.4	35.0	38.7
	9	2.44	4.98	7.69	10.6	13.6	16.8	20.1	23.4	26.9	30.4	34.0	37.7
	10	2.28	4.69	7.28	10.1	13.0	16.1	19.3	22.7	26.1	29.5	33.1	36.7
3	12	2.02	4.18	6.56	9.16	11.9	14.9	18.0	21.2	24.5	27.8	31.3	34.8
	14	1.80	3.76	5.95	8.38	11.0	13.8	16.7	19.8	23.0	26.3	29.6	33.1
	16	1.62	3.40	5.43	7.70	10.2	12.8	15.6	18.6	21.6	24.8	28.1	31.4
	18	1.47	3.10	4.99	7.11	9.42	11.9	14.6	17.4	20.4	23.5	26.7	29.9
	20	1.34	2.85	4.61	6.59	8.76	11.1	13.7	16.4	19.3	22.2	25.3	28.5
	24	1.15	2.45	3.99	5.73	7.67	9.82	12.2	14.6	17.3	20.1	23.0	26.0
	28	1.00	2.15	3.51	5.06	6.80	8.76	10.9	13.2	15.6	18.2	20.9	23.8
	32	0.88	1.91	3.13	4.52	6.11	7.89	9.83	11.9	14.2	16.6	19.2	21.8
	36	0.79	1.72	2.81	4.08	5.53	7.16	8.95	10.9	13.0	15.3	17.7	20.2
	2	3.74	7.47	11.2	15.0	18.9	22.8	26.7	30.6	34.5	38.5	42.4	46.4
	3	3.57	7.16	10.8	14.6	18.4	22.2	26.1	30.0	33.9	37.9	41.8	45.8
	4	3.38	6.82	10.4	14.1	17.8	21.7	25.5	29.4	33.3	37.3	41.2	45.1
	5	3.17	6.47	9.94	13.6	17.3	21.1	24.9	28.8	32.7	36.6	40.5	44.5
	6	2.97	6.14	9.52	13.1	16.7	20.5	24.3	28.2	32.1	36.0	39.9	43.8
	7	2.78	5.82	9.11	12.6	16.2	19.9	23.7	27.6	31.5	35.3	39.3	43.2
	8	2.60	5.52	8.73	12.1	15.7	19.4	23.2	27.0	30.8	34.7	38.6	42.5
	9	2.44	5.24	8.37	11.7	15.2	18.9	22.6	26.4	30.2	34.1	38.0	41.9
	10	2.28	4.98	8.03	11.3	14.8	18.4	22.1	25.8	29.7	33.5	37.4	41.3
6	12	2.02	4.51	7.41	10.6	14.0	17.5	21.1	24.8	28.5	32.3	36.2	40.1
	14	1.80	4.10	6.86	9.91	13.2	16.6	20.1	23.8	27.5	31.2	35.0	38.9
	16	1.62	3.76	6.37	9.29	12.4	15.8	19.2	22.8	26.5	30.2	33.9	37.7
	18	1.47	3.46	5.94	8.74	11.8	15.0	18.4	21.9	25.5	29.2	32.9	36.6
	20	1.34	3.21	5.56	8.23	11.2	14.3	17.6	21.0	24.6	28.2	31.9	35.6
	24	1.15	2.79	4.91	7.34	10.1	13.0	16.2	19.5	22.9	26.4	30.0	33.6
	28	1.00	2.47	4.38	6.61	9.13	11.9	14.9	18.1	21.4	24.7	28.2	31.8
	32	0.88	2.21	3.95	5.99	8.33	11.0	13.8	16.8	20.0	23.2	26.6	30.1
	36	0.79	2.00	3.58	5.46	7.65	10.1	12.8	15.7	18.7	21.9	25.1	28.5

Table 8-25 (cont.).
Coefficients C for Eccentrically Loaded Bolt Groups
Angle = 75°

$$C_{req} = \frac{P_u}{\phi r_n} \text{ or } \phi R_n = C \times \phi r_n$$

where

P_u = factored force, kips

ϕr_n = design strength per bolt, kips

ϕR_n = design strength of bolt group, kips

e = eccentricity of P_u with respect to centroid of bolt group, in. (not tabulated, may be determined by geometry.)

e_x = horizontal component of e, in.

s = bolt spacing, in.

C = coefficient tabulated below.

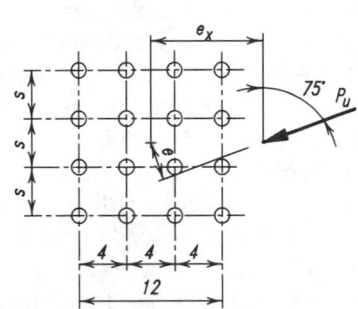

s, in.	e_x, in.	Number of bolts in one vertical row, n											
		1	2	3	4	5	6	7	8	9	10	11	12
3	2	3.89	7.75	11.6	15.5	19.3	23.1	26.9	30.8	34.6	38.5	42.3	46.2
	3	3.84	7.66	11.5	15.2	19.0	22.7	26.5	30.3	34.1	37.9	41.7	45.5
	4	3.79	7.54	11.3	15.0	18.7	22.4	26.1	29.8	33.5	37.3	41.0	44.8
	5	3.72	7.40	11.1	14.7	18.3	21.9	25.6	29.3	32.9	36.7	40.4	44.1
	6	3.65	7.25	10.8	14.4	17.9	21.5	25.1	28.7	32.4	36.1	39.8	43.5
	7	3.56	7.08	10.6	14.1	17.6	21.1	24.6	28.2	31.8	35.5	39.1	42.8
	8	3.47	6.90	10.3	13.7	17.2	20.6	24.1	27.7	31.3	34.9	38.5	42.2
	9	3.37	6.71	10.0	13.4	16.8	20.2	23.7	27.2	30.7	34.3	37.9	41.6
	10	3.27	6.52	9.77	13.1	16.4	19.8	23.2	26.7	30.2	33.7	37.3	41.0
	12	3.07	6.14	9.23	12.4	15.6	18.9	22.3	25.7	29.1	32.6	36.2	39.8
	14	2.87	5.76	8.71	11.8	14.9	18.1	21.4	24.7	28.1	31.6	35.1	38.7
	16	2.68	5.40	8.22	11.1	14.2	17.3	20.5	23.8	27.2	30.6	34.1	37.6
	18	2.50	5.07	7.76	10.6	13.5	16.6	19.7	23.0	26.3	29.7	33.1	36.6
	20	2.34	4.76	7.33	10.0	12.9	15.9	19.0	22.2	25.5	28.8	32.2	35.6
	24	2.06	4.23	6.57	9.10	11.8	14.7	17.6	20.7	23.9	27.1	30.4	33.8
	28	1.82	3.78	5.94	8.30	10.9	13.5	16.4	19.3	22.4	25.5	28.7	32.0
	32	1.63	3.41	5.41	7.61	10.0	12.6	15.3	18.1	21.0	24.1	27.2	30.4
	36	1.48	3.11	4.95	7.01	9.26	11.7	14.3	17.0	19.8	22.8	25.8	28.9
6	2	3.89	7.74	11.6	15.4	19.3	23.1	27.0	30.9	35.2	39.1	43.0	47.0
	3	3.84	7.64	11.4	15.2	19.0	22.8	26.6	30.5	34.4	38.3	42.2	46.1
	4	3.79	7.52	11.2	14.9	18.7	22.5	26.3	30.1	34.0	37.8	41.7	45.6
	5	3.72	7.38	11.0	14.7	18.4	22.1	25.9	29.7	33.6	37.4	41.3	45.2
	6	3.65	7.23	10.8	14.4	18.1	21.8	25.6	29.3	33.2	37.0	40.8	44.7
	7	3.56	7.07	10.6	14.2	17.8	21.5	25.2	29.0	32.8	36.6	40.4	44.3
	8	3.47	6.90	10.4	13.9	17.5	21.2	24.9	28.6	32.4	36.2	40.0	43.9
	9	3.37	6.73	10.1	13.6	17.2	20.8	24.5	28.3	32.0	35.8	39.6	43.5
	10	3.27	6.56	9.92	13.4	16.9	20.5	24.2	27.9	31.7	35.5	39.3	43.1
	12	3.07	6.21	9.48	12.9	16.4	19.9	23.6	27.3	31.0	34.7	38.5	42.3
	14	2.87	5.88	9.07	12.4	15.9	19.4	23.0	26.6	30.3	34.1	37.8	41.6
	16	2.68	5.57	8.67	11.9	15.4	18.8	22.4	26.0	29.7	33.4	37.1	40.9
	18	2.50	5.27	8.29	11.5	14.9	18.3	21.9	25.5	29.1	32.8	36.5	40.2
	20	2.34	4.99	7.94	11.1	14.4	17.8	21.3	24.9	28.5	32.2	35.8	39.6
	24	2.06	4.50	7.29	10.3	13.6	16.9	20.4	23.9	27.4	31.0	34.7	38.3
	28	1.82	4.08	6.73	9.67	12.8	16.1	19.4	22.9	26.4	30.0	33.6	37.2
	32	1.63	3.73	6.25	9.06	12.1	15.3	18.6	22.0	25.4	29.0	32.5	36.1
	36	1.48	3.43	5.82	8.51	11.4	14.5	17.8	21.1	24.5	28.0	31.5	35.1

ANCHOR RODS OR THREADED RODS

Cast-in-place anchor rods, illustrated in Figure 8-14, are generally made from unheaded rod material or headed bolt material. Drilled-in anchor rods, illustrated in Figure 8-15, are not normally used; their design is governed by manufacturer's specifications. Refer also to Cannon, Godfrey, and Moreadith (1981).

LRFD 8pecification Section A3.4 permits the use of unheaded rod material from the following ASTM specifications as anchor rods or threaded rods: A36, A193, A354, A449, A572, A588, and A687. Additionally, LRFD Specification Section A3.4 permits the use of headed bolts conforming to the provisions of LRFD Specification Section A3.3 for use as anchor rods. Headed bolts, however, are generally available only in lengths up to about eight inches. Furthermore, designations such as ASTM A325 and A490 apply only to bolts manufactured with a head and it is, therefore, improper to specify unheaded anchor rods or other similar threaded devices as ASTM A325 or A490.

The availability and strength of the aforementioned ASTM specifications for unheaded rod material and headed bolt material are summarized in Table 8-26. Suitable nuts may be selected from ASTM A563 or ASTM A194 grade 7. Because base plates typically have holes larger than oversized holes to allow for tolerances on the location of the anchor rod, washers are usually furnished from ASTM A36 steel plate; they may be round, square, or rectangular, are generally about $\frac{1}{2}$-in. thick, and generally have holes which are $\frac{1}{16}$-in. larger than the anchor rod diameter.

Minimum Edge Distance and Embedment Length

The recommendations of Shipp and Haninger (1983) for minimum anchor-rod (concrete) edge distance and embedment length for tensile forces, adopted from ACI 349, are summarized in Table 8-26. The edge distance requirement is intended to prevent blow-out of the side of the concrete foundation and is based on concrete with $f_c' = 3,000$ psi. For edge distance requirements for shear, refer to Shipp and Haninger (1983).

In addition to providing the recommended minimum embedment length, anchor rods must extend a distance above the foundation that is sufficient to permit full thread engagement of the nut; from RCSC Specification Section 2(b), "...the end of the [anchor rod] will be flush with or outside the face of the nut when properly installed."

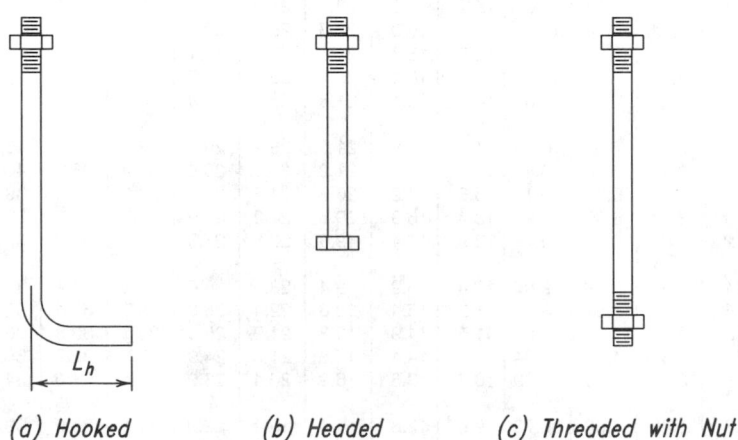

(a) Hooked   (b) Headed   (c) Threaded with Nut

Fig. 8-14. Typical cast-in-place anchor rods.

Note that it is seldom possible to fully tension anchor rods since the concrete usually cannot provide the necessary anchorage.

Welding to Anchor Rods

Though not typical, welds must sometimes be used in lieu of nuts to attach anchor rods to base plates. The use of weldable steels such as ASTM A36 or A572 is recommended for this purpose; anchor-rod material which is quenched and tempered should not be welded.

Hooked Anchor Rods

Hooked anchor rods should be used only for axially loaded columns to locate and prevent the displacement or overturning of columns due to erection loads or accidental collisions during erection. Additionally, high-strength steels are not recommended for use in hooked rods since bending with heat may materially affect their strength.

For the hooked rod of Figure 8-14a, the tensile force is resisted through bond development along the length and the mechanical anchorage of the hook. However, because smooth rods do not always form a reliable bond (due to oil used in threading among other things), the design of such anchor rods should be based upon the anchorage provided by the hook only. To prevent the anchor rod from pulling out and straightening, the hook should be designed to resist one-half the design tensile strength of the anchor rod ϕR_n,

where

$$\phi = 0.75$$
$$R_n = \phi_t F_u A_g$$

In the above equation, $\phi_t = 0.75$. From Fisher (1981), the bearing strength of the concrete is:

$$0.7 f_c' d L_h$$

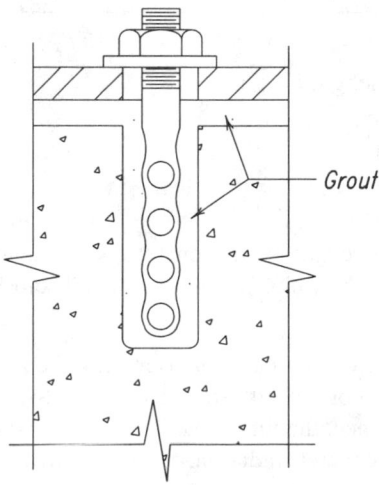

Fig. 8-15. Drilled-in anchor rods.

Table 8-26.
Anchor Rod Material Availability and Strength

Type	ASTM Design.	Material Type[b]	Grade	Diameter, d, in.	Proof Load	Min. Yield, ksi	Min. Tensile, ksi	Minimum Embdmt. Length, in.	Minimum Edge Dist., in.[e]
			Availability			**Strength**		**Minimum**	
Unheaded Rod Material (Only)	A36	C	—	to 8	—	36	58	12d	5d
	A572	HSLA	42	to 2	—	42	60	12d	5d
			50	to 6	—	50	65	17d	7d
	A588	HSLA, ACR	—	to 4	—	50	70	17d	7d
				over 4 to 5	—	46	67	17d	7d
				over 5 to 8	—	42	63	17d	7d
	A687	A, QT, NT	—	5/8 to 3	—	105	150[c]	19d	7d
Headed Bolt or Unheaded Rod Material	A354	A, QT	BD	1/4 to 2½	120	130	150	19d	7d
				over 2½ to 4	105	115	140	19d	7d
			BC	1/4 to 2½	105	109	125	17d	7d
				over 2½ to 4	95	99	115	17d	7d
	A449[d]	C, QT	—	1/4 to 1	85	92	120	17d	7d
				1⅛ to 1½	74	81	105	17d	7d
				1¾ to 3	55	58	90	17d	7d
Headed Bolt Mat. (Only)	A307	C	—	to 4	—	—	60	12d	5d
	A325[a,d]	C, QT	—	½ to 1	85	92	120	17d	7d
				1⅛ to 1½	74	81	105	17d	7d
	A490[a,d]	A, QT	—	½ to 1½	120	—	150	19d	7d

[a]Available with weathering (atmospheric corrosion resistance) characteristics comparable to ASTM A242 and A588 steels.
[b]A   = Alloy Steel
ACR = Atmospheric-Corrosion-Resistant Steel
C   = Carbon Steel
HSLA = High-Strength Low-Alloy Steel
NT   = Notch-Tough Steel (CVN 15 @ −20°F)
QT   = Quenched and Tempered Steel
[c]Maximum (ultimate tensile strength)
[d]Threaded rod material with properties meeting ASTM A325, A490, and A449 specifications may be obtained with the use of an appropriate steel (such as ASTM A193, grade B7), quenched and tempered after fabrication.
[e]Not less than 4 in.

Thus, the minimum hook length $L_{h\,min}$ is:

$$L_{h\,min} = \frac{\frac{\phi R_n}{2}}{0.7 f'_c d}$$

where f'_c is the specified strength of the concrete, ksi. The total embedded anchor rod length is then the hook length L_h plus the minimum embedment length from Table 8-26.

Headed Anchor Rods

When anchor rods are required for a calculated tensile force T_u, a more positive anchorage is formed when headed anchor rods, illustrated in Figure 8-14b, are used. With adequate embedment and edge distance, the limit state is either a tensile failure of the anchor rod or the pull-out of a cone of concrete radiating outward from the head (Marsh and Burdette, 1985a) as illustrated in Figure 8-16.

The design tensile strength of the anchor rod is ϕR_n,

where

$\phi = 0.75$

$R_n = \phi_t F_u A_g$

In the above equation, $\phi_t = 0.75$.

Using the projected surface area of the concrete cone and a limiting average stress on this area of $4\sqrt{f_c'}$, the minimum anchor rod length L_{min} is

$$L_{min} = \sqrt{\frac{A_{cp}}{3.14}}$$

where

$$A_{cp} = \frac{T_u}{4\phi_t \sqrt{f_c'}}$$

f_c' = specified strength of the concrete, psi

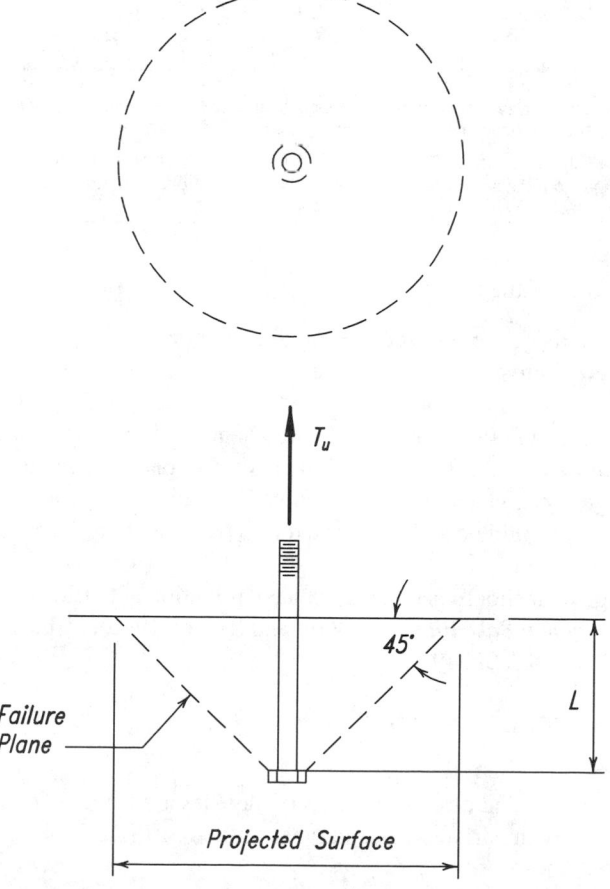

Fig. 8-16. Concrete cone subject to pull-out.

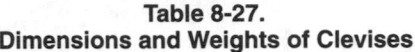

Table 8-27.
Dimensions and Weights of Clevises

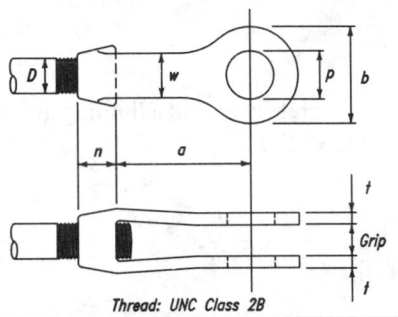

Grip=plate thickness + $\frac{1}{4}$ in.

Thread: UNC Class 2B

Clevis Number	Dimensions, in.							Weight, pounds	Design Strength ϕR_n*, kips
	Max. D	Max. p	b	n	a	w	t		
2	$\frac{5}{8}$	$\frac{3}{4}$	$1\frac{7}{16}$	$\frac{5}{8}$	$3\frac{7}{8}$	$1\frac{1}{16}$	$\frac{5}{16}$ $(+\frac{1}{32}, -0)$	1	5.25
$2\frac{1}{2}$	$\frac{7}{8}$	$1\frac{1}{2}$	$2\frac{1}{2}$	$1\frac{1}{8}$	4	$1\frac{1}{4}$	$\frac{5}{16}$ $(+\frac{1}{32}, -0)$	2	11.3
3	$1\frac{3}{8}$	$1\frac{3}{4}$	3	$1\frac{5}{16}$	5	$1\frac{1}{2}$	$\frac{1}{2}$ $(+\frac{1}{32}, -0)$	4	22.5
$3\frac{1}{2}$	$1\frac{1}{2}$	2	$3\frac{1}{2}$	$1\frac{5}{8}$	6	$1\frac{3}{4}$	$\frac{1}{2}$ $(+\frac{1}{32}, -0)$	6	27.0
4	$1\frac{3}{4}$	$2\frac{1}{4}$	4	$1\frac{3}{4}$	6	2	$\frac{1}{2}$ $(+\frac{1}{32}, -0)$	8	31.5
5	2	$2\frac{1}{2}$	5	$2\frac{1}{4}$	7	$2\frac{1}{2}$	$\frac{5}{8}$ $(+\frac{1}{16}, -0)$	16	56.4
6	$2\frac{1}{2}$	3	6	$2\frac{3}{4}$	8	3	$\frac{3}{4}$ $(+\frac{3}{32}, -0)$	26	81.0
7	3	$3\frac{3}{4}$	7	3	9	$3\frac{1}{2}$	$\frac{7}{8}$ $(+\frac{1}{8}, -0)$	36	103
8	4	4	8	4	10	4	$1\frac{1}{2}$ $(+\frac{1}{8}, -0)$	80	203

Notes:

Weights and dimensions of clevises are typical; products of all suppliers are essentially similar. User shall verify with the manufacturer that product meets design-strength specifications above.

*Tabulated design strengths for comparison with factored loads are based on ϕ=0.3. To determine safe working load (kips) for comparison with service loads, divide tabular design strength by 1.5. Safe working load, then, corresponds to a 5:1 factor of safety using maximum pin diameter.

T_u = tensile force in the anchor rod, kips

When the concrete cone intersects an edge of the pedestal or the cone from another anchor rod, the effective area of concrete is reduced; refer to the AISC Design Guide *Column Base Plates* (DeWolf and Ricker, 1990) and Marsh and Burdette (1985).

Marsh and Burdette (1985) showed that the head of the anchor rod usually provides sufficient anchorage and the use of an additional washer or plate does not add significantly to the anchorage. The nut and threading shown in Figure 8-14c is acceptable in lieu of a bolt head. The nut should be welded to the rod to prevent the rod from turning out when the top nut is tightened.

For the design of anchor rods for shear or a combination of tension and shear, see AISC Design Guide *Column Base Plates* (DeWolf and Ricker, 1990), Fisher (1981), Shipp and Haninger (1983), and ACI 349.

OTHER MECHANICAL FASTENERS

Clevises

Dimensions, weights, and design strengths of clevises are listed in Table 8-27. Compatability of clevises with various rods and pins is given in Table 8-28.

Turnbuckles

Dimensions, weights, and design strengths of turnbuckles are listed in Table 8-29.

Table 8-28.
Clevis Numbers Compatible with Various Rods and Pins

Dia. of Tap, in.	Diameter of Pin, in.															
	5/8	3/4	7/8	1	1¼	1½	1¾	2	2¼	2½	2¾	3	3¼	3½	3¾	4
5/8	2	2	2½	2½	2½	2½										
3/4	—	2½	2½	2½	2½	2½										
7/8	—	—	2½	2½	2½	2½										
1	—	—	—	3	3	3	3									
1¼	—	—	—	3	3	3	3	3½								
1⅜	—	—	—	3	3	3	3½	3½	4							
1½	—	—	—	3½	3½	3½	4	4	5							
1¾	—	—	—	—	4	4	5	5	5	5						
2	—	—	—	—	—	5	5	5	5	5	6	6				
2¼	—	—	—	—	—	—	—	6	6	6	6	6	7	7		
2½	—	—	—	—	—	—	—	6	6	6	7	7	7	7	7	
2¾	—	—	—	—	—	—	—	—	—	7	7	7	7	7	8	8
3	—	—	—	—	—	—	—	—	—	—	7	8	8	8	8	8
3¼	—	—	—	—	—	—	—	—	—	—	—	8	8	8	8	8
3½	—	—	—	—	—	—	—	—	—	—	—	8	8	8	8	8
3¾	—	—	—	—	—	—	—	—	—	—	—	8	8	8	8	8
4	—	—	—	—	—	—	—	—	—	—	—	8	8	8	8	8

Notes:
Tabular values assume that the net area of the clevis through the pin hole is greater than or equal to 125 percent of the net area of the rod, and is applicable to round rods without upset ends. For other net area ratios, the required clevis size may be calculated by reference to the dimensions tabulated in Tables 8-7 and 8-27.

Sleeve Nuts
Dimensions and weights of sleeve nuts are listed in Table 8-30.

Recessed-Pin Nuts
Dimensions and weights of recessed-pin nuts are listed in Table 8-31.

Cotter Pins
Dimensions and weights of cotter pins are listed in Table 8-32.

Table 8-29. Dimensions and Weights of Turnbuckles

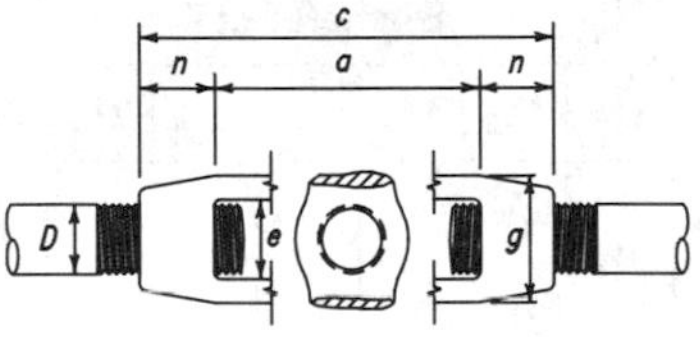

Threads: UNC and 4UN Class 2B

Diameter *D*, in.	Dimensions, in.					Weight (pounds) for Length *a*, in.						Design Strength, ϕR_n*, kips
	a	*n*	*c*	*e*	*g*	6	9	12	18	24	26	
3/8	6	9/16	7 1/8	9/16	1 1/32	0.41						1.80
1/2	6	3/4	7 1/2	11/16	1 5/16	0.75	0.80	1.00				3.30
5/8	6	29/32	7 13/16	13/16	1 1/2	1.00	1.38	1.50	2.43			5.25
3/4	6	1 1/16	8 1/8	15/16	1 23/32	1.45	1.63	2.13	3.06	4.25		7.80
7/8	6	1 7/32	8 7/16	1 3/32	1 7/8	1.85		2.83	4.20	5.43		10.8
1	6	1 3/8	8 3/4	1 9/32	2 1/32	2.60		3.20	4.40	6.85	10.0	14.0
1 1/8	6	1 9/16	9 1/8	1 13/32	2 9/32	2.72		4.70	6.10			17.4
1 1/4	6	1 3/4	9 1/2	1 9/16	2 17/32	3.58		4.70	7.13	11.3	13.1	22.8
1 3/8	6	1 15/16	9 7/8	1 11/16	2 3/4	4.50						26.1
1 1/2	6	2 1/8	10 1/4	1 27/32	3 1/32	5.50		8.00	9.13	16.8	19.4	31.5
1 5/8	6	2 1/4	10 1/2	1 31/32	3 9/32	7.50						36.8
1 3/4	6	2 1/2	11	2 1/8	3 9/16	9.50		15.3	16.0	19.5		42.5
1 7/8	6	2 3/4	11 1/2	2 3/8	4	11.5						55.8
2	6	2 3/4	11 1/2	2 3/8	4	11.5		15.3		27.5		55.8
2 1/4	6	3 3/8	12 3/4	2 11/16	4 5/8	18.0		35.3		43.5		72.0
2 1/2	6	3 3/4	13 1/2	3	5	23.3		33.6		42.4		90.0
2 3/4	6	4 1/8	14 1/4	3 1/4	5 5/8	31.5				54.0		113
3	6	4 1/2	15	3 5/8	6 1/8	39.5						145
3 1/4	6	5 1/4	16 1/2	3 7/8	6 3/4	60.5						183
3 1/2	6	5 1/4	16 1/2	3 7/8	6 3/4	60.5						183
3 3/4	6	6	18	4 5/8	8 1/2	95.0						252
4	6	6	18	4 5/8	8 1/2	95.0						252
4 1/4	9	6 3/4	22 1/2	5 1/4	9 3/4		152					351
4 1/2	9	6 3/4	22 1/2	5 1/4	9 3/4		152					351
4 3/4	9	6 3/4	22 1/2	5 1/4	9 3/4		152					351
5	9	7 1/2	24	6	10		200					442

Notes:

Weights and dimensions of turnbuckles are typical; products of all suppliers are essentially similar. User shall verify with the manufacturer that product meets design strength specifications above.

*Tabulated design strengths for comparison with factored loads are based on ϕ = 0.3. To determine safe working load (kips) for comparison with service loads, divide tabular design strength by 1.5. Safe working load, then, corresponds to a 5:1 factor of safety using maximum pin diameter.

Table 8-30.
Dimensions and Weights of Sleeve Nuts

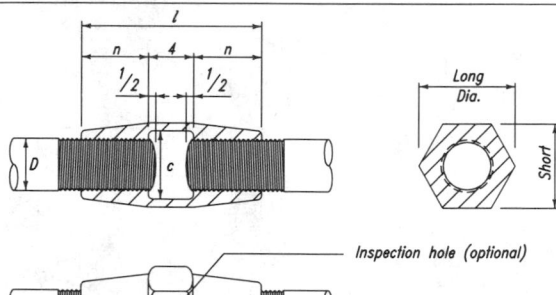

Thread: UNC and 4 UN Class 2B

Screw Dia. D, in.	Dimensions, in.					Weight, pounds
	Short Dia.	Long Dia.	Length *l*	Nut n	Clear c	
$\frac{3}{8}$	$\frac{11}{16}$	$\frac{25}{32}$	4	—	—	0.27
$\frac{7}{16}$	$\frac{25}{32}$	$\frac{7}{8}$	4	—	—	0.34
$\frac{1}{2}$	$\frac{7}{8}$	1	4	—	—	0.43
$\frac{9}{16}$	$\frac{15}{16}$	$1\frac{1}{16}$	5	—	—	0.64
$\frac{5}{8}$	$1\frac{1}{16}$	$1\frac{7}{32}$	5	—	—	0.93
$\frac{3}{4}$	$1\frac{1}{4}$	$1\frac{7}{16}$	5	—	—	1.12
$\frac{7}{8}$	$1\frac{7}{16}$	$1\frac{5}{8}$	7	$1\frac{7}{10}$	1	1.75
1	$1\frac{5}{8}$	$1\frac{13}{16}$	7	$1\frac{7}{16}$	$1\frac{1}{8}$	2.46
$1\frac{1}{8}$	$1\frac{13}{16}$	$2\frac{1}{16}$	$7\frac{1}{2}$	$1\frac{5}{8}$	$1\frac{1}{4}$	3.10
$1\frac{1}{4}$	2	$2\frac{1}{4}$	$7\frac{1}{2}$	$1\frac{5}{8}$	$1\frac{3}{8}$	4.04
$1\frac{3}{8}$	$2\frac{3}{16}$	$2\frac{1}{2}$	8	$1\frac{7}{8}$	$1\frac{1}{2}$	4.97
$1\frac{1}{2}$	$2\frac{3}{8}$	$2\frac{11}{16}$	8	$1\frac{7}{8}$	$1\frac{5}{8}$	6.16
$1\frac{5}{8}$	$2\frac{9}{16}$	$2\frac{15}{16}$	$8\frac{1}{2}$	$2\frac{1}{16}$	$1\frac{3}{4}$	7.36
$1\frac{3}{4}$	$2\frac{3}{4}$	$3\frac{1}{8}$	$8\frac{1}{2}$	$2\frac{1}{16}$	$1\frac{7}{8}$	8.87
$1\frac{7}{8}$	$2\frac{15}{16}$	$3\frac{5}{16}$	9	$2\frac{5}{16}$	2	10.4
2	$3\frac{1}{8}$	$3\frac{1}{2}$	9	$2\frac{5}{16}$	$2\frac{1}{8}$	12.2
$2\frac{1}{4}$	$3\frac{1}{2}$	$3\frac{15}{16}$	$9\frac{1}{2}$	$2\frac{1}{2}$	$2\frac{3}{8}$	16.2
$2\frac{1}{2}$	$3\frac{7}{8}$	$4\frac{3}{8}$	10	$2\frac{3}{4}$	$2\frac{5}{8}$	21.1
$2\frac{3}{4}$	$4\frac{1}{4}$	$4\frac{13}{16}$	$10\frac{1}{2}$	$2\frac{15}{16}$	$2\frac{7}{8}$	26.7
3	$4\frac{5}{8}$	$5\frac{1}{4}$	11	$3\frac{3}{16}$	$3\frac{1}{8}$	33.2
$3\frac{1}{4}$	5	$5\frac{5}{8}$	$11\frac{1}{2}$	$3\frac{3}{8}$	$3\frac{3}{8}$	40.6
$3\frac{1}{2}$	$5\frac{3}{8}$	6	12	$3\frac{5}{8}$	$3\frac{5}{8}$	49.1
$3\frac{3}{4}$	$5\frac{3}{4}$	$6\frac{3}{8}$	$12\frac{1}{2}$	$3\frac{13}{16}$	$3\frac{7}{8}$	58.6
4	$6\frac{1}{8}$	$6\frac{7}{8}$	13	$4\frac{1}{16}$	$4\frac{1}{8}$	69.2
$4\frac{1}{4}$	$6\frac{1}{2}$	$7\frac{1}{2}$	$13\frac{1}{2}$	$4\frac{3}{4}$	$4\frac{3}{8}$	75.0
$4\frac{1}{2}$	$6\frac{7}{8}$	$7\frac{15}{16}$	14	5	$4\frac{3}{4}$	90.0
$4\frac{3}{4}$	$7\frac{1}{4}$	$8\frac{3}{8}$	$14\frac{1}{2}$	$5\frac{1}{4}$	5	98.0
5	$7\frac{5}{8}$	$8\frac{7}{8}$	15	$5\frac{1}{2}$	$5\frac{1}{4}$	110
$5\frac{1}{4}$	8	$9\frac{1}{4}$	$15\frac{1}{2}$	$5\frac{3}{4}$	$5\frac{1}{2}$	122
$5\frac{1}{2}$	$8\frac{3}{8}$	$9\frac{3}{4}$	16	6	$5\frac{3}{4}$	142
$5\frac{3}{4}$	$8\frac{3}{4}$	$10\frac{1}{8}$	$16\frac{1}{2}$	$6\frac{1}{4}$	6	157
6	$9\frac{1}{8}$	$10\frac{5}{8}$	17	$6\frac{1}{2}$	$6\frac{1}{4}$	176

Notes:
Weights and dimensions of sleeve nuts are typical; products of all suppliers are essentially similar. User shall verify with the manufacturer that strengths of sleeve nut are greater than the corresponding connecting rod when the same material is used.

Table 8-31.
Dimensions and Weights of Recessed-Pin Nuts

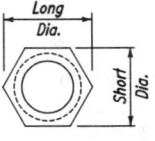

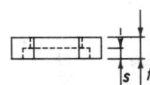

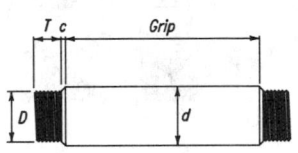

Material: Steel

Thread: 6 UN Class 2A/2B

Pin Dia. d, in.	Pin Dimensions, in.			Nut Dimensions, in.					Weight, pounds
	Thread		c	Thickness t	Diameter		Recess		
	D	T			Short Dia.	Long Dia.	Rough Dia.	s	
2, 2$\frac{1}{4}$	1$\frac{1}{2}$	1	$\frac{1}{8}$	$\frac{7}{8}$	3	3$\frac{3}{8}$	2$\frac{5}{8}$	$\frac{1}{4}$	1
2$\frac{1}{2}$, 2$\frac{3}{4}$	2	1$\frac{1}{8}$	$\frac{1}{8}$	1	3$\frac{5}{8}$	4$\frac{1}{8}$	3$\frac{1}{8}$	$\frac{1}{4}$	2
3, 3$\frac{1}{4}$, 3$\frac{1}{2}$	2$\frac{1}{2}$	1$\frac{1}{4}$	$\frac{1}{8}$	1$\frac{1}{8}$	4$\frac{3}{8}$	5	3$\frac{7}{8}$	$\frac{3}{8}$	3
3$\frac{3}{4}$, 4	3	1$\frac{3}{8}$	$\frac{1}{4}$	1$\frac{1}{4}$	4$\frac{7}{8}$	5$\frac{5}{8}$	4$\frac{3}{8}$	$\frac{3}{8}$	4
4$\frac{1}{4}$, 4$\frac{1}{2}$, 4$\frac{3}{4}$	3$\frac{1}{2}$	1$\frac{1}{2}$	$\frac{1}{4}$	1$\frac{3}{8}$	5$\frac{3}{4}$	6$\frac{5}{8}$	5$\frac{1}{4}$	$\frac{1}{2}$	5
5, 5$\frac{1}{4}$	4	1$\frac{5}{8}$	$\frac{1}{4}$	1$\frac{1}{2}$	6$\frac{1}{4}$	7$\frac{1}{4}$	5$\frac{3}{4}$	$\frac{1}{2}$	6
5$\frac{1}{2}$, 5$\frac{3}{4}$, 6	4$\frac{1}{2}$	1$\frac{3}{4}$	$\frac{1}{4}$	1$\frac{5}{8}$	7	8$\frac{1}{8}$	6$\frac{1}{2}$	$\frac{5}{8}$	8
6$\frac{1}{4}$, 6$\frac{1}{2}$	5	1$\frac{7}{8}$	$\frac{3}{8}$	1$\frac{3}{4}$	7$\frac{5}{8}$	8$\frac{7}{8}$	7	$\frac{5}{8}$	10
6$\frac{3}{4}$, 7	5$\frac{1}{2}$	2	$\frac{3}{8}$	1$\frac{7}{8}$	8$\frac{1}{8}$	9$\frac{3}{8}$	7$\frac{1}{2}$	$\frac{3}{4}$	12
7$\frac{1}{4}$, 7$\frac{1}{2}$	5$\frac{1}{2}$	2	$\frac{3}{8}$	1$\frac{7}{8}$	8$\frac{5}{8}$	10	8	$\frac{3}{4}$	14
7$\frac{3}{4}$, 8, 8$\frac{1}{4}$	6	2$\frac{1}{4}$	$\frac{3}{8}$	2$\frac{1}{8}$	9$\frac{3}{8}$	10$\frac{7}{8}$	8$\frac{3}{4}$	$\frac{3}{4}$	19
8$\frac{1}{2}$, 8$\frac{3}{4}$, 9	6	2$\frac{1}{4}$	$\frac{3}{8}$	2$\frac{1}{8}$	10$\frac{1}{4}$	11$\frac{7}{8}$	9$\frac{5}{8}$	$\frac{3}{4}$	24
9$\frac{1}{4}$, 9$\frac{1}{2}$	6	2$\frac{3}{8}$	$\frac{3}{8}$	2$\frac{1}{4}$	11$\frac{1}{4}$	13	10$\frac{5}{8}$	$\frac{3}{4}$	32
9$\frac{3}{4}$, 10	6	2$\frac{3}{8}$	$\frac{3}{8}$	2$\frac{1}{4}$	11$\frac{1}{4}$	13	10$\frac{5}{8}$	$\frac{3}{4}$	32

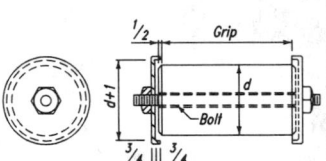

Typical Pin Cap Detail for Pins
over 10 in. in dia.
Dimensions shown are approximate

Notes:
Although nuts may be used on all sizes of pins as shown above, a detail similar to that shown at the left is preferrable for pin diameters over 10 inches. In this detail, the pin is held in place by a recessed cap at each end and secured by a bolt passing completely through the caps and pin. Suitable provisions must be made for attaching pilots and driving nuts.

Table 8-32.
Dimensions and Weights of Cotter Pins

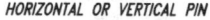
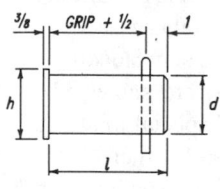
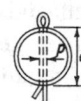
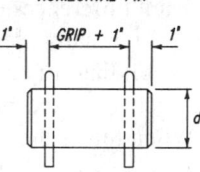

HORIZONTAL OR VERTICAL PIN

HORIZONTAL PIN

l = Length of pin, in.

Pin Diameter d, in.	Pins with Heads			Cotter		
	Head Diameter h, in.	Weight of One, pounds	Length c, in.	Diameter p, in.	Weight per 100, pounds
$1\frac{1}{4}$	$1\frac{1}{2}$	$0.19 + 0.35l$	$\frac{7}{8}$	$\frac{1}{4}$	2.64
$1\frac{1}{2}$	$1\frac{3}{4}$	$0.26 + 0.50l$	1	$\frac{1}{4}$	3.10
$1\frac{3}{4}$	2	$0.33 + 0.68l$	$1\frac{1}{8}$	$\frac{1}{4}$	3.50
2	$2\frac{3}{8}$	$0.47 + 0.89l$	$1\frac{1}{4}$	$\frac{3}{8}$	9.00
$2\frac{1}{4}$	$2\frac{5}{8}$	$0.58 + 1.13l$	$1\frac{3}{8}$	$\frac{3}{8}$	9.40
$2\frac{1}{2}$	$2\frac{7}{8}$	$0.70 + 1.39l$	$1\frac{1}{2}$	$\frac{3}{8}$	10.9
$2\frac{3}{4}$	$3\frac{1}{8}$	$0.82 + 1.68l$	$1\frac{5}{8}$	$\frac{3}{8}$	11.4
3	$3\frac{1}{2}$	$1.02 + 2.00l$	$1\frac{3}{4}$	$\frac{1}{2}$	28.5
$3\frac{1}{4}$	$3\frac{3}{4}$	$1.17 + 2.35l$	$1\frac{7}{8}$	$\frac{1}{2}$	28.5
$3\frac{1}{2}$	4	$1.34 + 2.73l$	$1\frac{7}{8}$	$\frac{1}{2}$	33.8
$3\frac{3}{4}$	$4\frac{1}{4}$	$1.51 + 3.13l$	$2\frac{1}{4}$	$\frac{1}{2}$	33.8

WELDED CONSTRUCTION

While AWS D1.1 is the traditional design specification for weld stresses in both buildings and bridges, AASHTO/AWS D1.5 also exists for dynamically loaded structures. There are significant differences between the two codes and, in the case of building structures, AWS D1.1 is normally used unless contract documents state otherwise.

Welds in building structures are predominantly designed for static loading. Some parts, however, such as crane runways and machinery supports, are subjected to dynamic loading. When this is the case, additional requirements and special joint details may be necessary. This may include reinforcing fillet welds at tee and corner joints, radius cuts on terminations of gusset type connections, radiographic or ultrasonic testing for quality control, or joint details in accordance with LRFD Specification Appendix K3. The contract documents should specifically enumerate these additional requirements when they are determined to be necessary.

Weldability of Steel

AWS has defined weldability as the capacity of a metal to be welded, under the fabrication conditions imposed, into a specific, suitably designed structure, and to perform satisfactorily in the intended service. AWS D1.1 is based on certain weldable grades of steel as listed therein by ASTM designation. It contains all of the steels permitted by LRFD Specification Section A3.1a.

The effect a steel's properties have upon its weldability relates to the reaction of the steel to the drastic heating and cooling cycle of welding. This weld quench can range from the practically instantaneous cooling of an accidental arc strike to the 10 minutes required to cool a high-heat-input electroslag weld. Due to the rapid cooling of the arc strike, the full-quench hardness for the carbon equivalent of the steel may be realized, resulting in brittleness and the potential for cracking. In contrast, the slower cooling rate of the electroslag weld may produce a more ductile and lower-strength metallurgical structure in the heat-affected zone (HAZ) of the base metal.

As they cool, welds develop residual shrinkage strains that can approach the yield strain as a limit; ductility and notch resistance are needed to accommodate these strains. Since chemical composition, grain size, and thickness affect ductility and notch resistance, they are the most important properties for weldability. These factors, discussed below, assume greater significance as the structure becomes large and must store greater elastic energy.

Table 8-33 summarizes several ASTM specifications and their requirements for the aforementioned properties. Note that there is a greater flexibility in grain size and carbon equivalents in these specifications for shapes, plates, and bars. Also, maximum tensile strength requirements are listed to exclude steels from the upper end of the chemical composition range which might require special welding procedures or weld repairs. In contrast, the requirements for structural tubing, pipe, sheet, and strip do not limit grain size or maximum tensile strength, but generally impose smaller limits on thickness.

Chemical analysis of a heat of steel is usually made during the processing as a control and upon completion after it has been tapped into a ladle. This heat analysis is used to compile a mill test report which also lists the customer's order number, steel grade, quantity and dimension of pieces shipped, and the results of any mechanical testing (tensile, flexural, Charpy impact, or other). This information may be obtained by request from the steel supplier when placing an order and is essential for good control of welded fabrication. It is imperative that the grade of steel to be welded is known since the proper welding procedure depends upon this information.

Table 8-33. ASTM Requirements for Properties Affecting Weldability of Steels					
ASTM Specification	Products Covered	Max. Carbon content, % by weight (heat analysis)	Max. tensile strength, ksi	Grain Size	Max. thickness, in.
A36	shapes	0.26	80	—*	none
	plates	0.25–0.29			
	bars	0.26–0.29			
A242	shapes, plates, bars	type 1, 0.15	none	—	4
A514	plates—quenched and tempered	varies among 13 grades, 0.14–0.21	130	fully killed, fine grain	6
A529	shapes, plates, bars	0.27	85	—	$\frac{1}{2}$
A572	shapes, plates, bars, sheet piling	varies among grades, 0.21–0.26	none	—*	Gr. 42: 6 Gr. 50: 4 Gr. 60, 65: $1\frac{1}{4}$
A588	shapes, plates, bars	varies among 5 grades, 0.15–0.20	none	fine grain	$F_y = 50$: 4 $F_y = 42$: 8
A852	plates	0.19	110	fine grain	4
A53 Grade B	tubing, pipe	0.30	none	—	2.344, 24 dia.
A500	tubing, pipe	Gr. A, B: 0.26 Gr. C: 0.23	none	—	$\frac{1}{2}$
A501	tubing, pipe	0.26	none	—	1
A618	tubing, pipe	Gr. Ia: 0.15 Gr. Ib: 0.20 Gr. II: 0.22 Gr. III: 0.23	none	—	$1\frac{1}{2}$
A570, Gr. 36, 50	sheet, strip	0.25	none	—	0.23
A606	sheet, strip	0.22	none	—	none
A607	sheet, strip	0.22–0.26	none	—	none

*Supplemental requirements can specify killed fine grain.

Chemical Composition

The most important element affecting weldability is carbon, however, the effect of other elements on weldability is related through a carbon equivalent formula. Weldability is enhanced as carbon equivalent decreases because the maximum hardness and consequent brittleness that a steel may reach after rapid liquid quenching from high temperature is directly proportional to the carbon equivalent. This relationship is illustrated in Figure 8-17 and is applicable to the surface in contact with the quench liquid where the quench rate is greatest.

Although no liquid is present in welding, the HAZ is subject to rapid cooling and consequent hardening by conduction of weld heat into the base metal. As the thickness of the section increases, so does the cooling rate, producing progressively harder and less ductile metallurgical constituents. Alloys such as Ni, Cr, and Mo in the steel permit hardening at slower cooling rates and at depths below the surface where the cooling rate is slower; pre-heat is the common remedy for reducing the cooling rate and hardness.

As the carbon content increases from 0.10 percent to 0.20 percent by weight, the maximum as-quenched hardness increases from 40 to 50 Rockwell C. Using the known hardness-strength relationship, it can be shown that the maximum as-quenched tensile strength increases from 180 to 260 ksi. Welding procedures are designed to keep weld quench rates far below these maximum rates. Also, electrodes are usually designed to deposit weld metal containing about 0.008 to 0.12 percent carbon to avoid cracking.

Grain Size

In general, weldability will be enhanced by steel with a finer grain size. As illustrated in Figure 8-18, grain size is a prime variable affecting the ductility and impact resistance for a wide variety of steel compositions.

The grain size of weld metal also varies and has a similar effect. Because they experience a slower cooling rate, high-heat-input welds show a larger grain size than the same process and electrode at a lower heat input. This is one reason the AWS D1.1 limits multi-pass SAW groove weld layers to a maximum size of ¼-in. Also, a subsequent pass will refine the grain of a previous pass.

Thickness

In general, as the thickness to be welded decreases, the weldability of the material is enhanced. Because of their greater mass, thick plates extract heat from and quench the weld more rapidly than thin plates with the identical weld. As a partial remedy, the plates may be pre-heated and held at temperatures of a few hundred degrees Fahrenheit for the welding operation. This pre-heat appreciably slows the quench rate and reduces weld hardness, as does post-heating.

As plate thickness increases, the notch impact resistance decreases as shown in Figure 8-19. This test was conducted on American Bureau of Shipping (ABS) class C ship plate in ¾-in., 1-in., 2-in., and 3-in. thicknesses using a severe crack-like notch in the ASTM A208 drop-weight test. The use of fine-grain steelmaking practice as specified by ASTM can improve notch toughness where required by the service of a particular structure.

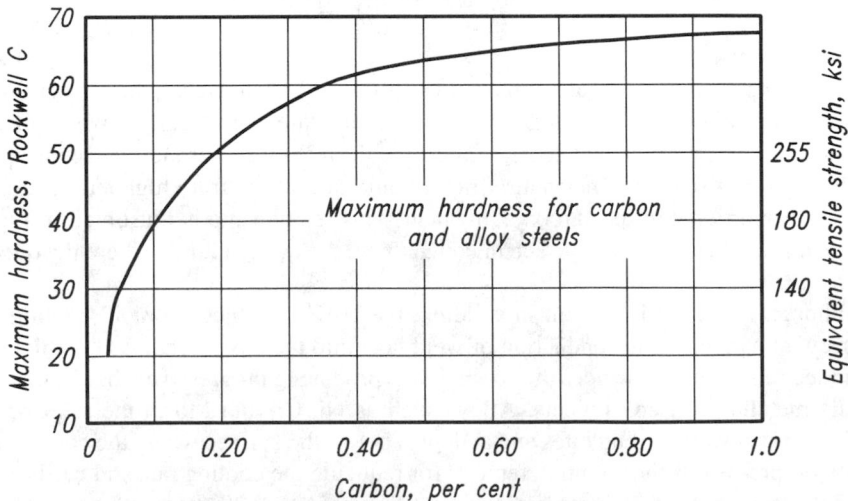

Fig. 8-17. *Influence of carbon content on the maximum hardness of steel as quenched (Stout and Doty, 1978), courtesy Welding Research Council.*

Structural Welding Materials and Processes

Filler metal and flux specifications are exclusively AWS specifications, having been removed from ASTM specifications. Additionally, AWS uses a coding system for consumable electrodes to designate the tensile strength and coating or flux combination. Since the coding for the several filler/flux combinations are consistent only with respect to the types of electrode used, it is very important that the applicable specifications be reviewed when specifying such welding requirements.

The welding processes discussed in this text are: shielded metal arc welding (SMAW), submerged arc welding (SAW), gas-metal arc welding (GMAW), flux-cored arc welding (FCAW), electroslag welding (ESW) and electrogas welding (EGW). Except for electroslag welding, each of these processes use electrical energy from an arc discharge between a steel-wire electrode and the base metal to provide heat for fusion. Electroslag welding uses a high-electrical-resistance molten-slag bath which occupies the entire joint. This slag melts both the electrode and the base metal.

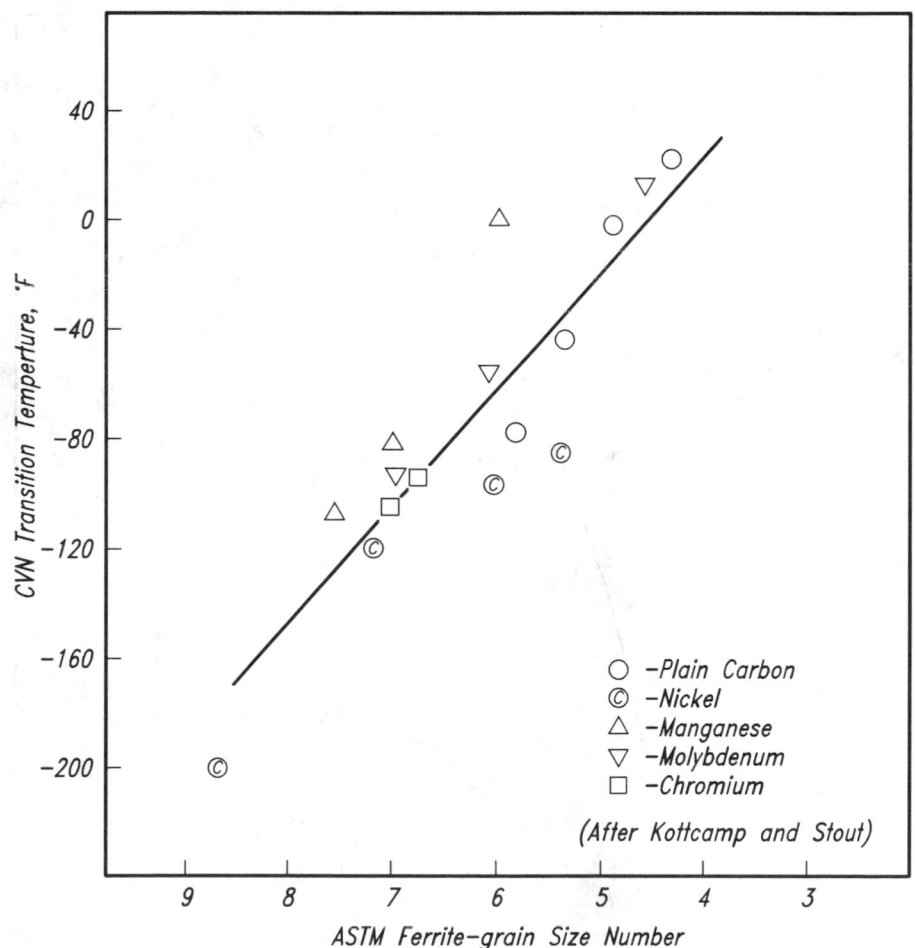

Fig. 8-18. *Effect of ferrite-grain size on CVN transitional temperature (Stout and Doty, 1978), courtesy Welding Research Council.*

Each of the aforementioned processes will be summarized here; a full description may be found in AWS (1978). Additionally, thermal cutting and air arc gouging will be discussed.

SMAW

There are two AWS Specifications for SMAW electrodes: AWS A5.1 and AWS A5.5. A condensation of the provisions of these specifications is given in Table 8-34.

AWS notation for SMAW electrodes is illustrated in Figure 8-20. This has also been extended to other processes. The welding positions noted in Figure 8-20 (flat, horizontal, vertical, and overhead) are illustrated in Figure 8-21. SMAW (stick) electrodes are made in a variety of low-carbon compositions. The extruded coatings contain aluminum, silicon, and other deoxidizers; the deposited weld is a mini-electric-furnace-killed steel with excellent ductility and resistance to cracking from weld shrinkage strains.

In the arc stream, moisture breaks down and liberates atomic hydrogen which is readily soluble in molten iron (Stout and Doty, 1978); see Figure 8-22. As the weld solidifies,

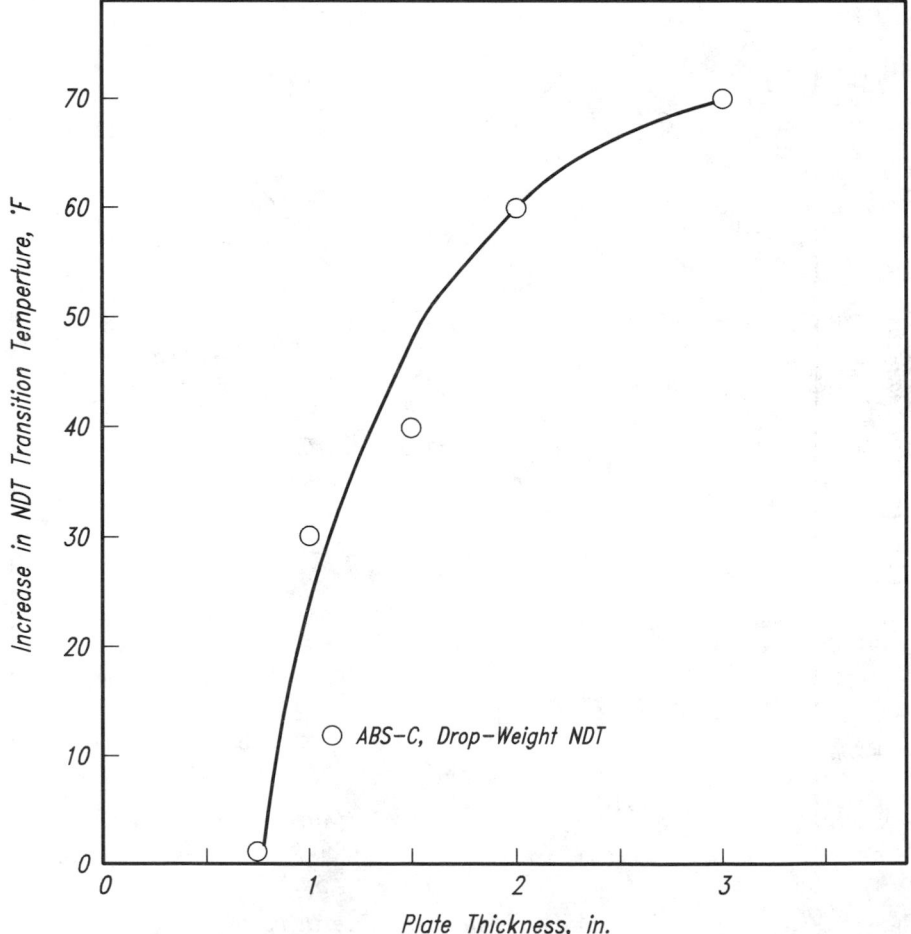

Fig. 8-19. Effect of plate thickness on the drop-weight NDT ductility transition temperature (Stout and Doty, 1978), courtesy Welding Research Council.

Table 8-34. Condensed AWS Specifications for SMAW Electrodes							
Electrode			**Min. Tensile Strength, ksi**	**Criteria for Composition of Deposited Weld Metal**	**Impact Test Criteria**		**Criteria for Radiographic Soundness**
Type	**AWS Spec.**	**Grades**			**Charpy V-Notch Test**	**Weld Metal Condition**	
Carbon Steel	A5.1	60	62	Not stipulated	Required for some grades only	As-welded	Stipulated for all but E6012, E6022
		70	72	Stipulated			
Low Alloy	A5.5	70	70	Stipulated (all grades)	Required for some grades only	Some as-welded, some stress-relieved	Stipulated for all grades
		80	80				
		90	90				
		100	100				
		110	110				
		120	120				
Note: A particular production welding condition may be more severe than the test conditions specified for the above.							

hydrogen becomes much less soluble and the atoms are rejected into voids where pairs combine to form a much less mobile molecular H_2. This molecular hydrogen can then exert pressure in lattice imperfections which is sufficient, when combined with weld shrinkage strains, to cause "fisheyes" or cracking in the weld material. This can be prevented by maintaining the moisture content of consumable electrodes below specified levels and through proper pre-heat.

E7015, E7016, E7018, and E7028 low-hydrogen electrodes have specially compounded and baked extruded coatings containing a limited moisture (hydrogen) content by weight. Coatings for the E70 electrode series can contain a maximum of 0.04 percent moisture, while the E120 electrode series is limited to only 0.015 percent. As the tensile strength of the base metal increases, electrodes with lower moisture content must be selected to avoid weld cracking. Since the electrode coating will absorb moisture when stored in damp or humid conditions, drying ovens near points of use in the shop are necessary for low-hydrogen electrodes.

ELECTRODE PROPERTIES					
E	70	1	6	—	A1
Electrode	70,000 psi min. tensile	Position code*	Coating characteristics**		Weld metal composition
*1 = All flat, vertical, overhead, and horizontal 2 = Flat and horizontal only **5, 6, 8 = Low hydrogen					

Fig. 8-20. AWS classification system for SMAW electrodes.

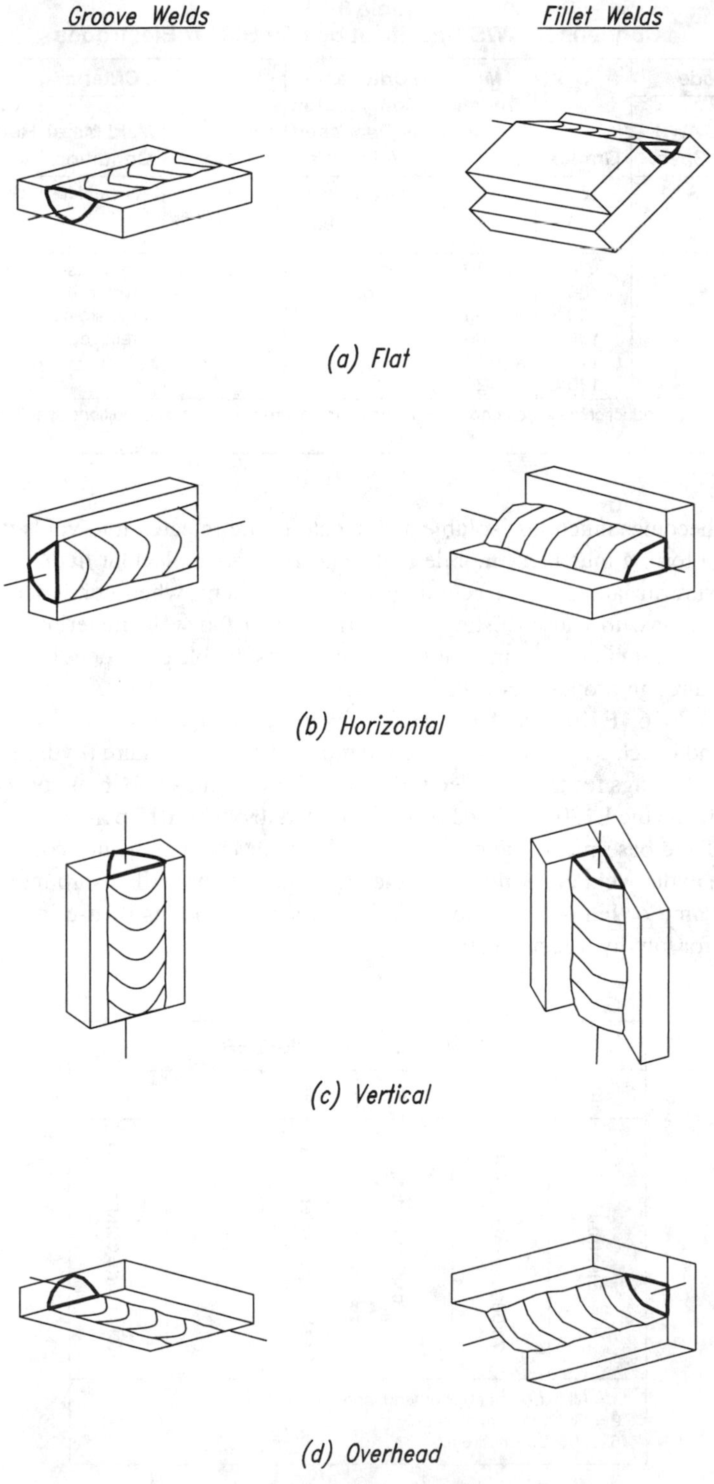

Groove Welds   Fillet Welds

(a) Flat

(b) Horizontal

(c) Vertical

(d) Overhead

Fig. 8-21. Welding positions.

The electrodes to be used with various base metals are shown in AWS D1.1 Table 4.1. Low-hydrogen electrodes are used with ASTM A572 and A588 steels among others. Filler metal matching the color of ASTM A588 steel is listed in AWS D1.1 Table 4.2.

SAW

The automatic and semi-automatic SAW processes provide consistent, high quality, and economical deposits which are particularly suitable for long welds. Their major limitation is that the work must be positioned to allow for near flat or horizontal welding.

In the SAW process, fluxes may be fused or agglomerated (finely powdered constituents bonded together with silicates), but are classified in AWS specifications only according to the weld metal properties produced in the standard specified weld tests. The applicable specifications are: AWS A5.17 and AWS A5.23.

AWS notation for SAW electrodes and fluxes is illustrated in Figure 8-23. Fluxes must be kept dry in storage to avoid an increase in moisture content and subsequent chance of hydrogen cracking in steels with higher yield strengths or highly restrained joints in thick members.

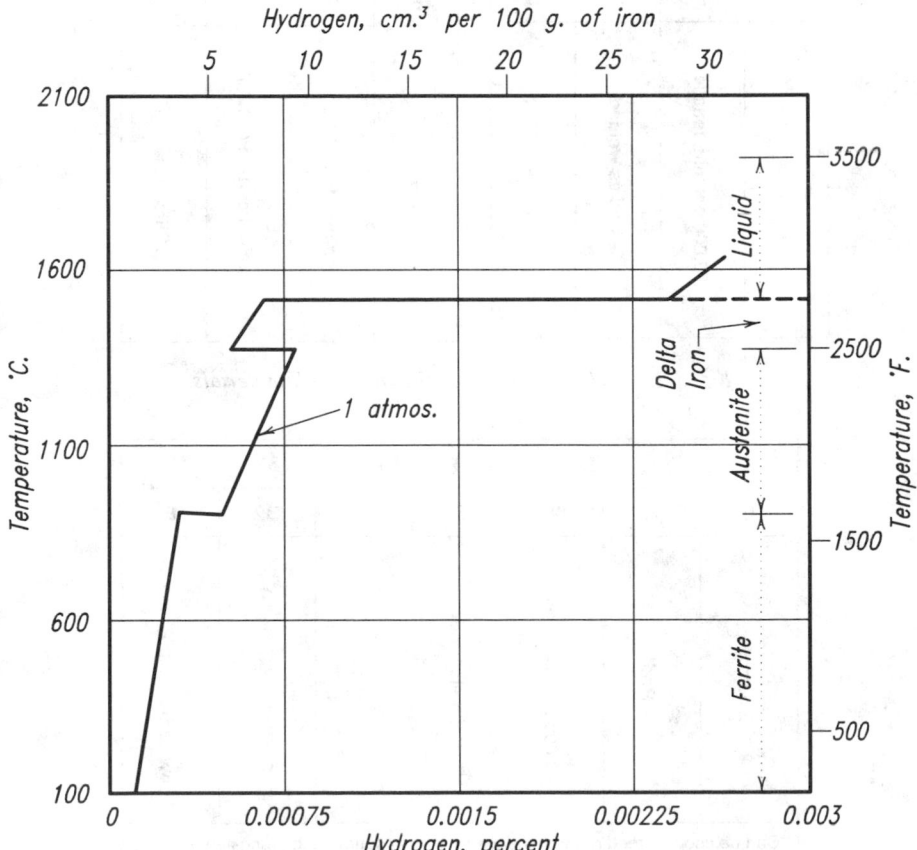

Fig. 8-22. Solubility of hydrogen in iron (Stout and Doty, 1978), courtesy Welding Research Council.

GMAW

The GMAW process can be used with mixtures of argon and two percent oxygen, argon and carbon dioxide, or pure carbon dioxide. While argon is inert, carbon dioxide can react with the weld metal and result in a reduction in ductility and impact properties at low temperatures. Despite this, 70 ksi electrodes have commonly been used with carbon dioxide gas with good results; a CVN 20 (20 ft-lb Charpy V-notch impact value) at $-20°F$ is specified in the AWS tests. Alloy electrodes producing up to 120 ksi minimum tensile strength with CVN 20 at $-60°F$, and three percent nickel electrodes producing 80 ksi minimum tensile strength with CVN 20 at $-100°F$ are available.

There are two AWS Specifications for GMAW electrodes: A5.18 and A5.28. Identification of these electrodes is illustrated in Figure 8-24.

FCAW

FCAW electrodes are made by forming a thin sheet strip into a U-shape and filling it with flux. After closing the tube, it is drawn to size as a continuous coil. AWS classifies these

FLUX CAPABILITY					ELECTRODE PROPERTIES			
F	7	A	6		E	M	12	K
Flux	70,000 psi min. tensile	Tested as welded	CVN 20 @ $-60°$ F		Electrode	Medium Mn (1.00% ±)	Nominal carbon (0.12%)	Silicon killed

Fig. 8-23. AWS classification system for SAW materials.

ELECTRODE PROPERTIES						
E	R	80	S	—	B2	L
Electrode	Rod*	80,000 psi min. tensile	Solid electrode		Cr (1¼%); Mo (½%)	Low carbon (0.05% max.)
*Can be used as feed rod with independent heat source (e.g., tungsten arc)						

Fig. 8-24. AWS classification system for GMAW electrodes.

electrodes according to: (1) whether or not carbon dioxide is used as a separate shielding gas; (2) suitability for either single or multiple pass applications; (3) the type of current; (4) the welding position; and, (5) the as-welded mechanical properties of the weld metal.

High weld-production rates may be attained with semi-automatic equipment which may be used in any position with the appropriate electrode. Where required by service conditions, flux-cored electrode grades can provide weld metal with CVN 20 impact values at temperatures in steps from 20°F to −100°F. Some of the deposits of the carbon steel electrodes will develop CVN 20 at −20°F, while the low alloy electrodes will develop CVN 20 at −100°F.

The applicable specifications are AWS A5.20 and AWS 5.29 (symbols are similar to AWS 5.20 with the addition of an alloy composition at the extreme right). The AWS classification system is illustrated in Figure 8-25.

ESW and EGW

With the ESW and EGW processes, 18-in. and greater thicknesses may be welded in one pass, using multiple electrodes, with the joint in a vertical plane. A single-electrode, semi-portable welding machine can join plates up to five inches thick. Furthermore, using either of these processes, it is possible to make girder flanges by welding mill-width plates and subsequently longitudinally cutting out three or more flange widths.

Note that AWS prohibits the use of these welding processes on quenched and tempered steels.

The composition of cored electrodes is based on weld-metal analysis, and the composition of solid electrodes is based on wire analysis. The coarse grains in the slow-cooled electroslag weld may make it difficult to test ultrasonically and the minimum size flaw detectable by RT is about 1½ percent of the thickness. This creates difficulty in the inspection of electroslag welding.

AWS A5.25 requires electrodes which contain nickel to provide CVN 15 impact values at either 0°F or −20°F. This specification is patterned after AWS A5.17 and A5.18 insofar as the electrodes are concerned; refer to Figure 8-26.

ELECTRODE PROPERTIES					
E	7X	T	1	—	2
Electrode	70,000 psi min. tensile	Tubular (flux cored)	Position code*		Usability code**

*0 = Flat and horizontal only
1 = All position
**2 = Single pass CO_2 shielded only

Fig. 8-25. AWS classification system for FCAW electrodes.

Thermal Cutting and Air-Arc Gouging

Thermally cut welding bevels are required to be smooth and free of notches or grooves in which weld slag may be trapped. Two cutting systems, oxy-fuel gas and plasma arc, are available. Oxy-fuel gas cutting may be used to cut almost any plate thickness commercially available except in stainless steel which must be plasma cut. Plasma arc cutting will cut thicknesses only up to about 1½-in., but is much faster than oxy-fuel gas cutting. This speed advantage increases as the plate thickness decreases; at a thickness of one inch, the cutting speed is over 300 percent faster with a water-injection plasma torch. The plasma arc cutting process, however, also leaves a slight taper in the cut as it descends.

If the plate being cut contains large discontinuities or non-metallic inclusions, turbulence may be created in the oxy-fuel cutting stream. As result, this may cause notches or gouges in the edge of the cut. The plasma arc stream is less susceptible to this as it moves with a higher velocity. Within the depth limits of the specifications, it is usually better practice to remove these by grinding than to weld repair and grind. Additionally, re-entrant thermal cuts should provide a smooth transition.

Carbon-air-arc gouging is a convenient method for removing weld defects, gouging the weld root to sound metal, or forming a U-groove on one side of a square butt joint. The carbon arc travels over the work and melts a weld-nugget-shaped area of the metal. This molten material is then blown away by a jet of compressed air, directed from the holder, parallel to the carbon electrode. Thus, air-arc gouging may be considered the opposite of welding in that each pass removes approximately one weld pass. Because the arc quench is similar in both air-arc gouging and welding, any pre-heat required for welding should also be used for air-arc gouging.

Inspection

The five most commonly used testing methods for welding inspection are: visual (VT), dye penetrant (DPT), magnetic particle (MT), radiographic (RT), and ultrasonic (UT). These methods are discussed in the following sections; refer also to AWS B1.0. Visual inspection is the most commonly specified procedure. Other, more stringent methods can

FLUX CAPABILITY					ELECTRODE PROPERTIES			
F	ES	7	2		E	W	T	1
Flux	Electroslag flux	70,000 psi min. tensile	CNV 15 @ 20°F		Electrode	Weld metal tested as deposited	70,000 psi min. tensile	CNV 15 @ 20°F

Fig. 8-26. AWS classification system for ESW materials.

add significant cost to the project and, therefore, should be specified only when essential to the integrity of the structure.

The engineer of record (EOR) must specify in the contract documents which type of weld inspection is required as well as the extent and application of each type. In the absence of instruction, AWS D1.1, paragraph 6.6.5 states that the fabricator or erector is responsible only for those weld discontinuities found by visual inspection. If additional inspection more stringent than visual is later required, the owner is normally responsible for the cost of weld repairs other than those identified by the visual inspection.

VT

Visual testing provides the most economical approach to checking weld quality. It is particularly good for inspecting single-pass welds, but is limited in that only surface imperfections may be detected. This type of inspection is especially effective when it includes both a check of the joint for accuracy and cleanliness before welding and an observation of the welding procedure. Acceptance criteria are specified in the AISC *Code of Standard Practice* and *Quality Criteria and Inspection Standards* (AISC, 1988), as well as AWS D1.1.

DPT

A red dye penetrant is applied to the work and penetrates any crack or crevice open to the surface. After removing excess dye, a white developer is applied. Where cracks are present, the red dye seeps through the developer, producing a visible red image. This process is summarized in Figure 8-27.

DPT may be used to detect tight cracks as long as they are open to the surface. Like VT, however, only surface cracks are detectable. Furthermore, deep weld ripples and scratches may give a false indication when DPT is used.

MT

A magnetizing current is introduced into the weldment to be inspected as shown in Figure 8-28. The magnetic field induced in the work will be distorted by any cracks, seams, inclusions, etc., located on or within approximately $\frac{1}{10}$-in. of the surface. A dry magnetic powder spread lightly on the surface will gather at such discontinuities, leaving a distinct mark. These magnetically held particles then show the size, location, and shape of the discontinuity.

This method will detect surface cracks filled with slag or contaminants which dye in DPT could not enter. Additionally, the powder may be picked up and preserved with clear

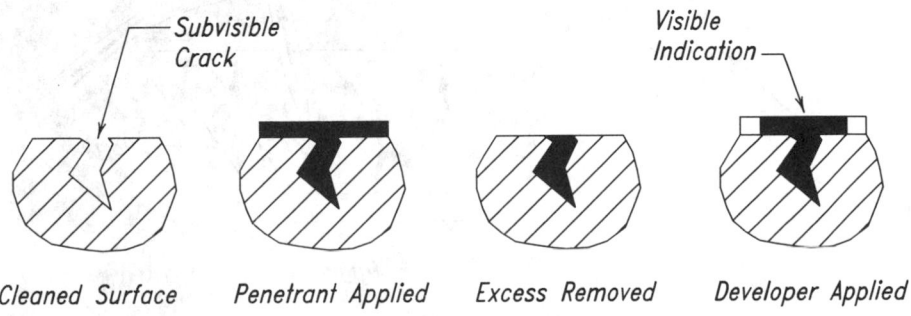

Fig. 8-27. Schematic diagram of DPT.

tape, providing accurate and detailed records of inspection results. However, this method requires relatively smooth surfaces and while cleanup is easy, demagnetization, when necessary, may not be.

RT

This method uses a radioactive source and an X-ray film process. RT can detect porosity, slag, voids, cracks, irregularities, and lack of fusion. To be detected, the imperfection must be oriented roughly parallel to the impinging radiation beam and occupy about 1½ percent of the metal thickness along that beam. The film negative provides a permanent record of the inspection.

Defects smaller than about 1½ percent of the metal thickness and defects not parallel to the beam may not register. RT of closed, inaccessible pipe joints is difficult to obtain and interpret and should be discouraged. Additionally, when the particle beam must penetrate varying thicknesses, as at fillets and tee or corner joints, RT is not readily interpreted and the resulting inspection may be less consistent. When this is the case, other inspection methods should be used. Other limitations of RT are that the required exposure time increases with material thickness and there is a worker hazard due to the radiation used in the method. The precautions for avoiding these hazards and the equipment and film costs make this method the most expensive inspection method.

UT

This process, illustrated in Figure 8-29, is analogous to radar and operates on a principle called pulse-echo. A short pulse of high-frequency sound is introduced into the metal. The reflection of this sound wave from the far end of the member and any voids encountered along the way may then be detected. Any reflections are displayed as pips on a display in which the horizontal grid represents the distance through the metal, and the vertical scale represents the area, and therefore the strength, of the reflecting surface. The point of origin of the sound wave can be readily moved around to check many orientations and can project the wave into the metal at angles of 90°, 70°, 60°, and 45°.

While UT can detect favorably oriented, flat discontinuities smaller than ¼₄-in. in carbon and low-alloy structural steels, austenitic stainless steels and extremely coarse-

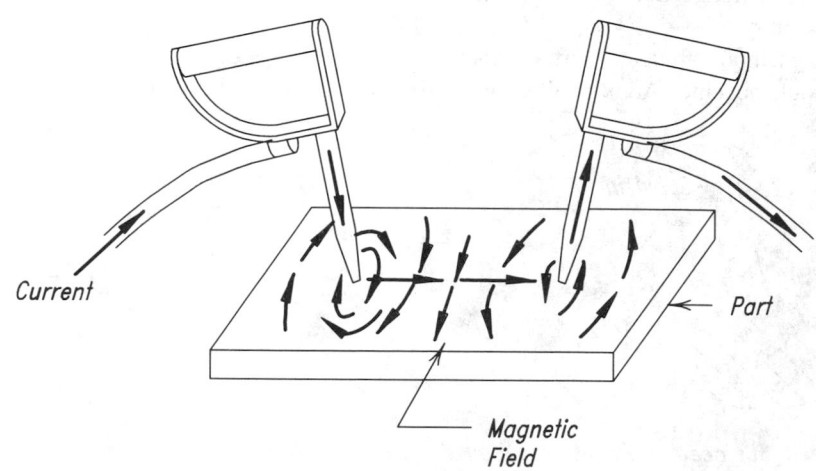

Fig. 8-28. Schematic diagram of MT.

grained steels such as electroslag weld metal are difficult to inspect. Also, certain joint geometry limits the use of UT and it is difficult to inspect members less than $\frac{5}{16}$-in. thick because there is a "dead area" at the origin of the sound wave.

The accuracy of UT depends upon the skill and training of the operator and frequent calibration of the instrument. ASNT has set training standards for UT operators. Despite the fact that UT is a more versatile, expedient, and economical inspection method than RT, it does not provide a permanent record like the X-ray negative in MT. Instead the operator must make a written record of discontinuity indications. For more information, see Krautkramer (1977) and Institute of Welding (1972).

Economical Considerations
On a weight basis, the cost of weld metal far exceeds the cost of any other material in a structure. Therefore, in addition to designing joints for the best welding position, significant economy can be achieved by selecting the proper weld type and an arrangement for the welds which requires a minimum amount of weld metal and the least amount of deposit time. Acceptance of prior qualification of welding procedures can also result in a more economical structure.

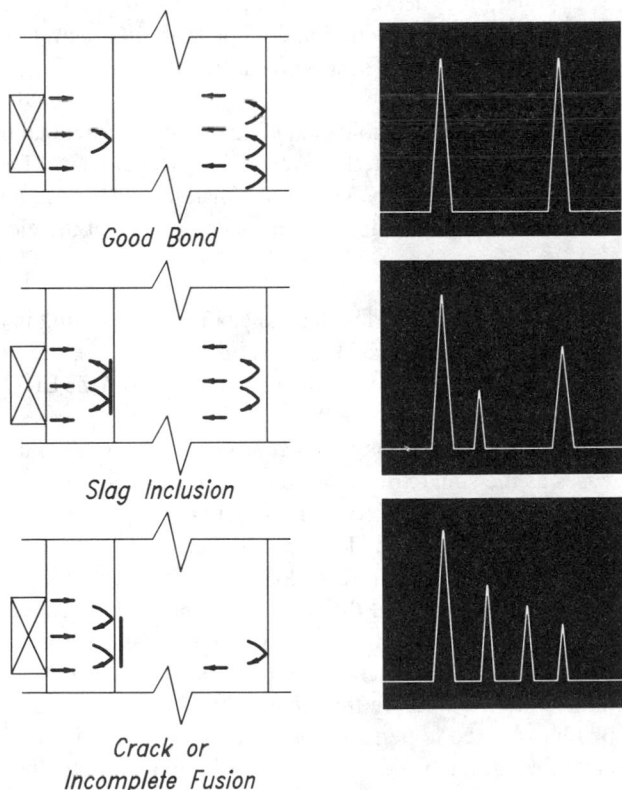

Good Bond

Slag Inclusion

*Crack or
Incomplete Fusion*

Fig. 8-29. Variations in UT reflections due to differences in acoustic properties caused by defects at the boundary.

Welding Position

When weld metal is deposited in the flat position, it can be deposited more quickly since gravity does not adversely affect the deposit. As a result, large electrodes and high currents may be used. In the vertical and overhead positions, electrode diameters above $5/32$-in. produce weld pools with surface tensions and arc forces which are unable to overcome the pull of gravity, causing the weld metal to run. Since the deposition rate in the flat position and in the horizontal position for single-pass fillet welds (not greater than $5/16$-in.) is approximately four times faster than that in the vertical or overhead position, there is strong economic incentive to design and position work for welding in the flat or horizontal position.

Weld Type

In general, in the flat position, the SAW, GMAW, or FCAW processes will be more economical than the SMAW process. However, the selection of the welding process should be left to the fabricator since the equipment and training of personnel will vary from one shop to another.

It is appropriate, though, for the designer to specify the type of weld to be used, e.g., fillet, groove, etc. The fillet weld will be most economical and should generally be selected instead of the groove weld in applications for which groove welds are not required. Additionally, fillet welds result in lesser distortion of the connected material. There are, however, situations, such as joints subjected to fatigue loading, in which the performance of the groove weld is superior. Complete-joint-penetration groove welds may incur the additional costs of non-destructive testing, backgouging, or backing bars; refer to Alexander (1991).

Fillet welds around the inside of a hole or slot require less weld metal than plug or slot welds of the same size. It should be noted, however, that the diameters of holes and widths of slots for fillet welds should be somewhat larger than those for plug and slot welds in metal of the same thickness to accommodate the necessary tilt of the electrode.

Weld Metal Volume

Welds which are oversized waste weld metal and labor time, resulting in an unnecessary increase in the cost of the connection. Thus, it is important to use the proper weld size required for strength or based upon the minimum weld size from the LRFD Specification and to not over-specify weld size.

While the strength of a fillet weld is in direct proportion to its size, the volume of the weld metal increases as the square of the weld size. Thus, a $5/8$-in. fillet weld is twice as strong as a $5/16$-in. fillet weld but also four times more costly. For this reason, it is more desirable to specify a smaller-sized and longer weld than a larger-sized and shorter weld.

In groove welds, double-bevel, double-V, double-J, and double-U welds are typically more economical than single welds of the same type since they use less weld metal. As an added benefit, the resulting symmetry results in less rotational distortion strain. Double welds, however, require more labor in edge preparation and proper cleaning of the weld root prior to commencing the weld on the second side. There may also be added cost if the piece must be repositioned to perform the weld on the second side. For this reason, many fabricators prefer a single weld in thicknesses up to about one inch.

Where single- or double-groove welds are to be used, bevel- and V-groove welds are usually less expensive since they may be flame cut; J- and U-groove welds are more expensive since they must be planed or air-arc gouged.

Deposit Time

Fillet welds sizes up to $\frac{5}{16}$-in. may be deposited in a single pass when deposited in the flat or horizontal position. Larger-size welds must be deposited in multiple passes which will require appreciably more time and weld metal. Thus, fillet welds sized not greater than $\frac{5}{16}$-in., where possible, will result in a significant savings in deposit time, weld material, and cost.

Prior Qualification of Procedures

Evidence of prior qualification of welding procedures, welders, welding operators, or tackers may be accepted at the discretion of the engineer of record (EOR). Fabricators certified in the AISC Quality Certification Program have the experience and documentation necessary to assure that the EOR could accept such prior qualifications (refer to Part 6 for a description of the AISC Quality Certification Program). Significant economic savings may be achieved by accepting such prior qualifications.

Minimizing Weld Repairs

Added cost in the form of weld repairs or replacement may be minimized if the designer considers the possibilities of lamellar tearing, fatigue cracking, notch development, and reduced impact toughness when designing welded connections.

Lamellar Tearing

A lamellar tear is a separation or crack in the base metal caused by through-thickness weld shrinkage strains. When steel is hot-rolled, sulphides or other inclusions are elongated to form microscopic platelets in the plane of the steel plate. These inclusions reduce the strength of the steel in the through-thickness direction below that in the longitudinal or transverse direction.

While special practices are available to produce low-sulphur steel which is resistant to lamellar tearing and ASTM A770 provides a testing method by which the through-thickness strength of the base metal may be measured, it is difficult to assure freedom from the possibility of lamellar tearing. Lamellar tearing is a phenomenon which can occur even in material with superior mechanical properties. Instead, the joint detail is most important in preventing lamellar tearing.

Some joint designs are inherently susceptible to lamellar tearing (AISC, 1973). For example, the complete-joint-penetration groove-welded tee joints in thick sections shown in Figure 8-30 can develop lamellar tears in the crossbar of the tee flange. Such tears can be detected with UT. Other susceptible joints are shown with improved details in Figures 8-31 and 8-32.

The probability of lamellar tearing may be minimized through good joint design and proper welding procedures. The joint design should minimize the weld size and, therefore, the resulting shrinkage strains. Additionally, the design should reduce the restraint which intensifies the local strains. The welding procedure should then establish a sequence to minimize component and internal restraint. Welding with low-hydrogen processes and effective pre-heat has also been shown to minimize lamellar tearing (Kaufmann, Pense, and Stout, 1981).

Fatigue Cracking

Because of their inherent rigidity, welded members are subjected to severe restrictions at service loads if subjected to the repeated variations in stress (fatigue loading). In a dynamically loaded structure, fatigue cracks at notches progress at a rate proportional to the stress range and to the number of stress cycles.

Gradual transitions of sections will help to alleviate these concentrations. The fatigue resistance of a butt weld in a tension member, for example, can be improved approximately 25 percent by grinding the weld reinforcement flush. Thus, any notches in the tension areas should be ground out. Additionally, all grinding should be done in the direction of the stress. Refer to LRFD Specification Appendix K3 for further information.

Notch Development

When subjected to lateral movement, a severe notch can result at locations of one-sided welds. For the fillet-welded joint subjected to lateral loading in Figure 8-33, the unwelded side has no strength in tension and a notch may form from the unwelded side. Using one fillet weld on each side will eliminate this condition. This is also true with partial-joint-penetration groove welds.

In the case of the backing bar of Figure 8-34a, the location of the tack welds may cause fatigue notches. An improved detail would be as shown in Figure 8-34b, where the backing bar is tack welded inside the groove. Any undercut would then be filled, or at least backed up, by the final weld joint. This is also applicable in the case of box members with corner backup. Note that backing bars should also be continuous throughout the length to avoid discontinuities at the base of the weld profile.

Impact Toughness

Different classifications of alloy electrodes and fluxes can produce welds with CVN 20 at selected temperatures between 0°F and −150°F.

Arc Strikes

Arc strikes may occur during welding procedures if the welding rod is lifted from the work while the current is on, or during magnetic particle testing if the magnetizing prod is lifted from the work while the current is on. As stated in *Quality Criteria and Inspection Standards* (AISC, 1988), arc strikes need not be removed in statically loaded structures.

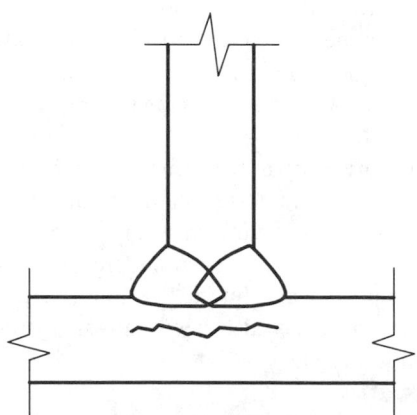

*Fig. 8-30. Lamellar tear resulting from shrinkage of large welds
in thick material under high restraint.*

Other Considerations in Welded Construction

Matching Electrodes

AWS D1.1 Table 4.1 lists matching electrodes for various steels by ASTM Specification and is referenced in LRFD Specification Table J2.5. Use of electrodes one strength-level higher than matching is permitted. Typical structural steel grades with F_y equal to 36 ksi and 50 ksi are normally welded with electrode material of 70 ksi nominal strength, indicated as E70XX for SMAW or its equivalent.

Welding Shapes from ASTM A6 Groups 4 and 5

When heavy shapes are spliced, extremely high shrinkage strains may develop in the base metal, inhibiting ductile deformation in the material and increasing the possibility of brittle fracture. Additionally, interior portions of heavy hot-rolled shapes and plates may contain a coarser grain structure and/or lower notch-toughness properties than other areas of the product.

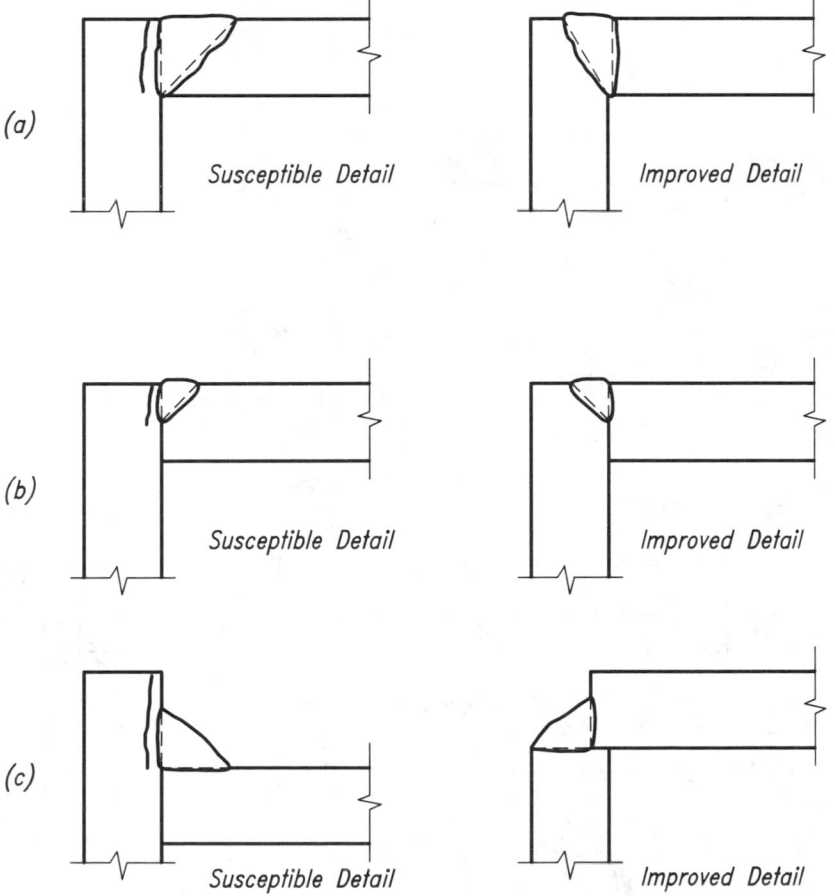

Fig. 8-31. Susceptibility to lamellar tearing can be reduced by careful detailing of welded connections.

LRFD Specification Sections A3.1c, J1.5, J1.6, J2.8, and M2.2 contain special material and fabrication requirements for ASTM A6 Groups 4 and 5 rolled shapes, shapes built-up from plates more than two inches thick, welded together to form the cross section, and shapes where the cross section is to be spliced by welding and subjected to primary tensile stress due to tension or flexure. These special requirements address notch toughness, access hole profiles, welding procedures, pre-heat, thermal cutting, grinding, and inspection requirements and are intended to minimize the possibility of cracking. The corresponding sections of the Commentary on the LRFD Specification provide further information, including alternative splice details and details for weld-access holes and beam copes.

Intersecting Welds and Triaxial Stresses

If a stiffener were to be welded into and around the corner as it meets two elements of a shape (i.e., the flange and web of a column), the welding arc would take the path of least resistance to the three plates meeting at the corner and a lack of fusion or slag pocket would result in that corner. In addition to creating a discontinuity, this would add to the weld shrinkage strains in that corner. Corners of stiffeners, then, should be clipped generously to preclude this problem.

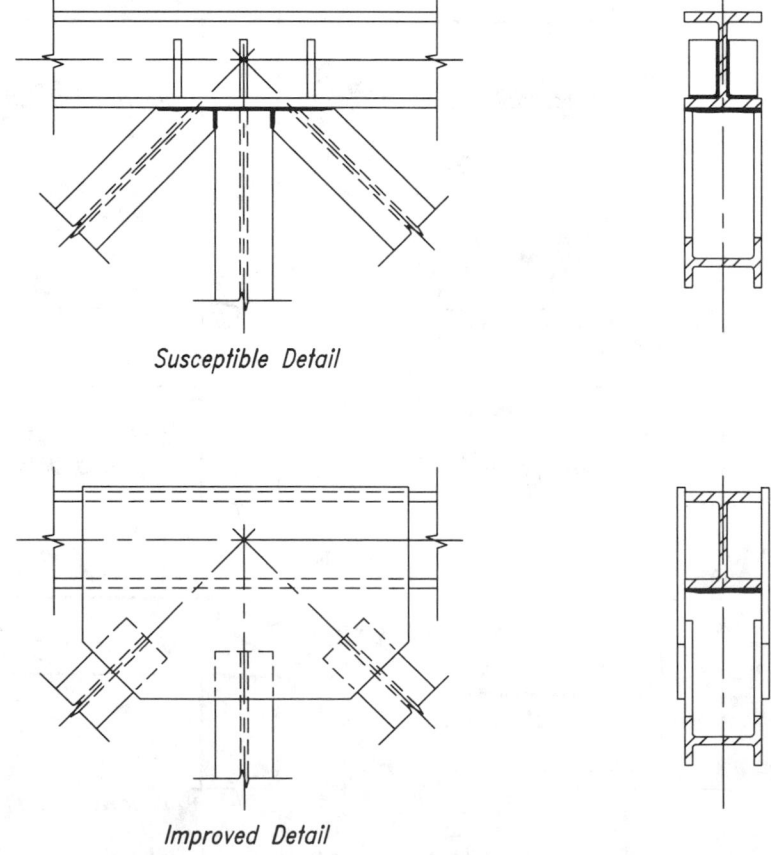

Susceptible Detail

Improved Detail

Figure 8-32.

In general, a ¾-in. clip will be adequate. In small stiffeners, where such a clip would remove a large portion of the effective area of the stiffener, and in shapes, the radii of which require a clip in excess of ¾-in., the clip dimension may be adjusted to suit conditions. For further information, see Butler, Pal, and Kulak (1972) and Blodgett (1980).

Painting Welded Connections

Paint is normally omitted in areas to be field welded. LRFD Specification Section M3.5 requires that, unless otherwise provided in the plans and specifications, surfaces within two inches of any field weld shall be free of materials that would prevent proper welding or produce objectionable fumes during welding. Since little is gained by an exhaustive identification of the small areas involved, most fabricators prefer to use the general note, "No paint on OSL of connection angles," where OSL stands for *outstanding leg*. This

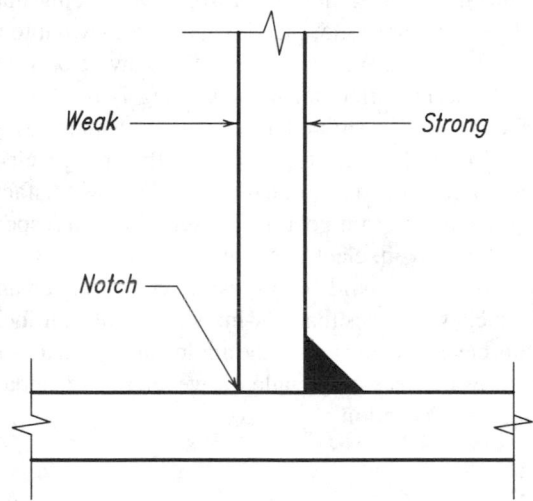

Fig. 8-33. One-sided fillet weld results in a severe notch.
A similar effect exists with a one-sided partial-penetration groove weld.

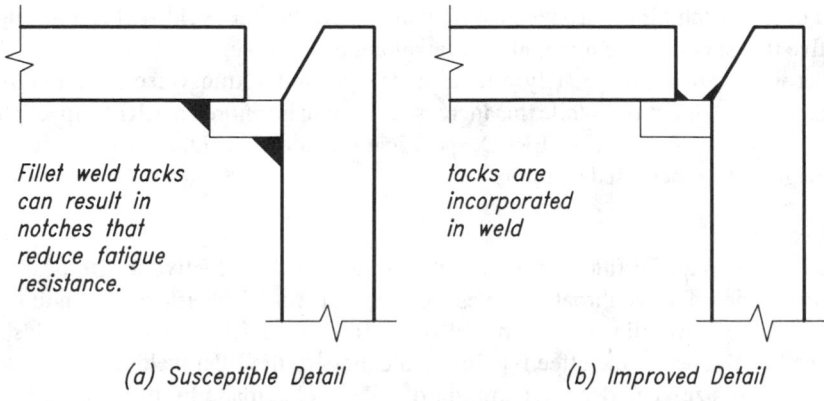

Fig. 8-34. Backing bar tack welds.

"no paint" requirement does not apply to shop welding where painting is normally done after the welds are made.

Clearances for Welding

Clearances are required to allow the welder to make proper welds. In the SMAW process, for example, the welder must hold an electrode, about ⅜-in. in diameter and 14 inches to 18 inches long, in full control, and in such a position that the far end of the rod is in near contact with the base metal. This welder must observe the weld through a protective window of very dark glass in a bulky protective hood. Furthermore, the welder must keep control of the stiff electrical cable which powers the welding process.

These conditions make welding difficult and it is imperative that other factors do not further hamper the welder. Ample room must be provided so that the welder or welding operator may manipulate the electrode and observe the weld as it is being deposited.

The preferred position of the electrode when welding in the horizontal position is in a plane forming 30° with the vertical side of the fillet weld being made. However, this angle, shown as angle x in Figure 8-35, may be varied somewhat to avoid contact with some projecting part of the work. A simple rule which may be used to provide adequate clearance for the electrode in horizontal fillet welding is that the clear distance to a projecting element should be at least one-half its height; distance $y/2$ in Figure 8-35b.

A special case of minimum clearance for welding with a straight electrode is illustrated in Figure 8-36. The 20° angle is the minimum which will allow satisfactory welding along the bottom of the angle and therefore governs the setback with respect to the end of the beam. If a ½-in. setback and ⅜-in. electrode diameter were used, the clearance between the angle and the beam flange could be no less than 1¼-in. for an angle with a leg dimension w of three inches, nor less than 1⅝-in. with a w of four inches. When it is not possible to provide this clearance, the end of the angle may be cut as noted by the optional cut in Figure 8-36 to allow the necessary angle. However, this secondary cut will increase the cost of fabricating the connection.

Fillet Welds

In Figure 8-37a, fillet welds A are loaded in longitudinal shear and fillet weld B is loaded in transverse shear. If the force R_u is increased to exceed the strength of the welds, rupture will occur on the planes of least resistance. As shown in Figure 8-37b, this is assumed to take place in the weld throat where the least cross-sectional area is present. Tests of fillet welds using matching electrodes have demonstrated that the weld will fail through its effective throat before the material will fail along the weld leg.

Fillet welds are approximately one-third stronger in the transverse direction than in the longitudinal direction. While this increased strength is ignored in LRFD Specification Section J2.4, the provisions of LRFD Specification Appendix J2.4 may be used to take advantage of this increased strength.

Effective Area

The effective area of a fillet weld A_w is the product of the effective length of the fillet weld times the effective throat thickness of the fillet weld. The effective length l of the fillet weld is the overall length of the full-sized fillet weld. Except for fillet welds made with the SAW process, the effective throat thickness of the fillet weld is $0.707w$, where w is the weld size. The deep penetration of fillet welds made by the SAW process is recognized in the LRFD Specification Section J2.2a wherein the effective throat thickness is considered to be equal to the weld size for ⅜-in. and smaller welds, and equal to the effective throat thickness plus 0.11 in. for fillet welds sizes over ⅜-in.

Minimum Effective Length

The minimum effective length of a fillet weld when used alone and not as a part of a continuing joint boundary (i.e., an end return or corner) must be greater than or equal to four times the nominal weld size. Thus, the shortest length of ⁵⁄₁₆-in. fillet weld which is permitted to be considered to transmit load is 1¼-in.

Conversely, regardless of the fillet-weld size used, the maximum effective size is limited to one-fourth the weld length. Intermittent fillet welds likewise are subject to this provision with the added requirement that the incremental length of weld must not be less than 1½-in; refer to LRFD Specification Section J2.2b.

Minimum Fillet Weld Size

When very small fillet-weld sizes are used, rapid cooling after welding creates internal stresses which, in turn, may lead to cracking of the weld. To preclude this, the minimum fillet-weld size is established in LRFD Specification Section J2.2b as a function of the thickness of the thicker of the parts joined. From this, if two ⅞-in. plates are joined, the minimum permissible fillet-weld size is ⁵⁄₁₆-in., even if a ¼-in. weld might provide adequate strength. Where different thicknesses are joined, the weld size need not exceed the thickness of the thinner part, unless a larger size is required for strength. If this is the case, adequate pre-heat must be provided to assure soundness of the weld.

Maximum Fillet-Weld Size

The maximum fillet-weld size on the edge of the material is limited in LRFD Specification Section J2.2b to the thickness of the element for material less than ¼-in. thick and ¹⁄₁₆-in. less than the thickness of the element for material greater than or equal to ¼-in. thick, unless the drawing is specially noted to build up the weld to achieve full throat size. This limitation recognizes that the exposed corner of the welded edge tends to melt

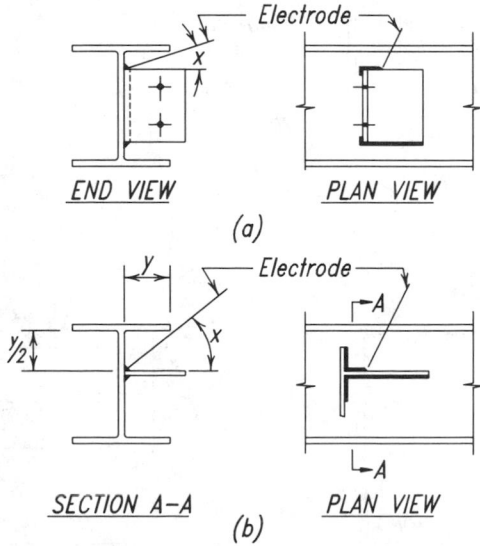

Fig. 8-35. Clearances for welding.

into the weld as illustrated in Figure 8-38, thereby reducing the leg dimension and the weld throat. Additionally, the toes of most rolled shapes do not have an ideal 90° corner. Thus the actual thickness of material at the weld is less than the nominal thickness t of the member. While the LRFD Specification permits the use of a larger weld size if the weld is built up to the full throat size, this is difficult to achieve.

End Returns

LRFD Specification Section J2.2b gives requirements on when fillet weld terminations must be returned around ends or sides. This is illustrated in Figure 8-39. Weld returns reinforce the effective weld where it is most highly stressed and, thus, inhibit cracking and progressive tearing throughout the length of the weld. Thus, they are required in fatigue applications and for connections which assume flexibility exists in the connected part or parts (e.g., the support legs of a double angle connection). If welds are not returned, they must terminate not less than the nominal weld size from the sides or ends.

Also, based upon LRFD Specification Section J2.2b, Figures 8-40 and 8-41 indicate examples where welds must be interrupted or should not be returned. In these instances, the welds, while in the same plane, lie on opposite sides of the contact surfaces. An attempt to weld around the corner will melt the corner material, creating a reduced thickness and notch. Furthermore, such welds cannot be made with a fully effective throat. Welding around such a corner should be avoided.

It is not recommended that weld be applied in the gap at the end of the beam web between the heels of the angles, as this reduces the flexibility of the connection angles. Furthermore, the setback of the beam web is not a controlled dimension as it may be used to account for the tolerance in length of the beam and may vary from zero in. to ½-in. or

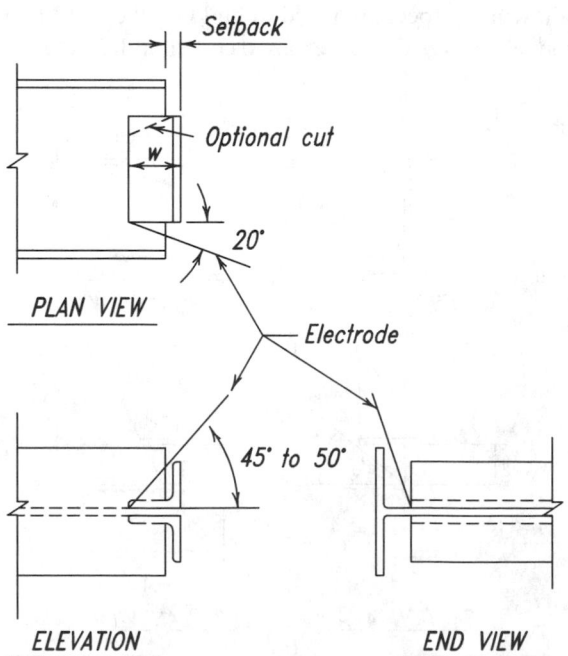

Fig. 8-36. Clearances for welding.

more. In any case, most beam webs are too thin for an effective minimum weld size to be applied along such an edge.

Fillet Welds in Holes or Slots

The recommended minimum hole diameters or slot widths for fillet welding are shown in Table 8-35. It is important to distinguish between plug or slot welds and fillet welds placed around the inside of a hole or slot. In the case of such fillet welds, the shear strength is the product of the effective throat thickness and the weld length measured along the line bisecting the throat area. If this effective area should exceed the area of the hole or slot, it cannot be considered to be a fillet weld and must be designed as a plug or slot weld.

Other Limitations on Fillet Welds

In concentrically loaded welded joints, the stresses are assumed to be uniformly distributed throughout the length of the welds. The design strength of a concentrically loaded fillet-weld group, then, is the sum of the design strengths of each weld in the group. LRFD Specification Section J1.8 provides that the center of gravity of a weld group should coincide with the gravity axis of an axially loaded member, or provision must be made for the resulting eccentricity. Certain welded members not subject to fatigue loading are excluded from this provision: "Eccentricity between gravity axes of such members...may be neglected in statically loaded members, but shall be considered in members subject to fatigue loading." This provision permits very significant cost savings in weld material

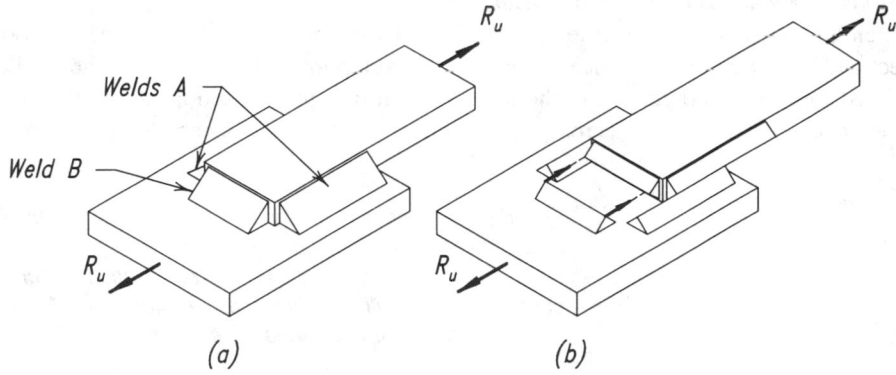

Fig. 8-37. Fillet welds.

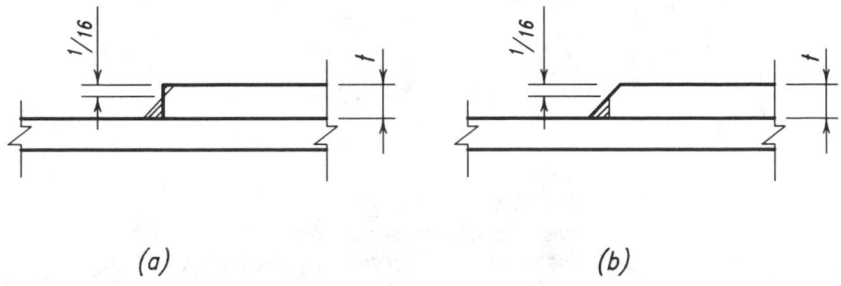

Figure 8-38.

and labor in the fabrication and erection of such statically loaded members as roof and floor trusses, bracing, etc.

Additionally, LRFD Specification Section J2.2b imposes other limitations on proportions of lap joints.

Minimum Shelf Dimensions

In Figure 8-42, the recommended minimum shelf dimensions for normal size SMAW fillet welds are summarized. This dimension is critical to the deposition of the weld. SAW fillet welds would require a greater shelf dimension to contain the flux, although this is sometimes provided by clamping auxiliary material to the member.

In Figure 8-43, the distance b must be large enough so that a full-size weld may be deposited on it. Select a gage that will permit enough clearance b to deposit an effective weld. The dimension b should be sufficient to accommodate the combined tolerances of the framing-angle length, the cope depth, and the beam mill over/underrun as well as the specified weld size.

Complete-Joint-Penetration Groove Welds

Assuming compliance with LRFD Specification Section J2, the design strength of complete-joint-penetration groove welds is equal to that of the base metal in all respects. Therefore, no allowance for the presence of such welds need be made in proportioning the connections of structural members for any type of static loading. Where members are of unequal cross section or different material strength, the strength of the complete-joint-penetration groove weld is limited to the strength of the weaker member.

Extension, Runoff, Backing, and Spacer Bars

When groove welds are used to splice plate girders and beams, LRFD Specification Section J7 requires that the splice be capable of developing the full strength of the smaller spliced section or 100 percent of the full section if the spliced sections are of the same size. To obtain a fully welded cross section, the termination at either end of the joint must

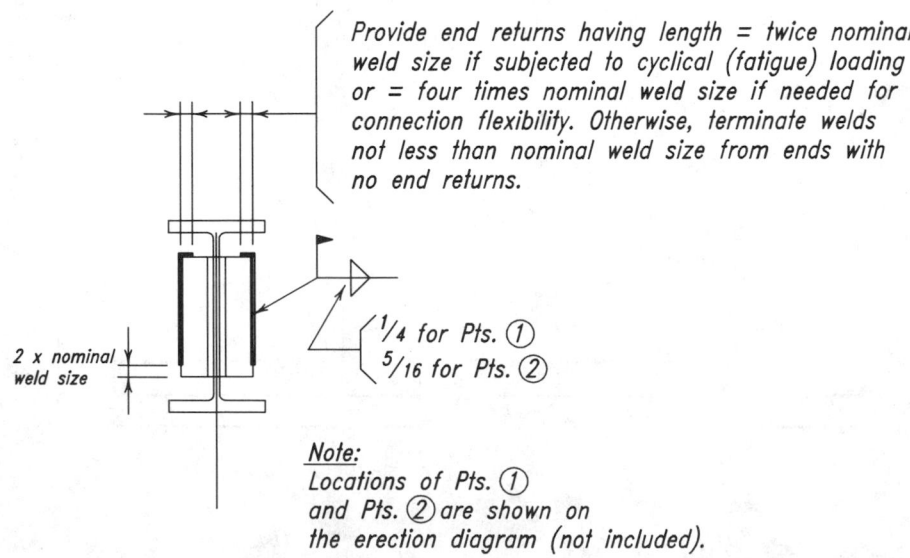

Provide end returns having length = twice nominal weld size if subjected to cyclical (fatigue) loading or = four times nominal weld size if needed for connection flexibility. Otherwise, terminate welds not less than nominal weld size from ends with no end returns.

$^1/_4$ for Pts. ①
$^5/_{16}$ for Pts. ②

2 x nominal weld size

Note:
Locations of Pts. ①
and Pts. ② are shown on
the erection diagram (not included).

Fig. 8-39. Weld returns.

be of sound weld metal. Extension or runoff bars are usually used to assure the soundness of the end of the weld. Frequently, the joint will require a backing or spacer bar which can be extended to serve as the extension or runoff bar.

Figure 8-44 demonstrates the application of extension, backing, and spacer bars in a splice or moment connection. Extension and backing bars should be of approved weldable material as specified in AWS D1.1, Section 8.2.4; spacer bars must be of the same material specification as the base metal. This can create a procurement problem since small tonnage requirements may make them difficult to obtain in the specified ASTM designation. Also indicated in Figure 8-44 is the use of a cover plate or seat angle for backing the weld.

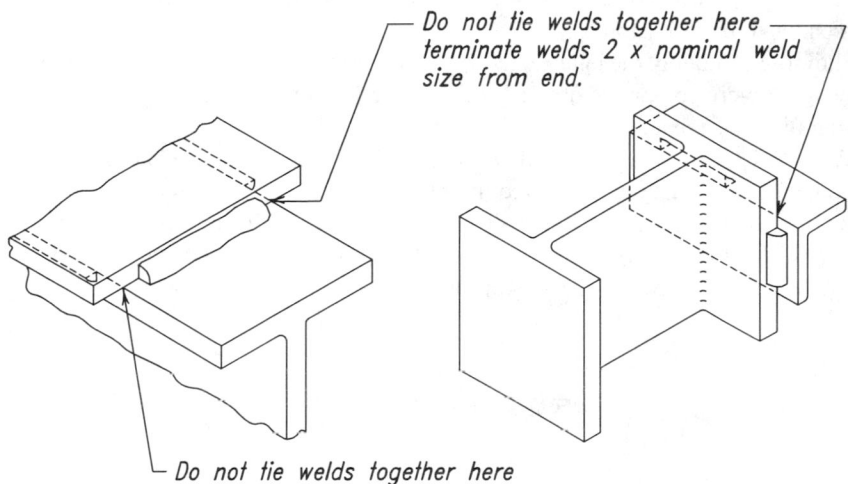

Fig. 8-40. Fillet welds on opposite sides of a common plane should not be continuous.

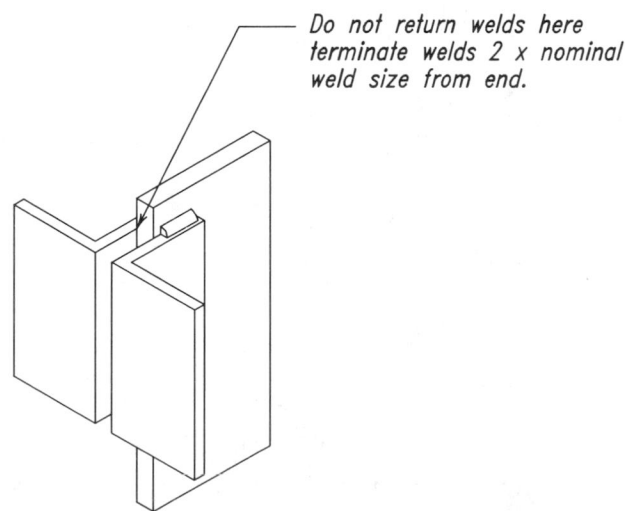

Fig. 8-41. Fillet welds should not be returned across thickness of material.

Table 8-35.
Recommended Minimum Hole Diameters or Slot Widths for Fillet Welding, in.

Plate Thickness, in.	Min. Diameter or Width, in.
$\frac{3}{16}$ and $\frac{1}{4}$	$1\frac{1}{16}$
$\frac{5}{16}$	$1\frac{3}{16}$
$\frac{3}{8}$	$1\frac{5}{16}$
$\frac{7}{16}$	$1\frac{1}{16}$
$\frac{1}{2}$	$1\frac{3}{16}$
$\frac{5}{8}$	$1\frac{5}{16}$

Shown in Figure 8-45 are flat-type extension bars, normally used with beveled grooves, and contour-type extension bars, normally used with J-grooves or U-grooves and shaped to follow the contour of the joint geometry. While the contour-type extension bar is shown as though it were comprised of two pieces, some fabricators might elect to mill the full contour in one piece and subsequently cut it to suit job requirements.

AWS D1.1, Section 3.12 states that runoff and extension bars need not be removed in statically loaded structures unless required by the engineer of record (EOR). Such might be the case where these bars would create an interference with other work. In dynamically and cyclically loaded structures, however, they must be removed and the welds made smooth and flush to the base metal abutting edges.

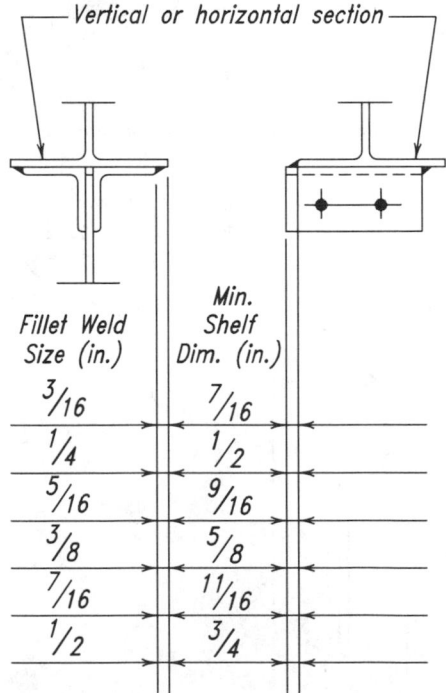

Fig. 8-42. *Recommended minimum shelf dimensions for SMAW fillet welds.*

According to AWS D1.1, Section 3.13, backing bars on groove-welded joints must be fully spliced to avoid stress concentrations or discontinuities and should be thoroughly fused with the weld metal. It is further required on dynamically loaded structures that the backing bars be removed and the surfaces finished smooth when they are transverse to the direction of stress. If this were the case for the flange splice of Figure 8-44, removal of the backing bars would be required and, therefore, the splice might be made more economically with another joint profile.

Weld Access Holes

The beam web is provided with an access hole or "rathole", as illustrated in Figure 8-44, to permit down-hand welding to the backing bars located below both the top and bottom flanges. The weld-access hole also provides increased relief from concentrated weld shrinkage strains and prevents the intersection or close juncture of welds in orthogonal directions. Weld-access holes should not be filled with weld metal since it is difficult to provide sound weld metal to fill such a void and doing so may introduce a state of triaxial stress under loading.

Partial-Joint-Penetration Groove Welds

Partial-joint-penetration groove welds are used primarily for welded compression splices, the connection of elements in heavy box sections and pedestals, and, in general, for joints where the stress to be transferred is substantially less than that which would require complete-joint-penetration groove welds. This type of weld is not, however, recom-

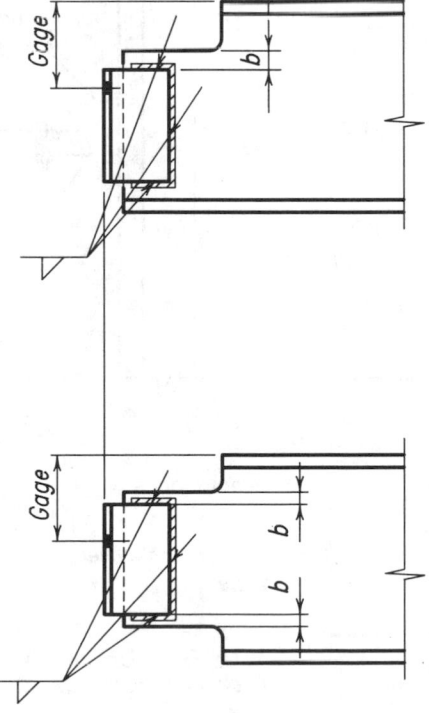

Figure 8-43

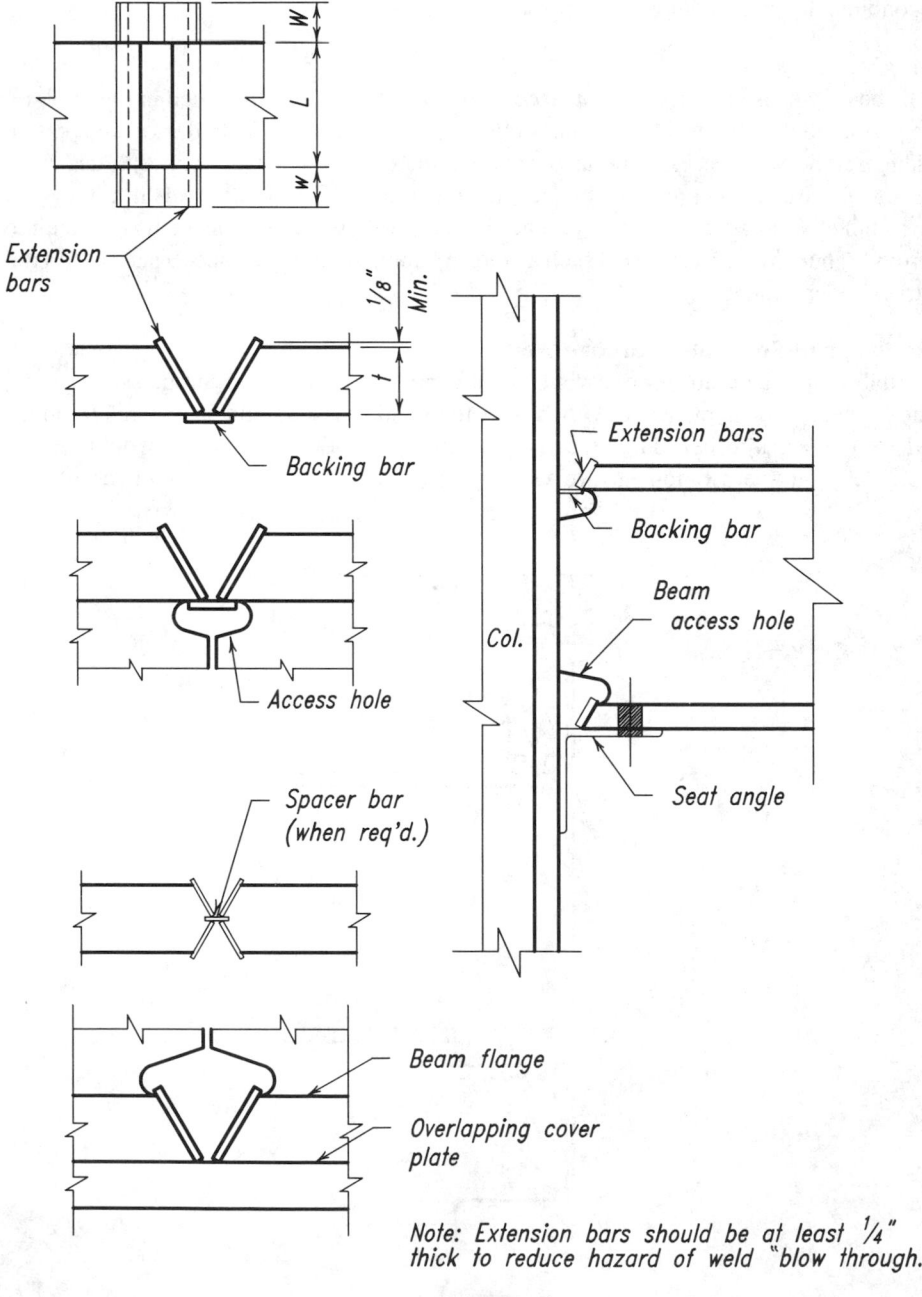

Extension bars

1/8" Min.

t

Backing bar

Access hole

Spacer bar (when req'd.)

Col.

Extension bars

Backing bar

Beam access hole

Seat angle

Beam flange

Overlapping cover plate

Note: Extension bars should be at least 1/4" thick to reduce hazard of weld "blow through."

Figure 8-44

mended in joints subject to dynamic or cyclical loading, except for joining the components of built-up members.

Effective Area

The effective area of a partial-joint-penetration groove weld A_w is the product of the effective length of the weld times the effective throat thickness of the fillet weld. These quantities are determined as follows.

The effective length is the width of the part joined. The effective throat thickness E is as determined from LRFD Specification Table J2.1, but not less than specified in LRFD Specification Table J2.3.

Nomenclature of partial-joint-penetration welds is shown in Figure 8-46. Note that the effective throat thickness shown is less than the dimensioned groove-weld size. AWS prequalified partial-joint-penetration welds establish for each joint an effective throat E as a function of the material thickness, weld-preparation size, or depth S. Thus, the design drawings should specify the effective weld length and the required effective throat. The shop drawings should then show the groove depth S and geometry which will provide for the specified effective throat E. Some fabricators may indicate both the weld size and the effective throat on the shop drawings to eliminate confusion.

The comments on "Extension, Runoff, Backing, and Spacer Bars" and "Weld Access Holes" for complete-joint-penetration groove welds also apply to partial-joint-penetration groove welds.

Intermittent Welds

In preparing the joint profile for intermittent partial-joint-penetration groove welds, a transition or "faring-in" of the joint at beginning and termination must be provided to ensure proper fusion with the base metal. The nominal angular value of this transition should be 45°as shown in Figure 8-46.

Flare Welds

A flare weld is a special case of the partial-joint-penetration groove weld wherein the convex surface of the connected part creates the joint preparation. This convexity may be the result of an edge preparation, but more often results when one (or both) joint component consists of a round rod or a shape with a rounded bend or corner radius created by bending or rolling as shown in Figure 8-47.

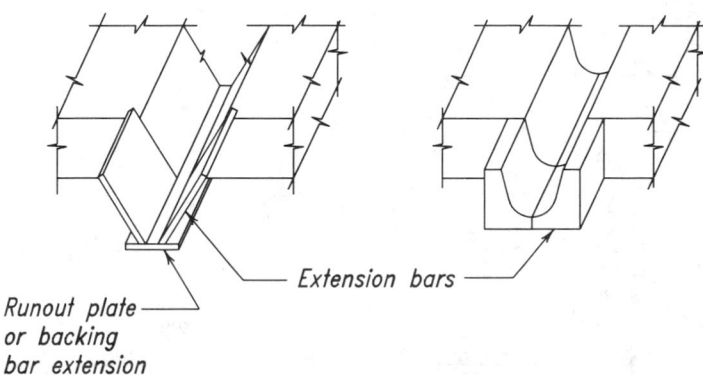

Runout plate or backing bar extension

Extension bars

Figure 8-45.

Effective Area

The effective area of a flare weld A_w is the product of the effective length of the weld times the effective throat thickness of the flare weld; the effective length is the width of the part joined and the effective throat thickness E is as determined from LRFD Specification Table J2.2.

Limitations

The deposition of effective weld metal to the bottom of the flare groove is very difficult because the welding arc short-circuits across the surfaces due to the sharp angular slopes. Thus, the quality of this weld is difficult to control; LRFD Specification Section J2.1a permits examination and adjustment of the weld strength based on random testing and special qualification.

Note that weldability of concrete reinforcing bars is not a part of ASTM specifications. In past experience, improperly welded concrete reinforcing bars have cracked and separated under no-load conditions. Typical deformed-type concrete reinforcing bars, such as ASTM A615, A616, and A617, are not produced to a controlled chemistry and their weldability must be carefully evaluated; refer to AWS D1.4.

Plug and Slot Welds

The use of plug and slot welds for stress transfer is limited to resisting shear loads in joint planes parallel to the faying surfaces. These welds should not be subjected to tensile stresses and are limited when subjected to stress reversal. Furthermore, some specifica-

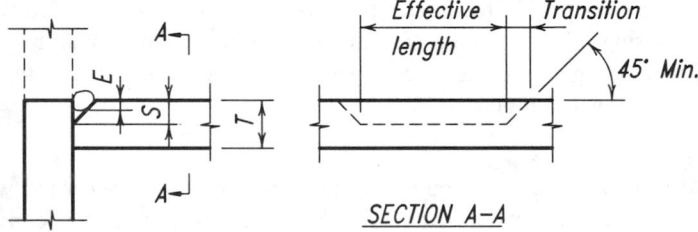

Fig. 8-46. Partial-joint-penetration groove weld nomenclature.

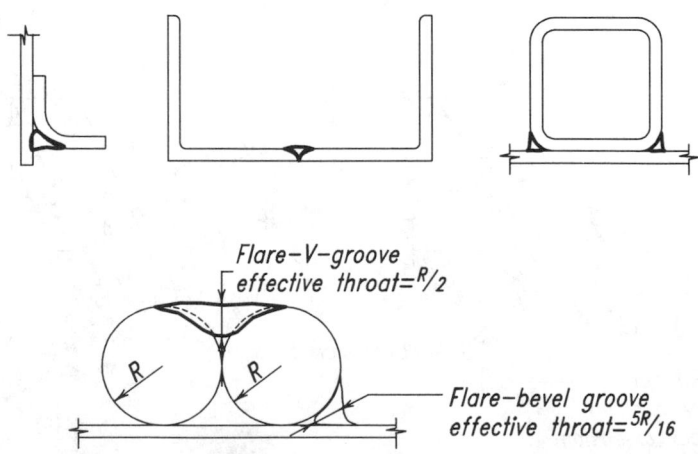

Fig. 8-47. Flare weld nomenclature.

tions do not permit their use as load-carrying welds. Because of these limitations, plug and slot welds are more frequently employed as stitch welds rather than as a means of primary stress transfer.

The effective area of a plug or slot weld A_w is the nominal cross-sectional area of the hole or slot. The proportions and spacing of holes and slots and the depth of weld are stipulated in LRFD Specification Section J2.3b and illustrated in Figure 8-48.

Design Strength of Welds

The design strength of welds is determined in accordance with LRFD Specification Sections J2.2 and J2.4. LRFD Specification requirements are based upon the provisions of AWS D1.1, except as noted in LRFD Specification Section J2.

For welds, the limit states of the weld-metal strength and the base-metal strength must be checked as applicable in LRFD Specification Table J2.5. These limit states assume that the matching electrode requirements of LRFD Specification Section J2.6 and Table J2.1 are met.

Weld Metal Design Strength

From LRFD Specification Section J2.4, the weld metal design strength is ϕR_n, where ϕ is a resistance factor from LRFD Specification Table J2.5 and:

$$R_n = F_w A_w$$

In the above equation,

$$F_w = 0.60F_{EXX}$$
A_w = effective area of the weld, in.^{2}

and ϕ is determined as follows:

For a fillet weld loaded in shear on its effective area,

$\phi = 0.75$;

For a complete-joint-penetration groove weld loaded in shear on its effective area,

$\phi = 0.80$;

For a partial-joint-penetration groove weld loaded in shear parallel to the axis of the weld,

$\phi = 0.75$;

For a partial-joint-penetration groove weld loaded in tension normal to the effective area,

$\phi = 0.80$;

For a plug or slot weld loaded in shear on its effective area,

$\phi = 0.75$.

Base Metal Design Strength

From LRFD Specification Section J2.4, the base metal design strength is ϕR_n, where ϕ is a resistance factor from LRFD Specification Table J2.5 and:

$$R_n = F_{BM} A_{BM}$$

In the above equation, A_{BM} is the cross-sectional area of the base metal. For a fillet weld loaded in tension or compression parallel to the axis of the weld,

$$\phi = 0.90$$
$$F_{BM} = F_y$$

For a complete-joint-penetration groove weld loaded in tension or compression normal to its effective area,

$$\phi = 0.90$$
$$F_{BM} = F_y$$

For a complete-joint-penetration groove weld loaded in shear on its effective area,

$$\phi = 0.90$$
$$F_{BM} = 0.60F_y$$

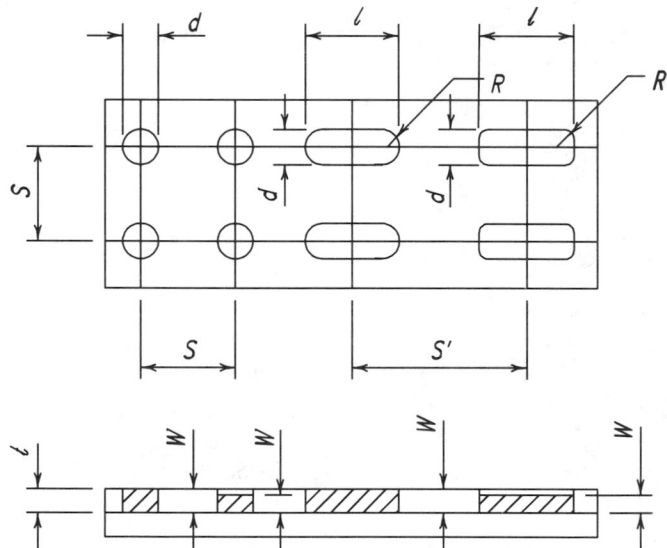

Plate thickness, in.	Min. hole dia. or slot width, d, in.	Hole and slot proportions, spacing and depth of weld	
$^3/_{16}$ & $^1/_4$	$^9/_{16}$	$d \geq (\ell + ^5/_{16})$, round to next higher odd $^1/_{16}$; also $d \leq 2^1/_4 W$	
$^5/_{16}$ & $^3/_8$	$^{11}/_{16}$	$S \geq 4d$ $S' \geq 2\ell$	Where $\ell \leq ^5/_8$, $W = \ell$
$^7/_{16}$ & $^1/_2$	$^{13}/_{16}$	$\ell \leq 10W$ $R = ^d/_2$	Where $\ell > ^5/_8$, $W = ^\ell/_2$ but, not less than $^5/_8$
$^9/_{16}$ & $^5/_8$	$^{15}/_{16}$	$R \geq \ell$	

Fig. 8-48. Plug and slot welds.

For a partial-joint-penetration groove weld loaded in tension or compression normal to its effective area or tension or compression parallel to the axis of the weld,

$\phi \quad = 0.90$

$F_{BM} = F_y$

Prequalified Welded Joints

AWS D1.1 contains provisions for prequalified welded joints which provide joint geometries, such as root openings, angles, and clearances, as illustrated in Figures 8-49 and 8-50, that will permit a qualified welder to deposit sound weld material. Thus, prequalified joints are concerned almost exclusively with the welding process as a method of joining metal and deal with welded joints only from fusion boundary to fusion boundary. The designer must satisfy all provisions of AWS D1.1 Sections 2, 3, and 4 before a joint is considered prequalified.

Prequalified welded joints are not, in themselves, adequate consideration of welded design details. To emphasize this, the AWS D1.1 Section 1.1 states: "...The use of prequalified joints is not intended as a substitute for engineering judgment with respect to the suitability of application of these joints to a weld assembly." The design and detailing for successful welded construction requires consideration of factors which include, but are not limited to, the magnitude, type, and distribution of forces to be transmitted, access, restraint against weld shrinkage, thickness of connected materials, residual stress, and distortion. Accordingly, the design and detailing must also satisfy the requirements of LRFD Specification Section J2.

The prequalified welded joints in Table 8-36 meet the requirements of the 1992 version of AWS D1.1 as well as the 1993 LRFD Specification. Because AWS D1.1 is revised every other year, designers and fabricators should verify this information with the latest issue of AWS D1.1. The designations such as B-L1a, B-U2, and B-P3 are those used in AWS standards. Note that lowercase letters, e.g., a, b, c, etc., are often used to differentiate between joints that would otherwise have the same joint designation. These prequalified welded joints are limited to those made by the SMAW, SAW, GMAW (except short circuit transfer), and FCAW procedures. Small deviations from dimensions, angles of grooves, and variation in depth of groove joints are permissible within the tolerances given.

In general, all fillet welds, whether illustrated or not, are prequalified, provided they conform to the requirements of AWS D1.1. Groove welds are classified using the conventions indicated in the tables. Welded joints other than those prequalified by AWS may be qualified, provided they are tested and qualified in accordance with AWS D1.1.

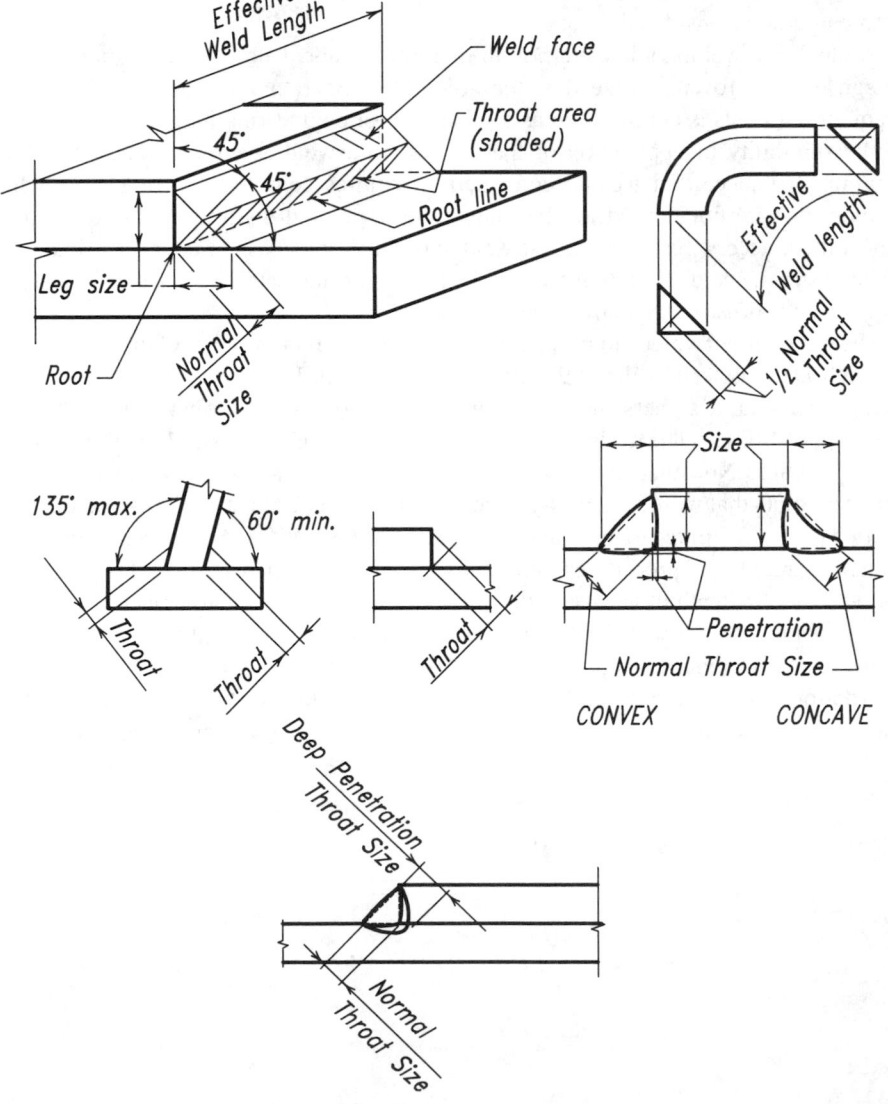

Fig. 8-49. Fillet weld nomenclature.

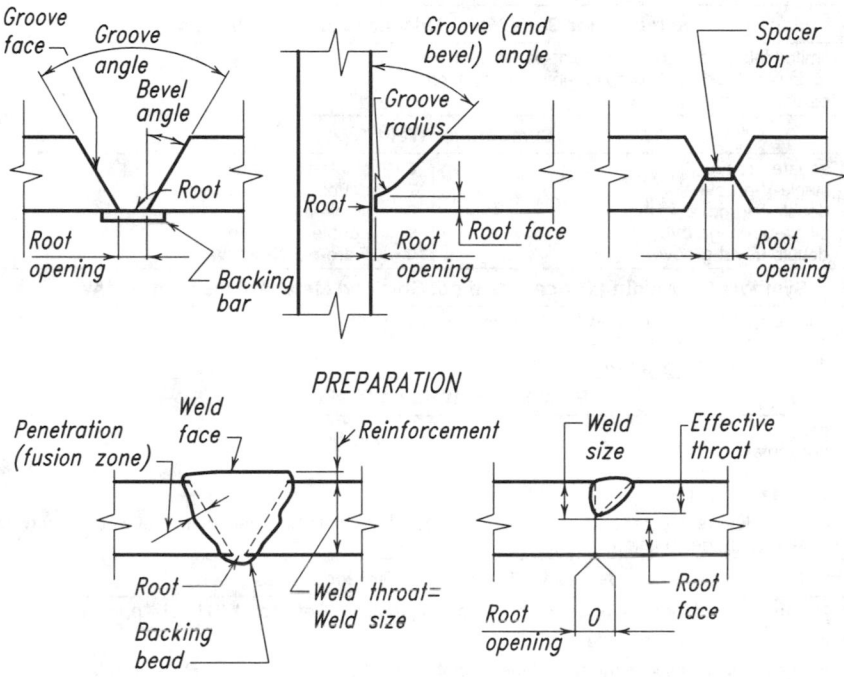

PREPARATION

COMPLETE–JOINT–PENETRATION   PARTIAL–JOINT–PENETRATION

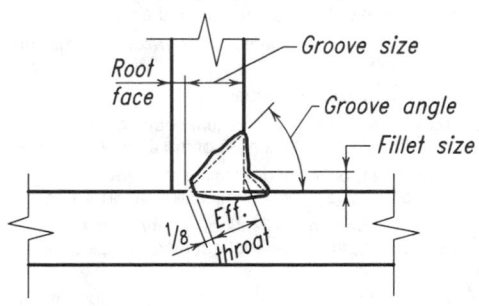

PARTIAL–JOINT–PENETRATION

(When Reinforcing Fillet
is Specified)

Fig. 8-50. Groove weld nomenclature.

Table 8-36.
Prequalified Welded Joints

Symbols for Joint Types

B	butt joint	BC	butt or corner joint
C	corner joint	TC	T- or corner joint
T	T-joint	BTC	butt, T-, or corner joint

Symbols for Base Metal Thickness and Penetration

L	limited thickness, complete-joint-penetration
U	unlimited thickness, complete-joint-penetration
P	partial-joint-penetration

Symbols for Weld Types

1	square-groove	6	single-U-groove
2	single-V-groove	7	double-U-groove
3	double-V-groove	8	single-J-groove
4	single-bevel-groove	9	double-J-groove
5	double-bevel-groove	10	Flare-bevel-groove

Symbols for Welding Processes if not Shielded Metal Arc welding (SMAW):

S	submerged arc welding SAW
G	gas metal arc welding GMAW
F	flux cored arc welding FCAW

Symbols for Welding Positions

F	flat
H	horizontal
V	vertical
OH	overhead

The lower case letters, e.g., a, b, c, d, etc., are used to differentiate between joints that would otherwise have the same joint designation.

Notes to Prequalified Welded Joints

A	Not prequalified for GMAW using short circuiting transfer. Refer to AWS D1.1 Appendix A.
B	Joints welded from one side only.
Br	Bridge applications limit the use of these joints to the horizontal position. Refer to AWS D1.1 Section 9.12.5.
C	Back gouge root to sound metal before welding second side.
E	Minimum effective throat (E) as shown in LRFD Specification Table J2.3; S as specified on drawings.
J	If fillet welds are used in buildings to reinforce groove welds in corner and T-joints, they shall be equal to $\frac{1}{4}T_1$, but need not exceed $\frac{3}{8}$-in. Groove welds in corner and T-joints in bridges shall be reinforced with fillet welds equal to $\frac{1}{4}T_1$, but not more than $\frac{3}{8}$-in.
J2	If fillet welds are used in buildings to reinforce groove welds in corner and T-joints, they shall be equal to $\frac{1}{4}T_1$, but not more than $\frac{3}{8}$-in.
L	Butt and T-joints are not prequalified for bridges.
M	Double-groove welds may have grooves of unequal depth, but the depth of the shallower groove shall be not less than one-fourth of the thickness of the thinner part joined.
Mp	Double-groove welds may have grooves of unequal depth, provided they conform to the limitations of Note E. Also, the effective throat (E), less any reduction, applies individually to each groove.
N	The orientation of the two members in the joints may vary from 135° to 180°, provided the basic joint configuration (groove angle, root face, root opening) remains the same and the design throat thickness is maintained.
Q	For corner and T-joints, the member orientation may be changed provided the groove dimensions are maintained as specified.
Q2	The member orientation may be changed provided the groove dimensions are maintained as specified.
R	The orientation of two members in the joint may vary from 45° to 135° for corner joints and from 45° to 90° for T-joints, provided the basic joint configuration (groove angle, root face, root opening) remains the same and the design throat thickness is maintained.
V	For corner joints, the ouside groove preparation may be in either or both members, provided the basic groove configuration is not changed and adequate edge distance is maintained to support the welding operations without excessive edge melting.

Table 8-36 (cont.).
Prequalified Welded Joints

Basic Weld Symbols

		Plug or Slot	Groove or Butt						
Back	Fillet		Square	V	Bevel	U	J	Flare V	Flare Bevel

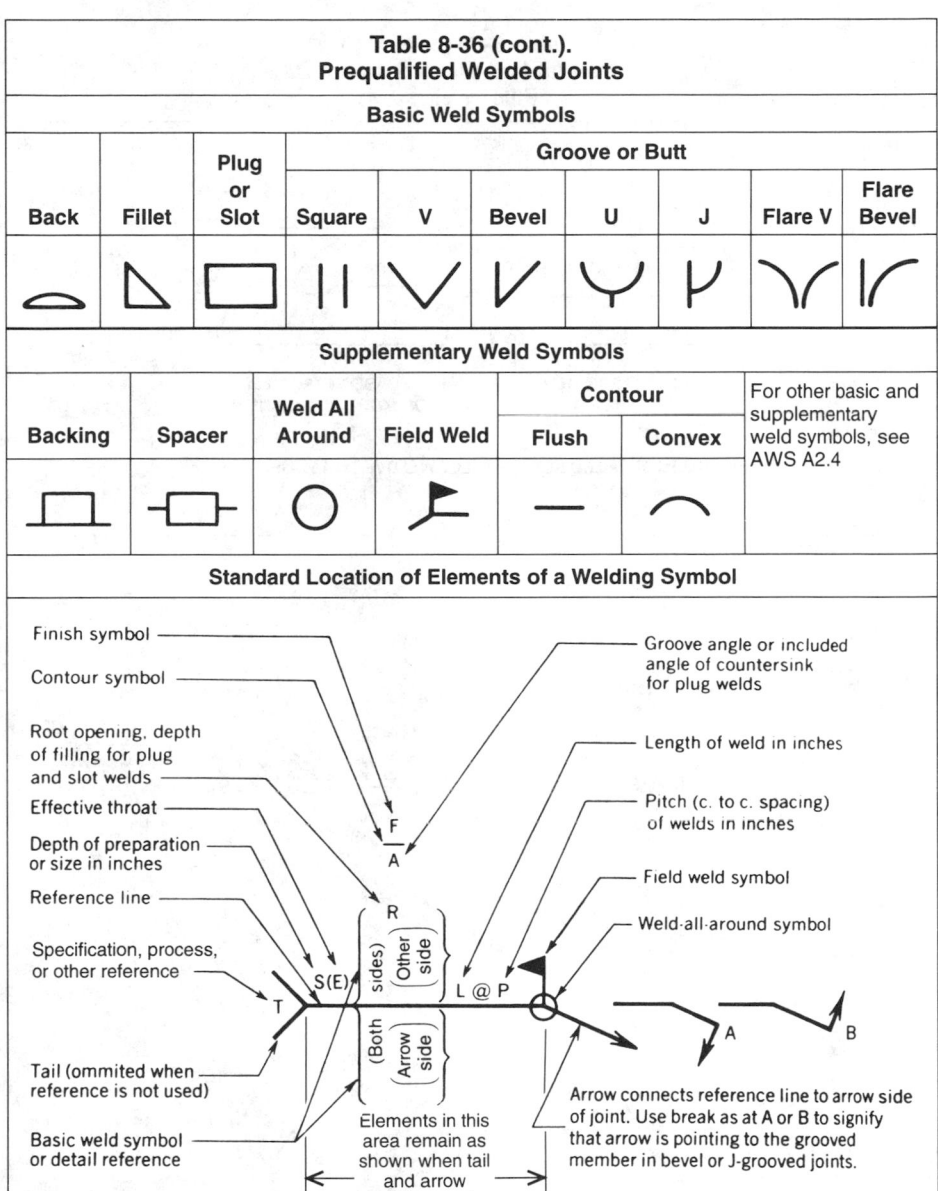

Supplementary Weld Symbols

		Weld All Around	Field Weld	Contour		For other basic and supplementary weld symbols, see AWS A2.4
Backing	Spacer			Flush	Convex	

Standard Location of Elements of a Welding Symbol

Finish symbol

Contour symbol

Root opening, depth of filling for plug and slot welds

Effective throat

Depth of preparation or size in inches

Reference line

Specification, process, or other reference

Tail (ommited when reference is not used)

Basic weld symbol or detail reference

Groove angle or included angle of countersink for plug welds

Length of weld in inches

Pitch (c. to c. spacing) of welds in inches

Field weld symbol

Weld-all-around symbol

F
A
R
S(E)
T
(sides)
(Other side)
(Both sides)
(Arrow side)
L @ P
A
B

Arrow connects reference line to arrow side of joint. Use break as at A or B to signify that arrow is pointing to the grooved member in bevel or J-grooved joints.

Elements in this area remain as shown when tail and arrow are reversed.

Note:

Size, weld symbol, length of weld, and spacing must read in that order, from left to right, along the reference line. Neither orientation of reference nor location of the arrow alters this rule.

The perpendicular leg of $\triangle$, V, P, $\mathsf{I\Gamma}$, weld symbols must be at left.

Arrow and other side welds are of the same size unless otherwise shown. Dimensions of fillet welds must be shown on both the arrow side and the other side symbol.

The point of the field weld symbol must point toward the tail.

Symbols apply between abrupt changes in direction of welding unless governed by the "all around" symbol or otherwise dimensioned.

These symbols do not explicitly provide for the case that frequently occurs in structural work, where duplicate material (such as stiffeners) occurs on the far side of a web or gusset plate. The fabricating industry has adopted this convention: that when the billing of the detail material discloses the existence of a member on the far side as well as on the near side, the welding shown for the near side shall be duplicated on the far side.

Table 8-36 (cont.).
Prequalified Welded Joints
Fillet Welds

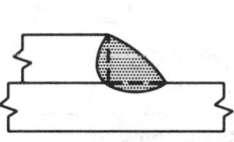
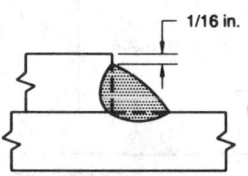

BASE METAL LESS THAN
1/4 in. THICK

(A)

BASE METAL 1/4 in.
OR MORE IN THICKNESS

(B)

MAXIMUM DETAILED SIZE OF FILLET WELD ALONG EDGES

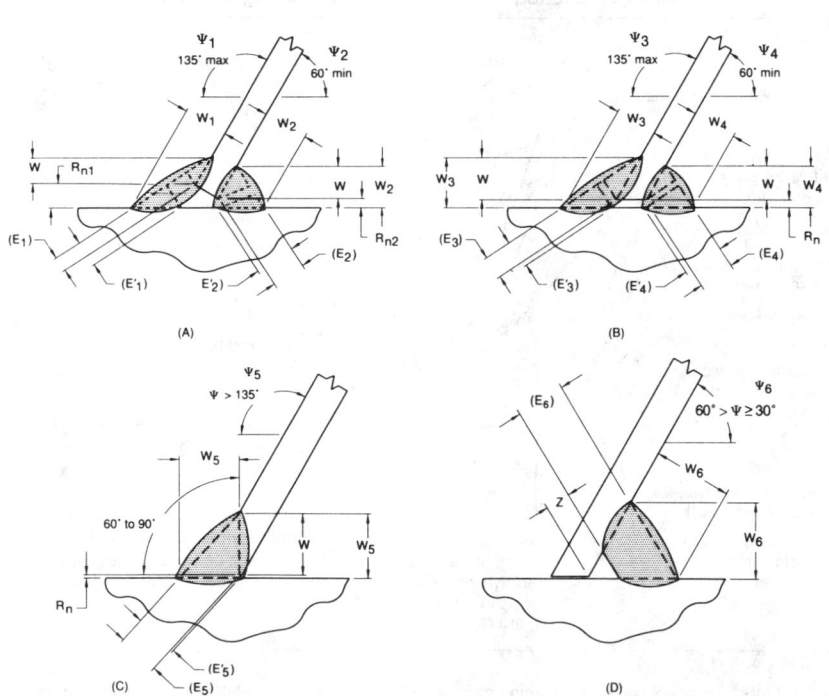

Notes:
1. E_n, $E_{n'}$ = effective throats dependent on magnitude of root opening R_n. See AWS D1.1 Section 3.3.1
   Subscript n represents 1, 2, 3, or 4.
2. t = thickness of thinner part.
3. Not prequalified for gas metal arc welding using short circuitry transfer. Refer to AWS D1.1.
4. Part (f), apply Z loss factor of AWS D1.1 Table 2.4 to determine effective thrust.
5. Part (f), not prequalified for angles under 30°. For welder qualfications, see AWS D1.1
   Table 10.5, Column 10.

*Angles smaller than 60° are permissible, however, if the weld is considered to be a partial-joint-penetration groove weld.

Table 8-36 (cont.).
Prequalified Welded Joints
Complete-Joint-Penetration Groove Welds

Square-groove weld (1)
Butt joint (B)
Corner Joint (C)

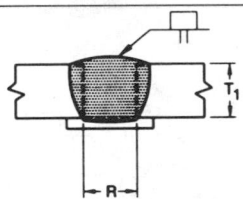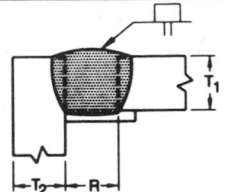

Welding Process	Joint Designation	Base Metal Thickness (U = unlimited)		Groove Preparation			Permitted Welding Positions	Gas Shielding for (FCAW)	Notes
		T_1	T_2	Root Opening	Tolerances As Detailed	As Fit Up			
SMAW	B-L1a	1/4 max	—	$R = T_1$	+1/16, −0	+1/4, −1/16	All	—	N
	C-L1a	1/4 max	U	$R = T_1$	+1/16, −0	+1/4, −1/16	All	—	—
GMAW FCAW	B-L1a-GF	3/8 max	—	$R = T_1$	+1/16, −0	+1/4, −1/16	All	Not Required	A, N

Square-groove weld (1)
Butt joint (B)

BACKGOUGE (EXCEPT B-L1-S)

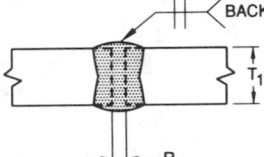

Welding Process	Joint Designation	Base Metal Thickness (U = unlimited)		Groove Preparation			Permitted Welding Positions	Gas Shielding for (FCAW)	Notes
		T_1	T_2	Root Opening	Tolerances As Detailed	As Fit Up			
SMAW	B-L1b	1/4 max	—	$R = T_1/2$	+1/16, −0	+1/6, −1/8	All	—	C, N
GMAW FCAW	B-L1b-GF	3/8 max	—	$R = 0$ to 1/8	+1/16, −0	+1/6, −1/8	All	Not Required	A, C, N
SAW	B-L1-S	3/8 max	—	$R = 0$	±0	+1/16, −0	F	—	N
SAW	B-L1a-S	5/8 max	—	$R = 0$	±0	+1/16, −0	F	—	C, N

Square-groove weld (1)
T-joint (T)
Corner joint (C)

BACKGOUGE

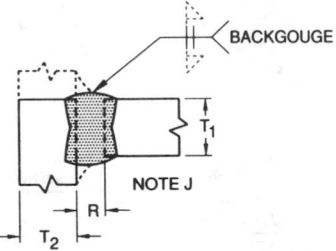

NOTE J

Welding Process	Joint Designation	Base Metal Thickness (U = unlimited)		Groove Preparation			Permitted Welding Positions	Gas Shielding for (FCAW)	Notes
		T_1	T_2	Root Opening	Tolerances As Detailed	As Fit Up			
SMAW	TC-L1b	1/4 max	U	$R = T_1/2$	+1/16, −0	+1/16, −1/8	All	—	C, J
GMAW FCAW	TC-L1-GF	3/8 max	U	$R = 0$ to 1/8	+1/16, −0	+1/16, −1/8	All	Not Required	A, C, J
SAW	TC-L1-S	3/8 max	U	$R = 0$	±0	+1/16, −0	F	—	J, C

Table 8-36 (cont.).
Prequalified Welded Joints
Complete-Joint-Penetration Groove Welds

Single-V-groove weld (2)
Butt joint (B)

Tolerances	
As Detailed	**As Fit Up**
$R = +^1/_{16}, -0$	$+^1/_4, -^1/_{16}$
$\alpha = +10°, -0°$	$+10°, -5°$

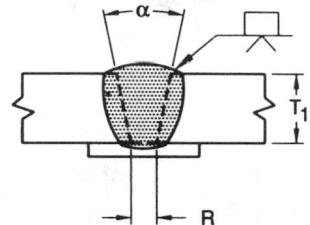

Welding Process	Joint Designation	Base Metal Thickness (U = unlimited)		Groove Preparation		Permitted Welding Positions	Gas Shielding for FCAW	Notes
		T_1	T_2	Root Opening	Groove Angle			
SMAW	B-U2a	U	—	$R = ^1/_4$	$\alpha = 45°$	All	—	N
				$R = ^3/_8$	$\alpha = 30°$	F, V, OH	—	N
				$R = ^1/_2$	$\alpha = 20°$	F, V, OH	—	N
GMAW FCAW	B-U2a-GF	U	—	$R = ^3/_{16}$	$\alpha = 30°$	F, V, OH	Required	A, N
				$R = ^3/_8$	$\alpha = 30°$	F, V, OH	Not req.	A, N
				$R = ^1/_4$	$\alpha = 45°$	F, V, OH	Not req.	A, N
SAW	B-L2a-S	2 max	—	$R = ^1/_4$	$\alpha = 30°$	F	—	N
SAW	B-U2-S	U	—	$R = ^5/_8$	$\alpha = 20°$	F	—	N

Single-V-groove weld (2)
Corner joint (C)

Tolerances	
As Detailed	**As Fit Up**
$R = +^1/_{16}, -0$	$+^1/_4, -^1/_{16}$
$\alpha = +10°, -0°$	$+10°, -5°$

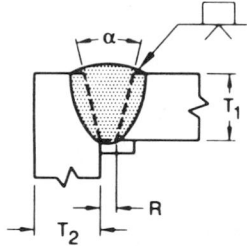

Welding Process	Joint Designation	Base Metal Thickness (U = unlimited)		Groove Preparation		Permitted Welding Positions	Gas Shielding for FCAW	Notes
		T_1	T_2	Root Opening	Groove Angle			
SMAW	C-U2a	U	U	$R = ^1/_4$	$\alpha = 45°$	All	—	Q
				$R = ^3/_8$	$\alpha = 30°$	F, V, OH	—	Q
				$R = ^1/_2$	$\alpha = 20°$	F, V, OH	—	Q
GMAW FCAW	C-U2a-GF	U	U	$R = ^3/_{16}$	$\alpha = 30°$	F, V, OH	Required	A
				$R = ^3/_8$	$\alpha = 30°$	F, V, OH	Not req.	A, Q
				$R = ^1/_4$	$\alpha = 45°$	F, V, OH	Not req.	A, Q
SAW	C-L2a-S	2 max	U	$R = ^1/_4$	$\alpha = 30°$	F	—	Q
SAW	C-U2-S	U	U	$R = ^5/_8$	$\alpha = 20°$	F	—	Q

Table 8-36 (cont.).
Prequalified Welded Joints
Complete-Joint-Penetration Groove Welds

Single-V-groove weld (2)
Butt joint (B)

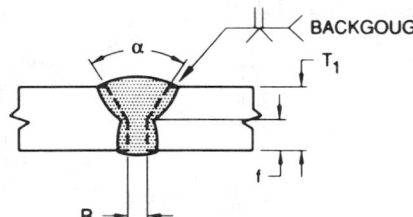

Welding Process	Joint Desig-nation	Base Metal Thickness (U = unlimited)		Groove Preparation			Permitted Welding Positions	Gas Shielding for FCAW	Notes
		T_1	T_2	Root Opening Root Face Groove Angle	Tolerances				
					As Detailed	As Fit Up			
SMAW	B-U2	U	—	$R = 0$ to $\frac{1}{8}$ $f = 0$ to $\frac{1}{8}$ $\alpha = 60°$	$+\frac{1}{16}, -0$ $+\frac{1}{16}, -0$ $+10°, -0°$	$+\frac{1}{16}, -\frac{1}{8}$ Not limited $+10°, -5°$	All	—	C, N
GMAW FCAW	B-U2-GF	U	—	$R = 0$ to $\frac{1}{8}$ $f = 0$ to $\frac{1}{8}$ $\alpha = 60°$	$+\frac{1}{16}, -0$ $+\frac{1}{16}, -0$ $+10°, -0°$	$+\frac{1}{16}, -\frac{1}{8}$ Not limited $+10°, -5°$	All	Not required	A, C, N
SAW	B-L2c-S	Over $\frac{1}{2}$ to 1	—	$R = 0, \alpha = 60°$ $f = \frac{1}{4}$ max	$R = \pm 0$ $f = +0, -f$ $\alpha = +10°, \ 0°$	$+\frac{1}{16}, -0$ $\pm\frac{1}{16}$ $+10°, -5°$	F	—	C, N
		Over 1 to $1\frac{1}{2}$	—	$R = 0, \alpha = 60°$ $f = \frac{1}{2}$ max					
		Over $1\frac{1}{2}$ to 2	—	$R = 0, \alpha = 60°$ $f = \frac{5}{8}$ max					

Single-V-groove weld (2)
Corner joint (C)

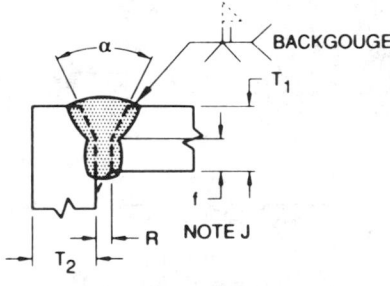

Welding Process	Joint Desig-nation	Base Metal Thickness (U = unlimited)		Groove Preparation			Permitted Welding Positions	Gas Shielding for FCAW	Notes
		T_1	T_2	Root Opening Root Face Groove Angle	Tolerances				
					As Detailed	As Fit Up			
SMAW	C-U2	U	U	$R = 0$ to $\frac{1}{8}$ $f = 0$ to $\frac{1}{8}$ $\alpha = 60°$	$+\frac{1}{16}, -0$ $+\frac{1}{16}, -0$ $+10°, -0°$	$+\frac{1}{16}, -\frac{1}{8}$ Not limited $+10°, -5°$	All	—	C, J, R
GMAW FCAW	C-U2-GF	U	U	$R = 0$ to $\frac{1}{8}$ $f = 0$ to $\frac{1}{8}$ $\alpha = 60°$	$+\frac{1}{16}, -0$ $+\frac{1}{16}, -0$ $+10°, -0°$	$+\frac{1}{16}, -\frac{1}{8}$ Not limited $+10°, -5°$	All	Not required	A, C, J, R
SAW	C-L2b-S	U	U	$R = 0$ $f = \frac{1}{4}$ max $\alpha = 60°$	± 0 $+0, -\frac{1}{4}$ $+10° -0°$	$+\frac{1}{16}, -0$ $\pm\frac{1}{16}$ $+10°, -5°$	F	—	C, J, R

Table 8-36 (cont.).
Prequalified Welded Joints
Complete-Joint-Penetration Groove Welds

Double-V-groove weld (3)
Butt joint (B)

Tolerances		
	As Detailed	As Fit Up
$R = \pm 0$		$+\frac{1}{4}, -0$
$f = \pm 0$		$+\frac{1}{16}, -0$
$\alpha = +10°, -0°$		$+10°, -5°$
Spacer SAW	± 0	$+\frac{1}{16}, -0$
SMAW	± 0	$+\frac{1}{8}, -0$

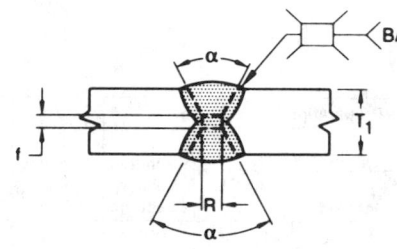

Welding Process	Joint Desig-nation	Base Metal Thickness (U = unlimited) T_1	T_2	Groove Preparation Root Opening	Root Face	Groove Angle	Permitted Welding Positions	Gas Shielding for (FCAW)	Notes
SMAW	B-U3a	U Spacer = $\frac{1}{8} \times R$	—	$R = \frac{1}{4}$	$f = 0$ to $\frac{1}{8}$	$\alpha = 45°$	All	—	C, M, N
				$R = \frac{3}{8}$	$f = 0$ to $\frac{1}{8}$	$\alpha = 30°$	F, V, OH	—	
				$R = \frac{1}{2}$	$f = 0$ to $\frac{1}{8}$	$\alpha = 20°$	F, V, OH	—	
SAW	B-U3a-S	U Spacer = $\frac{1}{4} \times R$	—	$R = \frac{5}{8}$	$f = 0$ to $\frac{1}{4}$	$\alpha = 20°$	F	—	C, M, N

Double-V-groove weld (3)
Butt joint (B)

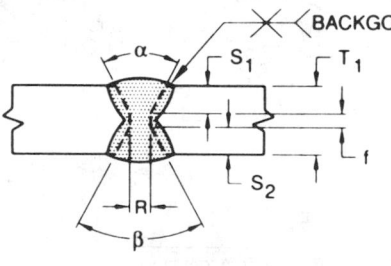

For B-U3c-S only	
T_1	S_1
Over to	
2 $2\frac{1}{2}$	$1\frac{3}{8}$
$2\frac{1}{2}$ 3	$1\frac{3}{4}$
3 $3\frac{5}{8}$	$2\frac{1}{8}$
$3\frac{5}{8}$ 4	$2\frac{3}{8}$
4 $4\frac{3}{4}$	$2\frac{3}{4}$
$4\frac{3}{4}$ $5\frac{1}{2}$	$3\frac{1}{4}$
$5\frac{1}{2}$ $6\frac{1}{4}$	$3\frac{3}{4}$

For $T_1 > 6\frac{1}{4}$, or $T_1 \leq 2$
$S_1 = \frac{2}{3}(T_1 - \frac{1}{4})$

Welding Process	Joint Desig-nation	Base Metal Thickness (U = unlimited) T_1	T_2	Groove Preparation Root Opening Root Face Groove Angle	Tolerances As Detailed	As Fit Up	Permitted Welding Positions	Gas Shielding for FCAW	Notes
SMAW	B-U3b	U	—	$R = 0$ to $\frac{1}{8}$	$+\frac{1}{16}, -0$	$+\frac{1}{16}, -\frac{1}{8}$	All	—	C, M, N
GMAW FCAW	B-U3-GF	U	—	$f = 0$ to $\frac{1}{8}$ $\alpha = \beta = 60°$	$+\frac{1}{16}, -0$ $+10°, -0°$	Not limited $+10°, -5°$	All	Not required	A, C, M, N
SAW	B-U3c-S	U	—	$R = 0$ $f = \frac{1}{4}$ min $\alpha = \beta = 60°$	$+\frac{1}{16}, -0$ $+\frac{1}{4}, -0$ $+10°, -0°$	$+\frac{1}{16}, -0$ $+\frac{1}{4}, -0$ $+10°, -5°$	F	—	C, M, N

To find S_1 see table above; $S_2 = T_1 - (S_1 + f)$

Table 8-36 (cont.).
Prequalified Welded Joints
Complete-Joint-Penetration Groove Welds

Single-bevel-groove weld (4)
Butt joint (B)

Tolerances		
As Detailed		**As Fit Up**
$R = +\frac{1}{16}, -0$		$+\frac{1}{4}, -\frac{1}{16}$
$\alpha = +10°, -0°$		$+10°, -5°$

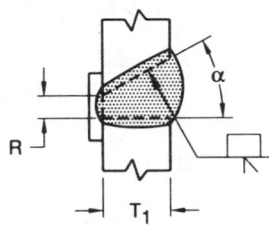

Welding Process	Joint Desig-nation	Base Metal Thickness (U = unlimited)		Groove Preparation		Permitted Welding Positions	Gas Shielding for FCAW	Notes
		T_1	T_2	Root Opening	Groove Angle			
SMAW	B-U4a	U	—	$R = \frac{1}{4}$	$\alpha = 45°$	All	—	Br, N
				$R = \frac{3}{8}$	$\alpha = 30°$	All	—	Br, N
GMAW FCAW	B-U4a-GF	U	—	$R = \frac{3}{16}$	$\alpha = 30°$	All	Required	A, Br, N
				$R = \frac{1}{4}$	$\alpha = 45°$	All	Not req.	A, Br, N
				$R = \frac{3}{8}$	$\alpha = 30°$	F	Not req.	A, Br, N

Single-bevel-groove weld (4)
T-joint (T)
Corner joint (C)

Tolerances		
As Detailed		**As Fit Up**
$R = +\frac{1}{16}, -0$		$+\frac{1}{4}, -\frac{1}{16}$
$\alpha = +10°, -0°$		$+10°, -5°$

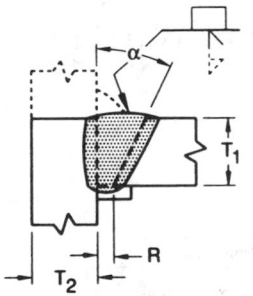

NOTE J
NOTE V

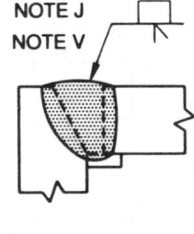

Welding Process	Joint Desig-nation	Base Metal Thickness (U = unlimited)		Groove Preparation		Permitted Welding Positions	Gas Shielding for FCAW	Notes
		T_1	T_2	Root Opening	Groove Angle			
SMAW	TC-U4a	U	U	$R = \frac{1}{4}$	$\alpha = 45°$	All	—	J, Q, V
				$R = \frac{3}{8}$	$\alpha = 30°$	F, V, OH	—	J, Q, V
GMAW FCAW	TC-U4a-GF	U	U	$R = \frac{3}{16}$	$\alpha = 30°$	All	Required	A, J, Q, V
				$R = \frac{3}{8}$	$\alpha = 30°$	F	Not req.	A, J, Q, V
				$R = \frac{1}{4}$	$\alpha = 45°$	All	Not req.	A, J, Q, V
SAW	TC-U4a-S	U	U	$R = \frac{3}{8}$	$\alpha = 30°$	F	—	J, Q, V
				$R = \frac{1}{4}$	$\alpha = 45°$			

Table 8-36 (cont.).
Prequalified Welded Joints
Complete-Joint-Penetration Groove Welds

Single-bevel-groove weld (4)
Butt joint (B)

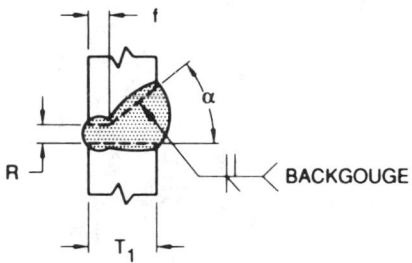

Welding Process	Joint Desig-nation	Base Metal Thickness (U = unlimited)		Groove Preparation			Permitted Welding Positions	Gas Shielding for FCAW	Notes
		T_1	T_2	Root Opening Root Face Groove Angle	Tolerances As Detailed	As Fit Up			
SMAW	B-U4b	U	—	$R = 0$ to $1/8$ $f = 0$ to $1/8$ $\alpha = 45°$	$+1/16, -0$ $+1/16, -0$ $+10°, -0°$	$+1/16, -1/8$ Not limited $+10°, -5°$	All	—	Br, C, N
GMAW FCAW	B-U4b-GF	U	—				All	Not required	A, Br, C, N

Single-bevel-groove weld (4)
T-joint (T)
Corner joint (C)

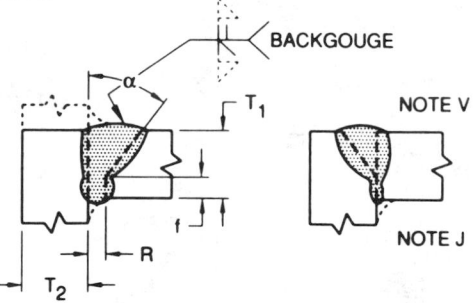

NOTE V

NOTE J

Welding Process	Joint Desig-nation	Base Metal Thickness (U = unlimited)		Groove Preparation			Permitted Welding Positions	Gas Shielding for FCAW	Notes
		T_1	T_2	Root Opening Root Face Groove Angle	Tolerances As Detailed	As Fit Up			
SMAW	TC-U4b	U	U	$R = 0$ to $1/8$ $f = 0$ to $1/8$ $\alpha = 45°$	$+1/16, -0$ $+1/16, -0$ $+10°, -0°$	$+1/16, -1/8$ Not limited $+10°, -5°$	All	—	C, J, R, V
GMAW FCAW	TC-U4b-GF	U	U				All	Not required	A, C, J, R, V
SAW	TC-U4b-S	U	U	$R = 0$ $f = 1/4$ max $\alpha = 60°$	±0 $+0, -1/8$ $+10°, -0°$	$+1/4, -0$ $\pm1/16$ $+10°, -5°$	F	—	C, J, R, V

Table 8-36 (cont.).
Prequalified Welded Joints
Complete-Joint-Penetration Groove Welds

Double-bevel-groove weld (5) Butt joint (B) T-joint (T) Corner joint (C)	Tolerances		
		As Detailed	As Fit Up
	R = ±0	+1/4, −0	
	f = 1/16, −0	±1/16	
	α = +10°, −0°	+10°, −5°	
	Spacer	+1/16, −0	+1/8, −0

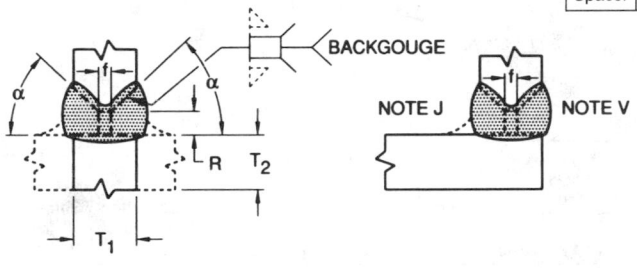

Welding Process	Joint Desig- nation	Base Metal Thickness (U = unlimited)		Groove Preparation			Permitted Welding Positions	Gas Shielding for FCAW	Notes
		T_1	T_2	Root Opening	Root Face	Groove Angle			
SMAW	B-U5b	U Spacer =1/8 × R	U	R = 1/4	f = 0 to 1/8	α = 45°	All	—	Br, C, M, N
	TC-U5a	U Spacer =1/4 × R	U	R = 1/4	f = 0 to 1/8	α = 45°	All	—	C, J, M, R, V
				R = 3/8	f = 0 to 1/8	α = 30°	F, OH	—	C, J, M, R, V

Double-bevel-groove weld (5)
Butt joint (B)

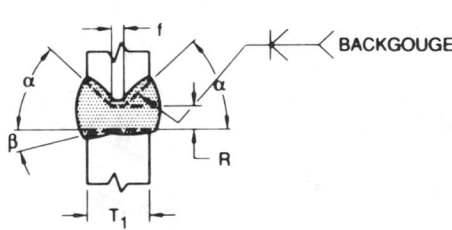

Welding Process	Joint Desig- nation	Base Metal Thickness (U = unlimited)		Groove Preparation			Permitted Welding Positions	Gas Shielding for FCAW	Notes
		T_1	T_2	Root Opening Root Face Groove Angle	Tolerances				
					As Detailed	As Fit Up			
SMAW	B-U5a	U	—	R = 0 to 1/8 f = 0 to 1/8 α = 45° β = 0° to 15°	+1/16, −0 +1/16, −0 α + β, +10°, −0°	+1/16, −1/8 Not limited α + β, +10°, −5°	All	—	Br, C, M, N
GMAW FCAW	B-U5-GF	U	—	R = 0 to 1/8 f = 0 to 1/8 α = 45° β = 0° to 15°	+1/16, −0 +1/16, −0 α + β = +10°, −0°	+1/16, −1/8 Not limited α + β = +10°, −5°	All	Not req.	A, Br, C, M, N

Table 8-36 (cont.).
Prequalified Welded Joints
Complete-Joint-Penetration Groove Welds

Double-bevel-groove weld (5)
T-joint (T)
Corner joint (C)

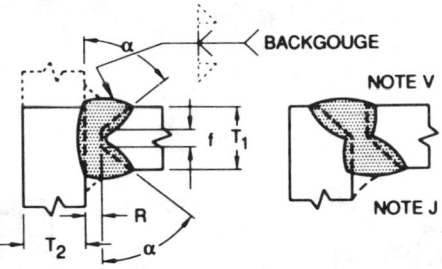

Welding Process	Joint Desig-nation	Base Metal Thickness (U = unlimited)		Groove Preparation			Permitted Welding Positions	Gas Shielding for (FCAW)	Notes
		T_1	T_2	Root Opening Root Face Groove Angle	Tolerances As Detailed	As Fit Up			
SMAW	TC-U5b	U	U	R = 0 to $1/8$	$+1/16, -0$	$+1/16, -1/8$	All	—	C, J, M,
				f = 0 to $1/8$	$+1/16, -0$	Not limited			R, V
GMAW FCAW	TC-U5-GF	U	U	α = 45°	$+10°, -0°$	$+10°, -5°$	All	Not req.	A, C, J, M, R, V
SAW	TC-U5-S	U	U	R = 0	±0	$+1/16, -0$	F	—	C, J, M,
				f = $3/16$ max	$+0, -3/16$	$±1/16$			R, V
				α = 60°	$+10°, -0°$	$+10°, -5°$			

Table 8-36 (cont.).
Prequalified Welded Joints
Complete-Joint-Penetration Groove Welds

Single-U-groove weld (6)
Butt joint (B)
Corner joint (C)

	Tolerances	
	As Detailed	As Fit Up
$R = +1/16, -0$		$+1/16, -1/8$
$\alpha = +10°, -0°$		$+10°, -5°$
$f = \pm 1/16$		Not limited
$r = +1/8, -0$		$+1/8, -0$

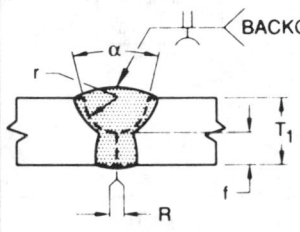 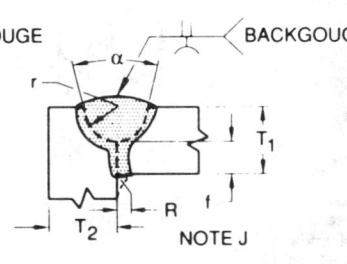

BACKGOUGE   BACKGOUGE   NOTE J

Welding Process	Joint Designation	Base Metal Thickness (U = unlimited)		Groove Preparation				Permitted Welding Positions	Gas Shielding for FCAW	Notes
		T_1	T_2	Root Opening	Groove Angle	Root Face	Groove Radius			
SMAW	B-U6	U	U	$R = 0$ to $1/8$	$\alpha = 45°$	$f = 1/8$	$r = 1/4$	All	—	C, N
				$R = 0$ to $1/8$	$\alpha = 20°$	$f = 1/8$	$r = 1/4$	F, OH	—	C, N
	C-U6	U	U	$R = 0$ to $1/8$	$\alpha = 45°$	$f = 1/8$	$r = 1/4$	All	—	C, J, R
				$R = 0$ to $1/8$	$\alpha = 20°$	$f = 1/8$	$r = 1/4$	F, OH	—	C, J, R
GMAW FCAW	B-U6-GF	U	U	$R = 0$ to $1/8$	$\alpha = 20°$	$f = 1/8$	$r = 1/4$	All	Not req.	A, C, N
	C-U6-GF	U	U	$R = 0$ to $1/8$	$\alpha = 20°$	$f = 1/8$	$r = 1/4$	All	Not req.	A, C, J, R

Double-U-groove weld (7)
Butt joint (B)

Tolerances				Tolerances	
For B-U7 and B-U7-GF				For B-U7-S	
As Detailed		As Fit Up		As Detailed	As Fit Up
$R = +1/16, -0$		$+1/16, -1/8$		$R = \pm 0$	$+1/16, -0$
$\alpha = +10°, -0°$		$+10°, -5°$		$\alpha = +0°, -1/4°$	$\pm 1/16$
$f = \pm 1/16, -0$		Not limited			
$r = +1/4, -0$		$\pm 1/16$			

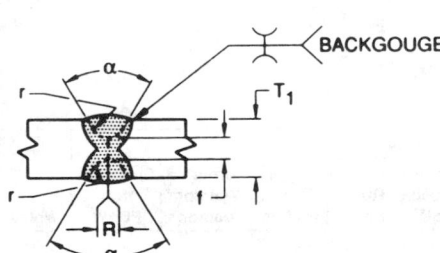

BACKGOUGE

Welding Process	Joint Designation	Base Metal Thickness (U = unlimited)		Groove Preparation				Permitted Welding Positions	Gas Shielding for FCAW	Notes
		T_1	T_2	Root Opening	Groove Angle	Root Face	Groove Radius			
SMAW	B-U7	U	—	$R = 0$ to $1/8$	$\alpha = 45°$	$f = 1/8$	$r = 1/4$	All	—	C, M, N
				$R = 0$ to $1/8$	$\alpha = 20°$	$f = 1/8$	$r = 1/4$	F, OH	—	C, M, N
GMAW FCAW	B-U7-GF	U	—	$R = 0$ to $1/8$	$\alpha = 20°$	$f = 1/8$	$r = 1/4$	All	Not required	A, C, M, N
SAW	B-U7-S	U	—	$R = 0$	$\alpha = 20°$	$f = 1/4$ max	$r = 1/4$	F	—	C, M, N

Table 8-36 (cont.).
Prequalified Welded Joints
Complete-Joint-Penetration Groove Welds

Single-J-groove weld (8)
Butt joint (B)

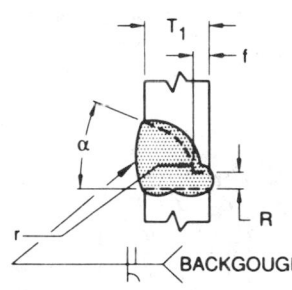

Tolerances	
As Detailed	**As Fit Up**
$R = +^1/_{16}, -0$	$+^1/_{16}, -^1/_8$
$\alpha = +10°, -0°$	$+10°, -5°$
$f = \pm^1/_{16}, -0$	Not limited
$r = +^1/_4, -0$	$\pm^1/_{16}$

BACKGOUGE

Welding Process	Joint Desig-nation	Base Metal Thickness (U = unlimited)		Groove Preparation				Permitted Welding Positions	Gas Shielding for FCAW	Notes
		T_1	T_2	Root Opening	Groove Angle	Root Face	Groove Radius			
SMAW	B-U8	U	—	$R = 0$ to $^1/_8$	$\alpha = 45°$	$f = ^1/_8$	$r = ^3/_8$	All	—	Br, C, N
GMAW FCAW	B-U8-GF	U	—	$R = 0$ to $^1/_8$	$\alpha = 30°$	$f = ^1/_8$	$r = ^3/_8$	All	Not required	A, Br, C, N

Single-J-groove weld (8)
T-joint (T)
Corner joint (C)

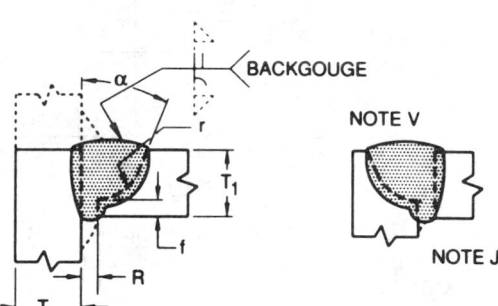

Tolerances	
As Detailed	**As Fit Up**
$R = +^1/_{16}, -0$	$+^1/_{16}, -^1/_8$
$\alpha = +10°, -0°$	$+10°, -5°$
$f = \pm^1/_{16}, -0$	Not limited
$r = +^1/_4, -0$	$\pm^1/_{16}$

BACKGOUGE

NOTE V

NOTE J

Welding Process	Joint Desig-nation	Base Metal Thickness (U = unlimited)		Groove Preparation				Permitted Welding Positions	Gas Shielding for FCAW	Notes
		T_1	T_2	Root Opening	Groove Angle	Root Face	Groove Radius			
SMAW	TC-U8a	U	U	$R = 0$ to $^1/_8$	$\alpha = 45°$	$f = ^1/_8$	$r = ^3/_8$	All	—	C, J, R, V
				$R = 0$ to $^1/_8$	$\alpha = 30°$	$f = ^1/_8$	$r = ^3/_8$	F, OH	—	C, J, R, V
GMAW FCAW	TC-U8a-GF	U	U	$R = 0$ to $^1/_8$	$\alpha = 30°$	$f = ^1/_8$	$r = ^3/_8$	All	Not required	A, C, J, R, V

Table 8-36 (cont.).
Prequalified Welded Joints
Complete-Joint-Penetration Groove Welds

Double-J-groove weld (9)
Butt joint (B)

Tolerances		
	As Detailed	As Fit Up
$R = +^1/_{16}, -0$		$+^1/_{16}, -^1/_8$
$\alpha = +10°, -0°$		$+10°, -5°$
$f = ±^1/_{16}, -0$		Not limited
$r = +^1/_8, -0$		$±^1/_{16}$

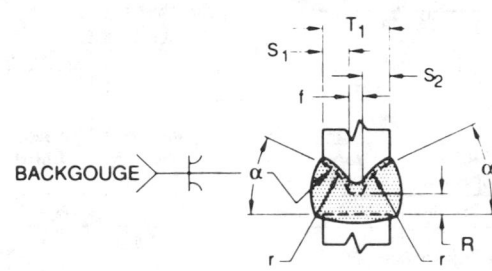

Welding Process	Joint Designation	Base Metal Thickness (U = unlimited)		Groove Preparation				Permitted Welding Positions	Gas Shielding for FCAW	Notes
		T_1	T_2	Root Opening	Groove Angle	Root Face	Groove Radius			
SMAW	B-U9	U	—	$R = 0$ to $^1/_8$	$\alpha = 45°$	$f = ^1/_8$	$r = ^3/_8$	All	—	Br, C, M, N
GMAW FCAW	B-U9-GF	U	—	$R = 0$ to $^1/_8$	$\alpha = 30°$	$f = ^1/_8$	$r = ^3/_8$	All	Not required	A, Br, C, M, N

Double-J-groove weld (9)
T-joint (T)
Corner joint (C)

Tolerances		
	As Detailed	As Fit Up
$R = +^1/_{16}, -0$		$+^1/_{16}, -^1/_8$
$\alpha = +10°, -0°$		$+10°, -5°$
$f = +^1/_{16}, -0$		Not limited
$r = +^1/_8, -0$		$±^1/_{16}$

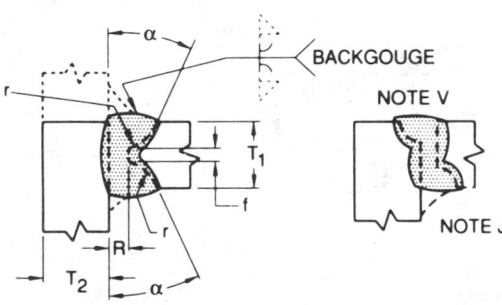

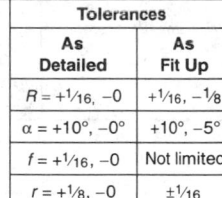

NOTE V

NOTE J

Welding Process	Joint Designation	Base Metal Thickness (U = unlimited)		Groove Preparation				Permitted Welding Positions	Gas Shielding for FCAW	Notes
		T_1	T_2	Root Opening	Groove Angle	Root Face	Groove Radius			
SMAW	TC-U9a	U	U	$R = 0$ to $^1/_8$	$\alpha = 45°$	$f = ^1/_8$	$r = ^3/_8$	All	—	C, J, M, R, V
				$R = 0$ to $^1/_8$	$\alpha = 30°$	$f = ^1/_8$	$r = ^3/_8$	F, OH	—	C, J, M, R, V
GMAW FCAW	TC-U9a-GF	U	U	$R = 0$ to $^1/_8$	$\alpha = 30°$	$f = ^1/_8$	$r = ^3/_8$	All	Not required	A, C, J, M, R, V

Table 8-36 (cont.).
Prequalified Welded Joints
Partial-Joint-Penetration Groove Welds

Square-groove weld (1)
Butt joint (B)

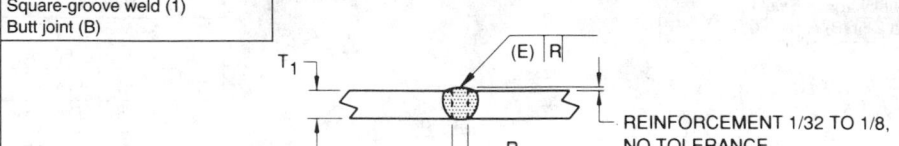

REINFORCEMENT 1/32 TO 1/8,
NO TOLERANCE

Welding Process	Joint Desig-nation	Base Metal Thickness (U = unlimited)		Groove Preparation			Permitted Welding Positions	Effective Throat (E)	Notes
		T_1	T_2	Root Opening	Tolerances				
					As Detailed	As Fit Up			
SMAW	B-P1a	1/8 max	—	$R = 0$ to $1/16$	$+1/16, -0$	$\pm 1/16$	All	$T_1 - 1/32$	B
	B-P1c	1/4 max	—	$R = \dfrac{T_1}{2}$ min	$+1/16, -0$	$\pm 1/16$	All	$\dfrac{T_1}{2}$	B

Square-groove weld (1)
Butt joint (B)

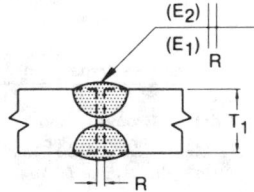

$E_1 + E_2$ must not exceed $\dfrac{3T_1}{4}$

Welding Process	Joint Desig-nation	Base Metal Thickness (U = unlimited)		Groove Preparation			Permitted Welding Positions	Effective Throat (E)	Notes
		T_1	T_2	Root Opening	Tolerances				
					As Detailed	As Fit Up			
SMAW	B-P1b	1/4 max	—	$R = \dfrac{T_1}{2}$	$\pm 1/16, -0$	$\pm 1/16$	All	$\dfrac{3T_1}{4}$	

Single-V-groove weld (2)
Butt joint (B)
Corner joint (C)

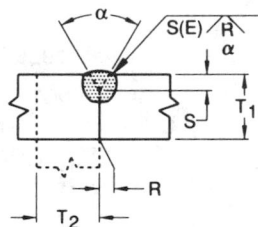

Welding Process	Joint Desig-nation	Base Metal Thickness (U = unlimited)		Groove Preparation			Permitted Welding Positions	Effective Throat (E)	Notes
		T_1	T_2	Root Opening Root Face Groove Angle	Tolerances				
					As Detailed	As Fit Up			
SMAW	BC-P2	1/4 min	U	$R = 0$ $f = 1/32$ min $\alpha - 60°$	$0, +1/16$ $+u, -0$ $+10°, -0°$	$+1/8, -1/16$ $\pm 1/16$ $+10°, -5°$	All	S	B, E, Q2
GMAW FCAW	BC-P2-GF	1/4 min	U	$R = 0$ $f = 1/8$ min $\alpha - 60°$	$0, +1/16$ $+u, -0$ $+10°, -0°$	$+1/8, -1/16$ $\pm 1/16$ $+10°, -5°$	All	S	A, B, E, Q2
SAW	BC-P2-S	7/16 min	U	$R = 0$ $f = 1/4$ min $\alpha - 60°$	± 0 $+u, -0$ $+10°, -0°$	$+1/16, -0*$ $\pm 1/16$ $+10°, -5°$	F	S	B, E, Q2

Table 8-36 (cont.).
Prequalified Welded Joints
Partial-Joint-Penetration Groove Welds

Double-V-groove weld (3)
Butt joint (B)

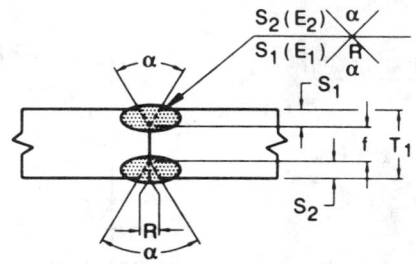

Welding Process	Joint Desig-nation	Base Metal Thickness (U = unlimited)		Groove Preparation			Permitted Welding Positions	Effective Throat (E)	Notes
		T_1	T_2	Root Opening Root Face Groove Angle	Tolerances As Detailed	As Fit Up			
SMAW	B-P3	½ min	—	$R = 0$ $f = ⅛$ min $\alpha = 60°$	$+\frac{1}{16}, -0$ $+u, -0$ $+10°, -0°$	$+⅛, -\frac{1}{16}$ $\pm\frac{1}{16}$ $+10°, -5°$	All	S	E, Mp, Q2
GMAW FCAW	B-P3-GF	½ min	—	$R = 0$ $f = ⅛$ min $\alpha = 60°$	$+\frac{1}{16}, -0$ $+u, -0$ $+10°, -0°$	$+⅛, -\frac{1}{16}$ $\pm\frac{1}{16}$ $+10°, -5°$	All	S	A, E, Mp, Q2
SAW	B-P3-S	¾ min	—	$R = 0$ $f = ¼$ min $\alpha = 60°$	± 0 $+u, -0$ $+10°, -0°$	$+\frac{1}{16}, -0^*$ $\pm\frac{1}{16}$ $+10°, -5°$	F	S	E, Mp, Q2

Table 8-36 (cont.).
Prequalified Welded Joints
Partial-Joint-Penetration Groove Welds

Single-bevel-groove weld (4)
Butt joint (B)
T-joint (T)
Corner joint (C)

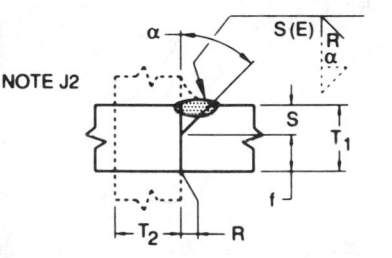

Welding Process	Joint Designation	Base Metal Thickness (U = unlimited)		Groove Preparation			Permitted Welding Positions	Effective Throat (E)	Notes
		T_1	T_2	Root Opening Root Face Groove Angle	Tolerances As Detailed	As Fit Up			
SMAW	BTC-P4	U	U	$R = 0$ $f = 1/8$ min $\alpha = 45°$	$+1/16, -0$ unlimited $+10°, -0°$	$+1/8, -1/16$ $\pm 1/16$ $+10°, -5°$	All	$S - 1/8$	B, E, J2, Q2, V
GMAW FCAW	BTC-P4-GF	$1/4$ min	U	$R = 0$ $f = 1/8$ min $\alpha = 45°$	$+1/16, -0$ unlimited* $+10°, -0°$	$+1/8, -1/16$ $\pm 1/16$ $+10°, -5°$	F, H V, OH	$\dfrac{S}{S - 1/8}$	A, B, E, J2, Q2, V
SAW	TC-P4-S	$7/16$ min	U	$R = 0$ $f = 1/4$ min $\alpha = 60°$	± 0 $+u, -0$ $+10°, -0°$	$+1/16, -0$ $\pm 1/16$ $+10°, -5°$	F	S	B, E, J2, Q2, V

Double-bevel-groove weld (5)
Butt joint (B)
T-joint (T)
Corner joint (C)

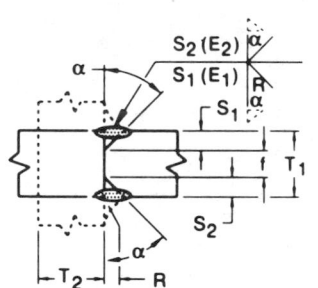
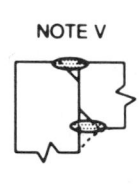

Welding Process	Joint Designation	Base Metal Thickness (U = unlimited)		Groove Preparation			Permitted Welding Positions	Effective Throat (E)	Notes
		T_1	T_2	Root Opening Root Face Groove Angle	Tolerances As Detailed	As Fit Up			
SMAW	BTC-P5	$5/16$ min	U	$R = 0$ $f = 1/8$ min $\alpha = 45°$	$+1/16, -0$ unlimited $+10°, -0°$	$+1/8, -1/16$ $\pm 1/16$ $+10°, -5°$	All	$(S - 1/8)$ $-1/4$	E, J2, L, Mp, Q2, V
GMAW FCAW	BTC-P5-GF	$1/2$ min	U	$R = 0$ $f = 1/8$ min $\alpha = 45°$	$+1/16, -0$ unlimited $+10°, -0°$	$+1/8, -1/16$ $\pm 1/16$ $+10°, -5°$	All	$(S_1 + S_2)$ $-1/4$	A, E, J2, L, Mp, Q2, V
SAW	TC-P5-S	$3/4$ min	U	$R = 0$ $f = 1/4$ min $\alpha = 60°$	± 0 unlimited $+10°, -0°$	$+1/16, -0$* $\pm 1/16$ $+10°, -5°$	F	$S_1 + S_2$	E, J2, L, Mp, Q2, V

*For flat and horizontal postiions $f = +u, -0$

Table 8-36 (cont.).
Prequalified Welded Joints
Partial-Joint-Penetration Groove Welds

Single-U-groove weld (6)
Butt joint (B)
Corner joint (C)

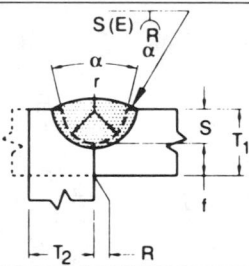

Welding Process	Joint Desig- nation	Base Metal Thickness (U = unlimited)		Groove Preparation			Permitted Welding Positions	Effective Throat (E)	Notes
		T_1	T_2	Root Opening Root Face Groove Radius Groove Angle	As Detailed (Tolerances)	As Fit Up (Tolerances)			
SMAW	BC-P6	1/4 min	U	$R = 0$ $f = 1/32$ min $r = 1/4$ $\alpha = 45°$	$+1/16, -0$ $+u, -0$ $+1/4, -0$ $+10°, -0°$	$+1/8, -1/16$ $\pm 1/16$ $\pm 1/16$ $+10°, -5°$	All	S	B, E, Q2
GMAW FCAW	BC-P6-GF	1/4 min	U	$R = 0$ $f = 1/8$ min $r = 1/4$ $\alpha = 20°$	$+1/16, -0$ $+u, -0$ $+1/4, -0$ $+10°, -0°$	$+1/8, -1/16$ $\pm 1/16$ $\pm 1/16$ $+10°, -5°$	All	S	A, B, E, Q2
SAW	BC-P6-S	7/16 min	U	$R = 0$ $f = 1/4$ min $r = 1/4$ $\alpha = 20°$	± 0 $+u, -0$ $+1/4, -0$ $+10°, -0°$	$+1/16, -0$ $\pm 1/16$ $\pm 1/16$ $+10°, -5°$	F	S	B, E, Q2

Double-U-groove weld (7)
Butt joint (B)

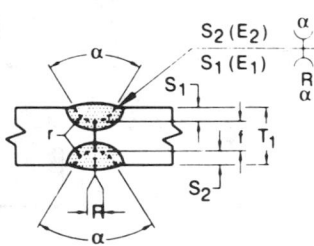

Welding Process	Joint Desig- nation	Base Metal Thickness (U = unlimited)		Groove Preparation			Permitted Welding Positions	Effective Throat (E)	Notes
		T_1	T_2	Root Opening Root Face Groove Radius Groove Angle	As Detailed (Tolerances)	As Fit Up (Tolerances)			
SMAW	B-P7	1/2 min	—	$R = 0$ $f = 1/8$ min $r = 1/4$ $\alpha = 45°$	$+1/16, -0$ $+u, -0$ $+1/4, -0$ $+10°, -0°$	$+1/8, -1/16$ $\pm 1/16$ $\pm 1/16$ $+10°, -5°$	All	$S_1 + S_2$	E, Mp, Q2
GMAW FCAW	B-P7-GF	1/2 min	—	$R = 0$ $f = 1/8$ min $r = 1/4$ $\alpha = 20°$	$+1/16, -0$ $+u, -0$ $+1/4, -0$ $+10°, -0°$	$+1/8, -1/16$ $\pm 1/16$ $\pm 1/16$ $+10°, -5°$	All	$S_1 + S_2$	A, E, Mp, Q2
SAW	B-P7-S	3/4 min	—	$R = 0$ $f = 1/4$ min $r = 1/4$ $\alpha = 20°$	± 0 $+u, -0$ $+1/4, -0$ $+10°, -0°$	$+1/16, -0$ $\pm 1/16$ $\pm 1/16$ $+10°, -5°$	F	$S_1 + S_2$	E, Mp, Q2

Table 8-36 (cont.).
Prequalified Welded Joints
Partial-Joint-Penetration Groove Welds

Single-J-groove weld (8)
Butt joint (B)
T-joint (T)
Corner joint (C)

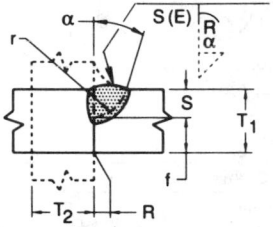

NOTE V

Welding Process	Joint Desig-nation	Base Metal Thickness (U = unlimited) T_1	T_2	Groove Preparation Root Opening Root Face Groove Radius Groove Angle	Tolerances As Detailed	As Fit Up	Permitted Welding Positions	Effective Throat (E)	Notes
SMAW	TC-P8*	1/4 min	U	$R = 0$ $f = 1/8$ min $r = 3/8$ $\alpha = 45°$	+1/16, -0 +u, -0 +1/4, -0 +10°, -0°	+1/8, -1/16 ±1/16 ±1/16 +10°, -5°	All	S	E, J2, Q2, V
SMAW	BC-P8**	1/4 min	U	$R = 0$ $f = 1/8$ min $r = 3/8$ $\alpha = 30°$	+1/16, -0 +u, -0 +1/4, -0 +10°, -0°	+1/8, -1/16 ±1/16 ±1/16 +10°, -5°	All	S	E, J2, Q2, V
GMAW FCAW	TC-P8-GF*	1/4 min	U	$R = 0$ $f = 1/8$ min $r = 3/8$ $\alpha = 45°$	+1/16, -0 +u, -0 +1/4, -0 +10°, -0°	+1/8, -1/16 ±1/16 ±1/16 +10°, -5°	All	S	A, E, J2, Q2, V
GMAW FCAW	BC-P8-GF**	1/4 min	U	$R = 0$ $f = 1/8$ min $r = 3/8$ $\alpha = 30°$	+1/16, -0 +u, -0 +1/4, -0 +10°, -0°	+1/8, -1/16 ±1/16 ±1/16 +10°, -5°	All	S	A, E, J2, Q2, V
SAW	TC-P8-S*	7/16 min	U	$R = 0$ $f = 1/4$ min $r = 1/2$ $\alpha = 45°$	±0 +u, -0 +1/4, -0 +10°, -0°	+1/16, -0 ±1/16 ±1/16 +10°, -5°	F	S	E, J2, Q2, V
SAW	C-P8-S**	7/16 min	U	$R = 0$ $f = 1/4$ min $r = 1/2$ $\alpha = 20°$	±0 +u, -0 +1/4, -0 +10°, -0°	+1/16, 0 ±1/16 ±1/16 +10°, -5°	F	S	E, J2, Q2, V

*Applies to inside corner joints.
**Applies to outside corner joints.

Table 8-36 (cont.).
Prequalified Welded Joints
Flare Welds

Double-J-groove weld (9)
Butt joint (B)
T-joint (T)
Corner joint (C)

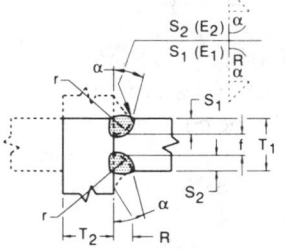

Welding Process	Joint Desig- nation	Base Metal Thickness (U = unlimited)		Groove Preparation			Permitted Welding Positions	Effective Throat (E)	Notes
		T_1	T_2	Root Opening Root Face Groove Radius Groove Angle	Tolerances As Detailed	As Fit Up			
SMAW	BTC-P9*	1/2 min	U	$R = 0$ $f = 1/8$ min $r = 3/8$ $\alpha = 45°$	$+1/16, -0$ $+u, -0$ $+1/4, -0$ $+10°, -0°$	$+1/8, -1/16$ $\pm 1/16$ $\pm 1/16$ $+10°, -5°$	All	$S_1 + S_2$	E, J2, Mp, Q2, V
GMAW FCAW	BTC-P9-GF**	1/2 min	U	$R = 0$ $f = 1/8$ min $r = 3/8$ $\alpha = 30°$	$+1/16, -0$ $+u, -0$ $+1/4, -0$ $+10°, -0°$	$+1/8, -1/16$ $\pm 1/16$ $\pm 1/16$ $+10°, -5°$	All	$S_1 + S_2$	A, J2, Mp, Q2, V
SAW	C-P9-S*	3/4 min	U	$R = 0$ $f = 1/4$ min $r = 1/2$ $\alpha = 45°$	± 0 $+u, -0$ $+1/4, -0$ $+10°, -0°$	$+1/16, -0$ $\pm 1/16$ $\pm 1/16$ $+10°, -5°$	F	$S_1 + S_2$	E, J2, Mp, Q2, V
SAW	C-P9-S**	3/4 min	U	$R = 0$ $f = 1/4$ min $r = 1/2$ $\alpha = 20°$	± 0 $+u, -0$ $+1/4, -0$ $+10°, -0°$	$-1/16, 0$ $\pm 1/16$ $\pm 1/16$ $+10°, 5°$	F	$S_1 + S_2$	E, J2, Mp, Q2, V
SAW	T-P9-S	3/4 min	U	$R = 0$ $f = 1/4$ min $r = 1/2$ $\alpha = 45°$	± 0 $+u, -0$ $+1/4, -0$ $+10°, -0°$	$+1/16, 0$ $\pm 1/16$ $\pm 1/16$ $+10°, -5°$	F	$S_1 + S_2$	E, J2, Mp, Q2

Single-J-groove weld (B)
Butt joint (B)
T-joint (T)
Corner joint (C)

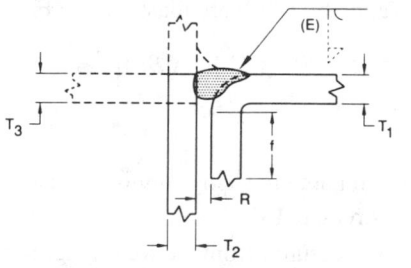 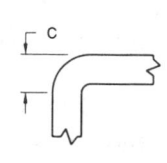

Welding Process	Joint Desig- nation	Base Metal Thickness (U = unlimited)			Groove Preparation			Permitted Welding Positions	Effective Throat (E)	Notes
		T_1	T_2	T_3	Root Opening Root Face Bend Radius	Tolerances As Detailed	As Fit Up			
SMAW	BTC-P10	3/16 min	U	T_1 min	$R = 0$ $f = 3/16$ min $C = \dfrac{3T_1}{2}$ min	$+1/16, -0$ $+U, -0$ $-0, +Not-$ Limited	$+1/8, -1/16$ $+U, -1/16$ $-0, +Not-$ Limited	All	$5/8 T_1$	J2, Q2, Z
GMAW FCAW	BTC-P10-GF	3/16 min	U	T_1 min	$R = 0$ $f = 3/16$ min $C = \dfrac{3T_1}{2}$ min	$+1/16, -0$ $+U, -0$ $-0, +Not-$ Limited	$+1/8, -1/16$ $+U, -1/16$ $-0, +Not-$ Limited	All	$5/8 T_1$	A, J2, Q2, Z
SAW	T-P10-S	1/2 min	1/2 min	N/A	$R = 0$ $f = 1/2$ min $C = \dfrac{3T_1}{2}$ min	± 0 $+U, -0$ $-0, +Not-$ Limited	$+1/16, -0$ $+U, -1/16$ $-0, +Not-$ Limited	F	$5/8 T_1$	J2, Q2, Z

*Applies to inside corner joints.
**Applies to outside corner joints.

ECCENTRICALLY LOADED WELD GROUPS

When the line of action of an applied load does not pass through the center of gravity (CG) of a weld group, the load is eccentric and results in a moment which must be considered in the design of the connection.

Eccentricity in the Plane of the Faying Surface

Eccentricity in the plane of the faying surface produces additional shear and the welds must then be designed to resist the combined effect of the direct shear from the applied load P_u and the additional shear from the induced moment $P_u e$. Two methods of analysis for this type of eccentricity will be discussed: (1) the instantaneous center of rotation method; and, (2) the elastic method.

Instantaneous Center of Rotation Method

Also known as the ultimate strength method (Crawford, 1968), this method considers the load-deformation relationship of each weld element as well as the variation in weld strength with respect to the direction of the applied force and, thus, more accurately predicts the ultimate strength of the eccentrically loaded connection (Butler, Pal, and Kulak, 1972). Eccentricity produces both a rotation about the centroid of the weld group and a translation of one connected element with respect to the other. The combined effect of this rotation and translation is equivalent to a rotation about a point defined as the instantaneous center of rotation (IC) as illustrated in Figure 8-51a. The location of the IC depends on the geometry of the weld group as well as the direction and point of application of the load. The individual resistance of each unit weld element is assumed to act on a line perpendicular to a ray passing through the instantaneous center and the centroid of that element, as illustrated in Figure 8-51b.

The load-deformation relationship of a single unit-weld element was originally given by Butler, Pal, and Kulak (1972) for E60 electrodes. New strength curves for E70 electrodes (Lesik and Kennedy, 1990) are illustrated in Figure 8-52, where:

$$R = 0.60F_{EXX}(1.0 + 0.50 \sin^{1.5}\theta) \, [p \, (1.9 - 0.9p)]^{0.3}$$

In the above equation,

R = shear force per unit area in a single unit-weld element at a deformation Δ, kips

F_{EXX} = weld electrode strength, ksi

θ = angle of loading measured from the weld longitudinal axis, degrees

p = ratio of element deformation to its deformation at maximum stress

Unlike the load-deformation relationship for bolts, strength and deformation of welds are dependent on the angle θ that the resultant elemental force makes with the axis of the weld element.

The critical weld element is usually the weld element farthest from the IC. While this may not always be the case, for the purpose of explanation, this will be assumed. The maximum deformation Δ_{max} may be determined as

$$\Delta_{max} = 1.087w \, (\theta + 6)^{-0.65} \le 0.17w$$

where w is the leg size of the weld and θ is expressed in degrees. The deformation of other weld elements is assumed to vary linearly with distance from the IC as,

$$\Delta = \frac{l_r}{l_{r\,max}} \Delta_{max}$$

More discussion of this method is contained in LRFD Specification Appendix J2.4 and its Commentary. These new provisions permit, for the first time, weld strength to exceed the $0.6F_{EXX}$ nominal value, which is the least strength applicable to longitudinally loaded $(\theta = 0°)$ elements. Load-deformation curves in Figure 8-52 for values of $\theta = 0°, 30°, 45°,$

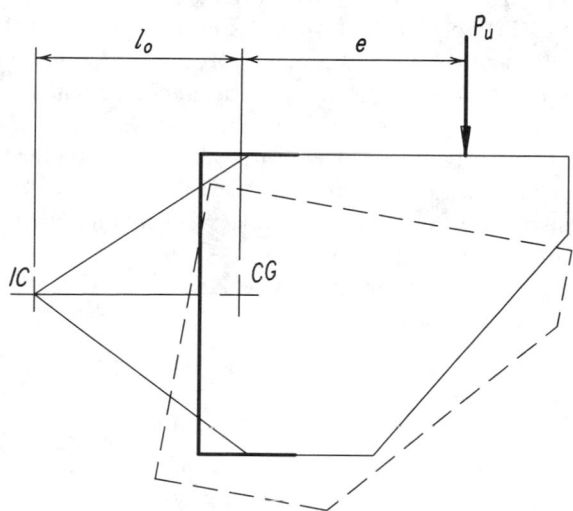

(a) Instantaneous center of rotation (IC)

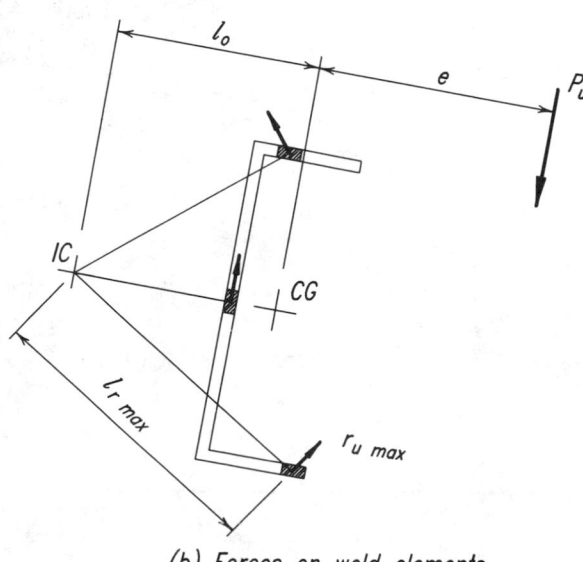

(b) Forces on weld elements

Fig. 8-51. Instantaneous center of rotation method.

60°, 75°, and 90° are shown relative to $R_o = 0.6F_{EXX}$. The ductility of the weld group is governed by Δ_{max} of the element that first reaches its limit. The total strength of all weld elements is the sum of the individual resistances of all welds in the group. If the correct location of the instantaneous center has been selected, the three equations of statics will be satisfied, i.e., $\Sigma F_x = 0$, $\Sigma F_y = 0$, $\Sigma M = 0$. Because of the non-linear nature of the requisite iterative solution, a minimum of twenty weld elements for the longest line segment is generally recommended for sufficient accuracy.

Tables 8-38 through 8-45 employ the instantaneous center of rotation method in accordance with LRFD Specification Appendix J2.4 for the weld patterns and eccentric conditions indicated and inclined loads at 0°, 15°, 30°, 45°, 60°, and 75°. Thus, unlike the First Edition LRFD Manual, tabulated values are not limited to a maximum weld nominal strength of $0.6F_{EXX}$. For some cases, significant increases of up to 50 percent of values tabulated previously are possible; many values reflect more moderate but, nevertheless, substantial increases on the order of 10 to 30 percent. The traditional and more conservative designs based upon a constant fillet weld nominal strength of $0.6F_{EXX}$ is also permitted, refer to AISC (1986).

For any of the weld group geometrics shown, the design strength of the eccentrically loaded weld group is ϕR_n, where

In the above equation,

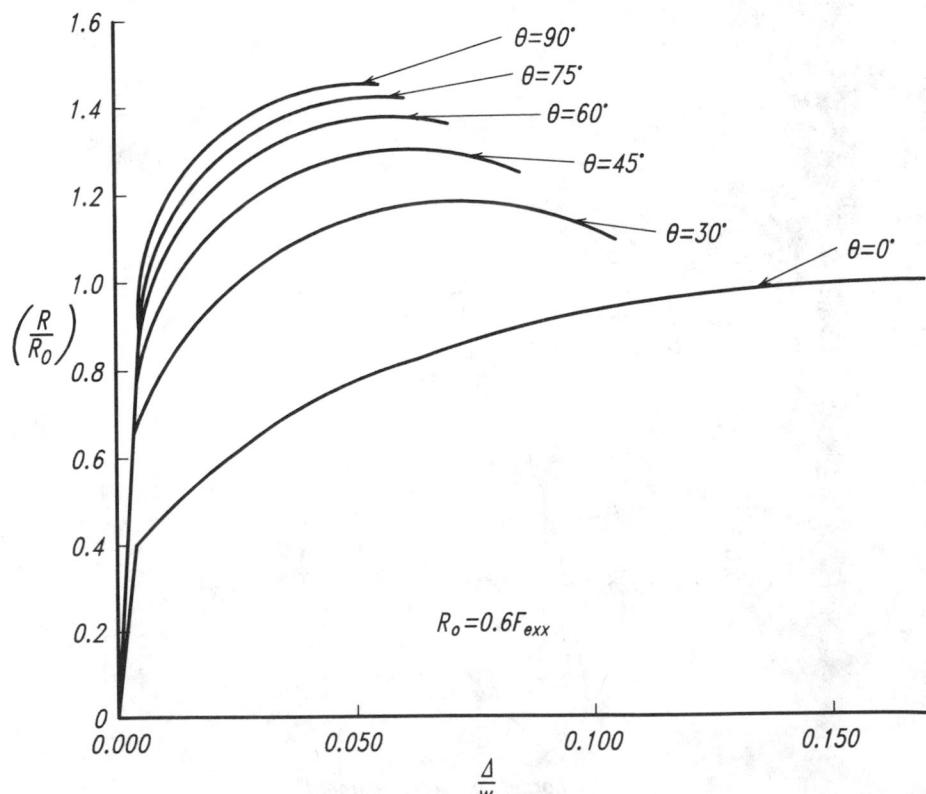

Fig. 8-52. Fillet weld strength as a function of force angle, θ.

C = tabular value (which includes $\phi = 0.75$)

$$\phi R_n = C C_1 D l$$

C_1 = electrode coefficient from Table 8-37 which adjusts tabular value, which is based on E70XX electrodes, for other electrodes. Note that this coefficient includes an additional reduction factor of 0.90 for E80 and E90 electrodes and 0.85 for E100 and E110; this accounts for the uncertainty of extrapolation to the higher strength electrodes.

D = number of sixteenths-of-an-inch in the weld size

l = length of the reference weld, in.

The first line in each table ($a = 0$) gives the design strength of a concentrically loaded weld group in accordance with LRFD Specification Appendix J2.4a. Linear interpolation within a given table between adjacent a and k values is permitted.

Figure C-J2.5 from LRFD Specification Commentary Section J2 indicates that, for equal-leg fillet welds, the area of the fusion surface is always larger than the leg dimension times the weld length. Therefore, the tabulated values are based upon the strength through the throat of the weld of

$$(0.75 \times 0.6 \times F_{EXX} \times 0.707 \times \frac{1}{16})$$

Tabulated values are valid for weld metal with a strength level equal to or matching the base material.

A convergence criterion of less than 0.5 percent unbalanced force was employed for the tabulated iterative solutions. Straight line interpolation between these angles may be significantly unconservative. Therefore, unless a direct analysis is performed, use only the values tabulated for the next lower angle. Since the coefficients in these tables were derived from physical tests with loading at ultimate strength levels, they should be used only for the weld patterns indicated and not in combination with any additional loading. In cases not treated by these tables, a special ultimate strength analysis is required if the instantaneous center of rotation method is to be used.

Example 8-3

Given: Refer to Figure 8-53. Determine the largest eccentric force P_u for which the design shear strength of the welds in the connection is adequate using the instantaneous center of rotation method. Use ⅜-in. fillet weld and 70 ksi electrode weld size.

A. Assume the load is vertical as illustrated in Figure 8-53 ($\theta = 0°$)

B. Assume the load acts at an angle of 75° with respect to vertical ($\theta = 75°$)

Solution A: $l = 10$ in.

$kl = 5$ in.

$k = 0.5$

From Table 8-42 with $\theta = 0°$, $x = 0.125$

Table 8-37.
Electrode Strength Coefficients

Electrode	F_{EXX} (ksi)	C_1
E60	60	0.857
E70	70	1.00
E80	80	1.03
E90	90	1.16
E100	100	1.21
E110	110	1.34

$xl + al = 10$ in.

$0.125(10$ in.$) + a$ $(10$ in.$) = 10$ in.

$a = 0.875$

By interpolation from Table 8-42 with $\theta = 0°$,

$C = 1.41$

Design shear strength

$\phi R_n = CC_1Dl$
$= 1.41(1.0)(6$ sixteenths$)(10$ in.$)$
$= 84.6$ kips

Comment: Note that this eccentricity has effectively reduced the shear strength of this weld group by 60 percent when compared with the eccentrically loaded case.

Solution B: From Solution A,

$k = 0.5$

$a = 0.875$

By interpolation from Table 8-42 with $\theta = 75°$,

$C = 2.59$

Design shear strength

$\phi R_n = CC_1Dl$
$= 2.59(1.0)(6$ sixteenths$)(10$ in.$)$
$= 155$ kips

Comment: In Solution B, the vertical component of the design strength is

$\phi R_n \sin75° = (155$ kips$)(0.966)$
$= 150$ kips

and the horizontal component of the design strength is

$$\phi R_n \cos 75° = (155 \text{ kips})(0.259)$$
$$= 40.1 \text{ kips}$$

Elastic Method

Alternatively, the elastic method may be used to analyze eccentrically loaded weld groups. It offers a simplified, conservative approach but does not render a consistent factor of safety and, in some cases, provides excessively conservative results. Furthermore, the elastic method ignores both the ductility of the weld group and the load redistribution which occurs. Refer to Higgins (1971).

In the elastic method, for a force applied parallel to the Y principle axis of the weld group, the eccentric force P_u is resolved into a force P_u acting through the center of gravity (CG) of the weld group and a moment $P_u e$ where e is the eccentricity. Each weld element is then assumed to support an equal share of the concentric force P_u, and a share of the eccentric moment $P_u e$ which is proportional to its distance from the CG. The weld most remote from the CG, then, is the most highly stressed. The resultant vectorial sum of these forces r_u is the required strength for the weld element.

The shear force per linear inch of weld due to the concentric force P_u is r_1,

where

$$r_1 = \frac{P_u}{l}$$

and l is the total length of the weld measured along the axis of each element.

The shear force per linear inch of weld due to the moment $P_u e$ varies with distance from the CG and will be maximum in the weld element which is most remote from the CG. The maximum shear due to the moment $P_u e$ is r_m,

where

$$r_m = \frac{P_u e c}{I_p}$$

In the above equation,

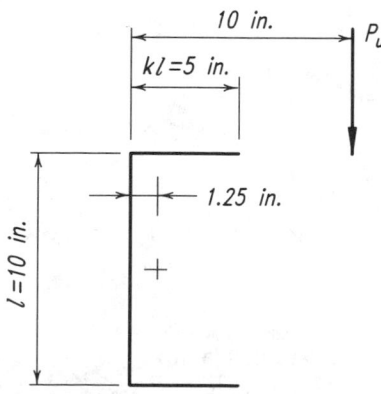

10 in.

P_u

$kl = 5$ in.

1.25 in.

$l = 10$ in.

Figure 8-53. Illustration for Example 8-3 and 8-4.

c = distance from CG to point on weld most remote from CG, in.

I_p = polar moment of inertia of the weld group, in.[4] per in.[2] ($I_p = I_x + I_y$). Refer to Figure 8-54. For section moduli and torsional constants of various welds treated as line elements, refer to Table 5 (page 7.4–7) of Blodgett (1966).

To determine the resultant force on the most highly stressed weld element, r_m must be resolved into vertical component r_2 and horizontal component r_3,

where

$$r_2 = \frac{P_u \, e c_x}{I_p}$$

$$r_3 = \frac{P_u \, e c_y}{I_p}$$

In the above equations, c_x and c_y are the horizontal and vertical components of the diagonal distance c. Thus, the resultant force is r_u,

where

$$r_u = \sqrt{(r_1 + r_2)^2 + (r_3)^2}$$

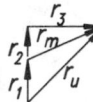

and the weld size must be chosen such that the design strength of the weld exceeds the required strength r_u.

For the more general case of an inclined eccentric force, i.e., not parallel to the Y principle axis of the bolt group, the effect of the X-direction component of the direct shear must also be included. Refer to Iwankiw (1987).

Example 8-4

Given: Refer to Example 8-3a. Recalculate the largest eccentric force P_u for which the design shear strength of the welds in the connection is adequate using the elastic method. Compare the result with that of Example 8-3a. Use ⅜-in. weld size, E70XX electrodes

$$I_p = 385 \text{ in.}^4 \text{ per in.}^2$$

Solution: Direct shear force per inch of weld

$$r_1 = \frac{P_u}{l}$$

$$= \frac{P_u}{20 \text{ in.}}$$

Additional shear force on weld due to eccentricity

$$r_2 = \frac{P_u \, e c_x}{I_p}$$

$$= \frac{P_u (8.75 \text{ in.}) (3.75 \text{ in.})}{385 \text{ in.}^4 \text{ per in.}^2}$$

$$= 0.0852 P_u$$

$$r_3 = \frac{P_u e c_y}{I_p}$$

$$= \frac{P_u(8.75 \text{ in.})(5 \text{ in.})}{385 \text{ in.}^4 \text{ per in.}^2}$$

$$= 0.114 P_u$$

Resultant shear force per inch of weld

$$r_u = \sqrt{(r_1 + r_2)^2 + (r_3)^2}$$

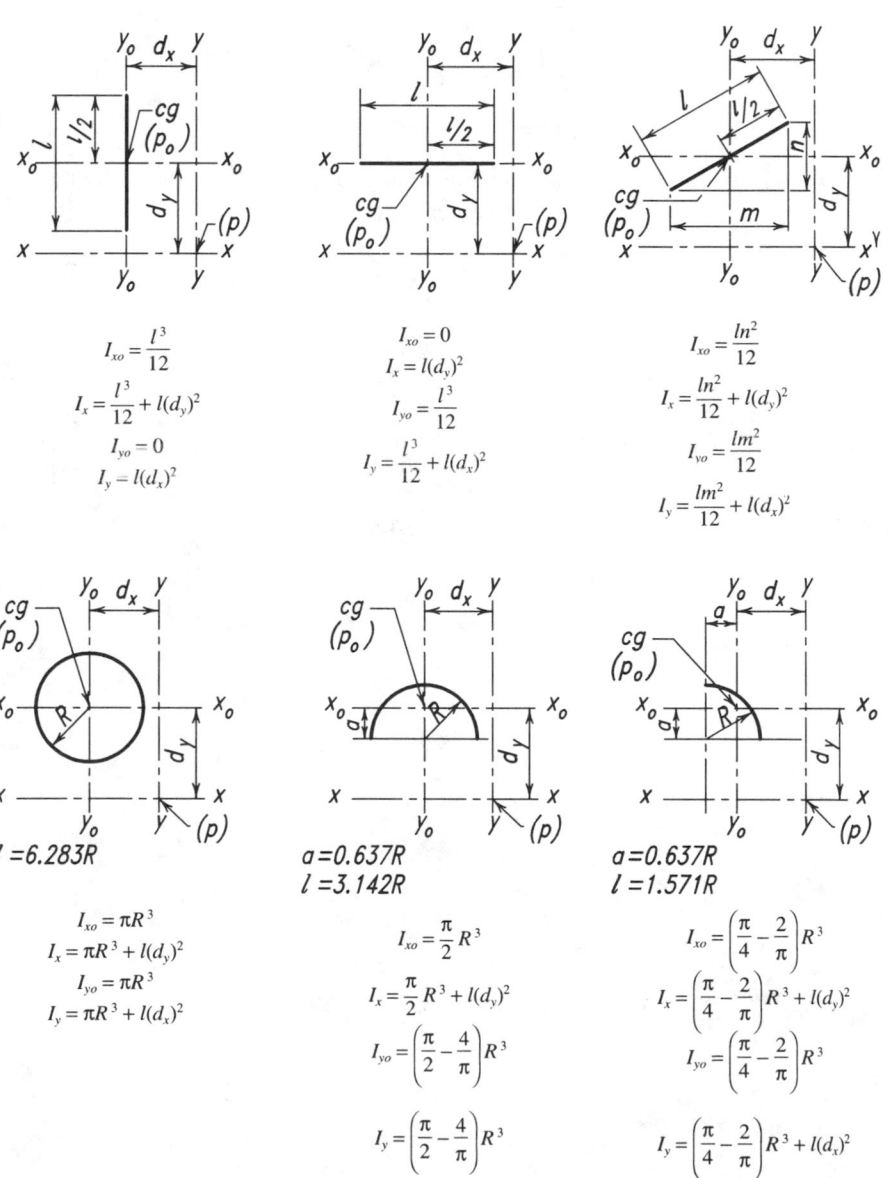

Fig. 8-54. Moments of inertia of various weld segments.

$$= \sqrt{\left(\frac{P_u}{20} + 0.0852 P_u\right)^2 + (0.114 P_u)^2}$$

$$= 0.177 P_u$$

Since r_u must be less than or equal to ϕr_n,

$$P_u \leq \frac{\phi r_n}{0.177}$$

$$\leq \frac{1.392 D}{0.177}$$

$$\leq \frac{1.392\ (6\ \text{sixteenths})}{0.177}$$

$$\leq 47.2\ \text{kips}$$

This is a 44 percent reduction in the strength predicted by the instantaneous center of rotation method in Example 8-3a.

Table 8-38.
Coefficients C for Eccentrically Loaded Weld Groups
Angle = 0°

$$\phi R_n = CC_1Dl \qquad C_{min} = \frac{P_u}{C_1Dl} \qquad D_{min} = \frac{P_u}{CC_1l} \qquad l_{min} = \frac{P_u}{CC_1D}$$

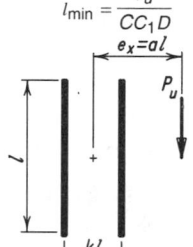

where

P_u = factored force, kips

D = number of sixteenths-of-an-inch in the fillet weld size

l = characteristic length of weld group, in.

a = e_x / l, in.

e_x = horizontal component of eccentricity of P_u with respect to centroid of weld group, in.

C = coefficient tabulated below which includes $\phi = 0.75$

C_1 = electode strength coefficient from Table 8-37 (1.0 for E70XX electrodes

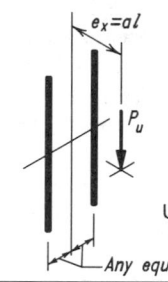

Special Case
(Load not in plane
of weld group)
Use C-values for k = 0

a	\multicolumn{17}{c}{k}
	0	0.1	0.2	0.3	0.4	0.5	0.6	0.7	0.8	0.9	1.0	1.2	1.4	1.6	1.8	2.0
0.00	2.78	2.78	2.78	2.78	2.78	2.78	2.78	2.78	2.78	2.78	2.78	2.78	2.78	2.78	2.78	2.78
0.10	2.78	2.78	2.78	2.78	2.78	2.77	2.75	2.74	2.73	2.71	2.70	2.67	2.64	2.61	2.59	2.78
0.15	2.75	2.75	2.75	2.74	2.73	2.71	2.70	2.69	2.67	2.66	2.64	2.63	2.60	2.58	2.55	2.53	2.50
0.20	2.64	2.63	2.63	2.62	2.60	2.59	2.58	2.57	2.56	2.55	2.54	2.52	2.50	2.48	2.46	2.44
0.25	2.48	2.48	2.48	2.47	2.47	2.46	2.46	2.45	2.45	2.44	2.44	2.43	2.41	2.40	2.39	2.38
0.30	2.32	2.32	2.32	2.32	2.33	2.33	2.33	2.33	2.33	2.33	2.33	2.33	2.33	2.32	2.32	2.31
0.40	2.00	2.00	2.01	2.03	2.05	2.07	2.08	2.10	2.11	2.12	2.14	2.15	2.16	2.17	2.18	2.18
0.50	1.72	1.72	1.74	1.77	1.80	1.83	1.86	1.89	1.91	1.93	1.95	1.99	2.01	2.03	2.05	2.06
0.60	1.50	1.50	1.52	1.55	1.59	1.63	1.67	1.71	1.74	1.77	1.79	1.84	1.87	1.90	1.92	1.94
0.70	1.32	1.32	1.34	1.38	1.42	1.47	1.51	1.55	1.59	1.62	1.65	1.71	1.75	1.79	1.81	1.84
0.80	1.17	1.18	1.20	1.24	1.28	1.33	1.38	1.42	1.46	1.50	1.53	1.59	1.64	1.68	1.71	1.74
0.90	1.05	1.06	1.08	1.12	1.17	1.22	1.27	1.31	1.35	1.39	1.43	1.49	1.54	1.59	1.62	1.66
1.00	0.957	0.963	0.986	1.02	1.07	1.12	1.17	1.21	1.26	1.29	1.33	1.40	1.45	1.50	1.54	1.58
1.20	0.806	0.812	0.835	0.872	0.916	0.963	1.01	1.06	1.10	1.14	1.17	1.24	1.30	1.35	1.40	1.44
1.40	0.695	0.701	0.724	0.758	0.799	0.844	0.889	0.932	0.973	1.01	1.05	1.12	1.18	1.23	1.28	1.32
1.60	0.611	0.616	0.638	0.670	0.708	0.750	0.792	0.833	0.873	0.911	0.947	1.01	1.07	1.13	1.17	1.22
1.80	0.544	0.550	0.570	0.600	0.635	0.674	0.714	0.753	0.791	0.828	0.863	0.928	0.987	1.04	1.09	1.13
2.00	0.491	0.496	0.515	0.542	0.576	0.612	0.650	0.687	0.723	0.758	0.792	0.855	0.912	0.964	1.01	1.05
2.20	0.447	0.452	0.470	0.495	0.526	0.560	0.596	0.631	0.665	0.699	0.731	0.792	0.848	0.899	0.945	0.988
2.40	0.410	0.415	0.431	0.455	0.484	0.516	0.550	0.583	0.616	0.648	0.679	0.738	0.792	0.842	0.887	0.929
2.60	0.379	0.384	0.399	0.421	0.448	0.478	0.510	0.542	0.573	0.604	0.634	0.691	0.743	0.791	0.836	0.877
2.80	0.352	0.357	0.371	0.392	0.417	0.446	0.476	0.506	0.536	0.565	0.594	0.649	0.699	0.746	0.790	0.830
3.00	0.329	0.333	0.347	0.366	0.390	0.417	0.446	0.474	0.503	0.531	0.559	0.611	0.661	0.706	0.748	0.788

Table 8-38 (cont.). Coefficients *C* for Eccentrically Loaded Weld Groups Angle = 15°

$\phi R_n = CC_1Dl$ $\quad C_{min} = \dfrac{P_u}{C_1Dl}$ $\quad D_{min} = \dfrac{P_u}{CC_1l}$ $\quad l_{min} = \dfrac{P_u}{CC_1D}$

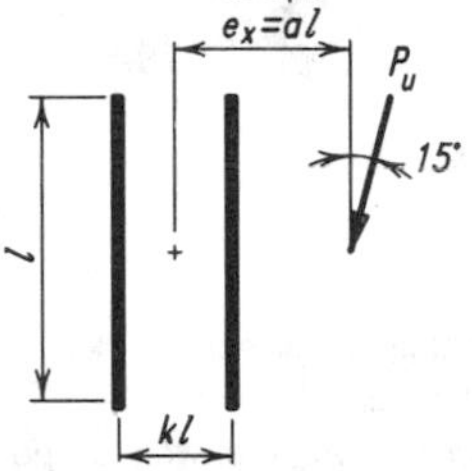

where

P_u = factored force, kips
D = number of sixteenths-of-an-inch in the fillet weld size
l = characteristic length of weld group, in.
a = e_x / l, in.
e_x = horizontal component of eccentricity of P_u with respect to centroid of weld group, in.
C = coefficient tabulated below which includes ϕ = 0.75
C_1 = electode strength coefficient from Table 8-37 (1.0 for E70XX electrodes)

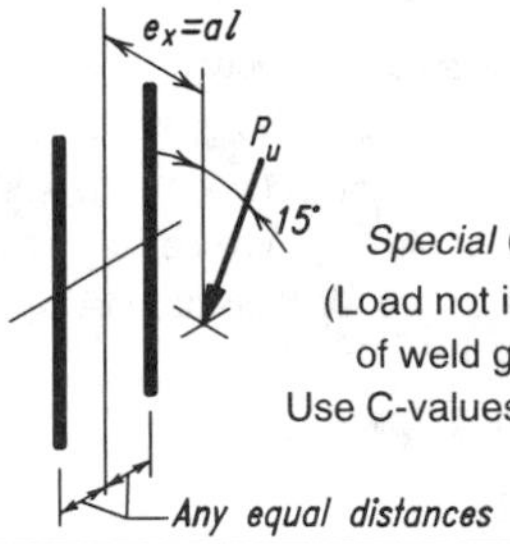

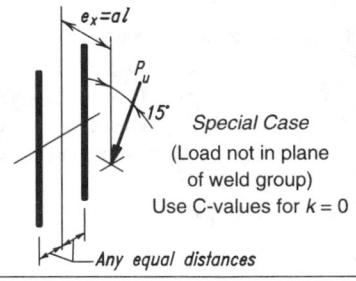

Special Case
(Load not in plane of weld group)
Use C-values for k = 0

	k															
a	**0**	**0.1**	**0.2**	**0.3**	**0.4**	**0.5**	**0.6**	**0.7**	**0.8**	**0.9**	**1.0**	**1.2**	**1.4**	**1.6**	**1.8**	**2.0**
0.00	2.97	2.97	2.97	2.97	2.97	2.97	2.97	2.97	2.97	2.97	2.97	2.97	2.97	2.97	2.97	2.97
0.10	2.84	2.84	2.84	2.83	2.82	2.82	2.81	2.80	2.80	2.79	2.78	2.77	2.75	2.74	2.73	2.72
0.15	2.76	2.76	2.75	2.75	2.74	2.73	2.72	2.72	2.71	2.70	2.70	2.69	2.68	2.67	2.66	2.65
0.20	2.63	2.63	2.63	2.62	2.62	2.62	2.61	2.61	2.61	2.61	2.60	2.60	2.59	2.59	2.58	2.58
0.25	2.48	2.48	2.48	2.48	2.49	2.49	2.49	2.49	2.49	2.50	2.50	2.50	2.51	2.51	2.51	2.51
0.30	2.32	2.32	2.32	2.33	2.34	2.35	2.36	2.37	2.38	2.39	2.39	2.41	2.42	2.43	2.43	2.43
0.40	2.01	2.01	2.02	2.04	2.06	2.09	2.12	2.14	2.16	2.18	2.19	2.22	2.25	2.27	2.28	2.30
0.50	1.74	1.74	1.76	1.78	1.82	1.86	1.89	1.93	1.96	1.99	2.01	2.05	2.09	2.12	2.15	2.17
0.60	1.52	1.52	1.54	1.57	1.62	1.66	1.70	1.75	1.78	1.82	1.85	1.90	1.95	1.99	2.02	2.05
0.70	1.34	1.35	1.37	1.40	1.45	1.50	1.54	1.59	1.63	1.67	1.70	1.77	1.82	1.87	1.90	1.94
0.80	1.20	1.20	1.22	1.26	1.31	1.36	1.41	1.46	1.50	1.54	1.58	1.65	1.71	1.76	1.80	1.84
0.90	1.08	1.08	1.11	1.15	1.19	1.24	1.30	1.34	1.39	1.43	1.47	1.54	1.60	1.66	1.70	1.74
1.00	0.979	0.985	1.01	1.05	1.09	1.15	1.20	1.25	1.29	1.33	1.37	1.45	1.51	1.57	1.62	1.66
1.20	0.826	0.832	0.856	0.893	0.938	0.987	1.04	1.09	1.13	1.17	1.21	1.29	1.35	1.41	1.46	1.51
1.40	0.714	0.719	0.743	0.778	0.820	0.866	0.913	0.960	1.00	1.05	1.09	1.16	1.23	1.28	1.34	1.39
1.60	0.628	0.633	0.656	0.688	0.727	0.770	0.815	0.859	0.901	0.941	0.980	1.05	1.12	1.18	1.23	1.28
1.80	0.560	0.566	0.587	0.617	0.653	0.693	0.735	0.777	0.817	0.855	0.893	0.963	1.03	1.09	1.14	1.19
2.00	0.506	0.511	0.530	0.558	0.592	0.630	0.669	0.708	0.746	0.783	0.819	0.887	0.949	1.01	1.06	1.11
2.20	0.461	0.466	0.484	0.510	0.541	0.577	0.613	0.650	0.687	0.722	0.757	0.822	0.882	0.938	0.989	1.04
2.40	0.423	0.428	0.445	0.469	0.499	0.532	0.566	0.601	0.636	0.670	0.703	0.765	0.824	0.878	0.928	0.974
2.60	0.391	0.396	0.412	0.434	0.462	0.493	0.526	0.559	0.591	0.624	0.656	0.716	0.772	0.825	0.873	0.918
2.80	0.363	0.368	0.383	0.404	0.430	0.460	0.491	0.522	0.553	0.584	0.614	0.672	0.727	0.778	0.825	0.869
3.00	0.339	0.344	0.358	0.378	0.403	0.430	0.460	0.489	0.519	0.549	0.578	0.634	0.686	0.736	0.781	0.824

Table 8-38 (cont.).
Coefficients C for Eccentrically Loaded Weld Groups
Angle = 30°

$$\phi R_n = CC_1 Dl \qquad C_{min} = \frac{P_u}{C_1 Dl} \qquad D_{min} = \frac{P_u}{CC_1 l} \qquad l_{min} = \frac{P_u}{CC_1 D}$$

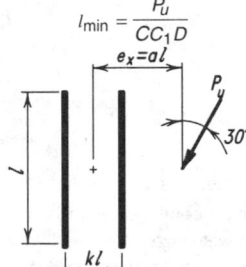

where

P_u = factored force, kips

D = number of sixteenths-of-an-inch
in the fillet weld size

l = characteristic length of weld group, in.

a = e_x / l, in.

e_x = horizontal component of eccentricity of
P_u with respect to centroid of weld group, in.

C = coefficient tabulated below which includes ϕ = 0.75

C_1 = electode strength coefficient from Table 8-37
(1.0 for E70XX electrodes)

Special Case
(Load not in plane
of weld group)
Use C-values for $k = 0$

Any equal distances

a	\multicolumn{17}{c}{k}
	0	0.1	0.2	0.3	0.4	0.5	0.6	0.7	0.8	0.9	1.0	1.2	1.4	1.6	1.8	2.0
0.00	3.28	3.28	3.28	3.28	3.28	3.28	3.28	3.28	3.28	3.28	3.28	3.28	3.28	3.28	3.28	3.28
0.10	3.03	3.03	3.04	3.04	3.04	3.05	3.05	3.06	3.06	3.06	3.07	3.07	3.07	3.06	3.06	3.05
0.15	2.87	2.87	2.87	2.87	2.88	2.88	2.88	2.89	2.90	2.90	2.91	2.92	2.94	2.94	2.95	2.95
0.20	2.72	2.73	2.73	2.73	2.74	2.74	2.75	2.76	2.76	2.77	2.78	2.79	2.81	2.82	2.83	2.84
0.25	2.57	2.57	2.57	2.58	2.59	2.61	2.62	2.63	2.64	2.66	2.67	2.68	2.70	2.72	2.73	2.74
0.30	2.41	2.41	2.42	2.43	2.45	2.47	2.49	2.51	2.53	2.54	2.56	2.59	2.61	2.63	2.64	2.66
0.40	2.11	2.11	2.12	2.14	2.17	2.20	2.24	2.28	2.30	2.33	2.35	2.40	2.43	2.46	2.49	2.51
0.50	1.84	1.85	1.86	1.89	1.93	1.98	2.02	2.06	2.10	2.13	2.17	2.22	2.27	2.31	2.34	2.37
0.60	1.62	1.63	1.65	1.68	1.73	1.78	1.83	1.88	1.92	1.96	2.00	2.07	2.12	2.17	2.21	2.25
0.70	1.44	1.45	1.47	1.51	1.56	1.61	1.66	1.72	1.77	1.81	1.85	1.92	1.99	2.04	2.09	2.13
0.80	1.30	1.30	1.33	1.37	1.42	1.47	1.52	1.58	1.63	1.68	1.72	1.80	1.87	1.93	1.98	2.02
0.90	1.17	1.18	1.20	1.24	1.29	1.35	1.41	1.46	1.51	1.56	1.61	1.69	1.76	1.82	1.88	1.92
1.00	1.07	1.08	1.10	1.14	1.19	1.25	1.30	1.36	1.41	1.46	1.51	1.59	1.66	1.73	1.78	1.84
1.20	0.907	0.913	0.939	0.979	1.03	1.08	1.14	1.19	1.24	1.29	1.33	1.42	1.49	1.56	1.62	1.68
1.40	0.786	0.792	0.817	0.855	0.901	0.951	1.00	1.05	1.10	1.15	1.20	1.28	1.35	1.42	1.49	1.54
1.60	0.693	0.699	0.723	0.759	0.801	0.848	0.897	0.946	0.993	1.04	1.08	1.16	1.24	1.31	1.37	1.42
1.80	0.619	0.625	0.648	0.681	0.721	0.765	0.811	0.857	0.901	0.945	0.987	1.07	1.14	1.21	1.27	1.32
2.00	0.559	0.565	0.587	0.617	0.655	0.696	0.740	0.783	0.825	0.866	0.907	0.984	1.05	1.12	1.18	1.24
2.20	0.510	0.516	0.536	0.564	0.599	0.638	0.679	0.720	0.761	0.800	0.838	0.912	0.981	1.04	1.10	1.16
2.40	0.469	0.474	0.493	0.520	0.552	0.589	0.628	0.667	0.705	0.743	0.779	0.850	0.917	0.978	1.04	1.09
2.60	0.433	0.439	0.456	0.481	0.512	0.546	0.583	0.620	0.657	0.693	0.728	0.795	0.860	0.920	0.975	1.03
2.80	0.403	0.408	0.424	0.448	0.477	0.509	0.544	0.579	0.614	0.649	0.683	0.748	0.809	0.867	0.922	0.972
3.00	0.376	0.382	0.397	0.419	0.446	0.477	0.510	0.543	0.577	0.610	0.642	0.705	0.764	0.821	0.873	0.923

Table 8-38 (cont.).
Coefficients C for Eccentrically Loaded Weld Groups
Angle = 45°

$$\phi R_n = CC_1 Dl \qquad C_{min} = \frac{P_u}{C_1 Dl} \qquad D_{min} = \frac{P_u}{CC_1 l} \qquad l_{min} = \frac{P_u}{CC_1 D}$$

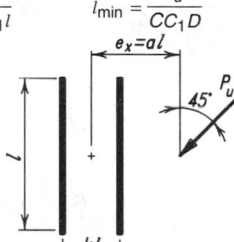

where

P_u = factored force, kips

D = number of sixteenths-of-an-inch in the fillet weld size

l = characteristic length of weld group, in.

a = e_x / l, in.

e_x = horizontal component of eccentricity of P_u with respect to centroid of weld group, in.

C = coefficient tabulated below which includes $\phi = 0.75$

C_1 = electode strength coefficient from Table 8-37 (1.0 for E70XX electrodes)

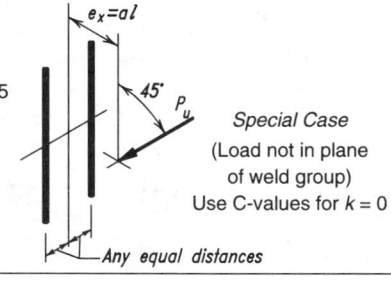

Special Case
(Load not in plane of weld group)
Use C-values for $k = 0$

—Any equal distances

a	\multicolumn{17}{c}{k}
	0	0.1	0.2	0.3	0.4	0.5	0.6	0.7	0.8	0.9	1.0	1.2	1.4	1.6	1.8	2.0
0.00	3.61	3.61	3.61	3.61	3.61	3.61	3.61	3.61	3.61	3.61	3.61	3.61	3.61	3.61	3.61	3.61
0.10	3.37	3.37	3.38	3.38	3.40	3.42	3.43	3.44	3.46	3.47	3.48	3.50	3.51	3.52	3.52	3.52
0.15	3.13	3.13	3.15	3.17	3.20	3.23	3.25	3.28	3.30	3.33	3.35	3.38	3.41	3.43	3.45	3.46
0.20	2.94	2.94	2.95	2.97	2.99	3.03	3.06	3.10	3.13	3.17	3.20	3.25	3.29	3.33	3.35	3.38
0.25	2.77	2.77	2.78	2.80	2.83	2.86	2.89	2.93	2.97	3.01	3.04	3.11	3.16	3.21	3.25	3.28
0.30	2.61	2.61	2.63	2.65	2.68	2.71	2.75	2.79	2.83	2.86	2.90	2.97	3.04	3.09	3.14	3.18
0.40	2.32	2.32	2.34	2.37	2.41	2.45	2.50	2.54	2.59	2.63	2.66	2.73	2.80	2.87	2.92	2.97
0.50	2.06	2.07	2.09	2.12	2.17	2.22	2.27	2.33	2.38	2.42	2.47	2.54	2.61	2.68	2.74	2.79
0.60	1.84	1.85	1.87	1.91	1.96	2.02	2.08	2.14	2.19	2.25	2.30	2.38	2.45	2.52	2.58	2.63
0.70	1.66	1.66	1.69	1.73	1.79	1.85	1.91	1.97	2.03	2.09	2.14	2.23	2.31	2.38	2.44	2.50
0.80	1.50	1.51	1.54	1.58	1.64	1.70	1.76	1.83	1.89	1.95	2.00	2.10	2.18	2.26	2.32	2.38
0.90	1.37	1.38	1.40	1.45	1.51	1.57	1.64	1.70	1.76	1.82	1.88	1.98	2.07	2.14	2.21	2.27
1.00	1.26	1.26	1.29	1.34	1.40	1.46	1.53	1.59	1.65	1.71	1.77	1.87	1.96	2.04	2.11	2.17
1.20	1.08	1.08	1.11	1.16	1.21	1.28	1.34	1.40	1.46	1.52	1.58	1.68	1.77	1.85	1.93	2.00
1.40	0.938	0.946	0.975	1.02	1.07	1.13	1.19	1.25	1.31	1.37	1.42	1.52	1.62	1.70	1.77	1.84
1.60	0.831	0.838	0.866	0.908	0.958	1.01	1.07	1.13	1.19	1.24	1.29	1.39	1.48	1.56	1.64	1.71
1.80	0.745	0.752	0.779	0.817	0.864	0.917	0.972	1.03	1.08	1.13	1.19	1.28	1.37	1.45	1.52	1.59
2.00	0.675	0.682	0.707	0.743	0.787	0.836	0.888	0.941	0.992	1.04	1.09	1.18	1.27	1.35	1.42	1.49
2.20	0.617	0.624	0.647	0.681	0.722	0.768	0.818	0.868	0.917	0.964	1.01	1.10	1.18	1.26	1.33	1.40
2.40	0.568	0.574	0.596	0.628	0.667	0.710	0.756	0.804	0.851	0.897	0.941	1.03	1.11	1.18	1.25	1.32
2.60	0.526	0.532	0.553	0.583	0.619	0.660	0.703	0.749	0.794	0.838	0.881	0.963	1.04	1.12	1.18	1.25
2.80	0.489	0.496	0.515	0.544	0.578	0.617	0.658	0.701	0.743	0.786	0.827	0.906	0.982	1.05	1.12	1.18
3.00	0.458	0.464	0.482	0.509	0.542	0.579	0.618	0.658	0.699	0.739	0.779	0.856	0.929	0.998	1.06	1.12

Table 8-38 (cont.).
Coefficients *C* for Eccentrically Loaded Weld Groups
Angle = 60°

$$\phi R_n = CC_1 Dl \qquad C_{min} = \frac{P_u}{C_1 Dl} \qquad D_{min} = \frac{P_u}{CC_1 l} \qquad l_{min} = \frac{P_u}{CC_1 D}$$

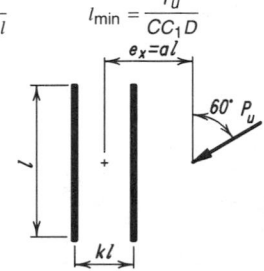

where

P_u = factored force, kips

D = number of sixteenths-of-an-inch in the fillet weld size

l = characteristic length of weld group, in.

a = e_x / l, in.

e_x = horizontal component of eccentricity of P_u with respect to centroid of weld group, in.

C = coefficient tabulated below which includes ϕ = 0.75

C_1 = electode strength coefficient from Table 8-37 (1.0 for E70XX electrodes)

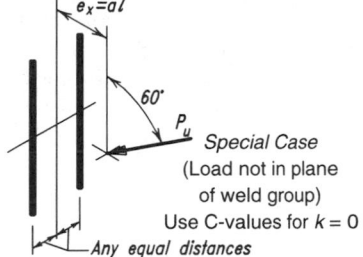

Special Case
(Load not in plane of weld group)
Use C-values for $k = 0$
Any equal distances

								k								
a	0	0.1	0.2	0.3	0.4	0.5	0.6	0.7	0.8	0.9	1.0	1.2	1.4	1.6	1.8	2.0
0.00	3.91	3.91	3.91	3.91	3.91	3.91	3.91	3.91	3.91	3.91	3.91	3.91	3.91	3.91	3.91	3.91
0.10	3.65	3.66	3.68	3.71	3.74	3.78	3.80	3.83	3.84	3.85	3.86	3.86	3.87	3.86	3.86	3.86
0.15	3.46	3.47	3.49	3.53	3.58	3.63	3.68	3.73	3.76	3.79	3.80	3.83	3.84	3.85	3.85	3.86
0.20	3.27	3.28	3.30	3.35	3.40	3.47	3.54	3.60	3.65	3.69	3.73	3.78	3.80	3.82	3.83	3.84
0.25	3.10	3.10	3.13	3.17	3.23	3.30	3.38	3.46	3.53	3.58	3.63	3.70	3.75	3.78	3.80	3.82
0.30	2.95	2.95	2.98	3.02	3.07	3.14	3.22	3.31	3.39	3.46	3.53	3.62	3.68	3.73	3.76	3.78
0.40	2.68	2.69	2.71	2.76	2.81	2.88	2.95	3.03	3.12	3.21	3.29	3.43	3.53	3.61	3.66	3.70
0.50	2.44	2.45	2.48	2.53	2.59	2.66	2.73	2.80	2.88	2.96	3.05	3.22	3.36	3.46	3.54	3.60
0.60	2.24	2.24	2.27	2.32	2.39	2.46	2.54	2.62	2.69	2.77	2.84	3.01	3.17	3.29	3.39	3.47
0.70	2.05	2.06	2.09	2.15	2.21	2.29	2.37	2.45	2.53	2.60	2.67	2.82	2.98	3.12	3.24	3.34
0.80	1.89	1.90	1.93	1.99	2.06	2.14	2.22	2.30	2.38	2.45	2.52	2.66	2.81	2.95	3.08	3.20
0.90	1.75	1.76	1.79	1.85	1.92	2.00	2.08	2.16	2.24	2.32	2.39	2.53	2.65	2.80	2.93	3.05
1.00	1.62	1.63	1.67	1.73	1.80	1.88	1.96	2.04	2.12	2.20	2.27	2.41	2.53	2.66	2.79	2.91
1.20	1.42	1.43	1.46	1.52	1.59	1.67	1.75	1.83	1.91	1.98	2.06	2.19	2.32	2.43	2.54	2.66
1.40	1.25	1.26	1.30	1.36	1.42	1.50	1.57	1.65	1.73	1.81	1.88	2.01	2.13	2.25	2.35	2.45
1.60	1.12	1.13	1.17	1.22	1.28	1.35	1.43	1.50	1.58	1.65	1.72	1.85	1.98	2.09	2.19	2.29
1.80	1.01	1.02	1.06	1.11	1.17	1.24	1.31	1.38	1.45	1.52	1.59	1.72	1.84	1.95	2.05	2.14
2.00	0.922	0.932	0.964	1.01	1.07	1.13	1.20	1.27	1.34	1.41	1.47	1.60	1.72	1.82	1.92	2.02
2.20	0.847	0.856	0.887	0.932	0.986	1.05	1.11	1.18	1.25	1.31	1.37	1.49	1.61	1.71	1.81	1.90
2.40	0.782	0.791	0.820	0.863	0.914	0.971	1.03	1.10	1.16	1.22	1.28	1.40	1.51	1.62	1.71	1.80
2.60	0.726	0.734	0.762	0.803	0.852	0.907	0.964	1.03	1.09	1.15	1.21	1.32	1.43	1.53	1.62	1.71
2.80	0.677	0.686	0.712	0.750	0.797	0.849	0.905	0.963	1.02	1.08	1.14	1.24	1.35	1.45	1.54	1.62
3.00	0.635	0.643	0.668	0.704	0.749	0.799	0.852	0.906	0.963	1.02	1.07	1.18	1.28	1.37	1.46	1.55

Table 8-38 (cont.).
Coefficients C for Eccentrically Loaded Weld Groups
Angle = 75°

$$\phi R_n = CC_1Dl \qquad C_{min} = \frac{P_u}{C_1Dl} \qquad D_{min} = \frac{P_u}{CC_1l} \qquad l_{min} = \frac{P_u}{CC_1D}$$

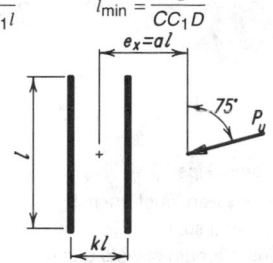

where

P_u = factored force, kips

D = number of sixteenths-of-an-inch in the fillet weld size

l = characteristic length of weld group, in.

a = e_x/l, in.

e_x = horizontal component of eccentricity of P_u with respect to centroid of weld group, in.

C = coefficient tabulated below which includes $\phi = 0.75$

C_1 = electode strength coefficient from Table 8-37 (1.0 for E70XX electrodes)

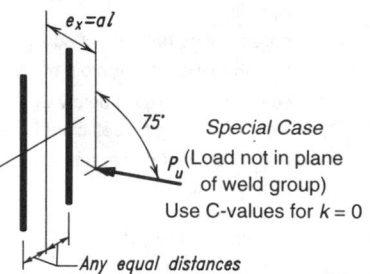

Special Case
P_u (Load not in plane of weld group)
Use C-values for $k = 0$

							k									
a	0	0.1	0.2	0.3	0.4	0.5	0.6	0.7	0.8	0.9	1.0	1.2	1.4	1.6	1.8	2.0
0.00	4.11	4.11	4.11	4.11	4.11	4.11	4.11	4.11	4.11	4.11	4.11	4.11	4.11	4.11	4.11	4.11
0.10	3.88	3.90	3.95	4.00	4.04	4.07	4.08	4.09	4.09	4.09	4.09	4.09	4.09	4.09	4.09	4.09
0.15	3.76	3.77	3.83	3.90	3.96	4.01	4.04	4.06	4.08	4.08	4.09	4.09	4.09	4.09	4.09	4.09
0.20	3.64	3.65	3.71	3.79	3.88	3.94	3.99	4.03	4.05	4.06	4.07	4.08	4.09	4.09	4.09	4.09
0.25	3.53	3.54	3.60	3.69	3.78	3.87	3.93	3.98	4.01	4.03	4.05	4.07	4.08	4.08	4.09	4.09
0.30	3.43	3.44	3.49	3.58	3.69	3.78	3.86	3.92	3.97	4.00	4.02	4.05	4.07	4.07	4.08	4.08
0.40	3.24	3.25	3.29	3.38	3.50	3.62	3.72	3.80	3.86	3.91	3.95	4.00	4.03	4.05	4.06	4.07
0.50	3.07	3.08	3.12	3.20	3.32	3.45	3.57	3.67	3.75	3.82	3.87	3.94	3.99	4.02	4.04	4.05
0.60	2.91	2.92	2.97	3.05	3.15	3.29	3.42	3.54	3.63	3.71	3.78	3.88	3.94	3.98	4.01	4.03
0.70	2.77	2.78	2.82	2.90	3.00	3.13	3.27	3.40	3.51	3.60	3.68	3.80	3.88	3.93	3.97	4.00
0.80	2.63	2.64	2.69	2.77	2.87	2.99	3.13	3.26	3.39	3.49	3.58	3.72	3.81	3.88	3.93	3.96
0.90	2.50	2.52	2.57	2.64	2.74	2.86	3.00	3.13	3.26	3.38	3.48	3.64	3.75	3.83	3.88	3.93
1.00	2.38	2.40	2.45	2.53	2.63	2.74	2.87	3.01	3.14	3.26	3.37	3.55	3.68	3.77	3.83	3.88
1.20	2.17	2.18	2.24	2.32	2.41	2.52	2.64	2.78	2.91	3.04	3.16	3.37	3.52	3.64	3.73	3.80
1.40	1.99	2.00	2.05	2.13	2.23	2.33	2.45	2.57	2.71	2.84	2.96	3.18	3.36	3.50	3.61	3.69
1.60	1.83	1.84	1.89	1.97	2.06	2.17	2.28	2.39	2.52	2.65	2.78	3.01	3.20	3.36	3.49	3.59
1.80	1.69	1.70	1.75	1.83	1.92	2.02	2.13	2.24	2.35	2.48	2.61	2.84	3.05	3.22	3.36	3.48
2.00	1.57	1.58	1.63	1.70	1.79	1.89	1.99	2.10	2.21	2.33	2.45	2.68	2.89	3.08	3.23	3.36
2.20	1.46	1.48	1.52	1.59	1.68	1.77	1.87	1.97	2.08	2.19	2.30	2.54	2.75	2.94	3.10	3.24
2.40	1.37	1.38	1.43	1.49	1.57	1.67	1.76	1.86	1.97	2.07	2.18	2.40	2.61	2.81	2.98	3.13
2.60	1.28	1.30	1.34	1.40	1.48	1.57	1.66	1.76	1.86	1.96	2.07	2.28	2.49	2.68	2.85	3.01
2.80	1.21	1.22	1.27	1.33	1.40	1.49	1.58	1.67	1.77	1.86	1.96	2.16	2.37	2.56	2.73	2.89
3.00	1.14	1.15	1.20	1.26	1.33	1.41	1.49	1.59	1.68	1.77	1.87	2.06	2.26	2.45	2.62	2.78

Table 8-39.
Coefficients *C* for Eccentrically Loaded Weld Groups
Angle = 0°

$$\phi R_n = CC_1 Dl \qquad C_{min} = \frac{P_u}{C_1 Dl} \qquad D_{min} = \frac{P_u}{CC_1 l} \qquad l_{min} = \frac{P_u}{CC_1 D}$$

where

P_u = factored force, kips

D = number of sixteenths-of-an-inch in the fillet weld size

l = characteristic length of weld group, in.

a = e_x / l, in.

e_x = horizontal component of eccentricity of P_u with respect to centroid of weld group, in.

C = coefficient tabulated below which includes $\phi = 0.75$

C_1 = electode strength coefficient from Table 8-37 (1.0 for E70XX electrodes)

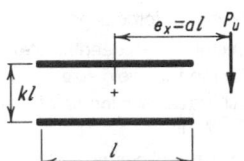

| a | \multicolumn{17}{c}{k} |
	0	0.1	0.2	0.3	0.4	0.5	0.6	0.7	0.8	0.9	1.0	1.2	1.4	1.6	1.8	2.0
0.00	4.18	4.18	4.18	4.18	4.18	4.18	4.18	4.18	4.18	4.18	4.18	4.18	4.18	4.18	4.18	4.18
0.10	3.24	3.27	3.36	3.48	3.61	3.73	3.83	3.91	3.97	4.01	4.05	4.09	4.12	4.13	4.14	4.15
0.15	2.92	2.95	3.03	3.15	3.29	3.43	3.56	3.68	3.77	3.84	3.90	3.98	4.04	4.07	4.09	4.11
0.20	2.65	2.68	2.75	2.85	2.99	3.15	3.30	3.43	3.55	3.65	3.73	3.85	3.93	3.99	4.03	4.06
0.25	2.41	2.44	2.50	2.60	2.73	2.88	3.04	3.19	3.33	3.44	3.54	3.70	3.81	3.89	3.95	3.99
0.30	2.20	2.23	2.29	2.39	2.50	2.64	2.81	2.96	3.11	3.24	3.36	3.54	3.68	3.79	3.86	3.92
0.40	1.86	1.88	1.95	2.03	2.14	2.26	2.39	2.55	2.71	2.86	2.99	3.22	3.40	3.55	3.66	3.74
0.50	1.60	1.62	1.68	1.76	1.85	1.96	2.08	2.21	2.36	2.51	2.66	2.91	3.12	3.29	3.43	3.55
0.60	1.40	1.42	1.47	1.54	1.63	1.73	1.84	1.96	2.08	2.22	2.36	2.63	2.86	3.05	3.21	3.34
0.70	1.24	1.26	1.30	1.37	1.45	1.54	1.64	1.75	1.86	1.98	2.11	2.38	2.61	2.81	2.99	3.14
0.80	1.11	1.13	1.17	1.23	1.30	1.39	1.48	1.58	1.68	1.79	1.90	2.15	2.39	2.60	2.78	2.95
0.90	1.00	1.02	1.06	1.11	1.18	1.26	1.35	1.44	1.53	1.63	1.73	1.96	2.19	2.40	2.59	2.76
1.00	0.914	0.929	0.965	1.02	1.08	1.15	1.23	1.31	1.40	1.49	1.59	1.79	2.02	2.22	2.41	2.59
1.20	0.777	0.789	0.821	0.866	0.920	0.984	1.05	1.13	1.20	1.28	1.36	1.53	1.72	1.92	2.11	2.28
1.40	0.674	0.685	0.713	0.753	0.802	0.856	0.918	0.982	1.05	1.12	1.19	1.34	1.50	1.68	1.85	2.02
1.60	0.594	0.604	0.629	0.665	0.709	0.759	0.812	0.871	0.931	0.993	1.06	1.19	1.33	1.48	1.64	1.80
1.80	0.531	0.541	0.563	0.595	0.635	0.680	0.729	0.780	0.836	0.892	0.950	1.07	1.20	1.33	1.47	1.62
2.00	0.481	0.489	0.509	0.538	0.574	0.616	0.661	0.708	0.757	0.810	0.862	0.971	1.09	1.21	1.33	1.47
2.20	0.439	0.446	0.465	0.492	0.524	0.562	0.604	0.647	0.693	0.740	0.789	0.890	0.994	1.10	1.22	1.34
2.40	0.404	0.410	0.427	0.452	0.483	0.517	0.555	0.596	0.638	0.681	0.727	0.820	0.917	1.02	1.12	1.23
2.60	0.374	0.379	0.396	0.419	0.447	0.479	0.514	0.552	0.592	0.632	0.674	0.760	0.850	0.941	1.04	1.14
2.80	0.348	0.353	0.368	0.390	0.416	0.446	0.479	0.514	0.551	0.589	0.628	0.709	0.793	0.878	0.966	1.06
3.00	0.325	0.330	0.344	0.364	0.389	0.417	0.448	0.481	0.516	0.552	0.588	0.664	0.742	0.822	0.904	0.989

Table 8-39 (cont.).
Coefficients C for Eccentrically Loaded Weld Groups
Angle = 15°

$$\phi R_n = CC_1 Dl \qquad C_{min} = \frac{P_u}{C_1 Dl} \qquad D_{min} = \frac{P_u}{CC_1 l} \qquad l_{min} = \frac{P_u}{CC_1 D}$$

where

P_u = factored force, kips

D = number of sixteenths-of-an-inch in the fillet weld size

l = characteristic length of weld group, in.

a = e_x / l, in.

e_x = horizontal component of eccentricity of P_u with respect to centroid of weld group, in.

C = coefficient tabulated below which includes $\phi = 0.75$

C_1 = electode strength coefficient from Table 8-37 (1.0 for E70XX electrodes)

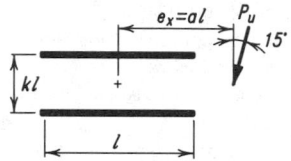

a	\multicolumn{16}{c}{k}
	0	0.1	0.2	0.3	0.4	0.5	0.6	0.7	0.8	0.9	1.0	1.2	1.4	1.6	1.8	2.0
0.00	4.11	4.11	4.11	4.11	4.11	4.11	4.11	4.11	4.11	4.11	4.11	4.11	4.11	4.11	4.11	4.11
0.10	3.29	3.30	3.34	3.43	3.55	3.66	3.76	3.83	3.89	3.94	3.97	4.02	4.04	4.06	4.07	4.07
0.15	2.97	2.99	3.03	3.11	3.22	3.35	3.48	3.59	3.68	3.76	3.82	3.90	3.96	4.00	4.02	4.04
0.20	2.70	2.71	2.76	2.84	2.94	3.07	3.20	3.34	3.45	3.55	3.64	3.76	3.85	3.91	3.95	3.98
0.25	2.46	2.48	2.53	2.61	2.71	2.82	2.95	3.09	3.22	3.34	3.44	3.61	3.72	3.81	3.87	3.91
0.30	2.25	2.27	2.32	2.40	2.50	2.61	2.73	2.87	3.00	3.13	3.25	3.44	3.58	3.69	3.77	3.83
0.40	1.91	1.93	1.98	2.05	2.15	2.25	2.37	2.49	2.62	2.75	2.87	3.10	3.29	3.44	3.56	3.65
0.50	1.65	1.66	1.71	1.79	1.87	1.97	2.08	2.19	2.30	2.42	2.55	2.79	2.99	3.17	3.32	3.44
0.60	1.44	1.46	1.50	1.57	1.65	1.75	1.85	1.95	2.06	2.17	2.28	2.51	2.72	2.91	3.08	3.22
0.70	1.28	1.29	1.34	1.40	1.48	1.56	1.66	1.76	1.85	1.96	2.06	2.27	2.48	2.67	2.85	3.00
0.80	1.15	1.16	1.20	1.26	1.33	1.41	1.50	1.59	1.69	1.78	1.88	2.07	2.27	2.45	2.63	2.79
0.90	1.04	1.05	1.09	1.14	1.21	1.29	1.37	1.45	1.54	1.63	1.72	1.90	2.08	2.26	2.43	2.59
1.00	0.945	0.958	0.993	1.05	1.11	1.18	1.26	1.34	1.42	1.50	1.59	1.75	1.92	2.10	2.26	2.42
1.20	0.803	0.814	0.846	0.892	0.946	1.01	1.08	1.15	1.22	1.30	1.37	1.52	1.67	1.82	1.97	2.12
1.40	0.697	0.707	0.735	0.776	0.825	0.881	0.942	1.01	1.07	1.14	1.21	1.34	1.47	1.61	1.74	1.87
1.60	0.615	0.624	0.649	0.686	0.731	0.781	0.835	0.894	0.954	1.01	1.08	1.20	1.32	1.44	1.56	1.68
1.80	0.550	0.559	0.581	0.614	0.655	0.701	0.751	0.802	0.858	0.913	0.968	1.08	1.19	1.30	1.41	1.52
2.00	0.497	0.505	0.526	0.556	0.593	0.635	0.681	0.729	0.778	0.830	0.881	0.985	1.09	1.19	1.29	1.39
2.20	0.454	0.461	0.480	0.508	0.542	0.580	0.623	0.667	0.713	0.760	0.808	0.904	0.999	1.10	1.19	1.28
2.40	0.418	0.424	0.442	0.467	0.499	0.534	0.573	0.615	0.657	0.700	0.745	0.834	0.924	1.01	1.10	1.19
2.60	0.387	0.392	0.409	0.433	0.462	0.495	0.531	0.570	0.609	0.650	0.691	0.775	0.859	0.942	1.03	1.11
2.80	0.360	0.365	0.381	0.403	0.430	0.461	0.494	0.530	0.568	0.606	0.644	0.724	0.802	0.881	0.959	1.03
3.00	0.336	0.341	0.356	0.377	0.402	0.431	0.463	0.496	0.532	0.568	0.604	0.678	0.753	0.827	0.900	0.972

Table 8-39 (cont.).
Coefficients C for Eccentrically Loaded Weld Groups
Angle = 30°

$$\phi R_n = CC_1 Dl \qquad C_{min} = \frac{P_u}{C_1 Dl} \qquad D_{min} = \frac{P_u}{CC_1 l} \qquad l_{min} = \frac{P_u}{CC_1 D}$$

where

P_u = factored force, kips

D = number of sixteenths-of-an-inch in the fillet weld size

l = characteristic length of weld group, in.

a = e_x / l, in.

e_x = horizontal component of eccentricity of P_u with respect to centroid of weld group, in.

C = coefficient tabulated below which includes ϕ = 0.75

C_1 = electrode strength coefficient from Table 8-37 (1.0 for E70XX electrodes)

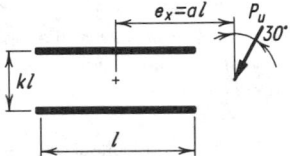

a	\multicolumn{17}{c}{k}
	0	0.1	0.2	0.3	0.4	0.5	0.6	0.7	0.8	0.9	1.0	1.2	1.4	1.6	1.8	2.0
0.00	3.91	3.91	3.91	3.91	3.91	3.91	3.91	3.91	3.91	3.91	3.91	3.91	3.91	3.91	3.91	3.91
0.10	3.37	3.38	3.40	3.45	3.50	3.56	3.62	3.67	3.71	3.75	3.77	3.81	3.83	3.84	3.85	3.85
0.15	3.07	3.07	3.10	3.14	3.20	3.27	3.35	3.43	3.50	3.56	3.61	3.69	3.74	3.77	3.79	3.81
0.20	2.82	2.83	2.85	2.89	2.94	3.01	3.09	3.17	3.26	3.35	3.42	3.54	3.62	3.68	3.72	3.75
0.25	2.60	2.61	2.63	2.68	2.73	2.80	2.87	2.94	3.03	3.12	3.21	3.36	3.48	3.56	3.62	3.67
0.30	2.40	2.41	2.44	2.49	2.55	2.62	2.69	2.76	2.84	2.92	3.01	3.18	3.32	3.43	3.51	3.57
0.40	2.06	2.07	2.10	2.16	2.22	2.30	2.38	2.46	2.54	2.61	2.69	2.83	2.99	3.14	3.25	3.35
0.50	1.80	1.80	1.84	1.89	1.97	2.04	2.13	2.21	2.29	2.36	2.44	2.57	2.70	2.85	2.98	3.10
0.60	1.58	1.59	1.62	1.68	1.75	1.83	1.91	2.00	2.07	2.15	2.23	2.36	2.49	2.61	2.74	2.86
0.70	1.41	1.42	1.45	1.51	1.58	1.66	1.74	1.82	1.90	1.97	2.05	2.18	2.30	2.42	2.53	2.64
0.80	1.26	1.27	1.31	1.37	1.43	1.51	1.59	1.66	1.74	1.82	1.89	2.03	2.15	2.26	2.36	2.46
0.90	1.15	1.15	1.19	1.25	1.31	1.38	1.46	1.53	1.61	1.68	1.75	1.88	2.01	2.12	2.22	2.32
1.00	1.05	1.06	1.09	1.14	1.20	1.27	1.35	1.42	1.49	1.57	1.63	1.76	1.88	1.99	2.09	2.19
1.20	0.891	0.900	0.932	0.979	1.03	1.10	1.17	1.23	1.30	1.37	1.43	1.56	1.67	1.78	1.88	1.97
1.40	0.774	0.783	0.812	0.855	0.906	0.963	1.02	1.09	1.15	1.21	1.27	1.39	1.50	1.60	1.70	1.79
1.60	0.684	0.692	0.719	0.757	0.805	0.857	0.913	0.972	1.03	1.09	1.14	1.26	1.36	1.46	1.55	1.64
1.80	0.612	0.620	0.644	0.679	0.723	0.771	0.823	0.876	0.931	0.985	1.04	1.14	1.24	1.33	1.42	1.51
2.00	0.554	0.562	0.584	0.616	0.655	0.700	0.749	0.799	0.848	0.899	0.949	1.05	1.14	1.23	1.31	1.39
2.20	0.506	0.513	0.534	0.563	0.600	0.641	0.686	0.733	0.779	0.826	0.874	0.965	1.05	1.14	1.22	1.30
2.40	0.465	0.472	0.491	0.519	0.553	0.591	0.633	0.676	0.720	0.764	0.808	0.896	0.979	1.06	1.14	1.21
2.60	0.431	0.437	0.455	0.481	0.512	0.548	0.587	0.628	0.669	0.711	0.752	0.835	0.915	0.992	1.07	1.14
2.80	0.401	0.406	0.423	0.448	0.477	0.511	0.547	0.585	0.625	0.664	0.704	0.782	0.858	0.931	1.00	1.07
3.00	0.375	0.380	0.396	0.419	0.447	0.478	0.513	0.548	0.586	0.623	0.661	0.734	0.808	0.878	0.946	1.01

Table 8-39 (cont.).
Coefficients C for Eccentrically Loaded Weld Groups
Angle = 45°

$$\phi R_n = CC_1 Dl \qquad C_{min} = \frac{P_u}{C_1 Dl} \qquad D_{min} = \frac{P_u}{CC_1 l} \qquad l_{min} = \frac{P_u}{CC_1 D}$$

where

P_u = factored force, kips

D = number of sixteenths-of-an-inch in the fillet weld size

l = characteristic length of weld group, in.

a = e_x / l, in.

e_x = horizontal component of eccentricity of P_u with respect to centroid of weld group, in.

C = coefficient tabulated below which includes $\phi = 0.75$

C_1 = electode strength coefficient from Table 8-37 (1.0 for E70XX electrodes)

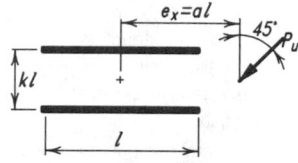

a	\multicolumn{17}{c}{k}
---	0	0.1	0.2	0.3	0.4	0.5	0.6	0.7	0.8	0.9	1.0	1.2	1.4	1.6	1.8	2.0
0.00	3.61	3.61	3.61	3.61	3.61	3.61	3.61	3.61	3.61	3.61	3.61	3.61	3.61	3.61	3.61	3.61
0.10	3.37	3.37	3.38	3.38	3.40	3.42	3.43	3.44	3.46	3.47	3.48	3.50	3.51	3.52	3.52	3.52
0.15	3.13	3.13	3.15	3.17	3.20	3.23	3.25	3.28	3.30	3.33	3.35	3.38	3.41	3.43	3.45	3.47
0.20	2.94	2.94	2.95	2.97	2.99	3.03	3.06	3.10	3.13	3.17	3.20	3.25	3.29	3.33	3.35	3.38
0.25	2.77	2.77	2.78	2.80	2.83	2.86	2.89	2.93	2.97	3.01	3.04	3.11	3.16	3.21	3.25	3.28
0.30	2.61	2.61	2.63	2.65	2.68	2.71	2.75	2.79	2.83	2.86	2.90	2.97	3.04	3.09	3.14	3.18
0.40	2.32	2.32	2.34	2.37	2.41	2.45	2.50	2.54	2.59	2.63	2.66	2.73	2.80	2.87	2.92	2.97
0.50	2.06	2.07	2.09	2.12	2.17	2.22	2.27	2.33	2.38	2.42	2.47	2.54	2.61	2.68	2.74	2.79
0.60	1.84	1.85	1.87	1.91	1.96	2.02	2.08	2.14	2.19	2.25	2.30	2.38	2.45	2.52	2.58	2.63
0.70	1.66	1.66	1.69	1.73	1.79	1.85	1.91	1.97	2.03	2.09	2.14	2.23	2.31	2.38	2.44	2.50
0.80	1.50	1.51	1.54	1.58	1.64	1.70	1.76	1.83	1.89	1.95	2.00	2.10	2.18	2.26	2.32	2.38
0.90	1.37	1.38	1.40	1.45	1.51	1.57	1.64	1.70	1.76	1.82	1.88	1.98	2.07	2.14	2.21	2.27
1.00	1.26	1.26	1.29	1.34	1.40	1.46	1.53	1.59	1.65	1.71	1.77	1.87	1.96	2.04	2.11	2.17
1.20	1.08	1.08	1.11	1.16	1.21	1.28	1.34	1.40	1.46	1.52	1.58	1.68	1.77	1.85	1.93	2.00
1.40	0.938	0.946	0.975	1.02	1.07	1.13	1.19	1.25	1.31	1.37	1.42	1.52	1.62	1.70	1.77	1.84
1.60	0.831	0.838	0.866	0.908	0.958	1.01	1.07	1.13	1.19	1.24	1.29	1.39	1.48	1.56	1.64	1.71
1.80	0.745	0.752	0.779	0.817	0.864	0.917	0.972	1.03	1.08	1.13	1.19	1.28	1.37	1.45	1.52	1.59
2.00	0.675	0.682	0.707	0.743	0.787	0.836	0.888	0.941	0.992	1.04	1.09	1.18	1.27	1.35	1.42	1.49
2.20	0.617	0.624	0.647	0.681	0.722	0.768	0.818	0.868	0.917	0.964	1.01	1.10	1.18	1.26	1.33	1.40
2.40	0.568	0.574	0.596	0.628	0.667	0.710	0.756	0.804	0.851	0.897	0.941	1.03	1.11	1.18	1.25	1.32
2.60	0.526	0.532	0.553	0.583	0.619	0.660	0.703	0.749	0.794	0.838	0.881	0.963	1.04	1.12	1.18	1.25
2.80	0.489	0.496	0.515	0.544	0.578	0.617	0.658	0.701	0.743	0.786	0.827	0.906	0.982	1.05	1.12	1.18
3.00	0.458	0.464	0.482	0.509	0.542	0.579	0.618	0.658	0.699	0.739	0.779	0.856	0.929	0.998	1.06	1.12

Table 8-39 (cont.).
Coefficients C for Eccentrically Loaded Weld Groups
Angle = 60°

$$\phi R_n = CC_1 Dl \qquad C_{min} = \frac{P_u}{C_1 Dl} \qquad D_{min} = \frac{P_u}{CC_1 l} \qquad l_{min} = \frac{P_u}{CC_1 D}$$

where

P_u = factored force, kips

D = number of sixteenths-of-an-inch in the fillet weld size

l = characteristic length of weld group, in.

a = e_x / l, in.

e_x = horizontal component of eccentricity of P_u with respect to centroid of weld group, in.

C = coefficient tabulated below which includes $\phi = 0.75$

C_1 = electode strength coefficient from Table 8-37 (1.0 for E70XX electrodes)

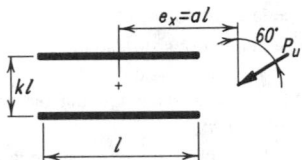

								k								
a	0	0.1	0.2	0.3	0.4	0.5	0.6	0.7	0.8	0.9	1.0	1.2	1.4	1.6	1.8	2.0
0.00	3.28	3.28	3.28	3.28	3.28	3.28	3.28	3.28	3.28	3.28	3.28	3.28	3.28	3.28	3.28	3.28
0.10	3.20	3.19	3.19	3.19	3.19	3.19	3.19	3.19	3.18	3.18	3.18	3.17	3.16	3.15	3.14	3.13
0.15	3.10	3.09	3.10	3.10	3.10	3.10	3.10	3.10	3.10	3.10	3.10	3.10	3.10	3.09	3.09	3.08
0.20	2.98	2.98	2.98	2.98	2.98	2.99	2.99	3.00	3.01	3.01	3.02	3.02	3.03	3.03	3.02	3.02
0.25	2.89	2.89	2.89	2.89	2.89	2.90	2.90	2.91	2.91	2.92	2.92	2.94	2.95	2.96	2.96	2.96
0.30	2.80	2.80	2.80	2.81	2.81	2.82	2.82	2.83	2.83	2.84	2.85	2.86	2.87	2.89	2.89	2.90
0.40	2.63	2.63	2.63	2.64	2.65	2.66	2.67	2.68	2.69	2.70	2.71	2.73	2.74	2.76	2.77	2.78
0.50	2.44	2.44	2.45	2.47	2.48	2.50	2.52	2.54	2.55	2.57	2.58	2.61	2.63	2.65	2.66	2.68
0.60	2.27	2.27	2.28	2.29	2.32	2.35	2.37	2.40	2.42	2.44	2.46	2.49	2.52	2.55	2.57	2.59
0.70	2.09	2.10	2.11	2.13	2.16	2.19	2.23	2.27	2.29	2.32	2.34	2.39	2.42	2.45	2.48	2.50
0.80	1.94	1.94	1.96	1.98	2.02	2.06	2.10	2.14	2.18	2.21	2.23	2.29	2.33	2.37	2.40	2.42
0.90	1.80	1.80	1.82	1.85	1.89	1.93	1.98	2.02	2.06	2.10	2.13	2.19	2.24	2.28	2.32	2.34
1.00	1.67	1.67	1.69	1.73	1.77	1.82	1.87	1.92	1.96	2.00	2.04	2.10	2.15	2.20	2.24	2.27
1.20	1.46	1.46	1.48	1.52	1.57	1.62	1.67	1.73	1.78	1.82	1.86	1.93	2.00	2.05	2.10	2.14
1.40	1.28	1.29	1.31	1.35	1.40	1.46	1.51	1.57	1.62	1.67	1.71	1.79	1.86	1.92	1.97	2.01
1.60	1.15	1.15	1.18	1.22	1.27	1.32	1.38	1.44	1.49	1.54	1.58	1.66	1.73	1.80	1.85	1.90
1.80	1.03	1.04	1.06	1.11	1.16	1.21	1.27	1.32	1.37	1.42	1.47	1.55	1.63	1.69	1.75	1.80
2.00	0.940	0.946	0.971	1.01	1.06	1.11	1.17	1.22	1.28	1.32	1.37	1.45	1.53	1.60	1.66	1.71
2.20	0.861	0.867	0.892	0.932	0.979	1.03	1.09	1.14	1.19	1.24	1.28	1.37	1.44	1.51	1.57	1.63
2.40	0.794	0.800	0.825	0.863	0.909	0.959	1.01	1.06	1.11	1.16	1.21	1.29	1.36	1.43	1.49	1.55
2.60	0.736	0.743	0.767	0.804	0.848	0.896	0.946	0.997	1.05	1.09	1.14	1.22	1.29	1.36	1.42	1.48
2.80	0.686	0.693	0.716	0.752	0.794	0.841	0.889	0.937	0.985	1.03	1.07	1.16	1.23	1.30	1.36	1.42
3.00	0.643	0.649	0.672	0.706	0.746	0.792	0.838	0.885	0.931	0.975	1.02	1.10	1.17	1.24	1.30	1.36

Table 8-39 (cont.).
Coefficients C for Eccentrically Loaded Weld Groups
Angle = 75°

$$\phi R_n = CC_1Dl \qquad C_{min} = \frac{P_u}{C_1Dl} \qquad D_{min} = \frac{P_u}{CC_1l} \qquad l_{min} = \frac{P_u}{CC_1D}$$

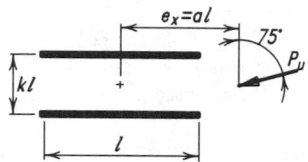

where

P_u = factored force, kips

D = number of sixteenths-of-an-inch in the fillet weld size

l = characteristic length of weld group, in.

a = e_x / l, in.

e_x = horizontal component of eccentricity of P_u with respect to centroid of weld group, in.

C = coefficient tabulated below which includes $\phi = 0.75$

C_1 = electode strength coefficient from Table 8-37 (1.0 for E70XX electrodes)

a	\multicolumn{17}{c}{k}
	0	0.1	0.2	0.3	0.4	0.5	0.6	0.7	0.8	0.9	1.0	1.2	1.4	1.6	1.8	2.0
0.00	2.97	2.97	2.97	2.97	2.97	2.97	2.97	2.97	2.97	2.97	2.97	2.97	2.97	2.97	2.97	2.97
0.10	2.86	2.86	2.87	2.87	2.88	2.88	2.88	2.89	2.89	2.89	2.89	2.88	2.87	2.85	2.84	2.55
0.15	2.89	2.89	2.89	2.89	2.89	2.89	2.89	2.89	2.88	2.88	2.87	2.86	2.85	2.83	2.82	2.81
0.20	2.88	2.88	2.88	2.88	2.88	2.87	2.87	2.86	2.86	2.85	2.85	2.84	2.82	2.81	2.80	2.78
0.25	2.87	2.87	2.87	2.87	2.86	2.86	2.85	2.85	2.84	2.84	2.83	2.81	2.80	2.79	2.78	2.76
0.30	2.86	2.86	2.86	2.85	2.85	2.84	2.84	2.83	2.82	2.82	2.81	2.80	2.78	2.77	2.76	2.74
0.40	2.84	2.83	2.83	2.82	2.82	2.81	2.80	2.79	2.79	2.78	2.77	2.76	2.74	2.73	2.72	2.71
0.50	2.79	2.79	2.78	2.77	2.77	2.76	2.76	2.75	2.74	2.73	2.73	2.71	2.70	2.69	2.68	2.67
0.60	2.74	2.73	2.73	2.72	2.72	2.71	2.70	2.70	2.69	2.68	2.68	2.67	2.66	2.65	2.64	2.63
0.70	2.66	2.66	2.66	2.66	2.65	2.65	2.64	2.64	2.64	2.63	2.63	2.62	2.61	2.61	2.60	2.59
0.80	2.59	2.59	2.59	2.58	2.58	2.58	2.58	2.58	2.58	2.58	2.58	2.57	2.57	2.56	2.56	2.56
0.90	2.51	2.51	2.51	2.51	2.51	2.51	2.51	2.51	2.52	2.52	2.52	2.52	2.52	2.52	2.52	2.52
1.00	2.42	2.42	2.43	2.43	2.43	2.44	2.44	2.45	2.45	2.46	2.46	2.47	2.48	2.48	2.48	2.48
1.20	2.25	2.25	2.25	2.27	2.28	2.30	2.31	2.32	2.33	2.34	2.35	2.37	2.38	2.39	2.40	2.40
1.40	2.08	2.08	2.09	2.11	2.13	2.15	2.18	2.19	2.21	2.23	2.24	2.27	2.29	2.31	2.32	2.33
1.60	1.92	1.93	1.94	1.96	1.99	2.02	2.05	2.08	2.10	2.12	2.14	2.17	2.20	2.22	2.24	2.26
1.80	1.78	1.78	1.80	1.83	1.86	1.89	1.93	1.96	1.99	2.02	2.04	2.08	2.12	2.15	2.17	2.19
2.00	1.65	1.66	1.67	1.70	1.74	1.78	1.82	1.86	1.89	1.92	1.95	2.00	2.04	2.07	2.10	2.12
2.20	1.54	1.54	1.56	1.60	1.64	1.68	1.72	1.76	1.80	1.83	1.86	1.92	1.96	2.00	2.03	2.06
2.40	1.44	1.44	1.46	1.50	1.54	1.59	1.63	1.68	1.71	1.75	1.78	1.84	1.89	1.93	1.97	2.00
2.60	1.35	1.35	1.37	1.41	1.45	1.50	1.55	1.60	1.64	1.67	1.71	1.77	1.82	1.87	1.91	1.94
2.80	1.26	1.27	1.29	1.33	1.37	1.42	1.47	1.52	1.56	1.60	1.64	1.71	1.76	1.81	1.85	1.88
3.00	1.19	1.20	1.22	1.26	1.30	1.35	1.40	1.45	1.50	1.54	1.58	1.64	1.70	1.75	1.79	1.83

Table 8-40.
Coefficients C for Eccentrically Loaded Weld Groups
Angle = 0°

$$\phi R_n = CC_1 Dl \qquad C_{min} = \frac{P_u}{C_1 Dl} \qquad D_{min} = \frac{P_u}{CC_1 l} \qquad l_{min} = \frac{P_u}{CC_1 D}$$

where

P_u = factored force, kips

D = number of sixteenths-of-an-inch in the fillet weld size

l = characteristic length of weld group, in.

a = e_x / l, in.

e_x = horizontal component of eccentricity of P_u with respect to centroid of weld group, in.

C = coefficient tabulated below which includes $\phi = 0.75$

C_1 = electrode strength coefficient from Table 8-37 (1.0 for E70XX electrodes)

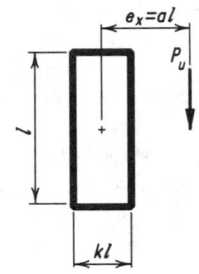

| a | \multicolumn{17}{c}{k} |
	0	0.1	0.2	0.3	0.4	0.5	0.6	0.7	0.8	0.9	1.0	1.2	1.4	1.6	1.8	2.0
0.00	2.78	3.20	3.62	4.04	4.45	4.87	5.29	5.71	6.12	6.54	6.96	7.80	8.63	9.47	10.3	11.1
0.10	2.78	3.07	3.42	3.78	4.15	4.53	4.91	5.30	5.69	6.08	6.47	7.25	8.03	8.82	9.61	10.4
0.15	2.75	3.05	3.37	3.71	4.06	4.42	4.78	5.15	5.52	5.89	6.27	7.02	7.78	8.54	9.31	10.1
0.20	2.64	2.95	3.25	3.57	3.91	4.25	4.59	4.94	5.30	5.66	6.02	6.75	7.49	8.23	8.98	9.74
0.25	2.48	2.79	3.10	3.40	3.72	4.04	4.38	4.71	5.06	5.40	5.75	6.46	7.18	7.91	8.65	9.39
0.30	2.32	2.61	2.92	3.22	3.52	3.83	4.15	4.47	4.80	5.14	5.48	6.17	6.88	7.59	8.32	9.05
0.40	2.00	2.26	2.54	2.83	3.12	3.41	3.71	4.01	4.32	4.64	4.96	5.62	6.30	6.99	7.70	8.42
0.50	1.72	1.95	2.20	2.47	2.75	3.03	3.31	3.59	3.89	4.19	4.50	5.13	5.78	6.46	7.14	7.84
0.60	1.50	1.70	1.92	2.17	2.44	2.70	2.97	3.24	3.52	3.80	4.09	4.70	5.33	5.98	6.64	7.32
0.70	1.32	1.50	1.70	1.93	2.17	2.43	2.68	2.93	3.20	3.47	3.75	4.33	4.93	5.56	6.20	6.86
0.80	1.17	1.33	1.52	1.73	1.95	2.20	2.43	2.67	2.92	3.18	3.45	4.00	4.58	5.18	5.80	6.44
0.90	1.05	1.20	1.37	1.56	1.77	2.00	2.23	2.45	2.69	2.93	3.18	3.71	4.26	4.84	5.44	6.05
1.00	0.957	1.09	1.24	1.42	1.62	1.83	2.05	2.26	2.49	2.72	2.96	3.45	3.98	4.53	5.10	5.69
1.20	0.806	0.916	1.05	1.21	1.38	1.57	1.76	1.95	2.15	2.36	2.57	3.02	3.50	4.00	4.53	5.08
1.40	0.695	0.790	0.908	1.05	1.20	1.37	1.54	1.72	1.89	2.08	2.27	2.68	3.12	3.58	4.06	4.57
1.60	0.611	0.694	0.800	0.923	1.06	1.21	1.37	1.53	1.69	1.86	2.03	2.40	2.80	3.23	3.67	4.15
1.80	0.544	0.619	0.714	0.825	0.950	1.09	1.23	1.37	1.52	1.68	1.84	2.18	2.54	2.93	3.35	3.79
2.00	0.491	0.558	0.645	0.746	0.860	0.984	1.12	1.25	1.38	1.53	1.67	1.99	2.33	2.69	3.08	3.49
2.20	0.447	0.509	0.588	0.680	0.785	0.898	1.02	1.14	1.27	1.40	1.54	1.83	2.14	2.48	2.84	3.22
2.40	0.410	0.467	0.540	0.625	0.721	0.827	0.939	1.05	1.17	1.29	1.42	1.69	1.99	2.30	2.64	3.00
2.60	0.379	0.431	0.499	0.578	0.667	0.765	0.869	0.977	1.09	1.20	1.32	1.57	1.85	2.15	2.46	2.80
2.80	0.352	0.401	0.464	0.538	0.621	0.712	0.809	0.911	1.01	1.12	1.23	1.47	1.73	2.01	2.31	2.63
3.00	0.329	0.375	0.434	0.503	0.580	0.666	0.757	0.853	0.949	1.05	1.16	1.38	1.62	1.89	2.17	2.47

Table 8-40 (cont.).
Coefficients C for Eccentrically Loaded Weld Groups
Angle = 15°

$$\phi R_n = CC_1 Dl \qquad C_{min} = \frac{P_u}{C_1 Dl} \qquad D_{min} = \frac{P_u}{CC_1 l} \qquad l_{min} = \frac{P_u}{CC_1 D}$$

where

P_u = factored force, kips

D = number of sixteenths-of-an-inch in the fillet weld size

l = characteristic length of weld group, in.

a = e_x / l, in.

e_x = horizontal component of eccentricity of P_u with respect to centroid of weld group, in.

C = coefficient tabulated below which includes $\phi = 0.75$

C_1 = electrode strength coefficient from Table 8-37 (1.0 for E70XX electrodes)

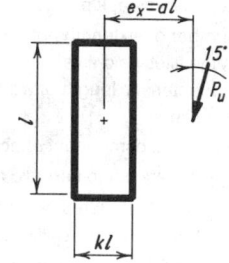

a								k								
	0	0.1	0.2	0.3	0.4	0.5	0.6	0.7	0.8	0.9	1.0	1.2	1.4	1.6	1.8	2.0
0.00	2.97	3.38	3.79	4.20	4.61	5.02	5.43	5.84	6.25	6.66	7.07	7.89	8.71	9.54	10.4	11.2
0.10	2.84	3.16	3.52	3.89	4.28	4.66	5.05	5.43	5.83	6.22	6.62	7.41	8.20	8.99	9.79	10.6
0.15	2.76	3.10	3.44	3.79	4.14	4.51	4.87	5.24	5.61	5.98	6.35	7.11	7.86	8.63	9.41	10.2
0.20	2.63	2.96	3.30	3.64	3.98	4.32	4.67	5.02	5.37	5.73	6.09	6.82	7.55	8.30	9.05	9.81
0.25	2.48	2.79	3.12	3.45	3.78	4.11	4.45	4.78	5.13	5.47	5.82	6.53	7.25	7.98	8.72	9.46
0.30	2.32	2.61	2.92	3.24	3.57	3.89	4.22	4.54	4.88	5.21	5.55	6.24	6.95	7.66	8.39	9.13
0.40	2.01	2.26	2.54	2.84	3.15	3.46	3.77	4.08	4.39	4.71	5.04	5.70	6.38	7.07	7.78	8.50
0.50	1.74	1.96	2.21	2.48	2.77	3.07	3.37	3.66	3.96	4.27	4.58	5.21	5.87	6.54	7.23	7.93
0.60	1.52	1.72	1.94	2.19	2.46	2.74	3.03	3.30	3.59	3.88	4.18	4.79	5.42	6.07	6.74	7.43
0.70	1.34	1.52	1.72	1.95	2.19	2.46	2.73	3.00	3.27	3.55	3.83	4.42	5.03	5.66	6.31	6.97
0.80	1.20	1.36	1.54	1.75	1.98	2.22	2.48	2.74	3.00	3.26	3.53	4.09	4.67	5.28	5.91	6.55
0.90	1.08	1.22	1.39	1.58	1.80	2.03	2.27	2.52	2.76	3.01	3.26	3.80	4.36	4.94	5.54	6.16
1.00	0.979	1.11	1.27	1.45	1.65	1.87	2.09	2.32	2.55	2.79	3.03	3.54	4.07	4.63	5.21	5.81
1.20	0.826	0.938	1.07	1.23	1.41	1.60	1.81	2.01	2.21	2.43	2.65	3.10	3.59	4.10	4.64	5.19
1.40	0.714	0.810	0.930	1.07	1.23	1.40	1.58	1.77	1.95	2.14	2.34	2.76	3.20	3.67	4.16	4.68
1.60	0.628	0.713	0.820	0.946	1.09	1.24	1.41	1.57	1.74	1.91	2.09	2.48	2.89	3.32	3.78	4.26
1.80	0.560	0.636	0.733	0.846	0.974	1.11	1.26	1.42	1.57	1.73	1.89	2.24	2.62	3.02	3.45	3.89
2.00	0.506	0.575	0.663	0.766	0.882	1.01	1.15	1.29	1.43	1.57	1.73	2.05	2.40	2.77	3.17	3.59
2.20	0.461	0.524	0.605	0.699	0.805	0.922	1.05	1.18	1.31	1.45	1.59	1.89	2.21	2.56	2.93	3.32
2.40	0.423	0.481	0.556	0.643	0.741	0.849	0.965	1.09	1.21	1.33	1.47	1.75	2.05	2.38	2.72	3.09
2.60	0.391	0.445	0.514	0.595	0.686	0.786	0.894	1.01	1.12	1.24	1.36	1.63	1.91	2.22	2.54	2.89
2.80	0.363	0.414	0.478	0.554	0.638	0.732	0.832	0.938	1.05	1.16	1.27	1.52	1.79	2.07	2.38	2.71
3.00	0.339	0.386	0.447	0.518	0.597	0.684	0.779	0.878	0.981	1.09	1.19	1.43	1.68	1.95	2.24	2.55

Table 8-40 (cont.).
Coefficients *C* for Eccentrically Loaded Weld Groups
Angle = 30°

$$\phi R_n = CC_1 Dl \qquad C_{min} = \frac{P_u}{C_1 Dl} \qquad D_{min} = \frac{P_u}{CC_1 l} \qquad l_{min} = \frac{P_u}{CC_1 D}$$

where

P_u = factored force, kips

D = number of sixteenths-of-an-inch in the fillet weld size

l = characteristic length of weld group, in.

a = e_x / l, in.

e_x = horizontal component of eccentricity of P_u with respect to centroid of weld group, in.

C = coefficient tabulated below which includes $\phi = 0.75$

C_1 = electrode strength coefficient from Table 8-37 (1.0 for E70XX electrodes)

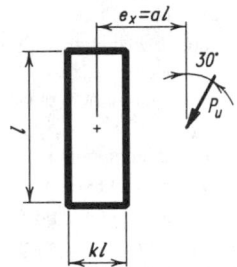

a	\multicolumn{17}{c}{k}
	0	0.1	0.2	0.3	0.4	0.5	0.6	0.7	0.8	0.9	1.0	1.2	1.4	1.6	1.8	2.0
0.00	3.28	3.67	4.06	4.45	4.84	5.23	5.62	6.01	6.40	6.79	7.18	7.96	8.74	9.53	10.3	11.1
0.10	3.03	3.45	3.85	4.24	4.62	5.00	5.38	5.76	6.14	6.52	6.90	7.67	8.44	9.22	10.0	10.8
0.15	2.87	3.25	3.63	4.02	4.40	4.77	5.15	5.52	5.89	6.26	6.64	7.39	8.15	8.93	9.70	10.5
0.20	2.72	3.07	3.43	3.80	4.17	4.53	4.89	5.25	5.61	5.97	6.34	7.09	7.84	8.60	9.37	10.2
0.25	2.57	2.88	3.22	3.57	3.93	4.29	4.65	4.99	5.34	5.69	6.05	6.77	7.52	8.27	9.04	9.81
0.30	2.41	2.70	3.02	3.35	3.70	4.05	4.40	4.75	5.09	5.43	5.77	6.48	7.21	7.95	8.71	9.48
0.40	2.11	2.36	2.64	2.95	3.27	3.60	3.94	4.28	4.61	4.94	5.27	5.95	6.65	7.36	8.09	8.85
0.50	1.84	2.07	2.32	2.60	2.90	3.21	3.53	3.86	4.19	4.50	4.83	5.48	6.16	6.85	7.56	8.28
0.60	1.62	1.83	2.06	2.31	2.59	2.88	3.18	3.50	3.82	4.12	4.43	5.07	5.73	6.40	7.09	7.79
0.70	1.44	1.63	1.84	2.07	2.33	2.60	2.89	3.19	3.50	3.79	4.09	4.70	5.33	5.99	6.66	7.35
0.80	1.30	1.46	1.65	1.87	2.11	2.37	2.64	2.93	3.22	3.50	3.79	4.38	4.98	5.61	6.26	6.93
0.90	1.17	1.32	1.50	1.70	1.93	2.18	2.43	2.70	2.98	3.25	3.52	4.08	4.67	5.28	5.90	6.55
1.00	1.07	1.21	1.37	1.56	1.78	2.01	2.25	2.51	2.77	3.02	3.28	3.82	4.38	4.97	5.57	6.20
1.20	0.907	1.03	1.17	1.34	1.53	1.73	1.95	2.18	2.41	2.65	2.88	3.37	3.89	4.44	5.00	5.59
1.40	0.786	0.890	1.02	1.17	1.34	1.52	1.72	1.93	2.14	2.34	2.56	3.01	3.49	3.99	4.52	5.07
1.60	0.693	0.785	0.901	1.04	1.19	1.36	1.53	1.72	1.91	2.10	2.30	2.72	3.16	3.62	4.12	4.63
1.80	0.619	0.703	0.808	0.931	1.07	1.22	1.38	1.55	1.73	1.90	2.08	2.47	2.88	3.31	3.77	4.25
2.00	0.559	0.635	0.731	0.844	0.970	1.11	1.26	1.41	1.57	1.74	1.91	2.26	2.64	3.05	3.48	3.93
2.20	0.510	0.579	0.668	0.771	0.887	1.01	1.15	1.30	1.45	1.60	1.75	2.08	2.44	2.82	3.22	3.64
2.40	0.469	0.533	0.614	0.710	0.818	0.935	1.06	1.20	1.34	1.48	1.62	1.93	2.27	2.62	3.00	3.40
2.60	0.433	0.493	0.569	0.658	0.758	0.867	0.984	1.11	1.24	1.37	1.51	1.80	2.11	2.45	2.80	3.18
2.80	0.403	0.458	0.529	0.613	0.706	0.808	0.918	1.03	1.16	1.28	1.41	1.68	1.98	2.29	2.63	2.99
3.00	0.376	0.428	0.495	0.573	0.661	0.757	0.860	0.969	1.08	1.20	1.32	1.58	1.86	2.16	2.48	2.82

Table 8-40 (cont.).
Coefficients C for Eccentrically Loaded Weld Groups
Angle = 45°

$$\phi R_n = CC_1 Dl \qquad C_{min} = \frac{P_u}{C_1 Dl} \qquad D_{min} = \frac{P_u}{CC_1 l} \qquad l_{min} = \frac{P_u}{CC_1 D}$$

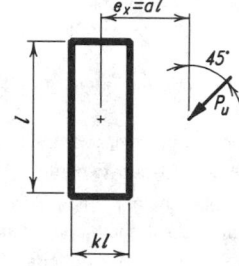

where

P_u = factored force, kips

D = number of sixteenths-of-an-inch
    in the fillet weld size

l = characteristic length of weld group, in.

a = e_x / l, in.

e_x = horizontal component of eccentricity of
    P_u with respect to centroid of weld group, in.

C = coefficient tabulated below which includes ϕ = 0.75

C_1 = electode strength coefficient from Table 8-37
    (1.0 for E70XX electrodes)

a	0	0.1	0.2	0.3	0.4	0.5	0.6	0.7	0.8	0.9	1.0	1.2	1.4	1.6	1.8	2.0
												k				
0.00	3.61	3.97	4.33	4.70	5.06	5.42	5.78	6.14	6.50	6.86	7.22	7.95	8.67	9.39	10.1	10.8
0.10	3.37	3.74	4.11	4.48	4.84	5.21	5.58	5.94	6.31	6.68	7.04	7.78	8.52	9.25	9.99	10.7
0.15	3.13	3.52	3.89	4.26	4.62	4.99	5.36	5.74	6.11	6.49	6.86	7.61	8.36	9.11	9.85	10.6
0.20	2.94	3.29	3.65	4.02	4.38	4.75	5.12	5.50	5.88	6.26	6.64	7.39	8.15	8.91	9.66	10.4
0.25	2.77	3.10	3.44	3.79	4.14	4.51	4.87	5.25	5.63	6.01	6.40	7.16	7.92	8.68	9.45	10.2
0.30	2.61	2.91	3.24	3.57	3.92	4.27	4.63	5.00	5.38	5.76	6.15	6.91	7.67	8.44	9.21	9.98
0.40	2.32	2.59	2.87	3.18	3.51	3.85	4.20	4.55	4.92	5.29	5.67	6.42	7.18	7.95	8.72	9.50
0.50	2.06	2.30	2.57	2.86	3.16	3.49	3.82	4.17	4.52	4.89	5.26	5.97	6.71	7.47	8.24	9.01
0.60	1.84	2.06	2.30	2.58	2.87	3.17	3.50	3.83	4.18	4.53	4.89	5.57	6.28	7.02	7.78	8.54
0.70	1.66	1.86	2.08	2.34	2.61	2.91	3.22	3.54	3.88	4.22	4.56	5.22	5.90	6.62	7.35	8.10
0.80	1.50	1.69	1.90	2.13	2.40	2.68	2.98	3.29	3.61	3.93	4.27	4.91	5.56	6.25	6.97	7.70
0.90	1.37	1.54	1.74	1.96	2.21	2.48	2.76	3.06	3.37	3.68	4.00	4.62	5.25	5.91	6.61	7.32
1.00	1.26	1.41	1.60	1.81	2.05	2.30	2.58	2.86	3.15	3.45	3.76	4.35	4.97	5.61	6.28	6.97
1.20	1.08	1.21	1.38	1.57	1.78	2.01	2.26	2.52	2.78	3.06	3.34	3.90	4.48	5.07	5.70	6.35
1.40	0.938	1.06	1.21	1.38	1.57	1.78	2.01	2.24	2.49	2.74	3.00	3.52	4.06	4.62	5.21	5.81
1.60	0.831	0.939	1.07	1.23	1.41	1.60	1.80	2.02	2.24	2.48	2.71	3.20	3.70	4.23	4.78	5.36
1.80	0.745	0.843	0.966	1.11	1.27	1.45	1.63	1.83	2.04	2.25	2.48	2.93	3.40	3.90	4.42	4.96
2.00	0.675	0.764	0.877	1.01	1.16	1.32	1.49	1.67	1.87	2.06	2.27	2.69	3.14	3.60	4.09	4.61
2.20	0.617	0.699	0.804	0.925	1.06	1.21	1.37	1.54	1.72	1.90	2.10	2.49	2.91	3.35	3.81	4.30
2.40	0.568	0.644	0.741	0.854	0.981	1.12	1.27	1.43	1.59	1.77	1.95	2.32	2.71	3.13	3.57	4.03
2.60	0.526	0.597	0.687	0.792	0.911	1.04	1.18	1.33	1.48	1.65	1.82	2.16	2.54	2.93	3.35	3.78
2.80	0.489	0.556	0.641	0.739	0.850	0.972	1.10	1.24	1.39	1.54	1.70	2.03	2.38	2.75	3.15	3.57
3.00	0.458	0.520	0.600	0.693	0.796	0.911	1.03	1.17	1.30	1.45	1.60	1.91	2.24	2.60	2.97	3.37

Table 8-40 (cont.).
Coefficients C for Eccentrically Loaded Weld Groups
Angle = 60°

$$\phi R_n = C C_1 D l \qquad C_{min} = \frac{P_u}{C_1 D l} \qquad D_{min} = \frac{P_u}{C C_1 l} \qquad l_{min} = \frac{P_u}{C C_1 D}$$

where

P_u = factored force, kips

D = number of sixteenths-of-an-inch in the fillet weld size

l = characteristic length of weld group, in.

a = e_x / l, in.

e_x = horizontal component of eccentricity of P_u with respect to centroid of weld group, in.

C = coefficient tabulated below which includes $\phi = 0.75$

C_1 = electode strength coefficient from Table 8-37 (1.0 for E70XX electrodes)

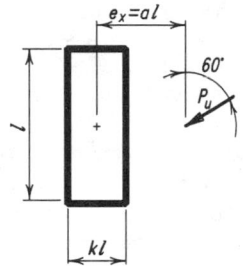

a	\multicolumn{17}{c}{k}
	0	0.1	0.2	0.3	0.4	0.5	0.6	0.7	0.8	0.9	1.0	1.2	1.4	1.6	1.8	2.0
0.00	3.91	4.23	4.56	4.89	5.22	5.54	5.87	6.20	6.53	6.85	7.18	7.84	8.49	9.15	9.80	10.5
0.10	3.65	3.97	4.30	4.64	4.99	5.35	5.70	6.05	6.40	6.74	7.08	7.75	8.42	9.08	9.74	10.4
0.15	3.46	3.78	4.11	4.45	4.80	5.17	5.53	5.90	6.26	6.61	6.97	7.66	8.34	9.02	9.69	10.4
0.20	3.27	3.60	3.92	4.26	4.61	4.97	5.35	5.72	6.09	6.46	6.82	7.54	8.24	8.93	9.61	10.3
0.25	3.10	3.42	3.74	4.07	4.41	4.78	5.15	5.53	5.92	6.29	6.67	7.40	8.12	8.83	9.52	10.2
0.30	2.95	3.25	3.57	3.89	4.23	4.59	4.97	5.35	5.74	6.12	6.50	7.25	7.98	8.70	9.41	10.1
0.40	2.68	2.96	3.26	3.57	3.90	4.24	4.61	4.99	5.38	5.77	6.16	6.93	7.69	8.43	9.16	9.87
0.50	2.44	2.70	2.98	3.29	3.61	3.95	4.31	4.68	5.06	5.45	5.84	6.61	7.37	8.12	8.87	9.60
0.60	2.24	2.47	2.74	3.04	3.35	3.69	4.04	4.40	4.77	5.15	5.54	6.31	7.08	7.83	8.58	9.31
0.70	2.05	2.28	2.53	2.81	3.12	3.45	3.79	4.14	4.50	4.87	5.25	6.02	6.79	7.55	8.30	9.04
0.80	1.89	2.10	2.34	2.62	2.92	3.23	3.56	3.91	4.26	4.62	4.99	5.75	6.51	7.27	8.02	8.76
0.90	1.75	1.95	2.18	2.44	2.73	3.04	3.36	3.69	4.03	4.38	4.74	5.49	6.24	6.99	7.74	8.48
1.00	1.62	1.81	2.03	2.29	2.56	2.86	3.17	3.49	3.82	4.17	4.52	5.24	5.98	6.72	7.46	8.20
1.20	1.42	1.59	1.79	2.02	2.28	2.55	2.84	3.15	3.46	3.78	4.11	4.80	5.50	6.21	6.93	7.66
1.40	1.25	1.41	1.59	1.81	2.04	2.30	2.57	2.85	3.14	3.45	3.76	4.41	5.07	5.75	6.44	7.16
1.60	1.12	1.26	1.43	1.63	1.85	2.09	2.34	2.60	2.88	3.16	3.46	4.08	4.69	5.34	6.01	6.70
1.80	1.01	1.14	1.30	1.48	1.69	1.91	2.14	2.39	2.65	2.92	3.20	3.78	4.36	4.97	5.61	6.27
2.00	0.922	1.04	1.19	1.36	1.55	1.75	1.97	2.20	2.45	2.70	2.97	3.52	4.08	4.65	5.26	5.90
2.20	0.847	0.956	1.09	1.25	1.43	1.62	1.83	2.04	2.27	2.51	2.76	3.28	3.82	4.37	4.94	5.55
2.40	0.782	0.884	1.01	1.16	1.33	1.51	1.70	1.91	2.12	2.35	2.58	3.08	3.59	4.11	4.66	5.24
2.60	0.726	0.821	0.943	1.08	1.24	1.41	1.59	1.78	1.99	2.20	2.42	2.89	3.38	3.88	4.40	4.96
2.80	0.677	0.767	0.881	1.01	1.16	1.32	1.49	1.67	1.87	2.07	2.28	2.73	3.20	3.68	4.18	4.70
3.00	0.635	0.719	0.827	0.952	1.09	1.24	1.40	1.58	1.76	1.95	2.15	2.58	3.03	3.49	3.97	4.47

Table 8-40 (cont.).
Coefficients C for Eccentrically Loaded Weld Groups
Angle = 75°

$$\phi R_n = CC_1 Dl \qquad C_{min} = \frac{P_u}{C_1 Dl} \qquad D_{min} = \frac{P_u}{CC_1 l} \qquad l_{min} = \frac{P_u}{CC_1 D}$$

where

P_u = factored force, kips

D = number of sixteenths-of-an-inch
in the fillet weld size

l = characteristic length of weld group, in.

a = e_x / l, in.

e_x = horizontal component of eccentricity of
P_u with respect to centroid of weld group, in.

C = coefficient tabulated below which includes $\phi = 0.75$

C_1 = electode strength coefficient from Table 8-37
(1.0 for E70XX electrodes)

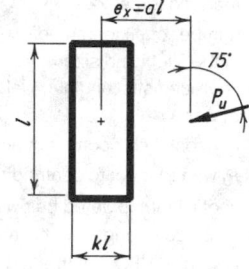

a	\multicolumn{17}{c}{k}
	0	**0.1**	**0.2**	**0.3**	**0.4**	**0.5**	**0.6**	**0.7**	**0.8**	**0.9**	**1.0**	**1.2**	**1.4**	**1.6**	**1.8**	**2.0**
0.00	4.11	4.40	4.70	5.00	5.29	5.59	5.89	6.18	6.48	6.78	7.07	7.67	8.26	8.85	9.45	10.0
0.10	3.88	4.17	4.49	4.81	5.14	5.45	5.76	6.07	6.36	6.66	6.96	7.53	8.11	8.69	9.27	9.84
0.15	3.76	4.04	4.36	4.69	5.03	5.36	5.69	6.00	6.31	6.61	6.91	7.50	8.09	8.67	9.25	9.83
0.20	3.64	3.92	4.23	4.57	4.92	5.26	5.60	5.92	6.24	6.55	6.86	7.46	8.05	8.64	9.23	9.81
0.25	3.53	3.80	4.11	4.45	4.81	5.16	5.50	5.84	6.16	6.48	6.80	7.41	8.01	8.60	9.19	9.78
0.30	3.43	3.70	4.00	4.34	4.70	5.05	5.40	5.75	6.08	6.41	6.72	7.35	7.96	8.56	9.16	9.75
0.40	3.24	3.51	3.80	4.14	4.49	4.86	5.22	5.57	5.91	6.25	6.57	7.22	7.85	8.46	9.07	9.67
0.50	3.07	3.34	3.63	3.95	4.31	4.67	5.04	5.40	5.75	6.10	6.43	7.08	7.71	8.34	8.96	9.57
0.60	2.91	3.17	3.46	3.78	4.13	4.49	4.86	5.23	5.60	5.95	6.30	6.96	7.60	8.23	8.84	9.46
0.70	2.77	3.02	3.31	3.63	3.97	4.33	4.70	5.07	5.44	5.80	6.16	6.84	7.50	8.13	8.75	9.36
0.80	2.63	2.87	3.16	3.48	3.81	4.17	4.53	4.91	5.28	5.65	6.01	6.72	7.39	8.04	8.66	9.27
0.90	2.50	2.74	3.02	3.34	3.67	4.02	4.38	4.75	5.13	5.50	5.87	6.58	7.27	7.93	8.57	9.20
1.00	2.38	2.62	2.89	3.20	3.53	3.87	4.23	4.60	4.98	5.35	5.72	6.45	7.15	7.83	8.48	9.11
1.20	2.17	2.39	2.66	2.96	3.28	3.61	3.96	4.32	4.69	5.06	5.44	6.18	6.90	7.60	8.28	8.93
1.40	1.99	2.20	2.45	2.74	3.05	3.38	3.71	4.06	4.42	4.79	5.16	5.91	6.64	7.36	8.06	8.74
1.60	1.83	2.03	2.27	2.54	2.84	3.16	3.49	3.82	4.17	4.53	4.90	5.64	6.39	7.12	7.83	8.53
1.80	1.69	1.88	2.11	2.37	2.66	2.96	3.28	3.61	3.95	4.29	4.65	5.39	6.13	6.87	7.59	8.30
2.00	1.57	1.75	1.97	2.22	2.49	2.79	3.10	3.41	3.74	4.08	4.43	5.15	5.88	6.62	7.34	8.07
2.20	1.46	1.63	1.84	2.08	2.34	2.63	2.92	3.23	3.55	3.88	4.22	4.92	5.64	6.37	7.10	7.83
2.40	1.37	1.53	1.73	1.96	2.21	2.48	2.77	3.07	3.37	3.69	4.02	4.71	5.42	6.14	6.87	7.59
2.60	1.28	1.44	1.63	1.85	2.09	2.35	2.62	2.91	3.21	3.52	3.84	4.51	5.20	5.91	6.63	7.36
2.80	1.21	1.36	1.54	1.75	1.98	2.23	2.50	2.78	3.07	3.36	3.67	4.32	5.00	5.70	6.41	7.13
3.00	1.14	1.28	1.46	1.66	1.88	2.12	2.38	2.65	2.92	3.22	3.52	4.14	4.80	5.49	6.19	6.91

Table 8-41.
Coefficients C for Eccentrically Loaded Weld Groups
Angle = 0°

$$\phi R_n = CC_1 Dl \qquad C_{min} = \frac{P_u}{C_1 Dl} \qquad D_{min} = \frac{P_u}{CC_1 l} \qquad l_{min} = \frac{P_u}{CC_1 D}$$

where

P_u = factored force, kips

D = number of sixteenths-of-an-inch
    in the fillet weld size

l = characteristic length of weld group, in.

a = e_x / l, in.

e_x = horizontal component of eccentricity of
    P_u with respect to centroid of weld group, in.

C = coefficient tabulated below which includes $\phi = 0.75$

C_1 = electode strength coefficient from Table 8-37
    (1.0 for E70XX electrodes)

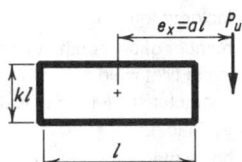

a	k															
	0	0.1	0.2	0.3	0.4	0.5	0.6	0.7	0.8	0.9	1.0	1.2	1.4	1.6	1.8	2.0
0.00	4.18	4.45	4.73	5.01	5.29	5.57	5.85	6.12	6.40	6.68	6.96	7.52	8.07	8.63	9.19	9.74
0.10	3.24	3.51	3.81	4.15	4.51	4.87	5.22	5.55	5.87	6.17	6.47	7.03	7.56	8.08	8.59	9.09
0.15	2.92	3.18	3.49	3.81	4.16	4.53	4.89	5.25	5.60	5.94	6.27	6.89	7.47	8.03	8.58	9.11
0.20	2.65	2.89	3.19	3.52	3.85	4.21	4.58	4.95	5.31	5.67	6.02	6.69	7.32	7.92	8.50	9.06
0.25	2.41	2.64	2.93	3.25	3.58	3.92	4.28	4.65	5.02	5.39	5.75	6.45	7.12	7.76	8.37	8.96
0.30	2.20	2.43	2.70	3.00	3.33	3.66	4.01	4.37	4.74	5.11	5.48	6.20	6.90	7.57	8.21	8.83
0.40	1.86	2.07	2.31	2.59	2.90	3.22	3.54	3.88	4.23	4.60	4.96	5.70	6.43	7.14	7.83	8.49
0.50	1.60	1.79	2.01	2.27	2.55	2.85	3.16	3.48	3.80	4.15	4.50	5.22	5.95	6.68	7.39	8.09
0.60	1.40	1.57	1.77	2.01	2.26	2.54	2.83	3.13	3.44	3.76	4.09	4.79	5.50	6.22	6.95	7.66
0.70	1.24	1.39	1.58	1.79	2.03	2.28	2.56	2.84	3.13	3.44	3.75	4.40	5.09	5.80	6.51	7.23
0.80	1.11	1.25	1.42	1.62	1.84	2.07	2.32	2.59	2.87	3.15	3.45	4.07	4.72	5.41	6.11	6.82
0.90	1.00	1.13	1.29	1.48	1.68	1.89	2.13	2.38	2.64	2.91	3.18	3.77	4.39	5.05	5.73	6.42
1.00	0.914	1.03	1.18	1.35	1.54	1.74	1.96	2.19	2.44	2.69	2.96	3.51	4.10	4.73	5.38	6.06
1.20	0.777	0.880	1.01	1.16	1.32	1.50	1.69	1.89	2.11	2.34	2.57	3.08	3.61	4.18	4.78	5.40
1.40	0.674	0.764	0.879	1.01	1.16	1.31	1.48	1.67	1.86	2.06	2.27	2.73	3.21	3.73	4.28	4.86
1.60	0.594	0.675	0.777	0.895	1.03	1.17	1.32	1.48	1.66	1.84	2.03	2.44	2.89	3.36	3.87	4.39
1.80	0.531	0.604	0.696	0.803	0.921	1.05	1.19	1.34	1.49	1.66	1.84	2.21	2.62	3.06	3.52	4.00
2.00	0.481	0.547	0.630	0.728	0.836	0.953	1.08	1.22	1.36	1.51	1.67	2.02	2.39	2.80	3.23	3.67
2.20	0.439	0.499	0.576	0.665	0.765	0.873	0.990	1.11	1.25	1.39	1.54	1.86	2.20	2.58	2.97	3.39
2.40	0.404	0.459	0.530	0.612	0.705	0.805	0.913	1.03	1.15	1.28	1.42	1.72	2.04	2.39	2.76	3.14
2.60	0.374	0.425	0.491	0.567	0.653	0.747	0.847	0.955	1.07	1.19	1.32	1.60	1.90	2.22	2.57	2.93
2.80	0.348	0.396	0.457	0.529	0.608	0.696	0.790	0.890	0.998	1.11	1.23	1.49	1.77	2.08	2.40	2.74
3.00	0.325	0.370	0.428	0.495	0.569	0.651	0.740	0.834	0.935	1.04	1.16	1.40	1.66	1.95	2.25	2.57

Table 8-41 (cont.).
Coefficients C for Eccentrically Loaded Weld Groups
Angle = 15°

$$\phi R_n = CC_1 Dl \qquad C_{min} = \frac{P_u}{C_1 Dl} \qquad D_{min} = \frac{P_u}{CC_1 l} \qquad l_{min} = \frac{P_u}{CC_1 D}$$

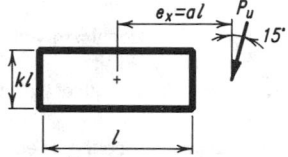

where

P_u = factored force, kips

D = number of sixteenths-of-an-inch in the fillet weld size

l = characteristic length of weld group, in.

a = e_x / l, in.

e_x = horizontal component of eccentricity of P_u with respect to centroid of weld group, in.

C = coefficient tabulated below which includes ϕ = 0.75

C_1 = electode strength coefficient from Table 8-37 (1.0 for E70XX electrodes)

								k								
a	0	0.1	0.2	0.3	0.4	0.5	0.6	0.7	0.8	0.9	1.0	1.2	1.4	1.6	1.8	2.0
0.00	4.11	4.40	4.70	5.00	5.29	5.59	5.89	6.18	6.48	6.78	7.07	7.67	8.26	8.85	9.45	10.0
0.10	3.29	3.56	3.85	4.19	4.55	4.91	5.26	5.61	5.95	6.29	6.62	7.26	7.88	8.49	9.09	9.69
0.15	2.97	3.24	3.53	3.85	4.20	4.56	4.93	5.30	5.66	6.01	6.35	7.01	7.65	8.27	8.89	9.51
0.20	2.70	2.95	3.24	3.56	3.89	4.25	4.62	4.99	5.37	5.73	6.09	6.78	7.45	8.08	8.71	9.31
0.25	2.46	2.70	2.98	3.29	3.63	3.97	4.33	4.70	5.08	5.45	5.82	6.54	7.23	7.90	8.54	9.17
0.30	2.25	2.48	2.75	3.05	3.38	3.71	4.06	4.43	4.80	5.17	5.55	6.29	7.01	7.70	8.36	9.01
0.40	1.91	2.12	2.36	2.65	2.95	3.27	3.61	3.95	4.30	4.67	5.04	5.78	6.52	7.25	7.95	8.64
0.50	1.65	1.83	2.06	2.32	2.60	2.90	3.22	3.54	3.88	4.22	4.58	5.31	6.04	6.78	7.51	8.23
0.60	1.44	1.61	1.82	2.06	2.32	2.60	2.89	3.20	3.52	3.84	4.18	4.88	5.60	6.33	7.06	7.78
0.70	1.28	1.43	1.62	1.84	2.08	2.34	2.62	2.91	3.21	3.51	3.83	4.49	5.19	5.90	6.62	7.34
0.80	1.15	1.29	1.46	1.67	1.89	2.13	2.38	2.65	2.93	3.23	3.53	4.15	4.82	5.50	6.21	6.92
0.90	1.04	1.17	1.33	1.52	1.72	1.95	2.19	2.44	2.70	2.98	3.26	3.86	4.49	5.15	5.83	6.52
1.00	0.945	1.07	1.22	1.39	1.59	1.79	2.02	2.25	2.50	2.76	3.03	3.60	4.19	4.83	5.48	6.15
1.20	0.803	0.908	1.04	1.19	1.36	1.54	1.74	1.95	2.17	2.40	2.65	3.16	3.70	4.27	4.86	5.48
1.40	0.697	0.789	0.907	1.04	1.19	1.35	1.53	1.72	1.91	2.12	2.34	2.80	3.30	3.82	4.35	4.92
1.60	0.615	0.698	0.803	0.924	1.06	1.21	1.36	1.53	1.71	1.90	2.09	2.51	2.97	3.45	3.93	4.45
1.80	0.550	0.625	0.719	0.829	0.951	1.08	1.23	1.38	1.54	1.71	1.89	2.28	2.69	3.13	3.58	4.06
2.00	0.497	0.565	0.651	0.752	0.863	0.984	1.12	1.26	1.40	1.56	1.73	2.08	2.47	2.87	3.29	3.73
2.20	0.454	0.516	0.595	0.687	0.790	0.902	1.02	1.15	1.29	1.43	1.59	1.91	2.27	2.65	3.04	3.45
2.40	0.418	0.475	0.548	0.633	0.728	0.832	0.943	1.06	1.19	1.32	1.47	1.77	2.10	2.45	2.82	3.20
2.60	0.387	0.440	0.508	0.586	0.675	0.772	0.875	0.987	1.11	1.23	1.36	1.65	1.96	2.28	2.63	2.99
2.80	0.360	0.409	0.473	0.546	0.629	0.719	0.817	0.921	1.03	1.15	1.27	1.54	1.83	2.14	2.46	2.80
3.00	0.336	0.383	0.442	0.511	0.589	0.674	0.765	0.863	0.967	1.08	1.19	1.44	1.72	2.01	2.31	2.63

Table 8-41 (cont.).
Coefficients C for Eccentrically Loaded Weld Groups
Angle = 30°

$$\phi R_n = CC_1 Dl \qquad C_{min} = \frac{P_u}{C_1 Dl} \qquad D_{min} = \frac{P_u}{CC_1 l} \qquad l_{min} = \frac{P_u}{CC_1 D}$$

where

P_u = factored force, kips

D = number of sixteenths-of-an-inch in the fillet weld size

l = characteristic length of weld group, in.

a = e_x / l, in.

e_x = horizontal component of eccentricity of P_u with respect to centroid of weld group, in.

C = coefficient tabulated below which includes $\phi = 0.75$

C_1 = electode strength coefficient from Table 8-37 (1.0 for E70XX electrodes)

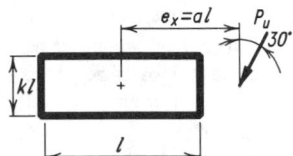

								k								
a	0	0.1	0.2	0.3	0.4	0.5	0.6	0.7	0.8	0.9	1.0	1.2	1.4	1.6	1.8	2.0
0.00	3.91	4.23	4.56	4.89	5.22	5.54	5.87	6.20	6.53	6.85	7.18	7.84	8.49	9.15	9.80	10.5
0.10	3.37	3.70	4.02	4.36	4.71	5.08	5.45	5.81	6.18	6.55	6.90	7.61	8.30	8.98	9.65	10.3
0.15	3.07	3.38	3.71	4.03	4.38	4.74	5.12	5.50	5.88	6.26	6.64	7.37	8.10	8.80	9.50	10.2
0.20	2.82	3.11	3.42	3.74	4.07	4.42	4.80	5.18	5.57	5.96	6.34	7.10	7.85	8.58	9.30	10.0
0.25	2.60	2.87	3.16	3.47	3.80	4.14	4.51	4.88	5.27	5.66	6.05	6.82	7.58	8.33	9.06	9.79
0.30	2.40	2.65	2.93	3.24	3.56	3.89	4.25	4.62	5.00	5.39	5.77	6.55	7.31	8.06	8.81	9.55
0.40	2.06	2.29	2.55	2.83	3.14	3.46	3.81	4.16	4.52	4.89	5.27	6.04	6.81	7.57	8.32	9.06
0.50	1.80	2.00	2.23	2.50	2.79	3.10	3.42	3.76	4.11	4.46	4.83	5.57	6.33	7.09	7.84	8.58
0.60	1.58	1.76	1.98	2.23	2.50	2.79	3.10	3.42	3.75	4.09	4.43	5.15	5.88	6.61	7.35	8.09
0.70	1.41	1.57	1.77	2.01	2.26	2.54	2.82	3.13	3.44	3.76	4.09	4.77	5.47	6.18	6.90	7.63
0.80	1.26	1.42	1.60	1.82	2.06	2.31	2.59	2.87	3.17	3.47	3.79	4.44	5.10	5.78	6.48	7.19
0.90	1.15	1.29	1.46	1.66	1.89	2.13	2.38	2.65	2.93	3.22	3.52	4.14	4.77	5.42	6.09	6.79
1.00	1.05	1.18	1.34	1.53	1.74	1.96	2.20	2.46	2.72	3.00	3.28	3.87	4.47	5.09	5.74	6.41
1.20	0.891	1.01	1.15	1.32	1.50	1.70	1.91	2.14	2.38	2.62	2.88	3.42	3.97	4.54	5.13	5.76
1.40	0.774	0.875	1.00	1.15	1.32	1.49	1.69	1.89	2.10	2.33	2.56	3.05	3.56	4.08	4.63	5.20
1.60	0.684	0.774	0.889	1.02	1.17	1.33	1.51	1.69	1.88	2.09	2.30	2.75	3.22	3.70	4.21	4.74
1.80	0.612	0.694	0.798	0.919	1.05	1.20	1.36	1.53	1.70	1.89	2.08	2.50	2.94	3.38	3.86	4.35
2.00	0.554	0.629	0.723	0.834	0.957	1.09	1.24	1.39	1.55	1.73	1.91	2.29	2.69	3.11	3.55	4.01
2.20	0.506	0.574	0.662	0.763	0.877	1.00	1.14	1.28	1.43	1.59	1.75	2.11	2.48	2.87	3.29	3.72
2.40	0.465	0.528	0.609	0.703	0.809	0.924	1.05	1.18	1.32	1.47	1.62	1.95	2.30	2.67	3.06	3.47
2.60	0.431	0.489	0.565	0.652	0.750	0.857	0.972	1.10	1.23	1.36	1.51	1.82	2.15	2.49	2.86	3.24
2.80	0.401	0.456	0.526	0.608	0.699	0.800	0.908	1.02	1.15	1.27	1.41	1.70	2.01	2.34	2.68	3.05
3.00	0.375	0.426	0.492	0.569	0.654	0.749	0.851	0.959	1.07	1.20	1.32	1.60	1.89	2.20	2.52	2.87

Table 8-41 (cont.).
Coefficients C for Eccentrically Loaded Weld Groups
Angle = 45°

$$\phi R_n = CC_1 Dl \qquad C_{min} = \frac{P_u}{C_1 Dl} \qquad D_{min} = \frac{P_u}{CC_1 l} \qquad l_{min} = \frac{P_u}{CC_1 D}$$

where

P_u = factored force, kips

D = number of sixteenths-of-an-inch
in the fillet weld size

l = characteristic length of weld group, in.

a = e_x / l, in.

e_x = horizontal component of eccentricity of
P_u with respect to centroid of weld group, in.

C = coefficient tabulated below which includes $\phi = 0.75$

C_1 = electode strength coefficient from Table 8-37
(1.0 for E70XX electrodes)

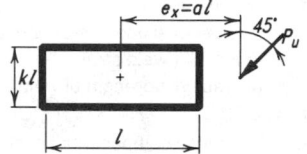

a	\multicolumn{17}{c}{k}
	0	0.1	0.2	0.3	0.4	0.5	0.6	0.7	0.8	0.9	1.0	1.2	1.4	1.6	1.8	2.0
0.00	3.61	3.97	4.33	4.70	5.06	5.42	5.78	6.14	6.50	6.86	7.22	7.95	8.67	9.39	10.1	10.8
0.10	3.37	3.74	4.11	4.48	4.84	5.21	5.58	5.94	6.31	6.68	7.04	7.78	8.52	9.25	9.98	10.7
0.15	3.13	3.52	3.89	4.26	4.62	4.99	5.36	5.74	6.11	6.49	6.86	7.61	8.36	9.11	9.85	10.6
0.20	2.94	3.29	3.65	4.02	4.38	4.75	5.12	5.50	5.88	6.26	6.64	7.39	8.15	8.91	9.67	10.4
0.25	2.77	3.10	3.44	3.79	4.14	4.51	4.87	5.25	5.63	6.01	6.40	7.16	7.92	8.68	9.45	10.2
0.30	2.61	2.91	3.24	3.57	3.92	4.27	4.63	5.00	5.38	5.76	6.15	6.91	7.67	8.44	9.21	9.98
0.40	2.32	2.59	2.87	3.18	3.51	3.85	4.20	4.55	4.92	5.29	5.67	6.42	7.18	7.95	8.72	9.50
0.50	2.06	2.30	2.57	2.86	3.16	3.49	3.82	4.17	4.52	4.89	5.26	5.97	6.71	7.47	8.24	9.01
0.60	1.84	2.06	2.30	2.58	2.87	3.17	3.50	3.83	4.18	4.53	4.89	5.57	6.28	7.02	7.78	8.54
0.70	1.66	1.86	2.08	2.34	2.61	2.91	3.22	3.54	3.88	4.22	4.56	5.22	5.90	6.62	7.35	8.10
0.80	1.50	1.69	1.90	2.13	2.40	2.68	2.98	3.29	3.61	3.93	4.27	4.91	5.56	6.25	6.97	7.70
0.90	1.37	1.54	1.74	1.96	2.21	2.48	2.76	3.06	3.37	3.68	4.00	4.62	5.25	5.91	6.61	7.32
1.00	1.26	1.41	1.60	1.81	2.05	2.30	2.58	2.86	3.15	3.45	3.76	4.35	4.97	5.61	6.28	6.97
1.20	1.08	1.21	1.38	1.57	1.78	2.01	2.26	2.52	2.78	3.06	3.34	3.90	4.48	5.07	5.70	6.35
1.40	0.938	1.06	1.21	1.38	1.57	1.78	2.01	2.24	2.49	2.74	3.00	3.52	4.06	4.62	5.21	5.81
1.60	0.831	0.939	1.07	1.23	1.41	1.60	1.80	2.02	2.24	2.48	2.71	3.20	3.70	4.23	4.78	5.36
1.80	0.745	0.843	0.966	1.11	1.27	1.45	1.63	1.83	2.04	2.25	2.48	2.93	3.40	3.90	4.42	4.96
2.00	0.675	0.764	0.877	1.01	1.16	1.32	1.49	1.67	1.87	2.06	2.27	2.69	3.14	3.60	4.09	4.61
2.20	0.617	0.699	0.804	0.925	1.06	1.21	1.37	1.54	1.72	1.90	2.10	2.49	2.91	3.35	3.81	4.30
2.40	0.568	0.644	0.741	0.854	0.981	1.12	1.27	1.43	1.59	1.77	1.95	2.32	2.71	3.13	3.57	4.03
2.60	0.526	0.597	0.687	0.792	0.911	1.04	1.18	1.33	1.48	1.65	1.82	2.16	2.54	2.93	3.35	3.78
2.80	0.489	0.556	0.641	0.739	0.850	0.972	1.10	1.24	1.39	1.54	1.70	2.03	2.38	2.75	3.15	3.57
3.00	0.458	0.520	0.600	0.693	0.796	0.911	1.03	1.17	1.30	1.45	1.60	1.91	2.24	2.60	2.97	3.37

Table 8-41 (cont.).
Coefficients *C* for Eccentrically Loaded Weld Groups
Angle = 60°

$$\phi R_n = CC_1 Dl \qquad C_{min} = \frac{P_u}{C_1 Dl} \qquad D_{min} = \frac{P_u}{CC_1 l} \qquad l_{min} = \frac{P_u}{CC_1 D}$$

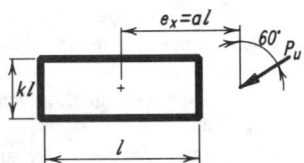

where

P_u = factored force, kips

D = number of sixteenths-of-an-inch
in the fillet weld size

l = characteristic length of weld group, in.

a = e_x / l, in.

e_x = horizontal component of eccentricity of
P_u with respect to centroid of weld group, in.

C = coefficient tabulated below which includes ϕ = 0.75

C_1 = electode strength coefficient from Table 8-37
(1.0 for E70XX electrodes)

								k								
a	0	0.1	0.2	0.3	0.4	0.5	0.6	0.7	0.8	0.9	1.0	1.2	1.4	1.6	1.8	2.0
0.00	3.28	3.67	4.06	4.45	4.84	5.23	5.62	6.01	6.40	6.79	7.18	7.96	8.74	9.53	10.3	11.1
0.10	3.20	3.59	3.98	4.37	4.76	5.14	5.53	5.92	6.30	6.69	7.08	7.86	8.63	9.41	10.2	11.0
0.15	3.10	3.50	3.90	4.29	4.67	5.05	5.43	5.82	6.20	6.58	6.96	7.74	8.51	9.29	10.1	10.9
0.20	2.98	3.38	3.79	4.17	4.56	4.94	5.31	5.69	6.07	6.45	6.83	7.59	8.36	9.13	9.91	10.7
0.25	2.89	3.27	3.66	4.04	4.43	4.80	5.17	5.55	5.92	6.29	6.67	7.43	8.19	8.96	9.74	10.5
0.30	2.80	3.16	3.54	3.92	4.29	4.66	5.02	5.39	5.76	6.13	6.50	7.25	8.01	8.77	9.55	10.3
0.40	2.63	2.96	3.30	3.66	4.02	4.38	4.74	5.09	5.44	5.80	6.16	6.89	7.64	8.40	9.17	9.94
0.50	2.44	2.74	3.06	3.40	3.75	4.10	4.46	4.80	5.14	5.49	5.84	6.54	7.28	8.02	8.78	9.56
0.60	2.27	2.54	2.84	3.16	3.49	3.83	4.18	4.53	4.86	5.20	5.54	6.23	6.93	7.66	8.42	9.18
0.70	2.09	2.35	2.63	2.93	3.25	3.58	3.92	4.26	4.59	4.92	5.25	5.93	6.63	7.34	8.07	8.82
0.80	1.94	2.18	2.44	2.73	3.03	3.35	3.68	4.01	4.34	4.66	4.99	5.65	6.34	7.04	7.75	8.49
0.90	1.80	2.02	2.27	2.54	2.83	3.14	3.46	3.78	4.11	4.43	4.74	5.40	6.07	6.76	7.46	8.18
1.00	1.67	1.88	2.11	2.37	2.65	2.95	3.26	3.58	3.90	4.21	4.52	5.16	5.82	6.50	7.19	7.90
1.20	1.46	1.64	1.85	2.09	2.35	2.62	2.91	3.21	3.52	3.82	4.11	4.73	5.36	6.02	6.69	7.38
1.40	1.28	1.45	1.64	1.86	2.10	2.35	2.63	2.91	3.20	3.48	3.76	4.35	4.96	5.58	6.23	6.89
1.60	1.15	1.30	1.47	1.67	1.89	2.13	2.39	2.66	2.93	3.19	3.46	4.01	4.60	5.20	5.82	6.47
1.80	1.03	1.17	1.33	1.51	1.72	1.95	2.19	2.44	2.69	2.94	3.20	3.72	4.28	4.85	5.45	6.07
2.00	0.940	1.06	1.21	1.39	1.58	1.79	2.01	2.25	2.49	2.72	2.97	3.47	3.99	4.55	5.12	5.72
2.20	0.861	0.974	1.11	1.27	1.46	1.65	1.86	2.09	2.31	2.53	2.76	3.24	3.74	4.27	4.82	5.40
2.40	0.794	0.899	1.03	1.18	1.35	1.54	1.73	1.94	2.15	2.36	2.58	3.03	3.52	4.02	4.55	5.11
2.60	0.736	0.834	0.956	1.10	1.26	1.43	1.62	1.82	2.02	2.22	2.42	2.86	3.32	3.80	4.31	4.84
2.80	0.686	0.778	0.893	1.03	1.18	1.34	1.52	1.70	1.89	2.08	2.28	2.69	3.13	3.60	4.08	4.59
3.00	0.643	0.729	0.837	0.964	1.11	1.26	1.43	1.61	1.79	1.97	2.15	2.55	2.97	3.41	3.88	4.37

Table 8-41 (cont.).
Coefficients C for Eccentrically Loaded Weld Groups
Angle = 75°

$$\phi R_n = CC_1 Dl \qquad C_{min} = \frac{P_u}{C_1 Dl} \qquad D_{min} = \frac{P_u}{CC_1 l} \qquad l_{min} = \frac{P_u}{CC_1 D}$$

where

P_u = factored force, kips

D = number of sixteenths-of-an-inch in the fillet weld size

l = characteristic length of weld group, in.

a = e_x / l, in.

e_x = horizontal component of eccentricity of P_u with respect to centroid of weld group, in.

C = coefficient tabulated below which includes ϕ = 0.75

C_1 = electode strength coefficient from Table 8-37 (1.0 for E70XX electrodes)

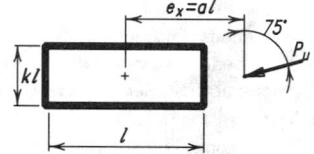

a	\multicolumn{16}{c}{k}															
	0	0.1	0.2	0.3	0.4	0.5	0.6	0.7	0.8	0.9	1.0	1.2	1.4	1.6	1.8	2.0
0.00	2.97	3.38	3.79	4.20	4.61	5.02	5.43	5.84	6.25	6.66	7.07	7.89	8.71	9.54	10.4	11.2
0.10	2.86	3.27	3.67	4.08	4.49	4.90	5.31	5.72	6.14	6.55	6.96	7.78	8.60	9.42	10.2	11.1
0.15	2.89	3.24	3.65	4.06	4.47	4.87	5.28	5.69	6.10	6.50	6.91	7.72	8.54	9.36	10.2	11.0
0.20	2.88	3.19	3.61	4.02	4.43	4.83	5.24	5.64	6.05	6.45	6.86	7.67	8.48	9.29	10.1	10.9
0.25	2.87	3.17	3.57	3.98	4.38	4.79	5.19	5.59	5.99	6.39	6.80	7.60	8.40	9.21	10.0	10.8
0.30	2.86	3.17	3.54	3.93	4.33	4.73	5.13	5.53	5.93	6.33	6.73	7.52	8.32	9.12	9.92	10.7
0.40	2.84	3.16	3.51	3.88	4.26	4.64	5.02	5.41	5.79	6.18	6.57	7.36	8.15	8.94	9.73	10.5
0.50	2.79	3.13	3.47	3.83	4.19	4.56	4.93	5.30	5.68	6.05	6.43	7.20	7.96	8.75	9.53	10.3
0.60	2.74	3.07	3.42	3.76	4.11	4.47	4.83	5.19	5.56	5.93	6.30	7.04	7.80	8.56	9.33	10.1
0.70	2.66	3.00	3.34	3.68	4.02	4.37	4.72	5.07	5.43	5.79	6.16	6.89	7.63	8.38	9.14	9.90
0.80	2.59	2.91	3.25	3.59	3.93	4.26	4.61	4.95	5.30	5.66	6.01	6.74	7.47	8.21	8.96	9.71
0.90	2.51	2.82	3.15	3.49	3.82	4.15	4.49	4.83	5.17	5.52	5.87	6.58	7.30	8.04	8.78	9.52
1.00	2.42	2.73	3.05	3.38	3.71	4.03	4.37	4.70	5.03	5.38	5.72	6.42	7.14	7.86	8.60	9.34
1.20	2.25	2.53	2.84	3.15	3.48	3.80	4.12	4.44	4.77	5.10	5.44	6.12	6.82	7.53	8.26	8.99
1.40	2.08	2.35	2.63	2.94	3.25	3.57	3.88	4.19	4.51	4.83	5.16	5.83	6.52	7.22	7.93	8.65
1.60	1.92	2.17	2.44	2.73	3.04	3.35	3.65	3.95	4.26	4.58	4.90	5.55	6.23	6.92	7.62	8.33
1.80	1.78	2.01	2.26	2.54	2.84	3.14	3.43	3.73	4.04	4.34	4.65	5.30	5.96	6.63	7.33	8.03
2.00	1.65	1.87	2.11	2.37	2.65	2.95	3.24	3.53	3.82	4.12	4.43	5.05	5.70	6.37	7.05	7.75
2.20	1.54	1.74	1.97	2.22	2.49	2.77	3.06	3.34	3.63	3.92	4.22	4.83	5.47	6.12	6.79	7.48
2.40	1.44	1.63	1.84	2.08	2.34	2.61	2.89	3.17	3.44	3.73	4.02	4.62	5.24	5.89	6.55	7.23
2.60	1.35	1.52	1.73	1.95	2.20	2.47	2.74	3.01	3.28	3.56	3.84	4.43	5.04	5.67	6.32	6.98
2.80	1.26	1.43	1.63	1.84	2.08	2.34	2.60	2.86	3.13	3.40	3.67	4.25	4.84	5.46	6.10	6.75
3.00	1.19	1.35	1.53	1.74	1.97	2.22	2.48	2.73	2.99	3.25	3.52	4.07	4.66	5.26	5.89	6.53

Table 8-42.
Coefficients C for Eccentrically Loaded Weld Groups
Angle = 0°

$$\phi R_n = CC_1 Dl \qquad C_{min} = \frac{P_u}{C_1 Dl} \qquad D_{min} = \frac{P_u}{CC_1 l} \qquad l_{min} = \frac{P_u}{CC_1 D}$$

where

P_u = factored force, kips

D = number of sixteenths-of-an-inch in the fillet weld size

l = characteristic length of weld group, in.

a = e_x / l, in.

e_x = horizontal component of eccentricity of P_u with respect to centroid of weld group, in.

C = coefficient tabulated below which includes $\phi = 0.75$

C_1 = electode strength coefficient from Table 8-37 (1.0 for E70XX electrodes)

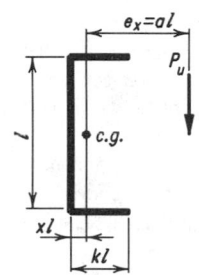

| | | | | | | | | | k | | | | | | | | |
a	0	0.1	0.2	0.3	0.4	0.5	0.6	0.7	0.8	0.9	1.0	1.2	1.4	1.6	1.8	2.0
0.00	1.39	1.81	2.28	2.65	3.06	3.48	3.90	4.32	4.73	5.15	5.57	6.40	7.24	8.07	8.91	9.74
0.10	1.39	1.71	2.09	2.48	2.88	3.28	3.69	4.10	4.51	4.92	5.33	6.16	6.99	7.82	8.65	9.48
0.15	1.37	1.69	2.05	2.43	2.81	3.20	3.60	4.00	4.40	4.80	5.21	6.02	6.84	7.65	8.47	9.28
0.20	1.32	1.63	1.98	2.33	2.70	3.08	3.46	3.84	4.23	4.62	5.01	5.80	6.58	7.38	8.17	8.97
0.25	1.24	1.56	1.88	2.22	2.57	2.93	3.29	3.65	4.03	4.40	4.77	5.53	6.30	7.07	7.84	8.62
0.30	1.16	1.46	1.77	2.09	2.42	2.76	3.10	3.45	3.81	4.16	4.53	5.26	6.00	6.75	7.51	8.27
0.40	0.998	1.27	1.55	1.84	2.13	2.43	2.74	3.06	3.38	3.71	4.04	4.73	5.43	6.14	6.87	7.61
0.50	0.860	1.09	1.35	1.61	1.87	2.14	2.41	2.70	3.00	3.30	3.61	4.25	4.92	5.60	6.30	7.01
0.60	0.748	0.952	1.17	1.41	1.65	1.89	2.14	2.40	2.67	2.95	3.24	3.84	4.47	5.13	5.80	6.49
0.70	0.659	0.838	1.04	1.25	1.46	1.68	1.91	2.15	2.40	2.66	2.93	3.50	4.09	4.72	5.36	6.03
0.80	0.586	0.746	0.922	1.11	1.31	1.51	1.72	1.94	2.17	2.42	2.67	3.20	3.77	4.36	4.98	5.62
0.90	0.527	0.671	0.829	1.00	1.18	1.37	1.56	1.77	1.98	2.21	2.44	2.95	3.48	4.05	4.64	5.25
1.00	0.478	0.609	0.752	0.909	1.08	1.25	1.43	1.62	1.82	2.03	2.25	2.73	3.23	3.77	4.33	4.92
1.20	0.403	0.512	0.633	0.766	0.910	1.06	1.22	1.39	1.56	1.75	1.94	2.36	2.81	3.29	3.80	4.34
1.40	0.348	0.441	0.546	0.661	0.787	0.922	1.06	1.21	1.36	1.53	1.70	2.08	2.48	2.92	3.38	3.86
1.60	0.305	0.387	0.479	0.581	0.692	0.813	0.938	1.07	1.21	1.36	1.51	1.85	2.21	2.61	3.03	3.48
1.80	0.272	0.345	0.427	0.518	0.618	0.727	0.840	0.958	1.09	1.22	1.36	1.66	1.99	2.35	2.74	3.15
2.00	0.245	0.311	0.385	0.467	0.558	0.657	0.760	0.868	0.983	1.10	1.23	1.51	1.81	2.14	2.50	2.88
2.20	0.223	0.283	0.350	0.425	0.508	0.599	0.694	0.793	0.897	1.01	1.13	1.38	1.66	1.97	2.30	2.65
2.40	0.205	0.260	0.321	0.390	0.467	0.551	0.639	0.729	0.826	0.929	1.04	1.27	1.53	1.82	2.12	2.46
2.60	0.189	0.240	0.297	0.360	0.431	0.509	0.591	0.675	0.765	0.860	0.961	1.18	1.42	1.69	1.98	2.29
2.80	0.176	0.223	0.276	0.335	0.401	0.474	0.550	0.628	0.712	0.801	0.895	1.10	1.33	1.57	1.85	2.14
3.00	0.164	0.208	0.257	0.313	0.375	0.443	0.514	0.588	0.666	0.749	0.838	1.03	1.24	1.48	1.73	2.01
x	0.000	0.008	0.029	0.056	0.089	0.125	0.164	0.204	0.246	0.289	0.333	0.424	0.516	0.610	0.704	0.800

Table 8-42 (cont.).
Coefficients C for Eccentrically Loaded Weld Groups
Angle = 15°

$$\phi R_n = CC_1 Dl \qquad C_{min} = \frac{P_u}{C_1 Dl} \qquad D_{min} = \frac{P_u}{CC_1 l} \qquad l_{min} = \frac{P_u}{CC_1 D}$$

where

P_u = factored force, kips

D = number of sixteenths-of-an-inch in the fillet weld size

l = characteristic length of weld group, in.

a = e_x / l, in.

e_x = horizontal component of eccentricity of P_u with respect to centroid of weld group, in.

C = coefficient tabulated below which includes $\phi = 0.75$

C_1 = electode strength coefficient from Table 8-37 (1.0 for E70XX electrodes)

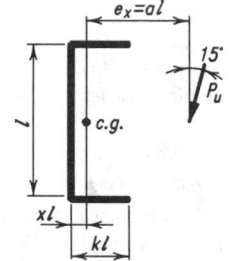

a	\multicolumn{17}{c}{k}																
	0	0.1	0.2	0.3	0.4	0.5	0.6	0.7	0.8	0.9	1.0	1.2	1.4	1.6	1.8	2.0
0.00	1.48	1.89	2.31	2.72	3.13	3.54	3.95	4.36	4.77	5.18	5.59	6.41	7.23	8.05	8.87	9.69
0.10	1.42	1.77	2.15	2.55	2.96	3.38	3.79	4.20	4.61	5.02	5.43	6.25	7.08	7.90	8.73	9.55
0.15	1.38	1.73	2.09	2.47	2.86	3.25	3.64	4.04	4.44	4.84	5.24	6.05	6.85	7.66	8.46	9.28
0.20	1.32	1.66	2.01	2.37	2.74	3.11	3.49	3.87	4.25	4.64	5.02	5.80	6.57	7.36	8.15	8.94
0.25	1.24	1.56	1.91	2.25	2.60	2.96	3.32	3.68	4.05	4.41	4.79	5.53	6.29	7.06	7.82	8.60
0.30	1.16	1.46	1.79	2.12	2.45	2.79	3.13	3.48	3.83	4.19	4.54	5.27	6.01	6.75	7.51	8.27
0.40	1.00	1.27	1.56	1.85	2.16	2.46	2.77	3.09	3.41	3.74	4.07	4.76	5.46	6.17	6.90	7.63
0.50	0.869	1.10	1.35	1.62	1.89	2.17	2.45	2.74	3.04	3.34	3.65	4.30	4.96	5.64	6.34	7.06
0.60	0.759	0.961	1.18	1.42	1.67	1.92	2.18	2.44	2.72	3.00	3.29	3.90	4.53	5.18	5.86	6.55
0.70	0.670	0.849	1.05	1.26	1.48	1.72	1.95	2.20	2.45	2.71	2.98	3.55	4.15	4.78	5.43	6.10
0.80	0.598	0.758	0.934	1.12	1.33	1.54	1.76	1.99	2.22	2.47	2.72	3.26	3.83	4.43	5.06	5.70
0.90	0.539	0.683	0.842	1.02	1.20	1.40	1.60	1.81	2.03	2.26	2.50	3.01	3.55	4.12	4.71	5.33
1.00	0.490	0.621	0.766	0.924	1.10	1.28	1.47	1.66	1.87	2.08	2.31	2.79	3.30	3.84	4.41	5.00
1.20	0.413	0.524	0.646	0.781	0.928	1.09	1.25	1.42	1.61	1.80	1.99	2.42	2.88	3.36	3.88	4.42
1.40	0.357	0.452	0.558	0.675	0.804	0.943	1.09	1.24	1.40	1.57	1.75	2.13	2.54	2.99	3.45	3.95
1.60	0.314	0.398	0.491	0.595	0.708	0.833	0.967	1.10	1.25	1.40	1.56	1.90	2.27	2.67	3.11	3.56
1.80	0.280	0.355	0.438	0.531	0.633	0.746	0.867	0.988	1.12	1.26	1.40	1.71	2.05	2.42	2.81	3.24
2.00	0.253	0.320	0.395	0.479	0.572	0.675	0.784	0.896	1.01	1.14	1.27	1.55	1.87	2.20	2.57	2.96
2.20	0.230	0.291	0.360	0.437	0.522	0.616	0.717	0.819	0.926	1.04	1.16	1.43	1.71	2.03	2.37	2.73
2.40	0.211	0.267	0.330	0.401	0.479	0.566	0.659	0.753	0.853	0.959	1.07	1.31	1.58	1.87	2.19	2.53
2.60	0.195	0.247	0.305	0.370	0.444	0.524	0.610	0.697	0.790	0.889	0.993	1.22	1.47	1.74	2.04	2.36
2.80	0.182	0.230	0.284	0.344	0.412	0.487	0.568	0.649	0.736	0.827	0.925	1.14	1.37	1.62	1.90	2.20
3.00	0.170	0.214	0.265	0.321	0.386	0.456	0.531	0.607	0.688	0.774	0.866	1.06	1.28	1.52	1.79	2.07
x	0.000	0.008	0.029	0.056	0.089	0.125	0.164	0.204	0.246	0.289	0.333	0.424	0.516	0.610	0.704	0.800

Table 8-42 (cont.).
Coefficients C for Eccentrically Loaded Weld Groups
Angle = 30°

$$\phi R_n = CC_1 Dl \qquad C_{min} = \frac{P_u}{C_1 Dl} \qquad D_{min} = \frac{P_u}{CC_1 l} \qquad l_{min} = \frac{P_u}{CC_1 D}$$

where

P_u = factored force, kips

D = number of sixteenths-of-an-inch in the fillet weld size

l = characteristic length of weld group, in.

a = e_x / l, in.

e_x = horizontal component of eccentricity of P_u with respect to centroid of weld group, in.

C = coefficient tabulated below which includes $\phi = 0.75$

C_1 = electode strength coefficient from Table 8-37 (1.0 for E70XX electrodes)

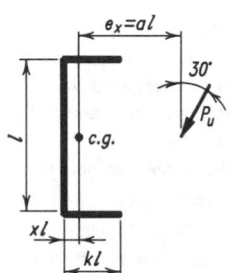

								k								
a	0	0.1	0.2	0.3	0.4	0.5	0.6	0.7	0.8	0.9	1.0	1.2	1.4	1.6	1.8	2.0
0.00	1.64	2.03	2.42	2.81	3.20	3.59	3.98	4.37	4.76	5.15	5.54	6.33	7.11	7.89	8.67	9.45
0.10	1.52	1.93	2.32	2.72	3.11	3.50	3.89	4.28	4.68	5.07	5.46	6.24	7.03	7.81	8.60	9.39
0.15	1.44	1.82	2.21	2.60	2.99	3.37	3.75	4.14	4.52	4.91	5.29	6.07	6.84	7.62	8.41	9.19
0.20	1.36	1.72	2.09	2.47	2.84	3.21	3.58	3.95	4.33	4.70	5.08	5.84	6.60	7.38	8.15	8.93
0.25	1.28	1.61	1.97	2.33	2.69	3.04	3.40	3.76	4.11	4.48	4.84	5.59	6.34	7.10	7.87	8.65
0.30	1.20	1.51	1.84	2.18	2.53	2.88	3.22	3.56	3.91	4.26	4.62	5.33	6.07	6.82	7.58	8.35
0.40	1.05	1.32	1.61	1.91	2.23	2.55	2.87	3.19	3.52	3.85	4.18	4.86	5.57	6.28	7.02	7.77
0.50	0.921	1.16	1.41	1.68	1.96	2.26	2.56	2.86	3.16	3.47	3.79	4.44	5.11	5.80	6.51	7.24
0.60	0.812	1.02	1.25	1.49	1.74	2.01	2.29	2.57	2.86	3.15	3.45	4.07	4.71	5.38	6.06	6.77
0.70	0.722	0.908	1.11	1.33	1.56	1.81	2.07	2.33	2.60	2.87	3.15	3.74	4.36	5.00	5.67	6.35
0.80	0.647	0.816	0.998	1.20	1.41	1.64	1.88	2.12	2.37	2.63	2.90	3.46	4.05	4.66	5.30	5.96
0.90	0.586	0.739	0.905	1.09	1.28	1.49	1.71	1.95	2.18	2.42	2.68	3.21	3.77	4.36	4.97	5.60
1.00	0.535	0.674	0.827	0.994	1.18	1.37	1.58	1.80	2.02	2.25	2.48	2.99	3.52	4.08	4.67	5.28
1.20	0.454	0.572	0.703	0.847	1.00	1.17	1.36	1.55	1.75	1.95	2.16	2.61	3.10	3.61	4.15	4.71
1.40	0.393	0.496	0.610	0.736	0.874	1.02	1.19	1.36	1.53	1.72	1.91	2.32	2.75	3.22	3.72	4.24
1.60	0.347	0.437	0.538	0.650	0.773	0.909	1.06	1.21	1.37	1.53	1.70	2.07	2.47	2.90	3.36	3.85
1.80	0.310	0.391	0.481	0.582	0.693	0.816	0.949	1.09	1.23	1.38	1.54	1.87	2.24	2.64	3.06	3.52
2.00	0.280	0.353	0.435	0.526	0.627	0.740	0.861	0.988	1.12	1.26	1.40	1.71	2.05	2.41	2.81	3.23
2.20	0.255	0.322	0.397	0.481	0.573	0.677	0.789	0.904	1.02	1.15	1.28	1.57	1.88	2.22	2.59	2.98
2.40	0.234	0.296	0.365	0.442	0.528	0.623	0.727	0.833	0.944	1.06	1.18	1.45	1.74	2.06	2.40	2.77
2.60	0.217	0.273	0.337	0.409	0.489	0.577	0.674	0.772	0.875	0.983	1.10	1.34	1.62	1.92	2.24	2.59
2.80	0.201	0.254	0.314	0.381	0.455	0.538	0.628	0.719	0.815	0.916	1.02	1.26	1.51	1.79	2.10	2.43
3.00	0.188	0.238	0.293	0.356	0.426	0.504	0.588	0.673	0.763	0.858	0.959	1.18	1.42	1.68	1.97	2.28
x	0.000	0.008	0.029	0.056	0.089	0.125	0.164	0.204	0.246	0.289	0.333	0.424	0.516	0.610	0.704	0.800

Table 8-42 (cont.).
Coefficients C for Eccentrically Loaded Weld Groups
Angle = 45°

$$\phi R_n = CC_1 Dl \qquad C_{min} = \frac{P_u}{C_1 Dl} \qquad D_{min} = \frac{P_u}{CC_1 l} \qquad l_{min} = \frac{P_u}{CC_1 D}$$

where

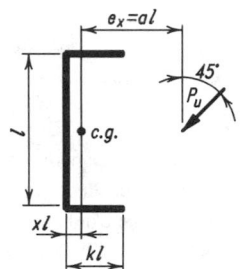

P_u = factored force, kips

D = number of sixteenths-of-an-inch in the fillet weld size

l = characteristic length of weld group, in.

a = e_x / l, in.

e_x = horizontal component of eccentricity of P_u with respect to centroid of weld group, in.

C = coefficient tabulated below which includes ϕ = 0.75

C_1 = electode strength coefficient from Table 8-37 (1.0 for E70XX electrodes)

| a | \multicolumn{16}{c}{k} |
	0	0.1	0.2	0.3	0.4	0.5	0.6	0.7	0.8	0.9	1.0	1.2	1.4	1.6	1.8	2.0
0.00	1.81	2.17	2.53	2.89	3.25	3.61	3.97	4.33	4.70	5.06	5.42	6.14	6.86	7.58	8.31	9.03
0.10	1.68	2.06	2.43	2.80	3.17	3.55	3.92	4.29	4.66	5.03	5.40	6.14	6.86	7.58	8.31	9.03
0.15	1.57	1.95	2.32	2.68	3.05	3.42	3.80	4.17	4.54	4.92	5.29	6.03	6.77	7.51	8.25	8.99
0.20	1.47	1.83	2.19	2.55	2.91	3.28	3.64	4.02	4.39	4.76	5.13	5.88	6.62	7.37	8.12	8.86
0.25	1.39	1.72	2.06	2.41	2.76	3.12	3.48	3.85	4.22	4.58	4.95	5.69	6.44	7.19	7.94	8.69
0.30	1.31	1.62	1.94	2.27	2.61	2.96	3.31	3.67	4.03	4.39	4.75	5.49	6.23	6.99	7.75	8.50
0.40	1.16	1.43	1.72	2.03	2.34	2.66	2.99	3.33	3.68	4.03	4.37	5.08	5.82	6.57	7.32	8.09
0.50	1.03	1.28	1.54	1.81	2.10	2.40	2.71	3.03	3.36	3.70	4.03	4.70	5.41	6.15	6.90	7.66
0.60	0.921	1.14	1.38	1.63	1.90	2.18	2.47	2.77	3.08	3.41	3.72	4.37	5.04	5.76	6.49	7.23
0.70	0.829	1.03	1.25	1.48	1.73	1.99	2.26	2.54	2.84	3.15	3.46	4.08	4.72	5.40	6.11	6.84
0.80	0.751	0.935	1.13	1.35	1.58	1.82	2.08	2.35	2.63	2.93	3.22	3.81	4.43	5.08	5.77	6.47
0.90	0.685	0.854	1.04	1.24	1.45	1.68	1.92	2.18	2.45	2.73	3.00	3.57	4.17	4.79	5.45	6.14
1.00	0.629	0.785	0.956	1.14	1.34	1.56	1.78	2.03	2.29	2.55	2.81	3.35	3.93	4.53	5.16	5.83
1.20	0.538	0.674	0.822	0.985	1.16	1.35	1.56	1.78	2.01	2.25	2.49	2.98	3.51	4.07	4.66	5.28
1.40	0.469	0.589	0.720	0.864	1.02	1.19	1.38	1.58	1.79	2.00	2.22	2.67	3.16	3.68	4.23	4.81
1.60	0.416	0.522	0.639	0.769	0.911	1.07	1.24	1.42	1.61	1.80	2.00	2.42	2.87	3.35	3.87	4.41
1.80	0.373	0.468	0.574	0.692	0.821	0.964	1.12	1.29	1.46	1.63	1.82	2.20	2.62	3.07	3.56	4.06
2.00	0.338	0.424	0.521	0.628	0.746	0.879	1.02	1.17	1.33	1.49	1.66	2.02	2.41	2.84	3.29	3.76
2.20	0.308	0.388	0.477	0.575	0.685	0.806	0.939	1.08	1.22	1.37	1.53	1.86	2.23	2.63	3.05	3.50
2.40	0.284	0.357	0.439	0.531	0.632	0.745	0.868	0.999	1.13	1.27	1.42	1.73	2.07	2.44	2.84	3.27
2.60	0.263	0.331	0.407	0.492	0.587	0.692	0.807	0.928	1.05	1.18	1.32	1.61	1.93	2.28	2.66	3.06
2.80	0.245	0.308	0.379	0.458	0.548	0.646	0.753	0.867	0.983	1.10	1.23	1.51	1.81	2.14	2.50	2.88
3.00	0.229	0.288	0.355	0.429	0.513	0.606	0.707	0.814	0.922	1.04	1.16	1.42	1.70	2.02	2.36	2.72
x	0.000	0.008	0.029	0.056	0.089	0.125	0.164	0.204	0.246	0.289	0.333	0.424	0.516	0.610	0.704	0.800

Table 8-42 (cont.).
Coefficients *C* for Eccentrically Loaded Weld Groups
Angle = 60°

$$\phi R_n = CC_1 Dl \qquad C_{min} = \frac{P_u}{C_1 Dl} \qquad D_{min} = \frac{P_u}{CC_1 l} \qquad l_{min} = \frac{P_u}{CC_1 D}$$

where

P_u = factored force, kips

D = number of sixteenths-of-an-inch in the fillet weld size

l = characteristic length of weld group, in.

a = e_x / l, in.

e_x = horizontal component of eccentricity of P_u with respect to centroid of weld group, in.

C = coefficient tabulated below which includes ϕ = 0.75

C_1 = electode strength coefficient from Table 8-37 (1.0 for E70XX electrodes)

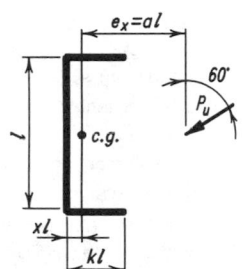

a	0	0.1	0.2	0.3	0.4	0.5	0.6	0.7	0.8	0.9	1.0	1.2	1.4	1.6	1.8	2.0
												k				
0.00	1.95	2.28	2.61	2.94	3.26	3.59	3.92	4.25	4.57	4.90	5.23	5.88	6.54	7.19	7.85	8.51
0.10	1.82	2.15	2.48	2.82	3.16	3.51	3.86	4.21	4.56	4.90	5.24	5.88	6.54	7.19	7.85	8.51
0.15	1.73	2.05	2.38	2.71	3.06	3.41	3.76	4.12	4.47	4.83	5.18	5.87	6.54	7.19	7.85	8.51
0.20	1.63	1.96	2.28	2.61	2.94	3.29	3.65	4.01	4.37	4.73	5.09	5.80	6.49	7.18	7.85	8.51
0.25	1.55	1.87	2.18	2.50	2.83	3.17	3.53	3.89	4.25	4.62	4.98	5.70	6.41	7.11	7.79	8.47
0.30	1.47	1.78	2.09	2.40	2.72	3.05	3.40	3.76	4.13	4.50	4.87	5.60	6.32	7.02	7.72	8.40
0.40	1.34	1.62	1.91	2.20	2.51	2.83	3.16	3.52	3.88	4.24	4.61	5.36	6.09	6.82	7.53	8.23
0.50	1.22	1.48	1.75	2.03	2.32	2.63	2.95	3.28	3.63	4.00	4.36	5.10	5.84	6.57	7.30	8.01
0.60	1.12	1.36	1.61	1.88	2.15	2.44	2.75	3.07	3.41	3.76	4.13	4.86	5.59	6.32	7.03	7.75
0.70	1.03	1.25	1.49	1.74	2.00	2.28	2.58	2.89	3.22	3.56	3.91	4.62	5.33	6.05	6.78	7.50
0.80	0.945	1.16	1.38	1.62	1.87	2.14	2.42	2.72	3.04	3.36	3.70	4.39	5.09	5.80	6.52	7.24
0.90	0.874	1.07	1.28	1.51	1.75	2.01	2.28	2.57	2.87	3.18	3.51	4.17	4.85	5.55	6.26	6.98
1.00	0.812	0.999	1.20	1.41	1.64	1.89	2.15	2.43	2.72	3.02	3.34	3.97	4.63	5.32	6.02	6.73
1.20	0.709	0.875	1.06	1.25	1.46	1.69	1.93	2.19	2.46	2.74	3.03	3.62	4.23	4.89	5.56	6.26
1.40	0.626	0.776	0.939	1.12	1.31	1.52	1.75	1.98	2.23	2.49	2.76	3.31	3.89	4.50	5.15	5.82
1.60	0.560	0.696	0.845	1.01	1.18	1.38	1.59	1.81	2.04	2.28	2.54	3.05	3.59	4.17	4.79	5.43
1.80	0.506	0.630	0.766	0.916	1.08	1.26	1.46	1.66	1.88	2.11	2.34	2.82	3.33	3.88	4.46	5.07
2.00	0.461	0.575	0.701	0.839	0.993	1.16	1.34	1.53	1.74	1.95	2.17	2.62	3.11	3.62	4.17	4.75
2.20	0.423	0.528	0.645	0.774	0.917	1.07	1.24	1.42	1.61	1.81	2.02	2.44	2.90	3.39	3.91	4.47
2.40	0.391	0.489	0.597	0.718	0.851	0.999	1.16	1.33	1.51	1.69	1.89	2.29	2.72	3.19	3.68	4.21
2.60	0.363	0.454	0.556	0.669	0.794	0.932	1.08	1.24	1.41	1.59	1.77	2.15	2.56	3.01	3.48	3.98
2.80	0.339	0.424	0.519	0.625	0.744	0.875	1.02	1.17	1.33	1.49	1.66	2.02	2.42	2.84	3.29	3.77
3.00	0.317	0.398	0.488	0.587	0.700	0.824	0.956	1.10	1.25	1.41	1.57	1.91	2.29	2.69	3.12	3.58
x	0.000	0.008	0.029	0.056	0.089	0.125	0.164	0.204	0.246	0.289	0.333	0.424	0.516	0.610	0.704	0.800

Table 8-42 (cont.).
Coefficients C for Eccentrically Loaded Weld Groups
Angle = 75°

$$\phi R_n = CC_1 Dl \qquad C_{min} = \frac{P_u}{C_1 Dl} \qquad D_{min} = \frac{P_u}{CC_1 l} \qquad l_{min} = \frac{P_u}{CC_1 D}$$

where

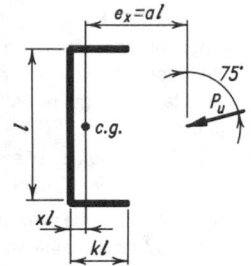

P_u = factored force, kips

D = number of sixteenths-of-an-inch in the fillet weld size

l = characteristic length of weld group, in.

a = e_x / l, in.

e_x = horizontal component of eccentricity of P_u with respect to centroid of weld group, in.

C = coefficient tabulated below which includes $\phi = 0.75$

C_1 = electode strength coefficient from Table 8-37 (1.0 for E70XX electrodes)

							k									
a	0	0.1	0.2	0.3	0.4	0.5	0.6	0.7	0.8	0.9	1.0	1.2	1.4	1.6	1.8	2.0
0.00	2.05	2.35	2.65	2.94	3.24	3.54	3.83	4.13	4.43	4.72	5.02	5.61	6.21	6.80	7.39	7.99
0.10	1.94	2.22	2.51	2.81	3.13	3.44	3.75	4.05	4.35	4.65	4.94	5.53	6.11	6.68	7.26	7.83
0.15	1.88	2.15	2.44	2.75	3.07	3.39	3.70	4.01	4.32	4.63	4.93	5.52	6.10	6.68	7.26	7.83
0.20	1.82	2.09	2.38	2.69	3.01	3.33	3.65	3.97	4.29	4.60	4.90	5.50	6.09	6.67	7.25	7.83
0.25	1.76	2.04	2.32	2.63	2.95	3.27	3.60	3.92	4.24	4.56	4.87	5.47	6.07	6.66	7.24	7.82
0.30	1.71	1.99	2.27	2.57	2.89	3.21	3.54	3.87	4.19	4.51	4.83	5.44	6.05	6.64	7.23	7.81
0.40	1.62	1.89	2.16	2.46	2.77	3.09	3.42	3.76	4.09	4.41	4.73	5.36	5.98	6.58	7.18	7.77
0.50	1.53	1.80	2.07	2.35	2.65	2.98	3.31	3.64	3.98	4.31	4.64	5.27	5.90	6.51	7.11	7.71
0.60	1.46	1.71	1.97	2.25	2.55	2.86	3.19	3.53	3.87	4.21	4.54	5.19	5.83	6.44	7.04	7.64
0.70	1.38	1.63	1.89	2.16	2.45	2.76	3.08	3.42	3.76	4.10	4.44	5.11	5.75	6.38	6.99	7.59
0.80	1.31	1.56	1.81	2.07	2.35	2.66	2.98	3.31	3.65	3.99	4.34	5.01	5.67	6.31	6.93	7.54
0.90	1.25	1.49	1.73	1.99	2.27	2.56	2.88	3.21	3.54	3.89	4.23	4.92	5.59	6.24	6.87	7.48
1.00	1.19	1.42	1.66	1.91	2.18	2.47	2.78	3.10	3.44	3.78	4.13	4.82	5.49	6.16	6.80	7.43
1.20	1.09	1.30	1.53	1.77	2.03	2.31	2.60	2.91	3.24	3.57	3.92	4.61	5.30	5.98	6.64	7.29
1.40	0.994	1.20	1.41	1.65	1.90	2.16	2.44	2.74	3.05	3.38	3.71	4.40	5.09	5.78	6.46	7.13
1.60	0.914	1.11	1.31	1.53	1.77	2.03	2.30	2.58	2.88	3.20	3.52	4.20	4.89	5.58	6.27	6.94
1.80	0.845	1.03	1.22	1.43	1.66	1.91	2.16	2.44	2.73	3.03	3.35	4.00	4.68	5.38	6.07	6.76
2.00	0.784	0.956	1.14	1.34	1.56	1.80	2.04	2.31	2.58	2.88	3.18	3.82	4.49	5.18	5.86	6.55
2.20	0.730	0.894	1.07	1.26	1.47	1.70	1.94	2.19	2.45	2.73	3.03	3.65	4.30	4.97	5.65	6.33
2.40	0.683	0.838	1.01	1.19	1.39	1.61	1.84	2.08	2.33	2.60	2.89	3.49	4.12	4.77	5.43	6.11
2.60	0.641	0.788	0.949	1.13	1.32	1.53	1.75	1.98	2.22	2.48	2.76	3.34	3.95	4.58	5.23	5.90
2.80	0.604	0.744	0.897	1.07	1.25	1.45	1.66	1.89	2.12	2.37	2.63	3.20	3.78	4.39	5.04	5.70
3.00	0.570	0.704	0.851	1.01	1.19	1.38	1.59	1.80	2.03	2.27	2.52	3.07	3.63	4.23	4.85	5.50
x	0.000	0.008	0.029	0.056	0.089	0.125	0.164	0.204	0.246	0.289	0.333	0.424	0.516	0.610	0.704	0.800

Table 8-43.
Coefficients *C* for Eccentrically Loaded Weld Groups
Angle = 0°

$$\phi R_n = CC_1 Dl \qquad C_{min} = \frac{P_u}{C_1 Dl} \qquad D_{min} = \frac{P_u}{CC_1 l} \qquad l_{min} = \frac{P_u}{CC_1 D}$$

where

P_u = factored force, kips

D = number of sixteenths-of-an-inch in the fillet weld size

l = characteristic length of weld group, in.

a = e_x / l, in.

e_x = horizontal component of eccentricity of P_u with respect to centroid of weld group, in.

C = coefficient tabulated below which includes $\phi = 0.75$

C_1 = electode strength coefficient from Table 8-37 (1.0 for E70XX electrodes)

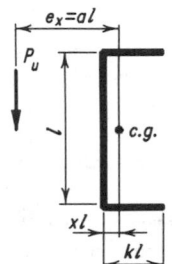

								k								
a	0	0.1	0.2	0.3	0.4	0.5	0.6	0.7	0.8	0.9	1.0	1.2	1.4	1.6	1.8	2.0
0.00	1.39	1.81	2.23	2.65	3.06	3.48	3.90	4.32	4.73	5.15	5.57	6.40	7.24	8.07	8.91	9.74
0.10	1.39	1.72	2.10	2.48	2.86	3.24	3.62	4.00	4.38	4.76	5.13	5.88	6.63	7.37	8.12	8.87
0.15	1.38	1.70	2.05	2.41	2.77	3.13	3.50	3.86	4.22	4.58	4.94	5.66	6.38	7.11	7.84	8.57
0.20	1.32	1.63	1.96	2.31	2.65	2.99	3.34	3.68	4.03	4.38	4.72	5.42	6.12	6.83	7.55	8.27
0.25	1.24	1.55	1.86	2.18	2.51	2.84	3.17	3.50	3.83	4.16	4.50	5.17	5.86	6.55	7.25	7.96
0.30	1.16	1.45	1.75	2.05	2.36	2.68	2.99	3.31	3.63	3.95	4.27	4.93	5.60	6.27	6.96	7.66
0.40	0.998	1.26	1.52	1.79	2.08	2.36	2.65	2.94	3.24	3.54	3.84	4.46	5.09	5.74	6.41	7.08
0.50	0.860	1.08	1.31	1.56	1.81	2.07	2.34	2.60	2.88	3.15	3.44	4.02	4.63	5.25	5.89	6.55
0.60	0.748	0.942	1.14	1.35	1.58	1.83	2.07	2.32	2.58	2.84	3.11	3.66	4.23	4.83	5.45	6.08
0.70	0.659	0.828	1.00	1.20	1.41	1.63	1.86	2.10	2.34	2.58	2.83	3.35	3.90	4.47	5.06	5.67
0.80	0.586	0.735	0.895	1.07	1.27	1.47	1.69	1.91	2.13	2.36	2.60	3.09	3.61	4.15	4.72	5.30
0.90	0.527	0.661	0.807	0.971	1.15	1.34	1.54	1.75	1.96	2.18	2.40	2.86	3.35	3.87	4.41	4.97
1.00	0.478	0.599	0.734	0.885	1.05	1.23	1.41	1.61	1.81	2.01	2.22	2.66	3.13	3.62	4.14	4.68
1.20	0.403	0.505	0.621	0.751	0.893	1.05	1.21	1.38	1.56	1.75	1.94	2.33	2.75	3.20	3.68	4.17
1.40	0.348	0.436	0.537	0.651	0.776	0.912	1.06	1.21	1.37	1.54	1.71	2.07	2.45	2.86	3.30	3.75
1.60	0.305	0.383	0.472	0.574	0.685	0.806	0.936	1.07	1.22	1.37	1.53	1.86	2.21	2.58	2.98	3.41
1.80	0.272	0.341	0.422	0.512	0.613	0.722	0.839	0.963	1.10	1.23	1.38	1.68	2.00	2.35	2.72	3.11
2.00	0.245	0.308	0.381	0.463	0.554	0.654	0.760	0.873	0.993	1.12	1.25	1.53	1.84	2.15	2.50	2.87
2.20	0.223	0.280	0.347	0.422	0.506	0.597	0.694	0.798	0.909	1.02	1.15	1.41	1.69	1.99	2.31	2.65
2.40	0.205	0.257	0.318	0.388	0.465	0.548	0.639	0.735	0.836	0.942	1.06	1.30	1.56	1.85	2.15	2.47
2.60	0.189	0.238	0.294	0.358	0.430	0.508	0.592	0.681	0.773	0.872	0.976	1.20	1.46	1.72	2.00	2.30
2.80	0.176	0.221	0.274	0.333	0.400	0.472	0.550	0.633	0.720	0.811	0.908	1.12	1.36	1.61	1.88	2.16
3.00	0.164	0.207	0.256	0.311	0.374	0.442	0.515	0.593	0.673	0.758	0.850	1.05	1.27	1.51	1.76	2.03
x	0.000	0.008	0.029	0.056	0.089	0.125	0.164	0.204	0.246	0.289	0.333	0.424	0.516	0.610	0.704	0.800

Table 8-43 (cont.).
Coefficients *C* for Eccentrically Loaded Weld Groups
Angle = 15°

$$\phi R_n = CC_1 Dl \qquad C_{min} = \frac{P_u}{C_1 Dl} \qquad D_{min} = \frac{P_u}{CC_1 l} \qquad l_{min} = \frac{P_u}{CC_1 D}$$

where

P_u = factored force, kips

D = number of sixteenths-of-an-inch in the fillet weld size

l = characteristic length of weld group, in.

a = e_x / l, in.

e_x = horizontal component of eccentricity of P_u with respect to centroid of weld group, in.

C = coefficient tabulated below which includes $\phi = 0.75$

C_1 = electode strength coefficient from Table 8-37 (1.0 for E70XX electrodes)

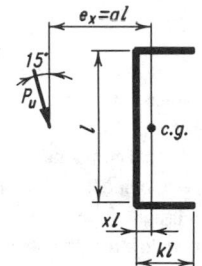

									k								
a	0	0.1	0.2	0.3	0.4	0.5	0.6	0.7	0.8	0.9	1.0	1.2	1.4	1.6	1.8	2.0
0.00	1.48	1.89	2.31	2.72	3.13	3.54	3.95	4.36	4.77	5.18	5.59	6.41	7.23	8.05	8.87	9.69
0.10	1.42	1.77	2.15	2.53	2.91	3.29	3.66	4.04	4.41	4.78	5.14	5.89	6.63	7.38	8.14	8.90
0.15	1.38	1.73	2.09	2.45	2.80	3.16	3.52	3.87	4.22	4.58	4.93	5.64	6.36	7.08	7.81	8.54
0.20	1.32	1.65	1.99	2.33	2.67	3.01	3.35	3.69	4.03	4.36	4.70	5.38	6.08	6.78	7.49	8.21
0.25	1.24	1.55	1.87	2.20	2.53	2.85	3.17	3.50	3.82	4.15	4.47	5.13	5.80	6.49	7.18	7.89
0.30	1.16	1.45	1.75	2.06	2.37	2.68	3.00	3.31	3.62	3.93	4.25	4.89	5.54	6.21	6.89	7.58
0.40	1.00	1.25	1.51	1.78	2.06	2.35	2.64	2.94	3.23	3.53	3.82	4.43	5.05	5.69	6.35	7.03
0.50	0.869	1.09	1.31	1.55	1.80	2.05	2.32	2.60	2.88	3.17	3.45	4.03	4.63	5.25	5.89	6.54
0.60	0.759	0.950	1.14	1.35	1.58	1.81	2.06	2.33	2.59	2.87	3.14	3.69	4.27	4.86	5.47	6.11
0.70	0.670	0.838	1.01	1.20	1.41	1.62	1.86	2.10	2.35	2.61	2.87	3.40	3.95	4.52	5.10	5.71
0.80	0.598	0.747	0.905	1.08	1.27	1.47	1.68	1.91	2.15	2.39	2.64	3.14	3.66	4.21	4.77	5.36
0.90	0.539	0.672	0.818	0.980	1.15	1.34	1.54	1.75	1.97	2.20	2.44	2.92	3.41	3.94	4.48	5.04
1.00	0.490	0.611	0.746	0.896	1.06	1.23	1.42	1.61	1.82	2.04	2.26	2.72	3.19	3.69	4.21	4.75
1.20	0.413	0.516	0.633	0.763	0.905	1.06	1.22	1.39	1.58	1.77	1.97	2.39	2.82	3.27	3.75	4.25
1.40	0.357	0.446	0.548	0.663	0.789	0.925	1.07	1.22	1.39	1.56	1.74	2.12	2.52	2.93	3.38	3.84
1.60	0.314	0.393	0.484	0.586	0.698	0.820	0.951	1.09	1.24	1.39	1.56	1.91	2.27	2.65	3.06	3.49
1.80	0.280	0.351	0.432	0.524	0.626	0.737	0.854	0.981	1.11	1.26	1.41	1.73	2.06	2.42	2.80	3.19
2.00	0.253	0.317	0.391	0.474	0.567	0.668	0.776	0.891	1.01	1.14	1.28	1.58	1.89	2.22	2.57	2.95
2.20	0.230	0.289	0.356	0.433	0.518	0.611	0.710	0.816	0.929	1.05	1.18	1.45	1.74	2.05	2.38	2.73
2.40	0.211	0.265	0.327	0.398	0.476	0.562	0.654	0.753	0.857	0.967	1.09	1.34	1.61	1.90	2.21	2.54
2.60	0.195	0.245	0.303	0.368	0.441	0.521	0.606	0.698	0.795	0.898	1.01	1.24	1.50	1.77	2.06	2.37
2.80	0.182	0.228	0.282	0.343	0.410	0.485	0.565	0.650	0.741	0.837	0.937	1.16	1.40	1.66	1.93	2.23
3.00	0.170	0.213	0.263	0.320	0.384	0.453	0.528	0.609	0.694	0.782	0.877	1.08	1.31	1.56	1.82	2.10
x	0.000	0.008	0.029	0.056	0.089	0.125	0.164	0.204	0.246	0.289	0.333	0.424	0.516	0.610	0.704	0.800

Table 8-43 (cont.).
Coefficients C for Eccentrically Loaded Weld Groups
Angle = 30°

$$\phi R_n = CC_1 Dl \qquad C_{min} = \frac{P_u}{C_1 Dl} \qquad D_{min} = \frac{P_u}{CC_1 l} \qquad l_{min} = \frac{P_u}{CC_1 D}$$

where

P_u = factored force, kips

D = number of sixteenths-of-an-inch
in the fillet weld size

l = characteristic length of weld group, in.

a = e_x / l, in.

e_x = horizontal component of eccentricity of
P_u with respect to centroid of weld group, in.

C = coefficient tabulated below which includes $\phi = 0.75$

C_1 = electode strength coefficient from Table 8-37
(1.0 for E70XX electrodes)

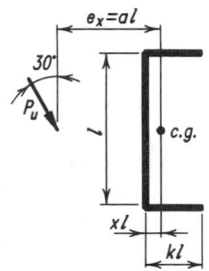

									k								
a	0	0.1	0.2	0.3	0.4	0.5	0.6	0.7	0.8	0.9	1.0	1.2	1.4	1.6	1.8	2.0
0.00	1.64	2.03	2.42	2.81	3.20	3.59	3.98	4.37	4.76	5.15	5.54	6.33	7.11	7.89	8.67	9.45
0.10	1.52	1.92	2.30	2.66	3.02	3.38	3.74	4.10	4.46	4.83	5.19	5.93	6.67	7.43	8.19	8.96
0.15	1.44	1.81	2.17	2.53	2.87	3.21	3.55	3.90	4.24	4.59	4.94	5.66	6.39	7.13	7.88	8.64
0.20	1.36	1.70	2.04	2.37	2.70	3.02	3.35	3.67	4.00	4.34	4.68	5.37	6.09	6.82	7.56	8.32
0.25	1.28	1.60	1.91	2.22	2.53	2.83	3.14	3.45	3.77	4.09	4.42	5.10	5.79	6.51	7.24	8.00
0.30	1.20	1.49	1.79	2.08	2.37	2.66	2.96	3.25	3.56	3.89	4.21	4.86	5.53	6.23	6.95	7.68
0.40	1.05	1.30	1.56	1.82	2.08	2.35	2.63	2.92	3.22	3.52	3.84	4.46	5.10	5.77	6.45	7.15
0.50	0.921	1.14	1.36	1.59	1.83	2.09	2.35	2.63	2.92	3.21	3.51	4.12	4.73	5.37	6.03	6.70
0.60	0.812	1.01	1.20	1.41	1.63	1.87	2.12	2.39	2.66	2.94	3.23	3.82	4.41	5.01	5.64	6.30
0.70	0.722	0.895	1.07	1.26	1.47	1.69	1.93	2.17	2.43	2.70	2.98	3.55	4.11	4.69	5.30	5.93
0.80	0.647	0.803	0.966	1.15	1.34	1.54	1.76	2.00	2.24	2.49	2.75	3.30	3.85	4.41	4.99	5.60
0.90	0.586	0.726	0.878	1.05	1.23	1.42	1.62	1.84	2.07	2.31	2.56	3.08	3.61	4.15	4.71	5.30
1.00	0.535	0.663	0.805	0.960	1.13	1.31	1.50	1.71	1.92	2.15	2.39	2.88	3.39	3.91	4.45	5.02
1.20	0.454	0.563	0.687	0.825	0.973	1.13	1.30	1.48	1.68	1.88	2.10	2.55	3.02	3.50	4.01	4.53
1.40	0.393	0.489	0.599	0.721	0.855	0.997	1.15	1.31	1.49	1.67	1.86	2.28	2.72	3.16	3.63	4.12
1.60	0.347	0.432	0.530	0.639	0.760	0.890	1.03	1.18	1.33	1.50	1.68	2.06	2.46	2.88	3.31	3.77
1.80	0.310	0.386	0.475	0.574	0.684	0.802	0.928	1.06	1.21	1.36	1.52	1.87	2.25	2.63	3.04	3.47
2.00	0.280	0.349	0.430	0.521	0.620	0.729	0.846	0.969	1.10	1.24	1.39	1.71	2.06	2.42	2.80	3.21
2.20	0.255	0.319	0.393	0.476	0.568	0.668	0.776	0.890	1.01	1.14	1.28	1.58	1.91	2.25	2.60	2.98
2.40	0.234	0.293	0.361	0.439	0.524	0.616	0.716	0.823	0.936	1.06	1.19	1.46	1.77	2.09	2.42	2.78
2.60	0.217	0.271	0.335	0.406	0.486	0.572	0.665	0.765	0.870	0.983	1.10	1.36	1.65	1.95	2.27	2.60
2.80	0.201	0.252	0.311	0.378	0.453	0.534	0.621	0.714	0.813	0.919	1.03	1.28	1.55	1.83	2.13	2.45
3.00	0.188	0.236	0.291	0.354	0.424	0.500	0.582	0.669	0.763	0.862	0.967	1.20	1.45	1.72	2.00	2.31
x	0.000	0.008	0.029	0.056	0.089	0.125	0.164	0.204	0.246	0.289	0.333	0.424	0.516	0.610	0.704	0.800

Table 8-43 (cont.).
Coefficients C for Eccentrically Loaded Weld Groups
Angle = 45°

$$\phi R_n = CC_1 Dl \qquad C_{min} = \frac{P_u}{C_1 Dl} \qquad D_{min} = \frac{P_u}{CC_1 l} \qquad l_{min} = \frac{P_u}{CC_1 D}$$

where

P_u = factored force, kips

D = number of sixteenths-of-an-inch
　　in the fillet weld size

l = characteristic length of weld group, in.

a = e_x / l, in.

e_x = horizontal component of eccentricity of
　　P_u with respect to centroid of weld group, in.

C = coefficient tabulated below which includes $\phi = 0.75$

C_1 = electode strength coefficient from Table 8-37
　　(1.0 for E70XX electrodes)

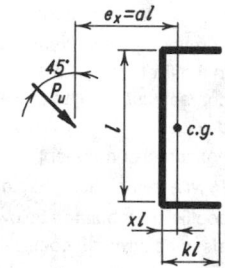

								k								
a	0	0.1	0.2	0.3	0.4	0.5	0.6	0.7	0.8	0.9	1.0	1.2	1.4	1.6	1.8	2.0
0.00	1.81	2.17	2.53	2.89	3.25	3.61	3.97	4.33	4.70	5.06	5.42	6.14	6.86	7.58	8.31	9.03
0.10	1.68	2.04	2.38	2.71	3.04	3.37	3.71	4.06	4.41	4.77	5.12	5.84	6.56	7.30	8.03	8.76
0.15	1.57	1.93	2.25	2.56	2.86	3.18	3.51	3.85	4.19	4.55	4.91	5.64	6.37	7.10	7.85	8.59
0.20	1.47	1.81	2.12	2.41	2.69	2.99	3.30	3.64	3.98	4.33	4.69	5.41	6.14	6.88	7.63	8.38
0.25	1.39	1.70	1.99	2.27	2.53	2.82	3.12	3.44	3.77	4.12	4.47	5.18	5.92	6.66	7.40	8.15
0.30	1.31	1.60	1.88	2.14	2.40	2.68	2.97	3.28	3.60	3.94	4.28	4.98	5.70	6.42	7.17	7.92
0.40	1.16	1.41	1.67	1.91	2.16	2.43	2.71	3.01	3.31	3.63	3.96	4.64	5.33	6.03	6.74	7.46
0.50	1.03	1.26	1.48	1.71	1.96	2.22	2.49	2.77	3.06	3.37	3.68	4.34	5.02	5.70	6.38	7.09
0.60	0.921	1.13	1.33	1.55	1.78	2.03	2.29	2.55	2.84	3.13	3.43	4.06	4.72	5.38	6.06	6.75
0.70	0.829	1.02	1.20	1.41	1.63	1.86	2.11	2.37	2.63	2.91	3.21	3.82	4.45	5.09	5.75	6.42
0.80	0.751	0.920	1.10	1.29	1.50	1.72	1.95	2.20	2.46	2.72	3.00	3.59	4.21	4.82	5.46	6.12
0.90	0.685	0.840	1.01	1.19	1.38	1.59	1.81	2.05	2.30	2.55	2.82	3.38	3.98	4.58	5.20	5.83
1.00	0.629	0.772	0.930	1.10	1.28	1.48	1.69	1.92	2.15	2.40	2.65	3.19	3.77	4.35	4.95	5.57
1.20	0.538	0.664	0.804	0.957	1.12	1.30	1.49	1.69	1.91	2.13	2.37	2.87	3.40	3.95	4.50	5.09
1.40	0.469	0.581	0.706	0.845	0.995	1.16	1.33	1.51	1.71	1.91	2.13	2.59	3.09	3.60	4.12	4.67
1.60	0.416	0.515	0.629	0.754	0.892	1.04	1.20	1.36	1.54	1.73	1.93	2.36	2.82	3.31	3.80	4.31
1.80	0.373	0.463	0.566	0.681	0.807	0.943	1.09	1.24	1.41	1.58	1.77	2.17	2.60	3.05	3.52	4.00
2.00	0.338	0.420	0.515	0.621	0.736	0.862	0.995	1.14	1.29	1.45	1.62	2.00	2.40	2.83	3.27	3.73
2.20	0.308	0.384	0.471	0.569	0.677	0.793	0.918	1.05	1.19	1.34	1.50	1.85	2.23	2.64	3.05	3.48
2.40	0.284	0.354	0.435	0.526	0.626	0.734	0.850	0.974	1.11	1.25	1.40	1.72	2.08	2.46	2.85	3.27
2.60	0.263	0.328	0.403	0.488	0.582	0.683	0.792	0.909	1.03	1.16	1.31	1.61	1.95	2.31	2.68	3.07
2.80	0.245	0.306	0.376	0.455	0.543	0.639	0.741	0.851	0.966	1.09	1.22	1.51	1.83	2.17	2.53	2.90
3.00	0.229	0.286	0.352	0.427	0.510	0.600	0.697	0.799	0.910	1.03	1.15	1.43	1.73	2.05	2.39	2.74
x	0.000	0.008	0.029	0.056	0.089	0.125	0.164	0.204	0.246	0.289	0.333	0.424	0.516	0.610	0.704	0.800

Table 8-43 (cont.).
Coefficients C for Eccentrically Loaded Weld Groups
Angle = 60°

$$\phi R_n = CC_1 Dl \qquad C_{min} = \frac{P_u}{C_1 Dl} \qquad D_{min} = \frac{P_u}{CC_1 l} \qquad l_{min} = \frac{P_u}{CC_1 D}$$

where

P_u = factored force, kips

D = number of sixteenths-of-an-inch in the fillet weld size

l = characteristic length of weld group, in.

a = e_x / l, in.

e_x = horizontal component of eccentricity of P_u with respect to centroid of weld group, in.

C = coefficient tabulated below which includes $\phi = 0.75$

C_1 = electode strength coefficient from Table 8-37 (1.0 for E70XX electrodes)

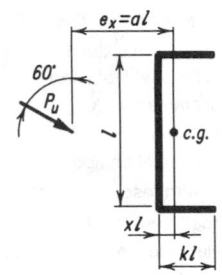

a	\multicolumn{16}{c}{k}
	0	0.1	0.2	0.3	0.4	0.5	0.6	0.7	0.8	0.9	1.0	1.2	1.4	1.6	1.8	2.0
0.00	1.95	2.28	2.61	2.94	3.26	3.59	3.92	4.25	4.57	4.90	5.23	5.88	6.54	7.19	7.85	8.51
0.10	1.83	2.13	2.42	2.72	3.03	3.36	3.69	4.03	4.37	4.71	5.05	5.73	6.41	7.08	7.74	8.40
0.15	1.73	2.03	2.30	2.58	2.88	3.19	3.52	3.86	4.20	4.55	4.89	5.59	6.27	6.96	7.63	8.31
0.20	1.63	1.93	2.19	2.45	2.74	3.04	3.36	3.69	4.03	4.38	4.73	5.43	6.13	6.82	7.51	8.19
0.25	1.55	1.84	2.09	2.34	2.62	2.92	3.23	3.55	3.88	4.22	4.56	5.26	5.97	6.67	7.37	8.06
0.30	1.47	1.76	2.00	2.25	2.52	2.81	3.12	3.43	3.76	4.09	4.42	5.11	5.80	6.51	7.22	7.93
0.40	1.34	1.60	1.84	2.08	2.34	2.62	2.91	3.22	3.54	3.87	4.20	4.87	5.54	6.23	6.91	7.61
0.50	1.22	1.46	1.69	1.93	2.18	2.45	2.73	3.04	3.35	3.67	3.99	4.66	5.33	6.01	6.69	7.37
0.60	1.12	1.34	1.56	1.79	2.04	2.30	2.57	2.86	3.16	3.48	3.80	4.46	5.14	5.81	6.49	7.16
0.70	1.03	1.23	1.44	1.67	1.91	2.16	2.42	2.70	3.00	3.30	3.62	4.27	4.94	5.62	6.30	6.98
0.80	0.945	1.14	1.34	1.55	1.79	2.03	2.29	2.56	2.84	3.14	3.45	4.08	4.75	5.42	6.10	6.79
0.90	0.874	1.06	1.25	1.45	1.68	1.91	2.16	2.42	2.70	2.98	3.28	3.91	4.56	5.24	5.91	6.60
1.00	0.812	0.983	1.17	1.37	1.58	1.81	2.05	2.30	2.57	2.84	3.13	3.75	4.39	5.05	5.73	6.41
1.20	0.709	0.862	1.03	1.21	1.41	1.62	1.84	2.08	2.33	2.59	2.86	3.44	4.06	4.70	5.36	6.03
1.40	0.626	0.766	0.922	1.09	1.27	1.46	1.67	1.89	2.13	2.37	2.63	3.18	3.76	4.38	5.02	5.66
1.60	0.560	0.688	0.831	0.988	1.16	1.34	1.53	1.74	1.95	2.18	2.43	2.94	3.50	4.08	4.70	5.32
1.80	0.506	0.623	0.756	0.901	1.06	1.23	1.41	1.60	1.80	2.02	2.25	2.73	3.26	3.82	4.41	5.00
2.00	0.461	0.569	0.692	0.828	0.975	1.13	1.30	1.48	1.67	1.88	2.09	2.55	3.05	3.58	4.14	4.71
2.20	0.423	0.524	0.638	0.765	0.902	1.05	1.21	1.38	1.56	1.75	1.95	2.39	2.86	3.37	3.90	4.45
2.40	0.391	0.484	0.591	0.710	0.840	0.979	1.13	1.29	1.46	1.64	1.83	2.24	2.69	3.17	3.69	4.21
2.60	0.363	0.451	0.551	0.662	0.785	0.916	1.06	1.21	1.37	1.54	1.72	2.11	2.54	3.00	3.49	3.99
2.80	0.339	0.421	0.515	0.620	0.736	0.861	0.994	1.14	1.29	1.45	1.62	2.00	2.40	2.85	3.31	3.79
3.00	0.317	0.395	0.484	0.584	0.693	0.812	0.939	1.07	1.22	1.37	1.53	1.89	2.28	2.70	3.15	3.60
x	0.000	0.008	0.029	0.056	0.089	0.125	0.164	0.204	0.246	0.289	0.333	0.424	0.516	0.610	0.704	0.800

Table 8-43 (cont.).
Coefficients C for Eccentrically Loaded Weld Groups
Angle = 75°

$$\phi R_n = CC_1 Dl \qquad C_{min} = \frac{P_u}{C_1 Dl} \qquad D_{min} = \frac{P_u}{CC_1 l} \qquad l_{min} = \frac{P_u}{CC_1 D}$$

where

P_u = factored force, kips

D = number of sixteenths-of-an-inch in the fillet weld size

l = characteristic length of weld group, in.

a = e_x / l, in.

e_x = horizontal component of eccentricity of P_u with respect to centroid of weld group, in.

C = coefficient tabulated below which includes $\phi = 0.75$

C_1 = electode strength coefficient from Table 8-37 (1.0 for E70XX electrodes)

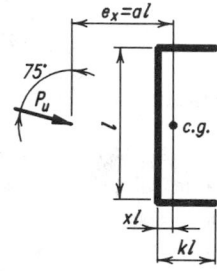

a	0	0.1	0.2	0.3	0.4	0.5	0.6	0.7	0.8	0.9	1.0	1.2	1.4	1.6	1.8	2.0
											k					
0.00	2.05	2.35	2.65	2.94	3.24	3.54	3.83	4.13	4.43	4.72	5.02	5.61	6.21	6.80	7.39	7.99
0.10	1.94	2.21	2.48	2.77	3.06	3.37	3.68	3.99	4.31	4.60	4.88	5.46	6.04	6.61	7.19	7.77
0.15	1.88	2.14	2.40	2.67	2.96	3.26	3.58	3.89	4.20	4.51	4.82	5.42	6.00	6.58	7.16	7.74
0.20	1.82	2.07	2.32	2.59	2.88	3.18	3.48	3.78	4.10	4.42	4.73	5.35	5.96	6.54	7.12	7.70
0.25	1.76	2.01	2.26	2.53	2.82	3.11	3.41	3.71	4.02	4.32	4.63	5.26	5.88	6.49	7.08	7.67
0.30	1.71	1.95	2.20	2.47	2.76	3.05	3.36	3.66	3.96	4.26	4.56	5.17	5.79	6.42	7.03	7.62
0.40	1.62	1.86	2.10	2.36	2.65	2.95	3.25	3.56	3.86	4.16	4.46	5.06	5.66	6.26	6.88	7.49
0.50	1.53	1.77	2.01	2.27	2.55	2.84	3.15	3.46	3.77	4.07	4.38	4.98	5.58	6.17	6.76	7.35
0.60	1.46	1.69	1.93	2.18	2.45	2.74	3.05	3.36	3.67	3.98	4.29	4.91	5.51	6.10	6.69	7.28
0.70	1.38	1.61	1.85	2.09	2.36	2.65	2.95	3.26	3.57	3.89	4.21	4.83	5.44	6.05	6.64	7.22
0.80	1.31	1.54	1.77	2.01	2.28	2.56	2.85	3.16	3.48	3.79	4.11	4.75	5.37	5.98	6.58	7.17
0.90	1.25	1.47	1.70	1.94	2.19	2.47	2.76	3.07	3.38	3.70	4.02	4.66	5.29	5.92	6.52	7.13
1.00	1.19	1.40	1.63	1.87	2.12	2.39	2.68	2.98	3.29	3.60	3.93	4.57	5.21	5.84	6.46	7.07
1.20	1.09	1.29	1.50	1.73	1.98	2.24	2.51	2.80	3.11	3.42	3.74	4.39	5.04	5.69	6.32	6.95
1.40	0.994	1.18	1.39	1.61	1.85	2.10	2.36	2.64	2.94	3.24	3.56	4.21	4.86	5.52	6.17	6.80
1.60	0.914	1.10	1.29	1.50	1.73	1.97	2.23	2.50	2.78	3.08	3.39	4.02	4.68	5.34	6.00	6.65
1.80	0.845	1.02	1.21	1.41	1.63	1.86	2.10	2.36	2.64	2.93	3.23	3.85	4.50	5.16	5.82	6.49
2.00	0.784	0.947	1.13	1.32	1.53	1.75	1.99	2.24	2.50	2.78	3.07	3.68	4.33	4.98	5.65	6.31
2.20	0.730	0.886	1.06	1.25	1.45	1.66	1.89	2.13	2.38	2.65	2.93	3.53	4.16	4.81	5.47	6.13
2.40	0.683	0.832	0.996	1.18	1.37	1.57	1.79	2.03	2.27	2.53	2.80	3.38	3.99	4.64	5.29	5.96
2.60	0.641	0.782	0.940	1.11	1.30	1.49	1.71	1.93	2.17	2.42	2.68	3.24	3.84	4.47	5.12	5.78
2.80	0.604	0.738	0.890	1.06	1.23	1.42	1.62	1.84	2.07	2.31	2.57	3.11	3.70	4.31	4.95	5.61
3.00	0.570	0.699	0.844	1.00	1.17	1.36	1.55	1.76	1.98	2.22	2.46	2.99	3.56	4.16	4.79	5.44
x	0.000	0.008	0.029	0.056	0.089	0.125	0.164	0.204	0.246	0.289	0.333	0.424	0.516	0.610	0.704	0.800

Table 8-44.
Coefficients C for Eccentrically Loaded Weld Groups
Angle = 0°

$$\phi R_n = CC_1 Dl \qquad C_{min} = \frac{P_u}{C_1 Dl} \qquad D_{min} = \frac{P_u}{CC_1 l} \qquad l_{min} = \frac{P_u}{CC_1 D}$$

where

P_u = factored force, kips

D = number of sixteenths-of-an-inch
in the fillet weld size

l = characteristic length of weld group, in.

a = e_x / l, in.

e_x = horizontal component of eccentricity of
P_u with respect to centroid of weld group, in.

C = coefficient tabulated below which includes ϕ = 0.75

C_1 = electode strength coefficient from Table 8-37
(1.0 for E70XX electrodes)

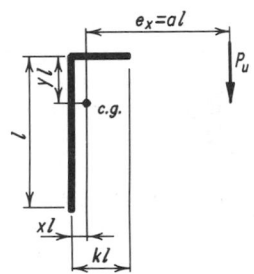

									k								
a	0	0.1	0.2	0.3	0.4	0.5	0.6	0.7	0.8	0.9	1.0	1.2	1.4	1.6	1.8	2.0
0.00	1.39	1.60	1.81	2.02	2.23	2.44	2.65	2.85	3.06	3.27	3.48	3.90	4.32	4.73	5.15	5.57
0.10	1.39	1.53	1.71	1.90	2.09	2.29	2.48	2.68	2.88	3.08	3.29	3.70	4.11	4.52	4.94	5.35
0.15	1.38	1.52	1.69	1.87	2.05	2.24	2.43	2.62	2.82	3.01	3.21	3.61	4.01	4.42	4.83	5.26
0.20	1.32	1.47	1.63	1.80	1.98	2.15	2.33	2.52	2.70	2.89	3.08	3.46	3.85	4.25	4.65	5.05
0.25	1.24	1.39	1.55	1.71	1.88	2.04	2.21	2.39	2.57	2.74	2.93	3.30	3.68	4.06	4.46	4.85
0.30	1.16	1.30	1.45	1.61	1.77	1.92	2.08	2.25	2.42	2.59	2.77	3.13	3.50	3.88	4.26	4.65
0.40	0.998	1.12	1.25	1.39	1.54	1.68	1.82	1.97	2.13	2.29	2.45	2.79	3.15	3.51	3.89	4.27
0.50	0.860	0.965	1.08	1.20	1.32	1.46	1.60	1.73	1.87	2.02	2.17	2.50	2.84	3.19	3.55	3.92
0.60	0.748	0.840	0.935	1.04	1.15	1.27	1.40	1.53	1.66	1.80	1.94	2.24	2.57	2.90	3.26	3.62
0.70	0.659	0.739	0.822	0.913	1.01	1.12	1.24	1.36	1.48	1.61	1.74	2.03	2.34	2.66	3.00	3.35
0.80	0.586	0.658	0.732	0.813	0.901	1.00	1.11	1.22	1.34	1.45	1.58	1.85	2.14	2.45	2.77	3.11
0.90	0.527	0.591	0.658	0.731	0.811	0.900	0.999	1.11	1.21	1.32	1.44	1.69	1.97	2.26	2.57	2.90
1.00	0.478	0.536	0.597	0.663	0.736	0.818	0.909	1.01	1.11	1.21	1.32	1.56	1.82	2.10	2.40	2.71
1.20	0.403	0.452	0.503	0.558	0.620	0.690	0.769	0.856	0.947	1.04	1.13	1.35	1.58	1.83	2.10	2.38
1.40	0.348	0.389	0.433	0.481	0.535	0.596	0.665	0.743	0.824	0.904	0.990	1.18	1.39	1.62	1.86	2.12
1.60	0.305	0.342	0.381	0.423	0.470	0.525	0.586	0.655	0.728	0.800	0.877	1.05	1.24	1.44	1.67	1.91
1.80	0.272	0.305	0.339	0.377	0.419	0.468	0.524	0.585	0.652	0.717	0.787	0.941	1.11	1.30	1.51	1.73
2.00	0.245	0.275	0.306	0.340	0.378	0.423	0.473	0.529	0.590	0.649	0.713	0.854	1.01	1.19	1.38	1.58
2.20	0.223	0.250	0.278	0.309	0.344	0.385	0.431	0.483	0.539	0.593	0.652	0.780	0.927	1.09	1.26	1.45
2.40	0.205	0.229	0.255	0.284	0.316	0.354	0.396	0.443	0.495	0.546	0.600	0.719	0.855	1.00	1.17	1.34
2.60	0.189	0.212	0.236	0.262	0.292	0.327	0.366	0.410	0.458	0.505	0.555	0.667	0.792	0.933	1.09	1.25
2.80	0.176	0.197	0.219	0.244	0.271	0.304	0.340	0.381	0.426	0.470	0.517	0.621	0.739	0.870	1.01	1.17
3.00	0.164	0.184	0.204	0.228	0.254	0.284	0.318	0.356	0.398	0.440	0.483	0.581	0.692	0.815	0.950	1.10
x	0.000	0.005	0.017	0.035	0.057	0.083	0.113	0.144	0.178	0.213	0.250	0.327	0.408	0.492	0.579	0.667
y	0.500	0.455	0.417	0.385	0.357	0.333	0.313	0.294	0.278	0.263	0.250	0.227	0.208	0.192	0.179	0.167

Table 8-44 (cont.).
Coefficients C for Eccentrically Loaded Weld Groups
Angle = ±15°

$$\phi R_n = CC_1 Dl \qquad C_{min} = \frac{P_u}{C_1 Dl} \qquad D_{min} = \frac{P_u}{CC_1 l} \qquad l_{min} = \frac{P_u}{CC_1 D}$$

where

P_u = factored force, kips

D = number of sixteenths-of-an-inch in the fillet weld size

l = characteristic length of weld group, in.

a = e_x / l, in.

e_x = horizontal component of eccentricity of P_u with respect to centroid of weld group, in.

C = coefficient tabulated below which includes ϕ = 0.75

C_1 = electode strength coefficient from Table 8-37 (1.0 for E70XX electrodes)

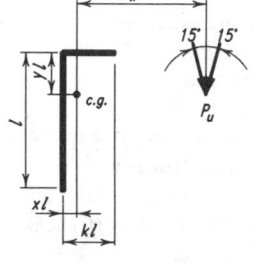

| | | | | | | | | | k | | | | | | | | |
a	0	0.1	0.2	0.3	0.4	0.5	0.6	0.7	0.8	0.9	1.0	1.2	1.4	1.6	1.8	2.0
0.00	1.48	1.69	1.89	2.10	2.31	2.51	2.72	2.92	3.13	3.33	3.54	3.95	4.36	4.77	5.18	5.59
0.10	1.42	1.56	1.73	1.90	2.09	2.28	2.47	2.67	2.88	3.09	3.30	3.73	4.17	4.59	5.02	5.45
0.15	1.38	1.53	1.69	1.85	2.03	2.20	2.39	2.57	2.76	2.95	3.15	3.54	3.95	4.37	4.80	5.23
0.20	1.32	1.48	1.63	1.78	1.94	2.11	2.28	2.46	2.64	2.82	3.00	3.38	3.77	4.16	4.57	4.97
0.25	1.24	1.39	1.55	1.69	1.85	2.00	2.16	2.33	2.50	2.67	2.85	3.21	3.59	3.97	4.36	4.76
0.30	1.16	1.31	1.46	1.60	1.74	1.89	2.04	2.20	2.36	2.53	2.70	3.05	3.42	3.79	4.18	4.57
0.40	1.00	1.13	1.27	1.41	1.53	1.66	1.80	1.94	2.09	2.25	2.41	2.74	3.09	3.45	3.83	4.21
0.50	0.869	0.978	1.09	1.22	1.34	1.46	1.59	1.72	1.85	2.00	2.15	2.47	2.80	3.15	3.51	3.88
0.60	0.759	0.854	0.953	1.06	1.18	1.29	1.41	1.53	1.65	1.79	1.93	2.23	2.55	2.89	3.24	3.60
0.70	0.670	0.753	0.840	0.936	1.04	1.15	1.26	1.37	1.48	1.61	1.74	2.03	2.33	2.65	2.99	3.34
0.80	0.598	0.672	0.749	0.834	0.927	1.03	1.13	1.23	1.34	1.46	1.58	1.85	2.14	2.45	2.78	3.12
0.90	0.539	0.605	0.675	0.750	0.834	0.925	1.02	1.12	1.23	1.33	1.45	1.71	1.98	2.28	2.59	2.91
1.00	0.490	0.550	0.613	0.681	0.758	0.842	0.933	1.03	1.12	1.23	1.34	1.58	1.84	2.12	2.42	2.73
1.20	0.413	0.464	0.517	0.574	0.639	0.712	0.791	0.878	0.964	1.06	1.15	1.36	1.60	1.86	2.13	2.41
1.40	0.357	0.401	0.446	0.496	0.552	0.616	0.686	0.764	0.841	0.921	1.01	1.20	1.41	1.64	1.89	2.15
1.60	0.314	0.352	0.392	0.436	0.485	0.542	0.605	0.675	0.745	0.818	0.897	1.07	1.26	1.47	1.70	1.94
1.80	0.280	0.314	0.350	0.389	0.433	0.484	0.541	0.604	0.668	0.735	0.806	0.962	1.14	1.33	1.54	1.77
2.00	0.253	0.283	0.315	0.351	0.391	0.437	0.489	0.546	0.606	0.666	0.731	0.875	1.04	1.21	1.41	1.62
2.20	0.230	0.258	0.287	0.319	0.356	0.398	0.446	0.499	0.554	0.609	0.669	0.801	0.950	1.11	1.29	1.49
2.40	0.211	0.237	0.263	0.293	0.327	0.366	0.410	0.458	0.510	0.561	0.616	0.738	0.877	1.03	1.20	1.38
2.60	0.195	0.219	0.243	0.271	0.302	0.338	0.379	0.424	0.472	0.520	0.571	0.685	0.814	0.957	1.12	1.28
2.80	0.182	0.203	0.226	0.252	0.281	0.314	0.352	0.394	0.439	0.484	0.532	0.639	0.759	0.894	1.04	1.20
3.00	0.170	0.190	0.211	0.235	0.262	0.294	0.329	0.368	0.411	0.453	0.498	0.598	0.711	0.838	0.977	1.13
x	0.000	0.005	0.017	0.035	0.057	0.083	0.113	0.144	0.178	0.213	0.250	0.327	0.408	0.492	0.579	0.667
y	0.500	0.455	0.417	0.385	0.357	0.333	0.313	0.294	0.278	0.263	0.250	0.227	0.208	0.192	0.179	0.167

Table 8-44 (cont.).
Coefficients C for Eccentrically Loaded Weld Groups
Angle = ±30°

$$\phi R_n = CC_1 Dl \qquad C_{min} = \frac{P_u}{C_1 Dl} \qquad D_{min} = \frac{P_u}{CC_1 l} \qquad l_{min} = \frac{P_u}{CC_1 D}$$

where

P_u = factored force, kips

D = number of sixteenths-of-an-inch in the fillet weld size

l = characteristic length of weld group, in.

a = e_x / l, in.

e_x = horizontal component of eccentricity of P_u with respect to centroid of weld group, in.

C = coefficient tabulated below which includes $\phi = 0.75$

C_1 = electode strength coefficient from Table 8-37 (1.0 for E70XX electrodes)

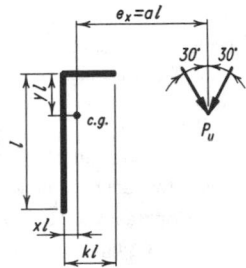

a	\multicolumn{17}{c}{k}
	0	0.1	0.2	0.3	0.4	0.5	0.6	0.7	0.8	0.9	1.0	1.2	1.4	1.6	1.8	2.0
0.00	1.64	1.83	2.03	2.22	2.42	2.61	2.81	3.01	3.20	3.40	3.59	3.98	4.37	4.76	5.15	5.54
0.10	1.52	1.68	1.85	2.03	2.21	2.39	2.58	2.77	2.97	3.16	3.36	3.77	4.18	4.59	5.00	5.41
0.15	1.44	1.60	1.76	1.91	2.08	2.25	2.42	2.60	2.78	2.97	3.16	3.55	3.95	4.37	4.79	5.20
0.20	1.36	1.51	1.67	1.83	1.98	2.14	2.30	2.47	2.64	2.82	3.00	3.37	3.76	4.16	4.57	4.98
0.25	1.28	1.43	1.58	1.73	1.88	2.03	2.18	2.34	2.51	2.68	2.85	3.21	3.59	3.98	4.37	4.78
0.30	1.20	1.34	1.49	1.63	1.78	1.92	2.06	2.22	2.37	2.54	2.71	3.06	3.43	3.81	4.20	4.60
0.40	1.05	1.18	1.31	1.44	1.58	1.71	1.84	1.98	2.13	2.28	2.44	2.78	3.13	3.50	3.88	4.27
0.50	0.921	1.03	1.15	1.27	1.40	1.52	1.64	1.77	1.91	2.06	2.21	2.53	2.87	3.23	3.60	3.98
0.60	0.812	0.910	1.02	1.13	1.25	1.36	1.47	1.59	1.72	1.86	2.00	2.31	2.64	2.98	3.34	3.71
0.70	0.722	0.811	0.908	1.01	1.12	1.22	1.33	1.44	1.56	1.69	1.83	2.12	2.43	2.77	3.12	3.48
0.80	0.647	0.729	0.815	0.908	1.01	1.11	1.21	1.31	1.43	1.55	1.67	1.95	2.25	2.58	2.91	3.26
0.90	0.586	0.660	0.736	0.820	0.912	1.01	1.10	1.20	1.31	1.42	1.54	1.81	2.10	2.41	2.73	3.07
1.00	0.535	0.601	0.671	0.747	0.831	0.920	1.01	1.11	1.21	1.31	1.43	1.68	1.96	2.25	2.56	2.89
1.20	0.454	0.510	0.568	0.632	0.704	0.783	0.868	0.952	1.04	1.14	1.24	1.47	1.72	1.99	2.28	2.58
1.40	0.393	0.441	0.492	0.547	0.610	0.680	0.756	0.835	0.916	1.00	1.10	1.30	1.53	1.77	2.04	2.32
1.60	0.347	0.389	0.433	0.482	0.537	0.600	0.668	0.742	0.814	0.893	0.978	1.16	1.37	1.60	1.84	2.10
1.80	0.310	0.347	0.387	0.430	0.480	0.536	0.599	0.666	0.733	0.805	0.882	1.05	1.24	1.45	1.67	1.92
2.00	0.280	0.314	0.349	0.389	0.433	0.485	0.542	0.604	0.666	0.732	0.802	0.958	1.13	1.33	1.54	1.76
2.20	0.255	0.286	0.318	0.354	0.395	0.442	0.495	0.552	0.610	0.670	0.735	0.880	1.04	1.22	1.42	1.63
2.40	0.234	0.263	0.292	0.325	0.363	0.406	0.455	0.508	0.562	0.618	0.679	0.813	0.963	1.13	1.31	1.51
2.60	0.217	0.243	0.270	0.301	0.335	0.376	0.421	0.471	0.521	0.573	0.630	0.754	0.896	1.05	1.22	1.41
2.80	0.201	0.226	0.251	0.280	0.312	0.349	0.392	0.438	0.486	0.535	0.588	0.704	0.836	0.984	1.15	1.32
3.00	0.188	0.211	0.235	0.261	0.292	0.327	0.366	0.410	0.455	0.501	0.550	0.660	0.784	0.924	1.08	1.24
x	0.000	0.005	0.017	0.035	0.057	0.083	0.113	0.144	0.178	0.213	0.250	0.327	0.408	0.492	0.579	0.667
y	0.500	0.455	0.417	0.385	0.357	0.333	0.313	0.294	0.278	0.263	0.250	0.227	0.208	0.192	0.179	0.167

Table 8-44 (cont.).
Coefficients C for Eccentrically Loaded Weld Groups
Angle = ±45°

$$\phi R_n = CC_1 Dl \qquad C_{min} = \frac{P_u}{C_1 Dl} \qquad D_{min} = \frac{P_u}{CC_1 l} \qquad l_{min} = \frac{P_u}{CC_1 D}$$

where

P_u = factored force, kips

D = number of sixteenths-of-an-inch in the fillet weld size

l = characteristic length of weld group, in.

a = e_x / l, in.

e_x = horizontal component of eccentricity of P_u with respect to centroid of weld group, in.

C = coefficient tabulated below which includes $\phi = 0.75$

C_1 = electode strength coefficient from Table 8-37 (1.0 for E70XX electrodes)

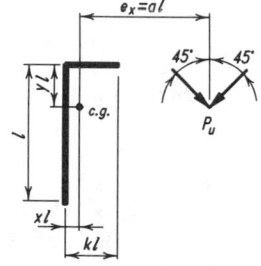

a	0	0.1	0.2	0.3	0.4	0.5	0.6	0.7	0.8	0.9	1.0	1.2	1.4	1.6	1.8	2.0
								k								
0.00	1.81	1.99	2.17	2.35	2.53	2.71	2.89	3.07	3.25	3.43	3.61	3.97	4.33	4.70	5.06	5.42
0.10	1.68	1.83	1.99	2.14	2.30	2.47	2.64	2.82	3.00	3.18	3.37	3.76	4.15	4.55	4.95	5.35
0.15	1.57	1.71	1.86	2.01	2.17	2.33	2.50	2.67	2.84	3.02	3.21	3.59	3.99	4.39	4.80	5.21
0.20	1.47	1.61	1.75	1.90	2.06	2.22	2.37	2.54	2.71	2.88	3.06	3.43	3.83	4.23	4.64	5.06
0.25	1.39	1.52	1.65	1.80	1.96	2.11	2.26	2.42	2.58	2.75	2.93	3.29	3.68	4.07	4.48	4.90
0.30	1.31	1.43	1.56	1.71	1.86	2.01	2.15	2.31	2.46	2.63	2.80	3.16	3.54	3.93	4.33	4.75
0.40	1.16	1.27	1.40	1.53	1.67	1.81	1.95	2.10	2.25	2.40	2.57	2.92	3.28	3.66	4.06	4.47
0.50	1.03	1.14	1.25	1.38	1.51	1.64	1.77	1.91	2.05	2.20	2.36	2.70	3.05	3.43	3.81	4.21
0.60	0.921	1.02	1.13	1.24	1.36	1.49	1.62	1.74	1.88	2.02	2.18	2.50	2.85	3.21	3.59	3.98
0.70	0.829	0.919	1.02	1.13	1.24	1.36	1.48	1.60	1.73	1.87	2.01	2.32	2.66	3.02	3.39	3.77
0.80	0.751	0.835	0.928	1.03	1.14	1.25	1.36	1.47	1.60	1.73	1.87	2.17	2.49	2.84	3.20	3.58
0.90	0.685	0.764	0.849	0.943	1.05	1.15	1.26	1.36	1.48	1.61	1.74	2.03	2.34	2.68	3.03	3.39
1.00	0.629	0.702	0.782	0.870	0.966	1.07	1.16	1.27	1.38	1.50	1.63	1.90	2.21	2.53	2.87	3.22
1.20	0.538	0.603	0.674	0.751	0.836	0.926	1.01	1.11	1.21	1.32	1.43	1.69	1.97	2.27	2.59	2.92
1.40	0.469	0.527	0.589	0.655	0.730	0.811	0.894	0.979	1.07	1.17	1.28	1.51	1.77	2.05	2.34	2.65
1.60	0.416	0.467	0.521	0.580	0.646	0.721	0.799	0.877	0.961	1.05	1.15	1.36	1.60	1.86	2.14	2.43
1.80	0.373	0.419	0.466	0.519	0.579	0.647	0.720	0.793	0.869	0.953	1.04	1.24	1.46	1.70	1.96	2.23
2.00	0.338	0.379	0.422	0.470	0.524	0.587	0.654	0.722	0.794	0.871	0.954	1.14	1.34	1.56	1.80	2.06
2.20	0.308	0.346	0.385	0.429	0.479	0.536	0.599	0.663	0.730	0.801	0.878	1.05	1.24	1.45	1.67	1.92
2.40	0.284	0.318	0.355	0.395	0.441	0.493	0.552	0.613	0.675	0.741	0.812	0.971	1.15	1.34	1.56	1.78
2.60	0.263	0.295	0.328	0.365	0.408	0.457	0.511	0.570	0.627	0.689	0.756	0.904	1.07	1.26	1.46	1.67
2.80	0.245	0.274	0.305	0.340	0.380	0.425	0.476	0.532	0.585	0.643	0.707	0.845	1.00	1.18	1.37	1.57
3.00	0.229	0.256	0.286	0.318	0.355	0.398	0.446	0.498	0.549	0.604	0.663	0.794	0.942	1.11	1.29	1.48
x	0.000	0.005	0.017	0.035	0.057	0.083	0.113	0.144	0.178	0.213	0.250	0.327	0.408	0.492	0.579	0.667
y	0.500	0.455	0.417	0.385	0.357	0.333	0.313	0.294	0.278	0.263	0.250	0.227	0.208	0.192	0.179	0.167

Table 8-44 (cont.).
Coefficients C for Eccentrically Loaded Weld Groups
Angle = ±60°

$$\phi R_n = CC_1 Dl \qquad C_{min} = \frac{P_u}{C_1 Dl} \qquad D_{min} = \frac{P_u}{CC_1 l} \qquad l_{min} = \frac{P_u}{CC_1 D}$$

where

P_u = factored force, kips

D = number of sixteenths-of-an-inch in the fillet weld size

l = characteristic length of weld group, in.

a = e_x / l, in.

e_x = horizontal component of eccentricity of P_u with respect to centroid of weld group, in.

C = coefficient tabulated below which includes ϕ = 0.75

C_1 = electode strength coefficient from Table 8-37 (1.0 for E70XX electrodes)

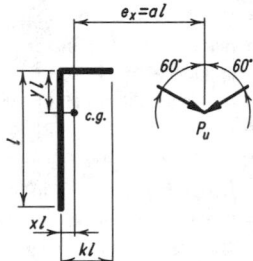

a	\multicolumn{17}{c}{k}
	0	0.1	0.2	0.3	0.4	0.5	0.6	0.7	0.8	0.9	1.0	1.2	1.4	1.6	1.8	2.0
---	---	---	---	---	---	---	---	---	---	---	---	---	---	---	---	---
0.00	1.95	2.12	2.28	2.44	2.61	2.77	2.94	3.10	3.26	3.43	3.59	3.92	4.25	4.57	4.90	5.23
0.10	1.83	1.94	2.07	2.20	2.35	2.51	2.68	2.85	3.03	3.21	3.40	3.78	4.17	4.55	4.90	5.23
0.15	1.73	1.84	1.96	2.10	2.25	2.40	2.57	2.74	2.92	3.10	3.28	3.67	4.07	4.47	4.86	5.23
0.20	1.63	1.74	1.87	2.00	2.15	2.31	2.47	2.64	2.81	2.99	3.18	3.56	3.96	4.37	4.78	5.16
0.25	1.55	1.66	1.78	1.92	2.06	2.22	2.38	2.55	2.72	2.89	3.07	3.45	3.85	4.27	4.68	5.08
0.30	1.47	1.58	1.70	1.83	1.98	2.13	2.29	2.46	2.63	2.80	2.98	3.35	3.75	4.16	4.58	5.00
0.40	1.34	1.44	1.56	1.69	1.83	1.98	2.13	2.29	2.45	2.62	2.80	3.16	3.55	3.96	4.38	4.81
0.50	1.22	1.32	1.43	1.56	1.69	1.83	1.99	2.14	2.30	2.46	2.63	2.99	3.37	3.77	4.19	4.61
0.60	1.12	1.21	1.32	1.44	1.57	1.71	1.85	2.00	2.15	2.31	2.48	2.83	3.21	3.60	4.01	4.43
0.70	1.03	1.12	1.22	1.33	1.46	1.59	1.73	1.88	2.02	2.18	2.34	2.68	3.05	3.44	3.85	4.26
0.80	0.945	1.03	1.13	1.24	1.36	1.49	1.63	1.76	1.90	2.05	2.21	2.55	2.91	3.29	3.69	4.09
0.90	0.874	0.958	1.05	1.16	1.27	1.40	1.53	1.66	1.79	1.94	2.09	2.42	2.77	3.15	3.53	3.93
1.00	0.812	0.893	0.983	1.08	1.19	1.31	1.44	1.56	1.69	1.83	1.98	2.30	2.65	3.01	3.39	3.79
1.20	0.709	0.783	0.865	0.957	1.06	1.17	1.29	1.40	1.52	1.65	1.79	2.09	2.42	2.76	3.13	3.51
1.40	0.626	0.695	0.771	0.855	0.948	1.05	1.16	1.26	1.38	1.50	1.63	1.91	2.22	2.55	2.89	3.25
1.60	0.560	0.623	0.693	0.771	0.857	0.952	1.05	1.15	1.25	1.36	1.49	1.75	2.04	2.35	2.68	3.03
1.80	0.506	0.564	0.629	0.701	0.781	0.868	0.957	1.05	1.15	1.25	1.37	1.61	1.89	2.18	2.50	2.83
2.00	0.461	0.515	0.575	0.642	0.716	0.798	0.880	0.964	1.06	1.16	1.26	1.49	1.75	2.03	2.33	2.64
2.20	0.423	0.473	0.530	0.590	0.659	0.735	0.813	0.891	0.978	1.07	1.17	1.39	1.63	1.90	2.18	2.48
2.40	0.391	0.438	0.489	0.545	0.608	0.680	0.755	0.829	0.910	0.997	1.09	1.30	1.53	1.78	2.05	2.33
2.60	0.363	0.407	0.454	0.506	0.565	0.632	0.704	0.774	0.850	0.932	1.02	1.22	1.43	1.67	1.93	2.20
2.80	0.339	0.380	0.424	0.472	0.527	0.590	0.658	0.726	0.797	0.875	0.958	1.14	1.35	1.57	1.82	2.08
3.00	0.317	0.356	0.397	0.442	0.494	0.553	0.618	0.683	0.750	0.824	0.903	1.08	1.27	1.49	1.72	1.97
x	0.000	0.005	0.017	0.035	0.057	0.083	0.113	0.144	0.178	0.213	0.250	0.327	0.408	0.492	0.579	0.667
y	0.500	0.455	0.417	0.385	0.357	0.333	0.313	0.294	0.278	0.263	0.250	0.227	0.208	0.192	0.179	0.167

Table 8-44 (cont.).
Coefficients C for Eccentrically Loaded Weld Groups
Angle = ±75°

$$\phi R_n = CC_1 Dl \qquad C_{min} = \frac{P_u}{C_1 Dl} \qquad D_{min} = \frac{P_u}{CC_1 l} \qquad l_{min} = \frac{P_u}{CC_1 D}$$

where

P_u = factored force, kips

D = number of sixteenths-of-an-inch
in the fillet weld size

l = characteristic length of weld group, in.

a = e_x / l, in.

e_x = horizontal component of eccentricity of
P_u with respect to centroid of weld group, in.

C = coefficient tabulated below which includes ϕ = 0.75

C_1 = electode strength coefficient from Table 8-37
(1.0 for E70XX electrodes)

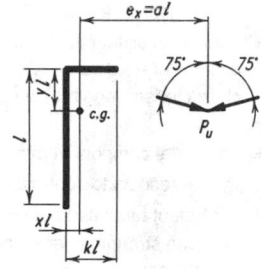

								k								
a	0	0.1	0.2	0.3	0.4	0.5	0.6	0.7	0.8	0.9	1.0	1.2	1.4	1.6	1.8	2.0
0.00	2.05	2.20	2.35	2.50	2.65	2.79	2.94	3.09	3.24	3.39	3.54	3.83	4.13	4.43	4.72	5.02
0.10	1.94	2.01	2.11	2.23	2.37	2.52	2.69	2.87	3.05	3.24	3.44	3.80	4.09	4.38	4.67	4.96
0.15	1.88	1.95	2.05	2.17	2.31	2.47	2.63	2.81	3.00	3.19	3.39	3.77	4.09	4.38	4.67	4.96
0.20	1.82	1.90	2.00	2.12	2.26	2.41	2.58	2.76	2.95	3.14	3.34	3.73	4.09	4.38	4.67	4.96
0.25	1.76	1.84	1.95	2.07	2.21	2.36	2.53	2.71	2.89	3.09	3.29	3.69	4.06	4.38	4.67	4.96
0.30	1.71	1.79	1.90	2.02	2.16	2.31	2.48	2.66	2.84	3.04	3.24	3.65	4.03	4.38	4.67	4.96
0.40	1.62	1.70	1.81	1.93	2.07	2.22	2.39	2.56	2.75	2.94	3.14	3.54	3.95	4.32	4.66	4.96
0.50	1.53	1.62	1.72	1.84	1.98	2.13	2.30	2.47	2.66	2.85	3.04	3.44	3.85	4.25	4.61	4.94
0.60	1.46	1.54	1.64	1.76	1.90	2.05	2.21	2.38	2.57	2.76	2.95	3.34	3.75	4.16	4.54	4.90
0.70	1.38	1.47	1.57	1.69	1.82	1.97	2.13	2.30	2.48	2.67	2.86	3.25	3.65	4.07	4.46	4.84
0.80	1.31	1.40	1.50	1.62	1.75	1.90	2.06	2.22	2.40	2.59	2.77	3.16	3.56	3.97	4.38	4.76
0.90	1.25	1.34	1.44	1.55	1.68	1.83	1.98	2.15	2.33	2.51	2.69	3.07	3.47	3.88	4.29	4.68
1.00	1.19	1.28	1.38	1.49	1.62	1.76	1.91	2.08	2.25	2.43	2.60	2.98	3.38	3.79	4.20	4.60
1.20	1.09	1.17	1.27	1.38	1.50	1.64	1.79	1.95	2.11	2.28	2.45	2.82	3.21	3.61	4.02	4.43
1.40	0.994	1.07	1.17	1.28	1.40	1.53	1.67	1.83	1.98	2.14	2.31	2.67	3.04	3.44	3.85	4.26
1.60	0.914	0.992	1.08	1.19	1.30	1.43	1.57	1.72	1.86	2.02	2.18	2.53	2.89	3.28	3.68	4.10
1.80	0.845	0.920	1.01	1.11	1.22	1.34	1.47	1.62	1.76	1.90	2.06	2.39	2.75	3.13	3.52	3.93
2.00	0.784	0.857	0.941	1.04	1.14	1.26	1.39	1.52	1.66	1.80	1.95	2.27	2.62	2.99	3.37	3.77
2.20	0.730	0.801	0.881	0.973	1.08	1.19	1.31	1.44	1.57	1.70	1.85	2.16	2.50	2.86	3.23	3.62
2.40	0.683	0.751	0.828	0.916	1.01	1.12	1.24	1.36	1.49	1.62	1.76	2.06	2.39	2.73	3.10	3.48
2.60	0.641	0.706	0.781	0.865	0.960	1.07	1.18	1.29	1.41	1.54	1.67	1.96	2.28	2.62	2.97	3.35
2.80	0.604	0.666	0.738	0.819	0.910	1.01	1.12	1.23	1.34	1.46	1.59	1.87	2.18	2.51	2.86	3.22
3.00	0.570	0.631	0.700	0.777	0.863	0.961	1.07	1.17	1.28	1.39	1.52	1.79	2.09	2.41	2.74	3.10
x	0.000	0.005	0.017	0.035	0.057	0.083	0.113	0.144	0.178	0.213	0.250	0.327	0.408	0.492	0.579	0.667
y	0.500	0.455	0.417	0.385	0.357	0.333	0.313	0.294	0.278	0.263	0.250	0.227	0.208	0.192	0.179	0.167

Table 8-45.
Coefficients C for Eccentrically Loaded Weld Groups
Angle = 0°

$$\phi R_n = CC_1 Dl \qquad C_{min} = \frac{P_u}{C_1 Dl} \qquad D_{min} = \frac{P_u}{CC_1 l} \qquad l_{min} = \frac{P_u}{CC_1 D}$$

where

P_u = factored force, kips

D = number of sixteenths-of-an-inch
   in the fillet weld size

l = characteristic length of weld group, in.

a = e_x / l, in.

e_x = horizontal component of eccentricity of
   P_u with respect to centroid of weld group, in.

C = coefficient tabulated below which includes $\phi = 0.75$

C_1 = electode strength coefficient from Table 8-37
   (1.0 for E70XX electrodes)

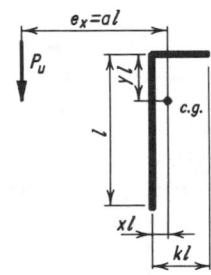

a	\multicolumn{17}{c}{k}
	0	0.1	0.2	0.3	0.4	0.5	0.6	0.7	0.8	0.9	1.0	1.2	1.4	1.6	1.8	2.0
---	---	---	---	---	---	---	---	---	---	---	---	---	---	---	---	---
0.00	1.39	1.60	1.81	2.02	2.23	2.44	2.65	2.85	3.06	3.27	3.48	3.90	4.32	4.73	5.15	5.57
0.10	1.39	1.55	1.74	1.93	2.12	2.31	2.49	2.66	2.83	2.99	3.14	3.45	3.76	4.08	4.42	4.76
0.15	1.38	1.53	1.71	1.88	2.06	2.22	2.38	2.54	2.69	2.83	2.98	3.28	3.59	3.91	4.24	4.58
0.20	1.32	1.47	1.63	1.79	1.94	2.09	2.23	2.37	2.52	2.67	2.81	3.12	3.42	3.75	4.08	4.41
0.25	1.24	1.38	1.52	1.66	1.80	1.94	2.07	2.21	2.35	2.51	2.66	2.96	3.27	3.59	3.91	4.25
0.30	1.16	1.29	1.42	1.54	1.67	1.79	1.92	2.06	2.20	2.35	2.51	2.82	3.12	3.44	3.76	4.10
0.40	0.998	1.11	1.22	1.32	1.42	1.54	1.66	1.80	1.94	2.09	2.24	2.55	2.85	3.16	3.48	3.81
0.50	0.860	0.958	1.05	1.14	1.24	1.34	1.46	1.58	1.72	1.86	2.01	2.31	2.61	2.91	3.22	3.54
0.60	0.748	0.833	0.913	0.998	1.09	1.18	1.29	1.41	1.54	1.67	1.81	2.10	2.40	2.69	2.99	3.31
0.70	0.659	0.733	0.805	0.884	0.967	1.06	1.15	1.26	1.38	1.51	1.64	1.92	2.21	2.50	2.79	3.09
0.80	0.586	0.652	0.719	0.791	0.868	0.951	1.04	1.14	1.25	1.37	1.50	1.77	2.05	2.32	2.61	2.90
0.90	0.527	0.586	0.648	0.715	0.786	0.864	0.949	1.04	1.15	1.26	1.38	1.63	1.90	2.17	2.44	2.73
1.00	0.478	0.532	0.589	0.651	0.717	0.791	0.870	0.957	1.05	1.16	1.27	1.51	1.77	2.03	2.29	2.57
1.20	0.403	0.448	0.497	0.551	0.609	0.674	0.744	0.820	0.904	0.997	1.10	1.31	1.55	1.80	2.04	2.30
1.40	0.348	0.387	0.430	0.476	0.528	0.586	0.648	0.716	0.790	0.874	0.963	1.16	1.37	1.60	1.83	2.07
1.60	0.305	0.340	0.378	0.419	0.466	0.518	0.573	0.634	0.702	0.776	0.856	1.03	1.23	1.44	1.66	1.88
1.80	0.272	0.303	0.337	0.374	0.416	0.463	0.514	0.569	0.630	0.697	0.771	0.932	1.11	1.31	1.51	1.72
2.00	0.245	0.273	0.304	0.338	0.376	0.419	0.465	0.516	0.572	0.633	0.700	0.848	1.02	1.20	1.39	1.58
2.20	0.223	0.249	0.277	0.308	0.343	0.382	0.425	0.472	0.523	0.579	0.641	0.778	0.932	1.10	1.28	1.47
2.40	0.205	0.228	0.254	0.283	0.315	0.351	0.391	0.434	0.482	0.534	0.591	0.718	0.861	1.02	1.18	1.36
2.60	0.189	0.211	0.235	0.261	0.291	0.325	0.362	0.402	0.446	0.495	0.548	0.667	0.800	0.945	1.10	1.27
2.80	0.176	0.196	0.218	0.243	0.271	0.302	0.337	0.375	0.416	0.461	0.511	0.622	0.747	0.882	1.03	1.18
3.00	0.164	0.183	0.204	0.227	0.253	0.282	0.315	0.350	0.389	0.431	0.478	0.582	0.700	0.825	0.962	1.11
x	0.000	0.005	0.017	0.035	0.057	0.083	0.113	0.144	0.178	0.213	0.250	0.327	0.408	0.492	0.579	0.667
y	0.500	0.455	0.417	0.385	0.357	0.333	0.313	0.294	0.278	0.263	0.250	0.227	0.208	0.192	0.179	0.167

Table 8-45 (cont.).
Coefficients C for Eccentrically Loaded Weld Groups
Angle = ±15°

$$\phi R_n = CC_1 Dl \qquad C_{min} = \frac{P_u}{C_1 Dl} \qquad D_{min} = \frac{P_u}{CC_1 l} \qquad l_{min} = \frac{P_u}{CC_1 D}$$

where

P_u = factored force, kips

D = number of sixteenths-of-an-inch in the fillet weld size

l = characteristic length of weld group, in.

a = e_x / l, in.

e_x = horizontal component of eccentricity of P_u with respect to centroid of weld group, in.

C = coefficient tabulated below which includes $\phi = 0.75$

C_1 = electode strength coefficient from Table 8-37 (1.0 for E70XX electrodes)

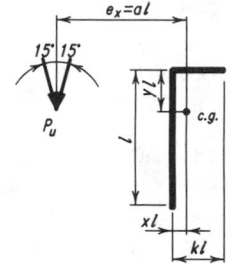

a							k									
	0	0.1	0.2	0.3	0.4	0.5	0.6	0.7	0.8	0.9	1.0	1.2	1.4	1.6	1.8	2.0
0.00	1.48	1.69	1.89	2.10	2.31	2.51	2.72	2.92	3.13	3.33	3.54	3.95	4.36	4.77	5.18	5.59
0.10	1.42	1.57	1.74	1.92	2.09	2.27	2.44	2.61	2.78	2.95	3.12	3.45	3.78	4.12	4.46	4.81
0.15	1.38	1.54	1.70	1.86	2.02	2.19	2.35	2.51	2.67	2.83	2.98	3.30	3.62	3.95	4.29	4.64
0.20	1.32	1.47	1.63	1.78	1.94	2.09	2.22	2.36	2.51	2.66	2.82	3.13	3.45	3.78	4.12	4.46
0.25	1.24	1.39	1.54	1.68	1.81	1.94	2.07	2.21	2.35	2.50	2.66	2.98	3.30	3.63	3.96	4.30
0.30	1.16	1.30	1.43	1.55	1.68	1.80	1.93	2.06	2.21	2.36	2.51	2.83	3.15	3.48	3.81	4.15
0.40	1.00	1.12	1.23	1.33	1.44	1.55	1.68	1.81	1.95	2.10	2.25	2.57	2.88	3.20	3.53	3.87
0.50	0.869	0.970	1.06	1.15	1.25	1.36	1.48	1.60	1.74	1.88	2.02	2.33	2.64	2.95	3.28	3.61
0.60	0.759	0.847	0.926	1.01	1.10	1.20	1.31	1.43	1.56	1.69	1.83	2.12	2.43	2.74	3.05	3.37
0.70	0.670	0.746	0.820	0.899	0.983	1.07	1.18	1.29	1.41	1.53	1.67	1.95	2.25	2.54	2.84	3.15
0.80	0.598	0.666	0.734	0.807	0.885	0.969	1.06	1.17	1.28	1.40	1.52	1.79	2.08	2.37	2.66	2.96
0.90	0.539	0.600	0.663	0.730	0.803	0.881	0.968	1.06	1.17	1.28	1.40	1.66	1.93	2.21	2.49	2.78
1.00	0.490	0.545	0.604	0.666	0.734	0.807	0.888	0.979	1.08	1.18	1.30	1.54	1.80	2.08	2.34	2.63
1.20	0.413	0.461	0.511	0.565	0.625	0.690	0.761	0.840	0.927	1.02	1.12	1.34	1.58	1.84	2.09	2.35
1.40	0.357	0.398	0.442	0.490	0.543	0.601	0.664	0.734	0.812	0.897	0.988	1.19	1.40	1.64	1.88	2.12
1.60	0.314	0.350	0.389	0.432	0.479	0.532	0.589	0.651	0.721	0.798	0.879	1.06	1.26	1.48	1.70	1.93
1.80	0.280	0.312	0.347	0.386	0.429	0.476	0.528	0.585	0.649	0.718	0.793	0.958	1.14	1.34	1.55	1.77
2.00	0.253	0.282	0.313	0.348	0.388	0.431	0.479	0.531	0.589	0.652	0.721	0.873	1.04	1.23	1.43	1.63
2.20	0.230	0.257	0.286	0.318	0.354	0.393	0.438	0.486	0.538	0.597	0.660	0.801	0.958	1.13	1.32	1.51
2.40	0.211	0.236	0.262	0.292	0.325	0.362	0.403	0.447	0.496	0.550	0.609	0.740	0.886	1.05	1.22	1.40
2.60	0.195	0.218	0.242	0.270	0.301	0.335	0.373	0.415	0.460	0.510	0.565	0.687	0.824	0.975	1.13	1.31
2.80	0.182	0.203	0.225	0.251	0.280	0.312	0.348	0.386	0.428	0.476	0.527	0.641	0.770	0.910	1.06	1.22
3.00	0.170	0.189	0.211	0.234	0.261	0.291	0.325	0.361	0.401	0.445	0.494	0.601	0.722	0.852	0.993	1.15
x	0.000	0.005	0.017	0.035	0.057	0.083	0.113	0.144	0.178	0.213	0.250	0.327	0.408	0.492	0.579	0.667
y	0.500	0.455	0.417	0.385	0.357	0.333	0.313	0.294	0.278	0.263	0.250	0.227	0.208	0.192	0.179	0.167

Table 8-45 (cont.).
Coefficients C for Eccentrically Loaded Weld Groups
Angle = ±30°

$$\phi R_n = CC_1Dl \qquad C_{min} = \frac{P_u}{C_1Dl} \qquad D_{min} = \frac{P_u}{CC_1l} \qquad l_{min} = \frac{P_u}{CC_1D}$$

where

P_u = factored force, kips

D = number of sixteenths-of-an-inch in the fillet weld size

l = characteristic length of weld group, in.

a = e_x / l, in.

e_x = horizontal component of eccentricity of P_u with respect to centroid of weld group, in.

C = coefficient tabulated below which includes $\phi = 0.75$

C_1 = electode strength coefficient from Table 8-37 (1.0 for E70XX electrodes)

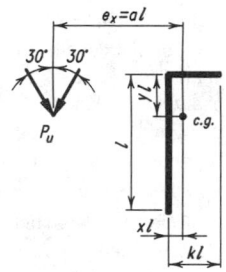

a	\multicolumn{17}{c}{k}															
	0	0.1	0.2	0.3	0.4	0.5	0.6	0.7	0.8	0.9	1.0	1.2	1.4	1.6	1.8	2.0
0.00	1.64	1.83	2.03	2.22	2.42	2.61	2.81	3.01	3.20	3.40	3.59	3.98	4.37	4.76	5.15	5.54
0.10	1.52	1.68	1.85	2.03	2.20	2.38	2.55	2.73	2.90	3.06	3.23	3.59	3.95	4.32	4.70	5.08
0.15	1.44	1.59	1.75	1.91	2.07	2.23	2.40	2.56	2.72	2.88	3.05	3.41	3.77	4.13	4.51	4.89
0.20	1.36	1.51	1.66	1.81	1.96	2.12	2.26	2.41	2.56	2.72	2.89	3.24	3.59	3.94	4.31	4.68
0.25	1.28	1.42	1.56	1.71	1.85	1.99	2.12	2.26	2.41	2.57	2.74	3.09	3.43	3.77	4.13	4.49
0.30	1.20	1.34	1.47	1.60	1.74	1.87	1.99	2.13	2.28	2.43	2.60	2.95	3.28	3.62	3.97	4.32
0.40	1.05	1.17	1.29	1.40	1.51	1.64	1.76	1.90	2.04	2.19	2.35	2.68	3.03	3.35	3.69	4.04
0.50	0.921	1.03	1.13	1.22	1.33	1.45	1.57	1.70	1.83	1.98	2.13	2.45	2.80	3.13	3.46	3.80
0.60	0.812	0.904	0.993	1.08	1.18	1.29	1.41	1.53	1.66	1.80	1.94	2.25	2.59	2.93	3.25	3.58
0.70	0.722	0.804	0.884	0.968	1.06	1.16	1.27	1.38	1.51	1.64	1.78	2.08	2.40	2.73	3.06	3.37
0.80	0.647	0.722	0.796	0.874	0.957	1.05	1.15	1.26	1.38	1.51	1.64	1.92	2.23	2.55	2.87	3.19
0.90	0.586	0.654	0.722	0.794	0.872	0.957	1.05	1.16	1.27	1.39	1.51	1.78	2.08	2.39	2.70	3.02
1.00	0.535	0.596	0.659	0.727	0.799	0.879	0.968	1.07	1.17	1.29	1.40	1.66	1.94	2.24	2.55	2.85
1.20	0.454	0.506	0.560	0.620	0.685	0.755	0.833	0.920	1.02	1.12	1.22	1.46	1.72	1.99	2.28	2.56
1.40	0.393	0.438	0.487	0.540	0.597	0.660	0.729	0.806	0.891	0.984	1.08	1.29	1.53	1.79	2.05	2.32
1.60	0.347	0.386	0.430	0.477	0.529	0.585	0.648	0.717	0.794	0.878	0.967	1.16	1.38	1.61	1.87	2.12
1.80	0.310	0.345	0.384	0.427	0.474	0.526	0.583	0.645	0.715	0.791	0.874	1.05	1.25	1.47	1.70	1.94
2.00	0.280	0.312	0.347	0.386	0.429	0.477	0.529	0.586	0.650	0.720	0.796	0.961	1.15	1.35	1.57	1.79
2.20	0.255	0.285	0.317	0.352	0.392	0.436	0.484	0.537	0.595	0.660	0.730	0.884	1.06	1.24	1.45	1.66
2.40	0.234	0.262	0.291	0.324	0.360	0.401	0.446	0.495	0.549	0.609	0.674	0.817	0.978	1.15	1.34	1.55
2.60	0.217	0.242	0.269	0.299	0.334	0.372	0.413	0.459	0.510	0.566	0.626	0.760	0.910	1.08	1.25	1.44
2.80	0.201	0.225	0.250	0.278	0.311	0.346	0.385	0.428	0.475	0.527	0.584	0.710	0.851	1.01	1.17	1.35
3.00	0.188	0.210	0.234	0.260	0.290	0.324	0.360	0.401	0.445	0.494	0.547	0.666	0.799	0.944	1.10	1.27
x	0.000	0.005	0.017	0.035	0.057	0.083	0.113	0.144	0.178	0.213	0.250	0.327	0.408	0.492	0.579	0.667
y	0.500	0.455	0.417	0.385	0.357	0.333	0.313	0.294	0.278	0.263	0.250	0.227	0.208	0.192	0.179	0.167

BOLTS, WELDS, AND CONNECTED ELEMENTS

Table 8-45 (cont.).
Coefficients C for Eccentrically Loaded Weld Groups
Angle = ±45°

$$\phi R_n = CC_1 Dl \qquad C_{min} = \frac{P_u}{C_1 Dl} \qquad D_{min} = \frac{P_u}{CC_1 l} \qquad l_{min} = \frac{P_u}{CC_1 D}$$

where

- P_u = factored force, kips
- D = number of sixteenths-of-an-inch in the fillet weld size
- l = characteristic length of weld group, in.
- a = e_x / l, in.
- e_x = horizontal component of eccentricity of P_u with respect to centroid of weld group, in.
- C = coefficient tabulated below which includes $\phi = 0.75$
- C_1 = electode strength coefficient from Table 8-37 (1.0 for E70XX electrodes)

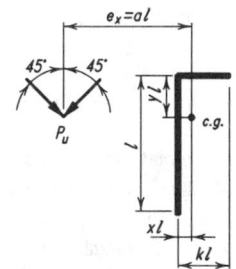

| a | \multicolumn{17}{c}{k} |
	0	0.1	0.2	0.3	0.4	0.5	0.6	0.7	0.8	0.9	1.0	1.2	1.4	1.6	1.8	2.0
0.00	1.81	1.99	2.17	2.35	2.53	2.71	2.89	3.07	3.25	3.43	3.61	3.97	4.33	4.70	5.06	5.42
0.10	1.68	1.83	1.99	2.15	2.32	2.50	2.67	2.85	3.02	3.19	3.37	3.74	4.10	4.48	4.85	5.21
0.15	1.57	1.71	1.86	2.02	2.18	2.36	2.53	2.69	2.86	3.03	3.21	3.58	3.95	4.33	4.71	5.08
0.20	1.47	1.60	1.74	1.88	2.04	2.21	2.38	2.54	2.71	2.88	3.06	3.42	3.80	4.18	4.56	4.94
0.25	1.39	1.51	1.64	1.77	1.92	2.07	2.24	2.41	2.58	2.75	2.93	3.29	3.66	4.04	4.42	4.80
0.30	1.31	1.43	1.55	1.67	1.81	1.95	2.11	2.28	2.45	2.62	2.80	3.16	3.54	3.91	4.29	4.66
0.40	1.16	1.27	1.38	1.49	1.63	1.77	1.92	2.07	2.23	2.39	2.57	2.92	3.27	3.64	4.03	4.41
0.50	1.03	1.13	1.23	1.35	1.48	1.61	1.74	1.88	2.03	2.19	2.36	2.72	3.06	3.40	3.76	4.14
0.60	0.921	1.01	1.11	1.22	1.34	1.46	1.58	1.72	1.86	2.01	2.18	2.52	2.87	3.21	3.55	3.91
0.70	0.829	0.911	1.00	1.11	1.21	1.33	1.45	1.58	1.71	1.86	2.01	2.35	2.70	3.04	3.37	3.71
0.80	0.751	0.828	0.915	1.01	1.11	1.22	1.33	1.45	1.58	1.72	1.87	2.19	2.53	2.88	3.21	3.54
0.90	0.685	0.757	0.839	0.927	1.02	1.12	1.22	1.34	1.47	1.60	1.74	2.05	2.38	2.73	3.05	3.38
1.00	0.629	0.696	0.773	0.854	0.938	1.03	1.14	1.25	1.36	1.49	1.63	1.92	2.24	2.58	2.91	3.23
1.20	0.538	0.598	0.666	0.735	0.810	0.892	0.985	1.09	1.20	1.31	1.43	1.70	2.00	2.31	2.64	2.96
1.40	0.469	0.523	0.582	0.644	0.712	0.786	0.868	0.960	1.06	1.16	1.28	1.52	1.80	2.09	2.40	2.71
1.60	0.416	0.464	0.516	0.572	0.633	0.700	0.774	0.857	0.948	1.05	1.15	1.38	1.63	1.90	2.19	2.50
1.80	0.373	0.416	0.463	0.513	0.570	0.631	0.699	0.775	0.858	0.947	1.04	1.25	1.49	1.74	2.02	2.31
2.00	0.338	0.377	0.419	0.466	0.518	0.574	0.636	0.705	0.782	0.865	0.954	1.15	1.37	1.60	1.86	2.14
2.20	0.308	0.344	0.383	0.426	0.474	0.526	0.583	0.647	0.718	0.795	0.878	1.06	1.26	1.49	1.73	1.99
2.40	0.284	0.317	0.353	0.392	0.436	0.485	0.538	0.598	0.664	0.736	0.812	0.983	1.17	1.38	1.61	1.85
2.60	0.263	0.293	0.327	0.363	0.405	0.450	0.500	0.555	0.617	0.684	0.756	0.916	1.09	1.29	1.50	1.73
2.80	0.245	0.273	0.304	0.338	0.377	0.420	0.467	0.518	0.576	0.638	0.707	0.857	1.03	1.21	1.41	1.62
3.00	0.229	0.256	0.285	0.317	0.353	0.393	0.437	0.486	0.540	0.599	0.663	0.805	0.964	1.14	1.33	1.53
x	0.000	0.005	0.017	0.035	0.057	0.083	0.113	0.144	0.178	0.213	0.250	0.327	0.408	0.492	0.579	0.667
y	0.500	0.455	0.417	0.385	0.357	0.333	0.313	0.294	0.278	0.263	0.250	0.227	0.208	0.192	0.179	0.167

Table 8-45 (cont.).
Coefficients C for Eccentrically Loaded Weld Groups
Angle = ±60°

$$\phi R_n = CC_1 Dl \qquad C_{min} = \frac{P_u}{C_1 Dl} \qquad D_{min} = \frac{P_u}{CC_1 l} \qquad l_{min} = \frac{P_u}{CC_1 D}$$

where

P_u = factored force, kips

D = number of sixteenths-of-an-inch in the fillet weld size

l = characteristic length of weld group, in.

a = e_x / l, in.

e_x = horizontal component of eccentricity of P_u with respect to centroid of weld group, in.

C = coefficient tabulated below which includes ϕ = 0.75

C_1 = electode strength coefficient from Table 8-37 (1.0 for E70XX electrodes)

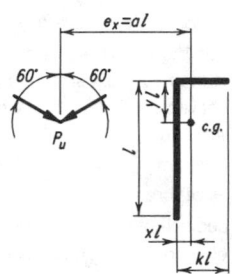

a	0	0.1	0.2	0.3	0.4	0.5	0.6	0.7	0.8	0.9	1.0	1.2	1.4	1.6	1.8	2.0
												k				
0.00	1.95	2.12	2.28	2.44	2.61	2.77	2.94	3.10	3.26	3.43	3.59	3.92	4.25	4.57	4.90	5.23
0.10	1.83	1.94	2.07	2.21	2.36	2.53	2.70	2.88	3.06	3.24	3.41	3.76	4.11	4.45	4.78	5.12
0.15	1.73	1.84	1.96	2.09	2.24	2.40	2.57	2.75	2.94	3.13	3.31	3.67	4.03	4.38	4.72	5.06
0.20	1.63	1.74	1.86	1.98	2.12	2.28	2.45	2.63	2.82	3.00	3.20	3.57	3.94	4.30	4.65	4.99
0.25	1.55	1.65	1.76	1.88	2.02	2.18	2.35	2.53	2.72	2.90	3.08	3.46	3.84	4.21	4.56	4.91
0.30	1.47	1.58	1.68	1.80	1.94	2.09	2.26	2.44	2.62	2.81	2.99	3.36	3.73	4.11	4.48	4.83
0.40	1.34	1.44	1.54	1.66	1.79	1.94	2.10	2.27	2.45	2.64	2.83	3.20	3.56	3.93	4.29	4.65
0.50	1.22	1.31	1.41	1.53	1.67	1.81	1.97	2.13	2.30	2.48	2.66	3.04	3.42	3.78	4.14	4.49
0.60	1.12	1.20	1.30	1.42	1.55	1.69	1.85	2.01	2.17	2.35	2.52	2.89	3.27	3.64	4.00	4.36
0.70	1.03	1.11	1.20	1.32	1.45	1.59	1.74	1.90	2.06	2.23	2.40	2.75	3.12	3.50	3.87	4.23
0.80	0.945	1.02	1.12	1.23	1.35	1.49	1.64	1.79	1.95	2.11	2.28	2.63	2.99	3.36	3.73	4.10
0.90	0.874	0.950	1.04	1.15	1.27	1.40	1.54	1.68	1.83	1.99	2.16	2.51	2.87	3.23	3.60	3.96
1.00	0.812	0.886	0.973	1.08	1.19	1.32	1.45	1.58	1.73	1.88	2.05	2.40	2.75	3.10	3.46	3.83
1.20	0.709	0.777	0.858	0.952	1.06	1.17	1.28	1.41	1.54	1.68	1.84	2.17	2.52	2.87	3.21	3.56
1.40	0.626	0.690	0.765	0.852	0.945	1.04	1.15	1.26	1.39	1.52	1.66	1.97	2.31	2.66	3.00	3.34
1.60	0.560	0.619	0.689	0.769	0.850	0.938	1.04	1.14	1.26	1.38	1.52	1.81	2.12	2.46	2.80	3.13
1.80	0.506	0.561	0.626	0.697	0.771	0.853	0.943	1.04	1.15	1.27	1.39	1.66	1.96	2.28	2.61	2.94
2.00	0.461	0.512	0.573	0.636	0.705	0.780	0.864	0.957	1.06	1.17	1.28	1.53	1.82	2.12	2.44	2.77
2.20	0.423	0.471	0.527	0.585	0.649	0.719	0.797	0.883	0.977	1.08	1.19	1.43	1.69	1.98	2.28	2.60
2.40	0.391	0.436	0.487	0.541	0.600	0.666	0.738	0.819	0.907	1.00	1.10	1.33	1.58	1.85	2.14	2.44
2.60	0.363	0.405	0.452	0.502	0.558	0.620	0.687	0.763	0.847	0.937	1.03	1.25	1.48	1.74	2.01	2.30
2.80	0.339	0.379	0.422	0.469	0.521	0.580	0.644	0.715	0.793	0.878	0.969	1.17	1.39	1.63	1.90	2.17
3.00	0.317	0.355	0.395	0.440	0.489	0.545	0.605	0.672	0.746	0.826	0.913	1.10	1.31	1.54	1.79	2.06
x	0.000	0.005	0.017	0.035	0.057	0.083	0.113	0.144	0.178	0.213	0.250	0.327	0.408	0.492	0.579	0.667
y	0.500	0.455	0.417	0.385	0.357	0.333	0.313	0.294	0.278	0.263	0.250	0.227	0.208	0.192	0.179	0.167

Table 8-45 (cont.).
Coefficients C for Eccentrically Loaded Weld Groups
Angle = ±75°

$$\phi R_n = CC_1 Dl \qquad C_{min} = \frac{P_u}{C_1 Dl} \qquad D_{min} = \frac{P_u}{CC_1 l} \qquad l_{min} = \frac{P_u}{CC_1 D}$$

where

P_u = factored force, kips

D = number of sixteenths-of-an-inch
in the fillet weld size

l = characteristic length of weld group, in.

a = e_x / l, in.

e_x = horizontal component of eccentricity of
P_u with respect to centroid of weld group, in.

C = coefficient tabulated below which includes ϕ = 0.75

C_1 = elecode strength coefficient from Table 8-37
(1.0 for E70XX electrodes)

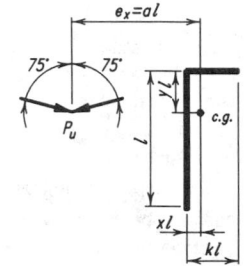

a	\multicolumn{16}{c}{k}
	0	0.1	0.2	0.3	0.4	0.5	0.6	0.7	0.8	0.9	1.0	1.2	1.4	1.6	1.8	2.0
0.00	2.05	2.20	2.35	2.50	2.65	2.79	2.94	3.09	3.24	3.39	3.54	3.83	4.13	4.43	4.72	5.02
0.10	1.94	2.01	2.08	2.20	2.34	2.49	2.65	2.81	2.97	3.13	3.28	3.59	3.93	4.25	4.57	4.88
0.15	1.88	1.94	2.03	2.15	2.29	2.44	2.61	2.77	2.94	3.10	3.25	3.56	3.87	4.20	4.52	4.84
0.20	1.82	1.89	1.97	2.09	2.24	2.39	2.56	2.73	2.90	3.07	3.23	3.53	3.84	4.15	4.48	4.80
0.25	1.76	1.83	1.92	2.04	2.19	2.35	2.52	2.69	2.86	3.03	3.20	3.51	3.81	4.11	4.42	4.75
0.30	1.71	1.78	1.87	2.00	2.14	2.30	2.47	2.65	2.82	3.00	3.17	3.49	3.79	4.09	4.39	4.70
0.40	1.62	1.69	1.79	1.91	2.05	2.21	2.38	2.56	2.74	2.92	3.10	3.43	3.75	4.05	4.35	4.65
0.50	1.53	1.61	1.70	1.83	1.97	2.12	2.29	2.47	2.66	2.84	3.03	3.38	3.70	4.02	4.32	4.61
0.60	1.46	1.53	1.63	1.75	1.89	2.05	2.21	2.39	2.57	2.76	2.95	3.31	3.65	3.97	4.28	4.58
0.70	1.38	1.46	1.56	1.68	1.82	1.97	2.14	2.31	2.49	2.68	2.87	3.25	3.60	3.93	4.24	4.55
0.80	1.31	1.39	1.49	1.61	1.75	1.90	2.06	2.24	2.41	2.60	2.79	3.17	3.54	3.88	4.20	4.51
0.90	1.25	1.33	1.43	1.54	1.68	1.83	1.99	2.16	2.34	2.52	2.71	3.10	3.47	3.82	4.16	4.47
1.00	1.19	1.27	1.37	1.48	1.62	1.77	1.93	2.10	2.27	2.45	2.64	3.03	3.41	3.76	4.10	4.43
1.20	1.09	1.16	1.26	1.37	1.50	1.64	1.80	1.97	2.14	2.31	2.50	2.87	3.26	3.64	3.99	4.34
1.40	0.994	1.07	1.16	1.27	1.40	1.54	1.69	1.85	2.01	2.19	2.36	2.73	3.12	3.50	3.87	4.23
1.60	0.914	0.987	1.08	1.18	1.30	1.44	1.58	1.74	1.90	2.07	2.24	2.59	2.97	3.36	3.74	4.11
1.80	0.845	0.915	1.00	1.11	1.22	1.35	1.49	1.64	1.80	1.95	2.11	2.45	2.82	3.21	3.61	3.99
2.00	0.784	0.852	0.937	1.04	1.15	1.27	1.40	1.55	1.70	1.84	2.00	2.33	2.69	3.06	3.46	3.85
2.20	0.730	0.797	0.878	0.972	1.08	1.20	1.33	1.46	1.60	1.74	1.89	2.21	2.56	2.92	3.31	3.71
2.40	0.683	0.748	0.825	0.915	1.02	1.13	1.26	1.39	1.52	1.65	1.80	2.10	2.44	2.80	3.17	3.56
2.60	0.641	0.704	0.778	0.865	0.963	1.07	1.19	1.32	1.44	1.57	1.71	2.00	2.33	2.68	3.05	3.42
2.80	0.604	0.664	0.736	0.819	0.913	1.02	1.13	1.25	1.37	1.49	1.62	1.91	2.23	2.57	2.93	3.30
3.00	0.570	0.628	0.698	0.777	0.867	0.966	1.07	1.18	1.30	1.42	1.55	1.82	2.13	2.46	2.81	3.18
x	0.000	0.005	0.017	0.035	0.057	0.083	0.113	0.144	0.178	0.213	0.250	0.327	0.408	0.492	0.579	0.667
y	0.500	0.455	0.417	0.385	0.357	0.333	0.313	0.294	0.278	0.263	0.250	0.227	0.208	0.192	0.179	0.167

Eccentricity Normal to the Plane of the Faying Surface

Figure 8-55 shows a bracket welded to a column flange. The eccentric load P_u can be resolved into a concentric force P_u at the faying surface of the connection and a moment $P_u e$ normal to the plane of the faying surface where e is the eccentricity. Each weld element is then assumed to support an equal share of the concentric force P_u, and the moment $P_u e$ is resisted by tension in the welds above the neutral axis and compression below the neutral axis.

In contrast to bolts, where the interaction of shear and tension must be considered, for welds, shear and tension may be combined vectorially for welds into a resultant shear. Thus, the solution of a weld loaded eccentrically normal to the plane of the faying surface is parallel to that discussed previously for welds loaded eccentrically in the plane of the faying surface; with the neutral axis assumed to be located at the CG of the weld group, this case is identical to that described previously for the elastic method.

CONSTRUCTION COMBINING BOLTS AND WELDS

In bearing-type connections in new construction, the rigidity of the welds prevents the initial joint slippage necessary to develop the strength of all the bolts in a connection that might combine both welds and bolts. Thus, bearing-type connections combining welds and bolts are permissible only if the design strength of the welds ϕR_n alone exceeds the required strength of the connection R_u. However, in situations where it can safely be assumed that joint slippage has occurred before welding is performed, welds may be used to reinforce existing bolted or riveted joints. Such is the case with structures previously in service. In this case, the design strength of the original bolt group may be used to carry the existing dead loads and the design strength of the welds need be adequate only to carry additional loads. Refer to LRFD Specification Section J1.9.

In slip-critical connections, since connection slip is neither expected nor required for the bolts to develop their strength, the design strengths of welds and high-strength bolts are additive. When high-strength bolts and welds are used together in a slip-critical connection, the bolts should preferably be fully tensioned before welding is performed. The design drawings should clearly indicate where this type of connection occurs.

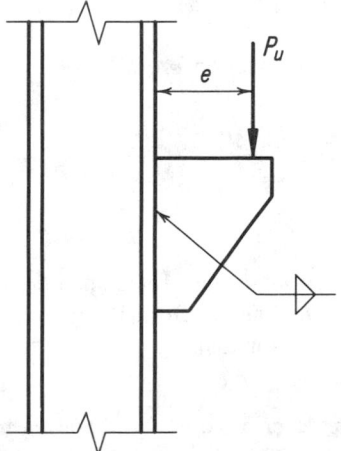

Fig. 8-55. Welds subjected to eccentricity normal to the plane of the faying surface.

CONNECTED ELEMENTS

Connected elements are the angles, plates, tees, gussets, and other connecting elements used in connections to transfer load from one structural member to another as well as the affected elements of the connected members.

Economical Considerations

Cost effective steel fabrication requires close cooperation between the designer, detailer, and fabricator. Effective communication and planning will allow the project to take full advantage of the strengths of all parties involved. Often, potential problems can be avoided through early consultation and good communication during the full life of a project.

Designs and details should be suited to the shop practices and standards of the fabricator. The resulting similarity throughout the project will further lend itself to the minimization of errors. For example, once gage lines conforming to standard machine set-ups are determined, they should be utilized as much as possible throughout any one job. Furthermore, it is desirable to keep the same bolt spacing throughout a project. Longitudinal spacing should preferably be three inches or a multiple of three inches, since most shops consider this to be standard.

At a minimum, gages and hole sizes on any one member should not be varied throughout the length of that member. This prevents unnecessary material re-handling and the need for multiple punching or drilling.

Design Strength of Connected Elements

The design strength of connecting elements is determined in accordance with the provisions of LRFD Specification Sections J4 and J5; the applicable limit states are shear yielding, shear rupture, block shear rupture, tension yielding, and tension rupture.

Shear Yielding

This limit state applies to the gross section of the connected element. From LRFD Specification Section J5.3, the design shear yielding strength is ϕR_n,

where

$\phi = 0.90$
$R_n = 0.60 F_y A_g$

Shear Rupture

This limit state applies to the net section From LRFD Specification Section J4.1, the design shear rupture strength is ϕR_n,

where

$\phi = 0.75$
$R_n = 0.60 F_u A_{nv}$

Table 8-46 gives the reduction in area for standard, oversized, short-slotted, and long-slotted holes in material thicknesses from $\frac{3}{16}$-in. to 1 in.; for other material thicknesses, multiply the tabular value for 1-in. thickness by the actual thickness.

Block Shear Rupture

The term block shear rupture describes a material tearing limit state which occurs in a combination of shear and tension. This phenomenon can occur at the end of a coped beam, shown in Figure 8-56, or at the end of a tension connection, shown in Figure 8-57. This

Table 8-46.
Reduction in Area for Holes, in.^{2}

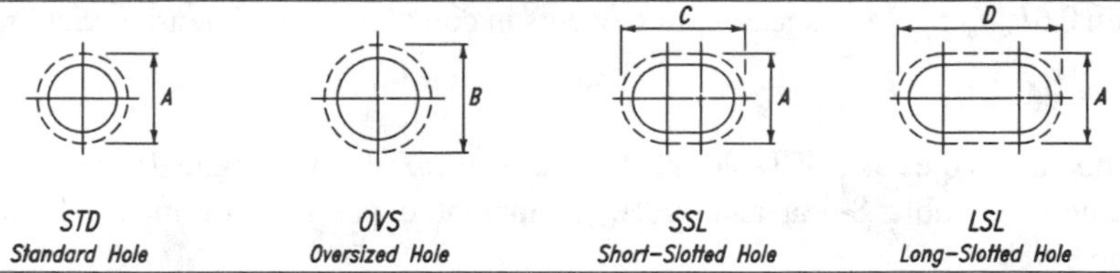

	STD	OVS	SSL	LSL
	Standard Hole	Oversized Hole	Short–Slotted Hole	Long–Slotted Hole

	$A \times t$							$B \times t$						
Thckns.	Bolt Diameter d_b, in.							Bolt Diameter d_b, in.						
t, in.	3/4	7/8	1	1⅛	1¼	1⅜	1½	3/4	7/8	1	1⅛	1¼	1⅜	1½
3/16	0.164	0.188	0.211	0.234	0.258	0.281	0.305	0.188	0.211	0.246	0.281	0.305	0.328	0.352
1/4	0.219	0.250	0.281	0.313	0.344	0.375	0.406	0.250	0.281	0.328	0.375	0.406	0.438	0.469
5/16	0.273	0.313	0.352	0.391	0.430	0.469	0.508	0.313	0.352	0.410	0.469	0.508	0.547	0.586
3/8	0.328	0.375	0.422	0.469	0.516	0.563	0.609	0.375	0.422	0.492	0.563	0.609	0.656	0.703
7/16	0.383	0.438	0.492	0.547	0.602	0.656	0.711	0.438	0.492	0.574	0.656	0.711	0.766	0.820
1/2	0.438	0.500	0.563	0.625	0.688	0.750	0.813	0.500	0.563	0.656	0.750	0.813	0.875	0.938
9/16	0.492	0.563	0.633	0.703	0.773	0.844	0.914	0.563	0.633	0.738	0.844	0.914	0.984	1.05
5/8	0.547	0.625	0.703	0.781	0.859	0.938	1.02	0.625	0.703	0.820	0.938	1.02	1.09	1.17
11/16	0.602	0.688	0.773	0.859	0.945	1.03	1.12	0.688	0.773	0.902	1.03	1.12	1.20	1.29
3/4	0.656	0.750	0.844	0.938	1.03	1.13	1.22	0.750	0.844	0.984	1.13	1.22	1.31	1.41
13/16	0.711	0.813	0.914	1.02	1.12	1.22	1.32	0.813	0.914	1.07	1.22	1.32	1.42	1.52
7/8	0.766	0.875	0.984	1.09	1.20	1.31	1.42	0.875	0.984	1.15	1.31	1.42	1.53	1.64
15/16	0.820	0.938	1.05	1.17	1.29	1.41	1.52	0.938	1.05	1.23	1.41	1.52	1.64	1.76
1	0.875	1.00	1.13	1.25	1.38	1.50	1.63	1.00	1.13	1.31	1.50	1.63	1.75	1.88

	$C \times t$							$D \times t$						
Thckns.	Bolt Diameter d_b, in.							Bolt Diameter d_b, in.						
t, in.	3/4	7/8	1	1⅛	1¼	1⅜	1½	3/4	7/8	1	1⅛	1¼	1⅜	1½
3/16	0.199	0.223	0.258	0.293	0.316	0.340	0.363	0.363	0.422	0.480	0.539	0.598	0.656	0.715
1/4	0.266	0.297	0.344	0.391	0.422	0.453	0.484	0.484	0.563	0.641	0.719	0.797	0.875	0.953
5/16	0.332	0.371	0.430	0.488	0.527	0.566	0.605	0.605	0.703	0.801	0.898	0.996	1.09	1.19
3/8	0.398	0.445	0.516	0.586	0.633	0.680	0.727	0.727	0.844	0.961	1.08	1.20	1.31	1.43
7/16	0.465	0.520	0.602	0.684	0.738	0.793	0.848	0.848	0.984	1.12	1.26	1.39	1.53	1.67
1/2	0.531	0.594	0.688	0.781	0.844	0.906	0.969	0.969	1.13	1.28	1.44	1.59	1.75	1.91
9/16	0.598	0.668	0.773	0.879	0.949	1.02	1.09	1.09	1.27	1.44	1.62	1.79	1.97	2.14
5/8	0.664	0.742	0.859	0.977	1.05	1.13	1.21	1.21	1.41	1.60	1.80	1.99	2.19	2.38
11/16	0.730	0.816	0.945	1.07	1.16	1.25	1.33	1.33	1.55	1.76	1.98	2.19	2.41	2.62
3/4	0.797	0.891	1.03	1.17	1.27	1.36	1.45	1.45	1.69	1.92	2.16	2.39	2.63	2.86
13/16	0.863	0.965	1.12	1.27	1.37	1.47	1.57	1.57	1.83	2.08	2.34	2.59	2.84	3.10
7/8	0.930	1.04	1.20	1.37	1.48	1.59	1.70	1.70	1.97	2.24	2.52	2.79	3.06	3.34
15/16	0.996	1.11	1.29	1.46	1.58	1.70	1.82	1.82	2.11	2.40	2.70	2.99	3.28	3.57
1	1.06	1.19	1.38	1.56	1.69	1.81	1.94	1.94	2.25	2.56	2.88	3.19	3.50	3.81

failure is usually the result of high reactions imposed on relatively thin material through a short connection.

The design block shear rupture strength is ϕR_n, where $\phi = 0.75$ and R_n is determined as follows.

For bolted connections, from LRFD Specification Section J4.3, when $F_u A_{nt} \geq 0.6 F_u A_{nv}$, shear yielding occurs in combination with tension rupture and,

$$R_n = 0.6 F_y A_{gv} + F_u A_{nt}$$

This case is the basis of Tables 8-47, where $\phi F_u A_{nt}$ is tabulated per inch of material thickness in Table 8-47a and $\phi(0.6F_y A_{gv})$ is tabulated per inch of material thickness in Table 8-47b. When $0.6F_u A_{nv} > F_u A_{nt}$, shear rupture occurs in combination with tension yielding and,

$$R_n = 0.6F_u A_{nv} + F_y A_{gt}$$

This case is the basis of Tables 8-48, where $\phi(0.6F_u A_{nv})$ is tabulated per inch of material thickness in Table 8-48a and $\phi F_y A_{gt}$ is tabulated per inch of material thickness in Table 8-48b.

For welded connections, block shear rupture is treated as for bolted connections; the only difference is that, in the absence of bolt holes, $A_{nv} = A_{gv}$ and $A_{nt} = A_{gt}$.

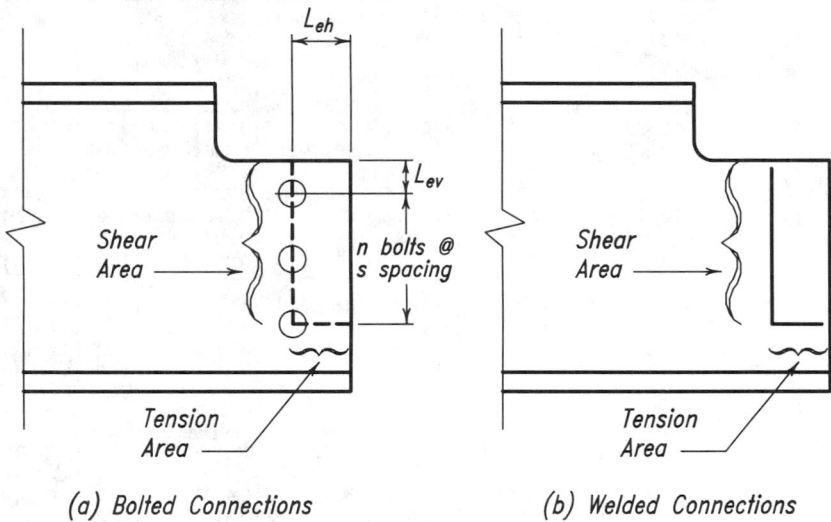

Fig. 8-56. Block shear rupture in coped beams.

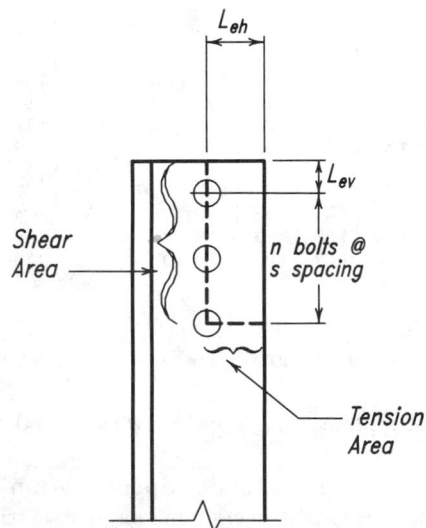

Fig. 8-57. Block shear rupture in ends of tension members.

Table 8-47a.
Block Shear Rupture
Tension Rupture Component per inch of thickness, $\phi[F_u A_{nt}] / t$, kips/in.

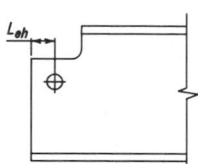

	F_u, ksi								
	58			65			70		
	Bolt Diameter d_b, in.			Bolt Diameter d_b, in.			Bolt Diameter d_b, in.		
L_{eh}, in.	$\frac{3}{4}$	$\frac{7}{8}$	1	$\frac{3}{4}$	$\frac{7}{8}$	1	$\frac{3}{4}$	$\frac{7}{8}$	1
1	24.5	21.8	19.0	27.4	24.4	21.3	29.5	26.3	23.0
$1\frac{1}{8}$	29.9	27.2	24.5	33.5	30.5	27.4	36.1	32.8	29.5
$1\frac{1}{4}$	35.3	32.6	29.9	39.6	36.6	33.5	42.7	39.4	36.1
$1\frac{3}{8}$	40.8	38.1	35.3	45.7	42.7	39.6	49.2	45.9	42.7
$1\frac{1}{2}$	46.2	43.5	40.8	51.8	48.8	45.7	55.8	52.5	49.2
$1\frac{5}{8}$	51.7	48.9	46.2	57.9	54.8	51.8	62.3	59.1	55.8
$1\frac{3}{4}$	57.1	54.4	51.7	64.0	60.9	57.9	68.9	65.6	62.3
$1\frac{7}{8}$	62.5	59.8	57.1	70.1	67.0	64.0	75.5	72.2	68.9
2	68.0	65.3	62.5	76.2	73.1	70.1	82.0	78.8	75.5
$2\frac{1}{4}$	78.8	76.1	73.4	88.4	85.3	82.3	95.2	91.9	88.6
$2\frac{1}{2}$	89.7	87.0	84.3	101	97.5	94.5	108	105	102
$2\frac{3}{4}$	101	97.9	95.2	113	110	107	121	118	115
3	111	109	106	125	122	119	135	131	128

Table 8-47b.
Block Shear Rupture
Shear Yielding Component per inch of thickness, $\phi[0.6F_yA_{gv}] / t$, kips/in.

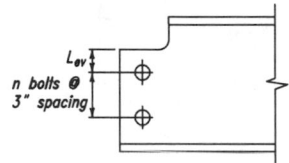

		F_y, ksi	
n	L_{ev}, in.	36	50
12	$1\frac{1}{4}$	555	771
	$1\frac{3}{8}$	557	773
	$1\frac{1}{2}$	559	776
	$1\frac{5}{8}$	561	779
	$1\frac{3}{4}$	563	782
	$1\frac{7}{8}$	565	785
	2	567	788
	$2\frac{1}{4}$	571	793
	$2\frac{1}{2}$	575	799
	$2\frac{3}{4}$	579	804
	3	583	810
11	$1\frac{1}{4}$	506	703
	$1\frac{3}{8}$	508	706
	$1\frac{1}{2}$	510	709
	$1\frac{5}{8}$	512	712
	$1\frac{3}{4}$	514	714
	$1\frac{7}{8}$	516	717
	2	518	720
	$2\frac{1}{4}$	522	726
	$2\frac{1}{2}$	527	731
	$2\frac{3}{4}$	531	737
	3	535	743
10	$1\frac{1}{4}$	458	636
	$1\frac{3}{8}$	460	638
	$1\frac{1}{2}$	462	641
	$1\frac{5}{8}$	464	644
	$1\frac{3}{4}$	466	647
	$1\frac{7}{8}$	468	650
	2	470	653
	$2\frac{1}{4}$	474	658
	$2\frac{1}{2}$	478	664
	$2\frac{3}{4}$	482	669
	3	486	675

Table 8-47b (cont.).
Block Shear Rupture
Shear Yielding Component per inch of thickness, $\phi[0.6F_yA_{gv}] / t$, kips/in.

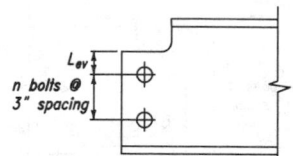

n	L_{ev}, in.	F_y, ksi	
		36	50
9	$1\frac{1}{4}$	409	568
	$1\frac{3}{8}$	411	571
	$1\frac{1}{2}$	413	574
	$1\frac{5}{8}$	415	577
	$1\frac{3}{4}$	417	579
	$1\frac{7}{8}$	419	582
	2	421	585
	$2\frac{1}{4}$	425	591
	$2\frac{1}{2}$	429	596
	$2\frac{3}{4}$	433	602
	3	437	608
8	$1\frac{1}{4}$	360	501
	$1\frac{3}{8}$	362	503
	$1\frac{1}{2}$	365	506
	$1\frac{5}{8}$	367	509
	$1\frac{3}{4}$	369	512
	$1\frac{7}{8}$	371	515
	2	373	518
	$2\frac{1}{4}$	377	523
	$2\frac{1}{2}$	381	529
	$2\frac{3}{4}$	385	534
	3	389	540
7	$1\frac{1}{4}$	312	433
	$1\frac{3}{8}$	314	436
	$1\frac{1}{2}$	316	439
	$1\frac{5}{8}$	318	442
	$1\frac{3}{4}$	320	444
	$1\frac{7}{8}$	322	447
	2	324	450
	$2\frac{1}{4}$	328	456
	$2\frac{1}{2}$	332	461
	$2\frac{3}{4}$	336	467
	3	340	473

Table 8-47b (cont.).
Block Shear Rupture
Shear Yielding Component per inch of thickness, $\phi[0.6F_yA_{gv}]/t$, kips/in.

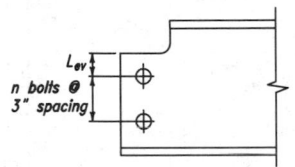

n	L_{ev}, in.	F_y, ksi	
		36	50
6	$1\frac{1}{4}$	263	366
	$1\frac{3}{8}$	265	368
	$1\frac{1}{2}$	267	371
	$1\frac{5}{8}$	269	374
	$1\frac{3}{4}$	271	377
	$1\frac{7}{8}$	273	380
	2	275	383
	$2\frac{1}{4}$	279	388
	$2\frac{1}{2}$	284	394
	$2\frac{3}{4}$	288	399
	3	292	405
5	$1\frac{1}{4}$	215	298
	$1\frac{3}{8}$	217	301
	$1\frac{1}{2}$	219	304
	$1\frac{5}{8}$	221	307
	$1\frac{3}{4}$	223	309
	$1\frac{7}{8}$	225	312
	2	227	315
	$2\frac{1}{4}$	231	321
	$2\frac{1}{2}$	235	326
	$2\frac{3}{4}$	239	332
	3	243	338
4	$1\frac{1}{4}$	166	231
	$1\frac{3}{8}$	168	233
	$1\frac{1}{2}$	170	236
	$1\frac{5}{8}$	172	239
	$1\frac{3}{4}$	174	242
	$1\frac{7}{8}$	176	245
	2	178	248
	$2\frac{1}{4}$	182	253
	$2\frac{1}{2}$	186	259
	$2\frac{3}{4}$	190	264
	3	194	270

Table 8-47b (cont.).
Block Shear Rupture
Shear Yielding Component per inch of thickness, $\phi[0.6F_yA_{gv}] / t$, kips/in.

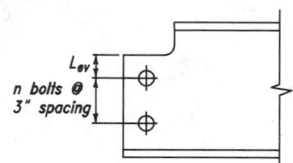

n	L_{ev}, in.	F_y, ksi	
		36	50
3	$1\frac{1}{4}$	117	163
	$1\frac{3}{8}$	119	166
	$1\frac{1}{2}$	122	169
	$1\frac{5}{8}$	124	172
	$1\frac{3}{4}$	126	174
	$1\frac{7}{8}$	128	177
	2	130	180
	$2\frac{1}{4}$	134	186
	$2\frac{1}{2}$	138	191
	$2\frac{3}{4}$	142	197
	3	146	203
2	$1\frac{1}{4}$	68.9	95.6
	$1\frac{3}{8}$	70.9	98.4
	$1\frac{1}{2}$	72.9	101
	$1\frac{5}{8}$	74.9	104
	$1\frac{3}{4}$	77.0	107
	$1\frac{7}{8}$	79.0	110
	2	81.0	113
	$2\frac{1}{4}$	85.1	118
	$2\frac{1}{2}$	89.1	124
	$2\frac{3}{4}$	93.2	129
	3	97.2	135

Table 8-48a.
Block Shear Rupture
Shear Rupture Component per inch of thickness, $\phi[0.6F_uA_{nv}]/t$, kips/in.

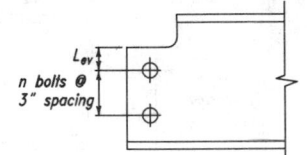

n	L_{ev}, in.	F_u, ksi								
		58			65			70		
		Bolt Diameter d_b, in.			Bolt Diameter d_b, in.			Bolt Diameter d_b, in.		
		$3/4$	$7/8$	1	$3/4$	$7/8$	1	$3/4$	$7/8$	1
12	$1^1/_4$	631	594	556	707	665	623	762	717	671
	$1^3/_8$	635	597	560	711	669	627	766	721	675
	$1^1/_2$	638	600	563	715	673	631	770	725	679
	$1^5/_8$	641	604	566	718	676	634	774	728	683
	$1^3/_4$	644	607	569	722	680	638	778	732	687
	$1^7/_8$	648	610	573	726	684	642	782	736	691
	2	651	613	576	729	687	645	786	740	695
	$2^1/_4$	657	620	582	737	695	653	793	748	703
	$2^1/_2$	664	626	589	744	702	660	801	756	711
	$2^3/_4$	670	633	595	751	709	667	809	764	719
	3	677	639	602	759	717	675	817	772	726
11	$1^1/_4$	576	542	507	645	607	569	695	654	612
	$1^3/_8$	579	545	511	649	611	572	699	658	616
	$1^1/_2$	582	548	514	653	614	576	703	662	620
	$1^5/_8$	586	551	517	656	618	580	707	665	624
	$1^3/_4$	589	555	520	660	622	583	711	669	628
	$1^7/_8$	592	558	524	664	625	587	715	673	632
	2	595	561	527	667	629	590	719	677	636
	$2^1/_4$	602	568	533	675	636	598	726	685	644
	$2^1/_2$	608	574	540	682	644	605	734	693	652
	$2^3/_4$	615	581	546	689	651	612	742	701	660
	3	622	587	553	697	658	620	750	709	667
10	$1^1/_4$	520	489	458	583	548	514	628	591	553
	$1^3/_8$	524	493	462	587	552	517	632	595	557
	$1^1/_2$	527	496	465	590	556	521	636	599	561
	$1^5/_8$	530	499	468	594	559	525	640	602	565
	$1^3/_4$	533	502	471	598	563	528	644	606	569
	$1^7/_8$	537	506	475	601	567	532	648	610	573
	2	540	509	478	605	570	536	652	614	577
	$2^1/_4$	546	515	484	612	578	543	660	622	585
	$2^1/_2$	553	522	491	620	585	550	667	630	593
	$2^3/_4$	560	529	498	627	592	558	675	638	600
	3	566	535	504	634	600	565	683	646	608

Table 8-48a (cont.).
Block Shear Rupture
Shear Rupture Component per inch of thickness, $\phi[0.6F_uA_{nv}]/t$, kips/in.

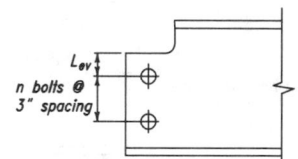

		F_u, ksi								
		58			65			70		
		Bolt Diameter d_b, in.			Bolt Diameter d_b, in.			Bolt Diameter d_b, in.		
n	L_{ev}, in.	$^3/_4$	$^7/_8$	1	$^3/_4$	$^7/_8$	1	$^3/_4$	$^7/_8$	1
9	$1^1/_4$	465	437	409	521	490	459	561	528	494
	$1^3/_8$	468	440	413	525	494	463	565	532	498
	$1^1/_2$	471	444	416	528	497	466	569	536	502
	$1^5/_8$	475	447	419	532	501	470	573	539	506
	$1^3/_4$	478	450	422	536	505	473	577	543	510
	$1^7/_8$	481	453	426	539	508	477	581	547	514
	2	484	457	429	543	512	481	585	551	518
	$2^1/_4$	491	463	436	550	519	488	593	559	526
	$2^1/_2$	498	470	442	558	527	495	600	567	534
	$2^3/_4$	504	476	449	565	534	503	608	575	541
	3	511	483	455	572	541	510	616	583	549
8	$1^1/_4$	409	385	361	459	431	404	494	465	435
	$1^3/_8$	413	388	364	463	435	408	498	469	439
	$1^1/_2$	416	392	367	466	439	411	502	473	443
	$1^5/_8$	419	395	370	470	442	415	506	476	447
	$1^3/_4$	422	398	374	473	446	419	510	480	451
	$1^7/_8$	426	401	377	477	450	422	514	484	455
	2	429	405	380	481	453	426	518	488	459
	$2^1/_4$	436	411	387	488	461	433	526	496	467
	$2^1/_2$	442	418	393	495	468	441	534	504	474
	$2^3/_4$	449	424	400	503	475	448	541	512	482
	3	455	431	406	510	483	455	549	520	490
7	$1^1/_4$	354	333	312	397	373	349	427	402	376
	$1^3/_8$	357	336	315	400	377	353	431	406	380
	$1^1/_2$	361	339	318	404	380	356	435	410	384
	$1^5/_8$	364	343	321	408	384	360	439	413	388
	$1^3/_4$	367	346	325	411	388	364	443	417	392
	$1^7/_8$	370	349	328	415	391	367	447	421	396
	2	374	352	331	419	395	371	451	425	400
	$2^1/_4$	380	359	338	426	402	378	459	433	408
	$2^1/_2$	387	365	344	433	410	386	467	441	415
	$2^3/_4$	393	372	351	441	417	393	474	449	423
	3	400	378	357	448	424	400	482	457	431

Table 8-48a (cont.).
Block Shear Rupture
Shear Rupture Component per inch of thickness, $\phi[0.6F_uA_{nv}]\,/\,t$, kips/in.

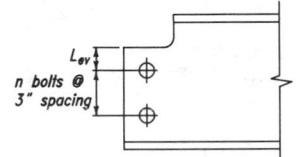

		F_u, ksi								
		58			**65**			**70**		
		Bolt Diameter d_b, in.			Bolt Diameter d_b, in.			Bolt Diameter d_b, in.		
n	L_{ev}, in.	$3/4$	$7/8$	1	$3/4$	$7/8$	1	$3/4$	$7/8$	1
6	$1^1/4$	299	281	263	335	314	294	360	339	317
	$1^3/8$	302	284	266	338	318	298	364	343	321
	$1^1/2$	305	287	269	342	322	302	368	347	325
	$1^5/8$	308	290	272	346	325	305	372	350	329
	$1^3/4$	312	294	276	349	329	309	376	354	333
	$1^7/8$	315	297	279	353	333	313	380	358	337
	2	318	300	282	356	336	316	384	362	341
	$2^1/4$	325	307	289	364	344	324	392	370	348
	$2^1/2$	331	313	295	371	351	331	400	378	356
	$2^3/4$	338	320	302	378	358	338	408	386	364
	3	344	326	308	386	366	346	415	394	372
5	$1^1/4$	243	228	214	272	256	239	293	276	258
	$1^3/8$	246	232	217	276	260	243	297	280	262
	$1^1/2$	250	235	220	280	263	247	301	284	266
	$1^5/8$	253	238	223	283	267	250	305	287	270
	$1^3/4$	256	241	227	287	271	254	309	291	274
	$1^7/8$	259	245	230	291	274	258	313	295	278
	2	263	248	233	294	278	261	317	299	282
	$2^1/4$	269	254	240	302	285	269	325	307	289
	$2^1/2$	276	261	246	309	293	276	333	315	297
	$2^3/4$	282	268	253	316	300	283	341	323	305
	3	289	274	259	324	307	291	348	331	313
4	$1^1/4$	188	176	165	210	197	185	226	213	199
	$1^3/8$	191	179	168	214	201	188	230	217	203
	$1^1/2$	194	183	171	218	205	192	234	221	207
	$1^5/8$	197	186	175	221	208	196	238	224	211
	$1^3/4$	201	189	178	225	212	199	242	228	215
	$1^7/8$	204	192	181	229	216	203	246	232	219
	2	207	196	184	232	219	207	250	236	222
	$2^1/4$	214	202	191	239	227	214	258	244	230
	$2^1/2$	220	209	197	247	234	221	266	252	238
	$2^3/4$	227	215	204	254	241	229	274	260	246
	3	233	222	210	261	249	236	282	268	254

Table 8-48a (cont.).
Block Shear Rupture
Shear Rupture Component per inch of thickness, $\phi[0.6F_uA_{nv}]/t$, kips/in.

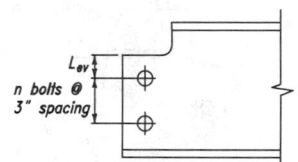

L_{ev}
n bolts @
3" spacing

		F_u, ksi								
		58			65			70		
		Bolt Diameter d_b, in.			Bolt Diameter d_b, in.			Bolt Diameter d_b, in.		
n	L_{ev}, in.	$3/4$	$7/8$	1	$3/4$	$7/8$	1	$3/4$	$7/8$	1
3	$1\frac{1}{4}$	132	124	116	148	139	130	159	150	140
	$1\frac{3}{8}$	135	127	119	152	143	133	163	154	144
	$1\frac{1}{2}$	139	131	122	155	146	137	167	158	148
	$1\frac{5}{8}$	142	134	126	159	150	141	171	161	152
	$1\frac{3}{4}$	145	137	129	163	154	144	175	165	156
	$1\frac{7}{8}$	148	140	132	166	157	148	179	169	159
	2	152	144	135	170	161	152	183	173	163
	$2\frac{1}{4}$	158	150	142	177	168	159	191	181	171
	$2\frac{1}{2}$	165	157	148	185	176	166	199	189	179
	$2\frac{3}{4}$	171	163	155	192	183	174	207	197	187
	3	178	170	161	199	190	181	215	205	195
2	$1\frac{1}{4}$	77	72	67	86	80	75	93	87	81
	$1\frac{3}{8}$	80	75	70	90	84	79	96	91	85
	$1\frac{1}{2}$	83	78	73	93	88	82	100	95	89
	$1\frac{5}{8}$	86	82	77	97	91	86	104	98	93
	$1\frac{3}{4}$	90	85	80	101	95	90	108	102	96
	$1\frac{7}{8}$	93	88	83	104	99	93	112	106	100
	2	96	91	86	108	102	97	116	110	104
	$2\frac{1}{4}$	103	98	93	115	110	104	124	118	112
	$2\frac{1}{2}$	109	104	100	122	117	112	132	126	120
	$2\frac{3}{4}$	116	111	106	130	124	119	140	134	128
	3	122	117	113	137	132	126	148	142	136

Table 8-48b.
Block Shear Rupture
Tension Yielding Component per inch of thickness
$$\phi[F_y A_{gt}] / t, \text{ kips/in.}$$

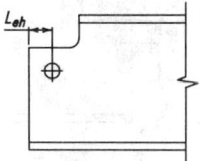

L_{eh}, in.	F_y, ksi	
	36	50
1	27.0	37.5
$1\frac{1}{8}$	30.4	42.2
$1\frac{1}{4}$	33.8	46.9
$1\frac{3}{8}$	37.1	51.6
$1\frac{1}{2}$	40.5	56.3
$1\frac{5}{8}$	43.9	60.9
$1\frac{3}{4}$	47.3	65.6
$1\frac{7}{8}$	50.6	70.3
2	54.0	75.0
$2\frac{1}{4}$	60.8	84.4
$2\frac{1}{2}$	67.5	93.8
$2\frac{3}{4}$	74.3	103
3	81.0	113

Tension Yielding
From LRFD Specification Section J5.2, the design tension yielding strength is ϕR_n,

where

$\phi = 0.90$
$R_n = F_y A_g$

Tension Rupture
From LRFD Specification Section J5.2, the design tension rupture strength is ϕR_n,

where

$\phi = 0.75$
$R_n = F_u A_n$

In the above equation, A_n is the net area not to exceed $0.85A_g$.

Table 8-46 gives the reduction in area for standard, oversized, short-slotted, and long-slotted holes in material thicknesses from $\frac{3}{16}$-in. to 1 in.; for material thicknesses not listed, multiply the tabular value for 1-in. thickness by the actual thickness.

Members with Copes, Blocks, or Cuts
When structural members frame together, a minimum clearance of $\frac{1}{2}$-in. should be provided, when possible. In cases where material removal is necessary to provide such a clearance, material may be removed by coping, blocking, or cutting as illustrated in Figures 8-58. Note the recommended practices for coping illustrated in Figure 8-59; the potential notch left by the first cut will occur in waste material which will subsequently be removed by the second cut. All re-entrant corners must be shaped notch-free per AWS D1.1 to a radius. An approximate minimum radius to which this corner must be shaped is $\frac{1}{2}$-in.

Material removal is costly, and should be avoided when possible. For example, the elevations of the tops of infill beams could be established at a sufficient distance below the tops of girders to clear the girder fillet. Alternatively, coping could be eliminated with a connection as illustrated in Figure 8-60; this detail also allows the use of a shorter beam length. When necessary, coping is usually the most economical method to remove material.

Copes, blocks, and cuts can significantly reduce the design strengths of members and may require web reinforcement; it may be more economical to use a heavier member than to provide such reinforcement. The design strength of the unreinforced coped member is determined from the limit states of flexural yielding, local buckling, and lateral torsional buckling, if applicable. Web reinforcement of coped beams is discussed in Part 9.

Flexural Yielding
The flexural yielding strength of a supported beam which is coped at the top and/or bottom is $\phi_b M_n$,

where

$\phi_b = 0.90$
$M_n = F_y S_{net}$

In the above equation, S_{net} is the net elastic section modulus, in.[3] Values of S_{net} are tabulated in Table 8-49.

The beam-end reaction R_u must be such that:

$$R_u \leq \frac{\phi_b M_n}{e}$$

where e is the distance from the face of the cope to the point of inflection of the beam, in. It is usually assumed that the point of inflection is located at the face of the supporting member and e is as shown in Figure 8-61. However, depending upon the connection type and stiffness and support condition, the point of inflection may move away from the face of the supporting member; when this is the case, a lesser value of e may be justified. In any case, the choice of e shown in Figure 8-61 will be conservative.

Local Web Buckling

For short copes no greater than the length of the connection angle(s), plate, or tee, local web buckling will generally not occur. If, however, the depth of the cope were such that

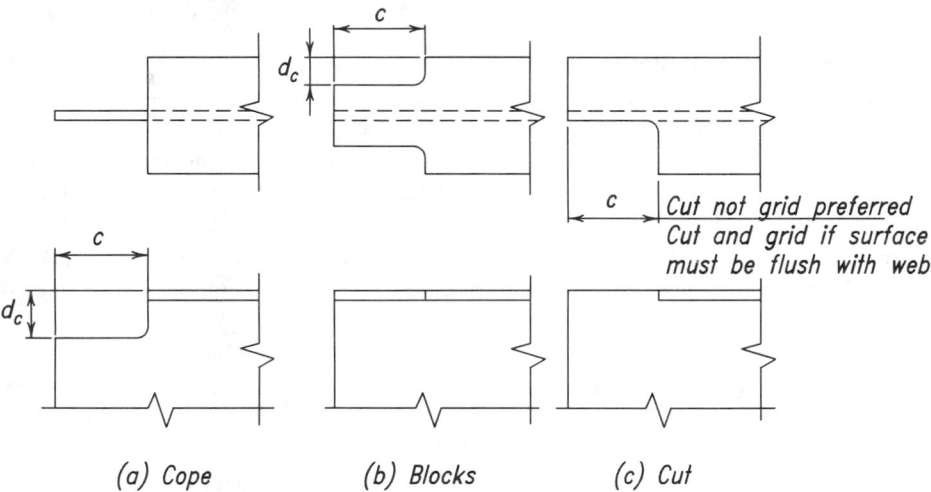

(a) Cope   (b) Blocks   (c) Cut

Fig. 8-58. Copes, blocks, and cuts.

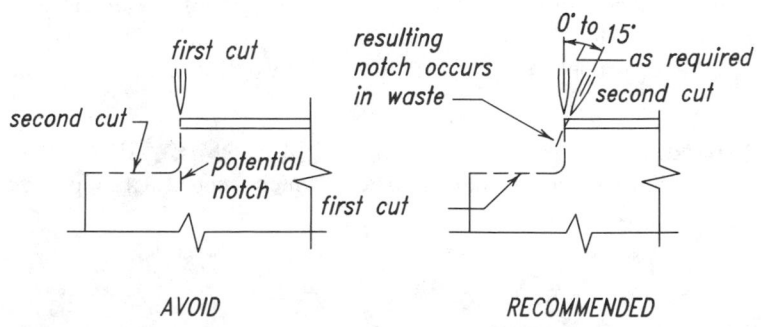

AVOID   RECOMMENDED

Fig. 8-59. Recommended coping practice.

$d_c > 0.2d$, the unreinforced web could buckle between the top of the cope and the beam flange if the beam web were thin.

In a reduced section, the design strength in local web buckling may be more critical than the design strength in flexural yielding. This design strength is critical at the compression zone of the web near the cope and is dependent on three parameters: (1) cope depth d_c; (2) cope length c; and (3) web thickness t_w. It should be noted that, for convenience, the dimension h_0 in Figure 8-61 is used instead of the more correct dimension h_1; this eliminates the detailed calculation required to locate the neutral axis of the coped beam. Alternatively, the dimension h_1 may be substituted for h_0 in the following local buckling calculations.

The beam end reaction R_u must be such that:

$$R_u \leq \frac{\phi F_{bc} S_{net}}{e}$$

where

S_{net} = elastic section modulus of the net section, in.^{3} from Table 8-49

e = distance from the end reaction to the face of the cope, in.

and ϕF_{bc} is determined as follows.

When a beam is coped at the top flange only, the design recommendations are based on the classical plate buckling formula with a k-factor based on three edges simply supported and one free edge. An additional factor f, which generally accounts for stress concentration at the cope, was developed to correlate with the coped beam buckling solutions (Cheng, et. al., 1984). From Figure 8-61, when the $c \leq 2d$ and $d_c \leq d/2$,

$$F_{cr} = \frac{\pi^2 E}{12(1-v^2)}\left(\frac{t_w}{h_o}\right)^2 fk$$

where

E = 29,000 ksi, modulus of elasticity of steel

v = 0.3, Poisson's ratio

f = plate buckling model adjustment factor

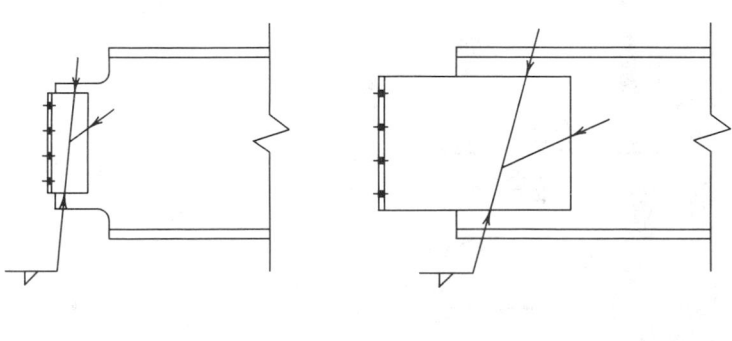

(a) Coping Required   (b) Coping Eliminated

Fig. 8-60. Minimizing coping requirements.

k = plate buckling coefficient

$h_o = d - d_c$, reduced beam depth, in.

Thus, the design buckling stress ϕF_{bc} for a beam coped at the top flange only is,

$$\phi F_{bc} = 23{,}590 \left(\frac{t_w}{h_o}\right)^2 fk$$

where f and k are determined from the following equations:

$$f = 2 \left(\frac{c}{d}\right) \text{ for } \frac{c}{d} \leq 1.0$$

$$f = 1 + \left(\frac{c}{d}\right) \text{ for } \frac{c}{d} > 1.0$$

$$k = 2.2 \left(\frac{h_o}{c}\right)^{1.65} \text{ for } \frac{c}{h_o} \leq 1.0$$

$$k = 2.2 \left(\frac{h_o}{c}\right) \text{ for } \frac{c}{h_o} > 1.0$$

When a beam is coped at both flanges, the design recommendations are based on the lateral buckling model with an adjustment factor f_d (Cheng, et al., 1984). From Figure 8-62, when at both flanges $c \leq 2d$ and $d_c \leq 0.2d$,

$$F_{cr} = 0.62 \pi E \frac{t_w^2}{ch_o} f_d$$

Thus, the design buckling stress ϕF_{bc} for a beam coped at both flanges is,

$$\phi F_{bc} = 50{,}840 \frac{t_w^2}{ch_o} f_d$$

and

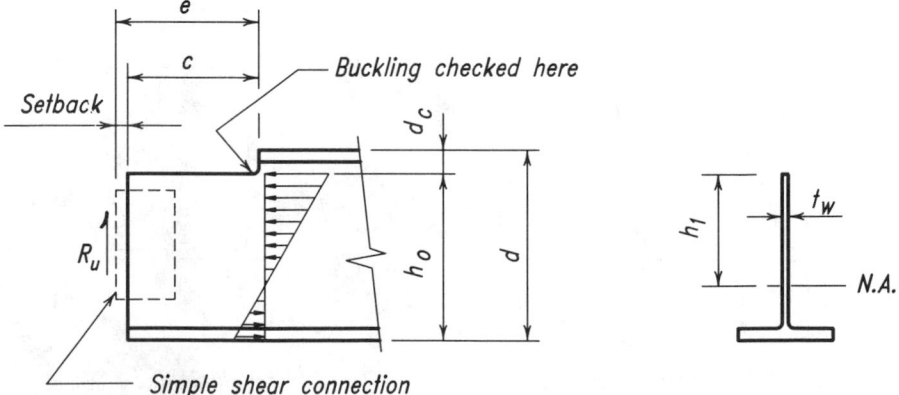

Fig. 8-61. *Local buckling of beam web coped at top flange only.*

$$f_d = 3.5 - 7.5\left(\frac{d_c}{d}\right)$$

where d_c is the larger of the top cope depth d_{ct} and the bottom cope depth d_{cb}.

Lateral Torsional Buckling

In laterally unbraced beams, copes, blocks, and cuts further reduce the out-of-plane rotational restraint. Cheng, et al. (1984) discusses the design strength of laterally unbraced coped beams. For laterally unbraced beams coped at the top only, this design strength may be determined with this information and the provisions of LRFD Specification Section F1.2. For laterally unbraced beams coped at the top and bottom, this design strength may be determined with this information and the provisions of LRFD Specification Appendix F1. A detailed discussion of this topic is beyond the scope of this text.

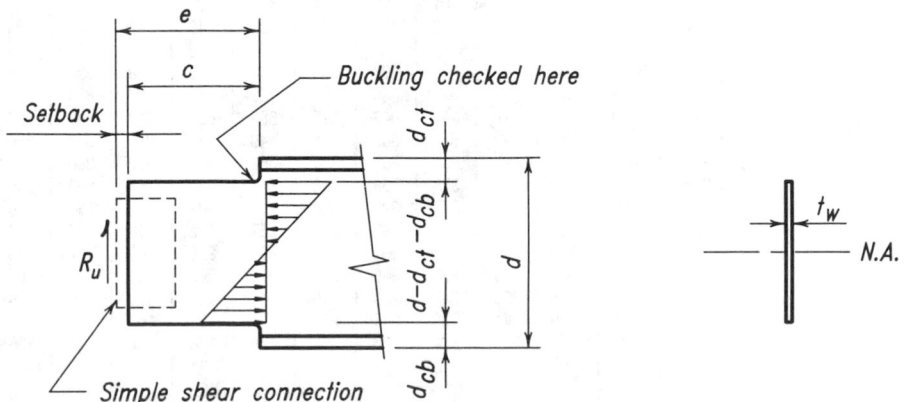

Fig. 8-62. *Local buckling of beam web coped at both flanges.*

Table 8-49.
Section Modulus of Coped W Shapes

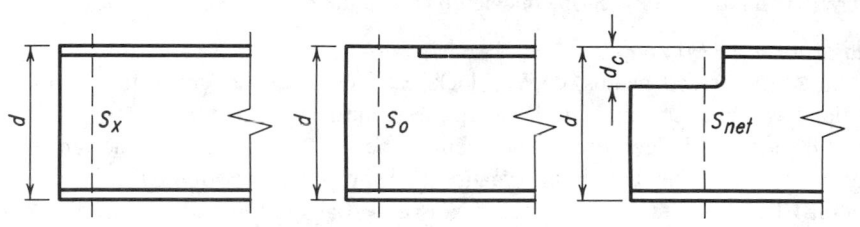

Designation	d in.	t_f in.	S_x in.^{3}	S_o in.^{3}	S_{net}, in.^{3} d_c, in. 2	3	4	5	6	7	8	9	10
W44×335	44.0	1.77	1410	492	451	431	411	392	373	355	337	320	303
×290	43.6	1.58	1240	417	382	365	348	332	316	300	285	270	255
×262	43.3	1.42	1120	374	342	327	312	297	283	269	255	241	228
×230	42.9	1.22	969	330	301	288	274	261	249	236	224	212	200
W40×593	43.0	3.23	2340	810	—	—	671	639	607	575	545	515	486
×503	42.1	2.76	1980	673	—	584	556	528	501	475	449	424	399
×431	41.3	2.36	1690	567	—	491	467	444	421	398	376	355	334
×372	40.6	2.05	1460	480	—	415	394	374	354	335	316	298	280
×321	40.1	1.77	1250	406	368	350	332	315	298	282	266	250	235
×297	39.8	1.65	1170	374	339	323	306	290	275	259	245	230	216
×277	39.7	1.58	1100	335	304	289	274	260	246	232	219	206	193
×249	39.4	1.42	992	299	271	258	245	232	219	207	195	183	172
×215	39.0	1.22	858	256	231	220	208	197	186	176	166	156	146
×199	38.7	1.07	769	247	224	213	202	191	180	170	160	150	141
×174	38.2	0.830	639	234	211	201	190	180	170	160	151	142	133
W40×466	42.4	2.95	1710	705	—	613	584	556	528	500	474	448	422
×392	41.6	2.52	1440	581	—	504	480	456	432	409	387	365	344
×331	40.8	2.13	1210	483	—	419	398	378	358	339	320	302	284
×278	40.2	1.81	1020	396	360	342	325	308	292	276	261	245	231
×264	40.0	1.73	971	371	337	321	305	289	274	259	244	230	216
×235	39.7	1.58	874	320	291	276	262	249	235	222	210	197	185
×211	39.4	1.42	785	286	259	246	234	221	209	198	186	175	165
×183	39.0	1.22	682	244	221	210	199	189	179	168	159	149	140
×167	38.6	1.03	599	234	212	201	191	181	171	161	152	143	134
×149	38.2	0.830	512	217	196	186	177	167	158	149	140	132	123
W36×848	42.5	4.53	3170	1094	—	—	903	858	813	770	728	687	647
×798	42.0	4.29	2980	1016	—	—	836	794	752	712	673	634	597
×650	40.5	3.54	2420	794	—	—	649	615	582	550	518	487	457
×527	39.3	2.91	1950	618	—	531	503	476	449	423	398	374	350
×439	38.3	2.44	1620	503	—	430	407	384	362	341	320	300	280
×393	37.8	2.20	1450	443	—	378	358	338	318	299	281	263	246
×359	37.4	2.01	1320	400	—	341	322	304	286	269	252	236	220
×328	37.1	1.85	1210	360	324	307	290	273	257	242	226	212	197
×300	36.7	1.68	1110	328	295	279	264	249	234	220	206	192	179
×280	36.5	1.57	1030	305	274	259	245	230	217	203	190	178	165
×260	36.3	1.44	953	285	256	242	228	215	202	190	177	166	154
×245	36.1	1.35	895	269	241	228	215	203	190	178	167	156	145
×230	35.9	1.26	837	253	227	214	202	190	179	168	157	146	136

Table 8-49 (cont.).
Section Modulus of Coped W Shapes

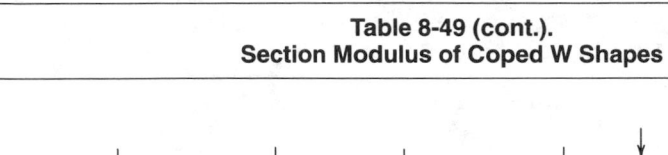

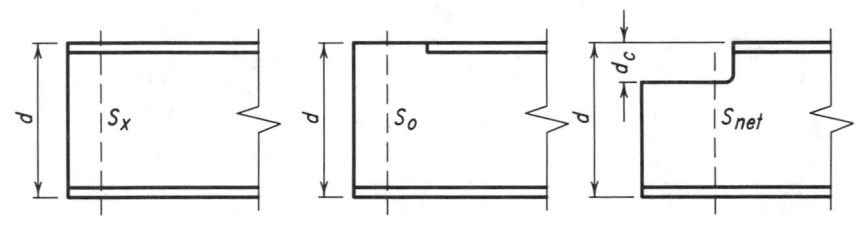

Designation	d in.	t_f in.	S_x in.^{3}	S_o in.^{3}	S_{net}, in.^{3} d_c, in.								
					2	3	4	5	6	7	8	9	10
W36×256	37.4	1.73	895	329	297	281	266	251	237	223	209	196	183
×232	37.1	1.57	809	295	266	251	238	224	211	199	186	174	163
×210	36.7	1.36	719	272	245	232	219	207	195	183	172	161	150
×194	36.5	1.26	664	249	224	212	201	189	178	167	157	146	137
×182	36.3	1.18	623	234	211	199	188	178	167	157	147	137	128
×170	36.2	1.10	580	218	196	185	175	165	155	146	137	128	119
×160	36.0	1.02	542	206	185	175	165	156	147	138	129	120	112
×150	35.9	0.940	504	195	176	166	157	148	139	130	122	114	106
×135	35.6	0.790	439	181	163	154	145	137	129	121	113	105	98.1
W33×354	35.6	2.09	1230	373	—	315	297	279	262	245	229	213	198
×318	35.2	1.89	1110	330	295	278	262	246	230	216	201	187	173
×291	34.8	1.73	1010	300	268	253	238	223	209	195	182	169	157
×263	34.5	1.57	917	268	239	226	212	199	186	174	162	151	139
×241	34.2	1.40	829	250	223	210	197	185	173	162	150	140	129
×221	33.9	1.28	757	230	205	193	181	170	159	148	138	128	118
×201	33.7	1.15	684	209	186	175	165	154	144	135	125	116	107
W33×169	33.8	1.22	549	191	170	161	151	141	132	124	115	107	98.6
×152	33.5	1.06	487	176	157	148	139	130	122	114	106	97.9	90.5
×141	33.3	0.960	448	165	147	139	130	122	114	106	98.8	91.6	84.6
×130	33.1	0.855	406	155	138	130	122	114	107	100	92.5	85.7	79.2
×118	32.9	0.740	359	143	128	120	113	106	98.6	91.9	85.4	79.1	73.0
W30×477	34.2	2.95	1530	475	—	398	374	350	327	305	283	262	242
×391	33.2	2.44	1250	378	—	315	295	276	257	239	222	205	188
×326	32.4	2.05	1030	305	—	254	237	221	206	191	177	163	150
×292	32.0	1.85	928	269	238	223	208	194	180	167	155	142	130
×261	31.6	1.65	827	240	212	198	185	172	160	148	137	126	115
×235	31.3	1.50	746	211	186	174	163	152	141	130	120	110	101
×211	30.9	1.32	663	192	170	159	148	138	128	118	109	100	91.2
×191	30.7	1.19	598	174	153	143	133	124	115	106	97.7	89.6	81.8
×173	30.4	1.07	539	158	139	130	121	112	104	96.1	88.4	81.0	73.9
W30×148	30.7	1.18	436	152	134	125	117	109	101	93.3	86.0	78.9	72.1
×132	30.3	1.00	380	139	123	115	107	99.3	92.1	85.1	78.3	71.8	65.5
×124	30.2	0.930	355	131	115	108	100	93.4	86.5	79.9	73.6	67.4	61.5
×116	30.0	0.850	329	124	109	102	95.3	88.6	82.1	75.8	69.7	63.9	58.2
×108	29.8	0.760	299	118	103	96.5	89.9	83.6	77.4	71.4	65.7	60.1	54.8
×99	29.7	0.670	269	110	96.4	90.0	83.9	77.9	72.1	66.5	61.1	56.0	51.0
×90	29.5	0.610	245	98.7	86.7	80.9	75.4	70.0	64.8	59.7	54.9	50.2	45.7

Table 8-49 (cont.).
Section Modulus of Coped W Shapes

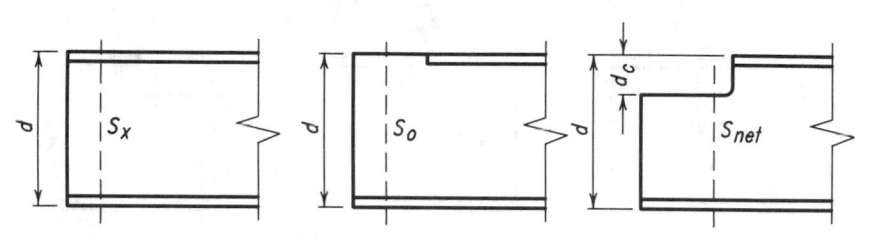

Designation	d in.	t_f in.	S_x in.^{3}	S_o in.^{3}	S_{net}, in.^{3}								
					d_c, in.								
					2	3	4	5	6	7	8	9	10
W27×539	32.5	3.54	1570	509	—	—	394	367	341	316	292	269	247
×448	31.4	2.99	1300	404	—	—	310	288	267	247	227	209	191
×368	30.4	2.48	1060	321	—	262	244	226	209	193	177	162	147
×307	29.6	2.09	884	259	—	211	196	181	167	154	141	128	116
×281	29.3	1.93	811	233	203	189	176	162	150	137	126	114	104
×258	29.0	1.77	742	212	185	172	159	147	136	124	114	103	93.3
×235	28.7	1.61	674	193	168	156	145	134	123	113	103	93.2	84.2
×217	28.4	1.50	624	174	152	141	130	120	111	101	92.3	83.7	75.5
×194	28.1	1.34	556	155	134	125	115	106	97.6	89.3	81.3	73.6	66.3
×178	27.8	1.19	502	145	126	117	108	100	91.5	83.6	76.1	68.8	61.9
×161	27.6	1.08	455	131	113	105	97.2	89.5	82.0	74.9	68.1	61.5	55.3
×146	27.4	0.975	411	118	102	95.0	87.7	80.7	74.0	67.5	61.3	55.3	49.7
W27×129	27.6	1.10	345	117	101	94.0	86.9	80.1	73.5	67.2	61.1	55.3	49.7
×114	27.3	0.930	299	106	91.6	84.9	78.4	72.2	66.2	60.5	54.9	49.6	44.6
×102	27.1	0.830	267	94.2	81.6	75.6	69.8	64.3	58.9	53.7	48.8	44.0	39.5
×94	26.9	0.745	243	88.0	76.2	70.6	65.1	59.9	54.9	50.1	45.4	41.0	36.8
×84	26.7	0.640	213	80.5	69.6	64.5	59.5	54.7	50.1	45.7	41.4	37.3	33.5
W24×492	29.7	3.54	1290	420	—	—	316	292	269	247	226	205	186
×408	28.5	2.99	1060	331	—	—	247	227	209	191	173	157	141
×335	27.5	2.48	864	261	—	209	193	177	162	147	133	120	108
×279	26.7	2.09	718	210	—	167	154	141	128	116	105	94.3	84.0
×250	26.3	1.89	644	184	158	146	134	123	112	101	91.2	81.7	72.6
×229	26.0	1.73	588	167	143	132	121	111	101	91.0	81.8	73.1	64.9
×207	25.7	1.57	531	149	127	117	107	98.0	89.0	80.4	72.2	64.4	57.0
×192	25.5	1.46	491	136	117	107	98.2	89.5	81.2	73.3	65.8	58.6	51.8
×176	25.2	1.34	450	124	106	97.6	89.4	81.4	73.8	66.5	59.6	53.0	46.8
×162	25.0	1.22	414	115	98.0	90.0	82.3	74.9	67.9	61.1	54.7	48.6	42.8
×146	24.7	1.09	371	104	88.5	81.2	74.2	67.5	61.1	54.9	49.1	43.6	38.3
×131	24.5	0.960	329	94.4	80.3	73.7	67.3	61.1	55.3	49.7	44.3	39.3	34.5
×117	24.3	0.850	291	84.4	71.7	65.7	60.0	54.5	49.2	44.2	39.4	34.8	30.5
×104	24.1	0.750	258	75.4	64.1	58.7	53.5	48.6	43.8	39.3	35.0	30.9	27.1
W24×103	24.5	0.980	245	82.9	70.7	64.9	59.3	53.9	48.8	43.9	39.2	34.8	30.6
×94	24.3	0.875	222	76.2	64.9	59.5	54.3	49.4	44.6	40.1	35.8	31.7	27.9
×84	24.1	0.770	196	68.3	58.0	53.2	48.6	44.1	39.8	35.8	31.9	28.2	24.8
×76	23.9	0.680	176	62.6	53.2	48.7	44.5	40.4	36.4	32.7	29.1	25.8	22.6
×68	23.7	0.585	154	57.5	48.8	44.7	40.8	37.0	33.4	29.9	26.6	23.5	20.6
W24×62	23.7	0.590	131	56.9	48.3	44.3	40.4	36.7	33.1	29.7	26.5	23.4	20.5
×55	23.6	0.505	114	51.1	43.4	39.7	36.2	32.9	29.7	26.6	23.7	20.9	18.3

Table 8-49 (cont.).
Section Modulus of Coped W Shapes

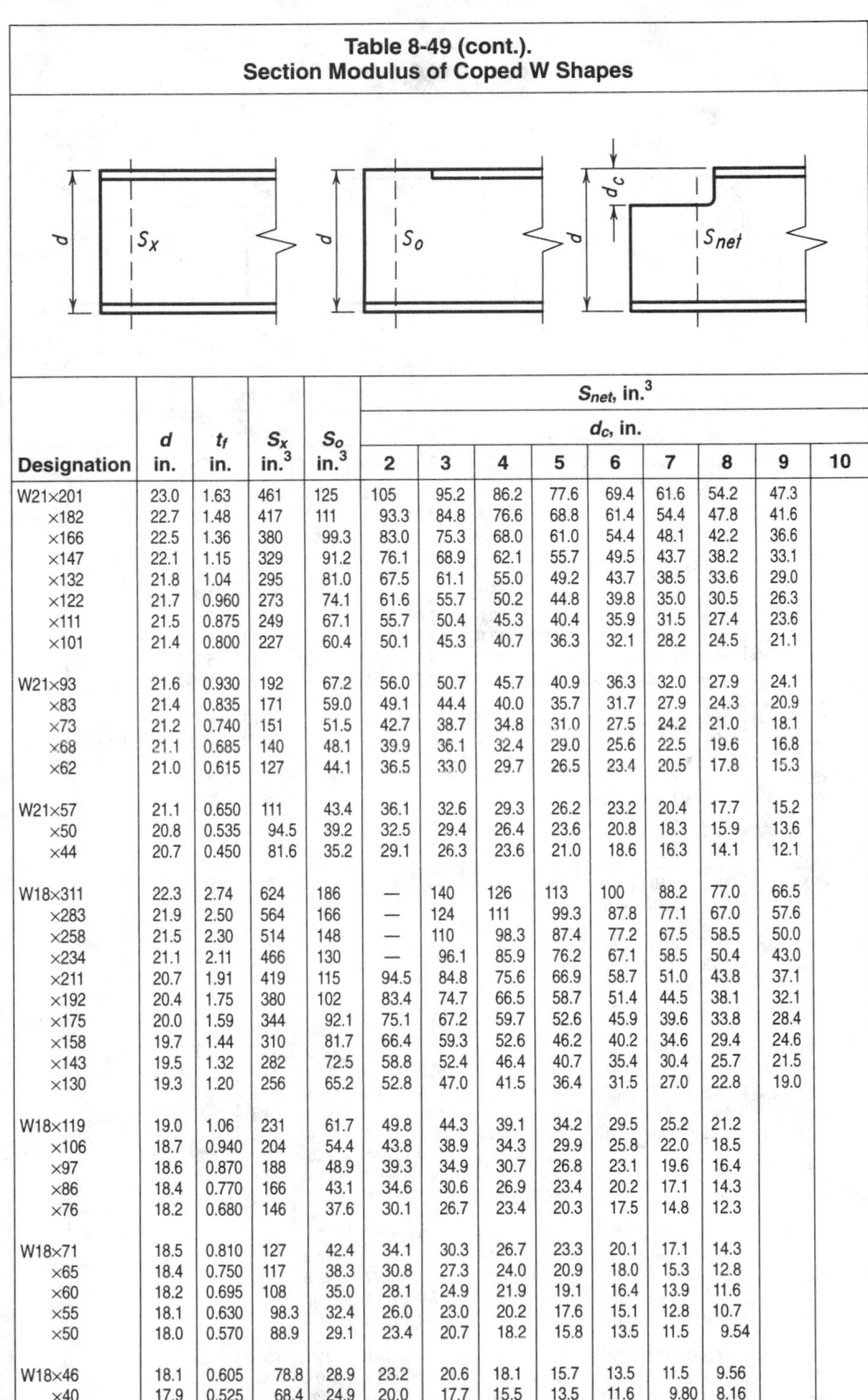

Designation	d in.	t_f in.	S_x in.^{3}	S_o in.^{3}	S_{net}, in.^{3} d_c, in.								
					2	3	4	5	6	7	8	9	10
W21×201	23.0	1.63	461	125	105	95.2	86.2	77.6	69.4	61.6	54.2	47.3	
×182	22.7	1.48	417	111	93.3	84.8	76.6	68.8	61.4	54.4	47.8	41.6	
×166	22.5	1.36	380	99.3	83.0	75.3	68.0	61.0	54.4	48.1	42.2	36.6	
×147	22.1	1.15	329	91.2	76.1	68.9	62.1	55.7	49.5	43.7	38.2	33.1	
×132	21.8	1.04	295	81.0	67.5	61.1	55.0	49.2	43.7	38.5	33.6	29.0	
×122	21.7	0.960	273	74.1	61.6	55.7	50.2	44.8	39.8	35.0	30.5	26.3	
×111	21.5	0.875	249	67.1	55.7	50.4	45.3	40.4	35.9	31.5	27.4	23.6	
×101	21.4	0.800	227	60.4	50.1	45.3	40.7	36.3	32.1	28.2	24.5	21.1	
W21×93	21.6	0.930	192	67.2	56.0	50.7	45.7	40.9	36.3	32.0	27.9	24.1	
×83	21.4	0.835	171	59.0	49.1	44.4	40.0	35.7	31.7	27.9	24.3	20.9	
×73	21.2	0.740	151	51.5	42.7	38.7	34.8	31.0	27.5	24.2	21.0	18.1	
×68	21.1	0.685	140	48.1	39.9	36.1	32.4	29.0	25.6	22.5	19.6	16.8	
×62	21.0	0.615	127	44.1	36.5	33.0	29.7	26.5	23.4	20.5	17.8	15.3	
W21×57	21.1	0.650	111	43.4	36.1	32.6	29.3	26.2	23.2	20.4	17.7	15.2	
×50	20.8	0.535	94.5	39.2	32.5	29.4	26.4	23.6	20.8	18.3	15.9	13.6	
×44	20.7	0.450	81.6	35.2	29.1	26.3	23.6	21.0	18.6	16.3	14.1	12.1	
W18×311	22.3	2.74	624	186	—	140	126	113	100	88.2	77.0	66.5	
×283	21.9	2.50	564	166	—	124	111	99.3	87.8	77.1	67.0	57.6	
×258	21.5	2.30	514	148	—	110	98.3	87.4	77.2	67.5	58.5	50.0	
×234	21.1	2.11	466	130	—	96.1	85.9	76.2	67.1	58.5	50.4	43.0	
×211	20.7	1.91	419	115	94.5	84.8	75.6	66.9	58.7	51.0	43.8	37.1	
×192	20.4	1.75	380	102	83.4	74.7	66.5	58.7	51.4	44.5	38.1	32.1	
×175	20.0	1.59	344	92.1	75.1	67.2	59.7	52.6	45.9	39.6	33.8	28.4	
×158	19.7	1.44	310	81.7	66.4	59.3	52.6	46.2	40.2	34.6	29.4	24.6	
×143	19.5	1.32	282	72.5	58.8	52.4	46.4	40.7	35.4	30.4	25.7	21.5	
×130	19.3	1.20	256	65.2	52.8	47.0	41.5	36.4	31.5	27.0	22.8	19.0	
W18×119	19.0	1.06	231	61.7	49.8	44.3	39.1	34.2	29.5	25.2	21.2		
×106	18.7	0.940	204	54.4	43.8	38.9	34.3	29.9	25.8	22.0	18.5		
×97	18.6	0.870	188	48.9	39.3	34.9	30.7	26.8	23.1	19.6	16.4		
×86	18.4	0.770	166	43.1	34.6	30.6	26.9	23.4	20.2	17.1	14.3		
×76	18.2	0.680	146	37.6	30.1	26.7	23.4	20.3	17.5	14.8	12.3		
W18×71	18.5	0.810	127	42.4	34.1	30.3	26.7	23.3	20.1	17.1	14.3		
×65	18.4	0.750	117	38.3	30.8	27.3	24.0	20.9	18.0	15.3	12.8		
×60	18.2	0.695	108	35.0	28.1	24.9	21.9	19.1	16.4	13.9	11.6		
×55	18.1	0.630	98.3	32.4	26.0	23.0	20.2	17.6	15.1	12.8	10.7		
×50	18.0	0.570	88.9	29.1	23.4	20.7	18.2	15.8	13.5	11.5	9.54		
W18×46	18.1	0.605	78.8	28.9	23.2	20.6	18.1	15.7	13.5	11.5	9.56		
×40	17.9	0.525	68.4	24.9	20.0	17.7	15.5	13.5	11.6	9.80	8.16		
×35	17.7	0.425	57.6	22.7	18.2	16.1	14.1	12.3	10.5	8.88	7.37		

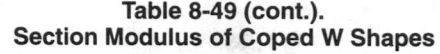

Table 8-49 (cont.).
Section Modulus of Coped W Shapes

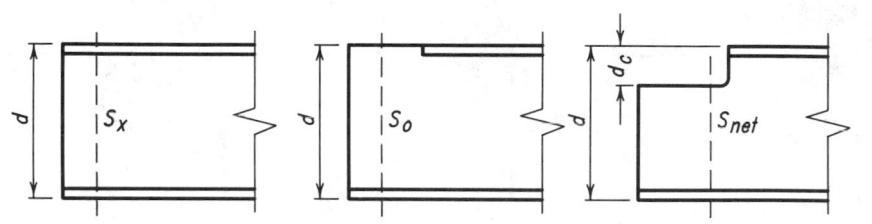

Designation	d in.	t_f in.	S_x in.^{3}	S_o in.^{3}	S_{net}, in.^{3} d_c, in.								
					2	3	4	5	6	7	8	9	10
W16×100	17.0	0.985	175	44.4	34.9	30.5	26.4	22.6	19.0	15.7			
×89	16.8	0.875	155	39.0	30.6	26.7	23.1	19.7	16.5	13.6			
×77	16.5	0.760	134	33.1	25.9	22.6	19.4	16.5	13.8	11.4			
×67	16.3	0.665	117	28.3	22.1	19.2	16.5	14.0	11.7	9.58			
W16×57	16.4	0.715	92.2	29.4	23.0	20.1	17.3	14.8	12.4	10.2			
×50	16.3	0.630	81.0	25.6	20.0	17.4	15.0	12.7	10.7	8.74			
×45	16.1	0.565	72.7	22.9	17.9	15.5	13.4	11.3	9.47	7.75			
×40	16.0	0.505	64.7	20.1	15.6	13.6	11.7	9.89	8.24	6.73			
×36	15.9	0.430	56.5	18.8	14.6	12.7	10.9	9.21	7.67	6.25			
W16×31	15.9	0.440	47.2	17.1	13.3	11.6	10.0	8.44	7.03	5.73			
×26	15.7	0.345	38.4	14.9	11.6	10.1	8.64	7.31	6.08	4.95			
W14×808	22.8	5.12	1400	451	—	—	—	—	244	216			
×730	22.4	4.91	1280	365	—	—	—	220	195	172			
×665	21.6	4.52	1150	317	—	—	—	187	165	144			
×605	20.9	4.16	1040	275	—	—	—	158	139	121			
×550	20.2	3.82	931	238	—	—	153	134	117	101			
×500	19.6	3.50	838	208	—	—	131	115	99.4	85.3			
×455	19.0	3.21	756	182	—	—	113	98.2	84.6	72.1			
W14×426	18.7	3.04	707	164	—	—	101	87.6	75.2	63.8			
×398	18.3	2.85	656	150	—	104	91.1	78.7	67.2	56.7			
×370	17.9	2.66	607	135	—	93.7	81.4	70.1	59.6	50.0			
×342	17.5	2.47	559	122	—	83.4	72.3	61.9	52.3	43.6			
×311	17.1	2.26	506	107	—	72.7	62.7	53.5	44.9	37.2			
×283	16.7	2.07	459	94.4	—	63.6	54.6	46.3	38.7	31.8			
×257	16.4	1.89	415	83.1	64.1	55.5	47.4	40.0	33.3	27.1			
×233	16.0	1.72	375	73.2	56.1	48.4	41.3	34.6	28.6	23.2			
×211	15.7	1.56	338	64.9	49.5	42.6	36.1	30.2	24.8	19.9			
×193	15.5	1.44	310	57.6	43.8	37.5	31.7	26.4	21.6	17.3			
×176	15.2	1.31	281	52.2	39.5	33.8	28.5	23.6	19.2	15.2			
×159	15.0	1.19	254	45.7	34.5	29.4	24.7	20.4	16.5	13.0			
×145	14.8	1.09	232	40.9	30.7	26.1	21.9	18.0	14.5	11.4			
W14×132	14.7	1.03	209	38.1	28.6	24.3	20.3	16.7	13.4				
×120	14.5	0.940	190	34.2	25.5	21.7	18.1	14.8	11.8				
×109	14.3	0.860	173	30.0	22.3	18.9	15.7	12.8	10.2				
×99	14.2	0.780	157	27.2	20.2	17.0	14.2	11.5	9.15				
×90	14.0	0.710	143	24.3	18.0	15.2	12.6	10.2	8.07				
W14×82	14.3	0.855	123	28.0	20.9	17.7	14.8	12.1	9.64				
×74	14.2	0.785	112	24.4	18.2	15.4	12.8	10.4	8.31				
×68	14.0	0.720	103	22.2	16.5	13.9	11.6	9.42	7.46				
×61	13.9	0.645	92.2	19.7	14.6	12.3	10.2	8.28	6.54				

Table 8-49 (cont.).
Section Modulus of Coped W Shapes

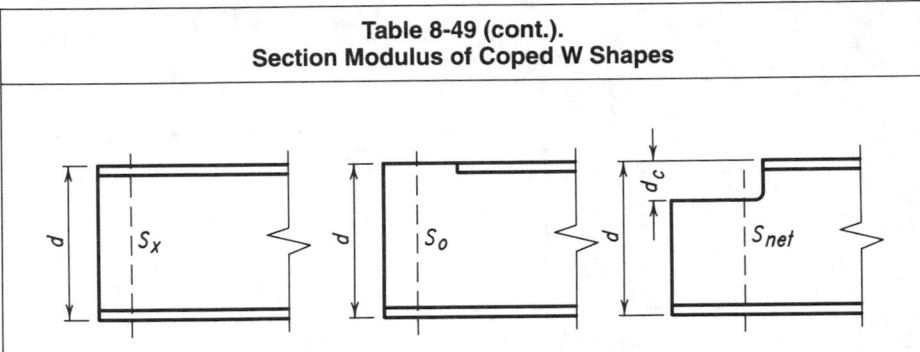

Designation	d in.	t_f in.	S_x in.^{3}	S_o in.^{3}	S_{net}, in.^{3} d_c, in.								
					2	3	4	5	6	7	8	9	10
W14×53	13.9	0.660	77.8	19.1	14.2	12.0	9.93	8.07	6.39				
×48	13.8	0.595	70.3	17.3	12.8	10.8	8.93	7.23	5.71				
×43	13.7	0.530	62.7	15.3	11.3	9.50	7.84	6.34	4.99				
W14×38	14.1	0.515	54.6	16.0	12.0	10.2	8.48	6.94	5.54				
×34	14.0	0.455	48.6	14.4	10.8	9.14	7.62	6.22	4.95				
×30	13.8	0.385	42.0	13.2	9.88	8.37	6.96	5.68	4.51				
W14×26	13.9	0.420	35.3	12.3	9.20	7.80	6.50	5.31	4.23				
×22	13.7	0.335	29.0	10.7	7.97	6.75	5.62	4.58	3.64				
W12×336	16.8	3.00	483	123	—	83.1	71.4	60.6	50.8				
×305	16.3	2.71	435	108	—	71.4	61.0	51.4	42.7				
×279	15.9	2.47	393	96.1	—	63.1	53.5	44.8	36.9				
×252	15.4	2.25	353	83.7	—	54.2	45.7	38.0	31.0				
×230	15.1	2.07	321	74.2	—	47.5	39.9	32.9	26.7				
×210	14.7	1.90	292	65.6	49.0	41.6	34.7	28.5	22.9				
×190	14.4	1.74	263	57.0	42.3	35.7	29.7	24.2	19.3				
×170	14.0	1.56	235	49.6	36.5	30.7	25.3	20.5	16.2				
×152	13.7	1.40	209	43.3	31.6	26.5	21.7	17.5	13.7				
×136	13.4	1.25	186	37.9	27.5	22.9	18.7	14.9	11.6				
×120	13.1	1.105	163	32.8	23.7	19.7	16.0	12.6	9.70				
×106	12.9	0.990	145	27.6	19.8	16.3	13.2	10.4					
×96	12.7	0.900	131	24.3	17.4	14.3	11.5	9.03					
×87	12.5	0.810	118	22.2	15.8	13.0	10.4	8.11					
×79	12.4	0.735	107	19.9	14.1	11.5	9.23	7.16					
×72	12.3	0.670	97.4	17.9	12.6	10.3	8.24	6.37					
×65	12.1	0.605	87.9	16.0	11.2	9.16	7.28	5.61					
W12×58	12.2	0.640	78.0	14.8	10.4	8.52	6.79	5.24					
×53	12.1	0.575	70.6	13.9	9.74	7.94	6.31	4.85					
W12×50	12.2	0.640	64.7	14.8	10.4	8.54	6.82	5.27					
×45	12.1	0.575	58.1	13.1	9.27	7.56	6.02	4.63					
×40	11.9	0.515	51.9	11.4	8.03	6.54	5.19	3.98					
W12×35	12.5	0.52	45.6	12.3	8.85	7.30	5.89	4.61					
×30	12.3	0.44	38.6	10.5	7.47	6.15	4.94	3.86					
×26	12.2	0.38	33.4	9.08	6.47	5.32	4.27	3.32					
W12×22	12.3	0.425	25.4	9.60	6.89	5.69	4.59	3.59					
×19	12.2	0.350	21.3	8.39	6.01	4.95	3.98	3.11					
×16	12.0	0.265	17.1	7.43	5.30	4.36	3.50	2.72					
×14	11.9	0.225	14.9	6.61	4.71	3.86	3.10	2.41					

Table 8-49 (cont.).
Section Modulus of Coped W Shapes

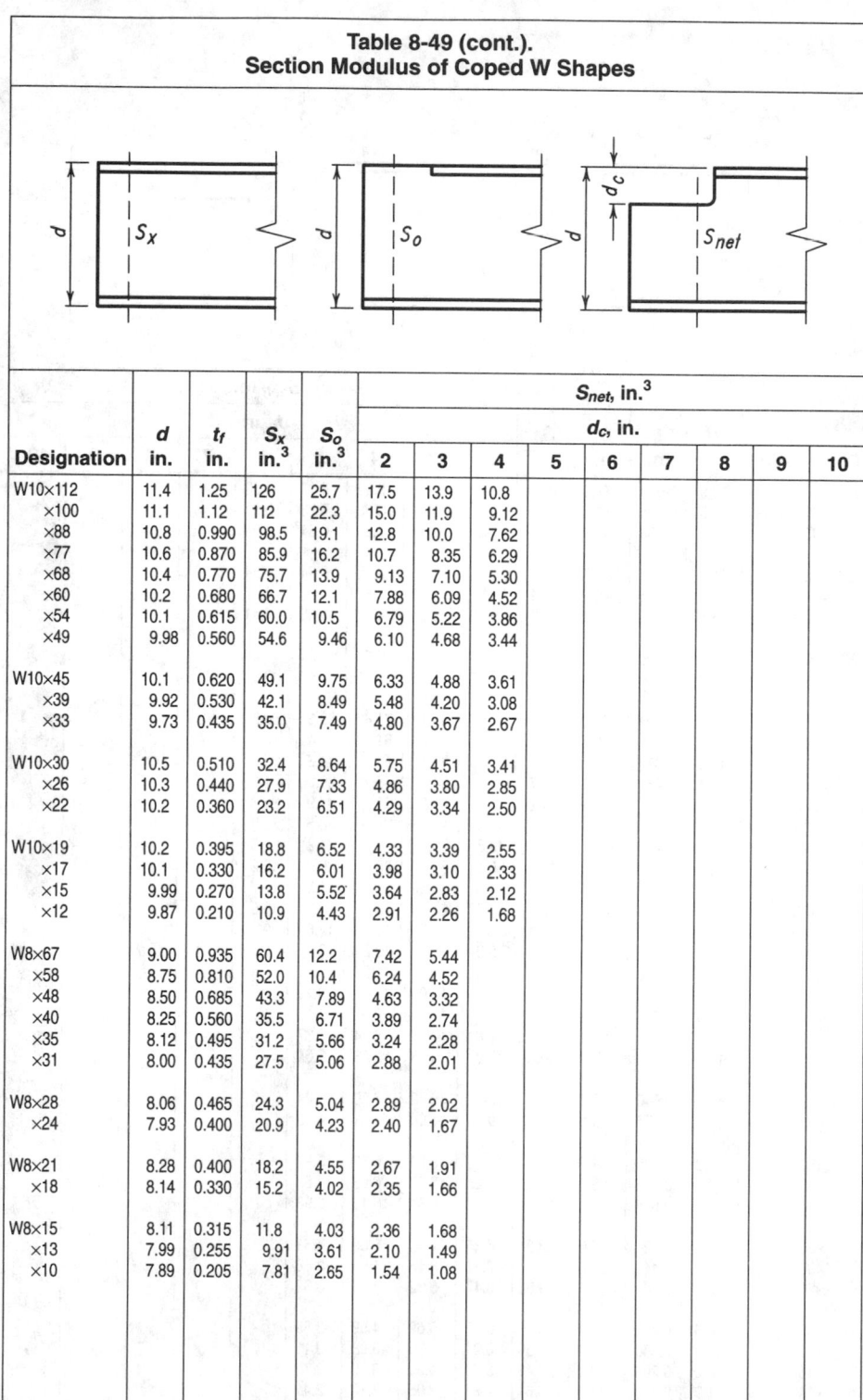

Designation	d in.	t_f in.	S_x in.^{3}	S_o in.^{3}	S_{net}, in.^{3}								
					d_c, in.								
					2	3	4	5	6	7	8	9	10
W10×112	11.4	1.25	126	25.7	17.5	13.9	10.8						
×100	11.1	1.12	112	22.3	15.0	11.9	9.12						
×88	10.8	0.990	98.5	19.1	12.8	10.0	7.62						
×77	10.6	0.870	85.9	16.2	10.7	8.35	6.29						
×68	10.4	0.770	75.7	13.9	9.13	7.10	5.30						
×60	10.2	0.680	66.7	12.1	7.88	6.09	4.52						
×54	10.1	0.615	60.0	10.5	6.79	5.22	3.86						
×49	9.98	0.560	54.6	9.46	6.10	4.68	3.44						
W10×45	10.1	0.620	49.1	9.75	6.33	4.88	3.61						
×39	9.92	0.530	42.1	8.49	5.48	4.20	3.08						
×33	9.73	0.435	35.0	7.49	4.80	3.67	2.67						
W10×30	10.5	0.510	32.4	8.64	5.75	4.51	3.41						
×26	10.3	0.440	27.9	7.33	4.86	3.80	2.85						
×22	10.2	0.360	23.2	6.51	4.29	3.34	2.50						
W10×19	10.2	0.395	18.8	6.52	4.33	3.39	2.55						
×17	10.1	0.330	16.2	6.01	3.98	3.10	2.33						
×15	9.99	0.270	13.8	5.52	3.64	2.83	2.12						
×12	9.87	0.210	10.9	4.43	2.91	2.26	1.68						
W8×67	9.00	0.935	60.4	12.2	7.42	5.44							
×58	8.75	0.810	52.0	10.4	6.24	4.52							
×48	8.50	0.685	43.3	7.89	4.63	3.32							
×40	8.25	0.560	35.5	6.71	3.89	2.74							
×35	8.12	0.495	31.2	5.66	3.24	2.28							
×31	8.00	0.435	27.5	5.06	2.88	2.01							
W8×28	8.06	0.465	24.3	5.04	2.89	2.02							
×24	7.93	0.400	20.9	4.23	2.40	1.67							
W8×21	8.28	0.400	18.2	4.55	2.67	1.91							
×18	8.14	0.330	15.2	4.02	2.35	1.66							
W8×15	8.11	0.315	11.8	4.03	2.36	1.68							
×13	7.99	0.255	9.91	3.61	2.10	1.49							
×10	7.89	0.205	7.81	2.65	1.54	1.08							

Other Elements in Connections

Shims

Shims are furnished to the erector for use in filling the spaces allowed for field clearance which might be present at connections such as simple shear connections, PR and FR moment connections, column base plates, and column splices. These shims, illustrated in Figure 8-63, may be either strip shims, with round punched holes, or finger shims, with slots cut through the edge. Whereas strip shims are less expensive to fabricate, finger shims may be laterally inserted and eliminate the need to remove erection bolts or pins already in place.

Finger shims, when inserted fully against the bolt shank, are acceptable for slip-critical connections and are not to be considered as an internal ply with the slotted hole determining the design strength of the connection. This is because less than 25 percent of the contact surface is lost and this is not enough to affect the performance of the joint.

Fillers

A filler is furnished to occupy spaces which will be present because of dimensional separations between elements of a connection across which load transfer occurs. Examples where fillers might be used are beams framing off center on a column and raised beams.

From LRFD Specification Section J6, fillers in welded connections and fillers thicker than ¾-in. in bolted bearing-type connections must be fully developed. In bolted bearing-type connections, fillers between ¼-in. and ¾-in. thick, inclusive, need not be developed, provided the design shear strength of the bolts is reduced by the factor

$$0.4(t - 0.25)$$

where t is the total thickness of the fillers up to ¾-in. In bolted slip-critical connections, fillers need not be fully developed.

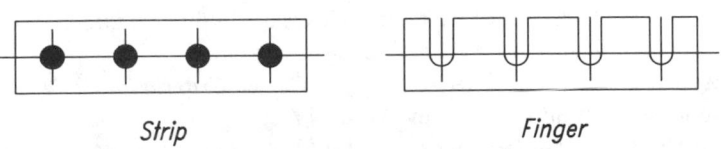

Strip   *Finger*

Fig. 8-63. Shims.

REFERENCES

Alexander, W. G., 1991, "Designing Longitudinal Welds for Bridge Members," *Engineering Journal*, Vol. 28, No. 1, (1st Qtr.), pp. 29–36, AISC, Chicago, IL.

American Concrete Institute, 1985, ACI 349 *Code Requirements for Nuclear Safety Related Concrete Structures*, Appendix B, ACI, Detroit, MI.

American Institute of Steel Construction, Inc., 1993, *Load and Resistance Factor Design Specification for Structural Steel Buildings*, AISC, Chicago, IL.

American Institute of Steel Construction, Inc., 1989, *Manual of Steel Construction—Allowable Stress Design*, 9th ed., AISC, Chicago, IL.

American Institute of Steel Construction, Inc., 1988, *Quality Criteria and Inspection Standards*, 3rd ed., AISC, Chicago, IL.

American Institute of Steel Construction, Inc., 1973, "Commentary on Highly Restrained Welded Connections," *Engineering Journal*, Vol. 10, No. 3, (3rd Qtr.), pp. 61–73, AISC, Chicago, IL.

American Welding Society, 1978, *Welding Handbook—Volume 2*, 7th ed., AWS, Miami, FL.

American Welding Society, 1977, *Guide for the Non-Destructive Inspection of Welds*, (AWS B1.0-77), AWS, Miami, FL.

Astaneh, A., 1985, "Procedure for Design and Analysis of Hanger-Type Connections," *Engineering Journal*, Vol. 22, No. 2, (2nd Qtr.), pp. 63–66, AISC, Chicago, IL.

Blodgett, O. W., 1966, *Design of Welded Structures*, James F. Lincoln Arc Welding Foundation, Cleveland, OH.

Blodgett, O. W., 1980, "Detailing to Achieve Practical Welded Fabrication," *Engineering Journal*, Vol. 17, No. 4, (4th Qtr.), pp. 106–119, AISC, Chicago, IL.

Bowman, M. D. and M. Betancourt, 1991, "Reuse of A325 and A490 High-Strength Bolts," *Engineering Journal*, Vol. 28, No. 3, (3rd Qtr.), pp. 110–118, AISC, Chicago, IL.

Butler, L. J., S. Pal, and G. L. Kulak, 1972, "Eccentrically Loaded Welded Connections," *Journal of the Structural Division*, Vol. 98, No. ST5, (May), pp. 989–1005, ASCE, New York, N.Y.

Cannon, R. W., D. A. Godfrey, and F. L. Moreadith, 1981, "Guide to the Design of Anchor Bolts and Other Steel Embedments," *Concrete International*, Vol. 3, No. 7, (July 1981), pp. 28–41, ACI, Detroit, MI..

Cheng, J. J., J. A. Yura, and C. P. Johnson, 1984, "Design and Behaviour of Coped Beams," Department of Civil Engineering, The University of Texas at Austin, Austin, TX.

Crawford, S. F. and G. L. Kulak, 1968, "Behavior of Eccentrically Loaded Bolted Connections," *Studies in Structural Engineering*, (No. 4), Department of Civil Engineering, Nova Scotia Technical College, Halifax, Nova Scotia.

DeWolf, J. T. and D. T. Ricker, 1990, *Column Base Plates*, AISC, Chicago, IL

Fisher, J. M., 1981, "Structural Details in Industrial Buildings," *Engineering Journal*, Vol. 18, No. 3, (3rd Qtr.), pp. 83–89, AISC, Chicago, IL.

Fisher, J. W. and J. H. A. Struik, 1974, *Guide to Design Criteria for Bolted and Riveted Joints*, John Wiley & Sons, Inc., New York, NY.

Grover, L., 1946, *Manual of Design for Arc Welded Steel Structures*, Air Reduction Sales Co., New York, NY.

Higgins, T. R., 1971, "Treatment of Eccentrically Loaded Connections in the AISC Manual," *Engineering Journal*, Vol. 8, No. 2, (April), pp. 52–54, AISC, Chicago, IL.

Institute of Welding, 1972, *Procedures and Recommendations for the Ultrasonic Testing of Butt Welds*, London, England.

Iwankiw, N. R., 1987, "Design for Eccentric and Inclined Loads on Bolt and Weld Groups," *Engineering Journal*, Vol. 24, No. 4, (4th Qtr.), pp. 164–171, AISC, Chicago, IL.

Kaufmann, J., A. W. Pense, and R. D. Stout, 1981, "An Evaluation of Factors Significant to Lamellar Tearing," *Welding Journal Research Supplement*, Vol. 60, No. 3, (March), AWS, Miami, FL.

Krautkramer, J., 1977, *Ultrasonic Testing of Materials*, 2nd. ed., Springer-Verlag, Berlin, West Germany.

Kulak, G. L., 1975, "Eccentrically Loaded Slip-Resistant Connections," *Engineering Journal*, Vol. 12, No. 2, (2nd Qtr.), pp. 52–55, AISC, Chicago, IL.

Lesik, D. F. and D. J. L. Kennedy, 1990, "Ultimate Strength of Fillet-Welded Connections Loaded in Plane," *Canadian Journal of Civil Engineering*, Vol. 17, No. 1, National Research Council of Canada, Ottawa, Canada.

Kulak, G. L. and Timler, 1984, "Tests on Eccentrically Loaded Fillet Welds," Department of Civil Engineering, University of Alberta, Edmonton, Canada.

Kulak, G. L., J. W. Fisher, and J. H. A. Struik, 1987, *Guide to Design Criteria for Bolted and Riveted Joints*, 2nd ed., John Wiley & Sons, New York, NY.

Marsh, M. L., and E. G. Burdette, 1985a, "Anchorage of Steel Building Components to Concrete," *Engineering Journal*, Vol. 22, No. 1, (1st Qtr.), pp. 33–39, AISC, Chicago, IL.

Marsh, M. L., and E. G. Burdette, 1985b, "Multiple Bolt Anchorages: Method for Determining the Effective Projected Area of Overlapping Stress Cones," *Engineering Journal*, Vol. 22, No. 1, (1st Qtr.), pp. 29–32, AISC, Chicago, IL.

Research Council on Structural Connections, 1988, *Load and Resistance Factor Design Specification for Structural Joints Using ASTM A325 or A490 Bolts*, AISC, Chicago, IL.

Shipp, J. G. and E. R. Haninger, 1983, "Design of Headed Anchor Bolts," *Engineering Journal*, Vol. 20, No. 2, (2nd Qtr.), pp. 58–69, AISC, Chicago. IL.

Stout, R. D. and W. D. Doty, 1978, *Weldability of Steels*, 3rd. ed., Welding Research Council, New York, NY

Thornton, W. A., 1985, "Prying Action—A General Treatment," *Engineering Journal*, Vol. 22, No. 2, (2nd Qtr.), pp. 67–75, AISC, Chicago, IL.

Tide, R. H. R., 1980, "Eccentrically Loaded Weld Groups—AISC Design Tables," *Engineering Journal*, Vol. 17, No. 4, (4th Qtr.), pp. 90–95, AISC, Chicago, IL.

PART 9

SIMPLE SHEAR AND PR MOMENT CONNECTIONS

SHEAR

OVERVIEW

Part 9 contains general information, design considerations, examples, and design aids for the design of simple shear connections, shear splices, PR moment connections, and special considerations in the aforementioned topics. It is based upon the provisions of the 1993 LRFD Specification. Supplementary information may also be found in the Commentary on the LRFD Specification.

Following are the general topics addressed.

SIMPLE SHEAR CONNECTIONS

The ends of members with simple shear connections are assumed to be unrestrained or free to rotate under load as illustrated in Figure 9-1. While simple shear connections do actually possess some rotational restraint, as illustrated by curve A in Figure 9-2, this small amount is usually neglected and the connection is idealized to be completely flexible. Accordingly, simple shear connections are sized only for the end reaction or shear R_u of the supported beam. Note that simple shear connections must provide flexibility to accommodate the required end rotation of the supported beam.

When members are designed with simple shear connections, provision must be made to stabilize the frame for gravity loads and also to resist lateral loads. A positive steel bracing system, such as X- or K-bracing, PR or FR construction, and concrete or masonry shear walls are three commonly used methods. PR moment connections (including flexible wind connections) are treated in this Part. FR moment connections are treated in Part 10. Bracing systems and connections are treated in Part 11. For the design of concrete or masonry shear walls, refer to ACI 318.

Considerations for Economical Simple Shear Connections

The AISC *Code of Standard Practice* states that, after the engineer of record (EOR) designs the structural members, the EOR may design and detail the connections or the EOR may have the fabricator develop the detailed configuration of the simple shear connections. In both cases, the fabricator must submit shop drawings for approval and verification that the EOR's design criteria and intent have been satisfied.

Regardless of which approach is taken, the AISC *Code of Standard Practice* states that the EOR is responsible for the adequacy of these connections. The fabricator is responsible for the accuracy of the detail dimensions, clearances, and general fit-up of the structural steel members and connecting materials for field assembly (refer to the AISC *Code of Standard Practice* Section 2 for definition of which items are and are not considered structural steel).

The latter approach is usually taken since there are economies inherent in allowing the fabricator to choose the most efficient connections for the fabricator's shop and erection processes. Whenever possible, the designer should give the fabricator and erector the flexibility to choose the connection types which offer the most economical shop fabrication and safest and most economical erection.

In taking this approach, however, some engineers of record specify general design criteria (e.g., one-half the total factored uniform load) from which the connections are to be developed without regard to the actual reactions. Thornton (1992) describes several of these practices and provides examples of the uneconomical and/or unsafe connections which can result from their use. Because of this, when the fabricator or detailer is to

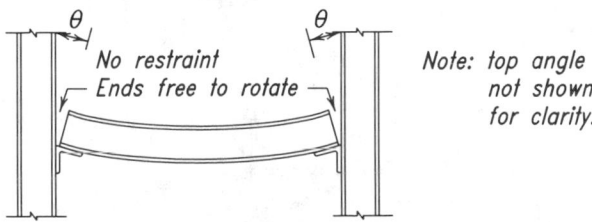

Figure 9.1. Illustration of simple shear connection.

develop the detailed configuration of the connections, the EOR must indicate the actual design reactions on the contract drawings or provide the fabricator with a method to accurately determine the required strength. In the absence of such information, connections will be selected to support one-half the total factored uniform load for the given beam, span, and grade of steel specified; no consideration will be given for the effects of any other loads unless specified on the contract drawings.

Comparing Two-Sided, Seated, and One-Sided Connections

Following is a general discussion of the advantages of two-sided, seated, and one-sided connections.

Two-sided connections, such as double-angle and shear end-plate connections, offer the following advantages: (1) suitability for use when the end reaction is large; (2) compactness (usually, the entire connection is contained within the flanges of the supported beam); and, (3) eccentricity perpendicular to the beam axis need not be considered for usual gages.

Unstiffened and stiffened seated connections offer the following advantages: (1) seats may be shop attached to the support, simplifying erection; (2) ample erection clearance is provided; (3) erection is fast and safe; and, (4) the bay length of the structure is easily maintained (seated connections may be preferable when maintaining bay length is a concern for repetitive bays of framing). Note that seated connections can cause erection

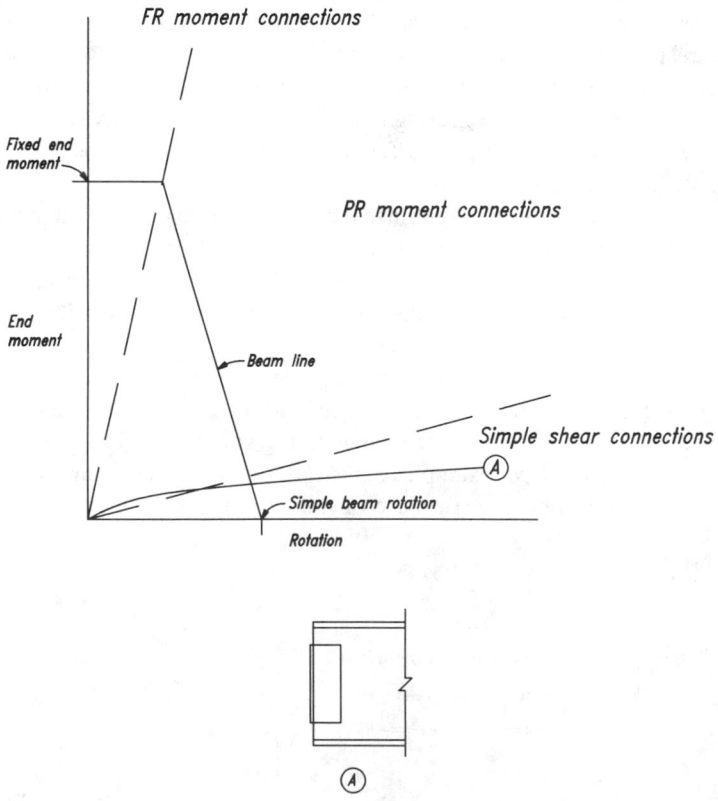

Figure 9-2. Simple shear connection behavior.

interference when floors are close, beams are deep, or seats protrude excessively from the column face; the practice of leaning or tilting the columns to erect a column-web connection is difficult, unsafe, and should always be avoided.

One-sided connections such as single-plate, single-angle, and tee connections offer the following advantages: (1) shop attachment of connecting materials to the support, simplifying shop fabrication and erection; (2) reduced material and shop labor requirements; and, (3) excellent safety during erection since double connections may be eliminated.

Erectability Considerations

In field-bolted connections, when beams or girders frame opposite each other and take the same open holes in the web of a column, as illustrated in Figure 9-3, the first member to be erected must be supported while the second member to be erected is brought into its final position. Note that hanging the beam on a partially inserted bolt or drift pin is dangerous; such a makeshift practice should not be attempted.

A temporary erection seat, usually an angle, is sometimes provided in the column web and located to clear the bottom flange of the supported member by approximately $\frac{3}{8}$-in. to accommodate mill, fabrication, and erection tolerances. The erection seat is sized and attached to the column web with sufficient bolts or welds to support the dead weight of the member, unless additional loading is indicated.

The sequence of erection is most important in determining the need for erection seats. If the erection sequence is known, the erection seat is provided on the side needing the support. If the erection sequence is not known, a seat can be provided on both sides of the column web. Erection seats may be reused at other locations, but are not generally required to be removed unless they create an interference, detract from the architectural appearance, or such removal is required in the contract documents.

In field-welded connections in which some means of temporary support must be provided until final welding is performed, temporary erection bolts are usually provided.

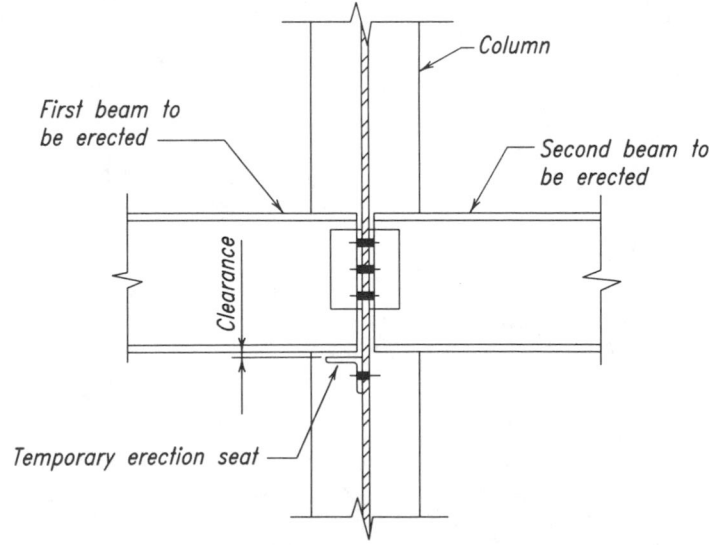

Figure 9-3. Erection seat.

Note that it is not necessary that these bolts be removed subsequent to final welding. Subject to the provisions of LRFD Specification Section J1.9, erection bolts may also serve as permanent attachment; refer to "Construction Combining Bolts and Welds" in Part 8.

Safety laws require that two bolts be placed for erection safety. As a general rule, then, two erection bolts are used for framing angles or similar connecting elements up to 12 inches long, four bolts are used for connecting elements up to 18 inches long, and six bolts are used for longer connecting elements. Additional erection bolts may be provided and serve two purposes: (1) they provide for the contingency of large temporary loads during erection; and, (2) they assist in pulling the connection angles up tightly against the web of the supporting beam prior to welding.

Some engineers prefer to locate erection bolts below the mid-depth of the connection; theoretically, this provides the greatest possible flexibility near the top of the connection, where the angles are expected to flex away from the supporting member. However, this practice does not ensure a close fit-up of the angle before welding. Other engineers prefer the more general practice of spacing the bolts equally along the length of the angles. In this latter case, the bolts are placed as closely as practical to the toes of the outstanding leg to provide greater flexibility.

Computer Software

CONXPRT is fully automated connection design software which provides for rapid design of economical simple shear connections. Based upon the AISC *Manual of Steel Construction, Volume II—Connections* and the engineering knowledge and experience of respected fabricators and design engineers, CONXPRT comes with preset guidelines, but can be modified to meet individual standards. It is menu-driven with a built-in shapes database and provides complete documentation of all design checks.

Double-Angle Connections

A double-angle connection is made with two angles, one on each side of the web of the beam to be supported, as illustrated in Figure 9-4. These angles may be bolted or welded to the supported beam as well as to the supporting member.

When the angles are welded to the support, adequate flexibility must be provided in the connection. As illustrated in Figure 9-4c, line welds are placed along the toes of the angles with a return at the top per LRFD Specification Section J2.2b. Note that welding across the entire top of the angles must be avoided as it would inhibit the flexibility and, therefore, the necessary end rotation of the connection; the performance of the resulting connection is unpredictable.

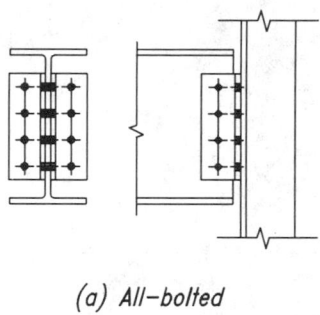

(a) All-bolted

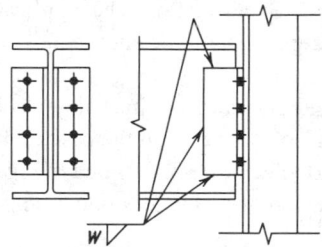

(b) Bolted/welded, angles welded to supported beam

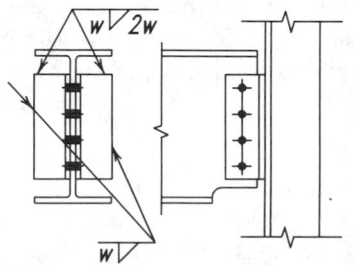

Note: weld returns on top of angles per LRFD Specification Section J2.2b.

(c) Bolted/welded, angles welded to support

Figure 9-4. Double-angle connections.

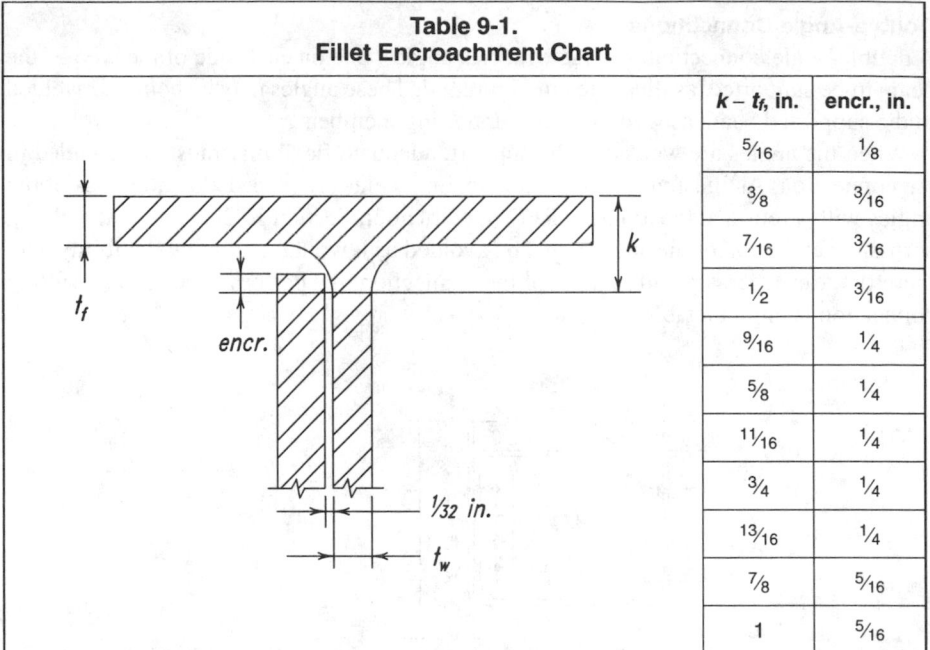

Table 9-1.
Fillet Encroachment Chart

$k - t_f$, in.	encr., in.
5/16	1/8
3/8	3/16
7/16	3/16
1/2	3/16
9/16	1/4
5/8	1/4
11/16	1/4
3/4	1/4
13/16	1/4
7/8	5/16
1	5/16

Design Checks

The design strengths of the bolts and/or welds and connected elements must be determined in accordance with the LRFD Specification; the applicable limit states are discussed in Part 8. In all cases, the design strength ϕR_n must equal or exceed the required strength R_u.

For usual gages of three inches and standard or short-slotted holes, eccentricity in double-angle connections may be neglected, except in the case of a double vertical row of bolts through the web of the supported beam, as illustrated in Figure 9-5. Eccentricity should always be considered in the design of welds for double-angle connections.

Recommended Angle Length and Thickness

To provide for stability during erection, it is recommended that the minimum angle length be one-half the T-dimension of the beam to be supported. The maximum length of the connection angles must be compatible with the T-dimension of an uncoped beam and the remaining web depth, exclusive of fillets, of a coped beam. Note that the angle may encroach on the fillet or fillets by 1/8-in. to 5/16-in., depending upon the radius of the fillets; refer to Table 9-1. To provide for flexibility, the maximum angle thickness for use with usual gages should be limited to 5/8-in.

Shop and Field Practices

Double-angle connections may be made to the webs of supporting girders and to the flanges of supporting columns. Because of bolting and welding clearances, double-angle connections may not be suitable for connections to the webs of W8 columns, unless gages are reduced or bolts are staggered, and may be impossible for W6 columns.

When framing to a girder web, both angles are usually shop attached to the web of the supported beam. When framing to a column web, both angles may be shop attached to the supported beam or to the column web. In the latter case, the bottom flange of the supported beam is coped to allow knifed erection (the beam web is lowered into place between the angles from above). Knifed erection requires that a total erection clearance of about ⅛-in. be provided between the angles as illustrated in Figure 9-6a. For bolted construction, this clearance may vary as gages will occur in minimum increments of ¹⁄₁₆-in. Shims must be furnished whenever measured clearances exceed ⅛-in.

When framing to a column flange, provision must be made for possible mill variation in the depth of the columns. If both angles are shop attached to the beam web, the beam length could be shortened to provide for mill overrun and shims could be furnished at the appropriate intervals to fill the resulting gaps or to provide for mill underrun; in general, shims are not required except for fairly long runs (i.e., six or more bays of framing). If both angles are shop attached to the column flange, the erected beam is knifed into place and play in the open holes usually furnishes the necessary adjustment to compensate for the mill variation in the columns; short slots can also be used.

Alternatively, in any of the aforementioned cases, one angle could be shop attached to the support and the other shipped loose. In this case, the spread between the outstanding legs should equal the decimal beam web thickness plus a clearance which will produce an opening to the next higher ¹⁄₁₆-in. increment, as illustrated in Figure 9-6b; short slots in the support-leg of the angle eliminate the need to provide for variations in web thickness. However, shipping one angle loose is not a desirable practice since it requires additional material handling as well as added erection costs and difficulty.

All-Bolted Double-Angle Connections

Tables 9-2 are design aids for all-bolted double-angle connections. Design strengths are tabulated for supported and supporting member material, as well as angle material with

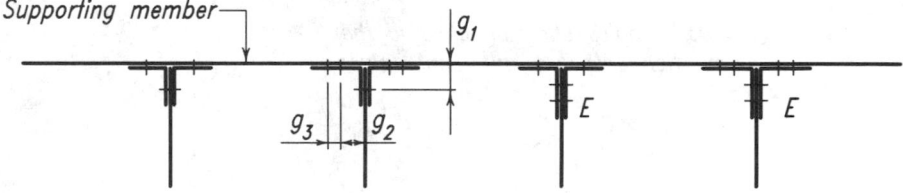

E indicates that eccentricity must
be considered in this leg.
Gages g_1, g_2, g_3 are usual gages
as shown below

	Usual gages* in angle legs, in.												
Leg	8	7	6	5	4	3½	3	2½	2	1¾	1½	1⅜	1¼	1
g_1	4½	4	3½	3	2½	2	1¾	1⅜	1⅛	1	⅞	⅞	¾	⅝
g_2	3	2½	2¼	2										
g_3	3	3	2½	1¾										

*Other gages are permitted to suit specific requirements subject to clearances and edge distance limitations.

Figure 9-5. Eccentricity in double-angle connections.

$F_y = 36$ ksi and $F_u = 58$ ksi and with $F_y = 50$ ksi and $F_u = 65$ ksi. All values, including slip-critical bolt design strengths, are for comparison with factored loads.

Tabulated bolt and angle design strengths consider the limit states of bolt shear, bolt bearing on the angles, shear yielding of the angles, shear rupture of the angles, and block shear rupture of the angles. Values are tabulated for 2 through 12 rows of ¾-in., ⅞-in, and 1 in. diameter A325 and A490 bolts at 3 in. spacing. For calculation purposes, angle edge distances L_{ev} and L_{eh} are assumed to be 1¼-in.

Tabulated beam web design strengths, per inch of web thickness, consider the limit state of bolt bearing on the beam web. For beams coped at the top flange only, the limit state of block shear rupture is also considered. Additionally, for beams coped at both the top and bottom flanges, the tabulated values consider the limit states of shear yielding

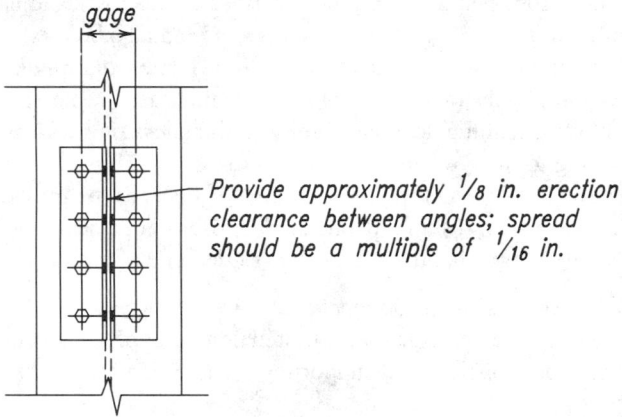

(a) Both angles shop attached to the column flange (beam knifed into place)

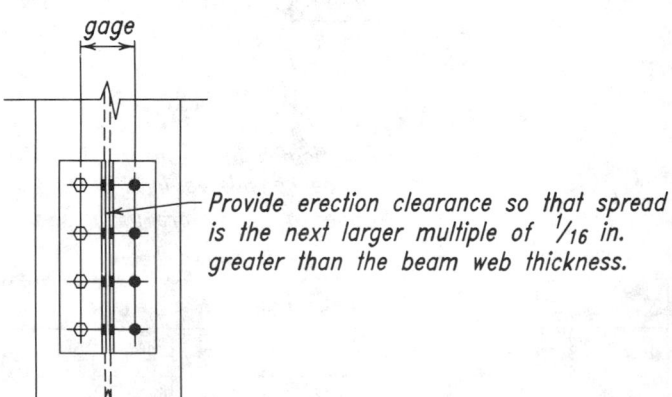

(b) One angle shop attached to the column flange, other shipped loose

Figure 9-6. Double-angle connection erection clearances.

and shear rupture of the beam web. Values are tabulated for beam web edge distances L_{eh} from $1\frac{1}{4}$-in. to 3 in. and for beam end distances L_{eh} of $1\frac{1}{2}$-in. and $1\frac{3}{4}$-in.; for calculation purposes, these end distances have been reduced to $1\frac{1}{4}$-in. and $1\frac{1}{2}$-in., respectively, to account for possible underrun in beam length. For coped members, the limit states of flexural yielding and local buckling must be checked independently. These limit states are discussed in Part 8; web reinforcement of coped members is treated in this Part under "Special Considerations".

Tabulated supporting member design strengths, per inch of flange or web thickness, consider the limit state of bolt bearing on the support.

Bolted/Welded Double-Angle Connections

Table 9-3 (see page 9-88) is a design aid arranged to permit substitution of welds for bolts in connections designed with Tables 9-2. Electrode strength is assumed to be 70 ksi. All values are for comparison with factored loads. Holes for erection bolts may be placed as required in angle legs that are to be field welded.

Welds A may be used in place of bolts through the supported-beam-web legs of the double angles or welds B may be used in place of bolts through the support legs of the double angles. Although it is permissible to use welds A and B from Table 9-3 in combination to obtain all-welded connections, it is recommended that such connections be chosen from Table 9-4. This table will allow increased flexibility in selection of angle lengths and connection strengths since Table 9-3 conforms to the bolt spacing and edge distance requirements for the bolted double-angle connections of Tables 9-2.

Weld design strengths are tabulated for the limit state of weld shear. Design strengths for welds A are determined by the instantaneous center of rotation method using Table 8-42 with $\theta = 0°$. Design strengths for welds B are determined by the elastic method. With the neutral axis assumed at one-sixth the depth of the angles measured downward and the tops of the angles in compression against each other through the beam web, the design strength of these welds is ϕR_n, where

$$\phi R_n = 2 \times \frac{1.392DL}{\sqrt{1 + \dfrac{12.96e^2}{L^2}}}$$

In the above equation, D is the number of sixteenths-of-an-inch in the weld size, L is the length of the connection angles, and e is the width of the leg of the connection angle attached to the support.

The tabulated minimum thicknesses of the supported beam web for welds A and the support for welds B match the shear yielding strength of these elements with the strength of the weld metal. Given the design shear yielding strength per unit length from LRFD Specification Section J5.3 as $0.9(0.60F_y t)$ and the weld strength constant (unit length design strength per $\frac{1}{16}$-in. weld size for 70 ksi electrodes) as 1.392 kips/in., the minimum supported beam web thickness for welds A (two lines of weld) is

$$t_{min} = \frac{D \times 1.392 \times 2}{0.9 \times 0.60F_y} = \frac{5.16D}{F_y}$$

where D is the number of sixteenths in the weld size. Similarly for welds B (one line of weld) the minimum supporting flange or web thickness is

$$t_{min} = \frac{2.58D}{F_y}$$

When welds line up on opposite sides of the support, the minimum thickness is the sum of the thicknesses required for each weld. In either case, when less than the minimum material thickness is present, the tabulated weld design strength must be reduced by the ratio of the thickness provided to the minimum thickness.

The minimum angle thickness when Table 9-3 is used is the weld size plus $\frac{1}{16}$-in. but not less than the angle thickness determined from Table 9-2. The angle length L must be as tabulated in Table 9-3. In general, $2L4\times3\frac{1}{2}$ will accommodate usual gages, with the 4 in. leg attached to the supporting member. Width of web legs in Case I may be optionally reduced from $3\frac{1}{2}$-in. to 3 in. Width of outstanding legs in Case II may be optionally reduced from 4 in. to 3 in. for values of L from $5\frac{1}{2}$ through $17\frac{1}{2}$-in.

All-Welded Double-Angle Connections

Table 9-4 (see page 9-89) is a design aid for all-welded double-angle connections. Electrode strength is assumed to be 70 ksi. All values are for comparison with factored loads. Holes for erection bolts may be placed as required in angle legs that are to be field welded.

Weld design strengths are tabulated for the limit state of weld shear. Design strengths for welds A are determined by the instantaneous center of rotation method using Table 8-42 with $\theta = 0°$. Design strengths for welds B are determined by the elastic method as discussed previously for bolted/welded double-angle connections.

The tabulated minimum thicknesses of the supported beam web for welds A and the support for welds B match the shear yielding strength of these elements with the strength of the weld metal and are determined as discussed previously for bolted/welded double angle connections. When welds line up on opposite sides of the support, the minimum thickness is the sum of the thicknesses required for each weld. When less than the minimum material thickness is present, the tabulated weld design strength must be reduced by the ratio of the thickness provided to the minimum thickness.

The minimum angle thickness when Table 9-4 is used must be equal to the weld size plus $\frac{1}{16}$-in. The angle length L must be as tabulated in Table 9-4. Use $2L4\times3$ for angle lengths greater than or equal to 18 in.; use $2L3\times3$ otherwise.

Example 9-1

Given: Refer to Figure 9-7. Use Table 9-2 to design an all-bolted double-angle connection for the W18×50 beam to W21×62 girder web connection.

$R_u = 60$ kips

W18×50

$t_w = 0.355$ in. $d = 17.99$ in.
$F_y = 50$ ksi, $F_u = 65$ ksi

top flange coped 2 in. deep by 4 in. long, $L_{ev} = 1\frac{1}{4}$-in., $L_{eh} = 1\frac{3}{4}$-in. (Assumed to be $1\frac{1}{2}$-in. for calculation purposes to account for possible underrun in beam lengths)

W21×62

$t_w = 0.400$ in.
$F_y = 50$ ksi, $F_u = 65$ ksi

Use ¾-in. diameter A325-N bolts in standard holes. Assume angle material with $F_y = 36$ ksi and $F_u = 58$ ksi.

Solution: *Design bolts and angles* (refer to Part 8)

From Table 9-2, for ¾-in. diameter A325-N bolts and angle material with $F_y = 36$ ksi and $F_u = 58$ ksi, select three rows of bolts and ¼-in. angle thickness.

$\phi R_n = 76.7$ kips > 60 kips   **o.k.**

Check supported beam web

From Table 9-2, for three rows of bolts, beam material with $F_y = 50$ ksi and $F_u = 65$ ksi, and $L_{ev} = 1\frac{1}{4}$-in. and $L_{eh} = 1\frac{3}{4}$-in. (Assumed to be 1½-in. for calculation purposes to account for possible underrun in beam lengths)

$\phi R_n = (195$ kips/in.$)(0.355$ in.$)$
      $= 69.2$ kips > 60 kips   **o.k.**

Check flexural yielding on the coped section (refer to Part 8)

From Table 8-49, $S_{net} = 23.4$ in.^{3}

$$\phi R_n = \frac{\phi F_y S_{net}}{e}$$

$$= \frac{0.9\,(50\text{ ksi})\,(23.4\text{ in.}^3)}{(4\text{ in.} + \tfrac{1}{2}\text{-in.})}$$

$$= 234 \text{ kips} > 60 \text{ kips}\quad \textbf{o.k.}$$

Check local web buckling at the cope (refer to Part 8)

$$\frac{c}{d} = \frac{4\text{ in.}}{17.99\text{ in.}} = 0.222$$

$$\frac{c}{h_o} = \frac{4\text{ in.}}{(17.99\text{ in.} - 2\text{ in.})} = 0.250$$

Since $\frac{c}{d} \leq 1.0$,

$$f = 2\left(\frac{c}{d}\right)$$

$$= 2(0.222)$$

$$= 0.444$$

Since $\frac{c}{h_o} \leq 1.0$,

$$k = 2.2 \left(\frac{h_o}{c}\right)^{1.65}$$

$$= 2.2 \left(\frac{1}{0.250}\right)^{1.65}$$

$$= 21.7$$

$$\phi F_{bc} = 23,590 \left(\frac{t_w}{h_o}\right)^2 fk$$

$$= 23,590 \left(\frac{0.355 \text{ in.}}{17.99 \text{ in.} - 2 \text{ in.}}\right)^2 (0.444)(21.7)$$

$$= 112 \text{ ksi}$$

$$\phi R_n = \frac{\phi F_{bc} S_{net}}{e}$$

$$= \frac{(112 \text{ ksi})(23.4 \text{ in.}^3)}{(4 \text{ in.} + \frac{1}{2}\text{-in.})}$$

$$= 582 \text{ kips} > 60 \text{ kips} \quad \textbf{o.k.}$$

Check supporting girder web

From Table 9-2, for three rows of bolts and girder material with $F_u = 65$ ksi,

$$\phi R_n = (527 \text{ kips/in.})(0.400 \text{ in.})$$

$$= 211 \text{ kips} > 60 \text{ kips} \quad \textbf{o.k.}$$

The connection, as summarized in Figure 9-7, is adequate.

Example 9-2

Given:

Refer to Figure 9-8. Use Table 9-2 to design an all-bolted double-angle connection for the W36×230 beam to W14×90 column-flange connection.

$R_u = 225$ kips

W36×230

$t_w = 0.760$ in.
$F_y = 50$ ksi, $F_u = 65$ ksi

W14×90

$t_f = 0.710$ in.
$F_y = 50$ ksi, $F_u = 65$ ksi

Use ¾-in. diameter A325-N bolts in standard holes. Assume angle material with $F_y = 36$ ksi and $F_u = 58$ ksi.

Solution:

Design bolts and angles

From Table 9-2, for ¾-in. diameter A325-N bolts and angle material with $F_y = 36$ ksi and $F_y = 58$ ksi, select eight rows of bolts and ⁵⁄₁₆-in. angle thickness.

$\phi R_n = 254$ kips > 225 kips   **o.k.**

Check supported beam web

From Table 9-2, for eight rows of bolts, beam material with $F_y = 50$ ksi and $F_u = 65$ ksi, and $L_{eh} = 1¾$-in.,

$\phi R_n = (702$ kips/in.$)(0.760$ in.$)$
$= 534$ kips > 225 kips   **o.k.**

Check supporting column flange

From Table 9-2, for eight rows of bolts and column material with $F_y = 50$ ksi and $F_u = 65$ ksi,

$\phi R_n = (1{,}404$ kips/in.$)(0.710$ in.$)$
$= 997$ kips > 225 kips   **o.k.**

Example 9-3

Given:

Refer to Example 9-1. Use Table 9-3 to substitute welds for bolts in the supported-beam-web legs of the double-angle connection (welds A).

Solution:

From Table 9-3, for three rows of bolts (an angle length of 8½-in.), a ³⁄₁₆-in. weld size provides $\phi R_n = 110$ kips. For beam web material with $F_y = 50$ ksi, the minimum web thickness is 0.31 in. Since $t_w = 0.355$ in. > 0.31 in., no reduction in the tabulated value is required.

$\phi R_n = 110$ kips > 60 kips   **o.k.**

Check minimum angle thickness

The minimum angle thickness for Table 9-3 is the weld size plus ¹⁄₁₆-in., but not less than the thickness determined from Table 9-2.

$t_{min} = ³⁄₁₆$-in. $+ ¹⁄₁₆$-in.
$= ¼$-in.

This thickness is equal to the thickness chosen previously from Table 9-2.

Example 9-4

Given:

Refer to Example 9-2. Use Table 9-3 to substitute welds for bolts in the support legs of the double-angle connection (welds B).

Solution:

From Table 9-3, for eight rows of bolts (an angle length of 23½-in.), a ⁵⁄₁₆-in. weld size provides $\phi R_n = 279$ kips. For beam web material with

$F_y = 50$ ksi, the minimum column flange thickness is 0.26 in. Since $t_f = 0.710$ in. > 0.26 in., no reduction of the tabulated value is required.

$\phi R_n = 279$ kips > 225 kips   **o.k.**

Check minimum angle thickness

The minimum angle thickness for Table 9-3 is the weld size plus $\frac{1}{16}$-in., but not less than the thickness determined from Table 9-2.

$t_{min} = \frac{5}{16}$-in. + $\frac{1}{16}$-in.

$\quad = \frac{3}{8}$-in.

Thus, the angle thickness must be increased to $\frac{3}{8}$-in. to accommodate the welded legs of the double-angle connection.

Example 9-5

Given:

Refer to Example 9-2. Use Table 9-4 to design an all-welded double-angle connection for the W36×230 beam to W14×90 column-flange connection.

Solution:

Design supported-beam-web angle leg welds (welds A)

From Table 9-4, for $L = 24$ in., a $\frac{3}{16}$-in. weld A size provides $\phi R_n = 248$ kips. For beam web material with $F_y = 50$ ksi, the minimum supported beam web thickness is 0.31 in. Since $t_w = 0.760$ in. > 0.31 in., no reduction of the tabulated value is required.

$\phi R_n = 248$ kips > 225 kips   **o.k.**

Design support angle leg welds (welds B)

From Table 9-4, for $L = 24$ in., a $\frac{1}{4}$-in. weld B size provides $\phi R_n = 229$ kips. For column flange material with $F_y = 50$ ksi, the minimum column flange thickness is 0.21 in. Since $t_f = 0.710$ in. > 0.21 in., no reduction of the tabulated value is required.

Check minimum angle thickness

The minimum angle thickness for Table 9-4 is the weld size plus $\frac{1}{16}$-in.

$t_{min} = \frac{1}{4}$-in. + $\frac{1}{16}$-in.

$\quad = \frac{5}{16}$-in.

Use 2L4×3×$\frac{5}{16}$.

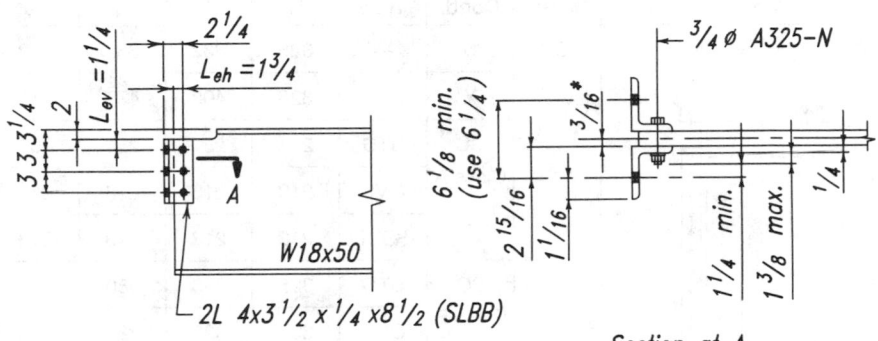

Section at A

* This dimension (see sketch, section at A) is determined to be one-half of the decimal web thickness rounded to the next higher $^1/_{16}$ in. Example: $0.355/2 = 0.1775$; use $^3/_{16}$ in. This will produce spacing of holes in the supporting beam slightly larger than detailed in the angles to permit spreading of angles (angles can be spread but not closed) at time of erection to supporting member. Alternatively, consider using horizontal slots in the support legs of the angles.

Fig. 9-7.

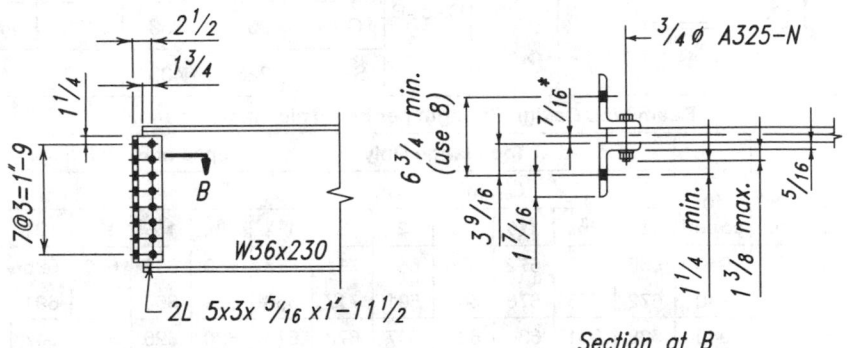

Section at B

* This dimemsion is one-half the decimal web thickness rounded to the next higher $^1/_{16}$ in., as in example 9-1.

Fig. 9-8.

$F_y = 36$ ksi
$F_u = 58$ ksi

Table 9-2.
All-Bolted Double-Angle Connections

3/4-in. Bolts	Bolt and Angle Design Strength, kips							
12 Rows	**ASTM**	**Thread**	**Hole**	**Angle Thickness, in.**			
W44	**Desig.**	**Cond.**	**Type**	**1/4**	**5/16**	**3/8**	**1/2**
	A325	N	—	326	382	382	382
		X	—	326	408	477	477
		SC Class A	STD	251	251	251	251
			OVS	213	213	213	213
			SSLT	213	213	213	213
		SC Class B	STD	326	380	380	380
			OVS	307	323	323	323
			SSLT	323	323	323	323
	A490	N	—	326	408	477	477
		X	—	326	408	489	596
		SC Class A	STD	313	313	313	313
			OVS	266	266	266	266
			SSLT	266	266	266	266
		SC Class B	STD	326	408	475	475
			OVS	307	383	403	403
			SSLT	326	403	403	403

Drawing labels: Varies, t, $11@3=33$, $2\frac{1}{4}$, L_{eh}, L_{ev}, $11@3=33$, L_{ev}

Beam Web Design Strength per Inch Thickness, kips/in.

			Coped at Top Flange Only						Coped at Both Flanges					
Hole	**L_{eh},***	**Un-**	**L_{ev}, in.**						**L_{ev}, in.**					
Type	**in.**	**coped**	**1 1/4**	**1 3/8**	**1 1/2**	**1 5/8**	**2**	**3**	**1 1/4**	**1 3/8**	**1 1/2**	**1 5/8**	**2**	**3**
STD	1 1/2	940	665	668	672	675	685	711	653	659	666	672	685	711
	1 3/4	940	672	675	678	682	691	717	653	659	666	672	691	717
OVS	1 1/2	940	628	631	634	637	647	673	613	620	626	633	647	673
	1 3/4	940	634	638	641	644	654	680	613	620	626	633	653	680
SSLT	1 1/2	940	665	668	672	675	685	711	653	659	666	672	685	711
	1 3/4	940	672	675	678	682	691	717	653	659	666	672	691	717

Support Design Strength per Inch Thickness, kips/in.	Notes: STD = Standard holes OVS = Oversized holes SSLT = Short-slotted holes transverse to direction of load	N = Threads included X = Threads excluded SC = Slip critical
1879	*Tabulated values include 1/4-in. reduction in end distance L_{eh} to account for possible underrun in beam length	

	F_y = 50 ksi
	F_u = 65 ksi

Table 9-2 (cont.).
All-Bolted Double-Angle Connections

¾-in Bolts	Bolt and Angle Design Strength, kips						
12 Rows	ASTM	Thread	Hole	Angle Thickness, in.			
W44	Desig.	Cond.	Type	¼	5/16	⅜	½
	A325	N	—	366	382	382	382
		X	—	366	457	477	477
		SC Class A	STD	251	251	251	251
			OVS	213	213	213	213
			SSLT	213	213	213	213
		SC Class B	STD	366	380	380	380
			OVS	323	323	323	323
			SSLT	323	323	323	323
	A490	N	—	366	457	477	477
		X	—	366	457	548	596
		SC Class A	STD	313	313	313	313
			OVS	266	266	266	266
			SSLT	266	266	266	266
		SC Class B	STD	366	457	475	475
			OVS	344	403	403	403
			SSLT	366	403	403	403

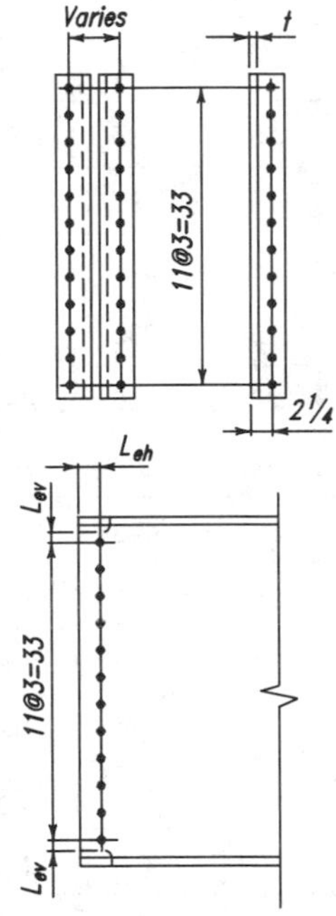

Beam Web Design Strength per Inch Thickness, kips/in.

			Coped at Top Flange Only						Coped at Both Flanges					
Hole Type	L_{eh},* in.	Un-coped	L_{ev}, in.						L_{ev}, in.					
			1¼	1⅜	1½	1⅝	2	3	1¼	1⅜	1½	1⅝	2	3
STD	1½	1053	754	758	762	765	776	806	731	739	746	753	775	806
	1¾	1053	764	767	771	775	786	815	731	739	746	753	775	815
OVS	1½	1053	712	716	720	723	734	764	687	695	702	709	731	764
	1¾	1053	722	725	729	733	744	773	687	695	702	709	731	773
SSLT	1½	1053	754	758	762	765	776	806	731	739	746	753	775	806
	1¾	1053	764	767	771	775	786	815	731	739	746	753	775	815

Support Design Strength per Inch Thickness, kips/in.	Notes: STD = Standard holes OVS = Oversized holes SSLT = Short-slotted holes transverse to direction of load	N = Threads included X = Threads excluded SC = Slip critical
2106	*Tabulated values include ¼-in. reduction in end distance L_{eh} to account for possible underrun in beam length.	

F_y = 36 ksi
F_u = 58 ksi

Table 9-2 (cont.).
All-Bolted Double-Angle Connections

3/4-in. Bolts		Bolt and Angle Design Strength, kips						
11 Rows		ASTM	Thread	Hole	Angle Thickness, in.			
W44, 40		Desig.	Cond.	Type	1/4	5/16	3/8	1/2
		A325	N	—	299	350	350	350
			X	—	299	373	437	437
			SC Class A	STD	230	230	230	230
				OVS	195	195	195	195
				SSLT	195	195	195	195
			SC Class B	STD	299	348	348	348
				OVS	281	296	296	296
				SSLT	296	296	296	296
		A490	N	—	299	373	437	437
			X	—	299	373	448	547
			SC Class A	STD	287	287	287	287
				OVS	244	244	244	244
				SSLT	244	244	244	244
			SC Class B	STD	299	373	435	435
				OVS	281	351	370	370
				SSLT	299	370	370	370

(diagram: Varies, t, 10@3=30, $2\frac{1}{4}$, L_{eh}, L_{ev}, 10@3=30)

Beam Web Design Strength per Inch Thickness, kips/in.

			Coped at Top Flange Only						Coped at Both Flanges					
Hole	L_{eh},*	Un-	L_{ev}, in.						L_{ev}, in.					
Type	in.	coped	1 1/4	1 3/8	1 1/2	1 5/8	2	3	1 1/4	1 3/8	1 1/2	1 5/8	2	3
STD	1 1/2	861	610	613	616	619	629	655	597	604	610	617	629	655
	1 3/4	861	616	620	623	626	636	662	597	604	610	617	636	662
OVS	1 1/2	861	575	579	582	585	595	621	561	568	574	581	595	621
	1 3/4	861	582	585	589	592	602	628	561	568	574	581	600	628
SSLT	1 1/2	861	610	613	616	619	629	655	597	604	610	617	629	655
	1 3/4	861	616	620	623	626	636	662	597	604	610	617	636	662

Support Design Strength per Inch Thickness, kips/in.	Notes: STD = Standard holes OVS = Oversized holes SSLT = Short-slotted holes transverse 　　　　to direction of load	N = Threads included X = Threads excluded SC = Slip critical
1723	*Tabulated values include 1/4-in. reduction in end distance L_{eh} to account for possible underrun in beam length.	

	F_y = 50 ksi
	F_u = 65 ksi

Table 9-2 (cont.).
All-Bolted Double-Angle Connections

¾-in. Bolts	Bolt and Angle Design Strength, kips						
11 Rows	ASTM Desig.	Thread Cond.	Hole Type	Angle Thickness, in.			
W44, 40				¼	⁵⁄₁₆	⅜	½
	A325	N	—	335	350	350	350
		X	—	335	418	437	437
		SC Class A	STD	230	230	230	230
			OVS	195	195	195	195
			SSLT	195	195	195	195
		SC Class B	STD	335	348	348	348
			OVS	296	296	296	296
			SSLT	296	296	296	296
	A490	N	—	335	418	437	437
		X	—	335	418	502	547
		SC Class A	STD	287	287	287	287
			OVS	244	244	244	244
			SSLT	244	244	244	244
		SC Class B	STD	335	418	435	435
			OVS	314	370	370	370
			SSLT	335	370	370	370

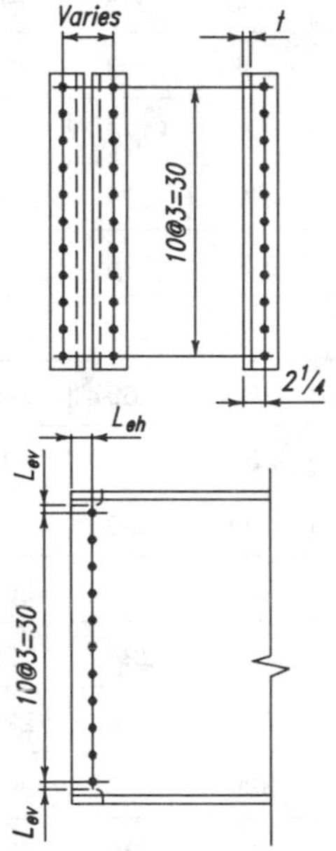

(Figures: connection diagrams showing "Varies", "t", "10@3=30", "2¼", "L_{eh}", "L_{ev}")

Beam Web Design Strength per Inch Thickness, kips/in.

Hole Type	L_{eh},* in.	Un-coped	Coped at Top Flange Only						Coped at Both Flanges					
			L_{ev}, in.						L_{ev}, in.					
			1¼	1⅜	1½	1⅝	2	3	1¼	1⅜	1½	1⅝	2	3
STD	1½	965	692	696	700	703	714	743	669	676	684	691	713	743
	1¾	965	702	705	709	713	724	753	669	676	684	691	713	753
OVS	1½	965	654	657	661	665	676	705	629	636	644	651	673	705
	1¾	965	663	667	671	674	685	714	629	636	644	651	673	714
SSLT	1½	965	692	696	700	703	714	743	669	676	684	691	713	743
	1¾	965	702	705	709	713	724	753	669	676	684	691	713	753

Support Design Strength per Inch Thickness, kips/in.	Notes: STD = Standard holes OVS = Oversized holes SSLT = Short-slotted holes transverse to direction of load	N = Threads included X = Threads excluded SC = Slip critical
1931	*Tabulated values include ¼-in. reduction in end distance L_{eh} to account for possible underrun in beam length.	

F_y = 36 ksi

F_u = 58 ksi

Table 9-2 (cont.).
All-Bolted Double-Angle Connections

¾-in. Bolt	Bolt and Angle Design Strength, kips						
10 Rows	**ASTM**	**Thread**	**Hole**	**Angle Thickness, in.**			
W44, 40, 36	**Desig.**	**Cond.**	**Type**	**¼**	**⁵⁄₁₆**	**⅜**	**½**
	A325	N	—	271	318	318	318
		X	—	271	338	398	398
		SC Class A	STD	209	209	209	209
			OVS	178	178	178	178
			SSLT	178	178	178	178
		SC Class B	STD	271	316	316	316
			OVS	254	269	269	269
			SSLT	269	269	269	269
	A490	N	—	271	338	398	398
		X	—	271	338	406	497
		SC Class A	STD	261	261	261	261
			OVS	222	222	222	222
			SSLT	222	222	222	222
		SC Class B	STD	271	338	396	396
			OVS	254	318	336	336
			SSLT	271	336	336	336

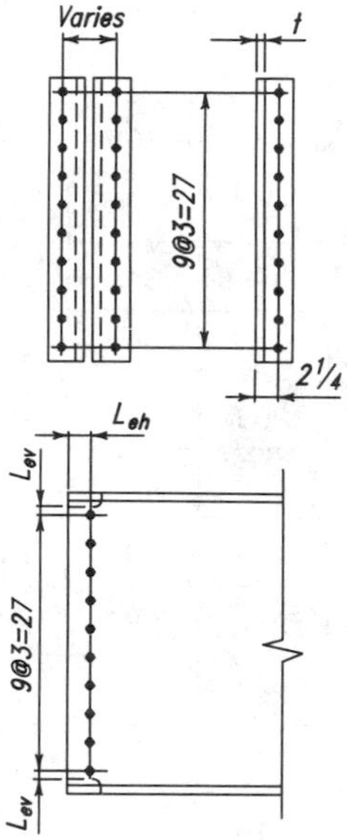

Drawing labels: Varies, t, $9@3=27$, $2¼$, L_{eh}, L_{ev}, $9@3=27$, L_{ev}

Beam Web Design Strength per Inch Thickness, kips/in.

			Coped at Top Flange Only						Coped at Both Flanges					
Hole	**L_{eh},***	**Un-**	**L_{ev}, in.**						**L_{ev}, in.**					
Type	**in.**	**coped**	**1¼**	**1⅜**	**1½**	**1⅝**	**2**	**3**	**1¼**	**1⅜**	**1½**	**1⅝**	**2**	**3**
STD	1½	783	554	557	561	564	574	600	542	548	555	561	574	600
	1¾	783	561	564	567	571	580	607	542	548	555	561	580	607
OVS	1½	783	523	526	530	533	543	569	509	515	522	529	543	569
	1¾	783	530	533	536	540	549	576	509	515	522	529	548	576
SSLT	1½	783	554	557	561	564	574	600	542	548	555	561	574	600
	1¾	783	561	564	567	571	580	607	542	548	555	561	580	607

Support Design Strength per Inch Thickness, kips/in.	Notes: STD = Standard holes OVS = Oversized holes SSLT = Short-slotted holes transverse to direction of load	N = Threads included X = Threads excluded SC = Slip critical
1566	*Tabulated values include ¼-in. reduction in end distance L_{eh} to account for possible underrun in beam length.	

	$F_y = 50$ ksi
	$F_u = 65$ ksi

Table 9-2 (cont.).
All-Bolted Double-Angle Connections

3/4-in. Bolts	Bolt and Angle Design Strength, kips						
10 Rows	ASTM	Thread	Hole	Angle Thickness, in.			
W44, 40, 36	Desig.	Cond.	Type	1/4	5/16	3/8	1/2

	A325	N	—	303	318	318	318
		X	—	303	379	398	398
		SC Class A	STD	209	209	209	209
			OVS	178	178	178	178
			SSLT	178	178	178	178
		SC Class B	STD	303	316	316	316
			OVS	269	269	269	269
			SSLT	269	269	269	269
	A490	N	—	303	379	398	398
		X	—	303	379	455	497
		SC Class A	STD	261	261	261	261
			OVS	222	222	222	222
			SSLT	222	222	222	222
		SC Class B	STD	303	379	396	396
			OVS	285	336	336	336
			SSLT	303	336	336	336

Beam Web Design Strength per Inch Thickness, kips/in.

			Coped at Top Flange Only						Coped at Both Flanges					
Hole	$L_{eh},$*	Un-	L_{ev}, in.						L_{ev}, in.					
Type	in.	coped	1 1/4	1 3/8	1 1/2	1 5/8	2	3	1 1/4	1 3/8	1 1/2	1 5/8	2	3
STD	1 1/2	878	630	634	637	641	652	681	607	614	622	629	651	681
	1 3/4	878	639	643	647	650	661	691	607	614	622	629	651	691
OVS	1 1/2	878	595	599	603	606	617	647	570	578	585	592	614	647
	1 3/4	878	605	608	612	616	627	656	570	578	585	592	614	656
SSLT	1 1/2	878	630	634	637	641	652	681	607	614	622	629	651	681
	1 3/4	878	639	643	647	650	661	691	607	614	622	629	651	691

Support Design Strength per Inch Thickness, kips/in.	Notes: STD = Standard holes OVS = Oversized holes SSLT = Short-slotted holes transverse to direction of load	N = Threads included X = Threads excluded SC = Slip critical
1755	*Tabulated values include 1/4-in. reduction in end distance L_{eh} to account for possible underrun in beam length.	

F_y = 36 ksi
F_u = 58 ksi

Table 9-2 (cont.).
All-Bolted Double-Angle Connections

¾-in. Bolts					Bolt and Angle Design Strength, kips				
9 Rows			ASTM	Thread	Hole	Angle Thickness, in.			
W44, 40, 36, 33			Desig.	Cond.	Type	¼	⁵⁄₁₆	⅜	½
			A325	N	—	243	286	286	286
				X	—	243	304	358	358
				SC Class A	STD	188	188	188	188
					OVS	160	160	160	160
					SSLT	160	160	160	160
				SC Class B	STD	243	285	285	285
					OVS	228	242	242	242
					SSLT	242	242	242	242
			A490	N	—	243	304	358	358
				X	—	243	304	365	447
				SC Class A	STD	235	235	235	235
					OVS	200	200	200	200
					SSLT	200	200	200	200
				SC Class B	STD	243	304	356	356
					OVS	228	285	303	303
					SSLT	243	303	303	303

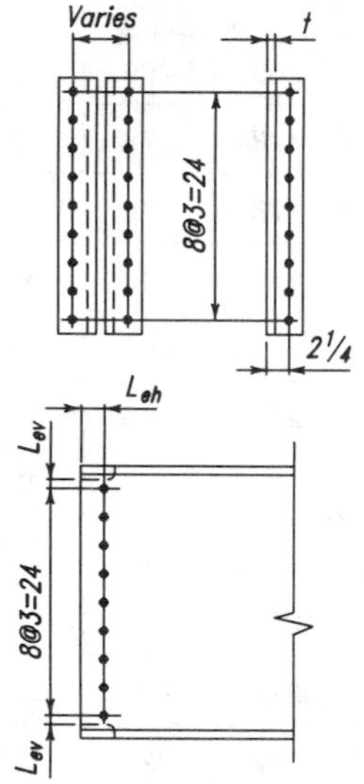

Beam Web Design Strength per Inch Thickness, kips/in.

			Coped at Top Flange Only						Coped at Both Flanges					
Hole	L_{eh},*	Un-	L_{ev}, in.						L_{ev}, in.					
Type	in.	coped	1¼	1⅜	1½	1⅝	2	3	1¼	1⅜	1½	1⅝	2	3
STD	1½	705	499	502	505	508	518	544	486	493	499	506	518	544
	1¾	705	505	509	512	515	525	551	486	493	499	506	525	551
OVS	1½	705	471	474	477	481	491	517	457	463	470	476	491	517
	1¾	705	478	481	484	487	497	523	457	463	470	476	496	523
SSLT	1½	705	499	502	505	508	518	544	486	493	499	506	518	544
	1¾	705	505	509	512	515	525	551	486	493	499	506	525	551

Support Design Strength per Inch Thickness, kips/in.	Notes:
	STD = Standard holes N = Threads included
	OVS = Oversized holes X = Threads excluded
	SSLT = Short-slotted holes transverse SC = Slip critical
	to direction of load
1409	*Tabulated values include ¼-in. reduction in end distance L_{eh} to account for possible underrun in beam length.

	$F_y = 50$ ksi
	$F_u = 65$ ksi

Table 9-2 (cont.).
All-Bolted Double-Angle Connections

$^3/_4$-in. Bolts	Bolt and Angle Design Strength, kips						
9 Rows	**ASTM**	**Thread**	**Hole**	\multicolumn{4}{c}{**Angle Thickness, in.**}

$^3/_4$-in. Bolts 9 Rows W44, 40, 36, 33	Bolt and Angle Design Strength, kips						
	ASTM Desig.	**Thread Cond.**	**Hole Type**	**Angle Thickness, in.**			
				$^1/_4$	$^5/_{16}$	$^3/_8$	$^1/_2$
	A325	N	—	272	286	286	286
		X	—	272	340	358	358
		SC Class A	STD	188	188	188	188
			OVS	160	160	160	160
			SSLT	160	160	160	160
		SC Class B	STD	272	285	285	285
			OVS	242	242	242	242
			SSLT	242	242	242	242
	A490	N	—	272	340	358	358
		X	—	272	340	409	447
		SC Class A	STD	235	235	235	235
			OVS	200	200	200	200
			SSLT	200	200	200	200
		SC Class B	STD	272	340	356	356
			OVS	256	303	303	303
			SSLT	272	303	303	303

Varies
t
$8@3=24$
$2^1/_4$
L_{eh}
L_{ev}
$8@3=24$
L_{ev}

Beam Web Design Strength per Inch Thickness, kips/in.

			Coped at Top Flange Only						Coped at Both Flanges					
			L_{ev}, in.						L_{ev}, in.					
Hole Type	**L_{eh},*** in.	**Un-coped**	$1^1/_4$	$1^3/_8$	$1^1/_2$	$1^5/_8$	2	3	$1^1/_4$	$1^3/_8$	$1^1/_2$	$1^5/_8$	2	3
STD	$1^1/_2$	790	568	572	575	579	590	619	545	552	559	567	589	619
	$1^3/_4$	790	577	581	585	588	599	628	545	552	559	567	589	628
OVS	$1^1/_2$	790	537	540	544	548	559	588	512	519	527	534	556	588
	$1^3/_4$	790	546	550	554	557	568	597	512	519	527	534	556	597
SSLT	$1^1/_2$	790	568	572	575	579	590	619	545	552	559	567	589	619
	$1^3/_4$	790	577	581	585	588	599	628	545	552	559	567	589	628

Support Design Strength per Inch Thickness, kips/in.	Notes: STD = Standard holes OVS = Oversized holes SSLT = Short-slotted holes transverse to direction of load	N = Threads included X = Threads excluded SC = Slip critical
1580	*Tabulated values include $^1/_4$-in. reduction in end distance L_{eh} to account for possible underrun in beam length.	

| F_y = 36 ksi |
| F_u = 58 ksi |

Table 9-2 (cont.).
All-Bolted Double-Angle Connections

³/₄-in. Bolts	Bot and Angle Design Strength, kips						

8 Rows	**ASTM**	**Thread**	**Hole**	**Angle Thickness, in.**			
W44, 40, 36, 33, 30	**Desig.**	**Cond.**	**Type**	**¹/₄**	**⁵/₁₆**	**³/₈**	**¹/₂**

	ASTM Desig.	Thread Cond.	Hole Type	¹/₄	⁵/₁₆	³/₈	¹/₂
	A325	N	—	215	254	254	254
		X	—	215	269	318	318
		SC Class A	STD	167	167	167	167
			OVS	142	142	142	142
			SSLT	142	142	142	142
		SC Class B	STD	215	253	253	253
			OVS	202	215	215	215
			SSLT	215	215	215	215
	A490	N	—	215	269	318	318
		X	—	215	269	323	398
		SC Class A	STD	209	209	209	209
			OVS	178	178	178	178
			SSLT	178	178	178	178
		SC Class B	STD	215	269	316	316
			OVS	202	253	269	269
			SSLT	215	269	269	269

Beam Web Design Strength per Inch Thickness, kips/in.

			Coped at Top Flange Only						Coped at Both Flanges					
			L_{ev}, in.						L_{ev}, in.					
Hole Type	**L_{eh},* in.**	**Un- coped**	**1¹/₄**	**1³/₈**	**1¹/₂**	**1⁵/₈**	**2**	**3**	**1¹/₄**	**1³/₈**	**1¹/₂**	**1⁵/₈**	**2**	**3**
STD	1¹/₂	626	443	446	450	453	463	489	431	437	444	450	463	489
	1³/₄	626	450	453	456	460	470	496	431	437	444	450	470	496
OVS	1¹/₂	626	419	422	425	429	438	464	405	411	418	424	438	464
	1³/₄	626	425	429	432	435	445	471	405	411	418	424	444	471
SSLT	1¹/₂	626	443	446	450	453	463	489	431	437	444	450	463	489
	1³/₄	626	450	453	456	460	470	496	431	437	444	450	470	496

Support Design Strength per Inch Thickness, kips/in.	Notes:
	STD = Standard holes N = Threads included
	OVS = Oversized holes X = Threads excluded
	SSLT = Short-slotted holes transverse SC = Slip critical
	to direction of load
1253	*Tabulated values include ¹/₄-in. reduction in end distance L_{eh} to account for possible underrun in beam length.

| | | | | F_y = 50 ksi |
| | | | | F_u = 65 ksi |

Table 9-2 (cont.).
All-Bolted Double-Angle Connections

¾-in. Bolts	Bolt and Angle Design Strength, kips						
8 Rows	ASTM	Thread	Hole	Angle Thickness, in.			
W44, 40, 36, 33, 30	Desig.	Cond.	Type	¼	⁵⁄₁₆	⅜	½
	A325	N	—	241	254	254	254
		X	—	241	302	318	318
		SC Class A	STD	167	167	167	167
			OVS	142	142	142	142
			SSLT	142	142	142	142
		SC Class B	STD	241	253	253	253
			OVS	215	215	215	215
			SSLT	215	215	215	215
	A490	N	—	241	302	318	318
		X	—	241	302	362	398
		SC Class A	STD	209	209	209	209
			OVS	178	178	178	178
			SSLT	178	178	178	178
		SC Class B	STD	241	302	316	316
			OVS	227	269	269	269
			SSLT	241	269	269	269

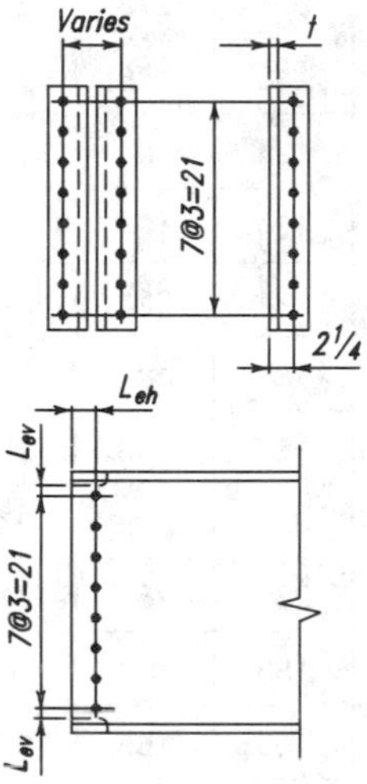

The diagram at left shows: *Varies*, t, $7@3=21$, $2\frac{1}{4}$, L_{eh}, L_{ev}, $7@3=21$, L_{ev}

Beam Web Design Strength per Inch Thickness, kips/in.

			Coped at Top Flange Only						Coped at Both Flanges					
			L_{ev}, in.						L_{ev}, in.					
Hole Type	L_{eh},* in.	Un-coped	1¼	1⅜	1½	1⅝	2	3	1¼	1⅜	1½	1⅝	2	3
STD	1½	702	506	509	513	517	528	557	483	490	497	505	527	557
	1¾	702	515	519	522	526	537	566	483	490	497	505	527	566
OVS	1½	702	478	482	486	489	500	530	453	461	468	475	497	530
	1¾	702	488	491	495	499	510	539	453	461	468	475	497	539
SSLT	1½	702	506	509	513	517	528	557	483	490	497	505	527	557
	1¾	702	515	519	522	526	537	566	483	490	497	505	527	566

Support Design Strength per Inch Thickness, kips/in.	Notes: STD = Standard holes N = Threads included OVS = Oversized holes X = Threads excluded SSLT = Short-slotted holes transverse SC = Slip critical to direction of load
1404	*Tabulated values include ¼-in. reduction in end distance L_{eh} to account for possible underrun in beam length.

$F_y = 36$ ksi
$F_u = 58$ ksi

Table 9-2 (cont.).
All-Bolted Double-Angle Connections

¾-in. Bolts	Bolt and Angle Design Strength, kips						
7 Rows	**ASTM Desig.**	**Thread Cond.**	**Hole Type**	**Angle Thickness, in.**			
W44, 40, 36, 33, 30, 27, 24 S24				¼	5/16	3/8	½
	A325	N	—	188	223	223	223
		X	—	188	234	278	278
		SC Class A	STD	146	146	146	146
			OVS	124	124	124	124
			SSLT	124	124	124	124
		SC Class B	STD	188	221	221	221
			OVS	176	188	188	188
			SSLT	188	188	188	188
	A490	N	—	188	234	278	278
		X	—	188	234	281	348
		SC Class A	STD	183	183	183	183
			OVS	155	155	155	155
			SSLT	155	155	155	155
		SC Class B	STD	188	234	277	277
			OVS	176	220	235	235
			SSLT	188	234	235	235

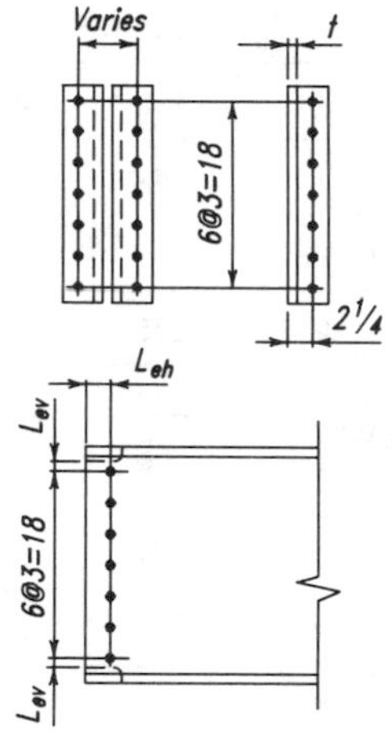

Figure labels: Varies, t, $6@3=18$, $2¼$, L_{eh}, L_{ev}, $6@3=18$, L_{ev}

Beam Web Design Strength per Inch Thickness, kips/in.

Hole Type	L_{eh},* in.	Un-coped	Coped at Top Flange Only						Coped at Both Flanges					
			L_{ev}, in.						L_{ev}, in.					
			1¼	1⅜	1½	1⅝	2	3	1¼	1⅜	1½	1⅝	2	3
STD	1½	548	388	391	394	398	407	433	375	382	388	395	407	433
	1¾	548	394	398	401	404	414	440	375	382	388	395	414	440
OVS	1½	548	367	370	373	376	386	412	352	359	365	372	386	412
	1¾	548	373	377	380	383	393	419	352	359	365	372	392	419
SSLT	1½	548	388	391	394	398	407	433	375	382	388	395	407	433
	1¾	548	394	398	401	404	414	440	375	382	388	395	414	440

Support Design Strength per Inch Thickness, kips/in.	Notes: STD = Standard holes　　　　　　N = Threads included
	OVS = Oversized holes　　　　　　　　X = Threads excluded
	SSLT = Short-slotted holes transverse　　SC = Slip critical
	to direction of load
1096	*Tabulated values include ¼-in. reduction in end distance L_{eh} to account for possible underrun in beam length.

	$F_y = 50$ ksi
	$F_u = 65$ ksi

Table 9-2 (cont.).
All-Bolted Double-Angle Connections

³⁄₄-in. Bolts		Bolt and Angle Design Strength, kips						
7 Rows		**ASTM Desig.**	**Thread Cond.**	**Hole Type**	**Angle Thickness, in.**			
W44, 40, 36, 33, 30, 27, 24 S24					¹⁄₄	⁵⁄₁₆	³⁄₈	¹⁄₂
		A325	N	—	210	223	223	223
			X	—	210	263	278	278
			SC Class A	STD	146	146	146	146
				OVS	124	124	124	124
				SSLT	124	124	124	124
			SC Class B	STD	210	221	221	221
				OVS	188	188	188	188
				SSLT	188	188	188	188
		A490	N	—	210	263	278	278
			X	—	210	263	315	348
			SC Class A	STD	183	183	183	183
				OVS	155	155	155	155
				SSLT	155	155	155	155
			SC Class B	STD	210	263	277	277
				OVS	197	235	235	235
				SSLT	210	235	235	235

Beam Web Design Strength per Inch Thickness, kips/in.

			Coped at Top Flange Only						Coped at Both Flanges					
			L_{ev}, in.						L_{ev}, in.					
Hole Type	L_{eh},* in.	Un-coped	1¹⁄₄	1³⁄₈	1¹⁄₂	1⁵⁄₈	2	3	1¹⁄₄	1³⁄₈	1¹⁄₂	1⁵⁄₈	2	3
STD	1¹⁄₂	614	444	447	451	455	466	495	420	428	435	442	464	495
	1³⁄₄	614	453	457	460	464	475	504	420	428	435	442	464	504
OVS	1¹⁄₂	614	420	423	427	431	442	471	395	402	410	417	439	471
	1³⁄₄	614	429	433	437	440	451	480	395	402	410	417	439	480
SSLT	1¹⁄₂	614	444	447	451	455	466	495	420	428	435	442	464	495
	1³⁄₄	614	453	457	460	464	475	504	420	428	435	442	464	504

Support Design Strength per Inch Thickness, kips/in.	Notes: STD = Standard holes N = Threads included
	OVS = Oversized holes X = Threads excluded
	SSLT = Short-slotted holes transverse SC = Slip critical
	to direction of load
1229	*Tabulated values include ¹⁄₄-in. reduction in end distance L_{eh} to account for possible underrun in beam length.

$F_y = 36$ ksi
$F_u = 58$ ksi

Table 9-2 (cont.).
All-Bolted Double-Angle Connections

¾-in. Bolts	Bolt and Angle Design Strength, kips						

6 Rows

W40, 36, 33, 30, 27, 24, 21
S24

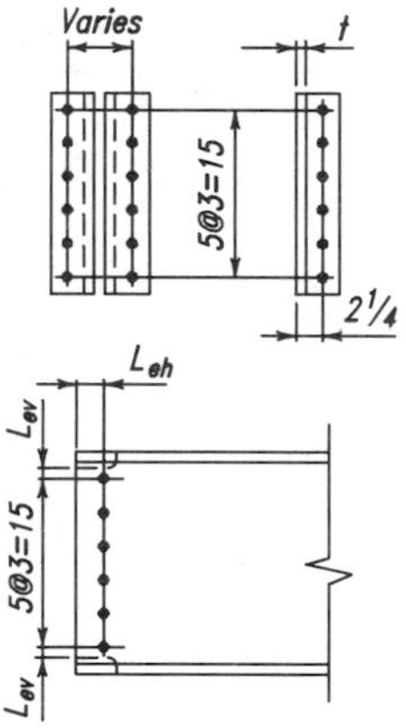

ASTM Desig.	Thread Cond.	Hole Type	Angle Thickness, in.			
			¼	⁵⁄₁₆	³⁄₈	½
A325	N	—	160	191	191	191
	X	—	160	200	239	239
	SC Class A	STD	125	125	125	125
		OVS	107	107	107	107
		SSLT	107	107	107	107
	SC Class B	STD	160	190	190	190
		OVS	150	161	161	161
		SSLT	160	161	161	161
A490	N	—	160	200	239	239
	X	—	160	200	240	298
	SC Class A	STD	157	157	157	157
		OVS	133	133	133	133
		SSLT	133	133	133	133
	SC Class B	STD	160	200	237	237
		OVS	150	188	202	202
		SSLT	160	200	202	202

Beam Web Design Strength per Inch Thickness, kips/in.

			Coped at Top Flange Only						Coped at Both Flanges					
Hole Type	L_{eh},* in.	Un- coped	L_{ev}, in.						L_{ev}, in.					
			1¼	1³⁄₈	1½	1⁵⁄₈	2	3	1¼	1³⁄₈	1½	1⁵⁄₈	2	3
STD	1½	470	332	336	339	342	352	378	320	326	333	339	352	378
	1¾	470	339	342	346	349	359	385	320	326	333	339	359	385
OVS	1½	470	314	318	321	324	334	360	300	307	313	320	334	360
	1¾	470	321	324	328	331	341	367	300	307	313	320	339	367
SSLT	1½	470	332	336	339	342	352	378	320	326	333	339	352	378
	1¾	470	339	342	346	349	359	385	320	326	333	339	359	385

Support Design Strength per Inch Thickness, kips/in.	Notes: STD = Standard holes　　　　　　N = Threads included OVS = Oversized holes　　　　　　　　X = Threads excluded SSLT = Short-slotted holes transverse　　SC = Slip critical 　　　　to direction of load
940	*Tabulated values include ¼-in. reduction in end distance L_{eh} to account for possible underrun in beam length.

	F_y = 50 ksi
	F_u = 65 ksi

Table 9-2 (cont.).
All-Bolted Double-Angle Connections

¾-in. Bolts	Bolt and Angle Design Strength, kips						
6 Rows	ASTM Desig.	Thread Cond.	Hole Type	Angle Thickness, in.			
W40, 36, 33, 30, 27, 24, 21 S24				¼	5/16	3/8	½
	A325	N	—	179	191	191	191
		X	—	179	224	239	239
		SC Class A	STD	125	125	125	125
			OVS	107	107	107	107
			SSLT	107	107	107	107
		SC Class B	STD	179	190	190	190
			OVS	161	161	161	161
			SSLT	161	161	161	161
	A490	N	—	179	224	239	239
		X	—	179	224	269	298
		SC Class A	STD	157	157	157	157
			OVS	133	133	133	133
			SSLT	133	133	133	133
		SC Class B	STD	179	224	237	237
			OVS	168	202	202	202
			SSLT	179	202	202	202

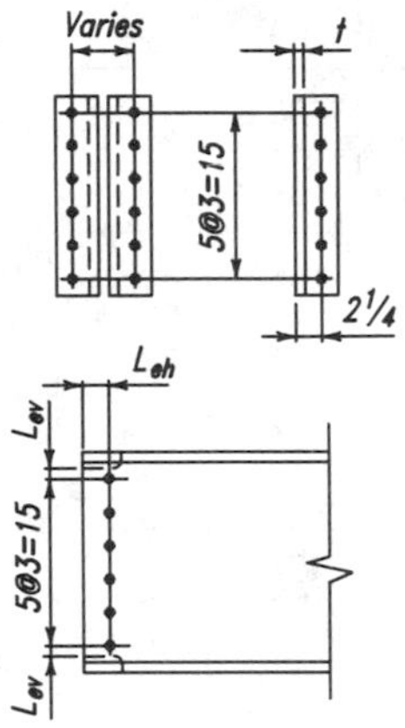

Varies, t, 5@3=15, 2¼, L_{eh}, L_{ev}, 5@3=15

Beam Web Design Strength per Inch Thickness, kips/in.

Hole Type	L_{eh},* in.	Un-coped	Coped at Top Flange Only						Coped at Both Flanges					
			L_{ev}, in.						L_{ev}, in.					
			1¼	1⅜	1½	1⅝	2	3	1¼	1⅜	1½	1⅝	2	3
STD	1½	527	381	385	389	392	403	433	358	366	373	380	402	433
	1¾	527	391	394	398	402	413	442	358	366	373	380	402	442
OVS	1½	527	361	365	369	372	383	413	336	344	351	358	380	413
	1¾	527	371	374	378	382	393	422	336	344	351	358	380	422
SSLT	1½	527	381	385	389	392	403	433	358	366	373	380	402	433
	1¾	527	391	394	398	402	413	442	358	366	373	380	402	442

Support Design Strength per Inch Thickness, kips/in.	Notes: STD = Standard holes N = Threads included OVS = Oversized holes X = Threads excluded SSLT = Short-slotted holes transverse SC = Slip critical to direction of load
1053	*Tabulated values include ¼-in. reduction in end distance L_{eh} to account for possible underrun in beam length.

$F_y = 36$ ksi

$F_u = 58$ ksi

Table 9-2 (cont.).
All-Bolted Double-Angle Connections

¾-in. Bolts	Bolt and Angle Design Strength, kips						
5 Rows	**ASTM Desig.**	**Thread Cond.**	**Hole Type**	**Angle Thickness, in.**			
				¼	⁵⁄₁₆	⅜	½
W30, 27, 24, 21, 18 S24, 20, 18 MC18	A325	N	—	132	159	159	159
		X	—	132	165	198	199
		SC Class A	STD	104	104	104	104
			OVS	88.8	88.8	88.8	88.8
			SSLT	88.8	88.8	88.8	88.8
		SC Class B	STD	132	158	158	158
			OVS	124	134	134	134
			SSLT	132	134	134	134
	A490	N	—	132	165	198	199
		X	—	132	165	198	249
		SC Class A	STD	131	131	131	131
			OVS	111	111	111	111
			SSLT	111	111	111	111
		SC Class B	STD	132	165	198	198
			OVS	124	155	168	168
			SSLT	132	165	168	168

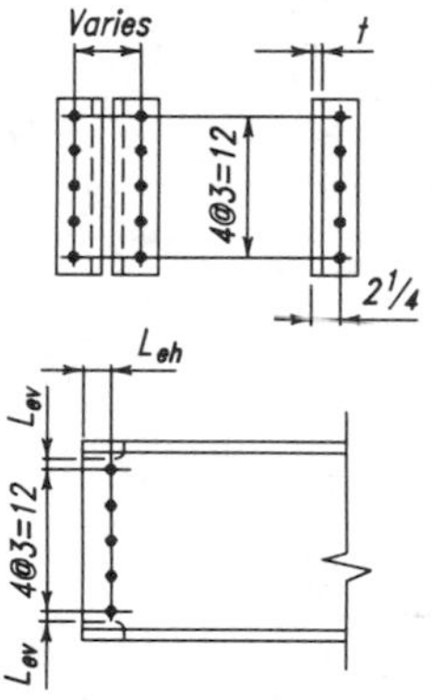

Varies, t, $4@3=12$, $2\frac{1}{4}$, L_{eh}, L_{ev}, $4@3=12$, L_{ev}

Beam Web Design Strength per Inch Thickness, kips/in.

Hole Type	$L_{eh,}$* in.	Un-coped	Coped at Top Flange Only						Coped at Both Flanges					
			L_{ev}, in.						L_{ev}, in.					
			1¼	1⅜	1½	1⅝	2	3	1¼	1⅜	1½	1⅝	2	3
STD	1½	392	277	280	283	287	296	322	264	271	277	284	296	322
	1¾	392	284	287	290	293	303	329	264	271	277	284	303	329
OVS	1½	392	262	265	269	272	282	308	248	254	261	268	282	308
	1¾	392	269	272	275	279	288	315	248	254	261	268	287	315
SSLT	1½	392	277	280	283	287	296	322	264	271	277	284	296	322
	1¾	392	284	287	290	293	303	329	264	271	277	284	303	329

Support Design Strength per Inch Thickness, kips/in.	Notes: STD = Standard holes N = Threads included OVS = Oversized holes X = Threads excluded SSLT = Short-slotted holes transverse SC = Slip critical to direction of load
783	*Tabulated values include ¼-in. reduction in end distance L_{eh} to account for possible underrun in beam length.

| | | | | $F_y = 50$ ksi |
| | | | | $F_u = 65$ ksi |

Table 9-2 (cont.).
All-Bolted Double-Angle Connections

¾-in. Bolts	Bolt and Angle Design Strength, kips						
5 Rows	**ASTM Desig.**	**Thread Cond.**	**Hole Type**	**Angle Thickness, in.**			
				¼	⁵⁄₁₆	⅜	½
W30, 27, 24, 21, 18 S24, 20, 18 MC18	A325	N	—	148	159	159	159
		X	—	148	185	199	199
		SC Class A	STD	104	104	104	104
			OVS	88.8	88.8	88.8	88.8
			SSLT	88.8	88.8	88.8	88.8
		SC Class B	STD	148	158	158	158
			OVS	134	134	134	134
			SSLT	134	134	134	134
	A490	N	—	148	185	199	199
		X	—	148	185	222	249
		SC Class A	STD	131	131	131	131
			OVS	111	111	111	111
			SSLT	111	111	111	111
		SC Class B	STD	148	185	198	198
			OVS	139	168	168	168
			SSLT	148	168	168	168

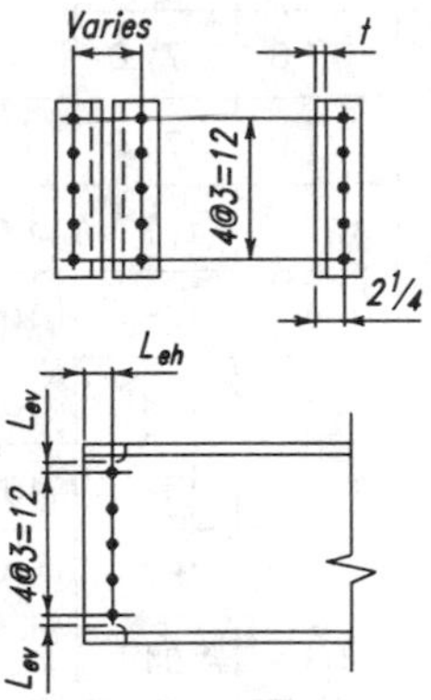

Beam Web Design Strength per Inch Thickness, kips/in.														
			Coped at Top Flange Only						**Coped at Both Flanges**					
			L_{ev}, in.						L_{ev}, in.					
Hole Type	L_{eh},* in.	**Un-coped**	1¼	1⅜	1½	1⅝	2	3	1¼	1⅜	1½	1⅝	2	3
STD	1½	439	319	323	327	330	341	370	296	303	311	318	340	370
	1¾	439	329	332	336	340	351	380	296	303	311	318	340	380
OVS	1½	439	303	306	310	314	325	354	278	285	293	300	322	354
	1¾	439	312	316	320	323	334	363	278	285	293	300	322	363
SSLT	1½	439	319	323	327	330	341	370	296	303	311	318	340	370
	1¾	439	329	332	336	340	351	380	296	303	311	318	340	380

Support Design Strength per Inch Thickness, kips/in.	Notes: STD = Standard holes OVS = Oversized holes SSLT = Short-slotted holes transverse to direction of load	N = Threads included X = Threads excluded SC = Slip critical
878	*Tabulated values include ¼-in. reduction in end distance L_{eh} to account for possible underrun in beam length.	

F_y = 36 ksi
F_u = 58 ksi

Table 9-2 (cont.).
All-Bolted Double-Angle Connections

¾-in. Bolts
4 Rows
W24, 21, 18, 16
S24, 20, 18, 15
C15
MC18

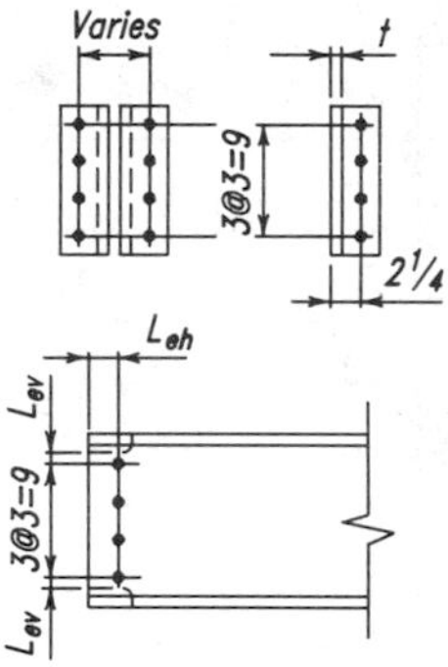

			Bolt and Angle Design Strength, kips			
ASTM Desig.	**Thread Cond.**	**Hole Type**	**Angle Thickness, in.**			
			¼	⁵⁄₁₆	⅜	½
A325	N	—	104	127	127	127
	X	—	104	131	157	159
	SC Class A	STD	83.5	83.5	83.5	83.5
		OVS	71.0	71.0	71.0	71.0
		SSLT	71.0	71.0	71.0	71.0
	SC Class B	STD	104	127	127	127
		OVS	97.9	108	108	108
		SSLT	104	108	108	108
A490	N	—	104	131	157	159
	X	—	104	131	157	199
	SC Class A	STD	104	104	104	104
		OVS	88.8	88.8	88.8	88.8
		SSLT	88.8	88.8	88.8	88.8
	SC Class B	STD	104	131	157	158
		OVS	97.9	122	134	134
		SSLT	104	131	134	134

Beam Web Design Strength per Inch Thickness, kips/in.

			Coped at Top Flange Only						Coped at Both Flanges					
Hole Type	L_{eh},* in.	Un-coped	L_{ev}, in.						L_{ev}, in.					
			1¼	1⅜	1½	1⅝	2	3	1¼	1⅜	1½	1⅝	2	3
STD	1½	313	221	225	228	231	241	267	209	215	222	228	241	267
	1¾	313	228	231	235	238	248	274	209	215	222	228	248	274
OVS	1½	313	210	213	216	220	230	256	196	202	209	215	230	256
	1¾	313	217	220	223	226	236	262	196	202	209	215	235	262
SSLT	1½	313	221	225	228	231	241	267	209	215	222	228	241	267
	1¾	313	228	231	235	238	248	274	209	215	222	228	248	274

Support Design Strength per Inch Thickness, kips/in.	Notes: STD = Standard holes N = Threads included OVS = Oversized holes X = Threads excluded SSLT = Short-slotted holes transverse SC = Slip critical to direction of load
626	*Tabulated values include ¼-in. reduction in end distance L_{eh} to account for possible underrun in beam length.

	$F_y = 50$ ksi
	$F_u = 65$ ksi

Table 9-2 (cont.).
All-Bolted Double-Angle Connections

¾-in. Bolts	Bolt and Angle Design Strength, kips						
4 Rows	ASTM Desig.	Thread Cond.	Hole Type	Angle Thickness, in.			
				¼	⁵⁄₁₆	⅜	½
W24, 21, 18, 16 S24, 20, 18, 15 C15 MC18	A325	N	—	117	127	127	127
		X	—	117	146	159	159
		SC Class A	STD	83.5	83.5	83.5	83.5
			OVS	71.0	71.0	71.0	71.0
			SSLT	71.0	71.0	71.0	71.0
		SC Class B	STD	117	127	127	127
			OVS	108	108	108	108
			SSLT	108	108	108	108
	A490	N	—	117	146	159	159
		X	—	117	146	176	199
		SC Class A	STD	104	104	104	104
			OVS	88.8	88.8	88.8	88.8
			SSLT	88.8	88.8	88.8	88.8
		SC Class B	STD	117	146	158	158
			OVS	110	134	134	134
			SSLT	117	134	134	134

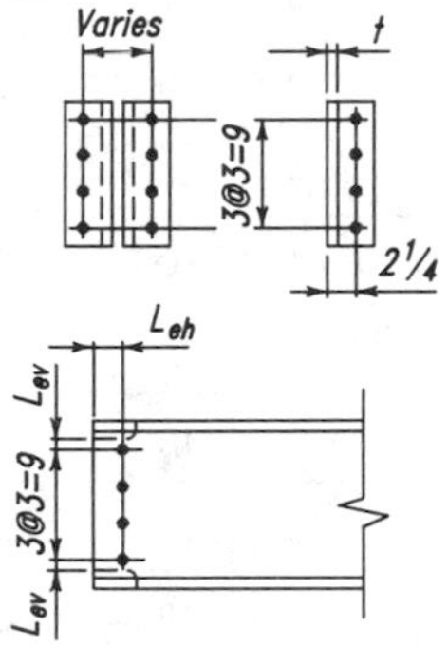

Beam Web Design Strength per Inch Thickness, kips/in.

Hole Type	L_{eh},* in.	Un-coped	Coped at Top Flange Only						Coped at Both Flanges					
			L_{ev}, in.						L_{ev}, in.					
			1¼	1⅜	1½	1⅝	2	3	1¼	1⅜	1½	1⅝	2	3
STD	1½	351	257	261	264	268	279	308	234	241	249	256	278	308
	1¾	351	266	270	274	277	288	318	234	241	249	256	278	318
OVS	1½	351	244	248	252	255	266	296	219	227	234	241	263	296
	1¾	351	254	257	261	265	276	305	219	227	234	241	263	305
SSLT	1½	351	257	261	264	268	279	308	234	241	249	256	278	308
	1¾	351	266	270	274	277	288	318	234	241	249	256	278	318

Support Design Strength per Inch Thickness, kips/in.	Notes: STD = Standard holes OVS = Oversized holes SSLT = Short-slotted holes transverse to direction of load	N = Threads included X = Threads excluded SC = Slip critical
702	*Tabulated values include ¼-in. reduction in end distance L_{eh} to account for possible underrun in beam length.	

F_y = 36 ksi
F_u = 58 ksi

Table 9-2 (cont.).
All-Bolted Double-Angle Connections

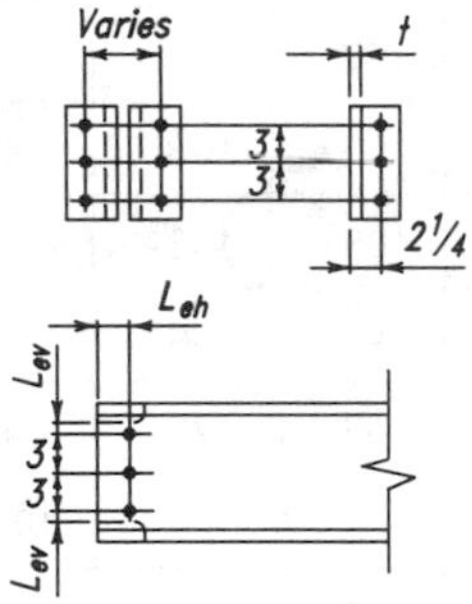

	Bolt and Angle Design Strength, kips						
¾-in. Bolts **3 Rows**	ASTM Desig.	Thread Cond.	Hole Type	Angle Thickness, in.			
				¼	⁵⁄₁₆	³⁄₈	½
W18, 16, 14, 12, 10*	A325	N	—	76.7	95.4	95.4	95.4
S18, 15, 12		X	—	76.7	95.8	115	119
C15, 12		SC Class A	STD	62.7	62.6	62.6	62.6
MC18, 13, 12			OVS	53.3	53.3	53.3	53.3
*Limited to W10×12, 15, 17, 19, 22, 26, 30.			SSLT	53.3	53.3	53.3	53.3
		SC Class B	STD	76.7	94.9	94.9	94.9
			OVS	71.8	80.7	80.7	80.7
			SSLT	76.7	80.7	80.7	80.7
	A490	N	—	76.7	95.8	115	119
		X	—	76.7	95.8	115	149
		SC Class A	STD	76.7	78.3	78.3	78.3
			OVS	66.6	66.6	66.6	66.6
			SSLT	66.6	66.6	66.6	66.6
		SC Class B	STD	76.7	95.8	115	119
			OVS	71.8	89.7	101	101
			SSLT	76.7	95.8	101	101

Beam Web Design Strength per Inch Thickness, kips/in.

			Coped at Top Flange Only						Coped at Both Flanges					
			L_{ev}, in.						L_{ev}, in.					
Hole Type	L_{eh},* in.	Un- coped	1¼	1³⁄₈	1½	1⁵⁄₈	2	3	1¼	1³⁄₈	1½	1⁵⁄₈	2	3
STD	1½	235	166	169	172	176	185	212	153	160	166	173	185	212
	1¾	235	173	176	179	182	192	218	153	160	166	173	192	218
OVS	1½	235	158	161	164	168	177	203	144	150	157	163	177	203
	1¾	235	164	168	171	174	184	210	144	150	157	163	183	210
SSLT	1½	235	166	169	172	176	185	212	153	160	166	173	185	212
	1¾	235	173	176	179	182	192	218	153	160	166	173	192	218

Support Design Strength per Inch Thickness, kips/in.	Notes: STD = Standard holes N = Threads included OVS = Oversized holes X = Threads excluded SSLT = Short-slotted holes transverse SC = Slip critical to direction of load
470	*Tabulated values include ¼-in. reduction in end distance L_{eh} to account for possible underrun in beam length.

	F_y = 50 ksi
	F_u = 65 ksi

Table 9-2 (cont.).
All-Bolted Double-Angle Connections

¾-in. Bolts	Bolt and Angle Design Strength, kips						
3 Rows	ASTM Desig.	Thread Cond.	Hole Type	Angle Thickness, in.			
				¼	⁵⁄₁₆	⅜	½
W18, 16, 14, 12, 10*	A325	N	—	85.9	95.4	95.4	95.4
S18, 15, 12							
C15, 12		X	—	85.9	107	119	119
MC18, 13, 12							
		SC Class A	STD	62.6	62.6	62.6	62.6
*Limited to W10×12, 15,			OVS	53.3	53.3	53.3	53.3
17, 19, 22, 26, 30.			SSLT	53.3	53.3	53.3	53.3
		SC Class B	STD	85.9	94.9	94.9	94.9
			OVS	80.4	80.7	80.7	80.7
			SSLT	80.7	80.7	80.7	80.7
	A490	N	—	85.9	107	119	119
		X	—	85.9	107	129	149
		SC Class A	STD	78.3	78.3	78.3	78.3
			OVS	66.6	66.6	66.6	66.6
			SSLT	66.6	66.6	66.6	66.6
		SC Class B	STD	85.9	107	119	119
			OVS	80.4	101	101	101
			SSLT	85.9	101	101	101

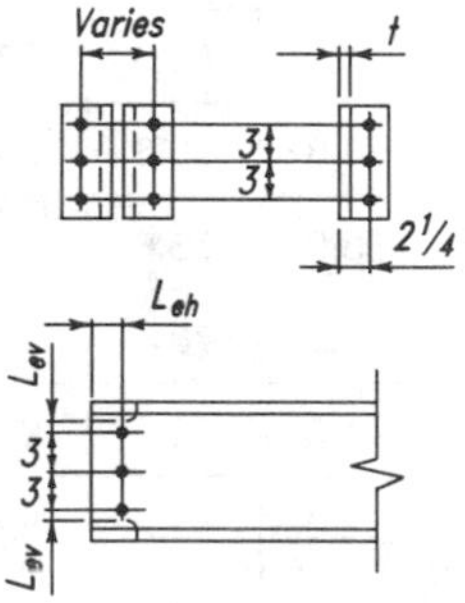

Beam Web Design Strength per Inch Thickness, kips/in.

Hole Type	$L_{eh},$* in.	Un-coped	Coped at Top Flange Only						Coped at Both Flanges					
			L_{ev}, in.						L_{ev}, in.					
			1¼	1⅜	1½	1⅝	2	3	1¼	1⅜	1½	1⅝	2	3
STD	1½	263	195	199	202	206	217	246	172	179	186	194	216	246
	1¾	263	204	208	212	215	226	256	172	179	186	194	216	256
OVS	1½	263	186	189	193	197	208	237	161	168	176	183	205	237
	1¾	263	195	199	203	206	217	246	161	168	176	183	205	246
SSLT	1½	263	195	199	202	206	217	246	172	179	186	194	216	246
	1¾	263	204	208	212	215	226	256	172	179	186	194	216	256

Support Design Strength per Inch Thickness, kips/in.	Notes: STD = Standard holes N = Threads included
	OVS = Oversized holes X = Threads excluded
	SSLT = Short-slotted holes transverse SC = Slip critical
	to direction of load
527	*Tabulated values include ¼-in. reduction in end distance L_{eh} to account for possible underrun in beam length.

F_y = 36 ksi
F_u = 58 ksi

Table 9-2 (cont.).
All-Bolted Double-Angle Connections

3/4-in. Bolts

2 Rows

W12, 10, 8
S12, 10, 8
C12, 10, 9, 8
MC13, 12, 10, 9, 8

		Bolt and Angle Design Strength, kips					
ASTM Desig.	**Thread Cond.**	**Hole Type**	**Angle Thickness, in.**			
			1/4	**5/16**	**3/8**	**1/2**
A325	N	—	48.9	61.2	63.6	63.6
	X	—	48.9	61.2	73.4	79.5
	SC Class A	STD	41.8	41.8	41.8	41.8
		OVS	35.5	35.5	35.5	35.5
		SSLT	35.5	35.5	35.5	35.5
	SC Class B	STD	48.9	61.2	63.3	63.3
		OVS	45.7	53.8	53.8	53.8
		SSLT	48.9	53.8	53.8	53.8
A490	N	—	48.9	61.2	73.4	79.5
	X	—	48.9	61.2	73.4	97.9
	SC Class A	STD	48.9	52.2	52.2	52.2
		OVS	44.4	44.4	44.4	44.4
		SSLT	44.4	44.4	44.4	44.4
	SC Class B	STD	48.9	61.2	73.4	79.1
		OVS	45.7	57.1	67.2	67.2
		SSLT	48.9	61.2	67.2	67.2

Beam Web Design Strength per Inch Thickness, kips/in.

			Coped at Top Flange Only						**Coped at Both Flanges**					
			L_{ev}, in.						L_{ev}, in.					
Hole Type	**L_{eh},* in.**	**Un-coped**	**1 1/4**	**1 3/8**	**1 1/2**	**1 5/8**	**2**	**3**	**1 1/4**	**1 3/8**	**1 1/2**	**1 5/8**	**2**	**3**
STD	1 1/2	157	110	114	117	120	130	156	97.9	104	111	117	130	156
	1 3/4	157	117	120	124	127	137	157	97.9	104	111	117	136	157
OVS	1 1/2	157	106	109	112	115	125	151	91.4	97.9	104	111	125	151
	1 3/4	157	112	116	119	122	132	157	91.4	97.9	104	111	131	157
SSLT	1 1/2	157	110	114	117	120	130	156	97.9	104	111	117	130	156
	1 3/4	157	117	120	124	127	137	157	97.9	104	111	117	136	157

Support Design Strength per Inch Thickness, kips/in.

313

Notes:
STD = Standard holes
OVS = Oversized holes
SSLT = Short-slotted holes transverse to direction of load

N = Threads included
X = Threads excluded
SC = Slip critical

*Tabulated values include 1/4-in. reduction in end distance L_{eh} to account for possible underrun in beam length.

	$F_y = 50$ ksi
	$F_u = 65$ ksi

Table 9-2 (cont.).
All-Bolted Double-Angle Connections

¾-in. Bolts	Bolt and Angle Design Strength, kips						
2 Rows				Angle Thickness, in.			
W12, 10, 8	ASTM Desig.	Thread Cond.	Hole Type	¼	⁵⁄₁₆	⅜	½
S12, 10, 8	A325	N	—	54.8	63.6	63.6	63.6
C12, 10, 9, 8							
MC13, 12, 10, 9, 8		X	—	54.8	68.6	79.5	79.5

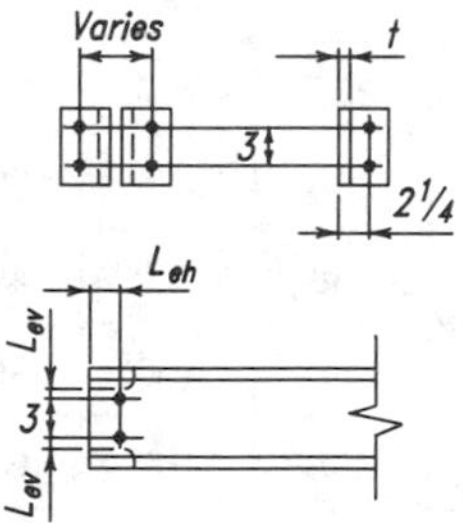

		SC Class A	STD	41.8	41.8	41.8	41.8
			OVS	35.5	35.5	35.5	35.5
			SSLT	35.5	35.5	35.5	35.5
		SC Class B	STD	54.8	63.3	63.3	63.3
			OVS	51.2	53.8	53.8	53.8
			SSLT	53.8	53.8	53.8	53.8
	A490	N	—	54.8	68.6	79.5	79.5
		X	—	54.8	68.6	82.3	99.4
		SC Class A	STD	52.2	52.2	52.2	52.2
			OVS	44.4	44.4	44.4	44.4
			SSLT	44.4	44.4	44.4	44.4
		SC Class B	STD	54.8	68.6	79.1	79.1
			OVS	51.2	64.0	67.2	67.2
			SSLT	54.8	67.2	67.2	67.2

Beam Web Design Strength per Inch Thickness, kips/in.

			Coped at Top Flange Only						Coped at Both Flanges					
			L_{ev}, in.						L_{ev}, in.					
Hole Type	L_{eh},* in.	Un-coped	1¼	1⅜	1½	1⅝	2	3	1¼	1⅜	1½	1⅝	2	3
STD	1½	176	133	136	140	144	155	176	110	117	124	132	154	176
	1¾	176	142	146	149	153	164	176	110	117	124	132	154	176
OVS	1½	176	127	131	135	138	149	176	102	110	117	124	146	176
	1¾	176	137	140	144	148	159	176	102	110	117	124	146	176
SSLT	1½	176	133	136	140	144	155	176	110	117	124	132	154	176
	1¾	176	142	146	149	153	164	176	110	117	124	132	154	176

Support Design Strength per Inch Thickness, kips/in.	Notes: STD = Standard holes N = Threads included OVS = Oversized holes X = Threads excluded SSLT = Short-slotted holes transverse SC = Slip critical to direction of load
351	*Tabulated values include ¼-in. reduction in end distance L_{eh} to account for possible underrun in beam length.

$F_y = 36$ ksi
$F_u = 58$ ksi

Table 9-2 (cont.).
All-Bolted Double-Angle Connections

$7/8$-in. Bolts 12 Rows W44		Bolt and Angle Design Strength, kips						
		ASTM Desig.	Thread Cond.	Hole Type	Angle Thickness, in.			
					$1/4$	$5/16$	$3/8$	$1/2$
		A325	N	—	307	383	460	520
			X	—	307	383	460	613
			SC Class A	STD	307	349	349	349
				OVS	286	297	297	297
				SSLT	297	297	297	297
			SC Class B	STD	307	383	460	520
				OVS	286	358	429	450
				SSLT	307	383	450	450
		A490	N	—	307	383	460	613
			X	—	307	383	460	613
			SC Class A	STD	307	383	439	439
				OVS	286	358	373	373
				SSLT	307	373	373	373
			SC Class B	STD	307	383	460	613
				OVS	286	358	429	565
				SSLT	307	383	460	565

Diagram labels: Varies, t, $11@3=33$, $2\frac{1}{4}$, L_{eh}, L_{ev}, $11@3=33$, L_{ev}

Beam Web Design Strength per Inch Thickness, kips/in.

Hole Type	L_{eh}, in.	Un-coped	Coped at Top Flange Only						Coped at Both Flanges					
			L_{ev}, in.						L_{ev}, in.					
			$1\frac{1}{4}$	$1\frac{3}{8}$	$1\frac{1}{2}$	$1\frac{5}{8}$	2	3	$1\frac{1}{4}$	$1\frac{3}{8}$	$1\frac{1}{2}$	$1\frac{5}{8}$	2	3
STD	$1\frac{1}{2}$	1096	628	631	634	637	647	673	613	620	626	633	647	673
	$1\frac{3}{4}$	1096	634	638	641	644	654	680	613	620	626	633	653	680
OVS	$1\frac{1}{2}$	1096	589	592	595	598	608	634	573	579	586	592	608	634
	$1\frac{3}{4}$	1096	595	599	602	605	615	641	573	579	586	592	612	641
SSLT	$1\frac{1}{2}$	1096	628	631	634	637	647	673	613	620	626	633	647	673
	$1\frac{3}{4}$	1096	634	638	641	644	654	680	613	620	626	633	653	680

Support Design Strength per Inch Thickness, kips/in.	Notes: STD = Standard holes N = Threads included OVS = Oversized holes X = Threads excluded SSLT = Short-slotted holes transverse SC = Slip critical to direction of load
2192	*Tabulated values include $1/4$-in. reduction in end distance L_{eh} to account for possible underrun in beam length.

F_y = 50 ksi	
F_u = 65 ksi	

Table 9-2 (cont.).
All-Bolted Double-Angle Connections

⁷⁄₈-in. Bolts	Bolt and Angle Design Strength, kips						
12 Rows	ASTM	Thread	Hole	Angle Thickness, in.			
W44	Desig.	Cond.	Type	¹⁄₄	⁵⁄₁₆	³⁄₈	¹⁄₂
	A325	N	—	344	430	516	520
		X	—	344	430	516	649
		SC Class A	STD	344	349	349	349
			OVS	297	297	297	297
			SSLT	297	297	297	297
		SC Class B	STD	344	430	516	520
			OVS	321	401	450	450
			SSLT	344	430	450	450
	A490	N	—	344	430	516	649
		X	—	344	430	516	687
		SC Class A	STD	344.	430	439	439
			OVS	321	373	373	373
			SSLT	344	373	373	373
		SC Class B	STD	344	430	516	649
			OVS	321	401	481	565
			SSLT	344	430	516	565

Beam Web Design Strength per Inch Thickness, kips/in.

			Coped at Top Flange Only						Coped at Both Flanges					
Hole	L_{eh},*	Un-	L_{ev}, in.						L_{ev}, in.					
Type	in.	coped	1¹⁄₄	1³⁄₈	1¹⁄₂	1⁵⁄₈	2	3	1¹⁄₄	1³⁄₈	1¹⁄₂	1⁵⁄₈	2	3
STD	1¹⁄₂	1229	712	716	720	723	734	764	687	695	702	709	731	764
	1³⁄₄	1229	722	725	729	733	744	773	687	695	702	709	731	773
OVS	1¹⁄₂	1229	669	672	676	680	691	720	642	649	656	664	686	720
	1³⁄₄	1229	678	682	685	689	700	729	642	649	656	664	686	729
SSLT	1¹⁄₂	1229	712	716	720	723	734	764	687	695	702	709	731	764
	1³⁄₄	1229	722	725	729	733	744	773	687	695	702	709	731	773

Support Design Strength per Inch Thickness, kips/in.	Notes:
	STD = Standard holes N = Threads included
	OVS = Oversized holes X = Threads excluded
	SSLT = Short-slotted holes transverse SC = Slip critical
	to direction of load
2457	*Tabulated values include ¹⁄₄-in. reduction in end distance L_{eh} to account for possible underrun in beam length.

$F_y = 36$ ksi
$F_u = 58$ ksi

Table 9-2 (cont.).
All-Bolted Double-Angle Connections

7⁄8-in. Bolts			Bolt and Angle Design Strength, kips						
11 Rows			**ASTM Desig.**	**Thread Cond.**	**Hole Type**	**Angle Thickness, in.**			
W44, 40						$1/4$	$5/16$	$3/8$	$1/2$

ASTM Desig.	Thread Cond.	Hole Type	$1/4$	$5/16$	$3/8$	$1/2$
A325	N	—	281	351	421	476
	X	—	281	351	421	561
	SC Class A	STD	281	320	320	320
		OVS	262	272	272	272
		SSLT	272	272	272	272
	SC Class B	STD	281	351	421	476
		OVS	262	327	393	412
		SSLT	281	351	412	412
A490	N	—	281	351	421	561
	X	—	281	351	421	561
	SC Class A	STD	281	351	402	402
		OVS	262	327	342	342
		SSLT	281	342	342	342
	SC Class B	STD	281	351	421	561
		OVS	262	327	393	518
		SSLT	281	351	421	518

Beam Web Design Strength per Inch Thickness, kips/in.

			Coped at Top Flange Only						Coped at Both Flanges					
Hole Type	**L_{eh},* in.**	**Un-coped**	**L_{ev}, in.**						**L_{ev}, in.**					
			$1\frac14$	$1\frac38$	$1\frac12$	$1\frac58$	2	3	$1\frac14$	$1\frac38$	$1\frac12$	$1\frac58$	2	3
STD	$1\frac12$	1005	575	579	582	585	595	621	561	568	574	581	595	621
	$1\frac34$	1005	582	585	589	592	602	628	561	568	574	581	600	628
OVS	$1\frac12$	1005	540	543	546	549	559	585	524	530	537	543	559	585
	$1\frac34$	1005	546	550	553	556	566	592	524	530	537	543	563	592
SSLT	$1\frac12$	1005	575	579	582	585	595	621	561	568	574	581	595	621
	$1\frac34$	1005	582	585	589	592	602	628	561	568	574	581	600	628

Support Design Strength per Inch Thickness, kips/in.	Notes: STD = Standard holes OVS = Oversized holes SSLT = Short-slotted holes transverse to direction of load	N = Threads included X = Threads excluded SC = Slip critical
2010	*Tabulated values include $1/4$-in. reduction in end distance L_{eh} to account for possible underrun in beam length.	

$F_y = 50$ ksi	
$F_u = 65$ ksi	

Table 9-2 (cont.).
All-Bolted Double-Angle Connections

$^7/_8$-in. Bolts	Bolt and Angle Design Strength, kips						
11 Rows	ASTM	Thread	Hole	Angle Thickness, in.			
W44, 40	Desig.	Cond.	Type	$^1/_4$	$^5/_{16}$	$^3/_8$	$^1/_2$
	A325	N	—	314	393	472	476
		X	—	314	393	472	595
		SC Class A	STD	314	320	320	320
			OVS	272	272	272	272
			SSLT	272	272	272	272
		SC Class B	STD	314	393	472	476
			OVS	294	367	412	412
			SSLT	314	393	412	412
	A490	N	—	314	393	472	595
		X	—	314	393	472	629
		SC Class A	STD	314	393	402	402
			OVS	294	342	342	342
			SSLT	314	342	342	342
		SC Class B	STD	314	393	472	595
			OVS	294	367	440	518
			SSLT	314	393	472	518

(Figure labels: Varies, t, $10@3=30$, $2^1/_4$, L_{eh}, L_{ev}, $10@3=30$, L_{ev})

Beam Web Design Strength per Inch Thickness, kips/in.

			Coped at Top Flange Only						Coped at Both Flanges					
Hole	L_{eh},*	Un-	L_{ev}, in.						L_{ev}, in.					
Type	in.	coped	$1^1/_4$	$1^3/_8$	$1^1/_2$	$1^5/_8$	2	3	$1^1/_4$	$1^3/_8$	$1^1/_2$	$1^5/_8$	2	3
STD	$1^1/_2$	1126	654	657	661	665	676	705	629	636	644	651	673	705
	$1^3/_4$	1126	663	667	671	674	685	714	629	636	644	651	673	714
OVS	$1^1/_2$	1126	614	618	621	625	636	665	587	594	602	609	631	665
	$1^3/_4$	1126	623	627	631	634	645	674	587	594	602	609	631	674
SSLT	$1^1/_2$	1126	654	657	661	665	676	705	629	636	644	651	673	705
	$1^3/_4$	1126	663	667	671	674	685	714	629	636	644	651	673	714

	Notes:
Support Design Strength per Inch Thickness, kips/in.	STD = Standard holes N = Threads included OVS = Oversized holes X = Threads excluded SSLT = Short-slotted holes transverse SC = Slip critical to direction of load
2252	*Tabulated values include $^1/_4$-in. reduction in end distance L_{eh} to account for possible underrun in beam length.

F_y = 36 ksi
F_u = 58 ksi

Table 9-2 (cont.).
All-Bolted Double-Angle Connections

⁷⁄₈-in. Bolts		Bolt and Angle Design Strength, kips						
10 Rows		**ASTM**	**Thread**	**Hole**	Angle Thickness, in.			
W44, 40, 36		**Desig.**	**Cond.**	**Type**	**¹⁄₄**	**⁵⁄₁₆**	**³⁄₈**	**¹⁄₂**
		A325	N	—	254	318	382	433
			X	—	254	318	382	509
			SC Class A	STD	254	291	291	291
				OVS	238	247	247	247
				SSLT	247	247	247	247
			SC Class B	STD	254	318	382	433
				OVS	238	297	356	375
				SSLT	254	318	375	375
		A490	N	—	254	318	382	509
			X	—	254	318	382	509
			SC Class A	STD	254	318	365	365
				OVS	238	297	311	311
				SSLT	254	311	311	311
			SC Class B	STD	254	318	382	509
				OVS	238	297	356	471
				SSLT	254	318	382	471

Figure labels: Varies, t, $2\frac{1}{4}$, $9@3=27$, L_{eh}, L_{ev}

Beam Web Design Strength per Inch Thickness, kips/in.

			Coped at Top Flange Only						Coped at Both Flanges					
Hole	**L_{eh},***	**Un-**	L_{ev}, in.						L_{ev}, in.					
Type	**in.**	**coped**	**1¹⁄₄**	**1³⁄₈**	**1¹⁄₂**	**1⁵⁄₈**	**2**	**3**	**1¹⁄₄**	**1³⁄₈**	**1¹⁄₂**	**1⁵⁄₈**	**2**	**3**
STD	1¹⁄₂	914	523	526	530	533	543	569	509	515	522	529	543	569
	1³⁄₄	914	530	533	536	540	549	576	509	515	522	529	548	576
OVS	1¹⁄₂	914	491	494	497	501	510	537	475	482	488	495	510	537
	1³⁄₄	914	498	501	504	507	517	543	475	482	488	495	514	543
SSLT	1¹⁄₂	914	523	526	530	533	543	569	509	515	522	529	543	569
	1³⁄₄	914	530	533	536	540	549	576	509	515	522	529	548	576

Support Design Strength per Inch Thickness, kips/in.	Notes: STD = Standard holes OVS = Oversized holes SSLT = Short-slotted holes transverse to direction of load	N = Threads included X = Threads excluded SC = Slip critical
1827	*Tabulated values include ¹⁄₄-in. reduction in end distance L_{eh} to account for possible underrun in beam length.	

$F_y = 50$ ksi	
$F_u = 65$ ksi	

Table 9-2 (cont.).
All-Bolted Double-Angle Connections

⁷⁄₈-in. Bolts	Bolt and Angle Design Strength, kips						
10 Rows	ASTM	Thread	Hole	Angle Thickness, in.			
W44, 40, 36	Desig.	Cond.	Type	¼	⁵⁄₁₆	³⁄₈	½
	A325	N	—	285	356	428	433
		X	—	285	356	428	541
		SC Class A	STD	285	291	291	291
			OVS	247	247	247	247
			SSLT	247	247	247	247
		SC Class B	STD	285	356	428	433
			OVS	266	333	375	375
			SSLT	285	356	375	375
	A490	N	—	285	356	428	541
		X	—	285	356	428	570
		SC Class A	STD	285	356	365	365
			OVS	266	311	311	311
			SSLT	285	311	311	311
		SC Class B	STD	285	356	428	541
			OVS	266	333	399	471
			SSLT	285	356	428	471

Beam Web Design Strength per Inch Thickness, kips/in.

			Coped at Top Flange Only						Coped at Both Flanges					
Hole	L_{eh},*	Un-	L_{ev}, in.						L_{ev}, in.					
Type	in.	coped	1¼	1³⁄₈	1½	1⁵⁄₈	2	3	1¼	1³⁄₈	1½	1⁵⁄₈	2	3
STD	1½	1024	595	599	603	606	617	647	570	578	585	592	614	647
	1¾	1024	605	608	612	616	627	656	570	578	585	592	614	656
OVS	1½	1024	559	563	567	570	581	610	532	540	547	554	576	610
	1¾	1024	569	572	576	580	591	620	532	540	547	554	576	620
SSLT	1½	1024	595	599	603	606	617	647	570	578	585	592	614	647
	1¾	1024	605	608	612	616	627	656	570	578	585	592	614	656

Support Design Strength per Inch Thickness, kips/in.	Notes: STD = Standard holes N = Threads included OVS = Oversized holes X = Threads excluded SSLT = Short-slotted holes transverse SC = Slip critical to direction of load
2048	*Tabulated values include ¼-in. reduction in end distance L_{eh} to account for possible underrun in beam length.

F_y = 36 ksi
F_u = 58 ksi

Table 9-2 (cont.).
All-Bolted Double-Angle Connections

⅞-in. Bolts	Bolt and Angle Design Strength, kips						
9 Rows	ASTM Desig.	Thread Cond.	Hole Type	Angle Thickness, in.			
W44, 40, 36, 33				¼	⁵⁄₁₆	⅜	½
	A325	N	—	228	285	343	390
		X	—	228	285	343	457
		SC Class A	STD	228	262	262	262
			OVS	213	223	223	223
			SSLT	223	223	223	223
		SC Class B	STD	228	285	343	390
			OVS	213	266	320	337
			SSLT	228	285	337	337
	A490	N	—	228	285	343	457
		X	—	228	285	343	457
		SC Class A	STD	228	285	329	329
			OVS	213	266	280	280
			SSLT	228	280	280	280
		SC Class B	STD	228	285	343	457
			OVS	213	266	320	424
			SSLT	228	285	343	424

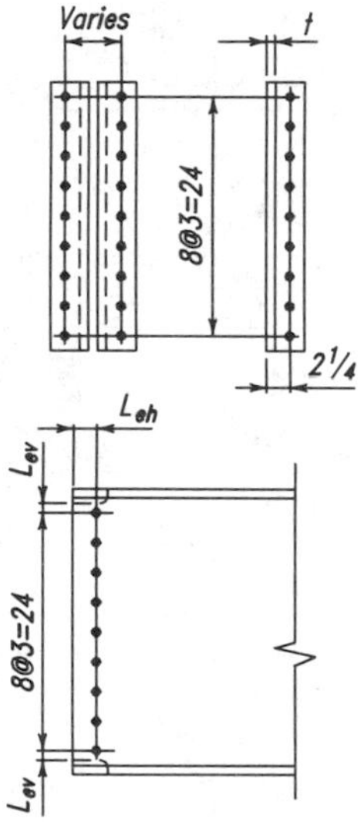

The diagram area shows: Varies, t, 8@3=24, 2¼, L_{eh}, L_{ev}, 8@3=24, L_{ev}.

Beam Web Design Strength per Inch Thickness, kips/in.

			Coped at Top Flange Only						Coped at Both Flanges					
			L_{ev}, in.						L_{ev}, in.					
Hole Type	L_{eh},* in.	Un-coped	1¼	1⅜	1½	1⅝	2	3	1¼	1⅜	1½	1⅝	2	3
STD	1½	822	471	474	477	481	491	517	457	463	470	476	491	517
	1¾	822	478	481	484	487	497	523	457	463	470	476	496	523
OVS	1½	822	442	445	449	452	462	488	426	433	439	446	462	488
	1¾	822	449	452	455	459	468	495	426	433	439	446	465	495
SSLT	1½	822	471	474	477	481	491	517	457	463	470	476	491	517
	1¾	822	478	481	484	487	497	523	457	463	470	476	496	523

Support Design Strength per Inch Thickness, kips/in.

1644

Notes:
STD = Standard holes
OVS = Oversized holes
SSLT = Short-slotted holes transverse to direction of load

N = Threads included
X = Threads excluded
SC = Slip critical

*Tabulated values include ¼-in. reduction in end distance L_{eh} to account for possible underrun in beam length.

	$F_y = 50$ ksi
	$F_u = 65$ ksi

Table 9-2 (cont.).
All-Bolted Double-Angle Connections

$\frac{7}{8}$-in. Bolts	Bolt and Angle Design Strength, kips						
9 Rows	**ASTM Desig.**	**Thread Cond.**	**Hole Type**	**Angle Thickness, in.**			
W44, 40, 36, 33				$\frac{1}{4}$	$\frac{5}{16}$	$\frac{3}{8}$	$\frac{1}{2}$

ASTM Desig.	Thread Cond.	Hole Type	$\frac{1}{4}$	$\frac{5}{16}$	$\frac{3}{8}$	$\frac{1}{2}$
A325	N	—	256	320	384	390
	X	—	256	320	384	487
	SC Class A	STD	256	262	262	262
		OVS	223	223	223	223
		SSLT	223	223	223	223
	SC Class B	STD	256	320	384	390
		OVS	239	299	337	337
		SSLT	256	320	337	337
A490	N	—	256	320	384	487
	X	—	256	320	384	512
	SC Class A	STD	256	320	329	329
		OVS	239	280	280	280
		SSLT	256	280	280	280
	SC Class B	STD	256	320	384	487
		OVS	239	299	358	424
		SSLT	256	320	384	424

Beam Web Design Strength per Inch Thickness, kips/in.

			Coped at Top Flange Only						Coped at Both Flanges					
			L_{ev}, in.						L_{ev}, in.					
Hole Type	L_{eh},* in.	**Un-coped**	$1\frac{1}{4}$	$1\frac{3}{8}$	$1\frac{1}{2}$	$1\frac{5}{8}$	2	3	$1\frac{1}{4}$	$1\frac{3}{8}$	$1\frac{1}{2}$	$1\frac{5}{8}$	2	3
STD	$1\frac{1}{2}$	921	537	540	544	548	559	588	512	519	527	534	556	588
	$1\frac{3}{4}$	921	546	550	554	557	568	597	512	519	527	534	556	597
OVS	$1\frac{1}{2}$	921	504	508	512	515	526	556	478	485	492	500	522	556
	$1\frac{3}{4}$	921	514	518	521	525	536	565	478	485	492	500	522	565
SSLT	$1\frac{1}{2}$	921	537	540	544	548	559	588	512	519	527	534	556	588
	$1\frac{3}{4}$	921	546	550	554	557	568	597	512	519	527	534	556	597

Support Design Strength per Inch Thickness, kips/in.	Notes: STD = Standard holes OVS = Oversized holes SSLT = Short-slotted holes transverse to direction of load	N = Threads included X = Threads excluded SC = Slip critical
1843	*Tabulated values include $\frac{1}{4}$-in. reduction in end distance L_{eh} to account for possible underrun in beam length.	

F_y = 36 ksi	
F_u = 58 ksi	

Table 9-2 (cont.).
All-Bolted Double-Angle Connections

7/8-in. Bolts				Bolt and Angle Design Strength, kips				
8 Rows	ASTM Desig.	Thread Cond.	Hole Type	Angle Thickness, in.			
W44, 40, 36, 33, 30				$\frac{1}{4}$	$\frac{5}{16}$	$\frac{3}{8}$	$\frac{1}{2}$
	A325	N	—	202	253	303	346
		X	—	202	253	303	405
		SC Class A	STD	202	233	233	233
			OVS	189	198	198	198
			SSLT	198	198	198	198
		SC Class B	STD	202	253	303	346
			OVS	189	236	283	300
			SSLT	202	253	300	300
	A490	N	—	202	253	303	405
		X	—	202	253	303	405
		SC Class A	STD	202	253	292	292
			OVS	189	236	249	249
			SSLT	202	249	249	249
		SC Class B	STD	202	253	303	405
			OVS	189	236	283	377
			SSLT	202	253	303	377

Drawing labels: Varies, t, $7@3=21$, $2\frac{1}{4}$, L_{eh}, L_{ev}, $7@3=21$, L_{ev}

Beam Web Design Strength per Inch Thickness, kips/in.

			Coped at Top Flange Only						Coped at Both Flanges					
Hole Type	L_{eh},* in.	Un-coped	L_{ev}, in.						L_{ev}, in.					
			$1\frac{1}{4}$	$1\frac{3}{8}$	$1\frac{1}{2}$	$1\frac{5}{8}$	2	3	$1\frac{1}{4}$	$1\frac{3}{8}$	$1\frac{1}{2}$	$1\frac{5}{8}$	2	3
STD	$1\frac{1}{2}$	731	419	422	425	429	438	464	405	411	418	424	438	464
	$1\frac{3}{4}$	731	425	429	432	435	445	471	405	411	418	424	444	471
OVS	$1\frac{1}{2}$	731	393	397	400	403	413	439	377	384	390	397	413	439
	$1\frac{3}{4}$	731	400	403	407	410	420	446	377	384	390	397	417	446
SSLT	$1\frac{1}{2}$	731	419	422	425	429	438	464	405	411	418	424	438	464
	$1\frac{3}{4}$	731	425	429	432	435	445	471	405	411	418	424	444	471

Support Design Strength per Inch Thickness, kips/in.	Notes: STD = Standard holes　　　　　　N = Threads included OVS = Oversized holes　　　　　　X = Threads excluded SSLT = Short-slotted holes transverse　SC = Slip critical 　　　　to direction of load
1462	*Tabulated values include $\frac{1}{4}$-in. reduction in end distance L_{eh} to account for possible underrun in beam length.

	$F_y = 50$ ksi
	$F_u = 65$ ksi

Table 9-2 (cont.).
All-Bolted Double-Angle Connections

$\frac{7}{8}$-in. Bolts	Bolt and Angle Design Strength, kips						
8 Rows	**ASTM Desig.**	**Thread Cond.**	**Hole Type**	**Angle Thickness, in.**			
W44, 40, 36, 33, 30				$\frac{1}{4}$	$\frac{5}{16}$	$\frac{3}{8}$	$\frac{1}{2}$
	A325	N	—	227	283	340	346
		X	—	227	283	340	433
		SC Class A	STD	227	233	233	233
			OVS	198	198	198	198
			SSLT	198	198	198	198
		SC Class B	STD	227	283	340	346
			OVS	211	264	300	300
			SSLT	227	283	300	300
	A490	N	—	227	283	340	433
		X	—	227	283	340	453
		SC Class A	STD	227	283	292	292
			OVS	211	249	249	249
			SSLT	227	249	249	249
		SC Class B	STD	227	283	340	433
			OVS	211	264	317	377
			SSLT	227	283	340	377

Beam Web Design Strength per Inch Thickness, kips/in.

			Coped at Top Flange Only						Coped at Both Flanges					
			L_{ev}, in.						L_{ev}, in.					
Hole Type	L_{eh},* **in.**	**Un-coped**	$1\frac{1}{4}$	$1\frac{3}{8}$	$1\frac{1}{2}$	$1\frac{5}{8}$	**2**	**3**	$1\frac{1}{4}$	$1\frac{3}{8}$	$1\frac{1}{2}$	$1\frac{5}{8}$	**2**	**3**
STD	$1\frac{1}{2}$	819	478	482	486	489	500	530	453	461	468	475	497	530
	$1\frac{3}{4}$	819	488	491	495	499	510	539	453	461	468	475	497	539
OVS	$1\frac{1}{2}$	819	450	453	457	461	472	501	423	430	438	445	467	501
	$1\frac{3}{4}$	819	459	463	466	470	481	510	423	430	438	445	467	510
SSLT	$1\frac{1}{2}$	819	478	482	486	489	500	530	453	461	468	475	497	530
	$1\frac{3}{4}$	819	488	491	495	499	510	539	453	461	468	475	497	539

Support Design Strength per Inch Thickness, kips/in.	Notes: STD = Standard holes OVS = Oversized holes SSLT = Short-slotted holes transverse to direction of load	N = Threads included X = Threads excluded SC = Slip critical
1638	*Tabulated values include $\frac{1}{4}$-in. reduction in end distance L_{eh} to account for possible underrun in beam length.	

F_y = 36 ksi
F_u = 58 ksi

Table 9-2 (cont.).
All-Bolted Double-Angle Connections

$\frac{7}{8}$-in. Bolts 7 Rows W44, 40, 36, 33, 30, 27, 24 S24			Bolt and Angle Design Strength, kips				
	ASTM Desig.	Thread Cond.	Hole Type	Angle Thickness, in.			
				$\frac{1}{4}$	$\frac{5}{16}$	$\frac{3}{8}$	$\frac{1}{2}$
	A325	N	—	176	220	264	303
		X	—	176	220	264	352
		SC Class A	STD	176	204	204	204
			OVS	164	173	173	173
			SSLT	173	173	173	173
		SC Class B	STD	176	220	264	303
			OVS	164	205	246	262
			SSLT	176	220	262	262
	A490	N	—	176	220	264	352
		X	—	176	220	264	352
		SC Class A	STD	176	220	256	256
			OVS	164	205	217	217
			SSLT	176	217	217	217
		SC Class B	STD	176	220	264	352
			OVS	164	205	246	329
			SSLT	176	220	264	329

Beam Web Design Strength per Inch Thickness, kips/in.

Hole Type	L_{eh},* in.	Un-coped	Coped at Top Flange Only						Coped at Both Flanges					
			L_{ev}, in.						L_{ev}, in.					
			$1\frac{1}{4}$	$1\frac{3}{8}$	$1\frac{1}{2}$	$1\frac{5}{8}$	2	3	$1\frac{1}{4}$	$1\frac{3}{8}$	$1\frac{1}{2}$	$1\frac{5}{8}$	2	3
STD	$1\frac{1}{2}$	639	367	370	373	376	386	412	352	359	365	372	386	412
	$1\frac{3}{4}$	639	373	377	380	383	393	419	352	359	365	372	392	419
OVS	$1\frac{1}{2}$	639	344	348	351	354	364	390	329	335	342	348	364	390
	$1\frac{3}{4}$	639	351	354	358	361	371	397	329	335	342	348	368	397
SSLT	$1\frac{1}{2}$	639	367	370	373	376	386	412	352	359	365	372	386	412
	$1\frac{3}{4}$	639	373	377	380	383	393	419	352	359	365	372	392	419

Support Design Strength per Inch Thickness, kips/in.	Notes:
1279	STD = Standard holes N = Threads included OVS = Oversized holes X = Threads excluded SSLT = Short-slotted holes transverse SC = Slip critical to direction of load *Tabulated values include $\frac{1}{4}$-in. reduction in end distance L_{eh} to account for possible underrun in beam length.

	F_y = 50 ksi
	F_u = 65 ksi

Table 9-2 (cont.).
All-Bolted Double-Angle Connections

7/8-in. Bolts	Bolt and Angle Design Strength, kips						
7 Rows	**ASTM Desig.**	**Thread Cond.**	**Hole Type**	**Angle Thickness, in.**			
W44, 40, 36, 33, 30, 27, 24 S24				$1/4$	$5/16$	$3/8$	$1/2$
	A325	N	—	197	247	296	303
		X	—	197	247	296	379
		SC Class A	STD	197	204	204	204
			OVS	173	173	173	173
			SSLT	173	173	173	173
		SC Class B	STD	197	247	296	303
			OVS	184	230	262	262
			SSLT	197	247	262	262
	A490	N	—	197	247	296	379
		X	—	197	247	296	395
		SC Class A	STD	197	247	256	256
			OVS	184	217	217	217
			SSLT	197	217	217	217
		SC Class B	STD	197	247	296	379
			OVS	184	230	276	329
			SSLT	197	247	296	329

Beam Web Design Strength per Inch Thickness, kips/in.

Hole Type	L_{eh},* in.	Un-coped	Coped at Top Flange Only						Coped at Both Flanges					
			L_{ev}, in.						L_{ev}, in.					
			$1^1/4$	$1^3/8$	$1^1/2$	$1^5/8$	2	3	$1^1/4$	$1^3/8$	$1^1/2$	$1^5/8$	2	3
STD	$1^1/2$	717	420	423	427	431	442	471	395	402	410	417	439	471
	$1^3/4$	717	429	433	437	440	451	480	395	402	410	417	439	480
OVS	$1^1/2$	717	395	399	402	406	417	446	368	376	383	390	412	446
	$1^3/4$	717	404	408	412	415	426	456	368	376	383	390	412	456
SSLT	$1^1/2$	717	420	423	427	431	442	471	395	402	410	417	439	471
	$1^3/4$	717	429	433	437	440	451	480	395	402	410	417	439	480

Support Design Strength per Inch Thickness, kips/in.	Notes: STD = Standard holes OVS = Oversized holes SSLT = Short-slotted holes transverse to direction of load	N = Threads included X = Threads excluded SC = Slip critical
1433	*Tabulated values include $1/4$-in. reduction in end distance L_{eh} to account for possible underrun in beam length.	

F_y = 36 ksi
F_u = 58 ksi

Table 9-2 (cont.).
All-Bolted Double-Angle Connections

7/8-in. Bolts

6 Rows

**W40, 36, 33, 30, 27, 24, 21
S24**

			Bolt and Angle Design Strength, kips				
ASTM Desig.	Thread Cond.	Hole Type	Angle Thickness, in.			
			1/4	5/16	3/8	1/2
A325	N	—	150	188	225	260
	X	—	150	188	225	300
	SC Class A	STD	150	175	175	175
		OVS	140	148	148	148
		SSLT	148	148	148	148
	SC Class B	STD	150	188	225	260
		OVS	140	175	210	225
		SSLT	150	188	225	225
A490	N	—	150	188	225	300
	X	—	150	188	225	300
	SC Class A	STD	150	188	219	219
		OVS	140	175	186	186
		SSLT	150	186	186	186
	SC Class B	STD	150	188	225	300
		OVS	140	175	210	280
		SSLT	150	188	225	282

Beam Web Design Strength per Inch Thickness, kips/in.

			Coped at Top Flange Only						Coped at Both Flanges					
			L_{ev}, in.						L_{ev}, in.					
Hole Type	L_{eh},* in.	Un-coped	1 1/4	1 3/8	1 1/2	1 5/8	2	3	1 1/4	1 3/8	1 1/2	1 5/8	2	3
STD	1 1/2	548	314	318	321	324	334	360	300	307	313	320	334	360
	1 3/4	548	321	324	328	331	341	367	300	307	313	320	339	367
OVS	1 1/2	548	296	299	302	305	315	341	280	286	293	299	315	341
	1 3/4	548	302	306	309	312	322	348	280	286	293	299	319	348
SSLT	1 1/2	548	314	318	321	324	334	360	300	307	313	320	334	360
	1 3/4	548	321	324	328	331	341	367	300	307	313	320	339	367

Support Design Strength per Inch Thickness, kips/in.	Notes: STD = Standard holes OVS = Oversized holes SSLT = Short-slotted holes transverse to direction of load	N = Threads included X = Threads excluded SC = Slip critical
1096	*Tabulated values include 1/4-in. reduction in end distance L_{eh} to account for possible underrun in beam length.	

	F_y = 50 ksi
	F_u = 65 ksi

Table 9-2 (cont.).
All-Bolted Double-Angle Connections

⅞-in. Bolts	Bolt and Angle Design Strength, kips						
6 Rows	**ASTM Desig.**	**Thread Cond.**	**Hole Type**	**Angle Thickness, in.**			
W40, 36, 33, 30, 27, 24, 21 S24				¼	⁵⁄₁₆	⅜	½
	A325	N	—	168	210	252	260
		X	—	168	210	252	325
		SC Class A	STD	168	175	175	175
			OVS	148	148	148	148
			SSLT	148	148	148	148
		SC Class B	STD	168	210	252	260
			OVS	157	196	225	225
			SSLT	168	210	225	225
	A490	N	—	168	210	252	325
		X	—	168	210	252	336
		SC Class A	STD	168	210	219	219
			OVS	157	186	186	186
			SSLT	168	186	186	186
		SC Class B	STD	168	210	252	325
			OVS	157	196	235	282
			SSLT	168	210	252	282

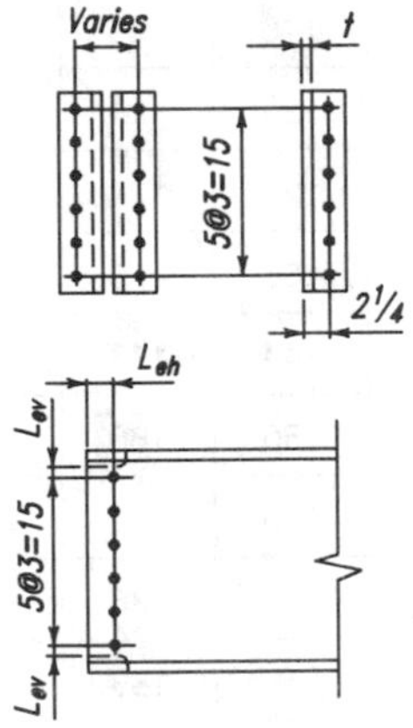

(diagram: Varies, t, 5@3=15, 2¼, L_{eh}, L_{ev}, 5@3=15, L_{ev})

Beam Web Design Strength per Inch Thickness, kips/in.														
			Coped at Top Flange Only						**Coped at Both Flanges**					
Hole Type	**L_{eh},* in.**	**Un-coped**	**L_{ev}, in.**						**L_{ev}, in.**					
			1¼	1⅜	1½	1⅝	2	3	1¼	1⅜	1½	1⅝	2	3
STD	1½	614	361	365	369	372	383	413	336	344	351	358	380	413
	1¾	614	371	374	378	382	393	422	336	344	351	358	380	422
OVS	1½	614	340	344	348	351	362	392	314	321	328	335	357	392
	1¾	614	350	353	357	361	372	401	314	321	328	335	357	401
SSLT	1½	614	361	365	369	372	383	413	336	344	351	358	380	413
	1¾	614	371	374	378	382	393	422	336	344	351	358	380	422

Support Design Strength per Inch Thickness, kips/in.	Notes: STD = Standard holes N = Threads included
	OVS = Oversized holes X = Threads excluded
	SSLT = Short-slotted holes transverse SC = Slip critical
	to direction of load
1229	*Tabulated values include ¼-in. reduction in end distance L_{eh} to account for possible underrun in beam length.

F_y = 36 ksi
F_u = 58 ksi

Table 9-2 (cont.).
All-Bolted Double-Angle Connections

$\frac{7}{8}$-in. Bolts

5 Rows

W30, 27, 24, 21, 18
S24, 20, 18
MC18

			Bolt and Angle Design Strength, kips				
ASTM Desig.	Thread Cond.	Hole Type	Angle Thickness, in.			
			$\frac{1}{4}$	$\frac{5}{16}$	$\frac{3}{8}$	$\frac{1}{2}$
A325	N	—	124	155	186	216
	X	—	124	155	186	248
	SC Class A	STD	124	145	145	145
		OVS	115	124	124	124
		SSLT	124	124	124	124
	SC Class B	STD	124	155	186	216
		OVS	115	144	173	187
		SSLT	124	155	186	187
A490	N	—	124	155	186	248
	X	—	124	155	186	248
	SC Class A	STD	124	155	183	183
		OVS	115	144	155	155
		SSLT	124	155	155	155
	SC Class B	STD	124	155	186	248
		OVS	115	144	173	231
		SSLT	124	155	186	235

Beam Web Design Strength per Inch Thickness, kips/in.

			Coped at Top Flange Only						Coped at Both Flanges					
Hole Type	$L_{eh,}$* in.	Un-coped	L_{ev}, in.						L_{ev}, in.					
			$1\frac{1}{4}$	$1\frac{3}{8}$	$1\frac{1}{2}$	$1\frac{5}{8}$	2	3	$1\frac{1}{4}$	$1\frac{3}{8}$	$1\frac{1}{2}$	$1\frac{5}{8}$	2	3
STD	$1\frac{1}{2}$	457	262	265	269	272	282	308	248	254	261	268	282	308
	$1\frac{3}{4}$	457	269	272	275	279	288	315	248	254	261	268	287	315
OVS	$1\frac{1}{2}$	457	247	250	253	257	266	293	231	238	244	251	266	293
	$1\frac{3}{4}$	457	254	257	260	263	273	299	231	238	244	251	270	299
SSLT	$1\frac{1}{2}$	457	262	265	269	272	282	308	248	254	261	268	282	308
	$1\frac{3}{4}$	457	269	272	275	279	288	315	248	254	261	268	287	315

Support Design Strength per Inch Thickness, kips/in.	Notes: STD = Standard holes OVS = Oversized holes SSLT = Short-slotted holes transverse to direction of load	N = Threads included X = Threads excluded SC = Slip critical
914	*Tabulated values include $\frac{1}{4}$-in. reduction in end distance L_{eh} to account for possible underrun in beam length.	

	F_y = 50 ksi
	F_u = 65 ksi

Table 9-2 (cont.).
All-Bolted Double-Angle Connections

$\frac{7}{8}$-in. Bolts	Bolt and Angle Design Strength, kips							
5 Rows	ASTM Desig.	Thread Cond.	Hole Type	Angle Thickness, in.			
				$\frac{1}{4}$	$\frac{5}{16}$	$\frac{3}{8}$	$\frac{1}{2}$
W30, 27, 24, 21, 18 S24, 20, 18 MC18	A325	N	—	139	174	208	216
		X	—	139	174	208	271
		SC Class A	STD	139	145	145	145
			OVS	124	124	124	124
			SSLT	124	124	124	124
		SC Class B	STD	139	174	208	216
			OVS	129	162	187	187
			SSLT	139	174	187	187
	A490	N	—	139	174	208	271
		X	—	139	174	208	278
		SC Class A	STD	139	174	183	183
			OVS	129	155	155	155
			SSLT	139	155	155	155
		SC Class B	STD	139	174	208	271
			OVS	129	162	194	235
			SSLT	139	174	208	235

Beam Web Design Strength per Inch Thickness, kips/in.														
			Coped at Top Flange Only						Coped at Both Flanges					
			L_{ev}, in.						L_{ev}, in.					
Hole Type	L_{eh},* in.	Un-coped	$1\frac{1}{4}$	$1\frac{3}{8}$	$1\frac{1}{2}$	$1\frac{5}{8}$	2	3	$1\frac{1}{4}$	$1\frac{3}{8}$	$1\frac{1}{2}$	$1\frac{5}{8}$	2	3
STD	$1\frac{1}{2}$	512	303	306	310	314	325	354	278	285	293	300	322	354
	$1\frac{3}{4}$	512	312	316	320	323	334	363	278	285	293	300	322	363
OVS	$1\frac{1}{2}$	512	286	289	293	297	308	337	259	266	273	281	303	337
	$1\frac{3}{4}$	512	295	299	302	306	317	346	259	266	273	281	303	346
SSLT	$1\frac{1}{2}$	512	303	306	310	314	325	354	278	285	293	300	322	354
	$1\frac{3}{4}$	512	312	316	320	323	334	363	278	285	293	300	322	363

Support Design Strength per Inch Thickness, kips/in.	Notes: STD = Standard holes OVS = Oversized holes SSLT = Short-slotted holes transverse to direction of load	N = Threads included X = Threads excluded SC = Slip critical
1024	*Tabulated values include $\frac{1}{4}$-in. reduction in end distance L_{eh} to account for possible underrun in beam length.	

| F_y = 36 ksi |
| F_u = 58 ksi |

Table 9-2 (cont.).
All-Bolted Double-Angle Connections

⁷⁄₈-in. Bolts 4 Rows				
W24, 21, 18, 16				
S24, 20, 18, 15				
C15				
MC18				

		Bolt and Angle Design Strength, kips					
ASTM Desig.	**Thread Cond.**	**Hole Type**	**Angle Thickness, in.**				
			¹⁄₄	⁵⁄₁₆	³⁄₈	¹⁄₂	
A325	N	—	97.9	122	147	173	
	X	—	97.9	122	147	196	
	SC Class A	STD	97.9	116	116	116	
		OVS	91.1	98.9	98.9	98.9	
		SSLT	97.9	98.9	98.9	98.9	
	SC Class B	STD	97.9	122	147	173	
		OVS	91.1	114	137	150	
		SSLT	97.9	122	147	150	
A490	N	—	97.9	122	147	196	
	X	—	97.9	122	147	196	
	SC Class A	STD	97.9	122	146	146	
		OVS	91.1	114	124	124	
		SSLT	97.9	122	124	124	
	SC Class B	STD	97.9	122	147	196	
		OVS	91.1	114	137	182	
		SSLT	97.9	122	147	188	

Beam Web Design Strength per Inch Thickness, kips/in.

Hole Type	L_{eh},* in.	Un- coped	Coped at Top Flange Only						Coped at Both Flanges					
			L_{ev}, in.						L_{ev}, in.					
			1¹⁄₄	1³⁄₈	1¹⁄₂	1⁵⁄₈	2	3	1¹⁄₄	1³⁄₈	1¹⁄₂	1⁵⁄₈	2	3
STD	1¹⁄₂	365	210	213	216	220	230	256	196	202	209	215	230	256
	1³⁄₄	365	217	220	223	226	236	262	196	202	209	215	235	262
OVS	1¹⁄₂	365	198	201	205	208	218	244	182	189	195	202	218	244
	1³⁄₄	365	205	208	211	215	224	250	182	189	195	202	221	250
SSLT	1¹⁄₂	365	210	213	216	220	230	256	196	202	209	215	230	256
	1³⁄₄	365	217	220	223	226	236	262	196	202	209	215	235	262

Support Design Strength per Inch Thickness, kips/in.	Notes: STD = Standard holes N = Threads included
	OVS = Oversized holes X = Threads excluded
	SSLT = Short-slotted holes transverse SC = Slip critical
	to direction of load
731	*Tabulated values include ¹⁄₄-in. reduction in end distance L_{eh} to account for possible underrun in beam length.

$F_y = 50$ ksi	
$F_u = 65$ ksi	

Table 9-2 (cont.).
All-Bolted Double-Angle Connections

$^{7}/_{8}$-in. Bolts	Bolt and Angle Design Strength, kips						
4 Rows	**ASTM Desig.**	**Thread Cond.**	**Hole Type**	**Angle Thickness, in.**			
				$^{1}/_{4}$	$^{5}/_{16}$	$^{3}/_{8}$	$^{1}/_{2}$
W24, 21, 18, 16 S24, 20, 18, 15 C15 MC18	A325	N	—	110	137	165	173
		X	—	110	137	165	216
		SC Class A	STD	110	116	116	116
			OVS	98.9	98.9	98.9	98.9
			SSLT	98.9	98.9	98.9	98.9
		SC Class B	STD	110	137	165	173
			OVS	102	128	150	150
			SSLT	110	137	150	150
	A490	N	—	110	137	165	216
		X	—	110	137	165	219
		SC Class A	STD	110	137	146	146
			OVS	102	124	124	124
			SSLT	110	124	124	124
		SC Class B	STD	110	137	165	216
			OVS	102	128	153	188
			SSLT	110	137	165	188

Beam Web Design Strength per Inch Thickness, kips/in.

			Coped at Top Flange Only						Coped at Both Flanges					
			L_{ev}, in.						L_{ev}, in.					
Hole Type	L_{eh},* in.	Un-coped	$1^{1}/_{4}$	$1^{3}/_{8}$	$1^{1}/_{2}$	$1^{5}/_{8}$	2	3	$1^{1}/_{4}$	$1^{3}/_{8}$	$1^{1}/_{2}$	$1^{5}/_{8}$	2	3
STD	$1^{1}/_{2}$	410	244	248	252	255	266	296	219	227	234	241	263	296
	$1^{3}/_{4}$	410	254	257	261	265	276	305	219	227	234	241	263	305
OVS	$1^{1}/_{2}$	410	231	235	238	242	253	282	204	211	219	226	248	282
	$1^{3}/_{4}$	410	240	244	248	251	262	292	204	211	219	226	248	292
SSLT	$1^{1}/_{2}$	410	244	248	252	255	266	296	219	227	234	241	263	296
	$1^{3}/_{4}$	410	254	257	261	265	276	305	219	227	234	241	263	305

Support Design Strength per Inch Thickness, kips/in.	Notes: STD = Standard holes N = Threads included OVS = Oversized holes X = Threads excluded SSLT = Short-slotted holes transverse SC = Slip critical to direction of load
819	*Tabulated values include $^{1}/_{4}$-in. reduction in end distance L_{eh} to account for possible underrun in beam length.

$F_y = 36$ ksi
$F_u = 58$ ksi

Table 9-2 (cont.).
All-Bolted Double-Angle Connections

⁷⁄₈-in. Bolts

3 Rows

W18, 16, 14, 12, 10*
S18, 15, 12
C15, 12
MC18, 13, 12

*Limited to W10×12, 15,
17, 19, 22, 26, 30

			Bolt and Angle Design Strength, kips				
ASTM Desig.	Thread Cond.	Hole Type	Angle Thickness, in.				
			¼	⁵⁄₁₆	³⁄₈	½	
A325	N	—	71.8	89.7	108	130	
	X	—	71.8	89.7	108	144	
	SC Class A	STD	71.8	87.3	87.3	87.3	
		OVS	66.7	74.2	74.2	74.2	
		SSLT	71.8	74.2	74.2	74.2	
	SC Class B	STD	71.8	89.7	108	130	
		OVS	66.7	83.4	100	112	
		SSLT	71.8	89.7	108	112	
A490	N	—	71.8	89.7	108	144	
	X	—	71.8	89.7	108	144	
	SC Class A	STD	71.8	89.7	108	110	
		OVS	66.7	83.4	93.2	93.2	
		SSLT	71.8	89.7	93.2	93.2	
	SC Class B	STD	71.8	89.7	108	144	
		OVS	66.7	83.4	100	133	
		SSLT	71.8	89.7	108	141	

Beam Web Design Strength per Inch Thickness, kips/in.

Hole Type	L_{eh},* in.	Un-coped	Coped at Top Flange Only						Coped at Both Flanges					
			L_{ev}, in.						L_{ev}, in.					
			1¼	1³⁄₈	1½	1⁵⁄₈	2	3	1¼	1³⁄₈	1½	1⁵⁄₈	2	3
STD	1½	274	158	161	164	168	177	203	144	150	157	163	177	203
	1¾	274	164	168	171	174	184	210	144	150	157	163	183	210
OVS	1½	274	149	153	156	159	169	195	133	140	146	153	169	195
	1¾	274	156	159	163	166	176	202	133	140	146	153	173	202
SSLT	1½	274	158	161	164	168	177	203	144	150	157	163	177	203
	1¾	274	164	168	171	174	184	210	144	150	157	163	183	210

Support Design Strength per Inch Thickness, kips/in.

548

Notes:
STD = Standard holes
OVS = Oversized holes
SSLT = Short-slotted holes transverse
      to direction of load

N = Threads included
X = Threads excluded
SC = Slip critical

*Tabulated values include ¼-in. reduction in end distance L_{eh} to account for possible underrun in beam length.

	F_y = 50 ksi
	F_u = 65 ksi

Table 9-2 (cont.).
All-Bolted Double-Angle Connections

7/8-in. Bolts	Bolt and Angle Design Strength, kips						
3 Rows	**ASTM Desig.**	**Thread Cond.**	**Hole Type**	**Angle Thickness, in.**			
				1/4	5/16	3/8	1/2
W18, 16, 14, 12, 10* S18, 15, 12 C15, 12 MC18, 13, 12	A325	N	—	80.4	101	121	130
		X	—	80.4	101	121	161
*Limited to W10×12, 15, 17, 19, 22, 26, 30		SC Class A	STD	80.4	87.3	87.3	87.3
			OVS	74.2	74.2	74.2	74.2
			SSLT	74.2	74.2	74.2	74.2
		SC Class B	STD	80.4	101	121	130
			OVS	74.7	93.4	112	112
			SSLT	80.4	101	112	112
	A490	N	—	80.4	101	121	161
		X	—	80.4	101	121	161
		SC Class A	STD	80.4	101	110	110
			OVS	74.7	93.2	93.2	93.2
			SSLT	80.4	93.2	93.2	93.2
		SC Class B	STD	80.4	101	121	161
			OVS	74.7	93.4	112	141
			SSLT	80.4	101	121	141

Beam Web Design Strength per Inch Thickness, kips/in.

Hole Type	L_{eh},* in.	Un-coped	Coped at Top Flange Only						Coped at Both Flanges					
			L_{ev}, in.						L_{ev}, in.					
			1 1/4	1 3/8	1 1/2	1 5/8	2	3	1 1/4	1 3/8	1 1/2	1 5/8	2	3
STD	1 1/2	307	186	189	193	197	208	237	161	168	176	183	205	237
	1 3/4	307	195	199	203	206	217	246	161	168	176	183	205	246
OVS	1 1/2	307	176	180	184	187	198	227	149	157	164	171	193	227
	1 3/4	307	186	189	193	197	208	237	149	157	164	171	193	237
SSLT	1 1/2	307	186	189	193	197	208	237	161	168	176	183	205	237
	1 3/4	307	195	199	203	206	217	246	161	168	176	183	205	246

Support Design Strength per Inch Thickness, kips/in.	Notes: STD = Standard holes OVS = Oversized holes SSLT = Short-slotted holes transverse to direction of load	N = Threads included X = Threads excluded SC = Slip critical
614	*Tabulated values include 1/4-in. reduction in end distance L_{eh} to account for possible underrun in beam length.	

$F_y = 36$ ksi
$F_u = 58$ ksi

Table 9-2 (cont.).
All-Bolted Double-Angle Connections

$7/8$-in. Bolts							
2 Rows							

	Bolt and Angle Design Strength, kips						
	ASTM Desig.	Thread Cond.	Hole Type	Angle Thickness, in.			
				$1/4$	$5/16$	$3/8$	$1/2$

W12, 10, 8
S12, 10, 8
C12, 10, 9, 8
MC13, 12, 10, 9, 8

Varies — t — $3\ddagger$ — $2\frac{1}{4}$ — L_{eh} — L_{ev} — $3\ddagger$ — L_{ev}

ASTM Desig.	Thread Cond.	Hole Type	$1/4$	$5/16$	$3/8$	$1/2$
A325	N	—	45.7	57.1	68.5	86.6
	X	—	45.7	57.1	68.5	91.4
	SC Class A	STD	45.7	57.1	58.2	58.2
		OVS	42.3	49.4	49.4	49.4
		SSLT	45.7	49.4	49.4	49.4
	SC Class B	STD	45.7	57.1	68.5	86.6
		OVS	42.3	52.9	63.4	74.9
		SSLT	45.7	57.1	68.5	74.9
A490	N	—	45.7	57.1	68.5	91.4
	X	—	45.7	57.1	68.5	91.4
	SC Class A	STD	45.7	57.1	68.5	73.1
		OVS	42.3	52.9	62.1	62.1
		SSLT	45.7	57.1	62.1	62.1
	SC Class B	STD	45.7	57.1	68.5	91.4
		OVS	42.3	52.9	63.4	84.6
		SSLT	45.7	57.1	68.5	91.4

Beam Web Design Strength per Inch Thickness, kips/in.

			Coped at Top Flange Only						Coped at Both Flanges					
Hole Type	L_{eh},* in.	Un-coped	L_{ev}, in.						L_{ev}, in.					
			$1\frac{1}{4}$	$1\frac{3}{8}$	$1\frac{1}{2}$	$1\frac{5}{8}$	2	3	$1\frac{1}{4}$	$1\frac{3}{8}$	$1\frac{1}{2}$	$1\frac{5}{8}$	2	3
STD	$1\frac{1}{2}$	183	106	109	112	115	125	151	91.4	97.9	104	111	125	151
	$1\frac{3}{4}$	183	112	116	119	122	132	158	91.4	97.9	104	111	131	158
OVS	$1\frac{1}{2}$	183	100	104	107	110	120	146	84.6	91.1	97.6	104	120	146
	$1\frac{3}{4}$	183	107	110	114	117	127	153	84.6	91.1	97.6	104	124	153
SSLT	$1\frac{1}{2}$	183	106	109	112	115	125	151	91.4	97.9	104	111	125	151
	$1\frac{3}{4}$	183	112	116	119	122	132	158	91.4	97.9	104	111	131	158

Support Design Strength per Inch Thickness, kips/in.	Notes:
365	STD = Standard holes　　　　　N = Threads included OVS = Oversized holes　　　　　X = Threads excluded SSLT = Short-slotted holes transverse　　SC = Slip critical 　　　　to direction of load *Tabulated values include $1/4$-in. reduction in end distance L_{eh} to account for possible underrun in beam length.

F_y = 50 ksi	
F_u = 65 ksi	

Table 9-2 (cont.).
All-Bolted Double-Angle Connections

$7/8$-in. Bolts	Bolt and Angle Design Strength, kips						
2 Rows	**ASTM Desig.**	**Thread Cond.**	**Hole Type**	**Angle Thickness, in.**			
				$1/4$	$5/16$	$3/8$	$1/2$
W12, 10, 8	A325	N	—	51.2	64.0	76.8	86.6
S12, 10, 8		X	—	51.2	64.0	76.8	102
C12, 10, 9, 8		SC Class A	STD	51.2	58.2	58.2	58.2
MC13, 12, 10, 9, 8			OVS	47.4	49.4	49.4	49.4
			SSLT	49.4	49.4	49.4	49.4
		SC Class B	STD	51.2	64.0	76.8	86.6
			OVS	47.4	59.2	71.1	74.9
			SSLT	51.2	64.0	74.9	74.9
	A490	N	—	51.2	64.0	76.8	102
		X	—	51.2	64.0	76.8	102
		SC Class A	STD	51.2	64.0	73.1	73.1
			OVS	47.4	59.2	62.1	62.1
			SSLT	51.2	62.1	62.1	62.1
		SC Class B	STD	51.2	64.0	76.8	102
			OVS	47.4	59.2	71.1	94.1
			SSLT	51.2	64.0	76.8	94.1

Beam Web Design Strength per Inch Thickness, kips/in.

			Coped at Top Flange Only						Coped at Both Flanges					
Hole Type	**L_{eh},* in.**	**Un-coped**	**L_{ev}, in.**						**L_{ev}, in.**					
			$1^1/4$	$1^3/8$	$1^1/2$	$1^5/8$	2	3	$1^1/4$	$1^3/8$	$1^1/2$	$1^5/8$	2	3
STD	$1^1/2$	205	127	131	135	138	149	179	102	110	117	124	146	179
	$1^3/4$	205	137	140	144	148	159	188	102	110	117	124	146	188
OVS	$1^1/2$	205	122	125	129	133	144	173	94.8	102	109	117	139	173
	$1^3/4$	205	131	135	138	142	153	182	94.8	102	109	117	139	182
SSLT	$1^1/2$	205	127	131	135	138	149	179	102	110	117	124	146	179
	$1^3/4$	205	137	140	144	148	159	188	102	110	117	124	146	188

Support Design Strength per Inch Thickness, kips/in.	Notes: STD = Standard holes N = Threads included
	OVS = Oversized holes X = Threads excluded
	SSLT = Short-slotted holes transverse SC = Slip critical
	to direction of load
410	*Tabulated values include $1/4$-in. reduction in end distance L_{eh} to account for possible underrun in beam length.

F_y = 36 ksi
F_u = 58 ksi

Table 9-2 (cont.).
All-Bolted Double-Angle Connections

1-in. Bolts	Bolt and Angle Design Strength, kips						
12 Rows	ASTM	Thread	Hole	Angle Thickness, in.			
W44	Desig.	Cond.	Type	$\frac{1}{4}$	$\frac{5}{16}$	$\frac{3}{8}$	$\frac{1}{2}$
	A325	N	—	286	358	429	573
		X	—	286	358	429	573
		SC Class A	STD	286	358	429	456
			OVS	258	323	387	388
			SSLT	286	358	388	388
		SC Class B	STD	286	358	429	573
			OVS	258	323	387	516
			SSLT	286	358	429	573
	A490	N	—	286	358	429	573
		X	—	286	358	429	573
		SC Class A	STD	286	358	429	573
			OVS	258	323	387	487
			SSLT	286	358	429	487
		SC Class B	STD	286	358	429	573
			OVS	258	323	387	516
			SSLT	286	358	429	573

Diagram labels: Varies, t, $11@3=33$, $2\frac{1}{2}$, L_{eh}, L_{ev}, $11@3=33$, L_{ev}

Beam Web Design Strength per Inch Thickness, kips/in.

			Coped at Top Flange Only						Coped at Both Flanges					
Hole	L_{eh},	Un-	L_{ev}, in.						L_{ev}, in.					
Type	in.	coped	$1\frac{1}{4}$	$1\frac{3}{8}$	$1\frac{1}{2}$	$1\frac{5}{8}$	2	3	$1\frac{1}{4}$	$1\frac{3}{8}$	$1\frac{1}{2}$	$1\frac{5}{8}$	2	3
STD	$1\frac{1}{2}$	1253	589	592	595	598	608	634	573	579	586	592	608	634
	$1\frac{3}{4}$	1253	595	599	602	605	615	641	573	579	586	592	612	641
OVS	$1\frac{1}{2}$	1253	534	538	541	544	554	580	516	523	529	536	554	580
	$1\frac{3}{4}$	1253	541	544	548	551	561	587	516	523	529	536	555	587
SSLT	$1\frac{1}{2}$	1253	589	592	595	598	608	634	573	579	586	592	608	634
	$1\frac{3}{4}$	1253	595	599	602	605	615	641	573	579	586	592	612	641

Support Design Strength per Inch Thickness, kips/in.	Notes:
	STD = Standard holes N = Threads included
	OVS = Oversized holes X = Threads excluded
	SSLT = Short-slotted holes transverse SC = Slip critical
	to direction of load
2506	*Tabulated values include $\frac{1}{4}$-in. reduction in end distance L_{eh} to account for possible underrun in beam length.

$F_y = 50$ ksi	
$F_u = 65$ ksi	

Table 9-2 (cont.).
All-Bolted Double-Angle Connections

1-in. Bolts	Bolt and Angle Design Strength, kips						
12 Rows	**ASTM Desig.**	**Thread Cond.**	**Hole Type**	**Angle Thickness, in.**			
W44				$1/4$	$5/16$	$3/8$	$1/2$
	A325	N	—	321	401	481	642
		X	—	321	401	481	642
		SC Class A	STD	321	401	456	456
			OVS	289	362	388	388
			SSLT	321	388	388	388
		SC Class B	STD	321	401	481	642
			OVS	289	362	434	579
			SSLT	321	401	481	588
	A490	N	—	321	401	481	642
		X	—	321	401	481	642
		SC Class A	STD	321	401	481	573
			OVS	289	362	434	487
			SSLT	321	401	481	487
		SC Class B	STD	321	401	481	642
			OVS	289	362	434	579
			SSLT	321	401	481	642

Diagram labels: Varies, t, $11@3=33$, $2\frac{1}{2}$, L_{eh}, L_{ev}, $11@3=33$, L_{ev}

Beam Web Design Strength per Inch Thickness, kips/in.

Hole Type	L_{eh},* in.	Un-coped	Coped at Top Flange Only						Coped at Both Flanges					
			L_{ev}, in.						L_{ev}, in.					
			$1\frac{1}{4}$	$1\frac{3}{8}$	$1\frac{1}{2}$	$1\frac{5}{8}$	2	3	$1\frac{1}{4}$	$1\frac{3}{8}$	$1\frac{1}{2}$	$1\frac{5}{8}$	2	3
STD	$1\frac{1}{2}$	1404	669	672	676	680	691	720	642	649	656	664	686	720
	$1\frac{3}{4}$	1404	678	682	685	689	700	729	642	649	656	664	686	729
OVS	$1\frac{1}{2}$	1404	608	612	615	619	630	659	579	586	593	601	622	659
	$1\frac{3}{4}$	1404	617	621	625	628	639	669	579	586	593	601	622	669
SSLT	$1\frac{1}{2}$	1404	669	672	676	680	691	720	642	649	656	664	686	720
	$1\frac{3}{4}$	1404	678	682	685	689	700	729	642	649	656	664	686	729

Support Design Strength per Inch Thickness, kips/in.	Notes:
	STD = Standard holes N = Threads included
	OVS = Oversized holes X = Threads excluded
	SSLT = Short-slotted holes transverse SC = Slip critical
	to direction of load
2808	*Tabulated values include $1/4$-in. reduction in end distance L_{eh} to account for possible underrun in beam length.

F_y = 36 ksi
F_u = 58 ksi

Table 9-2 (cont.).
All-Bolted Double-Angle Connections

1-in. Bolts	Bolt and Angle Design Strength, kips						
11 Rows	ASTM	Thread	Hole	Angle Thickness, in.			
W44, 40	Desig.	Cond.	Type	$\frac{1}{4}$	$\frac{5}{16}$	$\frac{3}{8}$	$\frac{1}{2}$
	A325	N	—	262	327	393	524
		X	—	262	327	393	524
		SC Class A	STD	262	327	393	418
			OVS	236	295	354	356
			SSLT	262	327	356	356
		SC Class B	STD	262	327	393	524
			OVS	236	295	354	472
			SSLT	262	327	393	524
	A490	N	—	262	327	393	524
		X	—	262	327	393	524
		SC Class A	STD	262	327	393	524
			OVS	236	295	354	446
			SSLT	262	327	393	446
		SC Class B	STD	262	327	393	524
			OVS	236	295	354	472
			SSLT	262	327	393	524

Diagram labels: Varies, t, 10@3=30, $2\frac{1}{2}$, L_{eh}, L_{ev}, 10@3=30, L_{ev}

Beam Web Design Strength per Inch Thickness, kips/in.

			Coped at Top Flange Only						Coped at Both Flanges					
			L_{ev}, in.						L_{ev}, in.					
Hole Type	L_{eh},* in.	Un-coped	$1\frac{1}{4}$	$1\frac{3}{8}$	$1\frac{1}{2}$	$1\frac{5}{8}$	2	3	$1\frac{1}{4}$	$1\frac{3}{8}$	$1\frac{1}{2}$	$1\frac{5}{8}$	2	3
STD	$1\frac{1}{2}$	1148	540	543	546	549	559	585	524	530	537	543	559	585
	$1\frac{3}{4}$	1148	546	550	553	556	566	592	524	530	537	543	563	592
OVS	$1\frac{1}{2}$	1148	490	494	497	500	510	536	472	479	485	492	510	536
	$1\frac{3}{4}$	1148	497	500	504	507	517	543	472	479	485	492	511	543
SSLT	$1\frac{1}{2}$	1148	540	543	546	549	559	585	524	530	537	543	559	585
	$1\frac{3}{4}$	1148	546	550	553	556	566	592	524	530	537	543	563	592

Support Design Strength per Inch Thickness, kips/in.	Notes: STD = Standard holes OVS = Oversized holes SSLT = Short-slotted holes transverse 　　　to direction of load	N = Threads included X = Threads excluded SC = Slip critical
2297	*Tabulated values include $\frac{1}{4}$-in. reduction in end distance L_{eh} to account for possible underrun in beam length.	

	F_y = 50 ksi
	F_u = 65 ksi

Table 9-2 (cont.).
All-Bolted Double-Angle Connections

1-in. Bolts	Bolt and Angle Design Strength, kips						
11 Rows	ASTM	Thread	Hole	Angle Thickness, in.			
W44, 40	Desig.	Cond.	Type	$\frac{1}{4}$	$\frac{5}{16}$	$\frac{3}{8}$	$\frac{1}{2}$

	ASTM Desig.	Thread Cond.	Hole Type	$\frac{1}{4}$	$\frac{5}{16}$	$\frac{3}{8}$	$\frac{1}{2}$
	A325	N	—	294	367	440	587
		X	—	294	367	440	587
		SC Class A	STD	294	367	418	418
			OVS	265	331	356	356
			SSLT	294	356	356	356
		SC Class B	STD	294	367	440	587
			OVS	265	331	397	529
			SSLT	294	367	440	539
	A490	N	—	294	367	440	587
		X	—	294	367	440	587
		SC Class A	STD	294	367	440	525
			OVS	265	331	397	446
			SSLT	294	367	440	446
		SC Class B	STD	294	367	440	587
			OVS	265	331	397	529
			SSLT	294	367	440	587

Beam Web Design Strength per Inch Thickness, kips/in.

Hole Type	L_{eh},* in.	Un-coped	Coped at Top Flange Only						Coped at Both Flanges					
			L_{ev}, in.						L_{ev}, in.					
			$1\frac{1}{4}$	$1\frac{3}{8}$	$1\frac{1}{2}$	$1\frac{5}{8}$	2	3	$1\frac{1}{4}$	$1\frac{3}{8}$	$1\frac{1}{2}$	$1\frac{5}{8}$	2	3
STD	$1\frac{1}{2}$	1287	614	618	621	625	636	665	587	594	602	609	631	665
	$1\frac{3}{4}$	1287	623	627	631	634	645	674	587	594	602	609	631	674
OVS	$1\frac{1}{2}$	1287	559	562	566	570	581	610	529	536	544	551	573	610
	$1\frac{3}{4}$	1287	568	572	575	579	590	619	529	536	544	551	573	619
SSLT	$1\frac{1}{2}$	1287	614	618	621	625	636	665	587	594	602	609	631	665
	$1\frac{3}{4}$	1287	623	627	631	634	645	674	587	594	602	609	631	674

Support Design Strength per Inch Thickness, kips/in.	Notes: STD = Standard holes OVS = Oversized holes SSLT = Short-slotted holes transverse to direction of load	N = Threads included X = Threads excluded SC = Slip critical
2574	*Tabulated values include $\frac{1}{4}$-in. reduction in end distance L_{eh} to account for possible underrun in beam length.	

F_y = 36 ksi
F_u = 58 ksi

Table 9-2 (cont.).
All-Bolted Double-Angle Connections

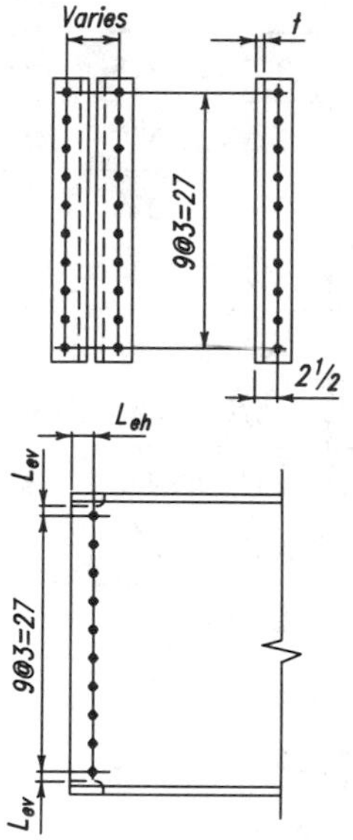

1-in. Bolts	Bolt and Angle Design Strength, kips						
10 Rows	**ASTM Desig.**	**Thread Cond.**	**Hole Type**	**Angle Thickness, in.**			
W44, 40, 36				1/4	5/16	3/8	1/2
	A325	N	—	238	297	356	475
		X	—	238	297	356	475
		SC Class A	STD	238	297	356	380
			OVS	214	268	321	323
			SSLT	238	297	323	323
		SC Class B	STD	238	297	356	475
			OVS	214	268	321	428
			SSLT	238	297	356	475
	A490	N	—	238	297	356	475
		X	—	238	297	356	475
		SC Class A	STD	238	297	356	475
			OVS	214	268	321	406
			SSLT	238	297	356	406
		SC Class B	STD	238	297	356	475
			OVS	214	268	321	428
			SSLT	238	297	356	475

Diagram labels: Varies, t, 9@3=27, 2½, L_{eh}, L_{ev}, 9@3=27

Beam Web Design Strength per Inch Thickness, kips/in.

Hole Type	L_{eh},* in.	Un-coped	Coped at Top Flange Only						Coped at Both Flanges					
			L_{ev}, in.						L_{ev}, in.					
			1¼	1⅜	1½	1⅝	2	3	1¼	1⅜	1½	1⅝	2	3
STD	1½	1044	491	494	497	501	510	537	475	482	488	495	510	537
	1¾	1044	498	501	504	507	517	543	475	482	488	495	514	543
OVS	1½	1044	446	450	453	456	466	492	428	435	441	448	466	492
	1¾	1044	453	456	460	463	473	499	428	435	441	448	467	499
SSLT	1½	1044	491	494	497	501	510	537	475	482	488	495	510	537
	1¾	1044	498	501	504	507	517	543	475	482	488	495	514	543

Support Design Strength per Inch Thickness, kips/in.	Notes: STD = Standard holes N = Threads included OVS = Oversized holes X = Threads excluded SSLT = Short-slotted holes transverse SC = Slip critical to direction of load
2088	*Tabulated values include ¼-in. reduction in end distance L_{eh} to account for possible underrun in beam length.

		F_y = 50 ksi
		F_u = 65 ksi

Table 9-2 (cont.).
All-Bolted Double-Angle Connections

1-in. Bolts	Bolt and Angle Design Strength, kips						

10 Rows	ASTM Desig.	Thread Cond.	Hole Type	Angle Thickness, in.			
W44, 40, 36				$\frac{1}{4}$	$\frac{5}{16}$	$\frac{3}{8}$	$\frac{1}{2}$
	A325	N	—	266	333	399	532
		X	—	266	333	399	532
		SC Class A	STD	266	333	380	380
			OVS	240	300	323	323
			SSLT	266	323	323	323
		SC Class B	STD	266	333	399	532
			OVS	240	300	360	480
			SSLT	266	333	399	490
	A490	N	—	266	333	399	532
		X	—	266	333	399	532
		SC Class A	STD	266	333	399	477
			OVS	240	300	360	406
			SSLT	266	333	399	406
		SC Class B	STD	266	333	399	532
			OVS	240	300	360	480
			SSLT	266	333	399	532

Drawing labels: Varies, t, $9@3=27$, $2\frac{1}{2}$, L_{eh}, L_{ev}, $9@3=27$, L_{ev}

Beam Web Design Strength per Inch Thickness, kips/in.

			Coped at Top Flange Only						Coped at Both Flanges					
			L_{ev}, in.						L_{ev}, in.					
Hole Type	L_{eh},* in.	Un-coped	$1\frac{1}{4}$	$1\frac{3}{8}$	$1\frac{1}{2}$	$1\frac{5}{8}$	2	3	$1\frac{1}{4}$	$1\frac{3}{8}$	$1\frac{1}{2}$	$1\frac{5}{8}$	2	3
STD	$1\frac{1}{2}$	1170	559	563	567	570	581	610	532	540	547	554	576	610
	$1\frac{3}{4}$	1170	569	572	576	580	591	620	532	540	547	554	576	620
OVS	$1\frac{1}{2}$	1170	509	513	516	520	531	560	480	487	494	502	524	560
	$1\frac{3}{4}$	1170	519	522	526	530	540	570	480	487	494	502	524	570
SSLT	$1\frac{1}{2}$	1170	559	563	567	570	581	610	532	540	547	554	576	610
	$1\frac{3}{4}$	1170	569	572	576	580	591	620	532	540	547	554	576	620

Support Design Strength per Inch Thickness, kips/in.	Notes: STD = Standard holes OVS = Oversized holes SSLT = Short-slotted holes transverse to direction of load	N = Threads included X = Threads excluded SC = Slip critical
2340	*Tabulated values include $\frac{1}{4}$-in. reduction in end distance L_{eh} to account for possible underrun in beam length.	

F_y = 36 ksi
F_u = 58 ksi

Table 9-2 (cont.).
All-Bolted Double-Angle Connections

1-in. Bolts	Bolt and Angle Design Strength, kips						
9 Rows	**ASTM Desig.**	**Thread Cond.**	**Hole Type**	**Angle Thickness, in.**			
W44, 40, 36, 33				$1/4$	$5/16$	$3/8$	$1/2$
	A325	N	—	213	266	320	426
		X	—	213	266	320	426
		SC Class A	STD	213	266	320	342
			OVS	192	240	288	291
			SSLT	213	266	291	291
		SC Class B	STD	213	266	320	426
			OVS	192	240	288	384
			SSLT	213	266	320	426
	A490	N	—	213	266	320	426
		X	—	213	266	320	426
		SC Class A	STD	213	266	320	426
			OVS	192	240	288	365
			SSLT	213	266	320	365
		SC Class B	STD	213	266	320	426
			OVS	192	240	288	384
			SSLT	213	266	320	426

The connection diagram shows dimensions: Varies, t, $8@3=24$, $2^1/2$, L_{eh}, L_{ev}, $8@3=24$, L_{ev}.

Beam Web Design Strength per Inch Thickness, kips/in.

Hole Type	L_{eh},* in.	Un-coped	Coped at Top Flange Only						Coped at Both Flanges					
			L_{ev}, in.						L_{ev}, in.					
			$1^1/4$	$1^3/8$	$1^1/2$	$1^5/8$	2	3	$1^1/4$	$1^3/8$	$1^1/2$	$1^5/8$	2	3
STD	$1^1/2$	940	442	445	449	452	462	488	426	433	439	446	462	488
	$1^3/4$	940	449	452	455	459	468	495	426	433	439	446	465	495
OVS	$1^1/2$	940	402	405	409	412	422	448	384	390	397	404	422	448
	$1^3/4$	940	409	412	415	419	428	455	384	390	397	404	423	455
SS:T	$1^1/2$	940	442	445	449	452	462	488	426	433	439	446	462	488
	$1^3/4$	940	449	452	455	459	468	495	426	433	439	446	465	495

Support Design Strength per Inch Thickness, kips/in.	Notes: STD = Standard holes N = Threads included
	OVS = Oversized holes X = Threads excluded
	SSLT = Short-slotted holes transverse SC = Slip critical
	to direction of load
1879	*Tabulated values include $1/4$-in. reduction in end distance L_{eh} to account for possible underrun in beam length.

	F_y = 50 ksi
	F_u = 65 ksi

Table 9-2 (cont.).
All-Bolted Double-Angle Connections

1-in. Bolts	Bolt and Angle Design Strength, kips						
9 Rows	**ASTM Desig.**	**Thread Cond.**	**Hole Type**	**Angle Thickness, in.**			
W44, 40, 36, 33				$1/4$	$5/16$	$3/8$	$1/2$
	A325	N	—	239	299	358	478
		X	—	239	299	358	478
		SC Class A	STD	239	299	342	342
			OVS	215	269	291	291
			SSLT	239	291	291	291
		SC Class B	STD	239	299	358	478
			OVS	215	269	323	430
			SSLT	239	299	358	441
	A490	N	—	239	299	358	478
		X	—	239	299	358	478
		SC Class A	STD	239	299	358	430
			OVS	215	269	323	365
			SSLT	239	299	358	365
		SC Class B	STD	239	299	358	478
			OVS	215	269	323	430
			SSLT	239	299	358	478

Beam Web Design Strength per Inch Thickness, kips/in.

Hole Type	L_{eh},* in.	Un-coped	Coped at Top Flange Only						Coped at Both Flanges					
			L_{ev}, in.						L_{ev}, in.					
			$1\frac{1}{4}$	$1\frac{3}{8}$	$1\frac{1}{2}$	$1\frac{5}{8}$	2	3	$1\frac{1}{4}$	$1\frac{3}{8}$	$1\frac{1}{2}$	$1\frac{5}{8}$	2	3
STD	$1\frac{1}{2}$	1053	504	508	512	515	526	556	478	485	492	500	522	556
	$1\frac{3}{4}$	1053	514	518	521	525	536	565	478	485	492	500	522	565
OVS	$1\frac{1}{2}$	1053	460	463	467	471	482	511	430	438	445	452	474	511
	$1\frac{3}{4}$	1053	469	473	476	480	491	520	430	438	445	452	474	520
SSLT	$1\frac{1}{2}$	1053	504	508	512	515	526	556	478	485	492	500	522	556
	$1\frac{3}{4}$	1053	514	518	521	525	536	565	478	485	492	500	522	565

Support Design Strength per Inch Thickness, kips/in.	Notes: STD = Standard holes OVS = Oversized holes SSLT = Short-slotted holes transverse to direction of load	N = Threads included X = Threads excluded SC = Slip critical
2106	*Tabulated values include $1/4$-in. reduction in end distance L_{eh} to account for possible underrun in beam length.	

F_y = 36 ksi
F_u = 58 ksi

Table 9-2 (cont.).
All-Bolted Double-Angle Connections

1-in. Bolts	Bolt and Angle Design Strength, kips						
8 Rows	ASTM	Thread	Hole	Angle Thickness, in.			
W44, 40, 36, 33, 30	Desig.	Cond.	Type	$\frac{1}{4}$	$\frac{5}{16}$	$\frac{3}{8}$	$\frac{1}{2}$
	A325	N	—	189	236	283	377
		X	—	189	236	283	377
		SC Class A	STD	189	236	283	304
			OVS	170	212	255	259
			SSLT	189	236	259	259
		SC Class B	STD	189	236	283	377
			OVS	170	212	255	340
			SSLT	189	236	283	377
	A490	N	—	189	236	283	377
		X	—	189	236	283	377
		SC Class A	STD	189	236	283	377
			OVS	170	212	255	325
			SSLT	189	236	283	325
		SC Class B	STD	189	236	283	377
			OVS	170	212	255	340
			SSLT	189	236	283	377

Beam Web Design Strength per Inch Thickness, kips/in.

Hole Type	L_{eh},* in.	Un-coped	Coped at Top Flange Only						Coped at Both Flanges					
			L_{ev}, in.						L_{ev}, in.					
			$1\frac{1}{4}$	$1\frac{3}{8}$	$1\frac{1}{2}$	$1\frac{5}{8}$	2	3	$1\frac{1}{4}$	$1\frac{3}{8}$	$1\frac{1}{2}$	$1\frac{5}{8}$	2	3
STD	$1\frac{1}{2}$	835	393	397	400	403	413	439	377	384	390	397	413	439
	$1\frac{3}{4}$	835	400	403	407	410	420	446	377	384	390	397	417	446
OVS	$1\frac{1}{2}$	835	358	361	365	368	378	404	340	346	353	359	378	404
	$1\frac{3}{4}$	835	365	368	371	375	384	410	340	346	353	359	379	410
SSLT	$1\frac{1}{2}$	835	393	397	400	403	413	439	377	384	390	397	413	439
	$1\frac{3}{4}$	835	400	403	407	410	420	446	377	384	390	397	417	446

Support Design Strength per Inch Thickness, kips/in.	Notes: STD = Standard holes N = Threads included OVS = Oversized holes X = Threads excluded SSLT = Short-slotted holes transverse SC = Slip critical to direction of load
1670	*Tabulated values include $\frac{1}{4}$-in. reduction in end distance L_{eh} to account for possible underrun in beam length.

$F_y = 50$ ksi	
$F_u = 65$ ksi	

Table 9-2 (cont.).
All-Bolted Double-Angle Connections

1-in. Bolts	Bolt and Angle Design Strength, kips						
8 Rows	ASTM	Thread	Hole	Angle Thickness, in.			
W44, 40, 36, 33, 30	Desig.	Cond.	Type	$\frac{1}{4}$	$\frac{5}{16}$	$\frac{3}{8}$	$\frac{1}{2}$
	A325	N	—	211	264	317	423
		X	—	211	264	317	423
		SC Class A	STD	211	264	304	304
			OVS	190	238	259	259
			SSLT	211	259	259	259
		SC Class B	STD	211	264	317	423
			OVS	190	238	286	381
			SSLT	211	264	317	392
	A490	N	—	211	264	317	423
		X	—	211	264	317	423
		SC Class A	STD	211	264	317	382
			OVS	190	238	286	325
			SSLT	211	264	317	325
		SC Class B	STD	211	264	317	423
			OVS	190	238	286	381
			SSLT	211	264	317	423

Figure labels: Varies, t, $7@3=21$, $2\frac{1}{2}$, L_{eh}, L_{ev}, $7@3=21$, L_{ev}

Beam Web Design Strength per Inch Thickness, kips/in.

			Coped at Top Flange Only						Coped at Both Flanges					
			L_{ev}, in.						L_{ev}, in.					
Hole Type	L_{eh},* in.	Un-coped	$1\frac{1}{4}$	$1\frac{3}{8}$	$1\frac{1}{2}$	$1\frac{5}{8}$	2	3	$1\frac{1}{4}$	$1\frac{3}{8}$	$1\frac{1}{2}$	$1\frac{5}{8}$	2	3
STD	$1\frac{1}{2}$	936	450	453	457	461	472	501	423	430	438	445	467	501
	$1\frac{3}{4}$	936	459	463	466	470	481	510	423	430	438	445	467	510
OVS	$1\frac{1}{2}$	936	410	414	418	421	432	461	381	388	395	403	425	461
	$1\frac{3}{4}$	936	420	423	427	431	442	471	381	388	395	403	425	471
SSLT	$1\frac{1}{2}$	936	450	453	457	461	472	501	423	430	438	445	467	501
	$1\frac{3}{4}$	936	459	463	466	470	481	510	423	430	438	445	467	510

Support Design Strength per Inch Thickness, kips/in.	Notes: STD = Standard holes OVS = Oversized holes SSLT = Short-slotted holes transverse to direction of load	N = Threads included X = Threads excluded SC = Slip critical
1872	*Tabulated values include $\frac{1}{4}$-in. reduction in end distance L_{eh} to account for possible underrun in beam length.	

F_y = 36 ksi
F_u = 58 ksi

Table 9-2 (cont.).
All-Bolted Double-Angle Connections

1-in. Bolts	Bolt and Angle Design Strength, kips						
7 Rows	ASTM Desig.	Thread Cond.	Hole Type	Angle Thickness, in.			
W44, 40, 36, 33, 30, 27, 24 S24				$\frac{1}{4}$	$\frac{5}{16}$	$\frac{3}{8}$	$\frac{1}{2}$
	A325	N	—	164	205	246	329
		X	—	164	205	246	329
		SC Class A	STD	164	205	246	266
			OVS	148	185	222	226
			SSLT	164	205	226	226
		SC Class B	STD	164	205	246	329
			OVS	148	185	222	296
			SSLT	164	205	246	329
	A490	N	—	164	205	246	329
		X	—	164	205	246	329
		SC Class A	STD	164	205	246	329
			OVS	148	185	222	284
			SSLT	164	205	246	284
		SC Class B	STD	164	205	246	329
			OVS	148	185	222	296
			SSLT	164	205	246	329

Diagram labels: Varies, t, $6@3=18$, $2\frac{1}{2}$, L_{eh}, L_{ev}, $6@3=18$, L_{ev}

Beam Web Design Strength per Inch Thickness, kips/in.

			Coped at Top Flange Only						Coped at Both Flanges					
Hole Type	L_{eh},* in.	Un-coped	L_{ev}, in.						L_{ev}, in.					
			$1\frac{1}{4}$	$1\frac{3}{8}$	$1\frac{1}{2}$	$1\frac{5}{8}$	2	3	$1\frac{1}{4}$	$1\frac{3}{8}$	$1\frac{1}{2}$	$1\frac{5}{8}$	2	3
STD	$1\frac{1}{2}$	731	344	348	351	354	364	390	329	335	342	348	364	390
	$1\frac{3}{4}$	731	351	354	358	361	371	397	329	335	342	348	368	397
OVS	$1\frac{1}{2}$	731	314	317	320	324	334	360	296	302	309	315	334	360
	$1\frac{3}{4}$	731	321	324	327	330	340	366	296	302	309	315	335	366
SSLT	$1\frac{1}{2}$	731	344	348	351	354	364	390	329	335	342	348	364	390
	$1\frac{3}{4}$	731	351	354	358	361	371	397	329	335	342	348	368	397

Support Design Strength per Inch Thickness, kips/in.	Notes: STD = Standard holes OVS = Oversized holes SSLT = Short-slotted holes transverse to direction of load	N = Threads included X = Threads excluded SC = Slip critical
1462	*Tabulated values include $\frac{1}{4}$-in. reduction in end distance L_{eh} to account for possible underrun in beam length.	

		F_y = 50 ksi
		F_u = 65 ksi

Table 9-2 (cont.).
All-Bolted Double-Angle Connections

1-in. Bolts 7 Rows W44, 40, 36, 33, 30, 27, 24 S24	Bolt and Angle Design Strength, kips						
	ASTM Desig.	Thread Cond.	Hole Type	Angle Thickness, in.			
				$1/4$	$5/16$	$3/8$	$1/2$
	A325	N	—	184	230	276	368
		X	—	184	230	276	368
		SC Class A	STD	184	230	266	266
			OVS	166	207	226	226
			SSLT	184	226	226	226
		SC Class B	STD	184	230	276	368
			OVS	166	207	249	331
			SSLT	184	230	276	343
	A490	N	—	184	230	276	368
		X	—	184	230	276	368
		SC Class A	STD	184	230	276	334
			OVS	166	207	249	284
			SSLT	184	230	276	284
		SC Class B	STD	184	230	276	368
			OVS	166	207	249	331
			SSLT	184	230	276	368

Diagram labels: Varies, t, $6@3=18$, $2\frac{1}{2}$, L_{eh}, L_{ev}, $6@3=18$, L_{ev}

Beam Web Design Strength per Inch Thickness, kips/in.

Hole Type	L_{eh},* in.	Un-coped	Coped at Top Flange Only						Coped at Both Flanges					
			L_{ev}, in.						L_{ev}, in.					
			$1\frac{1}{4}$	$1\frac{3}{8}$	$1\frac{1}{2}$	$1\frac{5}{8}$	2	3	$1\frac{1}{4}$	$1\frac{3}{8}$	$1\frac{1}{2}$	$1\frac{5}{8}$	2	3
STD	$1\frac{1}{2}$	819	395	399	402	406	417	446	368	376	383	390	412	446
	$1\frac{3}{4}$	819	404	408	412	415	426	456	368	376	383	390	412	456
OVS	$1\frac{1}{2}$	819	361	365	368	372	383	412	331	339	346	353	375	412
	$1\frac{3}{4}$	819	370	374	378	381	392	421	331	339	346	353	375	421
SSLT	$1\frac{1}{2}$	819	395	399	402	406	417	446	368	376	383	390	412	446
	$1\frac{3}{4}$	819	404	408	412	415	426	456	368	376	383	390	412	456

Support Design Strength per Inch Thickness, kips/in. 1638	Notes: STD = Standard holes OVS = Oversized holes SSLT = Short-slotted holes transverse to direction of load	N = Threads included X = Threads excluded SC = Slip critical
	*Tabulated values include $1/4$-in. reduction in end distance L_{eh} to account for possible underrun in beam length.	

$F_y = 36$ ksi

$F_u = 58$ ksi

Table 9-2 (cont.).
All-Bolted Double-Angle Connections

1-in. Bolts	Bolt and Angle Design Strength, kips						
6 Rows	**ASTM Desig.**	**Thread Cond.**	**Hole Type**	**Angle Thickness, in.**			
W40, 36, 33, 30, 27, 24, 21 S24				$1/4$	$5/16$	$3/8$	$1/2$
	A325	N	—	140	175	210	280
		X	—	140	175	210	280
		SC Class A	STD	140	175	210	228
			OVS	126	157	189	194
			SSLT	140	175	194	194
		SC Class B	STD	140	175	210	280
			OVS	126	157	189	252
			SSLT	140	175	210	280
	A490	N	—	140	175	210	280
		X	—	140	175	210	280
		SC Class A	STD	140	175	210	280
			OVS	126	157	189	243
			SSLT	140	175	210	243
		SC Class B	STD	140	175	210	280
			OVS	126	157	189	252
			SSLT	140	175	210	280

(Illustration: Varies, t, $5@3=15$, $2\frac{1}{2}$, L_{eh}, L_{ev}, $5@3=15$, L_{ev})

Beam Web Design Strength per Inch Thickness, kips/in.

			Coped at Top Flange Only						Coped at Both Flanges					
Hole Type	**L_{eh},* in.**	**Un-coped**	**L_{ev}, in.**						**L_{ev}, in.**					
			$1\frac{1}{4}$	$1\frac{3}{8}$	$1\frac{1}{2}$	$1\frac{5}{8}$	2	3	$1\frac{1}{4}$	$1\frac{3}{8}$	$1\frac{1}{2}$	$1\frac{5}{8}$	2	3
STD	$1\frac{1}{2}$	626	296	299	302	305	315	341	280	286	293	299	315	341
	$1\frac{3}{4}$	626	302	306	309	312	322	348	280	286	293	299	319	348
OVS	$1\frac{1}{2}$	626	270	273	276	280	289	315	252	258	265	271	289	315
	$1\frac{3}{4}$	626	277	280	283	286	296	322	252	258	265	271	291	322
SSLT	$1\frac{1}{2}$	626	296	299	302	305	315	341	280	286	293	299	315	341
	$1\frac{3}{4}$	626	302	306	309	312	322	348	280	286	293	299	319	348

Support Design Strength per Inch Thickness, kips/in.	Notes:
	STD = Standard holes N = Threads included
	OVS = Oversized holes X = Threads excluded
	SSLT = Short-slotted holes transverse SC = Slip critical
	to direction of load
1253	*Tabulated values include $1/4$-in. reduction in end distance L_{eh} to account for possible underrun in beam length.

$F_y = 50$ ksi	
$F_u = 65$ ksi	

Table 9-2 (cont.).
All-Bolted Double-Angle Connections

1-in. Bolts

6 Rows

W40, 36, 33, 30, 27, 24, 21
S24

ASTM Desig.	Thread Cond.	Hole Type	Angle Thickness, in.			
			$1/4$	$5/16$	$3/8$	$1/2$
A325	N	—	157	196	235	314
	X	—	157	196	235	314
	SC Class A	STD	157	196	228	228
		OVS	141	176	194	194
		SSLT	157	194	194	194
	SC Class B	STD	157	196	235	314
		OVS	141	176	211	282
		SSLT	157	196	235	294
A490	N	—	157	196	235	314
	X	—	157	196	235	314
	SC Class A	STD	157	196	235	286
		OVS	141	176	211	243
		SSLT	157	196	235	243
	SC Class B	STD	157	196	235	314
		OVS	141	176	211	282
		SSLT	157	196	235	314

Bolt and Angle Design Strength, kips

Beam Web Design Strength per Inch Thickness, kips/in.

Hole Type	L_{eh},* in.	Un-coped	Coped at Top Flange Only						Coped at Both Flanges					
			L_{ev}, in.						L_{ev}, in.					
			$1\frac{1}{4}$	$1\frac{3}{8}$	$1\frac{1}{2}$	$1\frac{5}{8}$	2	3	$1\frac{1}{4}$	$1\frac{3}{8}$	$1\frac{1}{2}$	$1\frac{5}{8}$	2	3
STD	$1\frac{1}{2}$	702	340	344	348	351	362	392	314	321	328	335	357	392
	$1\frac{3}{4}$	702	350	353	357	361	372	401	314	321	328	335	357	401
OVS	$1\frac{1}{2}$	702	311	315	319	322	333	363	282	289	297	304	326	363
	$1\frac{3}{4}$	702	321	324	328	332	343	372	282	289	297	304	326	372
SSLT	$1\frac{1}{2}$	702	340	344	348	351	362	392	314	321	328	335	357	392
	$1\frac{3}{4}$	702	350	353	357	361	372	401	314	321	328	335	357	401

Support Design Strength per Inch Thickness, kips/in.	Notes: STD = Standard holes N = Threads included OVS = Oversized holes X = Threads excluded SSLT = Short-slotted holes transverse SC = Slip critical to direction of load
1404	*Tabulated values include $1/4$-in. reduction in end distance L_{eh} to account for possible underrun in beam length.

F_y = 36 ksi
F_u = 58 ksi

Table 9-2 (cont.).
All-Bolted Double-Angle Connections

1-in. Bolts			Bolt and Angle Design Strength, kips					

5 rows

W30, 27, 24, 21, 18
S24, 20, 18
MC18

ASTM Desig.	Thread Cond.	Hole Type	Angle Thickness, in.			
			$1/4$	$5/16$	$3/8$	$1/2$
A325	N	—	115	144	173	231
	X	—	115	144	173	231
	SC Class A	STD	115	144	173	190
		OVS	104	130	156	162
		SSLT	115	144	162	162
	SC Class B	STD	115	144	173	231
		OVS	104	130	156	207
		SSLT	115	144	173	231
A490	N	—	115	144	173	231
	X	—	115	144	173	231
	SC Class A	STD	115	144	173	231
		OVS	104	130	156	203
		SSLT	115	144	173	203
	SC Class B	STD	115	144	173	231
		OVS	104	130	156	207
		SSLT	115	144	173	231

Beam Web Design Strength per Inch Thickness, kips/in.

Hole Type	$L_{eh,}$* in.	Un- coped	Coped at Top Flange Only						Coped at Both Flanges					
			L_{ev}, in.						L_{ev}, in.					
			$1 1/4$	$1 3/8$	$1 1/2$	$1 5/8$	2	3	$1 1/4$	$1 3/8$	$1 1/2$	$1 5/8$	2	3
STD	$1 1/2$	522	247	250	253	257	266	293	231	238	244	251	266	293
	$1 3/4$	522	254	257	260	263	273	299	231	238	244	251	270	299
OVS	$1 1/2$	522	226	229	232	236	245	271	207	214	221	227	245	271
	$1 3/4$	522	232	236	239	242	252	278	207	214	221	227	247	278
SSLT	$1 1/2$	522	247	250	253	257	266	293	231	238	244	251	266	293
	$1 3/4$	522	254	257	260	263	273	299	231	238	244	251	270	299

Support Design Strength per Inch Thickness, kips/in.	Notes: STD = Standard holes N = Threads included

Notes:
STD = Standard holes
OVS = Oversized holes
SSLT = Short-slotted holes transverse
    to direction of load

N = Threads included
X = Threads excluded
SC = Slip critical

Support Design Strength per Inch Thickness, kips/in.

1044

*Tabulated values include $1/4$-in. reduction in end distance L_{eh} to account for possible underrun in beam length.

	F_y = 50 ksi
	F_u = 65 ksi

Table 9-2 (cont.).
All-Bolted Double-Angle Connections

1-in. Bolts	Bolt and Angle Design Strength, kips						
5 Rows	ASTM Desig.	Thread Cond.	Hole Type	Angle Thickness, in.			
				$\frac{1}{4}$	$\frac{5}{16}$	$\frac{3}{8}$	$\frac{1}{2}$
W30, 27, 24, 21, 18 S24, 20, 18 MC18	A325	N	—	129	162	194	259
		X	—	129	162	194	259
		SC Class A	STD	129	162	190	190
			OVS	116	145	162	162
			SSLT	129	162	162	162
		SC Class B	STD	129	162	194	259
			OVS	116	145	174	233
			SSLT	129	162	194	245
	A490	N	—	129	162	194	259
		X	—	129	162	194	259
		SC Class A	STD	129	162	194	239
			OVS	116	145	174	203
			SSLT	129	162	194	203
		SC Class B	STD	129	162	194	259
			OVS	116	145	174	233
			SSLT	129	162	194	259

(Figure: diagram of double-angle connection, showing *Varies*, t, $4@3=12$, $2\frac{1}{2}$, L_{eh}, L_{ev}, $4@3=12$)

Beam Web Design Strength per Inch Thickness, kips/in.

Hole Type	L_{eh},* in.	Un-coped	Coped at Top Flange Only						Coped at Both Flanges					
			L_{ev}, in.						L_{ev}, in.					
			$1\frac{1}{4}$	$1\frac{3}{8}$	$1\frac{1}{2}$	$1\frac{5}{8}$	2	3	$1\frac{1}{4}$	$1\frac{3}{8}$	$1\frac{1}{2}$	$1\frac{5}{8}$	2	3
STD	$1\frac{1}{2}$	585	286	289	293	297	308	337	259	266	273	281	303	337
	$1\frac{3}{4}$	585	295	299	302	306	317	346	259	266	273	281	303	346
OVS	$1\frac{1}{2}$	585	262	266	269	273	284	313	233	240	247	254	276	313
	$1\frac{3}{4}$	585	271	275	279	282	293	323	233	240	247	254	276	323
SSLT	$1\frac{1}{2}$	585	286	289	293	297	308	337	259	266	273	281	303	337
	$1\frac{3}{4}$	585	295	299	302	306	317	346	259	266	273	281	303	346

Support Design Strength per Inch Thickness, kips/in.	Notes: STD = Standard holes OVS = Oversized holes SSLT = Short-slotted holes transverse to direction of load	N = Threads included X = Threads excluded SC = Slip critical
1170	*Tabulated values include $\frac{1}{4}$-in. reduction in end distance L_{eh} to account for possible underrun in beam length.	

F_y = 36 ksi
F_u = 58 ksi

Table 9-2 (cont.).
All-Bolted Double-Angle Connections

1-in. Bolts
4 Rows

W24, 21, 18, 16
S24, 20, 18, 15
C15
MC18

			Bolt and Angle Design Strength, kips				
ASTM Desig.	Thread Cond.	Hole Type	Angle Thickness, in.			
			$\frac{1}{4}$	$\frac{5}{16}$	$\frac{3}{8}$	$\frac{1}{2}$
A325	N	—	91.1	114	137	182
	X	—	91.1	114	137	182
	SC Class A	STD	91.1	114	137	152
		OVS	81.7	102	123	129
		SSLT	91.1	114	129	129
	SC Class B	STD	91.1	114	137	182
		OVS	81.7	102	123	163
		SSLT	91.1	114	137	182
A490	N	—	91.1	114	137	182
	X	—	91.1	114	137	182
	SC Class A	STD	91.1	114	137	182
		OVS	81.7	102	123	162
		SSLT	91.1	114	137	162
	SC Class B	STD	91.1	114	137	182
		OVS	81.7	102	123	163
		SSLT	91.1	114	137	182

Beam Web Design Strength per Inch Thickness, kips/in.

Hole Type	L_{eh},* in.	Un-coped	Coped at Top Flange Only						Coped at Both Flanges					
			L_{ev}, in.						L_{ev}, in.					
			$1\frac{1}{4}$	$1\frac{3}{8}$	$1\frac{1}{2}$	$1\frac{5}{8}$	2	3	$1\frac{1}{4}$	$1\frac{3}{8}$	$1\frac{1}{2}$	$1\frac{5}{8}$	2	3
STD	$1\frac{1}{2}$	418	198	201	205	208	218	244	182	189	195	202	218	244
	$1\frac{3}{4}$	418	205	208	211	215	224	250	182	189	195	202	221	250
OVS	$1\frac{1}{2}$	418	182	185	188	191	201	227	163	170	176	183	201	227
	$1\frac{3}{4}$	418	188	192	195	198	208	234	163	170	176	183	203	234
SSLT	$1\frac{1}{2}$	418	198	201	205	208	218	244	182	189	195	202	218	244
	$1\frac{3}{4}$	418	205	208	211	215	224	250	182	189	195	202	221	250

Support Design Strength per Inch Thickness, kips/in.	Notes:
835	STD = Standard holes N = Threads included OVS = Oversized holes X = Threads excluded SSLT = Short-slotted holes transverse SC = Slip critical to direction of load *Tabulated values include $\frac{1}{4}$-in. reduction in end distance L_{eh} to account for possible underrun in beam length.

F_y = 50 ksi	
F_u = 65 ksi	

Table 9-2 (cont.).
All-Bolted Double-Angle Connections

1-in. Bolts	Bolt and Angle Design Strength, kips						
4 Rows	**ASTM Desig.**	**Thread Cond.**	**Hole Type**	**Angle Thickness, in.**			
				$\frac{1}{4}$	$\frac{5}{16}$	$\frac{3}{8}$	$\frac{1}{2}$
W24, 21, 18, 16 S24, 20, 18, 15 C15 MC18	A325	N	—	102	128	153	204
		X	—	102	128	153	204
		SC Class A	STD	102	128	152	152
			OVS	91.6	114	129	129
			SSLT	102	128	129	129
		SC Class B	STD	102	128	153	204
			OVS	91.6	114	137	183
			SSLT	102	128	153	196
	A490	N	—	102	128	153	204
		X	—	102	128	153	204
		SC Class A	STD	102	128	153	191
			OVS	91.6	114	137	162
			SSLT	102	128	153	162
		SC Class B	STD	102	128	153	204
			OVS	91.6	114	137	183
			SSLT	102	128	153	204

Beam Web Design Strength per Inch Thickness, kips/in.

			Coped at Top Flange Only						Coped at Both Flanges					
Hole Type	**L_{eh},* in.**	**Un-coped**	**L_{ev}, in.**						**L_{ev}, in.**					
			$1\frac{1}{4}$	$1\frac{3}{8}$	$1\frac{1}{2}$	$1\frac{5}{8}$	2	3	$1\frac{1}{4}$	$1\frac{3}{8}$	$1\frac{1}{2}$	$1\frac{5}{8}$	2	3
STD	$1\frac{1}{2}$	468	231	235	238	242	253	282	204	211	219	226	248	282
	$1\frac{3}{4}$	468	240	244	248	251	262	292	204	211	219	226	248	292
OVS	$1\frac{1}{2}$	468	213	216	220	224	235	264	183	190	198	205	227	264
	$1\frac{3}{4}$	468	222	226	229	233	244	273	183	190	198	205	227	273
SSLT	$1\frac{1}{2}$	468	231	235	238	242	253	282	204	211	219	226	248	282
	$1\frac{3}{4}$	468	240	244	248	251	262	292	204	211	219	226	248	292

Support Design Strength per Inch Thickness, kips/in.	Notes: STD = Standard holes OVS = Oversized holes SSLT = Short-slotted holes transverse to direction of load	N = Threads included X = Threads excluded SC = Slip critical
936	*Tabulated values include $\frac{1}{4}$-in. reduction in end distance L_{eh} to account for possible underrun in beam length.	

AMERICAN INSTITUTE OF STEEL CONSTRUCTION

F_y = 36 ksi

F_u = 58 ksi

Table 9-2 (cont.).
All-Bolted Double-Angle Connections

1-in. Bolts	Bolt and Angle Design Strength, kips						
3 Rows	ASTM Desig.	Thread Cond.	Hole Type	Angle Thickness, in.			
				$\frac{1}{4}$	$\frac{5}{16}$	$\frac{3}{8}$	$\frac{1}{2}$
W18, 16, 14, 12, 10* **S18, 15, 12** **C15, 12** **MC18, 13, 12**	A325	N	—	66.7	83.4	100	133
		X	—	66.7	83.4	100	133
*Limited to W10×12, 15, 17, 19, 22, 26, 30		SC Class A	STD	66.7	83.4	100	114
			OVS	59.6	74.5	89.5	97.0
			SSLT	66.7	83.4	97.0	97.0
		SC Class B	STD	66.7	83.4	100	133
			OVS	59.6	74.5	89.5	119
			SSLT	66.7	83.4	100	133
	A490	N	—	66.7	83.4	100	133
		X	—	66.7	83.4	100	133
		SC Class A	STD	66.7	83.4	100	133
			OVS	59.6	74.5	89.5	119
			SSLT	66.7	83.4	100	122
		SC Class B	STD	66.7	83.4	100	133
			OVS	59.6	74.5	89.5	119
			SSLT	66.7	83.4	100	133

Figure illustration showing connection geometry with dimensions: Varies, t, 3, 3, 2½, L_{eh}, L_{ev}

Beam Web Design Strength per Inch Thickness, kips/in.

Hole Type	L_{eh},* in.	Un-coped	Coped at Top Flange Only						Coped at Both Flanges					
			L_{ev}, in.						L_{ev}, in.					
			$1\frac{1}{4}$	$1\frac{3}{8}$	$1\frac{1}{2}$	$1\frac{5}{8}$	2	3	$1\frac{1}{4}$	$1\frac{3}{8}$	$1\frac{1}{2}$	$1\frac{5}{8}$	2	3
STD	$1\frac{1}{2}$	313	149	153	156	159	169	195	133	140	146	153	169	195
	$1\frac{3}{4}$	313	156	159	163	166	176	202	133	140	146	153	173	202
OVS	$1\frac{1}{2}$	313	137	141	144	147	157	183	119	126	132	139	157	183
	$1\frac{3}{4}$	313	144	148	151	154	164	190	119	126	132	139	158	190
SSLT	$1\frac{1}{2}$	313	149	153	156	159	169	195	133	140	146	153	169	195
	$1\frac{3}{4}$	313	156	159	163	166	176	202	133	140	146	153	173	202

Support Design Strength per Inch Thickness, kips/in.

626

Notes:
STD = Standard holes
OVS = Oversized holes
SSLT = Short-slotted holes transverse
     to direction of load

N = Threads included
X = Threads excluded
SC = Slip critical

*Tabulated values include ¼-in. reduction in end distance L_{eh} to account for possible underrun in beam length.

$F_y = 50$ ksi	
$F_u = 65$ ksi	

Table 9-2 (cont.).
All-Bolted Double-Angle Connections

1-in. Bolts	Bolt and Angle Design Strength, kips						

1-in. Bolts 3 Rows W18, 16, 14, 12, 10* S18, 15, 12 C15, 12 C18, 13, 12 *Limited to W10×12, 15, 17, 19, 22, 26, 30	ASTM Desig.	Thread Cond.	Hole Type	Angle Thickness, in.			
				$\frac{1}{4}$	$\frac{5}{16}$	$\frac{3}{8}$	$\frac{1}{2}$
	A325	N	—	74.7	93.4	112	149
		X	—	74.7	93.4	112	149
		SC Class A	STD	74.7	93.4	112	114
			OVS	66.8	83.5	97.0	97.0
			SSLT	74.7	93.4	97.0	97.0
		SC Class B	STD	74.7	93.4	112	149
			OVS	66.8	83.5	100	134
			SSLT	74.7	93.4	112	147
	A490	N	—	74.7	93.4	112	149
		X	—	74.7	93.4	112	149
		SC Class A	STD	74.7	93.4	112	143
			OVS	66.8	83.5	100	122
			SSLT	74.7	93.4	112	122
		SC Class B	STD	74.7	93.4	112	149
			OVS	66.8	83.5	100	134
			SSLT	74.7	93.4	112	149

Varies — t — $3\frac{}{}$ $3\frac{}{}$ — $2\frac{1}{2}$

L_{eh}, L_{ev}, 3, 3, L_{ev}

Beam Web Design Strength per Inch Thickness, kips/in.

			Coped at Top Flange Only						Coped at Both Flanges					
			L_{ev}, in.						L_{ev}, in.					
Hole Type	L_{eh},* in.	Un- coped	$1\frac{1}{4}$	$1\frac{3}{8}$	$1\frac{1}{2}$	$1\frac{5}{8}$	2	3	$1\frac{1}{4}$	$1\frac{3}{8}$	$1\frac{1}{2}$	$1\frac{5}{8}$	2	3
STD	$1\frac{1}{2}$	351	176	180	184	187	198	227	149	157	164	171	193	227
	$1\frac{3}{4}$	351	186	189	193	197	208	237	149	157	164	171	193	237
OVS	$1\frac{1}{2}$	351	163	167	170	174	185	214	134	141	148	156	178	214
	$1\frac{3}{4}$	351	173	176	180	183	194	224	134	141	148	156	178	224
SSLT	$1\frac{1}{2}$	351	176	180	184	187	198	227	149	157	164	171	193	227
	$1\frac{3}{4}$	351	186	189	193	197	208	237	149	157	164	171	193	237

Support Design Strength per Inch Thickness, kips/in. 702	Notes: STD = Standard holes OVS = Oversized holes SSLT = Short-slotted holes transverse to direction of load *Tabulated values include $\frac{1}{4}$-in. reduction in end distance L_{eh} to account for possible underrun in beam length.	N = Threads included X = Threads excluded SC = Slip critical

$F_y = 36$ ksi
$F_u = 58$ ksi

Table 9-2 (cont.).
All-Bolted Double-Angle Connections

1-in. Bolts		Bolt and Angle Design Strength, kips						
2 Rows		ASTM Desig.	Thread Cond.	Hole Type	Angle Thickness, in.			
W12, 10, 8					$\frac{1}{4}$	$\frac{5}{16}$	$\frac{3}{8}$	$\frac{1}{2}$
S12, 10, 8 C12, 10, 9, 8		A325	N	—	42.3	52.9	63.4	84.6
MC13, 12, 10, 9, 8			X	—	42.3	52.9	63.4	84.6
			SC Class A	STD	42.3	52.9	63.4	76.1
				OVS	37.6	47.0	56.4	64.7
				SSLT	42.3	52.9	63.4	64.7
			SC Class B	STD	42.3	52.9	63.4	84.6
				OVS	37.6	47.0	56.4	75.2
				SSLT	42.3	52.9	63.4	84.6
		A490	N	—	42.3	52.9	63.4	84.6
			X	—	42.3	52.9	63.4	84.6
			SC Class A	STD	42.3	52.9	63.4	84.6
				OVS	37.6	47.0	56.4	75.2
				SSLT	42.3	52.9	63.4	81.1
			SC Class B	STD	42.3	52.9	63.4	84.6
				OVS	37.6	47.0	56.4	75.2
				SSLT	42.3	52.9	63.4	84.6

Beam Web Design Strength per Inch Thickness, kips/in.

Hole Type	L_{eh},* in.	Un-coped	Coped at Top Flange Only						Coped at Both Flanges					
			L_{ev}, in.						L_{ev}, in.					
			$1\frac{1}{4}$	$1\frac{3}{8}$	$1\frac{1}{2}$	$1\frac{5}{8}$	2	3	$1\frac{1}{4}$	$1\frac{3}{8}$	$1\frac{1}{2}$	$1\frac{5}{8}$	2	3
STD	$1\frac{1}{2}$	209	100	104	107	110	120	146	84.6	91.1	97.6	104	120	146
	$1\frac{3}{4}$	209	107	110	114	117	127	153	84.6	91.1	97.6	104	124	153
OVS	$1\frac{1}{2}$	209	93.4	96.7	99.9	103	113	139	75.2	81.7	88.2	94.7	113	139
	$1\frac{3}{4}$	209	100	103	107	110	120	146	75.2	81.7	88.2	94.7	114	146
SSLT	$1\frac{1}{2}$	209	100	104	107	110	120	146	84.6	91.1	97.6	104	120	146
	$1\frac{3}{4}$	209	107	110	114	117	127	153	84.6	91.1	97.6	104	124	153

Support Design Strength per Inch Thickness, kips/in.	Notes:
	STD = Standard holes N = Threads included
	OVS = Oversized holes X = Threads excluded
	SSLT = Short-slotted holes transverse SC = Slip critical
	to direction of load
418	*Tabulated values include $\frac{1}{4}$-in. reduction in end distance L_{eh} to account for possible underrun in beam length.

F_y = 50 ksi	
F_u = 65 ksi	

Table 9-2 (cont.).
All-Bolted Double-Angle Connections

1-in. Bolts	Bolt and Angle Design Strength, kips						
2 Rows	ASTM Desig.	Thread Cond.	Hole Type	Angle Thickness, in.			
				$1/4$	$5/16$	$3/8$	$1/2$
W12, 10, 8	A325	N	—	47.4	59.2	71.1	94.8
S12, 10, 8		X	—	47.4	59.2	71.1	94.8
C12, 10, 9, 8		SC Class A	STD	47.4	59.2	71.1	76.1
MC13, 12, 10, 9, 8			OVS	42.1	52.7	63.2	64.7
			SSLT	47.4	59.2	64.7	64.7
		SC Class B	STD	47.4	59.2	71.1	94.8
			OVS	42.1	52.7	63.2	84.2
			SSLT	47.4	59.2	71.1	94.8
	A490	N	—	47.4	59.2	71.1	94.8
		X	—	47.4	59.2	71.1	94.8
		SC Class A	STD	47.4	59.2	71.1	94.8
			OVS	42.1	52.7	63.2	81.1
			SSLT	47.4	59.2	71.1	81.1
		SC Class B	STD	47.4	59.2	71.1	94.8
			OVS	42.1	52.7	63.2	84.2
			SSLT	47.4	59.2	71.1	94.8

(Figure showing double-angle connection with dimensions: Varies, t, 3, $2\frac{1}{2}$, L_{eh}, L_{ev}, 3)

Beam Web Design Strength per Inch Thickness, kips/in.

Hole Type	L_{eh},* in.	Un-coped	Coped at Top Flange Only						Coped at Both Flanges					
			L_{ev}, in.						L_{ev}, in.					
			$1\frac{1}{4}$	$1\frac{3}{8}$	$1\frac{1}{2}$	$1\frac{5}{8}$	2	3	$1\frac{1}{4}$	$1\frac{3}{8}$	$1\frac{1}{2}$	$1\frac{5}{8}$	2	3
STD	$1\frac{1}{2}$	234	122	125	129	133	144	173	94.8	102	109	117	139	173
	$1\frac{3}{4}$	234	131	135	138	142	153	182	94.8	102	109	117	139	182
OVS	$1\frac{1}{2}$	234	114	117	121	125	136	165	84.2	91.6	98.9	106	128	165
	$1\frac{3}{4}$	234	123	127	130	134	145	174	84.2	91.6	98.9	106	128	174
SSLT	$1\frac{1}{2}$	234	122	125	129	133	144	173	94.8	102	109	117	139	173
	$1\frac{3}{4}$	234	131	135	138	142	153	182	94.8	102	109	117	139	182

Support Design Strength per Inch Thickness, kips/in.	Notes:
	STD = Standard holes N = Threads included
	OVS = Oversized holes X = Threads excluded
	SSLT = Short-slotted holes transverse SC = Slip critical
	to direction of load
468	*Tabulated values include $1/4$-in. reduction in end distance L_{eh} to account for possible underrun in beam length.

Table 9-3.
Combination Bolted/Welded Double-Angle Connections

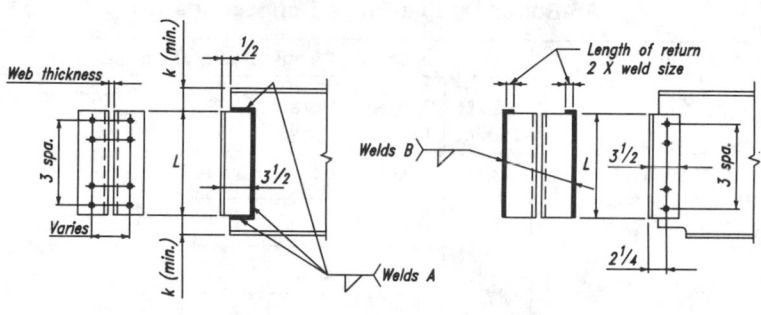

n	L	Weld Size, in.	ϕR_n, kips	Min. Web Thickness, in.		Weld Size, in.	ϕR_n, kips	Min. Support Thickness, in.	
				$F_y = 36$ ksi	$F_y = 50$ ksi			$F_y = 36$ ksi	$F_y = 50$ ksi
		Welds A (70 ksi)				**Welds B (70 ksi)**			
12	35½	5/16	593	0.72	0.52	3/8	550	0.43	0.31
		¼	475	0.57	0.41	5/16	458	0.36	0.26
		3/16	356	0.43	0.31	¼	366	0.29	0.21
11	32½	5/16	548	0.72	0.52	3/8	496	0.43	0.31
		¼	439	0.57	0.41	5/16	414	0.36	0.26
		3/16	329	0.43	0.31	¼	331	0.29	0.21
10	29½	5/16	506	0.72	0.52	3/8	443	0.43	0.31
		¼	405	0.57	0.41	5/16	369	0.36	0.26
		3/16	304	0.43	0.31	¼	295	0.29	0.21
9	26½	5/16	464	0.72	0.52	3/8	389	0.43	0.31
		¼	371	0.57	0.41	5/16	324	0.36	0.26
		3/16	278	0.43	0.31	¼	259	0.29	0.21
8	23½	5/16	423	0.72	0.52	3/8	335	0.43	0.31
		¼	338	0.57	0.41	5/16	279	0.36	0.26
		3/16	254	0.43	0.31	¼	223	0.29	0.21
7	20½	5/16	379	0.72	0.52	3/8	280	0.43	0.31
		¼	304	0.57	0.41	5/16	234	0.36	0.26
		3/16	228	0.43	0.31	¼	187	0.29	0.21
6	17½	5/16	334	0.72	0.52	3/8	226	0.43	0.31
		¼	267	0.57	0.41	5/16	188	0.36	0.26
		3/16	200	0.43	0.31	¼	150	0.29	0.21
5	14½	5/16	287	0.72	0.52	3/8	172	0.43	0.31
		¼	230	0.57	0.41	5/16	143	0.36	0.26
		3/16	172	0.43	0.31	¼	115	0.29	0.21
4	11½	5/16	237	0.72	0.52	3/8	120	0.43	0.31
		¼	190	0.57	0.41	5/16	100	0.36	0.26
		3/16	142	0.43	0.31	¼	79.9	0.29	0.21
3	8½	5/16	184	0.72	0.52	3/8	72.2	0.43	0.31
		¼	147	0.57	0.41	5/16	60.1	0.36	0.26
		3/16	110	0.43	0.31	¼	48.1	0.29	0.21
2	5½	5/16	125	0.72	0.52	3/8	32.8	0.43	0.31
		¼	100	0.57	0.41	5/16	27.3	0.36	0.26
		3/16	75.2	0.43	0.31	¼	21.9	0.29	0.21

Table 9-4.
All-Welded Double-Angle Connections

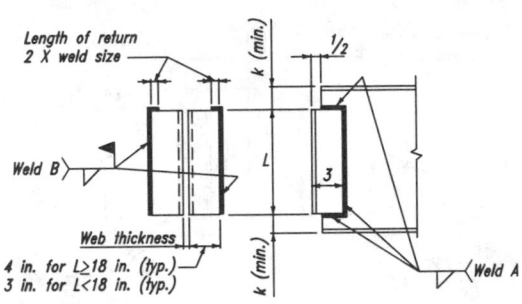

	Welds A (70 ksi)				Welds B (70 ksi)			
	Weld Size, in.	ϕR_n, kips	Min. Web Thickness, in.		Weld Size, in.	ϕR_n, kips	Min. Web Thickness, in.	
L			$F_y = 36$ ksi	$F_y = 50$ ksi			$F_y = 36$ ksi	$F_y = 50$ ksi
36	5/16	587	0.72	0.52	3/8	558	0.43	0.31
	1/4	469	0.57	0.41	5/16	465	0.36	0.26
	3/16	352	0.43	0.31	1/4	372	0.29	0.21
34	5/16	558	0.72	0.52	3/8	523	0.43	. 0.31
	1/4	446	0.57	0.41	5/16	436	0.36	0.26
	3/16	335	0.43	0.31	1/4	349	0.29	¯ 0.21
32	5/16	528	0.72	0.52	3/8	487	0.43	0.31
	1/4	422	0.57	0.41	5/16	406	0.36	0.26
	3/16	317	0.43	0.31	1/4	325	0.29	0.21
30	5/16	498	0.72	0.52	3/8	452	0.43	0.31
	1/4	398	0.57	0.41	5/16	376	0.36	0.26
	3/16	299	0.43	0.31	1/4	301	0.29	0.21
28	5/16	470	0.72	0.52	3/8	416	0.43	0.31
	1/4	376	0.57	0.41	5/16	347	0.36	0.26
	3/16	282	0.43	0.31	1/4	277	0.29	0.21
26	5/16	442	0.72	0.52	3/8	380	0.43	0.31
	1/4	364	0.57	0.41	5/16	317	0.36	0.26
	3/16	265	0.43	0.31	1/4	253	0.29	0.21
24	5/16	413	0.72	0.52	3/8	344	0.43	0.31
	1/4	330	0.57	0.41	5/16	286	0.36	0.26
	3/16	248	0.43	0.31	1/4	229	0.29	0.21
22	5/16	385	0.72	0.52	3/8	307	0.43	0.31
	1/4	308	0.57	0.41	5/16	256	0.36	0.26
	3/16	231	0.43	0.31	1/4	205	0.29	0.21
20	5/16	358	0.72	0.52	3/8	271	0.43	0.31
	1/4	286	0.57	0.41	5/16	226	0.36	0.26
	3/16	215	0.43	0.31	1/4	181	0.29	0.21
18	5/16	329	0.72	0.52	3/8	235	0.43	0.31
	1/4	264	0.57	0.41	5/16	196	0.36	0.26
	3/16	198	0.43	0.31	1/4	157	0.29	0.21

Table 9-4 (cont.).
All-Welded Double-Angle Connections

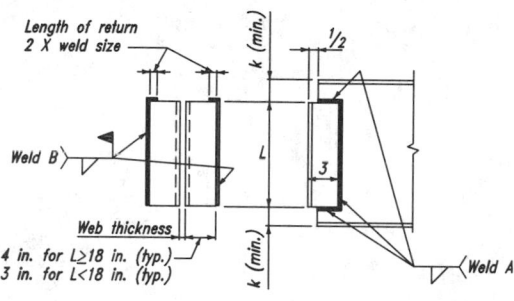

Length of return
2 X weld size

Weld B

Web thickness

4 in. for L≥18 in. (typ.)
3 in. for L<18 in. (typ.)

Weld A

L	Welds A (70 ksi)				Welds B (70 ksi)			
	Weld Size, In.	ϕR_n, kips	Min. Web Thickness, in.		Weld Size, in.	ϕR_n, kips	Min. Web Thickness, in.	
			$F_y = 36$ ksi	$F_y = 50$ ksi			$F_y = 36$ ksi	$F_y = 50$ ksi
16	5/16	299	0.72	0.52	3/8	222	0.43	0.31
	1/4	239	0.57	0.41	5/16	185	0.36	0.26
	3/16	180	0.43	0.31	1/4	148	0.29	0.21
14	5/16	269	0.72	0.52	3/8	185	0.43	0.31
	1/4	215	0.57	0.41	5/16	154	0.36	0.26
	3/16	161	0.43	0.31	1/4	123	0.29	0.21
12	5/16	236	0.72	0.52	3/8	149	0.43	0.31
	1/4	189	0.57	0.41	5/16	124	0.36	0.26
	3/16	142	0.43	0.31	1/4	99.3	0.29	0.21
10	5/16	203	0.72	0.52	3/8	113	0.43	0.31
	1/4	162	0.57	0.41	5/16	94.6	0.36	0.26
	3/16	122	0.43	0.31	1/4	75.7	0.29	0.21
9	5/16	185	0.72	0.52	3/8	96.2	0.43	0.31
	1/4	148	0.57	0.41	5/16	80.2	0.36	0.26
	3/16	111	0.43	0.31	1/4	64.2	0.29	0.21
8	5/16	168	0.72	0.52	3/8	79.5	0.43	0.31
	1/4	134	0.57	0.41	5/16	66.3	0.36	0.26
	3/16	101	0.43	0.31	1/4	53.0	0.29	0.21
7	5/16	149	0.72	0.52	3/8	63.6	0.43	0.31
	1/4	119	0.57	0.41	5/16	53.0	0.36	0.26
	3/16	89.5	0.43	0.31	1/4	42.4	0.29	0.21
6	5/16	130	0.72	0.52	3/8	48.7	0.43	0.31
	1/4	104	0.57	0.41	5/16	40.6	0.36	0.26
	3/16	77.8	0.43	0.31	1/4	32.4	0.29	0.21
5	5/16	111	0.72	0.52	3/8	35.1	0.43	0.31
	1/4	88.4	0.57	0.41	5/16	29.2	0.36	0.26
	3/16	66.3	0.43	0.31	1/4	23.4	0.29	0.21
4	5/16	90.8	0.72	0.52	3/8	23.2	0.43	0.31
	1/4	72.6	0.57	0.41	5/16	19.3	0.36	0.26
	3/16	54.5	0.43	0.31	1/4	15.5	0.29	0.21

Shear End-Plate Connections

A shear end-plate connection is made with a plate length less than the supported beam depth as illustrated in Figure 9-9. The end plate is always shop welded to the beam web with fillet welds on each side, but may be field bolted or welded to the supporting member. Welds connecting the end plate to the beam web should not be returned across the thickness of the beam web at the top or bottom of the end plate because of the danger of creating a notch in the beam web.

When the plate is welded to the support, adequate flexibility must be provided in the connection. Line welds are placed along the vertical edges of the plate with a return at the top per LRFD Specification Section J2.2b. Note that welding across the entire top of the plate must be avoided as it would inhibit the flexibility and, therefore, the necessary end rotation of the connection; the performance of the resulting connection is unpredictable.

The use of steels with F_y greater than 36 ksi for the end plate should be based on an engineering investigation that confirms that adequate flexibility will be provided. The strength and end-rotation characteristics of the shear end-plate connection will closely approximate that of the double-angle connection for similar thicknesses, gage lines, and length of connection.

Design Checks

The design strengths of the bolts and/or welds and connected elements must be determined in accordance with the LRFD Specification; the applicable limit states are discussed in Part 8. Note that the limit state of shear yielding of the beam web must be checked along the length of weld connecting the end plate to the beam web. In all cases, the design strength ϕR_n must equal or exceed the required strength R_u.

Recommended End-Plate Dimensions

End plates should be designed with a plate thickness between ¼-in. and ⅜-in., inclusive. The gage g should be between 3½-in. and 5½-in., inclusive, with top and bottom edge distances of 1¼-in.; lesser values of edge distance should be avoided.

Shop and Field Practices

Shear end-plate connections may be made to the flanges of supporting columns and to the webs of supporting girders. Because of bolting and welding clearances, shear end-plate connections may not be suitable for connections to the webs of W8 columns, unless gages are reduced, and may be impossible for W6 columns.

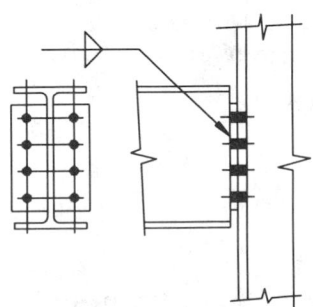

Figure 9-9. Shear end-plate connections.

When framing to a column flange, provision must be made for possible mill variation in the depth of the columns. The beam length could be shortened to provide for mill overrun and shims could be furnished at the appropriate intervals to fill the resulting gaps or to provide for mill underrun; in general shims are not required except for fairly long runs (i.e., six or more bays of framing).

Shear end-plate connections require close control in cutting the beam to the proper length and in squaring the beam ends such that both end plates are parallel. Additionally, any beam camber must not result in out-of-square end plates which make erection and field fit-up difficult.

Bolted/Welded Shear End-Plate Connections

Tables 9-5 are design aids for shear end-plate connections bolted to the supporting member and welded to the supported beam. Design strengths are tabulated for supported and supporting member material with $F_y = 36$ ksi and $F_u = 58$ ksi and with $F_y = 50$ ksi and $F_u = 65$ ksi. End-plate material is assumed to have $F_y = 36$ ksi and $F_u = 58$ ksi. Electrode strength is assumed to be 70 ksi. All values, including slip-critical bolt design strengths, are for comparison with factored loads.

Tabulated bolt and end-plate design strengths consider the limit states of bolt shear, bolt bearing on the end plate, shear yielding of the end plate, shear rupture of the end plate, and block shear rupture of the end plate. Values are included for 2 through 12 rows of $\frac{3}{4}$-in., $\frac{7}{8}$-in., and 1 in. diameter A325 and A490 bolts at 3 in. spacing. End-plate edge distances L_{ev} and L_{eh} are assumed to be $1\frac{1}{4}$-in.

Tabulated weld design strengths consider the limit state of weld shear assuming an effective weld length equal to the plate length minus twice the weld size. The tabulated minimum beam web thickness matches the shear yielding strength of the web material with the strength of the weld metal. As developed previously for double-angle connections,

$$t_{min} = \frac{5.16D}{F_y}$$

where D is the number of sixteenths-of-an-inch in the weld size. When less than the minimum material thickness is present, the tabulated weld design strength must be reduced by the ratio of the thickness provided to the minimum thickness.

Tabulated supporting member design strengths, per inch of flange or web thickness, consider the limit state of bolt bearing.

Example 9-6   Refer to Figure 9-10. Design a shear end-plate connection for the W18×50 beam to W21×62 girder web connection.

$R_u = 60$ kips

W18×50

$t_w = 0.355$ in. $d = 17.99$ in.

$F_y = 50$ ksi, $F_u = 65$ ksi

top flange coped 2 in. deep by $4\frac{1}{2}$-in. long

W21×62

$t_w = 0.400$ in.

$F_y = 50$ ksi, $F_u = 65$ ksi

Use ¾-in. diameter A325-N bolts in standard holes and 70 ksi electrodes. Assume plate material with $F_y = 36$ ksi and $F_u = 58$ ksi.

Solution:

Design bolts and end-plate

From Table 9-5, for ¾-in. diameter A325-N bolts and end-plate material with $F_y = 36$ ksi and $F_u = 58$ ksi, select three rows of bolts and ¼-in. plate thickness

$\phi R_n = 76.7$ kips > 60 kips **o.k.**

Check weld and beam web

From Table 9-5, for a ¼-in. weld size and three rows of bolts (an end-plate length of 8½-in.), a ¼-in. weld size provides $\phi R_n = 89.1$ kips. For beam web material with $F_y = 50$ ksi, the minimum web thickness is 0.41 in. Since $t_w = 0.355$ in. < 0.41 in. the tabular value must be reduced. Thus,

$$\phi R_n = 89.1 \text{ kips} \left(\frac{0.355 \text{ in.}}{0.41 \text{ in.}}\right)$$

$$= 77.1 \text{ kips} > 60 \text{ kips} \quad \textbf{o.k.}$$

Check flexural yielding on the coped section

From Table 8-49, $S_{net} = 23.4$ in.³

$$\phi R_n = \frac{0.9 F_y S_{net}}{e}$$

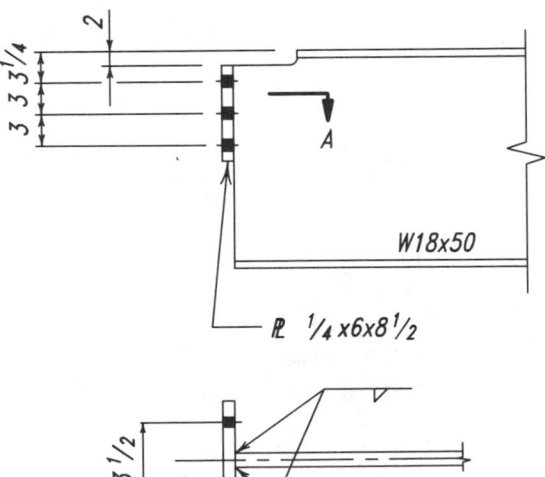

Section at A

Fig. 9-10.

$$= \frac{0.9 \,(50 \text{ ksi}) \,(23.4 \text{ in.}^4)}{(4\frac{1}{2}\text{-in.} + \frac{1}{4}\text{-in.})}$$

$$= 222 \text{ kips} > 60 \text{ kips} \quad \textbf{o.k.}$$

Check local web buckling at the cope

$$\frac{c}{d} = \frac{4\frac{1}{2}\text{-in.}}{17.99 \text{ in.}} = 0.250$$

$$\frac{c}{h_o} = \frac{4\frac{1}{2}\text{-in.}}{(17.99 \text{ in.} - 2 \text{ in.})} = 0.281$$

Since $\dfrac{c}{d} \le 1.0$,

$$f = 2\left(\frac{c}{d}\right)$$

$$= 2(0.250)$$

$$= 0.500$$

Since $\dfrac{c}{h_o} \le 1.0$,

$$k = 2.2 \left(\frac{h_o}{c}\right)^{1.65}$$

$$= 2.2 \left(\frac{1}{0.281}\right)^{1.65}$$

$$= 17.9$$

$$\phi F_{bc} = 23,590 \left(\frac{t_w}{h_o}\right)^2 fk$$

$$= 23,590 \left(\frac{0.355 \text{ in.}}{17.99 \text{ in.} - 2 \text{ in.}}\right)^2 (0.500)\,(17.9)$$

$$= 104 \text{ ksi}$$

$$\phi R_n = \frac{\phi F_{bc} S_{net}}{e}$$

$$= \frac{(104 \text{ ksi}) \,(23.4 \text{ in.}^3)}{(4\frac{1}{2}\text{- in.} + \frac{1}{4}\text{-in.})}$$

$$= 512 \text{ kips} > 60 \text{ kips} \quad \textbf{o.k.}$$

Check supporting girder web:

From Table 9-5, for three rows of bolts and girder material with $F_u = 65$ ksi,

$$\phi R_n = (527 \text{ kips/in.})(0.400 \text{ in.})$$

$$= 211 \text{ kips} > 60 \text{ kips} \quad \textbf{o.k.}$$

The connection, as summarized in Figure 9-10, is adequate.

F_y = 36 ksi				¾-in. Diameter Bolts	
F_y = 58 ksi				12 Rows	
				W44	

Table 9-5.
Bolted/Welded Shear End-Plate Connections

Bolt and End-Plate Design Strength, kips

ASTM Desig.	Thread Cond.	Hole Type	End-Plate Thickness, in.		
			¼	⁵⁄₁₆	⅜
A325	N	—	326	382	382
	X	—	326	408	477
	SC Class A	STD	251	251	251
		OVS	213	213	213
		SSLT	213	213	213
	SC Class B	STD	326	380	380
		OVS	307	323	323
		SSLT	323	323	323
A490	N	—	326	408	477
	X	—	326	408	489
	SC Class A	STD	313	313	313
		OVS	266	266	266
		SSLT	266	266	266
	SC Class B	STD	326	408	475
		OVS	307	383	403
		SSLT	326	403	403

Weld (70 ksi) and Beam Web Design Strength, kips

70 ksi Weld Size, in.	ϕR_n, kips	Minimum Beam Web Thickness, in.		Support Design Strength per Inch Thickness, kips/in.	
		F_y, ksi			
		36	50		
³⁄₁₆	293	0.43	0.31		
¼	390	0.57	0.41	F_u, ksi	
⁵⁄₁₆	485	0.72	0.52	58	65
⅜	580	0.86	0.62	1879	2106

STD = Standard holes
OVS = Oversized holes
SSLT = Short slotted holes transverse to direction of load

N = Threads included
X = Threads excluded
SC = Slip critical

¾-in. Diameter Bolts				$F_y = 36$ ksi	
11 Rows				$F_y = 58$ ksi	
W44, 40					

Table 9-5 (cont.).
Bolted/Welded Shear End-Plate Connections

Bolt and End-Plate Design Strength, kips

ASTM Desig.	Thread Cond.	Hole Type	End-Plate Thickness, in.		
			¼	⁵⁄₁₆	³⁄₈
A325	N	—	299	350	350
	X	—	299	373	437
	SC Class A	STD	230	230	230
		OVS	195	195	195
		SSLT	195	195	195
	SC Class B	STD	299	348	348
		OVS	281	296	296
		SSLT	296	296	296
A490	N	—	299	373	437
	X	—	299	373	448
	SC Class A	STD	287	287	287
		OVS	244	244	244
		SSLT	244	244	244
	SC Class B	STD	299	373	435
		OVS	281	351	370
		SSLT	299	370	370

Weld (70 ksi) and Beam Web Design Strength, kips

70 ksi Weld Size, in.	ϕR_n, kips	Minimum Beam Web Thickness, in.		Support Design Strength per Inch Thickness, kips/in.	
		F_y, ksi			
		36	50		
³⁄₁₆	268	0.43	0.31		
¼	356	0.57	0.41	F_u, ksi	
⁵⁄₁₆	444	0.72	0.52	58	65
³⁄₈	530	0.86	0.62	1723	1931

STD = Standard holes
OVS = Oversized holes
SSLT = Short slotted holes transverse to direction of load

N = Threads included
X = Threads excluded
SC = Slip critical

$F_y = 36$ ksi				¾-in. Diameter Bolts	
$F_y = 58$ ksi				10 Rows	
				W44, 40, 36	

Table 9-5 (cont.).
Bolted/Welded Shear End-Plate Connections

Bolt and End-Plate Design Strength, kips

ASTM Desig.	Thread Cond.	Hole Type	End-Plate Thickness, in.		
			¼	5/16	⅜
A325	N	—	271	318	318
	X	—	271	338	398
	SC Class A	STD	209	209	209
		OVS	178	178	178
		SSLT	178	178	178
	SC Class B	STD	271	316	316
		OVS	254	269	269
		SSLT	269	269	269
A490	N	—	271	338	398
	X	—	271	338	406
	SC Class A	STD	261	261	261
		OVS	222	222	222
		SSLT	222	222	222
	SC Class B	STD	271	338	396
		OVS	254	318	336
		SSLT	271	336	336

Weld (70 ksi) and Beam Web Design Strength, kips

70 ksi Weld Size, in.	ϕR_n, kips	Minimum Beam Web Thickness, in.		Support Design Strength per Inch Thickness, kips/in.	
		F_y, ksi			
		36	50		
3/16	243	0.43	0.31		
¼	323	0.57	0.41	F_u, ksi	
5/16	402	0.72	0.52	58	65
⅜	480	0.86	0.62	1566	1755

STD = Standard holes
OVS = Oversized holes
SSLT = Short slotted holes transverse to direction of load

N = Threads included
X = Threads excluded
SC = Slip critical

¾-in. Diameter Bolts		$F_y = 36$ ksi
9 Rows		$F_y = 58$ ksi
W44, 40, 36, 33		

Table 9-5 (cont.).
Bolted/Welded Shear End-Plate Connections

Bolt and End-Plate Design Strength, kips

ASTM Desig.	Thread Cond.	Hole Type	End-Plate Thickness, in.		
			¼	⁵⁄₁₆	⅜
A325	N	—	243	286	286
	X	—	243	304	358
	SC Class A	STD	188	188	188
		OVS	160	160	160
		SSLT	160	160	160
	SC Class B	STD	243	285	285
		OVS	228	242	242
		SSLT	242	242	242
A490	N	—	243	304	358
	X	—	243	304	365
	SC Class A	STD	235	235	235
		OVS	200	200	200
		SSLT	200	200	200
	SC Class B	STD	243	304	356
		OVS	228	285	303
		SSLT	243	303	303

Weld (70 ksi) and Beam Web Design Strength, kips

70 ksi Weld Size, in.	ϕR_n, kips	Minimum Beam Web Thickness, in.		Support Design Strength per Inch Thickness, kips/in.	
		F_y, ksi			
		36	50		
³⁄₁₆	218	0.43	0.31	F_u, ksi	
¼	290	0.57	0.41	58	65
⁵⁄₁₆	360	0.72	0.52	58	65
⅜	430	0.86	0.62	1409	1580

STD = Standard holes
OVS = Oversized holes
SSLT = Short slotted holes transverse
    to direction of load

N = Threads included
X = Threads excluded
SC = Slip critical

F_y = 36 ksi				¾-in. Diameter Bolts	
F_y = 58 ksi				**8 Rows**	
				W44, 40, 36, 33, 30	

Table 9-5 (cont.).
Bolted/Welded Shear End-Plate Connections

Bolt and End-Plate Design Strength, kips

ASTM Desig.	Thread Cond.	Hole Type	End-Plate Thickness, in.		
			¼	5⁄16	3⁄8
A325	N	—	215	254	254
	X	—	215	269	318
	SC Class A	STD	167	167	167
		OVS	142	142	142
		SSLT	142	142	142
	SC Class B	STD	215	253	253
		OVS	202	215	215
		SSLT	215	215	215
A490	N	—	215	269	318
	X	—	215	269	323
	SC Class A	STD	209	209	209
		OVS	178	178	178
		SSLT	178	178	178
	SC Class B	STD	215	269	316
		OVS	202	253	269
		SSLT	215	269	269

Weld (70 ksi) and Beam Web Design Strength, kips

70 ksi Weld Size, in.	ϕR_n, kips	Minimum Beam Web Thickness, in.		Support Design Strength per Inch Thickness, kips/in.	
		F_y, ksi			
		36	50		
3⁄16	193	0.43	0.31	F_u, ksi	
¼	256	0.57	0.41		
5⁄16	318	0.72	0.52	58	65
3⁄8	380	0.86	0.62	1253	1404

STD = Standard holes
OVS = Oversized holes
SSLT = Short slotted holes transverse
    to direction of load

N = Threads included
X = Threads excluded
SC = Slip critical

¾-in. Diameter Bolts			F_y = 36 ksi
7 Rows			F_y = 58 ksi
W44, 40, 36, 33, 30, 27, 24 S24			

Table 9-5 (cont.).
Bolted/Welded Shear End-Plate Connections

Bolt and End-Plate Design Strength, kips

ASTM Desig.	Thread Cond.	Hole Type	End-Plate Thickness, in.		
			¼	⁵⁄₁₆	⅜
A325	N	—	188	223	223
	X	—	188	234	278
	SC Class A	STD	146	146	146
		OVS	124	124	124
		SSLT	124	124	124
	SC Class B	STD	188	221	221
		OVS	176	188	188
		SSLT	188	188	188
A490	N	—	188	234	278
	X	—	188	234	281
	SC Class A	STD	183	183	183
		OVS	155	155	155
		SSLT	155	155	155
	SC Class B	STD	188	234	277
		OVS	176	220	235
		SSLT	188	234	235

Weld (70 ksi) and Beam Web Design Strength, kips

70 ksi Weld Size, in.	ϕR_n, kips	Minimum Beam Web Thickness, in.		Support Design Strength per Inch Thickness, kips/in.	
		F_y, ksi			
		36	50		
³⁄₁₆	168	0.43	0.31		
¼	223	0.57	0.41	F_u, ksi	
⁵⁄₁₆	277	0.72	0.52	58	65
⅜	330	0.86	0.62	1096	1229

STD = Standard holes
OVS = Oversized holes
SSLT = Short slotted holes transverse to direction of load

N = Threads included
X = Threads excluded
SC = Slip critical

F_y = 36 ksi			¾-in. Diameter Bolts		
F_y = 58 ksi			6 Rows		
			W40, 36, 33, 30, 27, 24, 21 S24		

Table 9-5 (cont.).
Bolted/Welded Shear End-Plate Connections

Bolt and End-Plate Design Strength, kips

ASTM Desig.	Thread Cond.	Hole Type	End-Plate Thickness, in.		
			¼	5⁄16	⅜
A325	N	—	160	191	191
	X	—	160	200	239
	SC Class A	STD	125	125	125
		OVS	107	107	107
		SSLT	107	107	107
	SC Class B	STD	160	190	190
		OVS	150	161	161
		SSLT	160	161	161
A490	N	—	160	200	239
	X	—	160	200	240
	SC Class A	STD	157	157	157
		OVS	133	133	133
		SSLT	133	133	133
	SC Class B	STD	160	200	237
		OVS	150	188	202
		SSLT	160	200	202

Weld (70 ksi) and Beam Web Design Strength, kips

70 ksi Weld Size, in.	ϕR_n, kips	Minimum Beam Web Thickness, in.		Support Design Strength per Inch Thickness, kips/in.	
		F_y, ksi			
		36	50		
3⁄16	143	0.43	0.31		
¼	189	0.57	0.41	F_u, ksi	
5⁄16	235	0.72	0.52	58	65
⅜	280	0.86	0.62	940	1053

STD = Standard holes
OVS = Oversized holes
SSLT = Short slotted holes transverse to direction of load

N = Threads included
X = Threads excluded
SC = Slip critical

¾-in. Diameter Bolts	$F_y = 36$ ksi
5 Rows	$F_y = 58$ ksi
W30, 27, 24, 21, 18 S24, 20, 18 MC18	

Table 9-5 (cont.).
Bolted/Welded Shear End-Plate Connections

Bolt and End-Plate Design Strength, kips

ASTM Desig.	Thread Cond.	Hole Type	End-Plate Thickness, in.		
			¼	⁵⁄₁₆	⅜
A325	N	—	132	159	159
	X	—	132	165	198
	SC Class A	STD	104	104	104
		OVS	88.8	88.8	88.8
		SSLT	88.8	88.8	88.8
	SC Class B	STD	132	158	158
		OVS	124	134	134
		SSLT	132	134	134
A490	N	—	132	165	198
	X	—	132	165	198
	SC Class A	STD	131	131	131
		OVS	111	111	111
		SSLT	111	111	111
	SC Class B	STD	132	165	198
		OVS	124	155	168
		SSLT	132	165	168

Weld (70 ksi) and Beam Web Design Strength, kips

70 ksi Weld Size, in.	ϕR_n, kips	Minimum Beam Web Thickness, in.		Support Design Strength per Inch Thickness, kips/in.	
		F_y, ksi			
		36	50		
³⁄₁₆	118	0.43	0.31	F_u, ksi	
¼	156	0.57	0.41	58	65
⁵⁄₁₆	193	0.72	0.52		
⅜	230	0.86	0.62	783	878

STD = Standard holes   N = Threads included
OVS = Oversized holes   X = Threads excluded
SSLT = Short slotted holes transverse   SC = Slip critical
     to direction of load

F_y = 36 ksi			¾-in. Diameter Bolts
F_y = 58 ksi			4 Rows
			W24, 21, 18, 16 S24, 20, 18, 15 C15 MC18

Table 9-5 (cont.).
Bolted/Welded Shear End-Plate Connections

Bolt and End-Plate Design Strength, kips

ASTM Desig.	Thread Cond.	Hole Type	End-Plate Thickness, in.		
			¼	⁵⁄₁₆	⅜
A325	N	—	104	127	127
	X	—	104	131	157
	SC Class A	STD	83.5	83.5	83.5
		OVS	71.0	71.0	71.0
		SSLT	71.0	71.0	71.0
	SC Class B	STD	104	127	127
		OVS	97.9	108	108
		SSLT	104	108	108
A490	N	—	104	131	157
	X	—	104	131	157
	SC Class A	STD	104	104	104
		OVS	88.8	88.8	88.8
		SSLT	88.8	88.8	88.8
	SC Class B	STD	104	131	157
		OVS	97.9	122	134
		SSLT	104	131	134

Weld (70 ksi) and Beam Web Design Strength, kips

70 ksi Weld Size, in.	ϕR_n, kips	Minimum Beam Web Thickness, in.		Support Design Strength per Inch Thickness, kips/in.	
		F_y, ksi			
		36	50		
³⁄₁₆	92.9	0.43	0.31	F_u, ksi	
¼	122	0.57	0.41		
⁵⁄₁₆	151	0.72	0.52	58	65
⅜	180	0.86	0.62	626	702

STD = Standard holes
OVS = Oversized holes
SSLT = Short slotted holes transverse to direction of load

N = Threads included
X = Threads excluded
SC = Slip critical

¾-in. Diameter Bolts				$F_y = 36$ ksi
3 Rows				$F_y = 58$ ksi

W18, 16, 14, 12, 10*
S18, 15, 12
C15, 12
MC18, 13, 12
*Limited to W10×12, 15, 17, 19, 22, 26, 30.

Table 9-5 (cont.).
Bolted/Welded Shear End-Plate Connections

Bolt and End-Plate Design Strength, kips

ASTM Desig.	Thread Cond.	Hole Type	End-Plate Thickness, in.		
			¼	⁵⁄₁₆	⅜
A325	N	—	76.7	95.4	95.4
	X	—	76.7	95.8	115
	SC Class A	STD	62.7	62.7	62.7
		OVS	53.3	53.3	53.3
		SSLT	53.3	53.3	53.3
	SC Class B	STD	76.7	94.9	94.9
		OVS	71.8	80.7	80.7
		SSLT	76.7	80.7	80.7
A490	N	—	76.7	95.8	115
	X	—	76.7	95.8	115
	SC Class A	STD	76.7	78.3	78.3
		OVS	66.6	66.6	66.6
		SSLT	66.6	66.6	66.6
	SC Class B	STD	76.7	95.8	115
		OVS	71.8	89.7	101
		SSLT	76.7	95.8	101

Weld (70 ksi) and Beam Web Design Strength, kips

70 ksi Weld Size, in.	ϕR_n, kips	Minimum Beam Web Thickness, in.		Support Design Strength per Inch Thickness, kips/in.	
		F_y, ksi			
		36	50		
³⁄₁₆	67.9	0.43	0.31		
¼	89.1	0.57	0.41	F_u, ksi	
⁵⁄₁₆	110	0.72	0.52	58	65
⅜	129	0.86	0.62	470	527

STD = Standard holes
OVS = Oversized holes
SSLT = Short slotted holes transverse to direction of load

N = Threads included
X = Threads excluded
SC = Slip critical

$F_y = 36$ ksi			3/4-in. Diameter Bolts		
$F_y = 58$ ksi			2 Rows		
			W12, 10, 8 S12, 10, 8 C12, 10, 9, 8 MC13, 12, 10, 9, 8		

Table 9-5 (cont.).
Bolted/Welded Shear End-Plate Connections

Bolt and End-Plate Design Strength, kips

ASTM Desig.	Thread Cond.	Hole Type	End-Plate Thickness, in.		
			1/4	5/16	3/8
A325	N	—	48.9	61.2	63.6
	X	—	48.9	61.2	73.4
	SC Class A	STD	41.8	41.8	41.8
		OVS	35.5	35.5	35.5
		SSLT	35.5	35.5	35.5
	SC Class B	STD	48.9	61.2	63.3
		OVS	45.7	53.8	53.8
		SSLT	48.9	53.8	53.8
A490	N	—	48.9	61.2	73.4
	X	—	48.9	61.2	73.4
	SC Class A	STD	48.9	52.2	52.2
		OVS	44.4	44.4	44.4
		SSLT	44.4	44.4	44.4
	SC Class B	STD	48.9	61.2	73.4
		OVS	45.7	57.1	67.2
		SSLT	48.9	61.2	67.2

Weld (70 ksi) and Beam Web Design Strength, kips

70 ksi Weld Size, in.	ϕR_n, kips	Minimum Beam Web Thickness, in.		Support Design Strength per Inch Thickness, kips/in.	
		F_y, ksi			
		36	50		
3/16	42.8	0.43	0.31	F_u, ksi	
1/4	55.7	0.57	0.41		
5/16	67.9	0.72	0.52	58	65
3/8	79.3	0.86	0.62	313	351

STD = Standard holes
OVS = Oversized holes
SSLT = Short slotted holes transverse to direction of load

N = Threads included
X = Threads excluded
SC = Slip critical

⁷⁄₈-in. Diameter Bolts			$F_y = 36$ ksi
12 Rows			$F_y = 58$ ksi
W44			

Table 9-5 (cont.).
Bolted/Welded Shear End-Plate Connections

Bolt and End-Plate Design Strength, kips

ASTM Desig.	Thread Cond.	Hole Type	End-Plate Thickness, in.		
			¼	⁵⁄₁₆	³⁄₈
A325	N	—	307	383	460
	X	—	307	383	460
	SC Class A	STD	307	349	349
		OVS	286	297	297
		SSLT	297	297	297
	SC Class B	STD	307	383	460
		OVS	286	358	429
		SSLT	307	383	450
A490	N	—	307	383	460
	X	—	307	383	460
	SC Class A	STD	307	383	439
		OVS	286	358	373
		SSLT	307	373	373
	SC Class B	STD	307	383	460
		OVS	286	358	429
		SSLT	307	383	460

Weld (70 ksi) and Beam Web Design Strength, kips

70 ksi Weld Size, in.	ϕR_n, kips	Minimum Beam Web Thickness, in.		Support Design Strength per Inch Thickness, kips/in.	
		F_y, ksi			
		36	50		
³⁄₁₆	293	0.43	0.31		
¼	390	0.57	0.41	F_u, ksi	
⁵⁄₁₆	485	0.72	0.52	58	65
³⁄₈	580	0.86	0.62	2192	2457

STD = Standard holes
OVS = Oversized holes
SSLT = Short slotted holes transverse to direction of load

N = Threads included
X = Threads excluded
SC = Slip critical

F_y = 36 ksi				$\frac{7}{8}$-in. Diameter Bolts	
F_y = 58 ksi				11 Rows	
				W44, 40	

Table 9-5 (cont.).
Bolted/Welded Shear End-Plate Connections

Bolt and End-Plate Design Strength, kips

ASTM Desig.	Thread Cond.	Hole Type	End-Plate Thickness, in.		
			$\frac{1}{4}$	$\frac{5}{16}$	$\frac{3}{8}$
A325	N	—	281	351	421
	X	—	281	351	421
	SC Class A	STD	281	320	320
		OVS	262	272	272
		SSLT	272	272	272
	SC Class B	STD	281	351	421
		OVS	262	327	393
		SSLT	281	351	412
A490	N	—	281	351	421
	X	—	281	351	421
	SC Class A	STD	281	351	402
		OVS	262	327	342
		SSLT	281	342	342
	SC Class B	STD	281	351	421
		OVS	262	327	393
		SSLT	281	351	421

Weld (70 ksi) and Beam Web Design Strength, kips

70 ksi Weld Size, in.	ϕR_n, kips	Minimum Beam Web Thickness, in.		Support Design Strength per Inch Thickness, kips/in.	
		F_y, ksi			
		36	50		
$\frac{3}{16}$	268	0.43	0.31		
$\frac{1}{4}$	356	0.57	0.41	F_u, ksi	
$\frac{5}{16}$	444	0.72	0.52	58	65
$\frac{3}{8}$	530	0.86	0.62	2010	2252

STD = Standard holes   N = Threads included
OVS = Oversized holes   X = Threads excluded
SSLT = Short slotted holes transverse   SC = Slip critical
       to direction of load

⅞-in. Diameter Bolts			F_y = 36 ksi
10 Rows			F_y = 58 ksi
W44, 40, 36			

Table 9-5 (cont.).
Bolted/Welded Shear End-Plate Connections

Bolt and End-Plate Design Strength, kips

ASTM Desig.	Thread Cond.	Hole Type	End-Plate Thickness, in.		
			¼	⁵⁄₁₆	⅜
A325	N	—	254	318	382
	X	—	254	318	382
	SC Class A	STD	254	291	291
		OVS	238	247	247
		SSLT	247	247	247
	SC Class B	STD	254	318	382
		OVS	238	297	356
		SSLT	254	318	375
A490	N	—	254	318	382
	X	—	254	318	382
	SC Class A	STD	254	318	365
		OVS	238	297	311
		SSLT	254	311	311
	SC Class B	STD	254	318	382
		OVS	238	297	356
		SSLT	254	318	382

Weld (70 ksi) and Beam Web Design Strength, kips

70 ksi Weld Size, in.	ϕR_n, kips	Minimum Beam Web Thickness, in.		Support Design Strength per Inch Thickness, kips/in.	
		F_y, ksi			
		36	50		
³⁄₁₆	243	0.43	0.31	F_u, ksi	
¼	323	0.57	0.41		
⁵⁄₁₆	402	0.72	0.52	58	65
⅜	480	0.86	0.62	1827	2048

STD = Standard holes
OVS = Oversized holes
SSLT = Short slotted holes transverse to direction of load
N = Threads included
X = Threads excluded
SC = Slip critical

F_y = 36 ksi				⅞-in. Diameter Bolts	
F_y = 58 ksi				9 Rows	
				W44, 40, 36, 33	

Table 9-5 (cont.).
Bolted/Welded Shear End-Plate Connections

Bolt and End-Plate Design Strength, kips

ASTM Desig.	Thread Cond.	Hole Type	End-Plate Thickness, in.		
			¼	⁵⁄₁₆	⅜
A325	N	—	228	285	343
	X	—	228	285	343
	SC Class A	STD	228	262	262
		OVS	213	223	223
		SSLT	223	223	223
	SC Class B	STD	228	285	343
		OVS	213	266	320
		SSLT	228	285	337
A490	N	—	228	285	343
	X	—	228	285	343
	SC Class A	STD	228	285	329
		OVS	213	266	280
		SSLT	228	280	280
	SC Class B	STD	228	285	343
		OVS	213	266	320
		SSLT	228	285	343

Weld (70 ksi) and Beam Web Design Strength, kips

70 ksi Weld Size, in.	ϕR_n, kips	Minimum Beam Web Thickness, in.		Support Design Strength per Inch Thickness, kips/in.	
		F_y, ksi			
		36	50		
³⁄₁₆	218	0.43	0.31		
¼	290	0.57	0.41	F_u, ksi	
⁵⁄₁₆	360	0.72	0.52	58	65
⅜	430	0.86	0.62	1644	1843

STD = Standard holes
OVS = Oversized holes
SSLT = Short slotted holes transverse to direction of load

N = Threads included
X = Threads excluded
SC = Slip critical

⅞-in. Diameter Bolts				$F_y = 36$ ksi	
8 Rows				$F_y = 58$ ksi	
W44, 40, 36, 33, 30					

Table 9-5 (cont.).
Bolted/Welded Shear End-Plate Connections

Bolt and End-Plate Design Strength, kips

ASTM Desig.	Thread Cond.	Hole Type	End-Plate Thickness, in.		
			¼	⁵⁄₁₆	⅜
A325	N	—	202	253	303
	X	—	202	253	303
	SC Class A	STD	202	233	233
		OVS	189	198	198
		SSLT	198	198	198
	SC Class B	STD	202	253	303
		OVS	189	236	283
		SSLT	202	253	300
A490	N	—	202	253	303
	X	—	202	253	303
	SC Class A	STD	202	253	292
		OVS	189	236	249
		SSLT	202	249	249
	SC Class B	STD	202	253	303
		OVS	189	236	283
		SSLT	202	253	303

Weld (70 ksi) and Beam Web Design Strength, kips

70 ksi Weld Size, in.	ϕR_n, kips	Minimum Beam Web Thickness, in.		Support Design Strength per Inch Thickness, kips/in.	
		F_y, ksi			
		36	50		
				F_u, ksi	
				58	65
³⁄₁₆	193	0.43	0.31		
¼	256	0.57	0.41		
⁵⁄₁₆	318	0.72	0.52		
⅜	380	0.86	0.62	1462	1638

STD = Standard holes
OVS = Oversized holes
SSLT = Short slotted holes transverse to direction of load

N = Threads included
X = Threads excluded
SC = Slip critical

F_y = 36 ksi			7/8-in. Diameter Bolts		
F_y = 58 ksi			7 Rows		
			W44, 40, 36, 33, 30, 27, 24 S24		

Table 9-5 (cont.).
Bolted/Welded Shear End-Plate Connections

Bolt and End-Plate Design Strength, kips

ASTM Desig.	Thread Cond.	Hole Type	End-Plate Thickness, in.		
			1/4	5/16	3/8
A325	N	—	176	220	264
	X	—	176	220	264
	SC Class A	STD	176	204	204
		OVS	164	173	173
		SSLT	173	173	173
	SC Class B	STD	176	220	264
		OVS	164	205	246
		SSLT	176	220	262
A490	N	—	176	220	264
	X	—	176	220	264
	SC Class A	STD	176	220	256
		OVS	164	205	217
		SSLT	176	217	217
	SC Class B	STD	176	220	264
		OVS	164	205	246
		SSLT	176	220	264

Weld (70 ksi) and Beam Web Design Strength, kips

70 ksi Weld Size, in.	ϕR_n, kips	Minimum Beam Web Thickness, in.		Support Design Strength per Inch Thickness, kips/in.	
		F_y, ksi			
		36	50		
3/16	168	0.43	0.31		
1/4	223	0.57	0.41	F_u, ksi	
5/16	277	0.72	0.52	58	65
3/8	330	0.86	0.62	1279	1433

STD = Standard holes
OVS = Oversized holes
SSLT = Short slotted holes transverse to direction of load

N = Threads included
X = Threads excluded
SC = Slip critical

$^7/_8$-in. Diameter Bolts				F_y = 36 ksi	
6 Rows				F_y = 58 ksi	
W40, 36, 33, 30, 27, 24, 21 S24					

Table 9-5 (cont.).
Bolted/Welded Shear End-Plate Connections

Bolt and End-Plate Design Strength, kips

ASTM Desig.	Thread Cond.	Hole Type	End-Plate Thickness, in.		
			$^1/_4$	$^5/_{16}$	$^3/_8$
A325	N	—	150	188	225
	X	—	150	188	225
	SC Class A	STD	150	175	175
		OVS	140	148	148
		SSLT	148	148	148
	SC Class B	STD	150	188	225
		OVS	140	175	210
		SSLT	150	188	225
A490	N	—	150	188	225
	X	—	150	188	225
	SC Class A	STD	150	188	219
		OVS	140	175	186
		SSLT	150	186	186
	SC Class B	STD	150	188	225
		OVS	140	175	210
		SSLT	150	188	225

Weld (70 ksi) and Beam Web Design Strength, kips

70 ksi Weld Size, in.	ϕR_n, kips	Minimum Beam Web Thickness, in.		Support Design Strength per Inch Thickness, kips/in.	
		F_y, ksi			
		36	50		
$^3/_{16}$	143	0.43	0.31	F_u, ksi	
$^1/_4$	189	0.57	0.41		
$^5/_{16}$	235	0.72	0.52	58	65
$^3/_8$	280	0.86	0.62	1096	1229

STD = Standard holes
OVS = Oversized holes
SSLT = Short slotted holes transverse to direction of load

N = Threads included
X = Threads excluded
SC = Slip critical

F_y = 36 ksi			**⅞-in. Diameter Bolts**		
F_y = 58 ksi			**5 Rows**		
			W30, 27, 24, 21, 18 **S24, 20, 18** **MC18**		

Table 9-5 (cont.).
Bolted/Welded Shear End-Plate Connections

Bolt and End-Plate Design Strength, kips

ASTM Desig.	Thread Cond.	Hole Type	End-Plate Thickness, in.		
			¼	⁵⁄₁₆	⅜
A325	N	—	124	155	186
	X	—	124	155	186
	SC Class A	STD	124	145	145
		OVS	115	124	124
		SSLT	124	124	124
	SC Class B	STD	124	155	186
		OVS	115	144	173
		SSLT	124	155	186
A490	N	—	124	155	186
	X	—	124	155	186
	SC Class A	STD	124	155	183
		OVS	115	144	155
		SSLT	124	155	155
	SC Class B	STD	124	155	186
		OVS	115	144	173
		SSLT	124	155	186

Weld (70 ksi) and Beam Web Design Strength, kips

70 ksi Weld Size, in.	ϕR_n, kips	Minimum Beam Web Thickness, in.		Support Design Strength per Inch Thickness, kips/in.	
		F_y, ksi			
		36	50		
³⁄₁₆	118	0.43	0.31		
¼	156	0.57	0.41	F_u, ksi	
⁵⁄₁₆	193	0.72	0.52	58	65
⅜	230	0.86	0.62	914	1024

STD = Standard holes　　　　　　　　　　　N = Threads included
OVS = Oversized holes　　　　　　　　　　　X = Threads excluded
SSLT = Short slotted holes transverse　　　SC = Slip critical
　　　　to direction of load

$7/8$-in. Diameter Bolts				$F_y = 36$ ksi	
4 Rows				$F_y = 58$ ksi	
W24, 21, 18, 16 S24, 20, 18, 15 C15 MC18					

Table 9-5 (cont.).
Bolted/Welded Shear End-Plate Connections

Bolt and End-Plate Design Strength, kips

ASTM Desig.	Thread Cond.	Hole Type	End-Plate Thickness, in.		
			$1/4$	$5/16$	$3/8$
A325	N	—	97.9	122	147
	X	—	97.9	122	147
	SC Class A	STD	97.9	116	116
		OVS	91.1	98.9	98.9
		SSLT	97.9	98.9	98.9
	SC Class B	STD	97.9	122	147
		OVS	91.1	114	137
		SSLT	97.9	122	147
A490	N	—	97.9	122	147
	X	—	97.9	122	147
	SC Class A	STD	97.9	122	146
		OVS	91.1	114	124
		SSLT	97.9	122	124
	SC Class B	STD	97.9	122	147
		OVS	91.1	114	137
		SSLT	97.9	122	147

Weld (70 ksi) and Beam Web Design Strength, kips

70 ksi Weld Size, in.	ϕR_n, kips	Minimum Beam Web Thickness, in.		Support Design Strength per Inch Thickness, kips/in.	
		F_y, ksi			
		36	50		
$3/16$	92.9	0.43	0.31	F_u, ksi	
$1/4$	122	0.57	0.41		
$5/16$	151	0.72	0.52	58	65
$3/8$	180	0.86	0.62	731	819

STD = Standard holes   N = Threads included
OVS = Oversized holes   X = Threads excluded
SSLT = Short slotted holes transverse   SC = Slip critical
    to direction of load

$F_y = 36$ ksi			**⅞-in. Diameter Bolts**		
$F_y = 58$ ksi			**3 Rows**		
			W18, 16, 14, 12, 10* S18, 15, 12 C15, 12 MC18, 13, 12 *Limited to W10×12, 15, 17, 19, 22, 26, 30		

Table 9-5 (cont.).
Bolted/Welded Shear End-Plate Connections

Bolt and End-Plate Design Strength, kips

ASTM Desig.	Thread Cond.	Hole Type	End-Plate Thickness, in.		
			¼	⁵⁄₁₆	⅜
A325	N	—	71.8	89.7	108
	X	—	71.8	89.7	108
	SC Class A	STD	71.8	87.3	87.3
		OVS	66.7	74.2	74.2
		SSLT	71.8	74.2	74.2
	SC Class B	STD	71.8	89.7	108
		OVS	66.7	83.4	100
		SSLT	71.8	89.7	108
A490	N	—	71.8	89.7	108
	X	—	71.8	89.7	108
	SC Class A	STD	71.8	89.7	108
		OVS	66.7	83.4	93.2
		SSLT	71.8	89.7	93.2
	SC Class B	STD	71.8	89.7	108
		OVS	66.7	83.4	100
		SSLT	71.8	89.7	108

Weld (70 ksi) and Beam Web Design Strength, kips

70 ksi Weld Size, in.	ϕR_n, kips	Minimum Beam Web Thickness, in.		Support Design Strength per Inch Thickness, kips/in.	
		F_y, ksi			
		36	50		
³⁄₁₆	67.9	0.43	0.31		
¼	89.1	0.57	0.41	F_u, ksi	
⁵⁄₁₆	110	0.72	0.52	58	65
⅜	129	0.86	0.62	548	614

STD = Standard holes	N = Threads included
OVS = Oversized holes	X = Threads excluded
SSLT = Short slotted holes transverse to direction of load	SC = Slip critical

⁷⁄₈-in. Diameter Bolts		$F_y = 36$ ksi
2 Rows		$F_y = 58$ ksi
W12, 10, 8 S12, 10, 8 C12, 10, 9, 8 MC13, 12, 10, 9, 8		

Table 9-5 (cont.).
Bolted/Welded Shear End-Plate Connections

Bolt and End-Plate Design Strength, kips

ASTM Desig.	Thread Cond.	Hole Type	End-Plate Thickness, in.		
			¹⁄₄	⁵⁄₁₆	³⁄₈
A325	N	—	45.7	57.1	68.5
	X	—	45.7	57.1	68.5
	SC Class A	STD	45.7	57.1	58.2
		OVS	42.3	49.5	49.5
		SSLT	45.7	49.5	49.5
	SC Class B	STD	45.7	57.1	68.5
		OVS	42.3	52.9	63.4
		SSLT	45.7	57.1	68.5
A490	N	—	45.7	57.1	68.5
	X	—	45.7	57.1	68.5
	SC Class A	STD	45.7	57.1	68.5
		OVS	42.3	52.9	62.1
		SSLT	45.7	57.1	62.1
	SC Class B	STD	45.7	57.1	68.5
		OVS	42.3	52.9	63.4
		SSLT	45.7	57.1	68.5

Weld (70 ksi) and Beam Web Design Strength, kips

70 ksi Weld Size, in.	ϕR_n, kips	Minimum Beam Web Thickness, in.		Support Design Strength per Inch Thickness, kips/in.	
		F_y, ksi			
		36	50		
³⁄₁₆	42.8	0.43	0.31	F_u, ksi	
¹⁄₄	55.7	0.57	0.41		
⁵⁄₁₆	67.9	0.72	0.52	58	65
³⁄₈	79.3	0.86	0.62	365	410

STD = Standard holes   N = Threads included
OVS = Oversized holes   X = Threads excluded
SSLT = Short slotted holes transverse SC = Slip critical
   to direction of load

F_y = 36 ksi			1-in. Diameter Bolts		
F_y = 58 ksi			12 Rows		
			W44		

Table 9-5 (cont.).
Bolted/Welded Shear End-Plate Connections

Bolt and End-Plate Design Strength, kips

ASTM Desig.	Thread Cond.	Hole Type	End-Plate Thickness, in.		
			$\frac{1}{4}$	$\frac{5}{16}$	$\frac{3}{8}$
A325	N	—	286	358	429
	X	—	286	358	429
	SC Class A	STD	286	358	429
		OVS	258	323	387
		SSLT	286	358	388
	SC Class B	STD	286	358	429
		OVS	258	323	387
		SSLT	286	358	429
A490	N	—	286	358	429
	X	—	286	358	429
	SC Class A	STD	286	358	429
		OVS	258	323	387
		SSLT	286	358	429
	SC Class B	STD	286	358	429
		OVS	258	323	387
		SSLT	286	358	429

Weld (70 ksi) and Beam Web Design Strength, kips

70 ksi Weld Size, in.	ϕR_n, kips	Minimum Beam Web Thickness, in. F_y, ksi		Support Design Strength per Inch Thickness, kips/in.	
		36	50		
$\frac{3}{16}$	293	0.43	0.31	F_u, ksi	
$\frac{1}{4}$	390	0.57	0.41	58	65
$\frac{5}{16}$	485	0.72	0.52		
$\frac{3}{8}$	580	0.86	0.62	2506	2808

STD = Standard holes
OVS = Oversized holes
SSLT = Short slotted holes transverse to direction of load

N = Threads included
X = Threads excluded
SC = Slip critical

1-in. Diameter Bolts		$F_y = 36$ ksi
11 Rows		$F_y = 58$ ksi
W44, 40		

Table 9-5 (cont.).
Bolted/Welded Shear End-Plate Connections

Bolt and End-Plate Design Strength, kips

ASTM Desig.	Thread Cond.	Hole Type	End-Plate Thickness, in.		
			$\frac{1}{4}$	$\frac{5}{16}$	$\frac{3}{8}$
A325	N	—	262	327	393
	X	—	262	327	393
	SC Class A	STD	262	327	393
		OVS	236	295	354
		SSLT	262	327	356
	SC Class B	STD	262	327	393
		OVS	236	295	354
		SSLT	262	327	393
A490	N	—	262	327	393
	X	—	262	327	393
	SC Class A	STD	262	327	393
		OVS	236	295	354
		SSLT	262	327	393
	SC Class B	STD	262	327	393
		OVS	236	295	354
		SSLT	262	327	393

Weld (70 ksi) and Beam Web Design Strength, kips

70 ksi Weld Size, in.	ϕR_n, kips	Minimum Beam Web Thickness, in.		Support Design Strength per Inch Thickness, kips/in.	
		F_y, ksi			
		36	50		
$\frac{3}{16}$	268	0.43	0.31	F_u, ksi	
$\frac{1}{4}$	356	0.57	0.41	58	65
$\frac{5}{16}$	444	0.72	0.52	2297	2574
$\frac{3}{8}$	530	0.86	0.62		

STD = Standard holes
OVS = Oversized holes
SSLT = Short slotted holes transverse to direction of load

N = Threads included
X = Threads excluded
SC = Slip critical

F_y = 36 ksi			1-in. Diameter Bolts		
F_y = 58 ksi			10 Rows		
			W44, 40, 36		

Table 9-5 (cont.).
Bolted/Welded Shear End-Plate Connections

Bolt and End-Plate Design Strength, kips

ASTM Desig.	Thread Cond.	Hole Type	End-Plate Thickness, in.		
			$\frac{1}{4}$	$\frac{5}{16}$	$\frac{3}{8}$
A325	N	—	238	297	356
	X	—	238	297	356
	SC Class A	STD	238	297	356
		OVS	214	268	321
		SSLT	238	297	323
	SC Class B	STD	238	297	356
		OVS	214	268	321
		SSLT	238	297	356
A490	N	—	238	297	356
	X	—	238	297	356
	SC Class A	STD	238	297	356
		OVS	214	268	321
		SSLT	238	297	356
	SC Class B	STD	238	297	356
		OVS	214	268	321
		SSLT	238	297	356

Weld (70 ksi) and Beam Web Design Strength, kips

70 ksi Weld Size, in.	ϕR_n, kips	Minimum Beam Web Thickness, in.		Support Design Strength per Inch Thickness, kips/in.	
		F_y, ksi			
		36	50		
$\frac{3}{16}$	243	0.43	0.31		
$\frac{1}{4}$	323	0.57	0.41	F_u, ksi	
$\frac{5}{16}$	402	0.72	0.52	58	65
$\frac{3}{8}$	480	0.86	0.62	2088	2340

STD = Standard holes	N = Threads included
OVS = Oversized holes	X = Threads excluded
SSLT = Short slotted holes transverse to direction of load	SC = Slip critical

1-in. Diameter Bolts			$F_y = 36$ ksi
9 Rows			$F_y = 58$ ksi
W44, 40, 36, 33			

Table 9-5 (cont.).
Bolted/Welded Shear End-Plate Connections

Bolt and End-Plate Design Strength, kips

ASTM Desig.	Thread Cond.	Hole Type	End-Plate Thickness, in.		
			$\frac{1}{4}$	$\frac{5}{16}$	$\frac{3}{8}$
A325	N	—	213	266	320
	X	—	213	266	320
	SC Class A	STD	213	266	320
		OVS	192	240	288
		SSLT	213	266	291
	SC Class B	STD	213	266	320
		OVS	192	240	288
		SSLT	213	266	320
A490	N	—	213	266	320
	X	—	213	266	320
	SC Class A	STD	213	266	320
		OVS	192	240	288
		SSLT	213	266	320
	SC Class B	STD	213	266	320
		OVS	192	240	288
		SSLT	213	266	320

Weld (70 ksi) and Beam Web Design Strength, kips

70 ksi Weld Size, in.	ϕR_n, kips	Minimum Beam Web Thickness, in.		Support Design Strength per Inch Thickness, kips/in.	
		F_y, ksi			
		36	50		
				F_u, ksi	
				58	65
$\frac{3}{16}$	218	0.43	0.31		
$\frac{1}{4}$	290	0.57	0.41		
$\frac{5}{16}$	360	0.72	0.52	58	65
$\frac{3}{8}$	430	0.86	0.62	1879	2106

STD = Standard holes   N   = Threads included
OVS = Oversized holes   X   = Threads excluded
SSLT = Short slotted holes transverse   SC = Slip critical
          to direction of load

F_y = 36 ksi			1-in. Diameter Bolts
F_y = 58 ksi			8 Rows
			W44, 40, 36, 33, 30

Table 9-5 (cont.).
Bolted/Welded Shear End-Plate Connections

Bolt and End-Plate Design Strength, kips

ASTM Desig.	Thread Cond.	Hole Type	End-Plate Thickness, in.		
			$1/4$	$5/16$	$3/8$
A325	N	—	189	236	283
	X	—	189	236	283
	SC Class A	STD	189	236	283
		OVS	170	212	255
		SSLT	189	236	259
	SC Class B	STD	189	236	283
		OVS	170	212	255
		SSLT	189	236	283
A490	N	—	189	236	283
	X	—	189	236	283
	SC Class A	STD	189	236	283
		OVS	170	212	255
		SSLT	189	236	283
	SC Class B	STD	189	236	283
		OVS	170	212	255
		SSLT	189	236	283

Weld (70 ksi) and Beam Web Design Strength, kips

70 ksi Weld Size, in.	ϕR_n, kips	Minimum Beam Web Thickness, in.		Support Design Strength per Inch Thickness, kips/in.	
		F_y, ksi			
		36	50		
$3/16$	193	0.43	0.31		
$1/4$	256	0.57	0.41	F_u, ksi	
$5/16$	318	0.72	0.52	58	65
$3/8$	380	0.86	0.62	1670	1872

STD = Standard holes
OVS = Oversized holes
SSLT = Short slotted holes transverse to direction of load

N = Threads included
X = Threads excluded
SC = Slip critical

1-in. Diameter Bolts			$F_y = 36$ ksi
7 Rows			$F_y = 58$ ksi
W44, 40, 36, 33, 30, 27, 24 S24			

Table 9-5 (cont.).
Bolted/Welded Shear End-Plate Connections

Bolt and End-Plate Design Strength, kips

ASTM Desig.	Thread Cond.	Hole Type	End-Plate Thickness, in.		
			$1/4$	$5/16$	$3/8$
A325	N	—	164	205	246
	X	—	164	205	246
	SC Class A	STD	164	205	246
		OVS	148	185	222
		SSLT	164	205	226
	SC Class B	STD	164	205	246
		OVS	148	185	222
		SSLT	164	205	246
A490	N	—	164	205	246
	X	—	164	205	246
	SC Class A	STD	164	205	246
		OVS	148	185	222
		SSLT	164	205	246
	SC Class B	STD	164	205	246
		OVS	148	185	222
		SSLT	164	205	246

Weld (70 ksi) and Beam Web Design Strength, kips

70 ksi Weld Size, in.	ϕR_n, kips	Minimum Beam Web Thickness, in.		Support Design Strength per Inch Thickness, kips/in.	
		F_y, ksi			
		36	50		
$3/16$	168	0.43	0.31	F_u, ksi	
$1/4$	223	0.57	0.41		
$5/16$	277	0.72	0.52	58	65
$3/8$	330	0.86	0.62	1462	1638

STD = Standard holes
OVS = Oversized holes
SSLT = Short slotted holes transverse to direction of load

N = Threads included
X = Threads excluded
SC = Slip critical

F_y = 36 ksi		1-in. Diameter Bolts		
F_y = 58 ksi		6 Rows		
		W40, 36, 30, 27, 24, 21 S24		

Table 9-5 (cont.).
Bolted/Welded Shear End-Plate Connections

Bolt and End-Plate Design Strength, kips

ASTM Desig.	Thread Cond.	Hole Type	End-Plate Thickness, in.		
			$\frac{1}{4}$	$\frac{5}{16}$	$\frac{3}{8}$
A325	N	—	140	175	210
	X	—	140	175	210
	SC Class A	STD	140	175	210
		OVS	126	157	189
		SSLT	140	175	194
	SC Class B	STD	140	175	210
		OVS	126	157	189
		SSLT	140	175	210
A490	N	—	140	175	210
	X	—	140	175	210
	SC Class A	STD	140	175	210
		OVS	126	157	189
		SSLT	140	175	210
	SC Class B	STD	140	175	210
		OVS	126	157	189
		SSLT	140	175	210

Weld (70 ksi) and Beam Web Design Strength, kips

70 ksi Weld Size, in.	ϕR_n, kips	Minimum Beam Web Thickness, in.		Support Design Strength per Inch Thickness, kips/in.	
		F_y, ksi			
		36	50		
$\frac{3}{16}$	143	0.43	0.31	F_u, ksi	
$\frac{1}{4}$	189	0.57	0.41		
$\frac{5}{16}$	235	0.72	0.52	58	65
$\frac{3}{8}$	280	0.86	0.62	1253	1404

STD = Standard holes	N = Threads included
OVS = Oversized holes	X = Threads excluded
SSLT = Short slotted holes transverse to direction of load	SC = Slip critical

1-in. Diameter Bolts		$F_y = 36$ ksi
5 Rows		$F_y = 58$ ksi
W30, 27, 24, 21, 18 S24, 20, 18 MC18		

Table 9-5 (cont.).
Bolted/Welded Shear End-Plate Connections

Bolt and End-Plate Design Strength, kips

ASTM Desig.	Thread Cond.	Hole Type	End-Plate Thickness, in.		
			$\frac{1}{4}$	$\frac{5}{16}$	$\frac{3}{8}$
A325	N	—	115	144	173
	X	—	115	144	173
	SC Class A	STD	115	144	173
		OVS	104	130	156
		SSLT	115	144	162
	SC Class B	STD	115	144	173
		OVS	104	130	156
		SSLT	115	144	173
A490	N	—	115	144	173
	X	—	115	144	173
	SC Class A	STD	115	144	173
		OVS	104	130	156
		SSLT	115	144	173
	SC Class B	STD	115	144	173
		OVS	104	130	156
		SSLT	115	144	173

Weld (70 ksi) and Beam Web Design Strength, kips

70 ksi Weld Size, in.	ϕR_n, kips	Minimum Beam Web Thickness, in.		Support Design Strength per Inch Thickness, kips/in.	
		F_y, ksi			
		36	50		
$\frac{3}{16}$	118	0.43	0.31	F_u, ksi	
$\frac{1}{4}$	156	0.57	0.41	58	65
$\frac{5}{16}$	193	0.72	0.52	58	65
$\frac{3}{8}$	230	0.86	0.62	1044	1170

STD = Standard holes
OVS = Oversized holes
SSLT = Short slotted holes transverse to direction of load

N = Threads included
X = Threads excluded
SC = Slip critical

$F_y = 36$ ksi				1-in. Diameter Bolts	
$F_y = 58$ ksi				4 Rows	
				W24, 21, 18, 16 S24, 20, 18, 15 C15 MC18	

Table 9-5 (cont.).
Bolted/Welded Shear End-Plate Connections

Bolt and End-Plate Design Strength, kips

ASTM Desig.	Thread Cond.	Hole Type	End-Plate Thickness, in.		
			$\frac{1}{4}$	$\frac{5}{16}$	$\frac{3}{8}$
A325	N	—	91.1	114	137
	X	—	91.1	114	137
	SC Class A	STD	91.1	114	137
		OVS	81.7	102	123
		SSLT	91.1	114	129
	SC Class B	STD	91.1	114	137
		OVS	81.7	102	123
		SSLT	91.1	114	137
A490	N	—	91.1	114	137
	X	—	91.1	114	137
	SC Class A	STD	91.1	114	137
		OVS	81.7	102	123
		SSLT	91.1	114	137
	SC Class B	STD	91.1	114	137
		OVS	81.7	102	123
		SSLT	91.1	114	137

Weld (70 ksi) and Beam Web Design Strength, kips

70 ksi Weld Size, in.	ϕR_n, kips	Minimum Beam Web Thickness, in.		Support Design Strength per Inch Thickness, kips/in.	
		F_y, ksi			
		36	50		
$\frac{3}{16}$	92.9	0.43	0.31		
$\frac{1}{4}$	122	0.57	0.41	F_u, ksi	
$\frac{5}{16}$	151	0.72	0.52	58	65
$\frac{3}{8}$	180	0.86	0.62	835	936

STD = Standard holes
OVS = Oversized holes
SSLT = Short slotted holes transverse
    to direction of load

N = Threads included
X = Threads excluded
SC = Slip critical

1-in. Diameter Bolts	F_y = 36 ksi
3 Rows	F_y = 58 ksi

W18, 16, 14, 12, 10*
S18, 15, 12
C15, 12
MC18, 13, 12
*Limited to W10×12, 15,
17, 19, 22, 26, 30

Table 9-5 (cont.).
Bolted/Welded Shear End-Plate Connections

Bolt and End-Plate Design Strength, kips

ASTM Desig.	Thread Cond.	Hole Type	End-Plate Thickness, in.		
			$^1/_4$	$^5/_{16}$	$^3/_8$
A325	N	—	66.7	83.4	100
	X	—	66.7	83.4	100
	SC Class A	STD	66.7	83.4	100
		OVS	59.6	74.6	89.5
		SSLT	66.7	83.4	97.0
	SC Class B	STD	66.7	83.4	100
		OVS	59.6	74.6	89.5
		SSLT	66.7	83.4	100
A490	N	—	66.7	83.4	100
	X	—	66.7	83.4	100
	SC Class A	STD	66.7	83.4	100
		OVS	59.6	74.6	89.5
		SSLT	66.7	83.4	100
	SC Class B	STD	66.7	83.4	100
		OVS	59.6	74.6	89.5
		SSLT	66.7	83.4	100

Weld (70 ksi) and Beam Web Design Strength, kips

70 ksi Weld Size, in.	ϕR_n, kips	Minimum Beam Web Thickness, in.		Support Design Strength per Inch Thickness, kips/in.	
		F_y, ksi			
		36	50	F_u, ksi	
$^3/_{16}$	67.9	0.43	0.31		
$^1/_4$	89.1	0.57	0.41		
$^5/_{16}$	110	0.72	0.52	58	65
$^3/_8$	129	0.86	0.62	626	702

STD = Standard holes
OVS = Oversized holes
SSLT = Short slotted holes transverse
to direction of load

N = Threads included
X = Threads excluded
SC = Slip critical

F_y = 36 ksi			1-in. Diameter Bolts		
F_y = 58 ksi			2 Rows		
			W12, 10, 8 S12, 10, 8 C12, 10, 9, 8 MC13, 12, 10, 9, 8		

Table 9-5 (cont.).
Bolted/Welded Shear End-Plate Connections

Bolt and End-Plate Design Strength, kips

ASTM Desig.	Thread Cond.	Hole Type	End-Plate Thickness, in.		
			$\frac{1}{4}$	$\frac{5}{16}$	$\frac{3}{8}$
A325	N	—	42.3	52.9	63.4
	X	—	42.3	52.9	63.4
	SC Class A	STD	42.3	52.9	63.4
		OVS	37.6	47.0	56.4
		SSLT	42.3	52.9	63.4
	SC Class B	STD	42.3	52.9	63.4
		OVS	37.6	47.0	56.4
		SSLT	42.3	52.9	63.4
A490	N	—	42.3	52.9	63.4
	X	—	42.3	52.9	63.4
	SC Class A	STD	42.3	52.9	63.4
		OVS	37.6	47.0	56.4
		SSLT	42.3	52.9	63.4
	SC Class B	STD	42.3	52.9	63.4
		OVS	37.6	47.0	56.4
		SSLT	42.3	52.9	63.4

Weld (70 ksi) and Beam Web Design Strength, kips

70 ksi Weld Size, in.	ϕR_n, kips	Minimum Beam Web Thickness, in.		Support Design Strength per Inch Thickness, kips/in.	
		F_y, ksi			
		36	50		
$\frac{3}{16}$	42.8	0.43	0.31	F_u, ksi	
$\frac{1}{4}$	55.7	0.57	0.41		
$\frac{5}{16}$	67.9	0.72	0.52	58	65
$\frac{3}{8}$	79.3	0.86	0.62	418	468

STD = Standard holes
OVS = Oversized holes
SSLT = Short slotted holes transverse to direction of load

N = Threads included
X = Threads excluded
SC = Slip critical

Unstiffened Seated Connections

An unstiffened seated connection is made with a seat angle and a top angle, as illustrated in Figure 9-11. These angles may be bolted or welded to the supported beam as well as to the supporting member. While the seat angle is assumed to carry the entire end reaction of the supported beam, the top angle must be placed as shown or in the optional side location for satisfactory performance and stability (Dalley and Roeder, 1989).

When the top angle is welded to the support and/or the supported beam, adequate flexibility must be provided in the connection. As illustrated in Figure 9-11b, line welds are placed along the toe of each angle leg. Note that welding along the sides of the vertical angle leg must be avoided as it would inhibit the flexibility and, therefore, the necessary end rotation of the connection; the performance of such a connection is unpredictable.

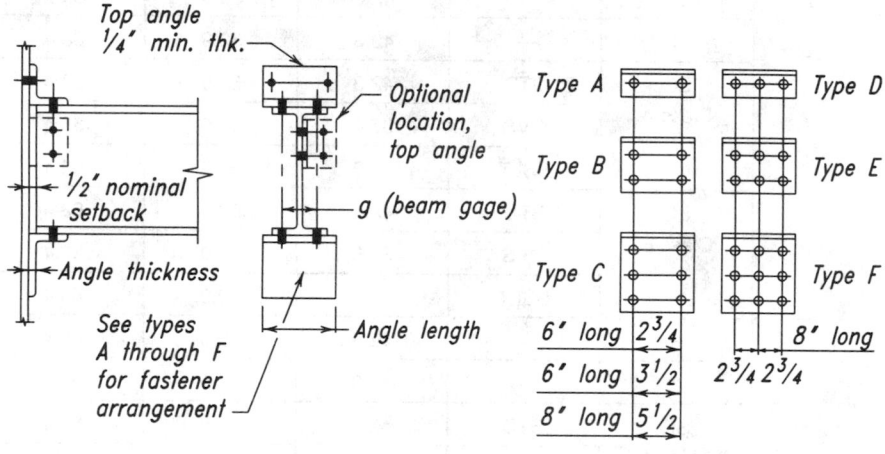

(a) All–bolted

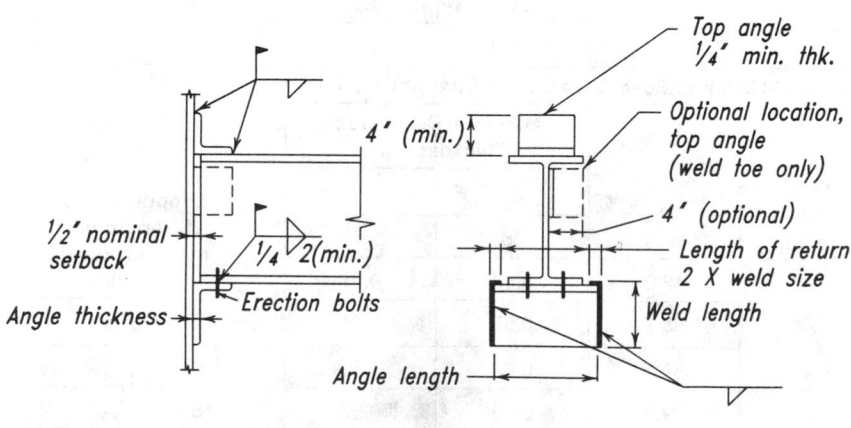

(b) All–welded

Figure 9-11. Unstiffened seated connections.

Refer to Garrett and Brockenbrough (1986) for the full design procedure for this connection.

Design Checks

The design strengths of the bolts and/or welds and connected elements must be determined in accordance with the LRFD Specification; the applicable limit states are discussed in Part 8. In all cases, the design strength ϕR_n must equal or exceed the required strength R_u.

Additionally, the strength of the supported beam web must be checked; the applicable limit states are local web yielding and web crippling. For local web yielding, from LRFD Specification Section K1.3, the design strength of the beam web is ϕR_n, where $\phi = 1.0$ and:

$$R_n = (2.5k + N)\, F_{yw} t_w$$

For any rolled beam shape, the design local web yielding strength may be determined from constants tabulated in the Factored Uniform Load Tables in Part 4. From these tables,

$$\phi R_n = \phi R_1 + N\,(\phi R_2)$$

where

$$\phi R_1 = \phi\,(2.5 k F_y t_w)$$
$$\phi R_2 = \phi\,(F_y t_w)$$

For web crippling, from LRFD Specification Section K1.4, the design strength of the beam web is ϕR_n, where $\phi = 0.75$ and, for $N / d \leq 0.2$:

$$R_n = 68 t_w^2 \left[1 + 3\left(\frac{N}{d}\right)\left(\frac{t_w}{t_f}\right)^{1.5} \right] \sqrt{\frac{F_{yw} t_f}{t_w}}$$

For $N / d > 0.2$:

$$R_n = 68 t_w^2 \left[1 + \left(\frac{4N}{d} - 0.2\right)\left(\frac{t_w}{t_f}\right)^{1.5} \right] \sqrt{\frac{F_{yw} t_f}{t_w}}$$

For any rolled beam shape, the design web crippling strength may be determined from constants tabulated in the Factored Uniform Load Tables in Part 4. From these tables, for $N / d \leq 0.2$:

$$\phi R_n = \phi R_3 + N\,(\phi R_4)$$

For $N / d > 0.2$:

$$\phi R_n = \phi R_5 + N\,(\phi R_6)$$

where

$$\phi R_3 = \phi\left(68 t_w^2 \sqrt{\frac{F_{yw} t_f}{t_w}}\right)$$

$$\phi R_4 = \phi \left[68 t_w^2 \left(\frac{3}{d} \right) \left(\frac{t_w}{t_f} \right)^{1.5} \sqrt{\frac{F_{yw} t_f}{t_w}} \right]$$

$$\phi R_5 = \phi \left[68 t_w^2 \left(1 - 0.2 \left(\frac{t_w}{t_f} \right)^{1.5} \right) \sqrt{\frac{F_{yw} t_f}{t_w}} \right]$$

$$\phi R_6 = \phi \left[68 t_w^2 \left(\frac{4}{d} \right) \left(\frac{t_w}{t_f} \right)^{1.5} \sqrt{\frac{F_{yw} t_f}{t_w}} \right]$$

Note that the beam design strength is tabulated in the Factored Uniform Load Table in Part 4 for $N = 3\frac{1}{4}$-in. (a 4-in. seat).

The top angle and its connections are not usually sized for any calculated strength requirement; a $\frac{1}{4}$-in. thick angle with a 4 in. vertical leg dimension will generally be adequate. It may be bolted with two bolts through each leg or welded with minimum-size welds to either the supported or the supporting members.

Shop and Field Practices

Unstiffened seated connections may be made to the webs and flanges of supporting columns. If adequate clearance exists, unstiffened seated connections may also be made to the webs of supporting girders.

To provide for overrun in beam length, the nominal setback for the beam end is $\frac{1}{2}$-in. To provide for underrun in beam length, this setback is assumed to be $\frac{3}{4}$-in. for calculation purposes.

The seat angle is usually shop attached to the support. Since the bottom flange typically establishes the plane of reference for seated connections, mill variation in beam depth may result in variation in the location of the top flange. Such variation is usually of no consequence with concrete slab and metal deck floors, but may be a concern when a grating or steel-plate floor is used. Thus, unless special care is required and the natural beam camber is controlled, the usual mill tolerances for member depth of $\frac{1}{8}$-in. to $\frac{1}{4}$-in. are ignored. However, when the top angle is shop attached to the supported beam and field bolted to the support, mill variation in beam depth must be considered. Slotted holes, as illustrated in Figure 9-12a, will accommodate both overrun and underrun in the beam depth and are the preferred method for economy and convenience to both the fabricator and erector. Alternatively, the angle could be shipped loose with clearance provided as shown in Figure 9-12b. When the top angle is to be field welded to the support, no provision for mill variation in the beam depth is necessary.

When the top angle is shop attached to the support, $\frac{1}{4}$-in. to $\frac{3}{8}$-in erection clearance must be provided as illustrated in Figure 9-12c. This range of clearances reflects the shop practice of most fabricators. Some fabricators supply shims for about twice the opening expected under the top angle in case of mill underrun in beam depth; others supply shims for openings as detailed and furnish additional shims only as required.

All-Bolted Unstiffened Seated Connections

Table 9-6 is a design aid for all-bolted unstiffened seats. Seat design strengths are tabulated, assuming a 4 in. outstanding leg, for angle material with $F_y = 36$ ksi and $F_u = 58$ ksi and beam material with $F_y = 36$ ksi and $F_u = 58$ ksi or with $F_y = 50$ ksi and $F_u = 65$

ksi. These tables will be conservative when used for angle material with F_y = 50 ksi and F_u = 65 ksi. All values are for comparison with factored loads.

Tabulated seat design strengths consider the limit states of shear yielding and flexural yielding of the outstanding angle leg and crippling of the beam web; the designer must independently check the design strength of the beam web in local yielding. Values are tabulated for a nominal beam setback of ½-in.; for calculation purposes, this setback is increased to ¾-in. to account for possible underrun in beam length.

Bolt design strengths are tabulated for the seat types illustrated in Figure 9-11a with ¾-in., ⅞-in., and 1 in. diameter A325 and A490 bolts. Vertical spacing of bolts and gages in seat angles may be arranged to suit conditions, provided they conform to the provisions of the LRFD Specification. Where thick angles are used, larger entering and tightening clearances may be required in the outstanding angle leg. The suitability of angle sizes and thicknesses for the seat types illustrated in Figure 9-11a are also listed.

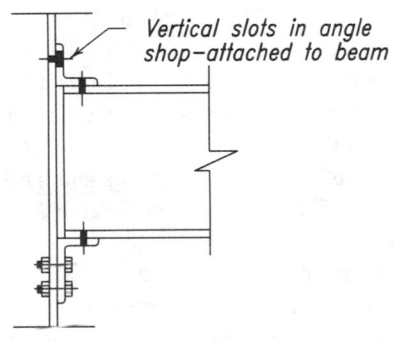

(a) Vertical slots

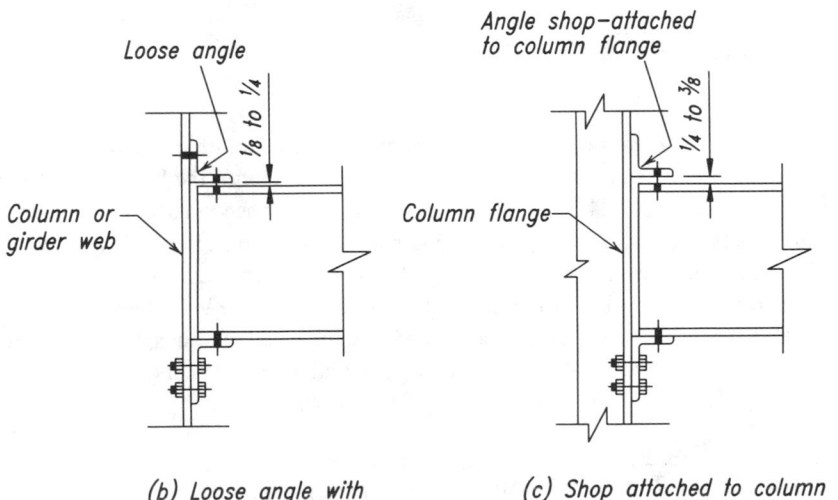

(b) Loose angle with clearance as shown   *(c) Shop attached to column flange with clearance as shown*

Figure 9-12. Providing for variation in beam depth with seated connections.

Bolted/Welded Unstiffened Seated Connections

Tables 9-6 and 9-7 may be used in combination to design unstiffened seated connections which are welded to the supporting member and bolted to the supported beam, or bolted to the supporting member and welded to the supported beam.

All-Welded Unstiffened Seated Connections

Table 9-7 is a design aid for all-welded unstiffened seats. Seat design strengths are tabulated, assuming either a 3½-in. or 4 in. outstanding leg (as indicated in the table), for angle material with $F_y = 36$ ksi and $F_u = 58$ ksi and beam material with $F_y = 36$ ksi and $F_u = 58$ ksi or with $F_y = 50$ ksi and $F_u = 65$ ksi. These tables will be conservative when used for angle material with $F_y = 50$ ksi and $F_u = 65$ ksi. Electrode strength is assumed to be 70 ksi. All values are for comparison with factored loads.

Tabulated seat design strengths consider the limit states of shear yielding and flexural yielding of the outstanding angle leg and crippling of the beam web; the designer must independently check the design strength of the beam web in local yielding. Values are tabulated for a nominal beam setback of ½-in.; for calculation purposes, this setback is increased to ¾-in. to account for possible underrun in beam length.

Weld design strengths are tabulated using the elastic method. The minimum and maximum angle thickness for each case is also tabulated. While these tabular values are based upon 70 ksi electrodes, they may be used for other electrodes, provided the tabular values are adjusted for the electrodes used (e.g., for 60 ksi electrodes, multiply the tabular values by 60/70 = 0.866, etc.) and the welds and base metal meet the required strength level provisions of LRFD Specification Section J2. Should combinations of material thickness and weld size selected from Table 9-7 exceed the limits set by LRFD Specification Section J2.2, increase the weld size or material thickness as required.

As can be seen from the following, reduction of the tabulated weld strength is not normally required when unstiffened seats line up on opposite sides of the supporting web. From Salmon and Johnson (1993), the design strength of the welds to the support is ϕR_n, where

$$\phi R_n = 2 \times \frac{1.392DL}{\sqrt{1 + \dfrac{20.25e^2}{L^2}}}$$

In the above equation, D is the number of sixteenths-of-an-inch in the weld size, L is the vertical leg dimension of the seat angle, and e is the eccentricity of the beam end reaction with respect to the weld lines. The term in the denominator which accounts for the eccentricity e increases the weld size far beyond what is required for shear alone, but with seats on both sides of the supporting member web, the forces due to eccentricity react against each other and have no effect on the web. Furthermore, as illustrated in Figure 9-13, there are actually two shear planes per weld; one at each weld toe and heel for a total of four shear planes. Thus, for an 8-in. long 7×4×¾ seat angle supporting a beam with $F_y = 36$ ksi and a web thickness of ⁹⁄₁₆-in. ($\phi R_n = 71.6$), the minimum support thickness would be

$$\frac{71.6}{0.9 \times 0.6 \times 36 \text{ ksi} \times 7 \text{ in.} \times 4 \text{ planes}} = 0.132 \text{ in.}$$

For the identical connection on both sides of the support, the minimum support thickness would be slightly larger than ¼-in. Thus, supporting web thickness is generally not a concern.

Example 9-7

Given:

Design an all-bolted unstiffened seated connection for a W16×50 beam to W14×90 column web connection

$R_u = 55$ kips

W16×50

$t_w = 0.380$ in. $d = 16.26$ in. $t_f = 0.630$ in.
$F_y = 50$ ksi, $F_u = 65$ ksi

W14×90

$t_w = 0.440$ in.
$F_y = 50$ ksi, $F_u = 65$ ksi

Use ⅞-in. diameter A325-N bolts in standard holes. Assume angle material with $F_y = 36$ ksi and $F_u = 58$ ksi.

Solution:

Design seat angle and bolts

Try 8 in. angle length with 5½-in. bolt gage. From Table 9-6, with $t_w = $ ⅜-in., a ¾-in. thick angle provides

$\phi R_n = 68.2$ kips > 55 kips   **o.k.**

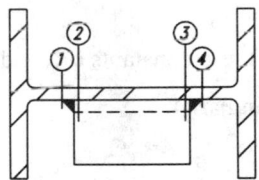

(a) Plan view

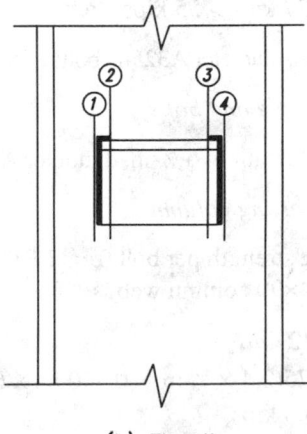

(b) Elevation

Figure 9-13. Shear planes for unstiffened seated connections.

and, for $\frac{7}{8}$-in. diameter A325-N bolts, connection type B (four bolts) provides

ϕR_n = 86.6 kips > 55 kips   **o.k.**

The table indicates a 6×4×¾ is available (4-in. OSL)

Check bolt bearing on the angle

The bearing strength per bolt from LRFD Specification Section J3.10 for the ¾-in. thick angle is

$\phi R_n = \phi\,(2.4dtF_u)$
$= 0.75\,(2.4 \times \frac{7}{8}\text{-in.} \times \frac{3}{4}\text{-in.} \times 58\text{ ksi})$
$= 68.5$ kips

Since this exceeds the strength of the bolts in single shear, bolt bearing is not critical.

Tabular values include check of local yielding strength of beam web **o.k.**

Check crippling of the beam web

$\dfrac{N}{d} = \dfrac{4\text{ in.}}{16.26\text{ in.}}$
$= 0.246$

Since $\dfrac{N}{d} > 0.2$, use constants ϕR_5 and ϕR_6 from the Factored Uniform Load Tables in Part 4,

where

$\phi R_n = \phi R_5 + N\,(\phi R_6)$
$= 60.8$ kips + 4 in.(7.73 kips/in.)
$= 91.7$ kips > 55 kips   **o.k.**

Use two $\frac{7}{8}$-in. diameter A325-N bolts to connect the beam to the seat angle.

Select top angle and bolts

Use L4×4×¼ with two $\frac{7}{8}$-in. diameter A325-N bolts through each leg.

Check supporting column

The bearing strength per bolt from LRFD Specification Section J.3.10 for the W14×90 column web is

$\phi R_n = \phi\,(2.4dtF_u)$
$= 0.75(2.4 \times \frac{7}{8}\text{-in.} \times 0.440\text{ in.} \times 65\text{ ksi})$
$= 45.0$ kips

Since this exceeds the strength of the bolts in single shear, bolt bearing is not critical.

Example 9-8

Given: Design an unstiffened seated connection for a W21×62 beam to W14×61 column flange connection.

$R_u = 55$ kips

W21×62

$t_w = 0.400$ in. $d = 20.99$ in. $t_f = 0.615$ in.
$F_y = 50$ ksi, $F_u = 65$ ksi

W14×61

$t_f = 0.645$ in.
$F_y = 50$ ksi, $F_u = 65$ ksi

Use ¾-in. diameter A325-N bolts in standard holes to connect the supported beam to the seat and top angles. Use 70 ksi electrode welds to connect the seat and top angles to the column flange. Assume angle material with $F_y = 36$ ksi and $F_u = 58$ ksi.

Solution: *Design seat angle and welds*

Try 8 in. angle length.

From Table 9-7, with $t_w \approx$ ⅜-in., a ¾-in. thick angle provides

$\phi R_n = 68.2$ kips > 55 kips   **o.k.**

and an 8×4 angle (4 in. OSL) with ⁵⁄₁₆-in. fillet welds provides

$\phi R_n = 66.8$ kips > 55 kips   **o.k.**

Check crippling of the beam web

$$\frac{N}{d} = \frac{4 \text{ in.}}{20.99 \text{ in.}}$$
$$= 0.191$$

Since $N/d \le 0.2$, use constants ϕR_3 and ϕR_4 from the Factored Uniform Load Tables in Part 4, where

$\phi R_n = \phi R_3 + N (\phi R_4)$
       $= 71.5$ kips $+ 4$ in.(5.36 kips/in.)
       $= 92.9$ kips > 55 kips   **o.k.**

Use two ¾-in. diameter A325-N bolts to connect the beam to the seat angle.

Select top angle, bolts, and welds

Use L4×4×¼ with two ¾-in. diameter A325-N bolts through the supported-beam leg of the angle. Use ³⁄₁₆-in. fillet weld along the toe of the angle (minimum size from LRFD Specification Table J2.4).

Table 9-6.
All-Bolted Unstiffened Seated Connections

4-in. Outstanding Angle Leg Design Strength, kips

Beam F_y, ksi	Angle Length, in.		6					8				
	Angle Thickns., in.	$3/8$	$1/2$	$5/8$	$3/4$	1	$3/8$	$1/2$	$5/8$	$3/4$	1
36	$3/16$	13.6	18.5	22.6	26.8	29.8	15.7	20.3	25.0	29.6	29.8
	$1/4$	15.7	23.3	30.0	34.9	43.3	18.1	26.5	32.6	38.1	43.3
	$5/16$	17.5	26.4	35.4	44.5	57.3	20.3	30.0	39.8	49.3	60.7
	$3/8$	19.2	29.2	39.5	49.9	70.3	22.2	33.1	44.3	55.6	75.0
	$7/16$	20.8	31.9	43.3	55.0	78.4	24.0	36.1	48.6	61.2	86.5
	$1/2$	22.2	34.4	47.1	59.9	85.8	25.6	38.9	52.6	66.5	94.4
	$9/16$	23.5	36.8	50.6	64.6	93.0	27.2	41.6	56.5	71.6	102
50	$3/16$	16.0	23.2	28.3	33.4	41.5	18.5	25.3	31.0	36.7	41.5
	$1/4$	18.5	28.0	37.7	44.3	56.6	21.3	31.8	41.2	47.9	60.2
	$5/16$	20.7	31.7	43.1	54.7	73.7	23.9	36.0	48.3	60.9	78.8
	$3/8$	22.6	35.2	48.3	61.5	88.2	26.1	39.9	53.9	68.2	96.9
	$7/16$	24.5	38.5	53.1	68.0	98.1	28.2	43.5	59.3	75.2	108
	$1/2$	26.1	41.7	57.8	74.3	108	30.2	47.0	64.3	82.0	118
	$9/16$	27.7	44.7	62.4	80.4	117	32.0	50.3	69.2	88.5	128

(left vertical label: Beam Web Thickness, in.)

Bolt Design Strength, kips									Available Angles		
Bolt Dia-meter, in.	ASTM Desig.	Thread Cond.	Connection Type from Figure 9-11a						Connec-tion Type	Angle Size	t, in.
			A	B	C	D	E	F			
$3/4$	A325	N	31.8	63.6	95.4	47.7	95.4	143	A, D	4×3	$3/8$–$1/2$
		X	39.8	79.5	119	59.6	119	179		$4\times3\frac{1}{2}$	$3/8$–$1/2$
	A490	N	39.8	79.5	119	59.6	119	179		4×4	$3/8$–$3/4$
		X	49.7	99.4	149	74.6	149	224	B, E	6×4	$3/8$–$3/4$
$7/8$	A325	N	43.3	86.6	130	64.9	130	195		7×4	$3/8$–$3/4$
		X	54.1	108	162	81.2	162	244		8×4	$1/2$–1
	A490	N	54.1	108	162	81.2	162	244	C, F[b]	8×4	$1/2$–1
		X	67.6	135	203	101	203	304	[b]Not suitable for use with 1-in. diameter bolts.		
1	A325	N	56.5	113	—	84.8	170	—			
		X	70.7	141	—	106	212	—			
	A490	N	70.7	141	—	106	212	—			
		X	88.4	177	—	133	265	—			

Table 9-7.
All-Welded Unstiffened Seated Connections

4-in. or 3½-in. Outstanding Angle Leg Design Strength, kips

Beam F_y, ksi	Angle Length, in.		6					8				
	Angle Thickns., in.		$\frac{3}{8}$	$\frac{1}{2}$	$\frac{5}{8}$	$\frac{3}{4}$	1	$\frac{3}{8}$	$\frac{1}{2}$	$\frac{5}{8}$	$\frac{3}{4}$	1
36	Beam Web Thickness, in.	$\frac{3}{16}$	13.6	18.5	22.6	26.8	29.8	15.7	20.3	25.0	29.6	29.8
		$\frac{1}{4}$	15.7	23.3	30.0	34.9	43.3	18.1	26.5	32.6	38.1	43.3
		$\frac{5}{16}$	17.5	26.4	35.4	44.5[a]	57.3	20.3	30.0	39.8	49.3[a]	60.7
		$\frac{3}{8}$	19.2	29.2	39.5	49.9[a]	70.3	22.2	33.1	44.3	55.6[a]	75.0
		$\frac{7}{16}$	20.8	31.9	43.3	55.0[a]	78.4[a]	24.0	36.1	48.6	61.2[a]	86.5
		$\frac{1}{2}$	22.2	34.4	47.1	59.9[a]	85.8[a]	25.6	38.9	52.6	66.5[a]	94.4[a]
		$\frac{9}{16}$	23.5	36.8	50.6	64.6[a]	93.0[a]	27.2	41.6	56.5	71.6[a]	102[a]
50		$\frac{3}{16}$	16.0	23.2	28.3	33.4	41.5	18.5	25.0	31.0	36.7	41.5
		$\frac{1}{4}$	18.5	28.0	37.7	44.3[a]	56.6	21.3	31.8	41.2	47.9	60.2
		$\frac{5}{16}$	20.7	31.7	43.1	54.7[a]	73.7	23.9	36.0	48.3	60.9[a]	78.8
		$\frac{3}{8}$	22.6	35.2	48.3	61.5[a]	88.2[a]	26.1	39.9	53.9	68.2[a]	96.9
		$\frac{7}{16}$	24.5	38.5	53.1	68.0[a]	98.1[a]	28.2	43.5	59.3	75.2[a]	108[a]
		$\frac{1}{2}$	26.1	41.7	57.8	74.3[a]	108[a]	30.2	47.0	64.3	82.0[a]	118[a]
		$\frac{9}{16}$	27.7	44.7	62.4	80.4[a]	117[a]	32.0	50.3	69.2	88.5[a]	128[a]

Weld (70 ksi) Design Strength, kips

70 ksi Weld Size, in.	Seat Angle Size (long leg vertical)				
	4×3½	5×3½	6×4	7×4	8×4
$\frac{1}{4}$	17.3	25.8	32.7	42.8	53.4
$\frac{5}{16}$	21.5	32.3	41.0	53.4	66.8
$\frac{3}{8}$	25.8	38.7	49.1	64.1	80.1
$\frac{7}{16}$	30.2	45.2	57.3	74.7	93.5
$\frac{1}{2}$	—	51.6	65.4	83.4	107
$\frac{5}{8}$	—	64.5	81.8	107	134
$\frac{11}{16}$	—	71.0	90.0	117	—
$\frac{3}{4}$	—	—	—	—	—

Available Angle Thickness, in.

	4×3½	5×3½	6×4	7×4	8×4
Minimum	$\frac{3}{8}$	$\frac{3}{8}$	$\frac{3}{8}$	$\frac{3}{8}$	$\frac{1}{2}$
Maximum	$\frac{1}{2}$	$\frac{3}{4}$	$\frac{3}{4}$	$\frac{3}{4}$	1

[a]Values apply only to angles with 4-in. outstanding leg.

Stiffened Seated Connections

A stiffened seated connection is made with a seat plate and stiffening element (e.g., a plate, pair of angles, or structural tee) and a top angle, as illustrated in Figure 9-14. The top angle may be bolted or welded to the supported beam as well as to the supporting

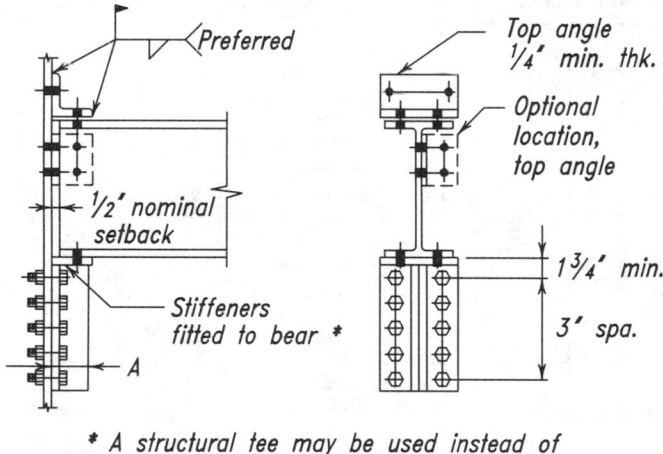

* A structural tee may be used instead of a pair of angles.

(a) All-bolted

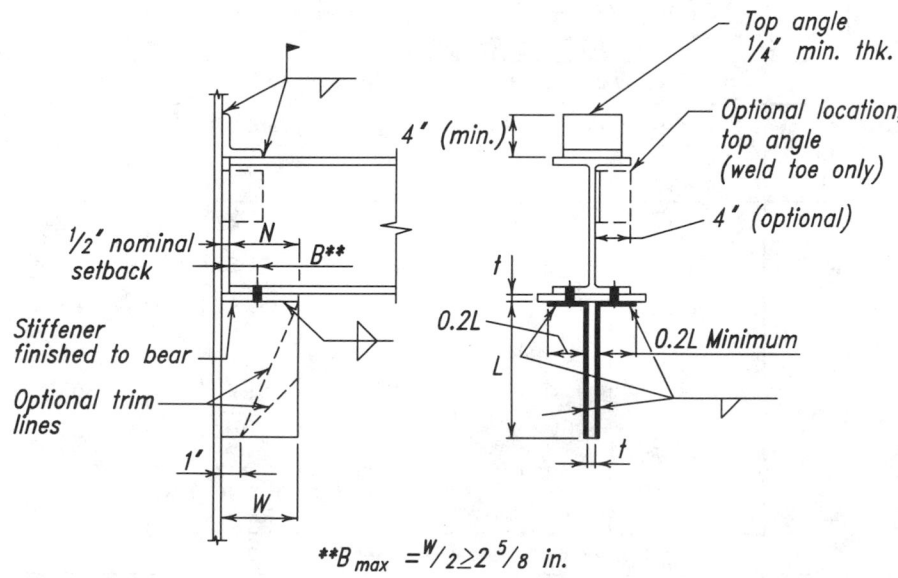

$$^{**}B_{max} = {}^W/_2 \geq 2\,{}^5/_8 \text{ in.}$$

(b) Bolted/welded

Figure 9-14. Stiffened seated connections.

member and the stiffening element may be bolted or welded to the support; the seat plate should be bolted to the supported beam as noted in the discussion (#2) below. While the stiffening element is assumed to carry the entire end reaction of the supported beam, the top angle must be placed as shown or in the optional side location for satisfactory performance and stability (Dalley and Roeder, 1989).

When the top angle is welded to the support and/or the supported beam, adequate flexibility must be provided in the connection. As illustrated in Figure 9-14b, line welds are placed along the toe of each angle leg. Note that welding along the sides of the vertical angle leg must be avoided as it inhibits the flexibility and, therefore, the necessary end rotation of the connection; the performance of such a connection is unpredictable.

Design Checks

The design strengths of the bolts and/or welds and connected elements must be determined in accordance with the LRFD Specification; the applicable limit states are discussed in Part 8. In all cases, the design strength ϕR_n must equal or exceed the required strength R_u.

Additionally, the strength of the supported beam web must be checked; the applicable limit states are local web yielding and web crippling. These design strengths may be determined as illustrated previously for unstiffened seated connections.

Stiffened seated connections such as the one shown in Figure 9-14b made to one side of the web of a supporting column may also need to be investigated for resistance to punching. In lieu of a more detailed analysis, Ellifrit and Sputo (1991) showed that punching will not be critical if the design parameters below and those summarized graphically in Figure 9-14b are met.

1. This simplified approach is applicable to the following column sections:

W14×43-730	W12×40-336	W10×33-112
W8×24-67	W6×20-25	W5×16-19

2. The supported beam must be bolted to the seat plate with ASTM A325 or A490 high-strength bolts to account for the prying action caused by rotation of the connection at ultimate load; welding the beam to the seat plate is not recommended because welds lack the required strength and ductility. The centerline of the bolts should be located no more than the greater of $W/2$ or $2\frac{5}{8}$-in. from the column web face.

3. For seated connections where $W = 8$ in. or $W = 9$ in. and $3\frac{1}{2}$-in. $< B \le W/2$, or where $W = 7$ in. and 3 in. $< B \le W/2$ for a W14×43 column, refer to Ellifrit and Sputo (1991). These limitations are summarized at the bottom of Table 9-9.

4. The top angle may be bolted or welded, but must have a minimum ¼-in. thickness.

5. The seat plate should not be welded to the column flange.

6. Except as noted, the maximum weld size for 70 ksi electrodes is limited to the column web thickness t_w for connections on one side of the web; for connections in line on both sides of a column web, the maximum weld size is $t_w/2$ for $F_y = 36$ ksi and $2t_w/3$ for $F_y = 50$ ksi. This approximately matches the shear yielding strength of the column web with the shear strength of the weld; as with unstiffened seated connections, the contribution of eccentricity to the required shear yielding strength is negligible.

The top angle and its connections are not usually sized for any calculated strength requirement; a ¼-in. thick angle with a 4-in. minimum vertical leg will usually be adequate. It may be bolted with two bolts through each leg or welded with minimum-size welds to either the supported or the supporting members.

Shop and Field Practices
The comments for unstiffened seated connections are equally applicable to stiffened seated connections.

All-Bolted Stiffened Seated Connections
Table 9-8 is a design aid for all-bolted stiffened seats. Stiffener design strengths are tabulated for stiffener material with F_y = 36 ksi and F_u = 58 ksi and with F_y = 50 ksi and F_y = 65 ksi. All values are for comparison with factored loads.

Tabulated values consider the limit state of bearing on the stiffening material. The designer must independently check the design strength of the beam web based upon the limit states of local web yielding and web crippling. Values are tabulated for a nominal beam setback of ½-in.; for calculation purposes, this setback is increased to ¾-in. to account for possible underrun in beam length.

Bolt design strengths are tabulated for two vertical rows of from three to seven ¾-in., ⅞-in., and 1 in. diameter ASTM A325 and A490 high-strength bolts based upon the limit state of bolt shear. Vertical spacing of fasteners in the stiffening element may be arranged to suit conditions, provided they conform to the provisions of the LRFD Specification.

Bolted/Welded Stiffened Seated Connections
Table 9-9 is a design aid for stiffened seated connections welded to the support and bolted to the supported beam. Electrode strength is assumed to be 70 ksi. All values are for comparison with factored loads.

Weld design strengths are tabulated using the elastic method. While these tabular values are based upon 70 ksi electrodes, they may be used for other electrodes, provided the tabular values are adjusted for the electrodes used (e.g., for 60 ksi electrodes, multiply the tabular values by 60/70 = 0.866, etc.) and the weld and base metal meet the provisions of LRFD Specification Section J2.

The thickness of the horizontal seat plate or tee flange should not be less than ⅜-in. If the seat and stiffener are composed of separate plates, finish the stiffener to bear against the seat. Welds connecting the two plates should have a strength not less than the horizontal welds to the support under the seat plate.

The designer must independently check the beam web for local web yielding and web crippling. The nominal beam setback of ½-in. should be assumed to be ¾-in. for calculation purposes to account for possible underrun in beam length.

The stiffener thickness may be conservatively determined as follows. When the stiffener has F_y = 36 ksi, the minimum stiffener thickness t for supported beams with unstiffened webs should not be less than t_w for supported beams with F_y = 36 ksi, and not less than $1.4t_w$ for supported beams with F_y = 50 ksi. For stiffener material with F_y = 50 ksi or greater, the minimum stiffener plate thickness t for supported beams with unstiffened webs should be the supported beam web thickness t_w multiplied by the ratio of F_y of the beam material to F_y of the stiffener material (e.g., F_y beam = 65 ksi, F_y stiffener = 50 ksi, $t = t_w \times 65/50$ minimum). Additionally, the minimum stiffener thickness t should be at least $2w$ for stiffener material with F_y = 36 ksi or $1.5w$ for stiffener material with F_y = 50 ksi, where w is the weld size for 70 ksi electrodes.

For stiffened seated connections in line on opposite sides of a column web with F_y = 36 ksi, select 70 ksi electrode weld size no greater than one-half the column web thickness

t_w; for column web material with $F_y = 50$ ksi, select 70 ksi electrode weld size no greater than two-thirds the column web thickness t_w. Should combinations of material thickness and weld size selected from Table 9-9 exceed the limits of LRFD Specification Section J2, increase the weld size or material thickness as required.

Example 9-9   Design a stiffened seated connection for a W21×68 beam to W14×90 column flange connection.

$R_u = 125$ kips

W21×68

$t_w = 0.430$ in. $d = 21.13$ in. $t_f = 0.685$ in.
$F_y = 50$ ksi, $F_u = 65$ ksi

W14×90

$t_f = 0.710$ in.
$F_y = 50$ ksi, $F_u = 65$ ksi

Use ¾-in. diameter A325-N bolts in standard holes to connect the supported beam to the seat plate and top angle. Use 70 ksi electrode welds to connect the stiffener and top angle to the column flange.

Solution: *Determine stiffener width W required for web crippling and local web yielding*

For web crippling, assume $N/d > 0.2$ and use constants ϕR_5 and ϕR_6 from the Factored Uniform Load Tables in Part 4.

$$W_{\min} = \frac{R_u - \phi R_5}{\phi R_6} + \text{setback}$$

$$= \frac{125 \text{ kips} - 75.8 \text{ kips}}{7.92 \text{ kips / in.}} + \tfrac{1}{2}\text{-in.}$$

$$= 6.71 \text{ in.}$$

For local web yielding, use constants ϕR_1 and ϕR_2 from the Factored Uniform Load Tables in Part 4.

$$W_{\min} = \frac{R_u - \phi R_1}{\phi R_2} + \text{setback}$$

$$= \frac{125 \text{ kips} - 77.3 \text{ kips}}{21.5 \text{ kips / in.}} + \tfrac{1}{2}\text{-in.}$$

$$= 2.72 \text{ in.}$$

The minimum stiffener width W for web crippling controls. To account for possible underrun in beam length, the minimum stiffener width should be increased by ¼-in. Thus, use $W = 7$ in.

Check assumption

$$\frac{N}{d} = \frac{7 \text{ in.}}{21.13 \text{ in.}}$$
$$= 0.331 > 0.2 \quad \textbf{o.k.}$$

Determine stiffener length L and stiffener to column flange weld size

From Table 9-9, a stiffener with $L = 15$ in. and ¼-in. weld size provides

$$\phi R_n = 139 \text{ kips} > 125 \text{ kips} \quad \textbf{o.k.}$$

Determine weld requirements for seat plate

Using ¼-in. fillet welds the minimum length of seat-plate-to-column-flange weld on each side of the stiffener is $0.2(L) = 3$ in. Use three inches of weld on each side of the stiffener. This also establishes the minimum weld between the seat plate and stiffener; use three inches of ¼-in. weld on both sides of the stiffener.

Determine seat plate dimensions

To accommodate two ¾-in. diameter A325-N bolts on a 5½-in. gage connecting the beam flange to the seat plate, a width of eight inches is adequate. This is greater than the width required to accommodate the seat-plate-to-column-flange welds.

Use PL⅜-in.×7 in.×8 in. for the seat plate.

Determine stiffener plate thickness

To develop the stiffener-to-seat-plate welds, the minimum stiffener thickness is

$$t_{min} = 2 \ (\text{¼-in.})$$
$$= \text{½-in.}$$

For a stiffener with $F_y = 36$ ksi and beam with $F_y = 50$ ksi, the minimum stiffener thickness is

$$t_{min} = 1.4 t_w$$
$$= 1.4 (0.430 \text{ in.})$$
$$= 0.602 \text{ in.}$$

The latter controls; use PL⅝-in.×7 in.×15 in. for the stiffener.

Select top angle, bolts, and welds

Use L4×4×¼ with two ¾-in. diameter A325-N bolts through the supported-beam leg of the angle. Use ⅛-in. fillet weld along the toe of the support leg of the angle (minimum size from LRFD Specification Table J2.4).

Example 9-10   Design a stiffened seated connection for a W21×68 beam to W14×90 column web connection.

$R_u = 125$ kips

W21×68

$t_w = 0.430$ $\qquad$ $d = 21.13$ in. $\qquad$ $t_f = 0.685$ in.
$F_y = 50$ ksi, $F_u = 65$ ksi

W14×90

$t_w = 0.440$
$F_y = 50$ ksi, $F_u = 65$ ksi

Use ¾-in. diameter A325-N bolts in standard holes to connect the supported beam to the seat plate and top angle. Use 70 ksi electrode welds to connect the stiffener and top angle to the column web. Assume angle material with $F_y = 36$ ksi and $F_u = 58$ ksi.

Solution: *Determine stiffener width W*

As calculated previously in Example 9-9, use $W = 7$ in.

Determine stiffener length L and stiffener to column web weld size

As calculated previously in Example 9-9, use $L = 15$ in. and ¼-in. weld size.

Determine weld requirements for seat plate

As calculated previously in Example 9-9, use three inches of ¼-in. weld on both sides of the seat plate for the seat-plate-to-column-web welds and for the seat-plate-to-stiffener welds.

Determine seat plate dimensions

For a column-web support, from Table 9-9, the maximum distance from the face to the support to the line of bolts between the beam flange and seat plate is 3½-in. The PL⅜-in.×7 in.×8 in. chosen previously in Example 9-9 will accommodate these bolts.

Determine stiffener plate thickness

As calculated previously in Example 9-9, use PL⅝-in.×7 in.×15 in.

Select top angle, bolts, and welds

Use L4×4×¼ with two ¾-in. diameter A325-N bolts through the supported-beam leg of the angle. Use 3/16-in. fillet weld along the toe of the support leg of the angle (minimum size from LRFD Specification Table J2.4).

Check column web

From Table 9-9, no limitation is placed on column web. Therefore, column web is **o.k.**

Table 9-8. All-Bolted Stiffened Seated Connections						
Stiffener Angle Design Strength, kips[a]						

Stiffener Material		$F_y = 36$ ksi $\phi R_n = 0.75 (1.8 \times 36) A_{pb}$			$F_y = 50$ ksi $\phi R_n = 0.75 (1.8 \times 50) A_{pb}$		
Stiffener Outstanding Leg A, in.[b]		$3\frac{1}{2}$	4	5	$3\frac{1}{2}$	4	5
Thickness of Stiffener Outstanding Legs, in.	$\frac{5}{16}$	83.5	98.7	129	116	137	179
	$\frac{3}{8}$	100	119	155	139	165	215
	$\frac{1}{2}$	134	158	207	186	219	287
	$\frac{5}{8}$	167	197	258	232	274	359
	$\frac{3}{4}$	201	237	310	278	329	430

Use minimum $\frac{3}{8}$-in. thick seat plate wide enough to extend beyond outstanding legs of stiffener.
[a]See LRFD Specification Sect. J8.
[b]Beam bearing length assumed $\frac{3}{4}$-in. less for calculation purposes.

Bolt Design Strength, kips

Bolt Diameter, in.	ASTM Desig.	Thread Cond.	Number of Bolts in One Vertical Row				
			3	4	5	6	7
$\frac{3}{4}$	A325	N	95.4	127	159	191	223
		X	119	159	199	239	278
	A490	N	119	159	199	239	278
		X	149	199	249	298	348
$\frac{7}{8}$	A325	N	130	173	216	260	303
		X	162	216	271	325	379
	A490	N	162	216	271	325	379
		X	203	271	338	406	474
1	A325	N	170	226	283	339	396
		X	212	283	353	424	495
	A490	N	212	283	353	424	495
		X	265	353	442	530	619

Table 9-9.
Bolted/Welded Stiffened Seated Connections

Stiffened Seat Design Strength, kips

L, in.	Width of Seat W, in.											
	4				5				6			
	70 ksi Weld Size, in.				70 ksi Weld Size, in.				70 ksi Weld Size, in.			
	$\frac{1}{4}$	$\frac{5}{16}$	$\frac{3}{8}$	$\frac{7}{16}$	$\frac{5}{16}$	$\frac{3}{8}$	$\frac{7}{16}$	$\frac{1}{2}$	$\frac{5}{16}$	$\frac{3}{8}$	$\frac{7}{16}$	$\frac{1}{2}$
6	34.0	42.5	51.1	59.6	35.2	42.2	49.3	56.3	29.9	35.9	41.9	47.8
7	44.9	56.1	67.3	78.6	46.9	56.2	65.6	75.0	40.1	48.1	56.1	64.1
8	56.7	70.8	85.0	99.2	59.8	71.7	83.7	95.6	51.4	61.7	72.0	82.2
9	69.2	86.5	104	121	73.7	88.5	103	118	63.8	76.6	89.3	102
10	82.3	103	123	144	88.5	106	124	142	77.2	92.6	108	123
11	95.8	120	144	168	104	125	146	167	91.3	110	128	146
12	110	137	165	192	120	144	168	192	106	127	149	170
13	124	155	186	217	137	164	192	219	122	146	170	195
14	138	173	207	242	154	185	216	246	138	165	193	220
15	152	191	229	267	171	206	240	274	154	185	216	247
16	167	209	250	292	189	227	265	302	171	205	240	274
17	181	227	272	318	207	248	290	331	188	226	264	301
18	196	245	294	343	225	270	315	360	206	247	288	329
19	211	263	316	369	243	291	340	388	223	268	313	357
20	225	281	338	394	261	313	365	417	241	289	337	386
21	240	300	359	419	279	335	391	446	259	311	362	414
22	254	318	381	445	297	357	416	476	277	332	388	443
23	269	336	403	470	315	378	442	505	295	354	413	472
24	283	354	425	495	334	400	467	534	313	376	438	501
25	297	372	446	520	352	422	492	563	331	397	464	530
26	312	390	468	546	370	444	518	592	349	419	489	559
27	326	408	489	571	388	466	543	621	368	441	515	588

Limitations for Connections to Column Webs

B	$2\frac{5}{8}$-in. max.	$2\frac{5}{8}$-in. max.	3 in. max.
	W12×40, W14×43 for L ≥ 9 in. limit weld ≤ $\frac{1}{4}$-in.		

Notes:
1. Values shown assume 70 ksi electrodes. For 60 ksi electrodes, multiply tabular values by 0.857, or enter table with 1.17 times the required strength R_u. For 80 ksi electrodes, multiply tabular values by 1.14, or enter table with 0.875 times the required strength R_u.
2. Tabulated values are valid for stiffeners with minimum thickness of

$$t_{min} = \frac{F_{y\,beam}}{F_{y\,stiffener}} \times t_w$$

   but not less than $2w$ for stiffeners with $F_y = 36$ ksi nor $1.5w$ for stiffeners with $F_y = 50$ ksi. In the above, t_w is the thickness of the unstiffened supported beam web and w is the nominal weld size.
3. Tabulated values may be limited by shear yielding of or bearing on the stiffener; refer to LRFD Specification Sections F2.2 and J8, respectively.

Table 9-9 (cont.).
Bolted/Welded Stiffened Seated Connections

Stiffener Design Strength, kips

Width of Seat W, in.

L, in.	7				8				9			
	70 ksi Weld Size, in.				70 ksi Weld Size, in.				70 ksi Weld Size, in.			
	$5/16$	$3/8$	$7/16$	$1/2$	$5/16$	$3/8$	$7/16$	$1/2$	$5/16$	$3/8$	$7/16$	$1/2$
11	81.0	97.2	113	130	72.5	87.1	116	145	65.6	78.7	105	131
12	94.7	114	133	151	85.1	102	136	170	77.1	92.5	123	154
13	109	131	153	174	98.3	118	157	197	89.3	107	143	179
14	124	149	174	198	112	135	180	224	102	123	164	204
15	139	167	195	223	127	152	203	253	116	139	185	232
16	155	186	217	249	142	170	227	283	130	156	208	260
17	172	206	240	275	157	189	251	314	144	173	231	289
18	188	226	264	301	173	208	277	346	159	191	255	319
19	205	246	287	329	189	227	303	378	175	210	280	350
20	223	267	312	356	206	247	329	411	191	229	305	381
21	240	288	336	384	222	267	356	445	207	248	331	413
22	258	309	361	412	240	287	383	479	223	268	357	446
23	275	330	385	440	257	308	411	514	240	288	384	480
24	293	352	410	469	274	329	439	548	257	308	411	513
25	311	373	435	498	292	350	467	584	274	329	438	548
26	329	395	461	526	309	371	495	619	291	349	466	582
27	347	417	486	555	327	393	524	655	308	370	494	617
28	365	438	511	584	345	414	552	690	326	391	522	652
29	383	460	537	613	363	436	581	726	344	412	550	687
30	402	482	562	643	381	457	610	762	362	434	578	723
31	420	504	588	672	399	479	639	799	379	455	607	759
32	438	526	613	701	417	501	668	835	397	477	636	795

Limitations for Connections to Column Webs

B	$3\frac{1}{2}$-in. max.	$3\frac{1}{2}$-in. max.	$3\frac{1}{2}$-in. max.
	W14×43, limit $B \leq 3$ in. See p. 9-139 "Design Checks", number 3	See p. 9-139 "Design Checks", number 3	See p. 9-139 "Design Checks", number 3

Notes:
1. Values shown assume 70 ksi electrodes. For 60 ksi electrodes, multiply tabular values by 0.857, or enter table with 1.17 times the required strength R_u. For 80 ksi electrodes, multiply tabular values by 1.14, or enter table with 0.875 times the required strength R_u.
2. Tabulated values are valid for stiffeners with minimum thickness of

$$t_{min} = \frac{F_{y\,beam}}{F_{y\,stiffener}} \times t_w$$

but not less than $2w$ for stiffeners with $F_y = 36$ ksi nor $1.5w$ for stiffeners with $F_y = 50$ ksi. In the above, t_w is the thickness of the unstiffened supported beam web and w is the nominal weld size.
3. Tabulated values may be limited by shear yielding of or bearing on the stiffener; refer to LRFD Specification Sections F2.2 and J8, respectively.

Single-Plate Connections
A single-plate connection is made with a plate as illustrated in Figure 9-15. The plate is always welded to the support on both sides of the plate and bolted to the supported member.

Design Checks
The design strengths of the bolts and/or welds and connected elements must be determined in accordance with the LRFD Specification; the applicable limit states are discussed in Part 8. In all cases, the design strength ϕR_n must equal or exceed the required strength R_u.

Eccentricity must be considered in the design of the single-plate connection; the bolts must be designed for the shear R_u and eccentric moment $R_u e_b$. The eccentricity on the bolts e_b depends upon the support condition present and whether standard or short-slotted holes are used in the plate (Astaneh et al., 1989).

A flexible support possesses relatively low rotational stiffness and permits the adjacent simply supported beam end rotation to be accommodated primarily through this supporting member's rotation. Such an end condition may exist with one-sided beam-to-girder-web connections or with deep beams connected to relatively light columns. For a flexible support with standard holes:

$$e_b = |(n-1) - a| \geq a$$

where a is the distance between the bolt line and weld line (see Figure 9-15), in., and n is the number of bolts.

For a flexible support with short-slotted holes:

$$e_b = \left| \frac{2n}{3} - a \right| \geq a$$

In contrast, a rigid support possesses relatively high rotational stiffness which constrains the adjacent simply supported beam end rotation to occur primarily within the end connection, such as a beam-to-column-flange connection or two concurrent beam-to-girder-web connections. For a rigid support with standard holes:

$$e_b = |(n-1) - a|$$

For a rigid support with short-slotted holes

$$e_b = \left| \frac{2n}{3} - a \right|$$

When the support condition is intermediate between flexible and rigid or cannot be readily classified as flexible or rigid, the larger value of e_b may conservatively be taken from the above equations.

For any combination of support condition and hole type, the 70 ksi electrode weld size should be equal to three-quarters of the plate thickness t_p for plate material with $F_y = 36$ ksi and $F_u = 58$ ksi. This ensures that the weld will not be the critical element in the connection, i.e., the plate yields before the weld yields.

The foregoing procedure is valid for single-plate connections with 2½-in. $\leq a \leq$ 3½-in.

Recommended Plate Length and Thickness

To provide for stability during erection, it is recommended that the minimum plate length be one-half the T-dimension of the beam to be supported. The maximum length of the plate must be compatible with the T-dimension of an uncoped beam and the remaining web depth, exclusive of fillets, of a coped beam. Note that the plate may encroach on the fillet or fillets by $\frac{1}{8}$-in. to $\frac{5}{16}$-in., depending upon the radius of the fillets; refer to Table 9-1. Note that if single-plate connections are used for laterally unsupported beams, for stability under service loading, the minimum depth connection as determined above should be increased by one row of bolts.

To prevent local buckling of the plate, the minimum plate thickness should be such that

$$t_{p\,min} = \frac{L}{64} \geq \frac{1}{4} \text{ in.}$$

where L is the length of the plate as illustrated in Figure 9-15. This minimum thickness is based on a simple conservative model which assumes that one-half the plate depth is subjected to uniform compression from flexure. Whereas usual local buckling limits are derived for long compression elements with plate aspect ratios approaching infinity, this case requires consideration of much shorter compression lengths using the aspect ratio

$$\frac{a}{\left(\frac{L}{2}\right)} = \frac{2a}{L}$$

and elastic plate-buckling theory for assumed simple and free edges in the direction of the flexural compression. The above minimum thickness is valid for A36 material only and $a \leq L/4$ for values of L between 12 in. and 27 in.; material specifications with higher yield strengths should not be used. The $\frac{1}{4}$-in. absolute minimum thickness is adequate for two- and three-bolt single plates with $a = 3$ in. Accordingly, Figure 9-15 lists the minimum plate thicknesses upon which Tables 9-10 are based. To provide for rotational ductility in the single plate, the maximum plate thickness should be such that

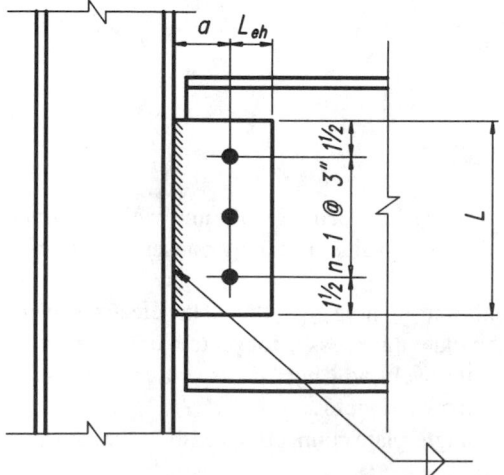

Minimum Plate Thickness	
n	$t_{p\,min}$, in.
2–5	$\frac{1}{4}$
6–7	$\frac{5}{16}$
8	$\frac{3}{8}$
9	$\frac{7}{16}$

Figure 9-15. Single-plate connections.

$$t_{p \, max} = \frac{d_b}{2} + \frac{1}{16} \text{ in. } \geq t_{p \, min}$$

where d_b is the bolt diameter, in.

Shop and Field Practices
Single-plate connections may be made to the webs of supporting girders and to the flanges of supporting columns. Because of bolting clearances, field-bolted single-plate connections may not be suitable for connections to the webs of supporting columns unless provision is made to extend the plate to locate the bolt line a sufficient distance beyond the column flanges. Such extension may require stiffening of the plate and the column web.

With the plate shop-attached to the support, side erection of the beam is permitted. Play in the open holes usually compensates for mill variation in column flange supports and other field adjustments. Thus, slotted holes are not normally required.

Bolted/Welded Single-Plate Connections
Tables 9-10 are design aids for single-plate connections welded to the support and bolted to the supported beam. Separate tables are included for supported and supporting member material with F_y = 36 ksi and F_u = 58 ksi and with F_y = 50 ksi and F_u = 65 ksi. Plate material is assumed to have F_y = 36 ksi and F_u = 58 ksi.

Tabulated bolt and plate design strengths consider the limit states of bolt shear, bolt bearing on the plate, shear yielding of the plate, shear rupture of the plate, block shear rupture of the plate, and weld shear. Values are tabulated for two through nine rows of $\frac{3}{4}$-in., $\frac{7}{8}$-in., and 1 in. diameter A325 and A490 bolts at three inches spacing. For calculation purposes, plate edge distances L_{ev} and L_{eh} are assumed to be $1\frac{1}{2}$-in. Weld sizes are tabulated equal to $\frac{3}{4}t_p$.

While the tabular values are based on a = 3 in., they may conservatively be used for values of a between $2\frac{1}{2}$-in. and 3 in. The tabulated values are valid for laterally supported beams, in steel and composite construction, all types of loading, snug-tightened and fully-tensioned bolts, and for supported and supporting members of all grades of steel.

Example 9-11

Given: Design a single-plate connection for a W16×50 beam to a W14×90 column flange.

R_u = 55 kips

W16×50

t_w = 0.380 in. d = 16.26 in. t_f = 0.630 in.
F_y = 50 ksi, F_u = 65 ksi

W14×90

t_f = 0.710 in.
F_y = 50 ksi, F_u = 65 ksi

Use $\frac{3}{4}$-in. diameter A325-N bolts in standard holes and 70 ksi electrode welds. Assume single plate material with F_y = 36 ksi and F_u = 58 ksi.

Solution: *Design bolts, single plate, and welds*

From Table 9-10, assuming the column provides a rigid support, for $\frac{3}{4}$-in. diameter A325-N bolts and single-plate material with $F_y = 36$ ksi and $F_u = 58$ ksi, select four rows of bolts, $\frac{1}{4}$-in. single-plate thickness, and $\frac{3}{16}$-in. fillet weld size.

$\phi R_n = 55.5$ kips > 55 kips   **o.k.**

Check supported beam web

From Table 9-2, for four rows of bolts, beam material with $F_y = 50$ ksi and $F_u = 65$ ksi, and $L_{ev} = 1\frac{1}{2}$-in. and $L_{eh} = 1\frac{1}{2}$-in. (Assumed to be $1\frac{1}{4}$-in. for calculation purposes to account for possible underrun in beam length),

$\phi R_n = (351$ kips/in.$)(0.380$ in.$)$
$= 133$ kips > 55 kips   **o.k.**

Example 9-12

Given: Design a single-plate connection for a W18×35 beam to a W21×62 girder-web.

$R_u = 40$ kips

W18×35

$t_w = 0.300$ in. $d = 17.70$ in.
$F_y = 50$ ksi, $F_u = 65$ ksi

top flange coped 2-in. deep by 3-in. long, $L_{ev} = 1\frac{1}{2}$-in., $L_{eh} = 1\frac{1}{2}$-in. (Assumed to be $1\frac{1}{4}$-in. for calculation purposes to account for possible underrun in beam length),

W21×62

$t_w = 0.400$ in.
$F_y = 50$ ksi, $F_u = 65$ ksi

Use $\frac{3}{4}$-in. diameter A325-N bolts in standard holes and 70 ksi electrode welds. Assume single-plate material with $F_y = 36$ ksi and $F_u = 58$ ksi.

Solution: *Design bolts, single plate, and welds*

From Table 9-10, assuming the girder provides a flexible support, for $\frac{3}{4}$-in. diameter A325-N bolts and single-plate material with $F_y = 36$ ksi and $F_u = 58$ ksi, select four rows of bolts, $\frac{1}{4}$-in. single-plate thickness, and $\frac{3}{16}$-in. fillet weld size.

$\phi R_n = 44.7$ kips > 40 kips   **o.k.**

Check supported beam web

From Table 9-2, for four rows of bolts, beam material with $F_y = 50$ ksi and $F_u = 65$ ksi, and $L_{ev} = 1\frac{1}{2}$-in. and $L_{eh} = 1\frac{1}{2}$-in. (Assumed to be $1\frac{1}{4}$-in.

for calculation purposes to account for possible underrun in beam length),

ϕR_n = (264 kips/in.)(0.300 in.)
= 79.2 kips > 40 kips   **o.k.**

Check flexural yielding of the coped section

From Table 8-49, S_{net} = 18.2 in.^{3}

$$\phi R_n = \frac{0.9 F_y S_{net}}{e}$$

$$= \frac{0.9 \, (50 \text{ ksi}) \, (18.2 \text{ in.}^3)}{3 \text{ in.} + 1\frac{3}{4}\text{-in.}}$$

= 172 kips > 40 kips   **o.k.**

Check local web buckling at the cope

$$\frac{c}{d} = \frac{3 \text{ in.}}{17.70 \text{ in.}} = 0.170$$

$$\frac{c}{h_o} = \frac{3 \text{ in.}}{17.70 \text{ in.} - 2 \text{ in.}} = 0.191$$

Since $\frac{c}{d} \le 1.0$,

$$f = 2 \left(\frac{c}{d} \right)$$
$$= 2(0.170)$$
$$= 0.340$$

Since $\frac{c}{h_o} \le 1.0$,

$$k = 2.2 \left(\frac{h_o}{c} \right)^{1.65}$$

$$= 2.2 \left(\frac{1}{0.191} \right)^{1.65}$$

$$= 33.8$$

$$\phi F_{bc} = 23{,}590 \left(\frac{t_w}{h_o} \right)^2 fk$$

$$= 23{,}590 \left(\frac{0.300 \text{ in.}}{17.70 \text{ in.} - 2 \text{ in.}} \right)^2 (0.340) \, (33.8)$$

$$= 99.0 \text{ ksi}$$

$$\phi R_n = \frac{\phi F_{bc} S_{net}}{e}$$

$$= \frac{(99.0 \text{ ksi}) \, (18.2 \text{ in.}^3)}{(3 \text{ in.} + 1\frac{3}{4}\text{-in.})}$$

= 379 kips > 40 kips   **o.k.**

¾-in. diameter bolts					**Table 9-10.**					
					Single-Plate Connections					
			Bolt, Weld, and Single-Plate Design Strength, kips							
	ASTM	**Thread**	**Support**	**Hole**	**Plate Thickness, in.**					
n	**Desig.**	**Cond.**	**Cond.**	**Type**	**¼**	**⁵⁄₁₆**	**³⁄₈**	**⁷⁄₁₆**	**½**	**⁹⁄₁₆**
9 (L = 27)	A325	N	Flexible	STD	—	—	—	115	—	—
				SSLT	—	—	—	130	—	—
			Rigid	STD	—	—	—	115	—	—
				SSLT	—	—	—	130	—	—
		X	Flexible	STD	—	—	—	144	—	—
				SSLT	—	—	—	162	—	—
			Rigid	STD	—	—	—	144	—	—
				SSLT	—	—	—	162	—	—
	A490	N	Flexible	STD	—	—	—	144	—	—
				SSLT	—	—	—	162	—	—
			Rigid	STD	—	—	—	144	—	—
				SSLT	—	—	—	162	—	—
		X	Flexible	STD	—	—	—	179	—	—
				SSLT	—	—	—	203	—	—
			Rigid	STD	—	—	—	179	—	—
				SSLT	—	—	—	203	—	—
8 (L = 24)	A325	N	Flexible	STD	—	—	106	106	—	—
				SSLT	—	—	113	113	—	—
			Rigid	STD	—	—	106	106	—	—
				SSLT	—	—	117	117	—	—
		X	Flexible	STD	—	—	132	132	—	—
				SSLT	—	—	142	142	—	—
			Rigid	STD	—	—	132	132	—	—
				SSLT	—	—	147	147	—	—
	A490	N	Flexible	STD	—	—	132	132	—	—
				SSLT	—	—	142	142	—	—
			Rigid	STD	—	—	132	132	—	—
				SSLT	—	—	147	147	—	—
		X	Flexible	STD	—	—	165	165	—	—
				SSLT	—	—	166	177	—	—
			Rigid	STD	—	—	165	165	—	—
				SSLT	—	—	166	183	—	—
7 (L = 21)	A325	N	Flexible	STD	—	96.4	96.4	96.4	—	—
				SSLT	—	96.4	96.4	96.4	—	—
			Rigid	STD	—	96.4	96.4	96.4	—	—
				SSLT	—	104	104	104	—	—
		X	Flexible	STD	—	120	120	120	—	—
				SSLT	—	120	120	120	—	—
			Rigid	STD	—	120	120	120	—	—
				SSLT	—	121	131	131	—	—
	A490	N	Flexible	STD	—	120	120	120	—	—
				SSLT	—	120	120	120	—	—
			Rigid	STD	—	120	120	120	—	—
				SSLT	—	121	131	131	—	—
		X	Flexible	STD	—	121	146	151	—	—
				SSLT	—	121	146	151	—	—
			Rigid	STD	—	121	146	151	—	—
				SSLT	—	121	146	163	—	—
Weld Size					**³⁄₁₆**	**¼**	**⁵⁄₁₆**	**³⁄₈**	**³⁄₈**	**⁷⁄₁₆**

STD = Standard holes   N = Threads included
SSLT = Short-slotted holes transverse   X = Threads excluded
       to direction of load

¾-in. diameter bolts

Table 9-10 (cont.).
Single-Plate Connections

Bolt, Weld, and Single-Plate Design Strength, kips

n	ASTM Desig.	Thread Cond.	Support Cond.	Hole Type	Plate Thickness, in.					
					¼	⁵⁄₁₆	⅜	⁷⁄₁₆	½	⁹⁄₁₆
6 (L = 18)	A325	N	Flexible	STD	—	79.2	79.2	79.2	—	—
				SSLT	—	79.2	79.2	79.2	—	—
			Rigid	STD	—	86.7	86.7	86.7	—	—
				SSLT	—	91.1	91.1	91.1	—	—
		X	Flexible	STD	—	99.0	99.0	99.0	—	—
				SSLT	—	99.0	99.0	99.0	—	—
			Rigid	STD	—	104	108	108	—	—
				SSLT	—	104	114	114	—	—
	A490	N	Flexible	STD	—	99.0	99.0	99.0	—	—
				SSLT	—	99.0	99.0	99.0	—	—
			Rigid	STD	—	104	108	108	—	—
				SSLT	—	104	114	114	—	—
		X	Flexible	STD	—	104	124	124	—	—
				SSLT	—	104	124	124	—	—
			Rigid	STD	—	104	125	135	—	—
				SSLT	—	104	125	142	—	—
5 (L = 15)	A325	N	Flexible	STD	62.0	62.0	62.0	62.0	—	—
				SSLT	62.0	62.0	62.0	62.0	—	—
			Rigid	STD	69.3	74.8	74.8	74.8	—	—
				SSLT	69.3	77.9	77.9	77.9	—	—
		X	Flexible	STD	69.3	77.5	77.5	77.5	—	—
				SSLT	69.3	77.5	77.5	77.5	—	—
			Rigid	STD	69.3	86.7	93.4	93.4	—	—
				SSLT	69.3	86.7	97.4	97.4	—	—
	A490	N	Flexible	STD	69.3	77.5	77.5	77.5	—	—
				SSLT	69.3	77.5	77.5	77.5	—	—
			Rigid	STD	69.3	86.7	93.4	93.4	—	—
				SSLT	69.3	86.7	97.4	97.4	—	—
		X	Flexible	STD	69.3	86.7	96.9	96.9	—	—
				SSLT	69.3	86.7	96.9	96.9	—	—
			Rigid	STD	69.3	86.7	104	117	—	—
				SSLT	69.3	86.7	104	121	—	—
4 (L = 12)	A325	N	Flexible	STD	44.7	44.7	44.7	44.7	—	—
				SSLT	44.7	44.7	44.7	44.7	—	—
			Rigid	STD	55.5	63.6	63.6	63.6	—	—
				SSLT	55.5	61.9	61.9	61.9	—	—
		X	Flexible	STD	55.0	55.9	55.9	55.9	—	—
				SSLT	55.0	55.9	55.9	55.9	—	—
			Rigid	STD	55.5	69.3	79.5	79.5	—	—
				SSLT	55.5	69.3	77.3	77.3	—	—
	A490	N	Flexible	STD	55.0	55.9	55.9	55.9	—	—
				SSLT	55.0	55.9	55.9	55.9	—	—
			Rigid	STD	55.5	69.3	79.5	79.5	—	—
				SSLT	55.5	69.3	77.3	77.3	—	—
		X	Flexible	STD	55.0	68.8	69.8	69.8	—	—
				SSLT	55.0	68.8	69.8	69.8	—	—
			Rigid	STD	55.5	69.3	83.2	97.1	—	—
				SSLT	55.5	69.3	83.2	96.7	—	—
Weld Size					³⁄₁₆	¼	⁵⁄₁₆	⅜	⅜	⁷⁄₁₆

STD = Standard holes
SSLT = Short-slotted holes transverse to direction of load

N = Threads included
X = Threads excluded

| 3/4-in. diameter bolts | | |

Table 9-10 (cont.).
Single-Plate Connections

Bolt, Weld, and Single-Plate Design Strength, kips

n	ASTM Desig.	Thread Cond.	Support Cond.	Hole Type	Plate Thickness, in.					
					1/4	5/16	3/8	7/16	1/2	9/16
3 (L = 9)	A325	N	Flexible	STD	27.8	27.8	27.8	27.8	—	—
				SSLT	27.8	27.8	27.8	27.8	—	—
			Rigid	STD	41.6	41.7	41.7	41.7	—	—
				SSLT	41.6	41.7	41.7	41.7	—	—
		X	Flexible	STD	34.3	34.8	34.8	34.8	—	—
				SSLT	34.3	34.8	34.8	34.8	—	—
			Rigid	STD	41.6	52.0	52.1	52.1	—	—
				SSLT	41.6	52.0	52.1	52.1	—	—
	A490	N	Flexible	STD	34.3	34.8	34.8	34.8	—	—
				SSLT	34.3	34.8	34.8	34.8	—	—
			Rigid	STD	41.6	52.0	52.1	52.1	—	—
				SSLT	41.6	52.0	52.1	52.1	—	—
		X	Flexible	STD	34.3	42.8	43.5	43.5	—	—
				SSLT	34.3	42.8	43.5	43.5	—	—
			Rigid	STD	41.6	52.0	62.4	65.1	—	—
				SSLT	41.6	52.0	62.4	65.1	—	—
2 (L = 6)	A325	N	Flexible	STD	14.0	14.0	14.0	14.0	—	—
				SSLT	14.0	14.0	14.0	14.0	—	—
			Rigid	STD	18.8	18.8	18.8	18.8	—	—
				SSLT	21.0	21.0	21.0	21.0	—	—
		X	Flexible	STD	17.2	17.5	17.5	17.5	—	—
				SSLT	17.2	17.5	17.5	17.5	—	—
			Rigid	STD	23.1	23.5	23.5	23.5	—	—
				SSLT	25.8	26.2	26.2	26.2	—	—
	A490	N	Flexible	STD	17.2	17.5	17.5	17.5	—	—
				SSLT	17.2	17.5	17.5	17.5	—	—
			Rigid	STD	23.1	23.5	23.5	23.5	—	—
				SSLT	25.8	26.2	26.2	26.2	—	—
		X	Flexible	STD	17.2	21.5	21.9	21.9	—	—
				SSLT	17.2	21.5	21.9	21.9	—	—
			Rigid	STD	23.1	28.9	29.3	29.3	—	—
				SSLT	25.8	32.3	32.8	32.8	—	—
Weld Size					3/16	1/4	5/16	3/8	3/8	7/16

STD = Standard holes   N = Threads included
SSLT = Short-slotted holes transverse   X = Threads excluded
         to direction of load

| Table 9-10 (cont.). Single-Plate Connections | | | | | $^7/_8$-in. diameter bolts | | | | | |

Bolt, Weld, and Single-Plate Design Strength, kips

n	ASTM Desig.	Thread Cond.	Support Cond.	Hole Type	Plate Thickness, in.					
					$^1/_4$	$^5/_{16}$	$^3/_8$	$^7/_{16}$	$^1/_2$	$^9/_{16}$
9 (L = 27)	A325	N	Flexible	STD	—	—	—	156	156	—
				SSLT	—	—	—	177	177	—
			Rigid	STD	—	—	—	156	156	—
				SSLT	—	—	—	177	177	—
		X	Flexible	STD	—	—	—	195	195	—
				SSLT	—	—	—	206	221	—
			Rigid	STD	—	—	—	195	195	—
				SSLT	—	—	—	206	221	—
	A490	N	Flexible	STD	—	—	—	195	195	—
				SSLT	—	—	—	206	221	—
			Rigid	STD	—	—	—	195	195	—
				SSLT	—	—	—	206	221	—
		X	Flexible	STD	—	—	—	206	235	—
				SSLT	—	—	—	206	235	—
			Rigid	STD	—	—	—	206	235	—
				SSLT	—	—	—	206	235	—
8 (L = 24)	A325	N	Flexible	STD	—	—	144	144	144	—
				SSLT	—	—	154	154	154	—
			Rigid	STD	—	—	144	144	144	—
				SSLT	—	—	157	160	160	—
		X	Flexible	STD	—	—	157	180	180	—
				SSLT	—	—	157	183	193	—
			Rigid	STD	—	—	157	180	180	—
				SSLT	—	—	157	183	200	—
	A490	N	Flexible	STD	—	—	157	180	180	—
				SSLT	—	—	157	183	193	—
			Rigid	STD	—	—	157	180	180	—
				SSLT	—	—	157	183	200	—
		X	Flexible	STD	—	—	157	183	209	—
				SSLT	—	—	157	183	209	—
			Rigid	STD	—	—	157	183	209	—
				SSLT	—	—	157	183	209	—
7 (L = 21)	A325	N	Flexible	STD	—	114	131	131	131	—
				SSLT	—	114	131	131	131	—
			Rigid	STD	—	114	131	131	131	—
				SSLT	—	114	137	142	142	—
		X	Flexible	STD	—	114	137	160	164	—
				SSLT	—	114	137	160	164	—
			Rigid	STD	—	114	137	160	164	—
				SSLT	—	114	137	160	178	—
	A490	N	Flexible	STD	—	114	137	160	164	—
				SSLT	—	114	137	160	164	—
			Rigid	STD	—	114	137	160	164	—
				SSLT	—	114	137	160	178	—
		X	Flexible	STD	—	114	137	160	183	—
				SSLT	—	114	137	160	183	—
			Rigid	STD	—	114	137	160	183	—
				SSLT	—	114	137	160	183	—
Weld Size					$^3/_{16}$	$^1/_4$	$^5/_{16}$	$^3/_8$	$^3/_8$	$^7/_{16}$

STD = Standard holes
SSLT = Short-slotted holes transverse to direction of load

N = Threads included
X = Threads excluded

⅞-in. diameter bolts

Table 9-10 (cont.).
Single-Plate Connections

Bolt, Weld, and Single-Plate Design Strength, kips

n	ASTM Desig.	Thread Cond.	Support Cond.	Hole Type	¼	5⁄16	⅜	7⁄16	½	9⁄16
6 (L = 18)	A325	N	Flexible	STD	—	97.9	108	108	108	—
				SSLT	—	97.9	108	108	108	—
			Rigid	STD	—	97.9	117	118	118	—
				SSLT	—	97.9	117	124	124	—
		X	Flexible	STD	—	97.9	117	135	135	—
				SSLT	—	97.9	117	135	135	—
			Rigid	STD	—	97.9	117	137	147	—
				SSLT	—	97.9	117	137	155	—
	A490	N	Flexible	STD	—	97.9	117	135	135	—
				SSLT	—	97.9	117	135	135	—
			Rigid	STD	—	97.9	117	137	147	—
				SSLT	—	97.9	117	137	155	—
		X	Flexible	STD	—	97.9	117	137	157	—
				SSLT	—	97.9	117	137	157	—
			Rigid	STD	—	97.9	117	137	157	—
				SSLT	—	97.9	117	137	157	—
5 (L = 15)	A325	N	Flexible	STD	65.3	81.6	84.4	84.4	84.4	—
				SSLT	65.3	81.6	84.4	84.4	84.4	—
			Rigid	STD	65.3	81.6	97.9	102	102	—
				SSLT	65.3	81.6	97.9	106	106	—
		X	Flexible	STD	65.3	81.6	97.9	106	106	—
				SSLT	65.3	81.6	97.9	106	106	—
			Rigid	STD	65.3	81.6	97.9	114	127	—
				SSLT	65.3	81.6	97.9	114	131	—
	A490	N	Flexible	STD	65.3	81.6	97.9	106	106	—
				SSLT	65.3	81.6	97.9	106	106	—
			Rigid	STD	65.3	81.6	97.9	114	127	—
				SSLT	65.3	81.6	97.9	114	131	—
		X	Flexible	STD	65.3	81.6	97.9	114	131	—
				SSLT	65.3	81.6	97.9	114	131	—
			Rigid	STD	65.3	81.6	97.9	114	131	—
				SSLT	65.3	81.6	97.9	114	131	—
4 (L = 12)	A325	N	Flexible	STD	52.2	60.8	60.8	60.8	60.8	—
				SSLT	52.2	60.8	60.8	60.8	60.8	—
			Rigid	STD	52.2	65.3	78.3	86.6	86.6	—
				SSLT	52.2	65.3	78.3	84.2	84.2	—
		X	Flexible	STD	52.2	65.3	76.0	76.0	76.0	—
				SSLT	52.2	65.3	76.0	76.0	76.0	—
			Rigid	STD	52.2	65.3	78.3	91.4	104	—
				SSLT	52.2	65.3	78.3	91.4	104	—
	A490	N	Flexible	STD	52.2	65.3	76.0	76.0	76.0	—
				SSLT	52.2	65.3	76.0	76.0	76.0	—
			Rigid	STD	52.2	65.3	78.3	91.4	104	—
				SSLT	52.2	65.3	78.3	91.4	104	—
		X	Flexible	STD	52.2	65.3	78.3	91.4	95.0	—
				SSLT	52.2	65.3	78.3	91.4	95.0	—
			Rigid	STD	52.2	65.3	78.3	91.4	104	—
				SSLT	52.2	65.3	78.3	91.4	104	—
Weld Size					3⁄16	¼	5⁄16	⅜	⅜	7⁄16

STD = Standard holes
SSLT = Short-slotted holes transverse to direction of load

N = Threads included
X = Threads excluded

							Plate Thickness, in.				

Table 9-10 (cont.).
Single-Plate Connections

7/8-in. diameter bolts

Bolt, Weld, and Single-Plate Design Strength, kips

n	ASTM Desig.	Thread Cond.	Support Cond.	Hole Type	1/4	5/16	3/8	7/16	1/2	9/16
3 (L = 9)	A325	N	Flexible	STD	37.9	37.9	37.9	37.9	37.9	—
				SSLT	37.9	37.9	37.9	37.9	37.9	—
			Rigid	STD	39.2	48.9	56.7	56.7	56.7	—
				SSLT	39.2	48.9	56.7	56.7	56.7	—
		X	Flexible	STD	39.2	47.4	47.4	47.4	47.4	—
				SSLT	39.2	47.4	47.4	47.4	47.4	—
			Rigid	STD	39.2	48.9	58.7	68.5	70.9	—
				SSLT	39.2	48.9	58.7	68.5	70.9	—
	A490	N	Flexible	STD	39.2	47.4	47.4	47.4	47.4	—
				SSLT	39.2	47.4	47.4	47.4	47.4	—
			Rigid	STD	39.2	48.9	58.7	68.5	70.9	—
				SSLT	39.2	48.9	58.7	68.5	70.9	—
		X	Flexible	STD	39.2	48.9	58.7	59.2	59.2	—
				SSLT	39.2	48.9	58.7	59.2	59.2	—
			Rigid	STD	39.2	48.9	58.7	68.5	78.3	—
				SSLT	39.2	48.9	58.7	68.5	78.3	—
2 (L = 6)	A325	N	Flexible	STD	19.0	19.0	19.0	19.0	19.0	—
				SSLT	19.0	19.0	19.0	19.0	19.0	—
			Rigid	STD	25.5	25.5	25.5	25.5	25.5	—
				SSLT	26.1	28.6	28.6	28.6	28.6	—
		X	Flexible	STD	20.1	23.8	23.8	23.8	23.8	—
				SSLT	20.1	23.8	23.8	23.8	23.8	—
			Rigid	STD	26.1	31.9	31.9	31.9	31.9	—
				SSLT	26.1	32.6	35.7	35.7	35.7	—
	A490	N	Flexible	STD	20.1	23.8	23.8	23.8	23.8	—
				SSLT	20.1	23.8	23.8	23.8	23.8	—
			Rigid	STD	26.1	31.9	31.9	31.9	31.9	—
				SSLT	26.1	32.6	35.7	35.7	35.7	—
		X	Flexible	STD	20.1	25.1	29.8	29.8	29.8	—
				SSLT	20.1	25.1	29.8	29.8	29.8	—
			Rigid	STD	26.1	32.6	39.2	39.9	39.9	—
				SSLT	26.1	32.6	39.2	44.6	44.6	—
Weld Size					3/16	1/4	5/16	3/8	3/8	7/16

STD = Standard holes
SSLT = Short-slotted holes transverse
to direction of load

N = Threads included
X = Threads excluded

1-in. diameter bolts

Table 9-10 (cont.).
Single-Plate Connections

Bolt, Weld, and Single-Plate Design Strength, kips

n	ASTM Desig.	Thread Cond.	Support Cond.	Hole Type	Plate Thickness, in.					
					1/4	5/16	3/8	7/16	1/2	9/16
9 (L = 27)	A325	N	Flexible	STD	—	—	—	192	204	204
				SSLT	—	—	—	192	220	231
			Rigid	STD	—	—	—	192	204	204
				SSLT	—	—	—	192	220	231
		X	Flexible	STD	—	—	—	192	220	247
				SSLT	—	—	—	192	220	247
			Rigid	STD	—	—	—	192	220	247
				SSLT	—	—	—	192	220	247
	A490	N	Flexible	STD	—	—	—	192	220	247
				SSLT	—	—	—	192	220	247
			Rigid	STD	—	—	—	192	220	247
				SSLT	—	—	—	192	220	247
		X	Flexible	STD	—	—	—	192	220	247
				SSLT	—	—	—	192	220	247
			Rigid	STD	—	—	—	192	220	247
				SSLT	—	—	—	192	220	247
8 (L = 24)	A325	N	Flexible	STD	—	—	146	171	188	188
				SSLT	—	—	146	171	195	201
			Rigid	STD	—	—	146	171	188	188
				SSLT	—	—	146	171	195	209
		X	Flexible	STD	—	—	146	171	195	220
				SSLT	—	—	146	171	195	220
			Rigid	STD	—	—	146	171	195	220
				SSLT	—	—	146	171	195	220
	A490	N	Flexible	STD	—	—	146	171	195	220
				SSLT	—	—	146	171	195	220
			Rigid	STD	—	—	146	171	195	220
				SSLT	—	—	146	171	195	220
		X	Flexible	STD	—	—	146	171	195	220
				SSLT	—	—	146	171	195	220
			Rigid	STD	—	—	146	171	195	220
				SSLT	—	—	146	171	195	220
7 (L = 21)	A325	N	Flexible	STD	—	107	128	149	171	171
				SSLT	—	107	128	149	171	171
			Rigid	STD	—	107	128	149	171	171
				SSLT	—	107	128	149	171	186
		X	Flexible	STD	—	107	128	149	171	192
				SSLT	—	107	128	149	171	192
			Rigid	STD	—	107	128	149	171	192
				SSLT	—	107	128	149	171	192
	A490	N	Flexible	STD	—	107	128	149	171	192
				SSLT	—	107	128	149	171	192
			Rigid	STD	—	107	128	149	171	192
				SSLT	—	107	128	149	171	192
		X	Flexible	STD	—	107	128	149	171	192
				SSLT	—	107	128	149	171	192
			Rigid	STD	—	107	128	149	171	192
				SSLT	—	107	128	149	171	192
Weld Size					3/16	1/4	5/16	3/8	3/8	7/16

STD = Standard holes
SSLT = Short-slotted holes transverse
       to direction of load

N = Threads included
X = Threads excluded

| | | | | 1-in. diameter bolts | | |

Table 9-10 (cont.).
Single-Plate Connections

Bolt, Weld, and Single-Plate Design Strength, kips

n	ASTM Desig.	Thread Cond.	Support Cond.	Hole Type	Plate Thickness, in.					
					1/4	5/16	3/8	7/16	1/2	9/16
6 (L = 18)	A325	N	Flexible	STD	—	91.5	110	128	141	141
				SSLT	—	91.5	110	128	141	141
			Rigid	STD	—	91.5	110	128	146	154
				SSLT	—	91.5	110	128	146	162
		X	Flexible	STD	—	91.5	110	128	146	165
				SSLT	—	91.5	110	128	146	165
			Rigid	STD	—	91.5	110	128	146	165
				SSLT	—	91.5	110	128	146	165
	A490	N	Flexible	STD	—	91.5	110	128	146	165
				SSLT	—	91.5	110	128	146	165
			Rigid	STD	—	91.5	110	128	146	165
				SSLT	—	91.5	110	128	146	165
		X	Flexible	STD	—	91.5	110	128	146	165
				SSLT	—	91.5	110	128	146	165
			Rigid	STD	—	91.5	110	128	146	165
				SSLT	—	91.5	110	128	146	165
5 (L = 15)	A325	N	Flexible	STD	61.0	76.3	91.5	107	110	110
				SSLT	61.0	76.3	91.5	107	110	110
			Rigid	STD	61.0	76.3	91.5	107	122	133
				SSLT	61.0	76.3	91.5	107	122	137
		X	Flexible	STD	61.0	76.3	91.5	107	122	137
				SSLT	61.0	76.3	91.5	107	122	137
			Rigid	STD	61.0	76.3	91.5	107	122	137
				SSLT	61.0	76.3	91.5	107	122	137
	A490	N	Flexible	STD	61.0	76.3	91.5	107	122	137
				SSLT	61.0	76.3	91.5	107	122	137
			Rigid	STD	61.0	76.3	91.5	107	122	137
				SSLT	61.0	76.3	91.5	107	122	137
		X	Flexible	STD	61.0	76.3	91.5	107	122	137
				SSLT	61.0	76.3	91.5	107	122	137
			Rigid	STD	61.0	76.3	91.5	107	122	137
				SSLT	61.0	76.3	91.5	107	122	137
4 (L = 12)	A325	N	Flexible	STD	48.8	61.0	73.2	79.5	79.5	79.5
				SSLT	48.8	61.0	73.2	79.5	79.5	79.5
			Rigid	STD	48.8	61.0	73.2	85.4	97.6	110
				SSLT	48.8	61.0	73.2	85.4	97.6	110
		X	Flexible	STD	48.8	61.0	73.2	85.4	97.6	99.3
				SSLT	48.8	61.0	73.2	85.4	97.6	99.3
			Rigid	STD	48.8	61.0	73.2	85.4	97.6	110
				SSLT	48.8	61.0	73.2	85.4	97.6	110
	A490	N	Flexible	STD	48.8	61.0	73.2	85.4	97.6	99.3
				SSLT	48.8	61.0	73.2	85.4	97.6	99.3
			Rigid	STD	48.8	61.0	73.2	85.4	97.6	110
				SSLT	48.8	61.0	73.2	85.4	97.6	110
		X	Flexible	STD	48.8	61.0	73.2	85.4	97.6	110
				SSLT	48.8	61.0	73.2	85.4	97.6	110
			Rigid	STD	48.8	61.0	73.2	85.4	97.6	110
				SSLT	48.8	61.0	73.2	85.4	97.6	110
Weld Size					3/16	1/4	5/16	3/8	3/8	7/16

STD = Standard holes
SSLT = Short-slotted holes transverse
to direction of load

N = Threads included
X = Threads excluded

1-in. diameter bolts

Table 9-10 (cont.).
Single-Plate Connections

Bolt, Weld, and Single-Plate Design Strength, kips

n	ASTM Desig.	Thread Cond.	Support Cond.	Hole Type	Plate Thickness, in.					
					1/4	5/16	3/8	7/16	1/2	9/16
3 (L = 9)	A325	N	Flexible	STD	36.6	45.8	49.5	49.5	49.5	49.5
				SSLT	36.6	45.8	49.5	49.5	49.5	49.5
			Rigid	STD	36.6	45.8	54.9	64.1	73.2	74.1
				SSLT	36.6	45.8	54.9	64.1	73.2	74.1
		X	Flexible	STD	36.6	45.8	54.9	61.9	61.9	61.9
				SSLT	36.6	45.8	54.9	61.9	61.9	61.9
			Rigid	STD	36.6	45.8	54.9	64.1	73.2	82.4
				SSLT	36.6	45.8	54.9	64.1	73.2	82.4
	A490	N	Flexible	STD	36.6	45.8	54.9	61.9	61.9	61.9
				SSLT	36.6	45.8	54.9	61.9	61.9	61.9
			Rigid	STD	36.6	45.8	54.9	64.1	73.2	82.4
				SSLT	36.6	45.8	54.9	64.1	73.2	82.4
		X	Flexible	STD	36.6	45.8	54.9	64.1	73.2	77.3
				SSLT	36.6	45.8	54.9	64.1	73.2	77.3
			Rigid	STD	36.6	45.8	54.9	64.1	73.2	82.4
				SSLT	36.6	45.8	54.9	64.1	73.2	82.4
2 (L = 6)	A325	N	Flexible	STD	23.0	24.9	24.9	24.9	24.9	24.9
				SSLT	23.0	24.9	24.9	24.9	24.9	24.9
			Rigid	STD	24.4	30.5	33.4	33.4	33.4	33.4
				SSLT	24.4	30.5	36.6	37.3	37.3	37.3
		X	Flexible	STD	23.0	28.7	31.1	31.1	31.1	31.1
				SSLT	23.0	28.7	31.1	31.1	31.1	31.1
			Rigid	STD	24.4	30.5	36.6	41.7	41.7	41.7
				SSLT	24.4	30.5	36.6	42.7	46.7	46.7
	A490	N	Flexible	STD	23.0	28.7	31.1	31.1	31.1	31.1
				SSLT	23.0	28.7	31.1	31.1	31.1	31.1
			Rigid	STD	24.4	30.5	36.6	41.7	41.7	41.7
				SSLT	24.4	30.5	36.6	42.7	46.7	46.7
		X	Flexible	STD	23.0	28.7	34.5	38.9	38.9	38.9
				SSLT	23.0	28.7	34.5	38.9	38.9	38.9
			Rigid	STD	24.4	30.5	36.6	42.7	48.8	52.1
				SSLT	24.4	30.5	36.6	42.7	48.8	54.9
Weld Size					3/16	1/4	5/16	3/8	3/8	7/16

STD = Standard holes　　　　　　　　N = Threads included
SSLT = Short-slotted holes transverse　　X = Threads excluded
　　　to direction of load

Single-Angle Connections

A single-angle connection is made with an angle on one side of the web of the beam to be supported, as illustrated in Figure 9-16. This angle is usually shop attached to the supporting member and may be bolted or welded to the supported beam as well as to the supporting member.

When the angle is welded to the support, adequate flexibility must be provided in the connection. As illustrated in Figure 9-16c, the weld is placed along the toe and across the bottom of the angle with a return at the top per LRFD Specification Section J2.2b. Note that welding across the entire top of the angle must be avoided as it would inhibit the flexibility and, therefore, the necessary end rotation of the connection; the performance of the resulting connection is unpredictable.

Design Checks

The design strengths of the bolts and/or welds and connected elements must be determined in accordance with the LRFD Specification; the applicable limit states are discussed in Part 8. In all cases, the design strength ϕR_n must equal or exceed the required strength R_u.

As illustrated in Figure 9-17, the effect of eccentricity should always be considered in the angle leg attached to the support. Additionally, eccentricity should be considered in the case of a double vertical row of bolts through the web of the supported beam or if the eccentricity exceeds three inches (2¾-in. gage + ¼-in. half web). Eccentricity should always be considered in the design of welds for single-angle connections.

Recommended Angle Length and Thickness

To provide for stability during erection, it is recommended that the mimimum angle length be one-half the T-dimension of the beam to be supported. The maximum length of the connection angles must be compatible with the T-dimension of an uncoped beam and the remaining web depth, exclusive of fillets, of a coped beam. Note that the angle may encroach on the fillet or fillets by ⅛-in. to ⁵⁄₁₆-in, depending upon the radius of the fillets; refer to Table 9-1.

A minimum angle thickness of ⅜-in. for ¾-in. and ⅞-in. diameter bolts, and ½-in. for 1 in. diameter bolts should be used. A 4×3 angle is normally selected for a single angle welded to the support with the 3 in. leg being the welded leg.

Shop and Field Practices

Single-angle connections may be made to the webs of supporting girders and to the flanges of supporting columns. Because of bolting and welding clearances, single-angle connections may not be suitable for connections to the webs of W8 columns, unless gages are reduced, and may be impossible for W6 columns.

When framing to a column flange, provision must be made for possible mill variation in the depth of the columns. Since the angle is usually shop attached to the column flange, play in the open holes or horizontal slots in the angle leg may be used to provide the necessary adjustment to compensate for the mill variation. Attaching the angle to the column flange offers the advantages of side erection of the beam and increased erection safety. Additionally, proper bay dimensions may be attained without the need for shims. These advantages are lost in the rare case that the angle is shop-attached to the supported beam web. The same is true for a girder web or truss support.

All-Bolted Single-Angle Connections
Table 9-11 is a design aid for all-bolted single-angle connections. The tabulated eccentrically loaded bolt group coefficients C are useful in determining the design strength ϕR_n, where

$$\phi R_n = C \times \phi r_n$$

In the above equation,

C = coefficient from Table 9-11

ϕr_n = the lesser of the design strength of one bolt in shear or bearing, kips

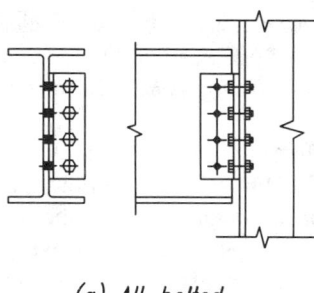

(a) All-bolted

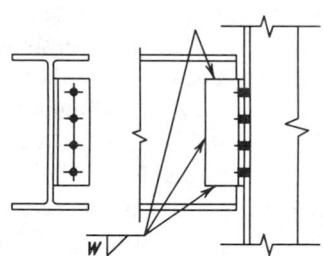

(b) Bolted/welded, angle welded to supported beam

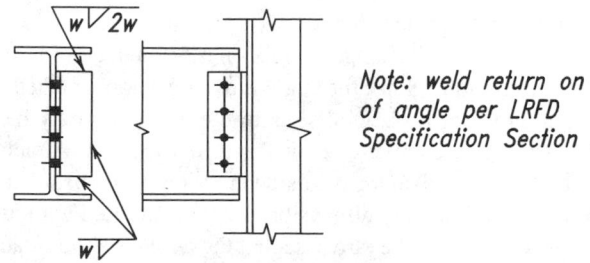

Note: weld return on top of angle per LRFD Specification Section J2.2b.

(c) Bolted/welded, angle welded to support

Figure 9-16. Single-angle connections.

Bolted/Welded Single-Angle Connections

Table 9-12 is a design aid for bolted/welded single angle connections. Electrode strength is assumed to be 70 ksi. All values are for comparison with factored loads. In the rare case where a single-angle connection must be field welded, erection bolts may be placed in the leg to be field welded.

Weld design strengths are determined by the instantaneous center of rotation method using Table 8-44 with $\theta = 0°$. The tabulated values assume a half-web thickness of ¼-in. and may be used conservatively for lesser half-web thicknesses; for half-web thicknesses greater than ¼-in., reduce the tabulated values proportionally to eight percent at a half-web thickness of ½-in. The tabulated minimum supporting flange or web thickness is the thicknesses which matches the strength of the support material to the strength of the weld material. In a manner similar to that illustrated previously for Tables 9-3, the minimum material thickness (for one line of weld) may be calculated as:

$$t = \frac{2.58D}{F_y}$$

where D is the number of sixteenths in the weld size. When welds line up on opposite sides of the support, the minimum thickness is the sum of the thicknesses required for each weld. In either case, when less than the minimum material thickness is present, the tabulated weld design strength should be multiplied by the ratio of the thickness provided to the minimum thickness.

Example 9-13

Given: Design an all-bolted single-angle connection (case I) for a W18×35 beam to W21×62 girder-web connection.

$R_u = 40$ kips

W18×35

$t_w = 0.300$ in. $d = 17.70$
$F_y = 50$ ksi, $F_u = 65$ ksi

top flange coped 2 in. deep by 4 in. long

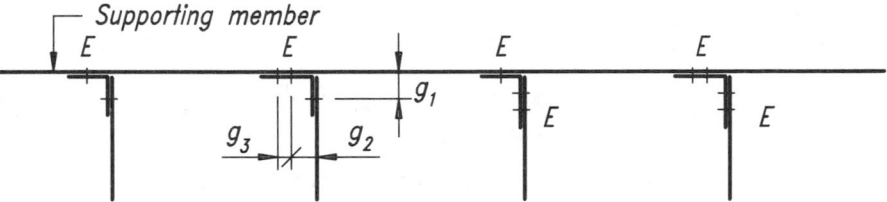

*E indicates that eccentricity must
be considered in this leg.
Gages g_1, g_2, and g_3 are usual gages
as shown in figure 9–5.*

Figure 9-17. Eccentricity in single-angle connections.

W21×62

$t_w = 0.400$ in.

$F_y = 50$ ksi, $F_u = 65$ ksi

Use ¾-in. diameter A325-N bolts in standard holes. Assume angle material with $F_y = 36$ ksi and $F_u = 58$ ksi.

Solution: *Design bolts and single angle*

Since half-web dimension of W18×35 is less than ¼-in., tabular values in Table 9-11 may conservatively be used. Bolt shear is more critical than bolt bearing; thus, $\phi r_n = 15.9$ kips

$$C_{min} = \frac{R_u}{\phi r_n}$$

$$= \frac{40 \text{ kips}}{15.9 \text{ kips / bolt}}$$

$$= 2.52$$

From Table 9-11, try a four-bolt connection with a ⅜-in. thick angle

$C = 3.07 > 2.52$ **o.k.**

Check shear yielding of the angle

$\phi R_n = 0.9 \ (0.6 F_y A_g)$

$\quad\quad = 0.9[0.6 \times 36 \text{ ksi} \ (11½\text{-in.} \times ⅜\text{-in.})]$

$\quad\quad = 83.8$ kips > 40 kips   **o.k.**

Check shear rupture of the angle

$\phi R_n = 0.75 \ (0.6 F_u A_n)$

$\quad\quad = 0.75[0.6 \times 58 \text{ ksi} \ (11½\text{-in.} \times ⅜\text{-in.} - 4 \times 0.875 \text{ in.} \times ⅜ \text{ in.})]$

$\quad\quad = 78.3$ kips > 40 kips   **o.k.**

Check block shear rupture of the angle

From Tables 8-47 and 8-48, with $L_{eh} = L_{ev} = 1¼$-in., $0.6 F_u A_{nv} > F_u A_{nt}$. Thus,

$\phi R_n = \phi \ [0.6 F_u A_{nv} + F_y A_{gt}]$

From Tables 8-48a and 8-48b,

$\phi R_n = (188 \text{ kips/in.} + 33.8 \text{ kips/in.}) ⅜\text{-in.}$

$\quad\quad = 83.2$ kips > 40 kips   **o.k.**

Check flexure of the support-leg of the angle

The required strength M_u is

$M_u = R_u e$

$\quad\quad = 40 \text{ kips} \times 2¼\text{-in.}$

$\quad\quad = 90$ in.-kips

For flexural yielding

$$\phi M_n = \phi F_y S_x$$

$$= 0.9 \, (36 \text{ ksi}) \left[\frac{(\frac{3}{8}\text{-in.}) \, (11\frac{1}{2}\text{-in.})^2}{6} \right]$$

$$= 268 \text{ in.-kips} > 90 \text{ in.-kips} \quad \textbf{o.k.}$$

For flexural rupture using general equation from Table 12-1 (bracket plates),

$$\phi M_n = \phi F_u S_{net}$$

$$= 0.75(58 \text{ ksi}) \left[\frac{\frac{3}{8}\text{-in.}}{6} \left((11\frac{1}{2} \text{ -in.})^2 - \frac{(3\text{in.})^2(4)(4^2 - 1)(0.875 \text{ in.})}{11\frac{1}{2}\text{-in.}} \right) \right]$$

$$= 248 \text{ in.-kips} > 90 \text{ in.-kips} \quad \textbf{o.k.}$$

Check the supported beam web

From Table 9-2, for four rows of bolts, beam material with $F_y = 50$ ksi and $F_u = 65$ ksi, and $L_{ev} = 1\frac{1}{4}$-in. and $L_{eh} = 1\frac{1}{2}$-in. (Assumed to be $1\frac{1}{4}$-in. for calculation purposes to provide for possible underrun in beam length)

$$\phi R_n = (257 \text{ kips/in.})(0.300 \text{ in.})$$

$$= 77.1 \text{ kips} > 40 \text{ kips} \quad \textbf{o.k.}$$

Check flexural yielding on the coped section

From Table 8-49, $S_{net} = 18.2$ in.^{3}

$$\phi R_n = \frac{0.9 F_y S_{net}}{e}$$

$$= \frac{0.9 \, (50 \text{ ksi}) \, (18.2 \text{ in.}^3)}{(4 \text{ in.} + \frac{1}{2}\text{-in.})}$$

$$= 182 \text{ kips} > 40 \text{ kips} \quad \textbf{o.k.}$$

Check local web buckling at the cope

$$\frac{c}{d} = \frac{4 \text{ in.}}{17.70 \text{ in.}} = 0.226$$

$$\frac{c}{h_o} = \frac{4 \text{ in.}}{(17.70 \text{ in.} - 2 \text{ in.})} = 0.255$$

Since $\frac{c}{d} \le 1.0$,

$$f = 2\left(\frac{c}{d}\right)$$

$$= 2(0.226)$$

$$= 0.452$$

Since $\frac{c}{h_o} \le 1.0$,

$$k = 2.2 \left(\frac{h_o}{c}\right)^{1.65}$$

$$= 2.2 \left(\frac{1}{0.255} \right)^{1.65}$$

$$= 21.0$$

$$\phi F_{bc} = 23,590 \left(\frac{t_w}{h_o} \right)^2 fk$$

$$= 23,590 \left(\frac{0.300 \text{ in.}}{17.70 \text{ in.} - 2 \text{ in.}} \right)^2 (0.452)(21.0)$$

$$= 81.8 \text{ ksi}$$

$$\phi R_n = \frac{\phi F_{bc} S_{net}}{e}$$

$$= \frac{(81.8 \text{ ksi})(18.2 \text{ in.}^3)}{(4 \text{ in.} + \frac{1}{2}\text{-in.})}$$

$$= 331 \text{ kips} > 40 \text{ kips} \quad \textbf{o.k.}$$

Check supporting girder web

From Table 9-2 for four rows of bolts and girder material with $F_u = 65$ ksi. Taking half the tabulated value,

$$\phi R_n = \frac{1}{2}(702 \text{ kips/in.})(0.400)$$

$$= 140 \text{ kips} > 40 \text{ kips} \quad \textbf{o.k.}$$

Example 9-14   Design a single-angle connection for a W16×50 beam to W14×90 column flange connection.

$R_u = 55$ kips

W16×50

$t_w = 0.380$ in. $d = 16.26$ in. $t_f = 0.630$ in.
$F_y = 50$ ksi, $F_u = 65$ ksi

W14×90

$t_f = 0.710$
$F_y = 50$ ksi, $F_u = 65$ ksi

Use ¾-in. diameter A325-N bolts to connect the supported beams to the single angle. Use 70 ksi electrode welds to connect the single angle to the column flange. Assume angle material with $F_y = 36$ ksi and $F_u = 58$ ksi.

Solution: *Design single angle, bolts, and welds*

Since half-web dimension of W16×50 is less than ¼-in., tabular values in Table 9-12 may conservatively be used.

From Table 9-12, try a four-bolt single angle (L4×3×⅜).

$$\phi R_n = 63.6 \text{ kips} > 55 \text{ kips} \quad \textbf{o.k.}$$

Also from Table 9-12, with a $\frac{3}{16}$-in. fillet weld size

$\phi R_n = 56.6$ kips > 55 kips   **o.k.**

Use four-bolt single-angle (L4×3×⅜)

Check supported beam web

The bearing strength of the beam web per bolt is

$$\phi r_n = \phi \, (2.4 d t F_u)$$
$$= 0.75(2.4 \times \frac{3}{4}\text{-in.} \times 0.380 \text{ in.} \times 65 \text{ ksi})$$
$$= 33.3 \text{ kips}$$

Since this exceeds the single shear strength per bolt, bolt bearing on the beam web is not critical.

Check support

From Table 9-12, the minimum support thickness for the $\frac{3}{16}$-in. welds is 0.31 in.

$t_w = 0.710 > 0.31$ **o.k.**

Table 9-11.
All-Bolted Single-Angle Connections

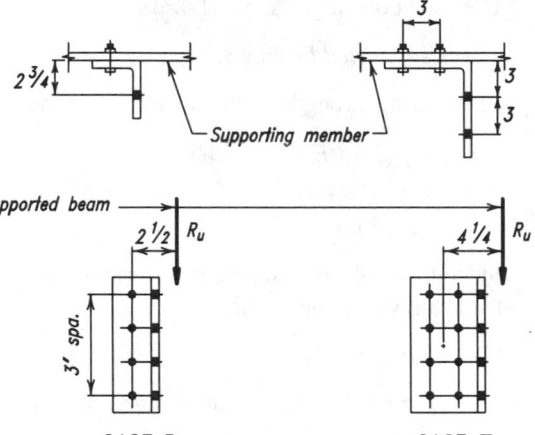

Note: standard holes in support leg of angle

Eccentrically Loaded Bolt Group Coefficients, C

Number of Bolts in One Vertical Row, n	Case I	Case II
12	11.4	21.5
11	10.4	19.4
10	9.37	17.3
9	8.35	15.1
8	7.32	13.0
7	6.27	10.8
6	5.22	8.70
5	4.15	6.63
4	3.07	4.70
3	1.99	2.94
2	1.03	1.61
1	—	0.518

$\phi R_n = C \times \phi r_n$

where

C = coefficient from Table above
ϕr_n = design strength of one bolt in shear or bearing, kips/bolt

Notes:
For eccentricities less than or equal to those shown above, tabulated values may be used.
For greater eccentricities, coefficient C should be recalculated from Table 8-18 or Table 8-19.
Connection may be bearing-type or slip-critical.

Table 9-12.
Bolted/Welded Single-Angle Connections

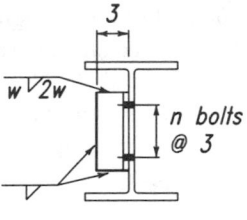 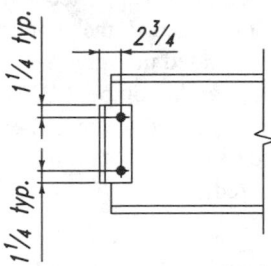

Number of Bolts in One Vertical Row	A325-N Bolt Shear Strength, kips		Angle Size	Angle Length	Weld (70 ksi)		Minimum t_w of Supporting Member with Angles Both Sides of Web	
	$3/4$-in.	$7/8$-in.	$(F_y = 36\ ksi)$	in.	Design Strength, kips	Size in.	$F_y = 36$ ksi	$F_y = 50$ ksi
12	191	260		35½	270	$5/16$	0.72	0.52
					216	$1/4$	0.57	0.41
					162	$3/16$	0.43	0.31
11	175	238		32½	247	$5/16$	0.72	0.52
					198	$1/4$	0.57	0.41
					148	$3/16$	0.43	0.31
10	159	217		29½	227	$5/16$	0.72	0.52
					182	$1/4$	0.57	0.41
					136	$3/16$	0.43	0.31
9	143	195		26½	205	$5/16$	0.72	0.52
					164	$1/4$	0.57	0.41
					123	$3/16$	0.43	0.31
8	127	173	L4×3×$3/8$	23½	185	$5/16$	0.72	0.52
					150	$1/4$	0.57	0.41
					111	$3/16$	0.43	0.31
7	111	152		20½	164	$5/16$	0.72	0.52
					131	$1/4$	0.57	0.41
					98.4	$3/16$	0.43	0.31
6	95.4	130		17½	141	$5/16$	0.72	0.52
					113	$1/4$	0.57	0.41
					84.5	$3/16$	0.43	0.31
5	79.5	108		14½	118	$5/16$	0.72	0.52
					94.5	$1/4$	0.57	0.41
					70.9	$3/16$	0.43	0.31
4	63.6	86.6		11½	94.3	$5/16$	0.72	0.52
					75.4	$1/4$	0.57	0.41
					56.6	$3/16$	0.43	0.31
3	47.7	64.9		8½	68.9	$5/16$	0.72	0.52
					55.1	$1/4$	0.57	0.41
					41.3	$3/16$	0.43	0.31
2	31.8	43.3		5½	42.1	$5/16$	0.72	0.52
					33.7	$1/4$	0.57	0.41
					25.2	$3/16$	0.43	0.31

Notes:
Gage in angle leg attached to beam web as well as leg width may be decreased. 3-in. welded leg may not be increased or decreased.

Tabulated weld design strengths are based on a $1/4$-in. half web for the supported member. Smaller half webs will result in these values being conservative. For half webs over $1/4$-in., weld values must be reduced proportionally to 8% for a $1/2$-in. half web or recalculated.

When the beam web thickness of the supporting member is less than the minimum and single-angle connections are back to back, either stagger the angles, or multiply the weld design strength by the ratio of the actual web thickness to the tabulated minimum thickness to determine the reduced weld design strength.

Tee Connections

A tee connection is made with a structural tee as illustrated in Figure 9-18. The tee may be bolted or welded to the supported beam as well as to the supporting member.

When the tee is welded to the support, adequate flexibility must be provided in the connection. As illustrated in Figure 9-18b, line welds are placed along the toes of the tee flange with a return at the top per LRFD Specification Section J2.2b. Note that welding across the entire top of the tee must be avoided as it would inhibit the flexibility and, therefore, the necessary end rotation of the connection; the performance of the resulting connection is unpredictable.

Design Checks

The design strengths of the bolts and/or welds and connected elements must be determined in accordance with the LRFD Specification; the applicable limit states are discussed in Part 8. In all cases, the design strength ϕR_n must equal or exceed the required strength R_u.

When the tee is welded to the support and bolted to the supported beam, for ductility in the tee connection, the 70 ksi weld size w must be such that

$$w_{min} = 0.0158 \frac{F_y t_f^2}{b} \left(\frac{b^2}{L^2} + 2 \right)$$

but need not exceed $\frac{3}{4}t_s$. In the above equation, t_f is the thickness of the tee flange, t_s is the thickness of the tee stem, and b and L are as illustrated in Figure 9-19.

For a tee bolted to the support and bolted or welded to the supported beam, the minimum diameter for bolts through the tee flange for ductility must be such that

$$d_{b \, min} = 0.163 t_f \sqrt{\frac{F_y}{b} \left(\frac{b^2}{L^2} + 2 \right)}$$

but need not exceed $0.69\sqrt{t_s}$. Additionally, to provide for rotational ductility when the tee stem is bolted to the supported beam, the maximum tee stem thickness should be such that

$$t_{s \, max} = \frac{d_b}{2} + \frac{1}{16} \text{ in.}$$

When the tee stem is welded to the supported beam, there is no perceived ductility problem for this weld.

In either case, eccentricity must be considered in the design of the tee connection. For a flexible support, the bolts or welds attaching the tee flange to the support must be designed for the shear R_u; the bolts through the tee stem must be designed for the shear R_u and the eccentric moment $R_u a$ where a is the distance from the face of the support to the centroid of the bolt group through the tee stem. For a rigid support, the bolts or welds attaching the tee flange to the support must be designed for the shear R_u and the eccentric moment $R_u a$; the bolts through the tee stem must be designed for the shear R_u.

---

*Note this value has been increased by $\frac{1}{4}$-in. to account for possible underrun in beam length.

Recommended Tee Length and Flange and Web Thicknesses

To provide for stability during erection, it is recommended that the mimimum tee length be one-half the *T*-dimension of the beam to be supported. The maximum length of the tee must be compatible with the *T*-dimension of an uncoped beam and the remaining web depth, exclusive of fillets, of a coped beam. Note that the tee may encroach on the fillet or fillets by ⅛-in. to 5⁄16-in, depending upon the radius of the fillets; refer to Table 9-1.

The flange thickness of tees used in simple shear connections should be held to a minimum to permit the flexure necessary to accommodate the end rotation of the beam.

Shop and Field Practices

Tee connections may be made to the webs of supporting girders and to the flanges of supporting columns. Because of bolting and welding clearances, tee connections may not be suitable for connections to the webs of **W8** columns, unless gages are reduced, and may be impossible for **W6** columns.

When framing to a column flange, provision must be made for possible mill variation in the depth of the columns. If the tee is shop attached to the column flange, play in the open holes usually furnishes the necessary adjustment to compensate for the mill variation. This approach offers the advantage of side erection of the beam. Alternatively, if the tee is shop attached to the supported beam web, the beam length could be shortened to provide for mill overrun and shims could be furnished at the appropriate intervals to fill the resulting gaps or to provide for mill underrun.

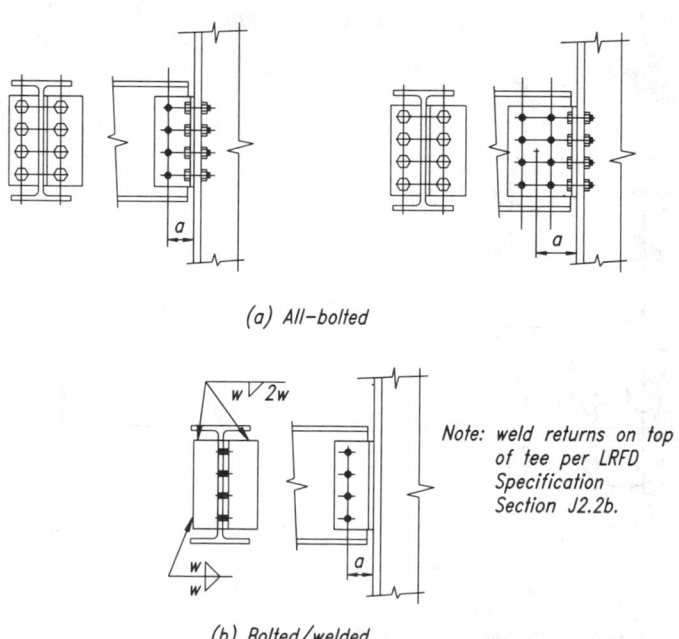

(a) All–bolted

Note: weld returns on top
of tee per LRFD
Specification
Section J2.2b.

(b) Bolted/welded

Figure 9-18. Tee Connections.

When a single vertical row of bolts is used in a tee stem, a 4-in. or 5-in. stem is required to accommodate the end distance of the supported beam and possible overrun/underrun in beam length. A double vertical row of bolts will require a 7-in. or 8-in. tee stem. There is no maximum limit on l_h for the tee stem.

Example 9-15

Given:
Design an all-bolted tee connection for a W16×50 beam to a W14×90 column flange.

$R_u = 55$ kips

W16×50

$t_w = 0.380$ in. $d = 16.26$ in. $t_f = 0.630$ in.
$F_y = 50$ ksi, $F_u = 65$ ksi

W14×90

$t_f = 0.710$ in.
$F_y = 50$ ksi, $F_u = 65$ ksi

Use ¾-in. diameter A325-N bolts in standard holes. Assume the tee has $F_y = 50$ ksi and $F_u = 65$ ksi.

Solution:
Try WT5×22.5 ($d = 5.050$ in., $b_f = 8.020$ in., $t_f = 0.620$ in., $t_s = 0.350$ in., $k_1 = {}^{11}\!/_{16}$-in.) with a four-bolt connection ($L = 11½$-in.) and $L_{eh} = 1¼$-in.

Check limitation on tee stem thickness

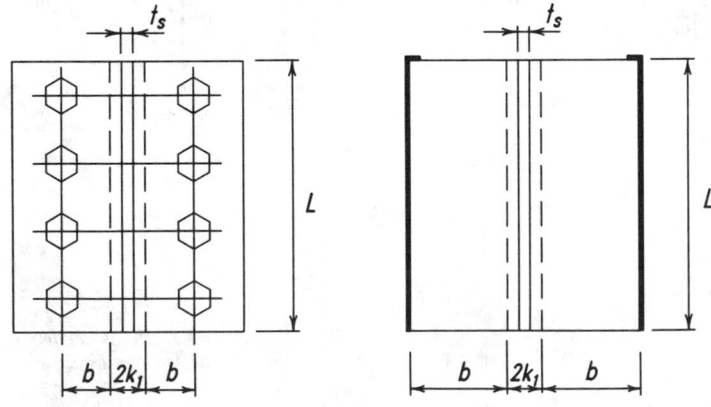

(a) Bolted flange   (b) Welded flange

Figure 9-19. Illustration of variables for tee connectins.

$$t_{s\,max} = \frac{d_b}{2} + \frac{1}{16}\text{-in.}$$

$$= \frac{\frac{3}{4}\text{-in.}}{2} + \frac{1}{16}\text{-in.}$$

$$= 0.438 \text{ in.} > 0.350 \text{ in.} \quad \textbf{o.k.}$$

Check limitation on bolt diameter for bolts through tee flange

Assuming a 5½-in. gage,

$$b = \frac{g - 2k_1}{2}$$

$$= \frac{5\frac{1}{2}\text{-in.} - 2\,(^{11}/_{16}\text{-in.})}{2}$$

$$= 2.06 \text{ in.}$$

$$d_{b\,min} = 0.163t_f\sqrt{\frac{F_y}{b}\left(\frac{b^2}{L^2} + 2\right)} \le 0.69\sqrt{t_s}$$

$$= 0.163(0.620 \text{ in.})\sqrt{\frac{50 \text{ ksi}}{2.06 \text{ in.}}\left(\frac{(2.06 \text{ in.})^2}{(11\frac{1}{2}\text{-in.})^2} + 2\right)} \le 0.69\sqrt{0.350 \text{ in.}}$$

$$= 0.710 \text{ in.} \le 0.408 \text{ in.}$$

$$= 0.408 \text{ in.}$$

Since $d_b = \frac{3}{4}$-in. $> d_{b\,min} = 0.408$ in.,   **o.k.**

Check bolt group through beam web for shear and bearing

$$a = d - L_{eh}$$

$$= 5.050 \text{ in.} - 1\frac{1}{4}\text{-in.}$$

$$= 3.80 \text{ in.}$$

Assuming the column provides a rigid support,

$$e_b = 0$$

Since bolt shear is more critical than bolt bearing, $\phi r_n = 15.9$ kips, Thus,

$$\phi R_n = n \times \phi r_n$$

$$= 4 \text{ bolts} \times 15.9 \text{ kips}$$

$$= 63.6 \text{ kips} > 55 \text{ kips} \quad \textbf{o.k.}$$

Check shear yielding of the tee stem

$$\phi R_n = 0.9\,(0.6F_y A_g)$$

$$= 0.9[0.6 \times 50 \text{ ksi } (\,11\frac{1}{2}\text{-in.} \times 0.350 \text{ in.})]$$

$$= 109 \text{ kips} > 55 \text{ kips} \quad \textbf{o.k.}$$

Check shear rupture of the tee stem

$$\phi R_n = 0.75\,(0.6F_u A_n)$$

$$= 0.75[0.6 \times 65 \text{ ksi } (11\frac{1}{2}\text{-in.} - 4 \times 0.875 \text{ in.})(0.350 \text{ in.})]$$

$$= 81.9 \text{ kips} > 55 \text{ kips} \quad \textbf{o.k.}$$

Check block shear rupture of the tee stem

From Tables 8-47 and 8-48, with $L_{eh} = L_{ev} = 1\frac{1}{4}$-in., $0.6F_u A_{nv} > F_u A_{nt}$. Thus,

$$\phi R_n = \phi[0.6F_u A_{nv} + F_y A_{gt}]$$

From Tables 8-48a and 8-48b,

$$\phi R_n = (210 \text{ kips/in.} + 46.9 \text{ kips/in.})(0.350 \text{ in.})$$
$$= 89.9 \text{ kips} > 55 \text{ kips} \quad \textbf{o.k.}$$

Check bolt group through support for shear and bearing

Calculate tensile force per bolt r_{ut}.

$$2r_{ut}[2 \times (1.5 \text{ in.} + 4.5 \text{ in.})] = R_u e$$

$$r_{ut} = \frac{55 \text{ kips}(5.050 \text{ in.} - 1\frac{1}{4}\text{-in.})}{2 \text{ bolts } (12 \text{ in.})}$$
$$= 8.71 \text{ kips/bolt}$$

Check design strength of bolts for tension-shear interaction

$$r_{uv} = \frac{55 \text{ kips}}{8 \text{ bolts}}$$
$$= 6.88 \text{ kips/bolt} < 15.9 \text{ kips/bolt} \quad \textbf{o.k.}$$
$$F_t = 117 \text{ ksi} - 1.9f_v \leq 90 \text{ ksi}$$
$$= 117 \text{ ksi} - 1.9\left(\frac{6.88 \text{ kips / bolt}}{0.4418 \text{ in.}^2}\right) \leq 90 \text{ ksi}$$
$$= 87.4 \text{ ksi}$$
$$\phi r_n = \phi F_t A_b$$
$$= 0.75(87.4 \text{ ksi})(0.4418 \text{ in.}^2)$$
$$= 29.0 \text{ kips/bolt} > 8.71 \text{ kips/bolt} \quad \textbf{o.k.}$$

Check bearing strength at bolt holes

With $L_e = 1\frac{1}{4}$-in. and $s = 3$ in., the bearing strength of the tee flange exceeds the single shear strength of the bolts. Therefore, bearing strength is **o.k.**

Check prying action

$$b = \frac{g - t_s}{2}$$
$$= \frac{5\frac{1}{2}\text{-in.} - 0.350 \text{ in.}}{2}$$
$$= 2.58 \text{ in.}^2$$
$$a = \frac{b_f - g}{2}$$

$$= \frac{8.020 \text{ in.} - 5\frac{1}{2}\text{-in.}}{2}$$

$$= 1.26 \text{ in.}$$

Since $a = 1.26$ in. is less than $1.25b = 3.23$ in., use $a = 1.26$ in. for calculation purposes.

$$b' = b - d/2$$

$$= 2.58 \text{ in.} - \frac{\frac{3}{4}\text{-in.}}{2}$$

$$= 2.21 \text{ in.}$$

$$a' = a + d/2$$

$$= 1.26 \text{ in.} + \frac{\frac{3}{4}\text{-in.}}{2}$$

$$= 1.64 \text{ in.}$$

$$\rho = \frac{b'}{a'}$$

$$= \frac{2.21 \text{ in.}}{1.64 \text{ in.}}$$

$$= 1.35$$

$$\beta = \frac{1}{\rho}\left(\frac{\phi r_n}{r_{ut}} - 1\right)$$

$$= \frac{1}{1.35}\left(\frac{29.0 \text{ kips / bolt}}{8.71 \text{ kips / bolt}} - 1\right)$$

$$= 1.72$$

Since $\beta \geq 1$, set $\alpha' = 1.0$

$$p = \frac{11\frac{1}{2}\text{-in.}}{4 \text{ bolts}}$$

$$= 2.88 \text{ in./bolt}$$

$$\delta = 1 - \frac{d'}{p}$$

$$= 1 - \frac{\frac{13}{16}\text{-in.}}{2.88 \text{ in.}}$$

$$= 0.718$$

$$t_{req} = \sqrt{\frac{4.44 r_{ut} b'}{p F_y (1 + \delta \alpha')}}$$

$$= \sqrt{\frac{4.44(8.71 \text{ kips / bolt})(2.21 \text{ in.})}{(2.88 \text{ in. / bolt})(50 \text{ ksi})(1 + (0.718)(1.0)]}}$$

$$= 0.588 \text{ in.} < 0.620 \text{ in.} \quad \textbf{o.k.}$$

Similarly, checks of the tee flange for shear yielding, shear rupture, and block shear will show that the tee flange is **o.k.**

Check the supported beam web

From Table 9-2, for four rows of $\frac{3}{4}$-in. diameter bolts and an uncoped beam with $F_y = 50$ ksi and $F_u = 65$ ksi,

$$\phi R_n = (351 \text{ kips/in.})(0.380 \text{ in.})$$
$$= 133 \text{ kips} > 55 \text{ kips} \quad \textbf{o.k.}$$

Check the supporting column flange

From Table 9-2, for four rows of $\frac{3}{4}$-in. diameter bolts with $F_y = 50$ ksi and $F_u = 65$ ksi,

$$\phi R_n = (702 \text{ kips/in.})(0.710 \text{ in.})$$
$$= 498 \text{ kips} > 55 \text{ kips} \quad \textbf{o.k.}$$

Example 9-16

Given:

Redesign the tee connecton of Example 9-15 to be bolted to the supported beam and welded to the support for a factored end reaction $R_u = 37$ kips.

Solution:

Try WT5×22.5 ($d = 5.050$ in., $b_f = 8.020$ in., $t_f = 0.620$ in., $t_s = 0.350$ in., $k_1 = \frac{11}{16}$-in.) with a four-bolt connection ($L = 11\frac{1}{2}$-in.) and $L_{eh} = 1\frac{1}{4}$-in.

Check limitation on tee stem thickness

$$t_{s\,max} = \frac{d_b}{2} + \frac{1}{16}\text{-in.}$$

$$= \frac{\frac{3}{4}\text{-in.}}{2} + \frac{1}{16}\text{-in.}$$

$$= 0.438 \text{ in.} > 0.350 \text{ in.} \quad \textbf{o.k.}$$

Design the welds connecting the tee flange to the column flange

This connection is inherently flexible because the welds are at the toes of the WT flanges. This is true independent of the rigidity of the support. Therefore, it is recommended that this connection be designed with a flexible support condition; any rigidity-induced weld forces are accounted for by the minimum required weld size w_{min}. Thus,

$$b = \frac{b_f - 2k_1}{2}$$

$$= \frac{8.020 \text{ in.} - 2\,(\frac{11}{16}\text{-in.})}{2}$$

$$= 3.32 \text{ in.}$$

$$w_{min} = 0.0158 \frac{F_y t_f^2}{b}\left(\frac{b^2}{L^2} + 2\right) \le \frac{3}{4} t_s$$

$$= 0.0158 \frac{(50 \text{ ksi})(0.620 \text{ in.})^2}{3.32 \text{ in.}}\left(\frac{(3.32 \text{ in.})^2}{(11\frac{1}{2}\text{-in.})^2} + 2\right) \le \frac{3}{4}\,(0.350 \text{ in.})$$

$$= 0.191 \text{ in.} \le 0.263 \text{ in.}$$

$$= 0.191 \text{ in.}$$

Try $\frac{1}{4}$-in. fillet welds.

$\phi R_n = 1.392Dl$
$= 1.392(4 \text{ sixteenths})(2 \times 11\frac{1}{2}\text{-in.})$
$= 128 \text{ kips} > 37 \text{ kips}$ **o.k.**

Use $\frac{1}{4}$-in. fillet welds.

Check stem side of connection

Since the connection is flexible, the tee stem and bolts must be designed for both the shear and the eccentric moment e_b where

$e_b = a$
$= 3.80 \text{ in.}$

Thus the tee stem and bolts must be designed for $R_u = 37$ kips and $R_u e_b = 141$ in.-kips.

Check bolt group through beam web for shear and bearing

From Table 8-18 for $\theta = 0°$ with s = 3 in., $e_x = e_b = 3.80$ in., and $n = 4$ bolts,

$C = 2.45$

and, since bolt shear is more critical than bolt bearing,

$\phi R_n = C \times \phi r_n$
$= 2.45(15.9 \text{ kips/bolt})$
$= 39.0 \text{ kips} > 37 \text{ kips}$ **o.k.**

Check flexure on the tee stem

For flexural yielding,

$\phi M_n = \phi F_y S_x$
$= 0.9(50 \text{ ksi})\dfrac{(0.350 \text{ in.})(11\frac{1}{2}\text{-in.})^2}{6}$
$= 347 \text{ in.-kips} > 141 \text{ in.-kips}$ **o.k.**

For flexural rupture (see Table 12-1),

$S_{net} = \dfrac{t}{6}\left[d^2 - \dfrac{s^2 n(n^2 - 1)(d_b + 0.125 \text{ in.})}{d} \right]$
$= \dfrac{0.350 \text{ in.}}{6}\left[(11\frac{1}{2}\text{-in.})^2 - \dfrac{(3 \text{ in.})^2(4)(4^2 - 1)(0.875 \text{ in.})}{11\frac{1}{2}\text{-in.}} \right]$
$= 5.32 \text{ in.}^3$
$\phi M_n = \phi F_u S_{net}$
$= 0.75(65 \text{ ksi})(5.32 \text{ in.}^3)$
$= 259 \text{ in.-kips} > 141 \text{ in.-kips}$ **o.k.**

Check shear yielding of the tee stem

$\phi R_n = 0.9(0.6F_y A_g)$

$$= 0.9[0.6 \times 50 \text{ ksi } (11\tfrac{1}{2}\text{-in.} \times 0.350 \text{ in.})]$$
$$= 109 \text{ kips} > 37 \text{ kips} \quad \textbf{o.k.}$$

Check shear rupture of the tee stem

$$\phi R_n = 0.75(0.6 F_u A_n)$$
$$= 0.75[0.6 \times 65 \text{ ksi } (11\tfrac{1}{2}\text{-in.} - 4 \times 0.875 \text{ in.})(0.350 \text{ in.})]$$
$$= 81.9 \text{ kips} > 37 \text{ kips} \quad \textbf{o.k.}$$

Check block shear rupture of the tee stem

From Tables 8-47 and 8-48, with $L_{eh} = L_{ev} = 1\tfrac{1}{4}$-in., $0.6 F_u A_{nv} > F_u A_{nt}$. Thus,

$$\phi R_n = \phi[0.6 F_u A_{nv} + F_y A_{gt}]$$

From Tables 8-48a and 8-48b,

$$\phi R_n = (210 \text{ kips/in.} + 46.9 \text{ kips/in.})(0.350 \text{ in.})$$
$$= 89.9 \text{ kips} > 37 \text{ kips} \quad \textbf{o.k.}$$

Check supported beam web

From Tables 9-2, for four rows of $\tfrac{3}{4}$-in. diameter bolts and an uncoped beam with $F_y = 50$ ksi and $F_u = 65$ ksi,

$$\phi R_n = (351 \text{ kips/in.})(0.380 \text{ in.})$$
$$= 133 \text{ kips} > 37 \text{ kips} \quad \textbf{o.k.}$$

Check supporting column flange

From Table 9-3, for beam web material with $F_y = 50$ ksi, the minimum support thickness is 0.26 in. for $\tfrac{1}{4}$-in. fillet welds

$$t_f = 0.710 \text{ in.} > 0.26 \text{ in.} \quad \textbf{o.k.}$$

SHEAR SPLICES

Shear splices are usually made with a single plate, as shown in Figure 9-20a, or two plates, as shown in Figures 9-20b and 9-20c. When a highly flexible splice is desired, the splice utilizing four normal framing angles, shown in Figure 9-21, is especially useful. These shear splices may be made by bolting and/or welding.

The design strengths of the bolts and/or welds and connected elements must be determined in accordance with the LRFD Specification; the applicable limit states are discussed in Part 8. In all cases, the design strength ϕR_n must equal or exceed the required strength R_u.

Eccentricity must be considered in the design of shear splices except all-bolted framing-angle-type shear splices as illustrated in Figure 9-5. When the splice is symmetrical, as shown for the bolted splice in Figure 9-20a, each side of the splice is equally restrained regardless of the relative flexibility of the spliced members. Accordingly, as illustrated in Figure 9-22, the eccentricity of the shear to the center of gravity of either bolt group is equal to half the distance between the centroids of the bolt groups and each bolt group must be designed for the shear R_u and one-half the eccentric moment $R_u e$ (Kulak and Green, 1990). This principle is also applicable to symmetrical welded splices.

When the splice is not symmetrical, as shown in Figures 9-20b and 9-20c, one side of the splice will possess a higher degree of rigidity. For the splice shown in Figure 9-20b, the right side is more rigid because the stiffness of the weld group exceeds the stiffness

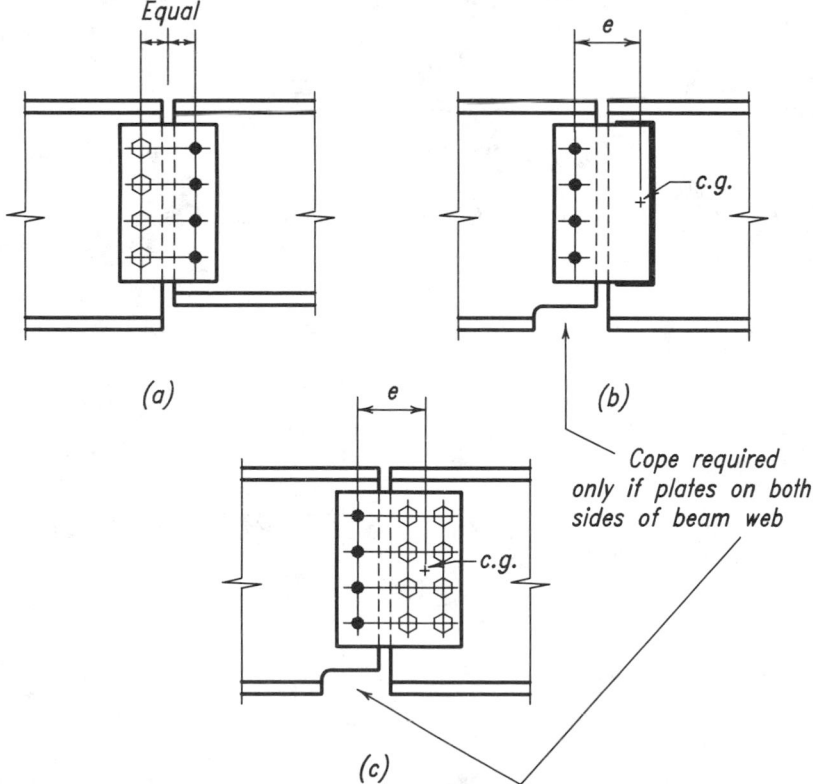

Figure 9-20. Shear splices utilizing plates.

of the bolt group, even if the bolts are fully tensioned. Also, for the splice shown in Figure 9-20c, the right side is more rigid since there are two vertical rows of bolts while the left side has only one. In these cases, it is conservative to design the side with the higher rigidity for the shear R_u and the full eccentric moment $R_u e$; the side with the lower rigidity is then designed for the shear R_u only. This principle is independent of the relative flexibility of the spliced members.

Some splices, such as those which occur at expansion joints, require special attention and are beyond the scope of this Manual.

Example 9-17

Given: Design an all-bolted single-plate shear splice between a W24×55 beam and W24×68 beam.

$$R_u = 60 \text{ kips}$$

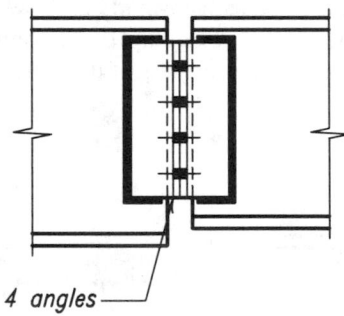

4 angles

Figure 9-21. Shear splice utilizing angles.

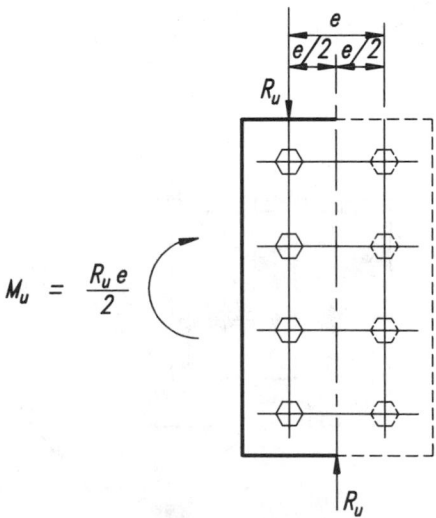

$$M_u = \frac{R_u e}{2}$$

Figure 9-22. Eccentricity in symmetrical shear splices.

W24×55

$t_w = 0.395$ in.
$F_y = 50$ ksi, $F_u = 65$ ksi

W24×68

$t_w = 0.415$ in.
$F_y = 50$ ksi, $F_u = 65$ ksi

Use ⅞-in. diameter A325-N bolts with five inches between vertical bolt rows. Assume plate material with $F_y = 36$ ksi and $F_u = 58$ ksi.

Solution: *Design bolt groups*

Using a symmetrical splice, each bolt group will carry one-half the eccentric moment. Thus, the eccentricity on each bolt group $e = 2\frac{1}{2}$-in.

For bolt shear, $\phi r_n = 21.6$ kips/bolt. For bearing on the web of the W24×55, $\phi r_n = 40.4$ kips/bolt. Since bolt shear is more critical,

$$C_{min} = \frac{R_u}{\phi r_n}$$
$$= \frac{60 \text{ kips}}{21.6 \text{ kips / bolt}}$$
$$= 2.78$$

From Table 8-18 with $\theta = 0°$ and $e_x = 2\frac{1}{2}$-in., a four-bolt connection provides

$C = 3.07 > 2.78$ **o.k.**

Design splice plate

Try PL⅜-in.×8 in.×10½-in.

Check bolt bearing on plate

$$\phi R_n = C(2.4dtF_u)$$
$$= 3.07(2.4 \times ⅞\text{-in.} \times ⅜\text{-in.} \times 58 \text{ ksi})$$
$$= 140 \text{ kips} > 60 \text{ kips}   \textbf{o.k.}$$

Check flexure of the plate

$$M_u = \frac{R_u e}{2}$$
$$= \frac{60 \text{ kips} \times 5\text{-in.}}{2}$$
$$= 150 \text{ in.-kips}$$

For flexural yielding,

$$\phi M_u = \phi F_y S_x$$
$$= 0.9 (36 \text{ ksi}) \left[\frac{⅜\text{-in. } (12 \text{ in.})^2}{6} \right]$$

$$= 292 \text{ in.-kips} > 150 \text{ in.-kips} \quad \textbf{o.k.}$$

For flexural rupture (with Table 12-1),

$$\phi M_n = \phi F_u S_{net}$$
$$= 0.75(58 \text{ ksi})(6.19 \text{ in.}^3)$$
$$= 269 \text{ in.-kips} > 150 \text{ in.-kips} \quad \textbf{o.k.}$$

Check shear yielding of the plate

$$\phi R_n = \phi \, (0.6 F_y A_g)$$
$$= 0.9(0.6 \times 36 \text{ ksi})(12 \text{ in.} \times \tfrac{3}{8}\text{-in.})$$
$$= 87.5 \text{ kips} > 60 \text{ kips} \quad \textbf{o.k.}$$

Check shear rupture of the plate

$$\phi R_n = \phi \, (0.6 F_u A_n)$$
$$= 0.75(0.6 \times 58 \text{ ksi})(12 \text{ in.} - 4 \times 1 \text{ in.}) \, \tfrac{3}{8}\text{-in.}$$
$$= 78.3 \text{ kips} > 60 \text{ kips} \quad \textbf{o.k.}$$

Check block shear rupture of the plate

From Tables 8-47 and 8-48, with four $\tfrac{7}{8}$-in. diameter bolts and $L_{ev} = L_{eh} = 1\tfrac{1}{2}$-in., $0.6 F_u A_{gt} > F_y A_{gt}$. Thus,

$$\phi R_n = \phi \, [0.6 F_u A_{nv} + F_y A_{gt}]$$
$$= (183 \text{ kips/in.} + 40.5 \text{ kips/in.}) \, \tfrac{3}{8}\text{-in.}$$
$$= 83.8 \text{ kips} > 60 \text{ kips} \quad \textbf{o.k.}$$

Use PL$\tfrac{3}{8}$-in. $\times$ 8 in. $\times$ 10$\tfrac{1}{2}$-in.

Example 9-18

Given: Refer to Figure 9-23. Design a single-plate shear splice between a W16×31 beam and W16×50 beam (not illustrated)

$$R_u = 50 \text{ kips}$$

W16×31

$$t_w = 0.275 \text{ in.}$$
$$F_y = 50 \text{ ksi}, \; F_u = 65 \text{ ksi}$$

W16×50

$$t_w = 0.380$$
$$F_y = 50 \text{ ksi}, \; F_u = 65 \text{ ksi}$$

Use $\tfrac{3}{4}$-in. diameter A325-N bolts through the web of the W16×50 and 70 ksi electrode welds to the web of the W16×31. Assume plate material with $F_y = 36$ ksi and $F_u = 58$ ksi.

Solution: *Design weld group*

Since splice is unsymmetrical and the weld group is more rigid, it will be designed for the full eccentric moment.

Assume PL⅜-in.×8 in.×1′-0 as illustrated in Figure 9-23.

$$k = \frac{kl}{l} = \frac{3\frac{1}{2}\text{-in.}}{12 \text{ in.}} = 0.292$$

By interpolation from Table 8-42 with $\theta = 0°$, $x = 0.0538$ and $xl = 0.646$ in.

$$al = 6\frac{1}{2}\text{-in.} - 0.646 \text{ in.}$$
$$= 5.85 \text{ in.}$$

Thus $a = \dfrac{al}{l} = \dfrac{5.85 \text{ in.}}{12 \text{ in.}} = 0.488$ in.

By interpolation, $C = 1.61$ and the required weld size is

$$D_{req} = \frac{R_u}{CC_1l}$$
$$= \frac{50 \text{ kips}}{(1.61)(1.0)(12 \text{ in.})}$$
$$= 2.59 \rightarrow 3 \text{ sixteenths}$$

From LRFD Specification Table J2.4, the minimum weld size is ³⁄₁₆-in. Use ³⁄₁₆-in. weld size.

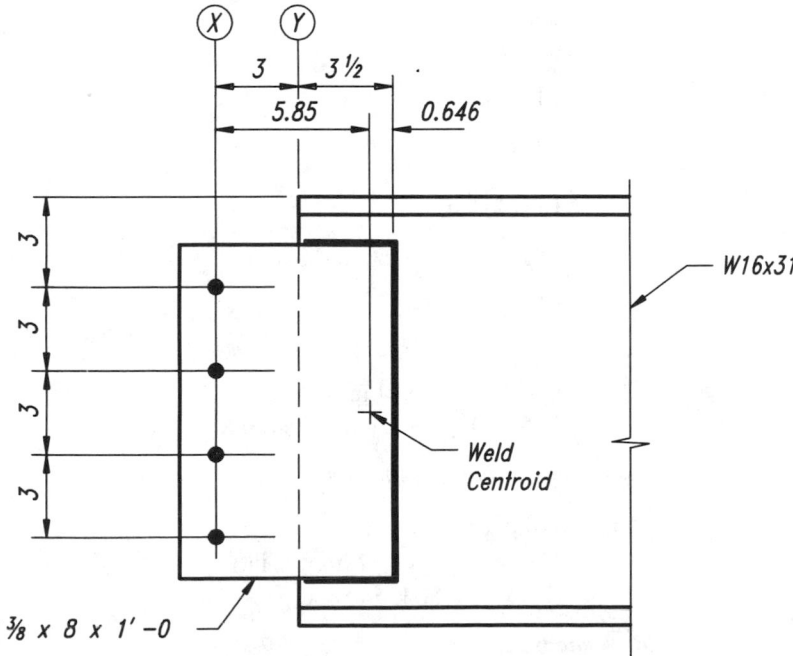

Figure 9-23. Shear splice for Example 9-18.

Check shear yielding of beam web (W16×31)

$$t_{min} = \frac{2.58D}{F_y}$$

$$= \frac{2.58 \ (3 \ \text{sixteenths})}{50 \ \text{ksi}}$$

$$= 0.154 < 0.275 \ \text{in.} \quad \textbf{o.k.}$$

Design bolt group

Since the weld group was designed for the full eccentric moment, the bolt group will be designed for shear only.

For bolt shear $\phi r_n = 15.9$ kips/bolt. For bearing on the ⅜-in. thick single plate, $\phi r_n = 29.4$ kips/bolt. Since bolt shear is more critical,

$$n_{min} = \frac{R_u}{\phi r_n}$$

$$= \frac{50 \ \text{kips}}{15.9 \ \text{kips} / \text{bolt}}$$

$$= 3.14 \rightarrow 4 \ \text{bolts}$$

Design single plate

As before, try PL⅜-in.×8 in.×1'-0.

Check flexure of the plate

$$M_u = R_u e$$

$$= 50 \ \text{kips} \ (3 \ \text{in.})$$

$$= 150 \ \text{in.-kips}$$

For flexural yielding

$$\phi M_n = \phi F_y S_x$$

$$= 0.9 \ (36 \ \text{ksi}) \left[\frac{⅜\text{-in.} \ (12 \ \text{in.})^2}{6} \right]$$

$$= 292 \ \text{in.-kips} > 150 \ \text{in.-kips} \quad \textbf{o.k.}$$

For flexural rupture (with Table 12-1),

$$\phi M_n = \phi F_u s_{net}$$

$$= 0.75(58 \ \text{ksi})(6.54 \ \text{in.}^3)$$

$$= 285 \ \text{in.-kips} > 150 \ \text{in.-kips} \quad \textbf{o.k.}$$

Check shear yielding of the plate

$$\phi R_n = \phi \ (0.6 F_y A_g)$$

$$= 0.9(0.6 \times 36 \ \text{ksi})(12 \ \text{in.} \times ⅜\text{-in.})$$

$$= 87.5 \ \text{kips} > 50 \ \text{kips} \quad \textbf{o.k.}$$

Check shear rupture of the plate

$$\phi R_n = \phi \ (0.6 F_u A_n)$$

$$= 0.75(0.6 \times 58 \text{ ksi})(12 \text{ in.} - 4 \times 1 \text{ in.}) \, \%\text{-in.}$$
$$= 78.3 \text{ kips} > 50 \text{ kips} \quad \textbf{o.k.}$$

Check block shear rupture of the plate

From Tables 8-47 and 8-48, with four $\%$-in. diameter bolts and $L_{ev} = L_{eh} = 1\frac{1}{2}$-in., $0.6F_u A_{nv} > F_y A_{gt}$. Thus,

$$\phi R_n = \phi \, [0.6F_u A_{nv} + F_y A_{gt}]$$
$$= (194 \text{ kips/in.} + 40.5 \text{ kips/in.})\%\text{-in.}$$
$$= 87.9 \text{ kips} > 50 \text{ kips} \quad \textbf{o.k.}$$

Use PL$\%$-in.$\times$8 in.$\times$1$'$-0

SPECIAL CONSIDERATIONS FOR SIMPLE SHEAR CONNECTIONS

Web Reinforcement of Coped Beams

The design strength of coped beams based on the limit state of flexural yielding, local buckling, and lateral torsional buckling was discussed previously in Part 8. When the strength of a reduced section is inadequate, the designer has two basic options: (1) select a different section to eliminate the need for reinforcement; or (2) provide reinforcement to increase the strength of the inadequate section. In spite of the increase in material cost, the former may be the most economical option due to the appreciable labor cost of adding stiffeners and/or doublers.

When the original section must be reinforced, Figure 9-24 illustrates several reinforcing details which may be useful. The doubler plate illustrated in Figure 9-24a and the longitudinal stiffener illustrated in Figure 9-24b are used with rolled sections where $h/t_w \le 60$. The combination of longitudinal and transverse stiffeners shown in Figure 9-24c is required for thin-webbed plate-girders, where $h/t_w > 60$.

Doubler Plates

When a doubler plate is used to stiffen the web of a coped beam, the required doubler plate thickness $t_{d\,req}$ is determined by substituting the quantity $(t_w + t_{d\,req})$ for t_w in the calculations of the design strength ϕR_n. Design checks for flexural yielding and local web buckling are then made as discussed previously in Part 8. To prevent local crippling of the beam web, the doubler plate must be extended at least a distance d_c (depth of cope) beyond the cope as illustrated in Figure 9-24a.

Longitudinal Stiffeners

When longitudinal stiffening is used to stiffen the web of a coped beam, the stiffening elements must be proportioned to meet the width-thickness ratios specified in LRFD Specification Table B5.1. The stiffened section must then be checked for flexural yielding; local web buckling need not be checked. To prevent local crippling of the beam web, longitudinal stiffeners must be extended a distance d_c beyond the cope as illustrated in Figure 9-24b.

Combination Longitudinal and Transverse Stiffening

When longitudinal and transverse stiffening is used in combination to stiffen a coped plate girder, the stiffening elements must be proportioned to meet the width-thickness ratios specified in LRFD Specification Table B5.1. The stiffened section must then be checked for flexural yielding; local web buckling need not be checked. To prevent local crippling of the beam web, longitudinal stiffeners must be extended a distance $c/3$ beyond the cope as illustrated in Figure 9-24c.

Example 9-19

Given: For a W21×62 (t_w = 0.400 in., d = 20.99 in., F_y = 50 ksi, F_u = 65 ksi) coped 8-in. deep by 9-in. long at the top flange only:

A. calculate the design strength of the beam end considering the limit states of flexural yielding and local buckling assuming e = 9½-in.

B. determine the alternative W21 that would eliminate the need for stiffening for a required strength of 95 kips

C. design a doubler plate to stiffen the W21×62 for a required strength of 95 kips

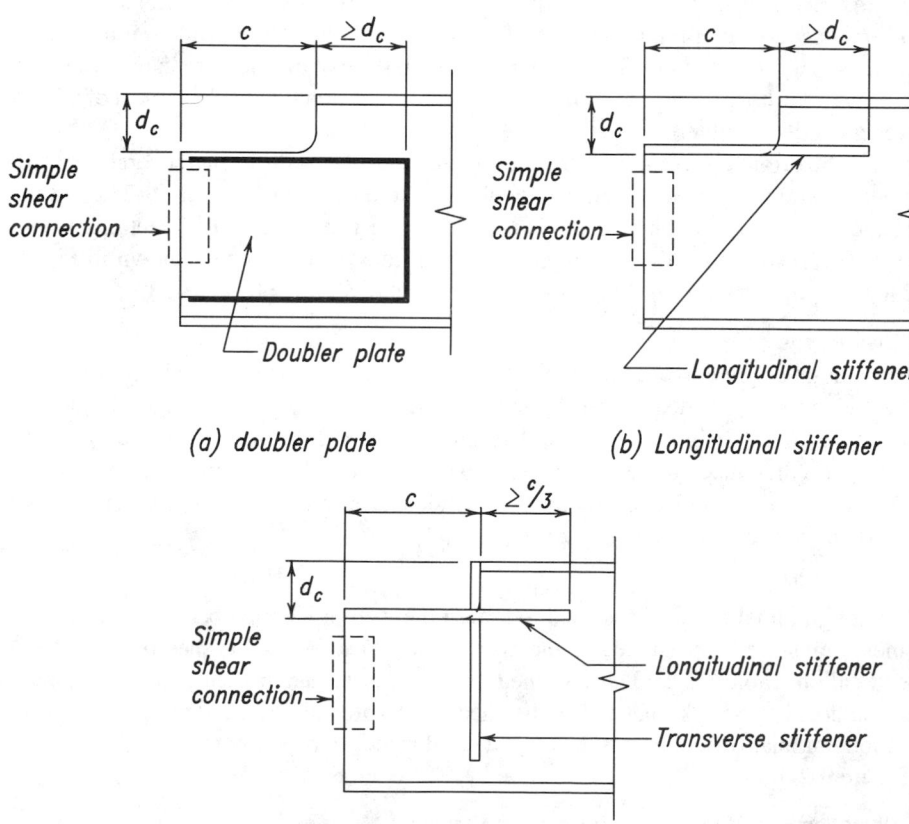

(a) doubler plate   *(b) Longitudinal stiffener*

(c) Combination longitudinal and transverse stiffeners

Figure 9-24. *Stiffening for coped member ends.*

D. design longitudinal stiffening for the W21×62 for a required strength of 95 kips

Solution A: *Check flexural yielding*

From Table 8-49, $S_{net} = 17.8$ in.^{3}

$$\phi R_n = \frac{\phi F_y S_{net}}{e}$$

$$= \frac{0.9 \ (50 \text{ ksi}) \ (17.8 \text{ in.}^3)}{9\frac{1}{2}\text{-in.}}$$

$$= 84.3 \text{ kips}$$

Check local buckling

Verify parameters

$$2d = 42.0 \text{ in.}$$

$$d/2 = 10.5 \text{ in.}$$

Since $c \le 2d$ and $d_c \le d/2$, procedure from Part 8 may be used.

$$\frac{c}{d} = \frac{9 \text{ in.}}{20.99 \text{ in.}} = 0.429$$

$$\frac{c}{h_o} = \frac{9 \text{ in.}}{20.99 \text{ in.} - 8 \text{ in.}} = 0.693$$

Since $c/d \le 1.0$,

$$f = 2\left(\frac{c}{d}\right)$$

$$= 2(0.429)$$

$$= 0.858$$

Since $c/h_o \le 1.0$,

$$k = 2.2\left(\frac{h_o}{c}\right)^{1.65}$$

$$= 2.2\left(\frac{1}{0.693}\right)^{1.65}$$

$$= 4.03$$

For a top cope only, the critical buckling stress is

$$\phi F_{bc} = 23,590\left(\frac{t_w}{h_o}\right)^2 fk$$

$$= 23,590\left(\frac{0.400 \text{ in.}}{20.99 \text{ in.} - 8 \text{ in.}}\right)^2 (0.858) \ (4.03)$$

$$= 77.3 \text{ ksi}$$

and the design strength is

$$\phi R_n = \frac{\phi F_{bc} S_{net}}{e}$$

$$= \frac{(77.3 \text{ ksi}) (17.8 \text{ in.}^3)}{9\frac{1}{2}\text{-in.}}$$

$$= 145 \text{ kips}$$

The design strength of the coped W21×62 is controlled by flexural yielding where

$$\phi R_n = 84.3 \text{ kips}$$

Solution B: If the required strength R_u were 95 kips, the W21×62 would be inadequate due to the limit state of flexural yielding. The required net elastic section modulus S_{req} would be

$$S_{req} = \frac{R_u e}{\phi F_y}$$

$$= \frac{95 \text{ kips} (9\frac{1}{2}\text{-in.})}{0.9 (50 \text{ ksi})}$$

$$= 20.1 \text{ in.}^3$$

From Table 8-49, a W21×73 with an 8-in. deep cope provides

$$S_{net} = 21.0 \text{ in.}^3 > 20.1 \text{ in.}^3 \quad \textbf{o.k.}$$

Check local buckling

Since the W21×62 provided $\phi R_n = 145$ kips > 95 kips for the limit state of local buckling, local buckling is not critical for the W21×73 with a 8-in. deep cope.

Solution C: *Design doubler plate*

From Solutions A and B, the doubler plate must provide for 95 kips − 84.3 kips = 10.7 kips. Conservatively ignoring the effect of the Ad^2 term in computing the section modulus, the required section modulus for the doubler plate is

$$S_{req} = \frac{(R_u - \phi R_{n \, beam}) \, e}{\phi F_y}$$

$$= \frac{(95 \text{ kips} - 84.3 \text{ kips}) \, 9\frac{1}{2}\text{-in.}}{0.9 (36 \text{ ksi})}$$

$$= 3.14 \text{ in.}^3$$

For an 8-in. deep plate,

$$t_{req} = \frac{6 S_{req}}{d^2}$$

$$= \frac{6 (3.14 \text{ in.}^3)}{(8 \text{ in.})^2}$$

= 0.294 in.

Thus, since the doubler plate must extend at least d_c beyond the cope, use PL$\frac{5}{16}$-in.×8-in.×1′-5

Solution D: *Design longitudinal stiffeners*

Try PL $\frac{1}{4}$-in.×4 in. slotted to fit over beam web, $F_y = 50$ ksi. The neutral axis is located 4.40 in. from the bottom flange (8.84 in. from the top of the stiffener) and the elastic section modulus of the reinforced section is as follows:

	I_o (in.^{4})	Ad^2 (in.^{4})	$I_o + Ad^2$ (in.^{4})
Stiffener	0.00521	76.0	76.0
W21×62 web	63.2	28.6	91.8
W21×62 bottom flange	0.160	84.9	85.1
			$I_x = 253$ in.^{4}

$$S_{net} = \frac{I_x}{c}$$
$$= \frac{253 \text{ in.}^4}{8.84 \text{ in.}}$$
$$= 28.6 \text{ in.}^3$$

and the design strength of the section is

$$\phi R_n = \frac{\phi F_y S_{net}}{e}$$
$$= \frac{0.9 \, (50 \text{ ksi}) \, (28.6 \text{ in.}^3)}{9\frac{1}{2}\text{-in.}}$$
$$= 136 \text{ kips} > 95 \text{ kips} \quad \textbf{o.k.}$$

Thus, since the longitudinal stiffening must extend at least d_c beyond the cope,

Use PL $\frac{1}{4}$-in.×4 in.×1′-5.

Example 9-20

Given: For a W21×62 ($t_w = 0.400$ in., $d = 20.99$ in., $F_y = 50$ ksi, $F_u = 65$ ksi) coped 3-in. deep by 7-in. long at the top flange and 4-in. deep by 7-in. long at the bottom flange. Calculate the design strength of the beam end considering the limit states of flexural yielding and local buckling assuming $e = 7\frac{1}{2}$-in.

Solution: *Check flexural yielding*

$$S_{net} = \frac{t_w h_o^2}{6}$$
$$= \frac{(0.400 \text{ in.}) \, (20.99 \text{ in.} - 3 \text{ in.} - 4 \text{ in.})^2}{6}$$
$$= 13.1 \text{ in.}^3$$

$$\phi R_n = \frac{\phi F_y S_{net}}{e}$$

$$= \frac{0.9 \,(50 \text{ ksi}) \,(13.1 \text{ in.}^3)}{7\frac{1}{2}\text{-in.}}$$

$$= 78.6 \text{ kips}$$

Check local buckling

Verify parameters

$$2d \quad = 42.0 \text{ in.}$$
$$0.2d = 4.20 \text{ in.}$$

Since, for each cope, $c \le 2d$ and $d_c \le 0.2d$, procedure from Part 8 may be used.

$$f_d = 3.5 - 7.5 \left(\frac{d_c}{d} \right)$$

$$= 3.5 - 7.5 \left(\frac{4 \text{ in.}}{20.99 \text{ in.}} \right)$$

$$= 2.07$$

For the doubly coped beam, the critical stress is

$$\phi F_{bc} = 50,840 \left(\frac{t_w^2}{ch_o} \right) f_d$$

$$= 50,840 \left[\frac{(0.400 \text{ in.})^2}{(7 \text{ in.}) \,(20.99 \text{ in.} - 3 \text{ in.} - 4 \text{ in.})} \right] (2.07)$$

$$= 172 \text{ ksi}$$

and the design strength is

$$\phi R_n = \frac{\phi F_{bc} S_{net}}{e}$$

$$= \frac{(172 \text{ ksi}) \,(13.1 \text{ in.}^3)}{7\frac{1}{2}\text{-in.}}$$

$$= 300 \text{ kips}$$

Simple Shear Connections at Stiffened Column-Web Locations

Stiffeners are obstacles to direct connections to column web. Figure 9-25a illustrates a seat angle welded to the toes of the column flanges; Figure 9-25d shows a vertical plate extended beyond the column flanges. Figures 9-25b and 9-25c offer two additional options for framing at locations of diagonal stiffeners; these should be examined carefully as they may create erection problems. Additionally, the deep cope of Figure 9-25c may significantly reduce the design strength of the beam at the end connection. Alternatively, the bottom transverse stiffener could be extended to serve as a seat plate with a bearing stiffener provided to distribute the beam reaction.

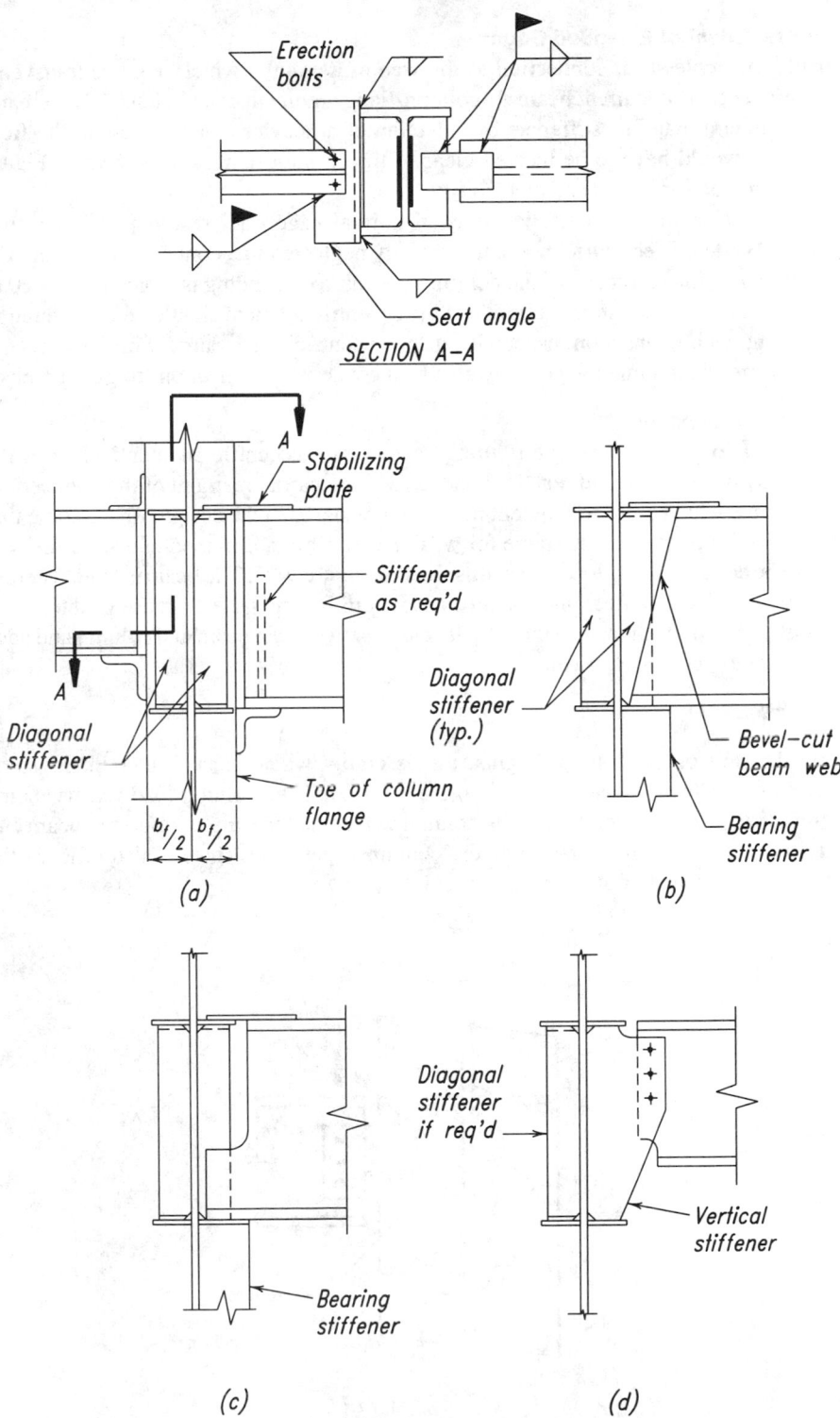

Figure 9-25. *Simple shear connections at stiffened column-web locations.*

Eccentric Effect of Extended Gages

Consider a simple shear connection to the web of a column which requires transverse stiffeners for two concurrent beam-to-column-flange moment connections. If it were not possible to eliminate the stiffeners by selection of a heavier column section, the field connection would have to be located clear of the column flanges, as shown in Figure 9-26, to provide for access and erectability.

The extension of the connection beyond normal gage lines results in an eccentric moment. While this eccentric moment is usually neglected in a connection framing to a column flange, the resistance of the column to weak-axis bending is typically only 20 to 50 percent of that in the strong axis. Thus the eccentric moment should be considered in this column-web connection, especially if the eccentricity e is large. Similarly, eccentricities larger than normal gages may also be a concern in connections to girder webs.

Column-Web Supports

There are two components contributing to the total eccentric moment: (1) $R_u e$ the eccentricity of the beam end reaction; and (2) M_{pr} the partial restraint of the connection. To determine what eccentric moment must be considered in the design, first assume that the column is part of a braced frame for weak-axis bending, is pinned-ended with $K = 1$, and will be concentrically loaded, as illustrated in Figure 9-27. The beam is loaded before the column and will deflect under load as shown in Figure 9-27. Because of the partial restraint of the connection, a couple M_{pr} develops between the beam and column and adds to the eccentric couple $R_u e$. Thus,

$$M_{con} = R_u e + M_{pr}$$

As the loading of the column begins, the assembly will deflect further in the same direction under load, as indicated in Figure 9-28, until the column load reaches some magnitude P_{sbr} when the rotation of the column will equal the simply supported beam end rotation. At this load, the rotation of the column negates M_{pr} since it also relieves the partial restraint effect of the connection and,

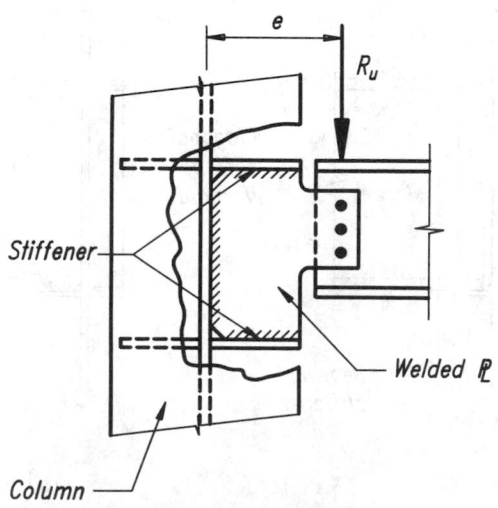

Figure 9-26. Eccentric effect of extended gages.

$$M_{con} = R_u e$$

As the column load is increased above P_{sbr}, the column rotation exceeds the simply supported beam end rotation and a moment M'_{pr} results such that

$$M_{con} = R_u e - M'_{pr}$$

Note that the partial restraint of the connection now actually stabilizes the column and reduces its effective length factor K below the originally assumed value of 1. Thus, since

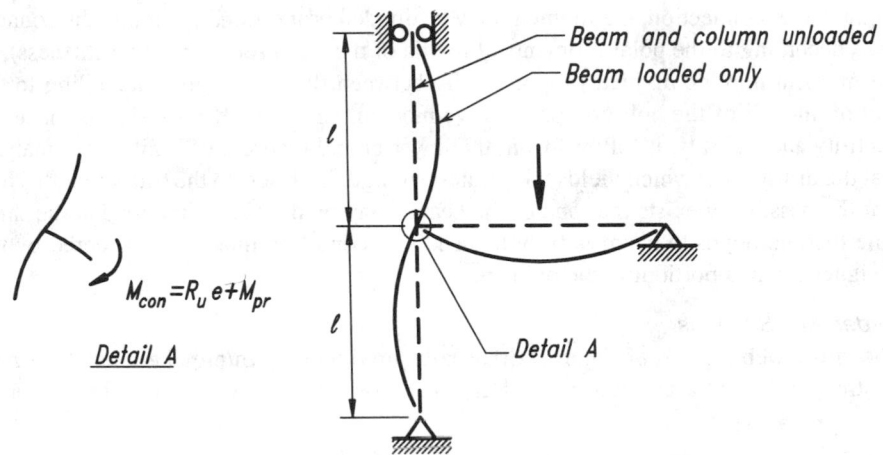

Figure 9-27. *Illustration of beam, column, and connection behavior under loading of beam only.*

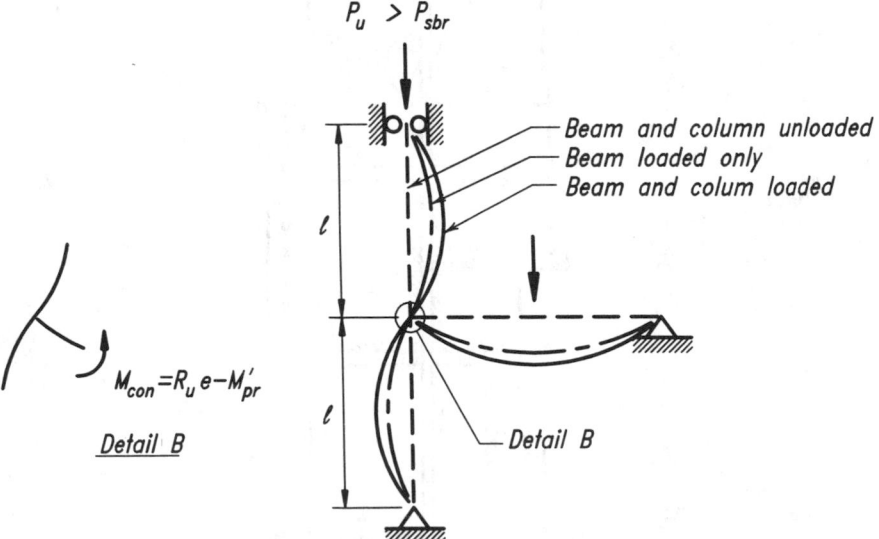

Figure 9-28. *Illustration of beam, column, and connection behavior under loading of beam and column.*

M'_{pr} must be greater than zero, it must also be true that $R_u e > M_{con}$. It is therefore conservative to design the connection for the shear R_u and the eccentric moment $R_u e$.

The welds connecting the plate to the supporting column web should be designed to resist the full shear R_u only; the top and bottom plate-to-stiffener welds have minimal strength normal to their length, are not assumed to carry any calculated force, and may be of minimum size in accordance with LRFD Specification Section J2.

If simple shear connections frame to both sides of the column web as illustrated in Figure 9-29, each connection should be designed for its respective shear R_{u1} and R_{u2}, and the eccentric moment $|R_{u2}e_2 - R_{u1}e_1|$ may be apportioned between the two simple shear connections as the designer sees fit; the total eccentric moment may be assumed to act on the larger connection, the moment may be divided proportionally among the connections according to the polar moments of inertia of the bolt groups (relative stiffness), or the moment may be divided proportionally between the connections according to the section moduli of the bolt groups (relative moment strength). If provision is made for ductility and stability, it follows from the lower bound theorem of limit states analysis that the distribution which yields the greatest strength is closest to the true strength. Note that the possibility exists that one of the beams may be devoid of live load at the same time that the opposite beam is fully loaded. This condition must be considered by the designer when apportioning the moment.

Girder-Web Supports

The girder-web support of Figure 9-30 usually provides only minimal torsional stiffness or strength. When larger-than-normal gages are used, the end rotation of the supported

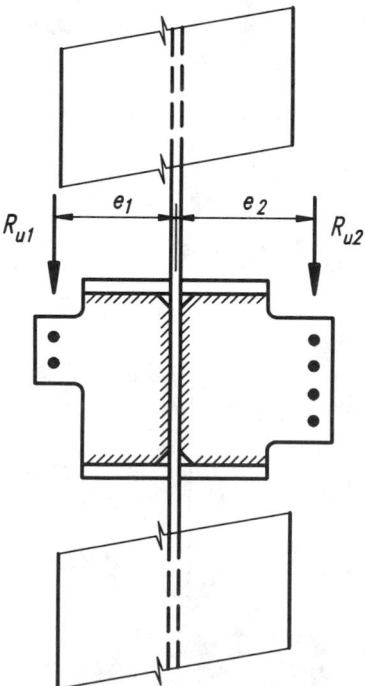

Figure 9-29. Columns subjected to dual eccentric moments.

beam will usually be accommodated through rotation of the girder support. It follows that the bolt group should be designed to resist both the shear R_u and the eccentric moment $R_u e$. The beam end reaction will then be carried through to the center of the supporting girder web.

The welds connecting the plate to the supporting girder web should be designed to resist the shear R_u only; the top and bottom plate-to-girder-flange welds have minimal strength normal to their length, are not assumed to carry any calculated force, and may be of minimum size in accordance with LRFD Specification Section J2.

Similarly, for the girder illustrated in Figure 9-31 supporting two eccentric reactions, each connection should be designed for its respective shear R_{u1} and R_{u2}, and the eccentric moment $|R_{u2}e_2 - R_{u1}e_1|$ may be apportioned between the two simple shear connections as the designer sees fit.

Alternative Treatment of Eccentric Moment
In the foregoing treatment of eccentric moments with column- and girder-web supports, it is possible to design the support (instead of the connection) for the eccentric moment $R_u e$. Additionally, when metal deck is used with puddle welds or self-tapping screws , the

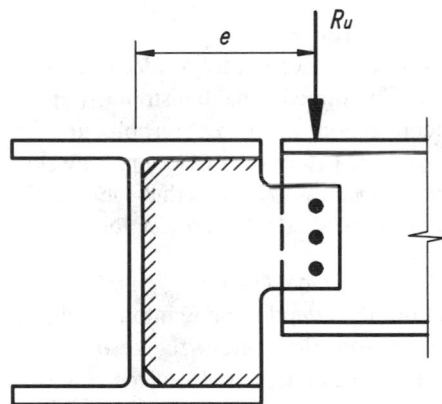

Figure 9-30. Eccentric moments on girder-web supports.

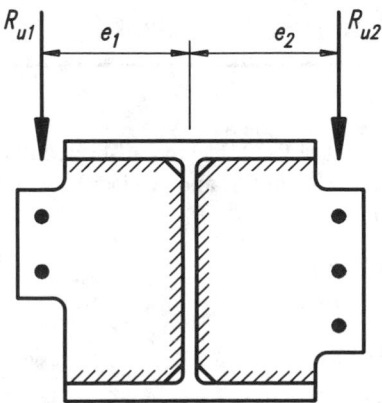

Figure 9-31. Girders subjected to dual eccentric moments.

metal deck tends to reduce relative movement between the two members and thus will tend to carry all or some of the eccentric moment. In these cases, the connection may be designed for the shear R_u only or the shear R_u and a reduced eccentric moment.

Simple Shear Connections for Large End Reactions

In general, large end reactions necessitate the use of double-angle connections since the design strengths and limitations of other simple shear connections may preclude their use. Such cases as this may be encountered with deep beams, heavily loaded beams on short spans, or girders supporting concentrated loads located near the connections.

In bolted construction, large end reactions may necessitate the use of A490 bolts and or bolts of diameter larger than one inch. In welded construction, beams having end reactions greater than the strengths tabulated usually will require connections with larger weld sizes, since the length of welds is restricted by the depth of the beam. In either of these cases, connection angles thicker than the recommended maximum ⅝-in. thickness may be required. Past experience has proven that adequate flexibility is obtained if the width of the outstanding angle leg dimension is increased by one inch for each additional sixteenth of an inch in angle thickness. The availability of angles of suitable size and thickness should be considered in establishing a final design.

Double Connections

When beams frame opposite each other and are welded to the web of the supporting girder or column, there are usually no dimensional constraints imposed on one connection by the presence of the other connection unless erection bolts are common to each connection. When the connections are bolted to the web of the supporting column or girder, however, the close proximity of the connections requires that some or all fasteners be common to both connections. This is known as a double connection.

Supported Beams of Different Nominal Depths

When beams of different nominal depths frame into a double connection, care must be taken to avoid interference from the bottom flange of the shallower beam with the entering and tightening clearances for the bolts of the connection for the deeper beam. Access to the bolts which will support the deeper beam may be provided by coping or blocking the bottom flange of the shallower beam. Alternatively, stagger may be used to favorably position the bolts around the bottom flange of the shallower beam.

Example 9-21

Given: Refer to Figure 9-32. Design all-bolted double-angle connections for the W12×40 beam (A) and W21×50 beam (B) to W30×99 girder-web connection.

$R_{uA} = 25$ kips
$R_{uB} = 110$ kips

W12×40

$t_w = 0.295$ in., $d = 11.94$ in.
$F_y = 50$ ksi, $F_u = 65$ ksi

top and bottom flanges coped 2-in. deep by 5-in. long (bottom cope allows for entering and tightening bolts through support)

W21×50

t_w = 0.380 in., d = 20.83 in.
F_y = 50 ksi, F_u = 65 ksi

top flange (only) coped 2-in. deep by 5-in. long

W30×99

t_w = 0.520 in., d = 29.65 in.
F_y = 50 ksi, F_u = 65 ksi

Use ¾-in. diameter A325-N bolts in standard holes. Assume angle material with F_y = 36 ksi and F_u = 58 ksi

Solution: *Design bolts and angles for* W12×40 *(beam A)*

From Table 9-2, for ¾-in. diameter A325-N bolts and angle material with F_y = 36 ksi and F_u = 58 ksi, select two rows of bolts and ¼-in. angle thickness

ϕR_n = 48.9 > 25 kips **o.k.**

Check supported beam web (beam A)

From Table 9-2, for two rows of bolts and beam material with F_y = 50 ksi and F_u = 65 ksi, and L_{ev} = 1¼-in. and L_{eh} = 1½-in. (assumed to be 1¼-in. for calculation purposes to account for possible underrun in beam length)

ϕR_n = (110 kips/in.)(0.295 in.)
$\qquad$ = 32.5 kips > 25 kips **o.k.**

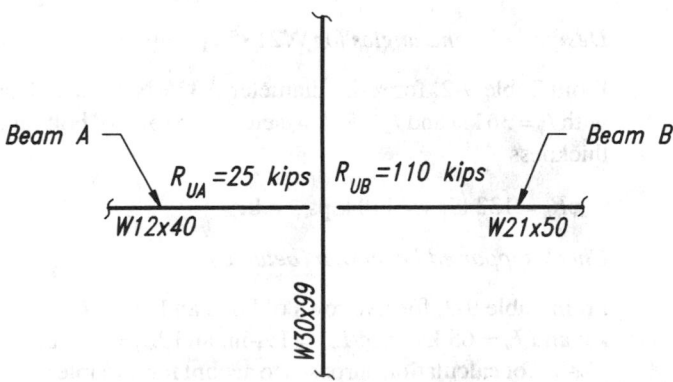

PART PLAN

Figure 9-32. Illustration for Example 9-21.

Check flexural yielding of the coped sections (beam A)

$$S_{net} = \frac{t_w h_o^2}{6}$$

$$= \frac{(0.295 \text{ in.})(11.94 \text{ in.} - 2 \text{ in.} - 2 \text{ in.})^2}{6}$$

$$= 3.10 \text{ in.}^3$$

$$\phi R_n = \frac{\phi F_y S_{net}}{e}$$

$$= \frac{0.9 \ (50 \text{ ksi}) \ (3.10 \text{ in.}^3)}{5 \text{ in.} + \frac{1}{2}\text{-in.}}$$

$$= 25.4 \text{ kips} > 25 \text{ kips} \quad \textbf{o.k.}$$

Check local buckling at the cope (beam A)

$$f_d = 3.5 - 7.5 \left(\frac{d_c}{d} \right)$$

$$= 3.5 - 7.5 \left(\frac{2 \text{ in.}}{11.94 \text{ in.}} \right)$$

$$= 2.24$$

$$\phi F_{bc} = 50,840 \ \frac{t_w^2}{ch_o} f_d$$

$$= 50,840 \left[\frac{(0.295 \text{ in.})^2}{(5 \text{ in.}) \ (11.94 \text{ in.} - 2 \text{ in.} - 2 \text{ in.})} \right] (2.24)$$

$$= 250 \text{ ksi}$$

$$\phi R_n = \frac{\phi F_{bc} S_{net}}{e}$$

$$= \frac{(250 \text{ ksi})(3.10 \text{ in.}^3)}{5 \text{ in.} + \frac{1}{2}\text{-in.}}$$

$$= 141 \text{ kips} > 25 \text{ kips} \quad \textbf{o.k.}$$

Design bolts and angles for W21×50 *(beam B)*

From Table 9-2, for $\frac{3}{4}$-in. diameter A325-N bolts and angle material with $F_y = 36$ ksi and $F_u = 58$ ksi, select five rows of bolts and $\frac{1}{4}$-in. angle thickness.

$$\phi R_n = 132 \text{ kips} > 110 \text{ kips} \quad \textbf{o.k.}$$

Check supported beam web (beam B)

From Table 9-2, for five rows of bolts and beam material with $F_y = 50$ ksi and $F_u = 65$ ksi, and $L_{ev} = 1\frac{1}{4}$-in. and $L_{eh} = 1\frac{1}{2}$-in. (assumed to be $1\frac{1}{4}$-in. for calculation purposes to acount for possible underrun in beam length)

$$\phi R_n = (319 \text{ kips/in.})(0.380 \text{ in.})$$

$$= 121 \text{ kips} > 110 \text{ kips} \quad \textbf{o.k.}$$

Check flexural yielding of the coped section (beam B)

From Table 8-49, $S_{net} = 32.5$ in.^{3}

$$\phi R_n = \frac{\phi F_y S_{net}}{e}$$

$$= \frac{0.9\,(50\,\text{ksi})\,(32.5\,\text{in.}^3)}{5\,\text{in.} + 1\tfrac{1}{2}\text{-in.}}$$

$$= 266\,\text{kips} > 110\,\text{kips}$$

Check local web buckling at the cope (beam B)

$$\frac{c}{d} = \frac{5\,\text{in.}}{20.83\,\text{in.}} = 0.240$$

$$\frac{c}{h_o} = \frac{5\,\text{in.}}{20.83\,\text{in.} - 2\,\text{in.}} = 0.266$$

Since $c/d \leq 1.0$,

$$f = 2\left(\frac{c}{d}\right)$$

$$= 2(0.240)$$

$$= 0.480$$

Since $c/h_o \leq 1.0$,

$$k = 2.2\left(\frac{h_o}{c}\right)^{1.65}$$

$$= 2.2\left(\frac{1}{0.266}\right)^{1.65}$$

$$= 19.6$$

$$\phi F_{bc} = 23{,}590\left(\frac{t_w}{h_o}\right)^2 fk$$

$$= 23{,}590\left(\frac{0.380\,\text{in.}}{20.83\,\text{in.} - 2\,\text{in.}}\right)^2 (0.480)\,(19.6)$$

$$= 90.4\,\text{ksi}$$

$$\phi R_n = \frac{\phi F_{bc} S_{net}}{e}$$

$$= \frac{(90.4\,\text{ksi})\,(32.5\,\text{in.}^3)}{5\,\text{in.} + \tfrac{1}{2}\text{-in.}}$$

$$= 534\,\text{kips} > 110\,\text{kips} \quad \textbf{o.k.}$$

Check supporting girder web

The required bearing strength per bolt is maximum for the bolts which are common to both connections. From beam A, each bolt transmits one-fourth of 25 kips or 6.25 kips/bolt. From beam B, each bolt transmits one-tenth of 110 kips or 11.0 kips. Thus,

$$r_u = 6.25 \text{ kips/bolt} + 11.0 \text{ kips/bolt}$$
$$= 17.3 \text{ kips/bolt}$$

From LRFD Specification Section J3.10, the design bearing strength per bolt is

$$\phi r_n = \phi (2.4 dt F_u)$$
$$= 0.75(2.4 \times \text{¾-in.} \times 0.520 \text{ in.} \times 65 \text{ ksi})$$
$$= 45.6 \text{ kips/bolt} > 17.3 \text{ kips} \quad \textbf{o.k.}$$

Supported Beams Offset Laterally

Frequently, beams do not frame exactly opposite each other, but are offset slightly as illustrated in Figure 9-33. Several connection configurations are possible, depending on the offset dimension.

If the offset were equal to the gage on the support, the connection could be designed with all bolts on the same gage lines as shown in Figure 9-33b and the angles arranged as shown in Figure 9-33d. If the offset were less than the gage on the support, staggering the bolts as shown in Figure 9-33c would reduce the required gage and the angles could be arranged as shown in Figure 9-33c. In any case, each bolt transmits an equal share of its beam reaction(s) to the supporting member. Once the geometry of the connection has been determined, the distribution of the forces is patterned after that in the design of a typical connection. For normal gages, eccentricity may be ignored in this type of connection.

Example 9-22

Given: For the all-bolted double-angle connection design of Example 9-1, suppose that two such connections were made back to back for beams with an offset. Determine the design changes necessary to accommodate an offset of 6 in.

Solution: Since the offset dimension (6 in.) is approximately equal to the gage on the support from Example 9-1 (6¼-in.), use a connection configuration similar to that illustrated in Figure 9-33d. All aspects of these connections than are unchanged with the exception of the middle vertical row of bolts (through both connections) which now carry their proportional share of the reaction of both connections.

Check supporting girder web

The required bearing strength per bolt is

$$r_u = \frac{2 \text{ connections} \times 60 \text{ kips / connection}}{6 \text{ bolts}}$$
$$= 20 \text{ kips/bolt}$$

From LRFD Specification Section J3.10, the design strength per bolt is

$$\begin{aligned}\phi r_n &= \phi\,(2.4dtF_u)\\ &= 0.75(2.4 \times \tfrac{3}{4}\text{-in.} \times 0.400\text{ in.} \times 65\text{ ksi})\\ &= 35.1\text{ kips/bolt} > 20\text{ kips/bolt}\quad \textbf{o.k.}\end{aligned}$$

W18x50
W16x45
65K
D
W16x45
65K
E
E
PART PLAN
(Beam flush top)
(a)
℄ Beam B
℄ Beam A
D
4½
4½
1¼
1¼
3
3@3 =9
A
B
F
F
SECTION E – E
Bolts on same gage
(b)
℄ Beam B
℄ Beam A
D
3
4½
3@3 =9
3@3 =9
A
B
F
F
SECTION E – E
Bolts staggered
(c)
D
SECTION F – F
Bolts on same gage
(d)
D
SECTION F – F
Bolts staggered
(e)

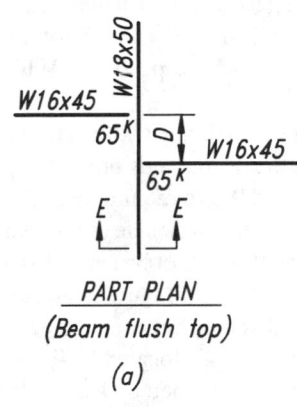

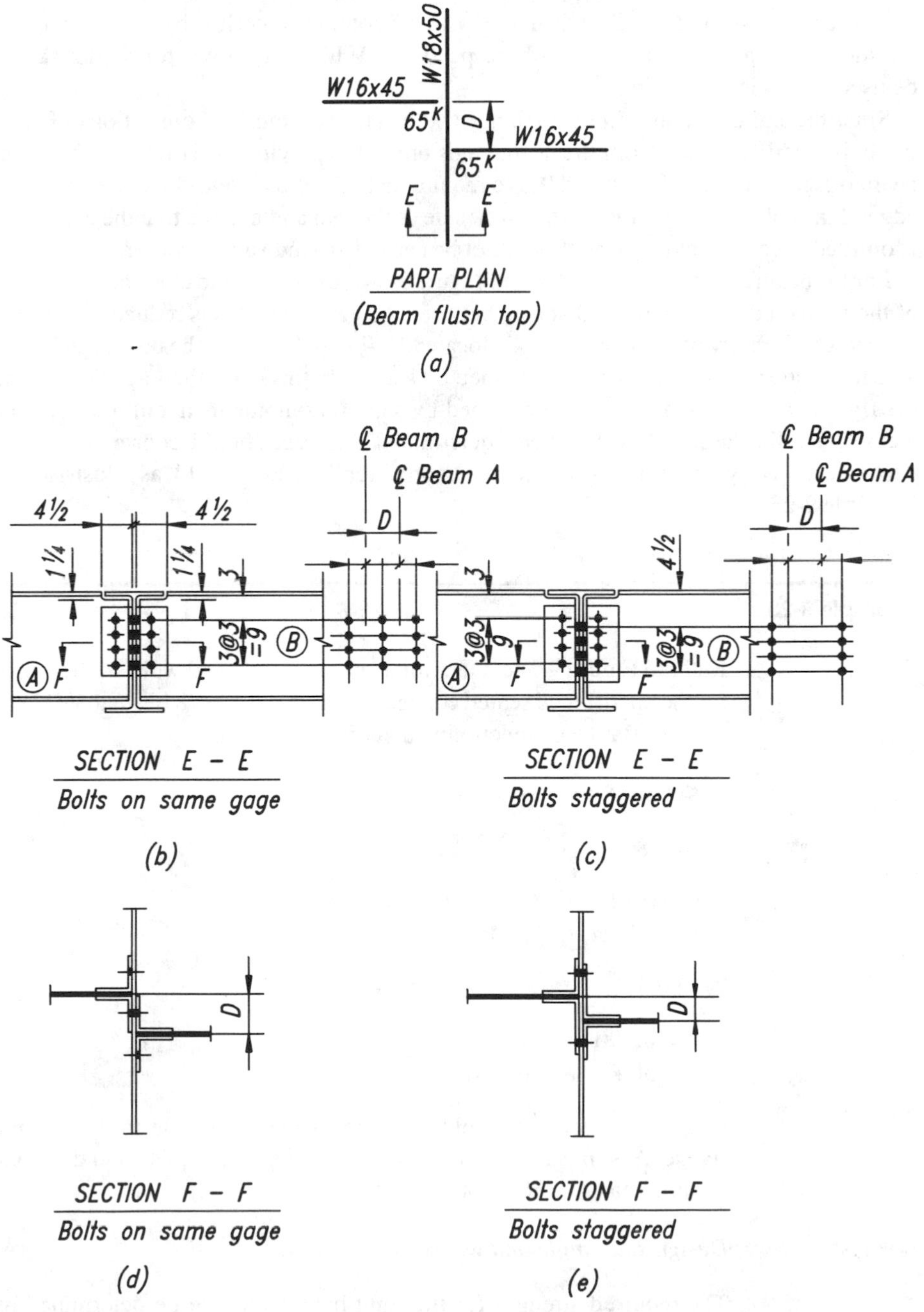

Figure 9-33. Offset beams.

Beams Offset From Column Centerline

Framing to the Column Flange from the Strong Axis

As illustrated in Figure 9-34, beam-to-column-flange connections offset from the column centerline may be supported on a typical welded seat, stiffened or unstiffened, provided the welds for the seat can be spaced approximately equally on either side of the beam centerline. Two such seats offset from the W12×65 column centerline by 2¼-in. and 3¼-in. are shown in Figures 9-34a and 9-34b, respectively. While not shown, top angles should be used with this connection.

Since the entire seat fits within the flange width of the column, the connection of Figure 9-34a is readily selected from the design aids presented previously. However, the larger beam offsets in Figures 9-34b and 9-34c require that one of the welds be made along the edge of the column flange against the back side of the seat angle. Note that the end return is omitted because weld returns should not be carried around such a corner.

For the beam offset of 5½-in. shown in Figure 9-34c, the seat angle overhangs the edge of the beam and the horizontal distance between the vertical welds is reduced to 3½-in.; the center of gravity of the weld group is located 1¼-in. to the left of the beam centerline. The force on each weld may be determined by statics. In this case, the larger force is in the right-hand weld and may be determined by summing moments about the left hand weld. Once the larger force has been determined, the seat should conservatively be designed to carry twice the force in the more highly loaded weld as illustrated in Example 9-23.

Example 9-23

Given: Refer to Figure 9-34c. Determine the seat angle and weld size required for the unstiffened seated connection for the W14×48 beam to W12×65 column-flange connection with a offset of 5½-in.

$R_u = 30$ kips

W14×48

$t_w = 0.340$ in., $d = 13.79$ in., $t_f = 0.595$ in.
$F_y = 50$ ksi, $F_u = 65$ ksi

W12×65

$t_w = 0.390$
$F_y = 50$ ksi, $F_u = 65$ ksi

Use 70 ksi electrode welds to connect the seat angle to the column flange. Assume a 4 in. outstanding angle leg is adequate and the angle material has $F_y = 36$ ksi and $F_u = 58$ ksi.

Solution: *Design seat angle and welds*

The required strength for the right-hand weld can be determined by summing moments about the left-hand weld.

$$R_{uR} = \frac{(30 \text{ kips}) (3 \text{ in.})}{3\frac{1}{2}\text{-in.}}$$
$$= 25.7 \text{ kips}$$

Selecting the welds on both sides of the seat to resist this force, the total required strength would be 51.4 kips.

From Table 9-7, with $t_w \approx \frac{5}{16}$-in., a $\frac{3}{4}$-in. seat angle thickness provides

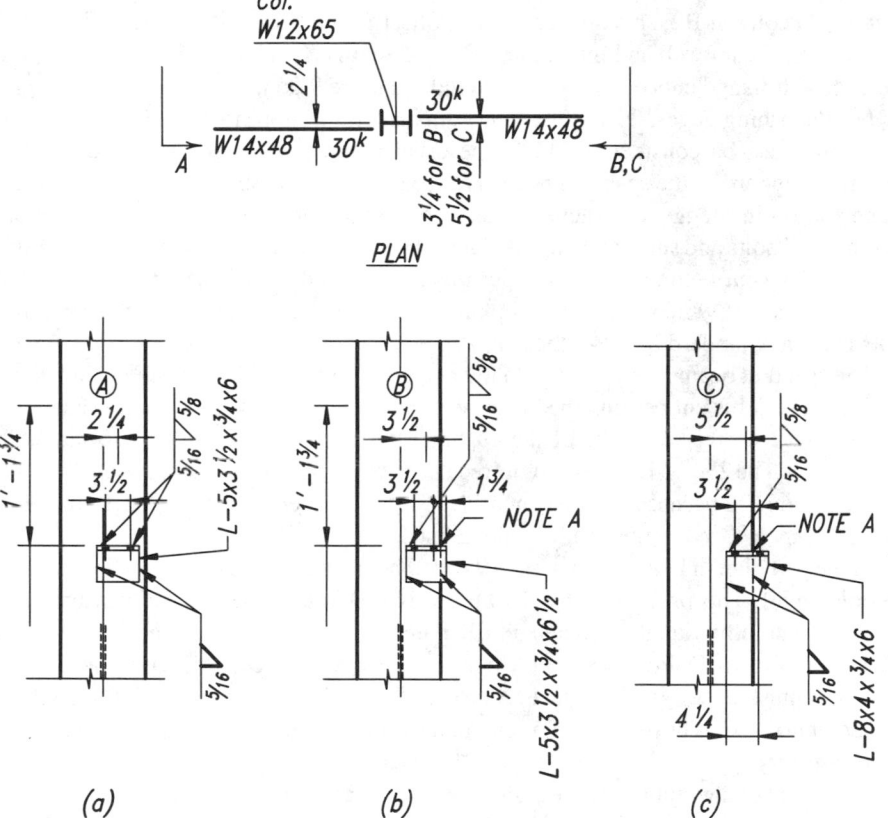

NOTE A

End return is omitted because the AWS Code does not permit weld returns to be carried around the corner formed by the column flange toe and seat angle heel.

NOTE B

Beam and top angle not shown for clarity.

Figure 9-34. Offset beam-to-column-flange connections.

$$\phi R_n = 54.7 \text{ kips} > 51.4 \text{ kips} \quad \textbf{o.k.}$$

and an 8×4 angle with $\frac{5}{16}$-in. fillet weld provides

$$\phi R_n = 66.8 \text{ kips} > 51.4 \text{ kips} \quad \textbf{o.k.}$$

Use L8×4× $\frac{3}{4}$× 6 in. for the seat angle.

Framing to Column Flange from the Weak Axis

Spandrel beams X and Y in the part plan shown in Figure 9-35 are offset $4\frac{1}{8}$-in. from the centerline of column C1, permitting the beam web to be connected directly to the column flange. At column B2, spandrel beam X is offset five inches and requires a $\frac{7}{8}$-in. filler between the beam web and the column flange. Beams X and Y are both plain-punched beams, with flange cuts on one side as noted in Figure 9-35a, Section F-F.

In establishing gages, the requirements of other connections to the column at adjacent locations must be considered. While the usual flange gage is $3\frac{1}{2}$-in. for the W8×28 columns supporting the spandrel beams, for beams Z, the combination of a 4-in. column gage and $1\frac{1}{2}$-in. stagger of fasteners is used to provide entering and tightening clearance for the field bolts and sufficient edge distance on the column flange as illustrated in Figure 9-35b. The 4-in. column gage also permits a $1\frac{1}{2}$-in. edge distance at the ends of the spandrel beams, which will accommodate the normal length tolerance of $\pm\frac{1}{4}$-in. as specified in "Standard Mill Practice" in Part 1.

The spandrel beams are shown with the notation "Cut and Grind Flush FS" in Sections E-E and F-F. This cut permits the beam web to lie flush against the column flange. The uncut flange on the near side of the spandrel beam contributes to the stiffness of the connection. The $2\frac{1}{2}×\frac{7}{8}$-in. filler is required between the spandrel beam web and the flange of the column B2 because of the $\frac{7}{8}$-in. offset. Since the filler in Section E-E, Figure 9-35a is thicker than $\frac{3}{4}$-in., it must be fully developed.

In the part plan in Figure 9-36a, the W16×40 beam is offset $6\frac{1}{4}$-in. from the centerline of column D1. This prevents the web of the W16×40 from being placed flush against the side of the column flange. A plate and filler are used to connect the beam to the column flange, as shown in Figure 9-36b. Such a connection is eccentric and one group of fasteners must be designed for the eccentricity. Lack of space on the inner flange face of the column requires development of the moment induced by the eccentricity in the beam web fasteners.

To minimize the number of field fasteners, the plate in this case is shop bolted to the beam and field bolted to the column. A careful check must be made to ensure that the beam can be erected without interference from fittings on the column web. Some fabricators would elect to shop attach the plate to the column to eliminate possible interferences and permit use of plain-punched beams. Additionally, if the column were a heavy section, the fabricator may elect to shop weld the plate to the column to avoid drilling the thick flanges. The welding of this plate to the column creates a much stiffer connection and the design should be modified to recognize the increased rigidity.

If the centerline of the W16 were offset $6\frac{1}{16}$-in. from line 1, it would be possible to cope or cut the flanges flush top and bottom and frame the web directly to the column flange with details similar to those shown in Figure 9-35. This type of framing also provides a connection with more rigidity than normally contemplated in simple construction. A coped connection of this type would create a bending moment at the root of the cope which might require reinforcement of the beam web.

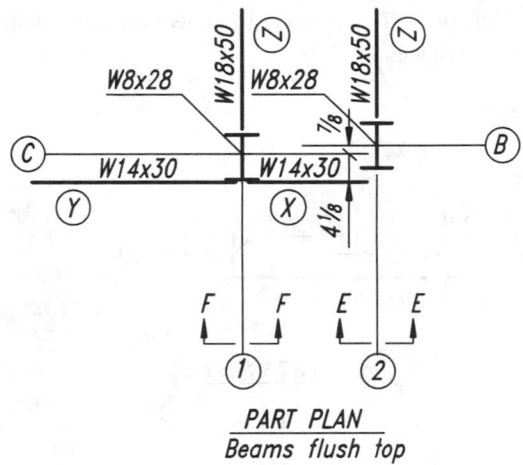

PART PLAN
Beams flush top

Part Column Details
C1 and B2

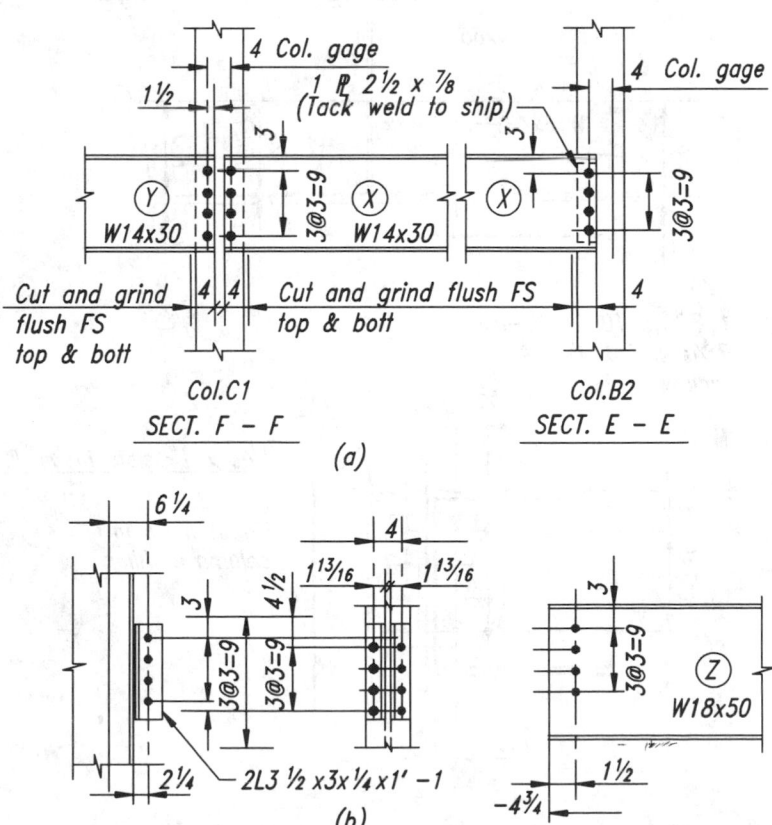

Figure 9-35. Offset beam-to-column-flange connections.

One method frequently adopted to avoid moment transfer to the column because of beam connection rigidity is to use slotted holes and a bearing connection to provide some flexibility. The slotted holes would be provided in the connection plate only and would be in the field connection only. These slotted connections also would accommodate fabrication and erection tolerances.

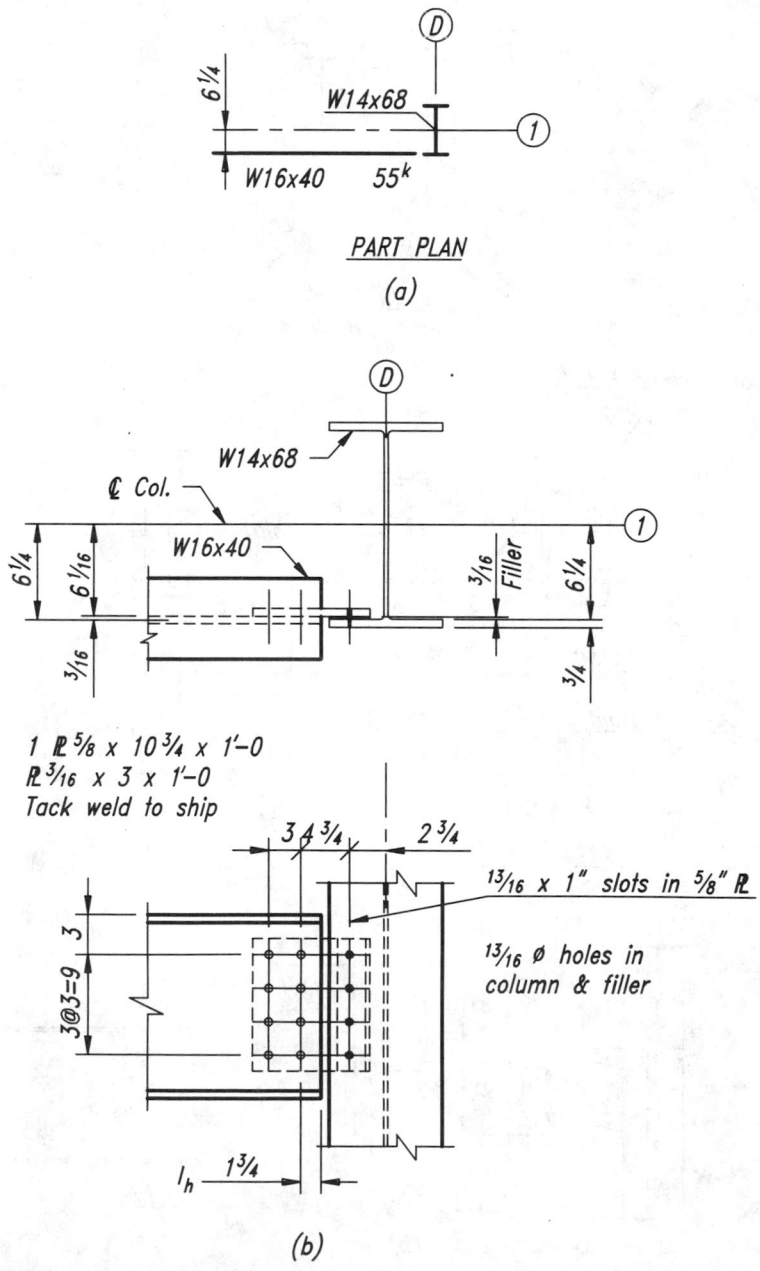

PART PLAN

(a)

1 ℝ ⅝ x 10 ¾ x 1'-0
ℝ 3/16 x 3 x 1'-0
Tack weld to ship

13/16 x 1" slots in ⅝" ℝ

13/16 Ø holes in
column & filler

(b)

Figure 9-36. Offset beam-to-column-flange connections.

The type of connection detailed in Figure 9-36 is similar to a coped beam and should be checked for buckling as illustrated in Parts 8 and 9. The following differences are apparent and should be recognized in the analysis:

1. The effective length of equivalent "cope" is longer by the amount of end distance to the first bolt gage line.

2. There is an inherent eccentricity due to the beam web and plate thickness. The ordinary web and plate thicknesses normally will not require an analysis for this condition, since the inelastic rotation allowed by the LRFD Specification will relieve this secondary moment effect. Two plates may sometimes be required to counter this eccentricity when dimensions are significant.

3. The connection plate can be made of sufficient thickness as required for bending or buckling stresses with a minimum thickness of $\frac{3}{8}$-in.

Example 9-24

Given: Refer to Figure 9-36. Design the connection between the W16×40 beam and W14×68 column flange.

$R_u = 55$ kips

W16×40

$t_w = 0.305$ in., $d = 16.01$ in.
$F_y = 50$ ksi, $F_u = 65$ ksi

W14×68

$t_f = 0.720$ in.
$F_y = 50$ ksi, $F_u = 65$ ksi

Use $\frac{3}{4}$-in. diameter A325-N bolts in standard holes except use short-slotted holes in plate for bolts through the column flange.

Solution: *Design bolts connecting beam web to plate*

For bolt shear, $\phi r_n = 15.9$ kips/bolt. For bolt bearing on the beam web, $\phi r_n = 26.8$ kips/bolt. Since bolt shear is more critical,

$$C_{min} = \frac{R_u}{\phi r_n}$$
$$= \frac{55 \text{ kips}}{15.9 \text{ kips / bolt}}$$
$$= 3.45$$

From Table 8-19 with $\theta = 0°$ and an eccentricity of $6\frac{1}{4}$-in. as shown in Figure 9-36, a four row by two vertical row bolt group provides

$C = 3.59 > 3.45$ **o.k.**

Design bolts connecting plate to column flange

Try one vertical row of four $\frac{3}{4}$-in. diameter A325-N bolts.

Check bolt shear

$$\phi R_n = \phi(F_v A_v) \times n$$

$$= \left[0.75 \times 48 \text{ ksi} \times \frac{\pi}{4} (\text{3/4-in.})^2 \right] \times 4 \text{ bolts}$$

$$= 63.6 \text{ kips} > 55 \text{ kips} \quad \textbf{o.k.}$$

Check bolt bearing on column flange

$$\phi R_n = \phi(2.4 dt F_u) \times n$$
$$= 0.75(2.4 \times \text{3/4-in.} \times 0.720 \text{ in.} \times 65 \text{ ksi}) \times 4 \text{ bolts}$$
$$= 253 \text{kips} > 55 \text{ kips} \quad \textbf{o.k.}$$

Design connection plate

Try PL5/8-in. ×12 in.

Check flexural strength of the plate

The required strength is

$$M_u = R_u e$$
$$= (55 \text{ kips})(4\text{3/4-in.})$$
$$= 261 \text{ in.-kips}$$

For flexural yielding

$$\phi M_n = \phi F_y S_x$$

$$= 0.9(36 \text{ ksi}) \left[\frac{\text{5/8-in. } (12 \text{ in.})^2}{6} \right]$$

$$= 486 \text{ in.-kips} > 261 \text{ in.-kips} \quad \textbf{o.k.}$$

For flexural rupture (from Table 12-1),

$$\phi M_n = \phi F_u S_{net}$$
$$= 0.75(58 \text{ ksi})(11.0 \text{ in.}^3)$$
$$= 479 \text{ in.-kips} > 261 \text{ in.-kips} \quad \textbf{o.k.}$$

Check shear yielding of the plate

$$\phi R_n = \phi(0.6F_y) A_g$$
$$= 0.9(0.6 \times 36 \text{ ksi})(12 \text{ in.} \times \text{5/8-in.})$$
$$= 146 \text{ kips} > 55 \text{ kips} \quad \textbf{o.k.}$$

Check shear rupture of the plate

$$\phi R_n = \phi(0.6F_u) A_n$$
$$= 0.75(0.6 \times 58 \text{ ksi})(12 \text{ in.} - 4 \times 0.875 \text{ in.})\text{5/8-in.}$$
$$= 139 \text{ kips} > 55 \text{ kips} \quad \textbf{o.k.}$$

Check block shear rupture of the plate

From Table 8-47 and 8-48 with $n = 4$, $L_{ev} = 1\text{1/2-in.}$, $L_{eh} = 1\text{1/4-in.}$ $0.6F_u A_{nv} > F_u A_{nt}$. Thus,

$$\phi R_n = \phi[0.6F_u A_{nv} + F_y A_{gt}]$$

From Table 8-48a and 8-48b,

$$\phi R_n = (194 \text{ kips/in.} + 33.8 \text{ kips/in.})\frac{5}{8}\text{-in.}$$
$$= 142 \text{ kips} > 55 \text{ kips} \quad \textbf{o.k.}$$

Check local buckling of the plate

This check is analogous to the local buckling check for doubly coped beams as illustrated previously in Parts 8 and 9 where $c = 6$ in. and $d_c = 1\frac{1}{2}$-in. at both the top and bottom flanges.

$$f_d = 3.5 - 7.5\left(\frac{d_c}{d}\right)$$
$$= 3.5 - 7.5\left(\frac{1\frac{1}{2}\text{-in.}}{16.01 \text{ in.}}\right)$$
$$= 2.80$$

$$\phi F_{bc} = 50,840\left(\frac{t_w^2}{ch_o}\right)f_d$$
$$= 50,840\left[\frac{(\frac{5}{8}\text{-in.})^2}{(6 \text{ in.})(16.01 \text{ in.} - 1.5 \text{ in.} - 1.5 \text{ in.})}\right](2.80)$$
$$= 712 \text{ ksi}$$

$$\phi R_n = \frac{\phi F_{bc} S_{net}}{e}$$
$$= \frac{(712 \text{ ksi})(11.0 \text{ in.}^3)}{4\frac{3}{4}\text{-in.}}$$
$$= 1,650 \text{ kips} > 55 \text{ kips} \quad \textbf{o.k.}$$

Framing to the Column Web

If the offset of the beam from the centerline of the column web is small enough that the connection may still be centered on or under the supported beam, no special considerations need be made. However, when the offset of the beam is too large to permit the centering of the connection under the beam as in Figure 9-37, it may be necessary to consider the effect of eccentricity in the fastener group.

The offset of the beam in Figure 9-37 requires that the top and bottom flanges be blocked to provide erection clearance at the column flange. Since only half of each flange, then, remains in which to punch holes, a 6-in. outstanding leg is used for both the seat and top angles of these connections; this permits the use of two field bolts to each of the seat and top angles, as required for safety reasons.

Example 9-25

Given: Refer to Figure 9-38. Design the seat angle and weld size required for the unstiffened seated connection for the W16×45 beam to W12×53 "column-web" connection.

$$R_u = 30 \text{ kips}$$

W16×45

$t_w = 0.345$ in.
$F_y = 50$ ksi, $F_u = 65$ ksi

W12×53

$t_f = 0.575$ in., $d = 12.06$ in.
$F_y = 50$ ksi, $F_u = 65$ ksi

Use 70 ksi electrode welds to connect the seat angle to the column-flange toes. Assume a 4-in. outstanding angle leg is adequate and the angle material has $F_y = 36$ ksi and $F_u = 58$ ksi.

Solution: *Design seat angle and welds*

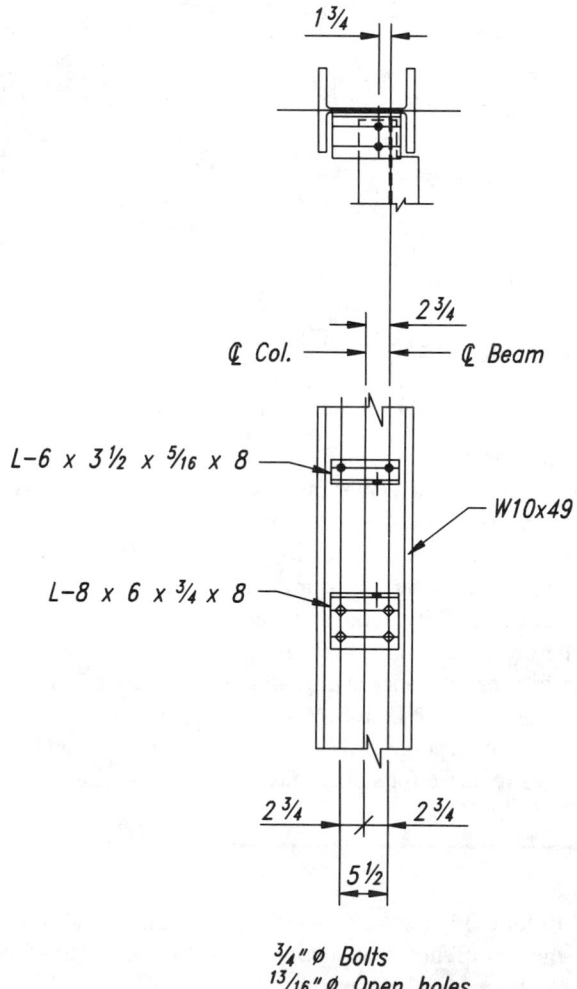

Figure 9-37. Offset beam-to-column-web connections.

The required strength for the left-hand weld can be determined by summing moments about the right-hand weld.

$$R_{uL} = \frac{(30\,\text{kips})(10\tfrac{1}{2}\text{-in.})}{(12.06\,\text{in.})}$$
$$= 26.1\,\text{kips}$$

Selecting the welds on both sides of the seat to resist this force, the total required strength would be 52.2 kips.

From Table 9-7, with $t_w \approx \tfrac{5}{16}$-in., a $\tfrac{3}{4}$-in. seat angle thickness provides

$$\phi R_n = 54.7\,\text{kips} > 52.2\,\text{kips} \quad \textbf{o.k.}$$

and an 8×4 angle with $\tfrac{5}{16}$-in. fillet welds provides

$$\phi R_n = 66.8\,\text{kips} > 52.2\,\text{kips} \quad \textbf{o.k.}$$

Connections for Raised Beams

When raised beams are connected to column flanges or webs, there is usually no special consideration required. However, when the support is a girder, the differing tops of steel may preclude the use of typical connections. Figure 9-39 shows several typical details

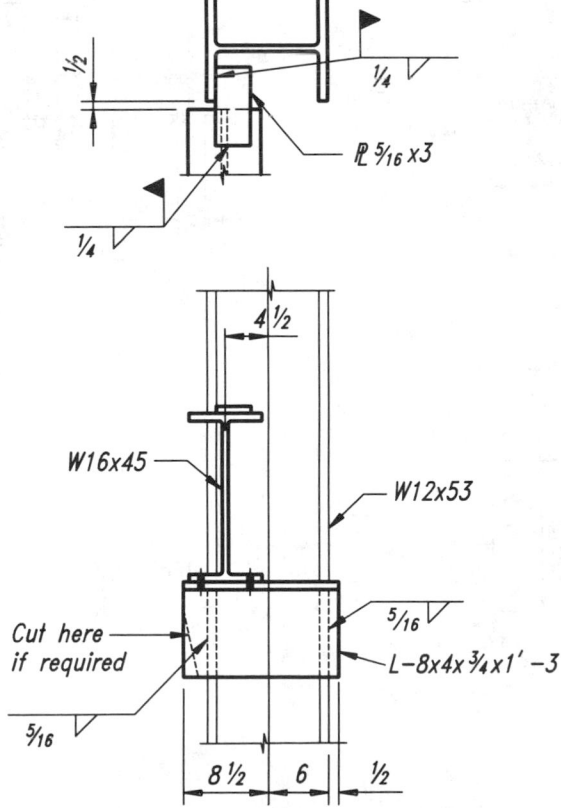

Figure 9-38. Illustration for Example 9-25.

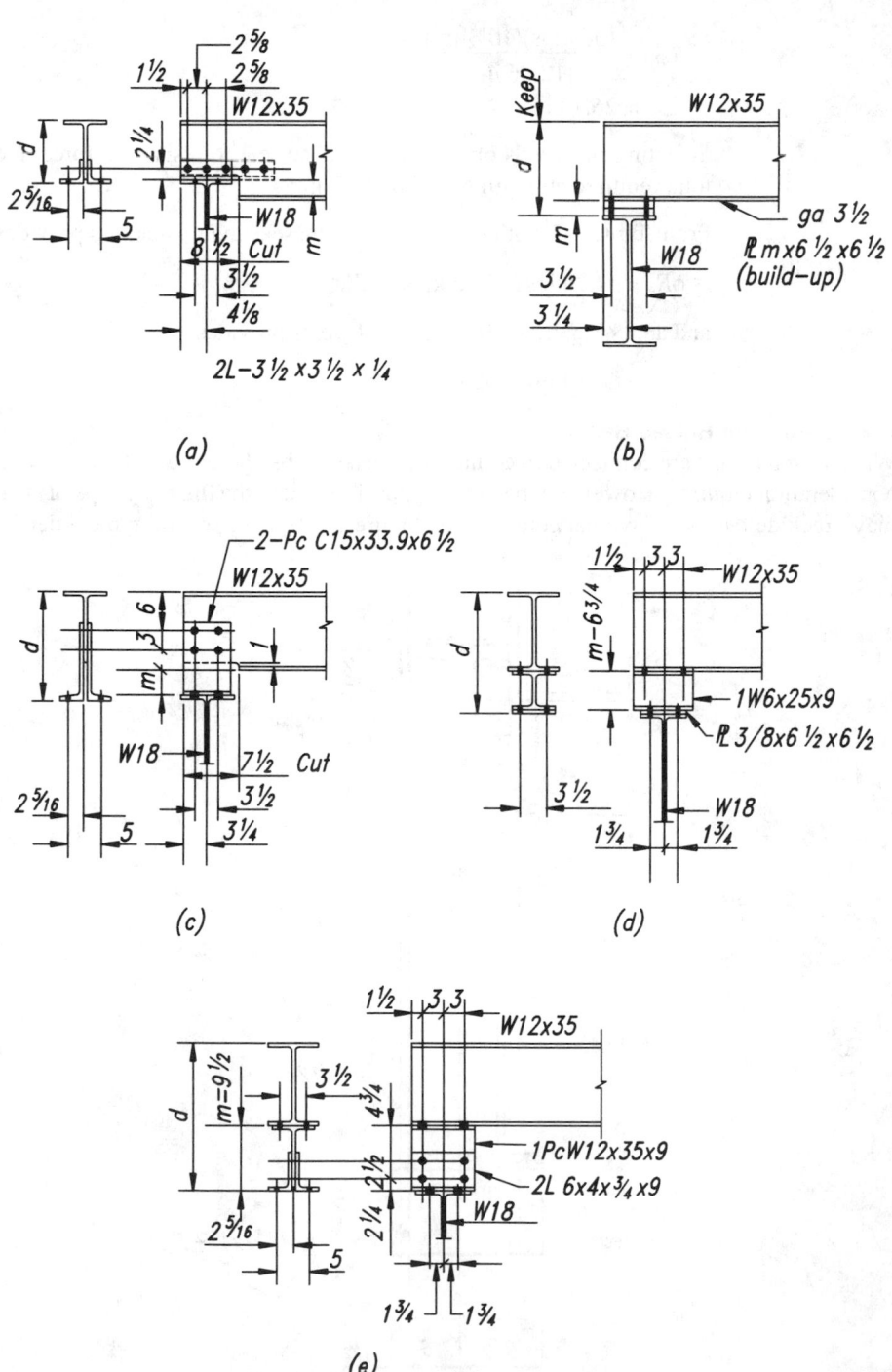

Figure 9-39. *Typical bolted raised beam connections.*

commonly used for such cases in bolted construction. Figure 9-40 shows several typical details commonly used in welded construction.

In Figure 9-39a, since the top of the W12×35 is located somewhat less than 12 inches above the top of the W18 supporting beam, a double-angle connection is used. This connection would be designed for the beam reaction and the shop bolts would be governed by double shear or bearing, just as if they were located in a vertical position. However, the field bolts are not required to carry any calculated force under gravity loading.

The maximum permissible distance m depends on the beam reaction, since the web remaining after the bottom cope must provide sufficient area to resist the vertical shear as well as the bending moment which would be critical at the end of the cope. The beam can be reinforced by extending the angles beyond the cope and adding additional shop bolts for development. The angle size and/or thickness can be increased to gain shear area or section modulus, if required. The effect of any eccentricity would be a matter of judgment, but could be neglected for small dimensions.

When this connection is used for flexure or for dynamic or cyclical loading, the web is subjected to high stress concentrations at the end of the cope, and it is good practice to extend the angles as shown in Figure 9-39a by the dashed lines to add at least two additional web fasteners.

Figure 9-39b covers the case where the bottom flange of the W12×35 is located a few inches above the top of the W18. The beam bears directly upon fillers and is connected to the W18 by four field bolts which are not required to transmit a calculated gravity load. If the distance m exceeds the thickest plate which can be punched, two or more plates may be used. Even though the fillers in this case need only be 6½-in. square, the amount of material required increases rapidly as m increases. If m exceeds 2 or 3 in., another type of detail may be more economical.

The detail shown in Figure 9-39c is used frequently when m is up to 6 or 7 in. The load on the shop bolts in this case is no greater than that in Figure 9-39a. However, to provide more lateral stiffness, the fittings are cut from a 15 in. channel and are detailed to overlap the beam web sufficiently to permit four shop bolts on two gage lines.

A stool or pedestal, cut from a rolled shape, can be used with or without fillers to provide for the necessary m-distance as in Figure 9-39d. A pair of connection angles and a tee will also serve a similar purpose, as shown in Figure 9-39e. To provide adequate strength to carry the beam end reaction and to provide lateral stiffness, the web thickness of the pedestal in each of these cases should be at least as thick as the member being supported.

In Figure 9-40a, welded framing angles are substituted for the bolted angles of Figure 9-39a. In Figure 9-40b, a single horizontal plate is shown replacing the pair of framing angles; this results in a savings in material and the amount of shop welding. In this case, particular care must be taken in cutting the beam web and positioning the plate at right angles to the beam web. For this reason, if only a few connections of this type are to be made, some fabricators prefer to use the angles as in Figure 9-40a. If sufficient duplication were available to warrant making a simple jig to position the plate during welding, the solution of Figure 9-40b may be economical.

Figure 9-40c shows a tee centered on the beam web and welded to the bottom flange of the beam. The tee stem thickness should not be less than the beam web thickness. The welded solutions shown in Figures 9-40d and 9-40e are capable of providing good lateral

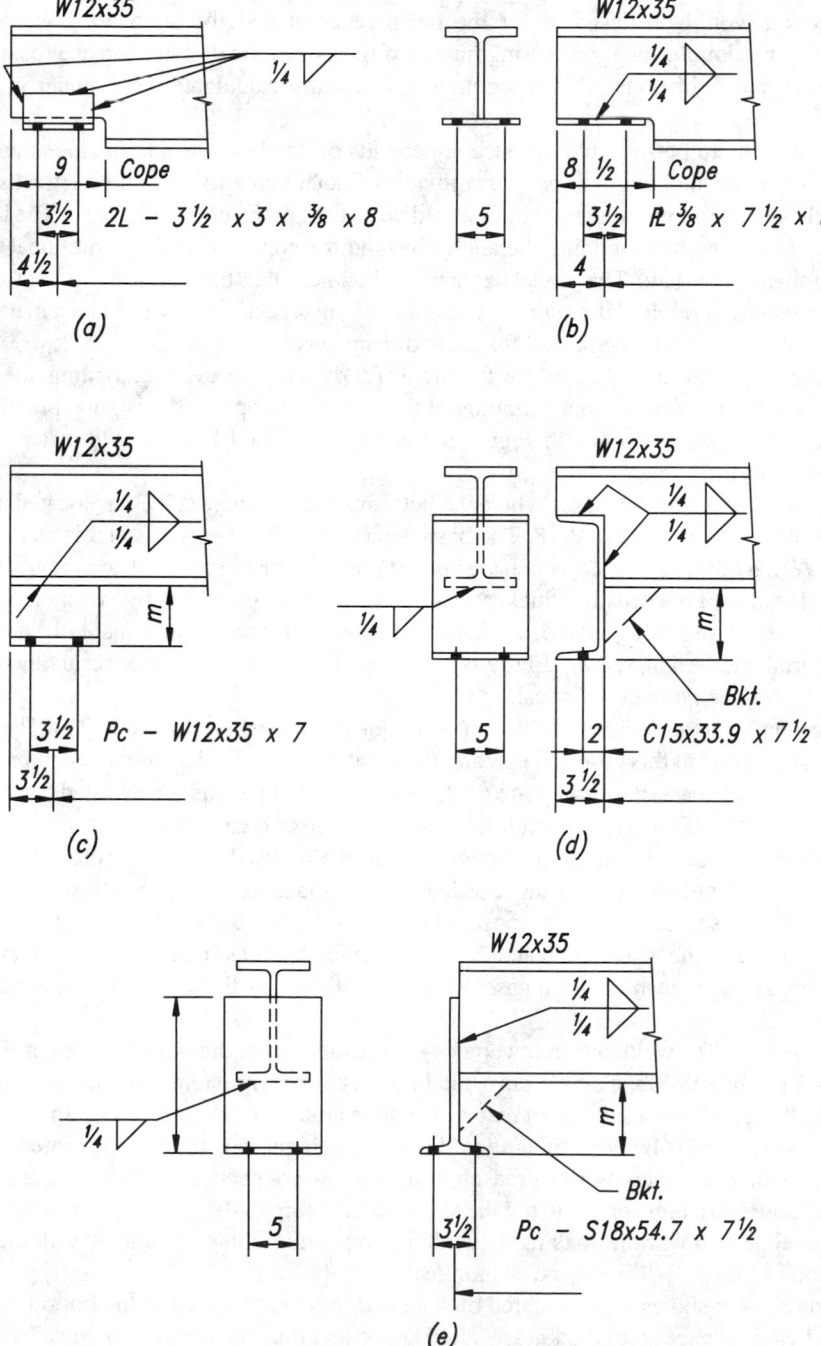

Figure 9-40. Typical welded raised beam connections.

stiffness. The latter two types also permit end rotation as the beam deflects under load. However, if the *m* distance exceeds 3 or 4 in., it is advisable to shop weld a triangular bracket plate at one end of the beam, as indicated by the dashed lines, to prevent the beam from deflecting along its longitudinal axis.

Other equally satisfactory details may be devised to meet the needs of connections for raised beams. They will vary depending on the size of the supported beam and the distance *m*. When using this type of connection where the load is transmitted through bearing, the provisions of LRFD Specification Sections K1.3 and K1.4 must be satisfied for both the supported and supporting members. For the detail of Figure 9-40b, since the rolled fillet has been removed by the cut, the value of *k* would be taken as the thickness of the plate plus the fillet weld size.

LRFD Specification Section B6 requires stability and restraint against rotation about the beam's longitudinal axis. This provision is most easily accomplished with a floor on top of the supported beam. In the absence of a floor, the top flange may be supported by a strut or bracket attached to the supporting member. When the beam is encased in a wall, this stability may also be provided with wall anchors; refer to "Wall Anchors" in Part 12.

This discussion has considered that the field bolts which attach the beam to the pedestal or support beam, are subject to no calculated load. It is important, however, to recognize that when the beam deflects about its neutral axis, a tensile force can be exerted on the outside bolts. The intensity of this tensile force is a function of the dimension *d* indicated in Figure 9-39, the span length of the supported member, and the beam stiffness. If these forces are large, high-strength bolts should be used and the connection analyzed for the effects of prying action.

Raised beam connections such as these are used frequently as equipment or machinery supports where it is important to maintain a true and level surface or elevation. When this tolerance becomes important, the dimension *d* should be noted "keep" to advise the fabricator of this importance, as shown in Figure 9-39b. Since the supporting beam is subject to certain camber/deflection tolerances, it also may be appropriate to furnish shim packs between the connection and the supporting member.

Connections for Tubular and Pipe Members
Several typical connections for tubular and pipe members are illustrated in Figure 9-41. For more information, refer to Palmer (1990), Sherman and Ales (1991), Sherman and Herlache (1988), and Ricker (1985).

Non-Rectangular Simple Shear Connections
It is often necessary to design connections for beams which do not frame into a support orthogonally. Such a beam may be inclined with respect to the supporting member in various directions. Depending upon the relative angular position which a beam assumes, the connection may be classified among three categories: skewed, sloped, or canted. These conditions are illustrated in Figure 9-42 for beam-to-girder web connections; the same descriptions apply to beam-to-column flange and web connections. Additionally, beams may be oriented in a combination of any or all of these conditions. For any condition of skewed, sloped, or canted framing, the single-plate connection is generally the simplest and most economical of those illustrated in this text.

Skewed Connections
A beam is said to be skewed when its flanges are parallel to the flanges of the supporting beam, but the webs incline to each other. The angle of skew A appears in Figure 9-42a

and represents the horizontal bevel to which the fittings must be bent or set, or the direction of gage lines on a seated connection.

When the skew angle is less than 15° (3 in 12 slope), a pair of double angles can be bent inward or outward to make the connection as shown in Figure 9-43. While bent angle sections are usually drawn as bending in a straight line from the heel, rolled angles will tend to bend about the root of the fillet (dimension k in Manual Part 1). This produces a significant jog in the leg alignment, which is magnified by the amount of bend. Above this angle of skew, it becomes impractical to bend rolled angles.

For skews approximately greater than 5° (1 in 12 slope), a pair of bent plates, shown in Figure 9-44, may be a more practical solution. Bent plates are not subject to the deformation problem described for bent angles, but the radius and direction of the bend must be considered to avoid cracking during the cold-bending operation.

Bent plates exhibit better ductility when bent perpendicular to the rolling direction and are, therefore, less likely to crack. Whenever possible, bent connection plates should be billed with the width dimension parallel to the bend line. The length of the plate is measured on its mid-thickness, without regard to the radius of the bend. While this will

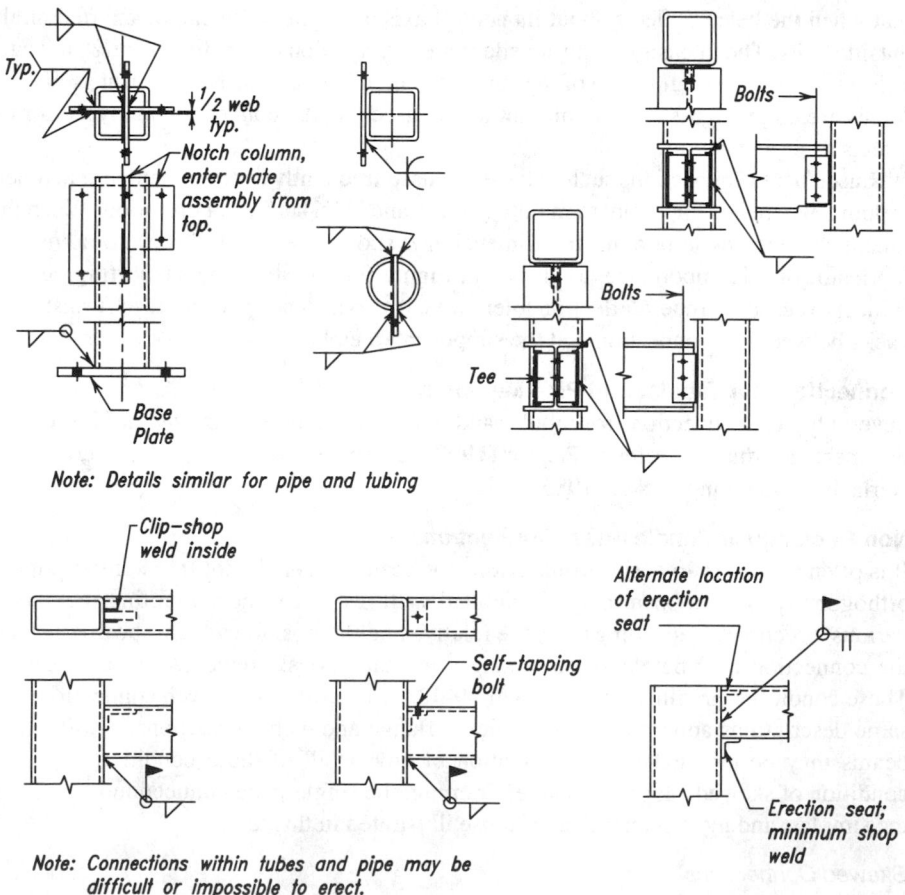

Figure 9-41. Typical connections for tubular and pipe members.

provide a plate which is slightly longer than necessary, this will be corrected when the bend is laid out to the proper radius prior to fabrication.

Table 9-13 gives the generally accepted minimum inside-bending radius for plate thickness t for various grades of steel. Values are for bend lines transverse to the direction of final rolling. When bend lines are parallel to the direction of final rolling, the tabular values may have to be approximately doubled. When bend lines are longer than 36 inches, all radii may have to be increased if problems in bending are encountered.

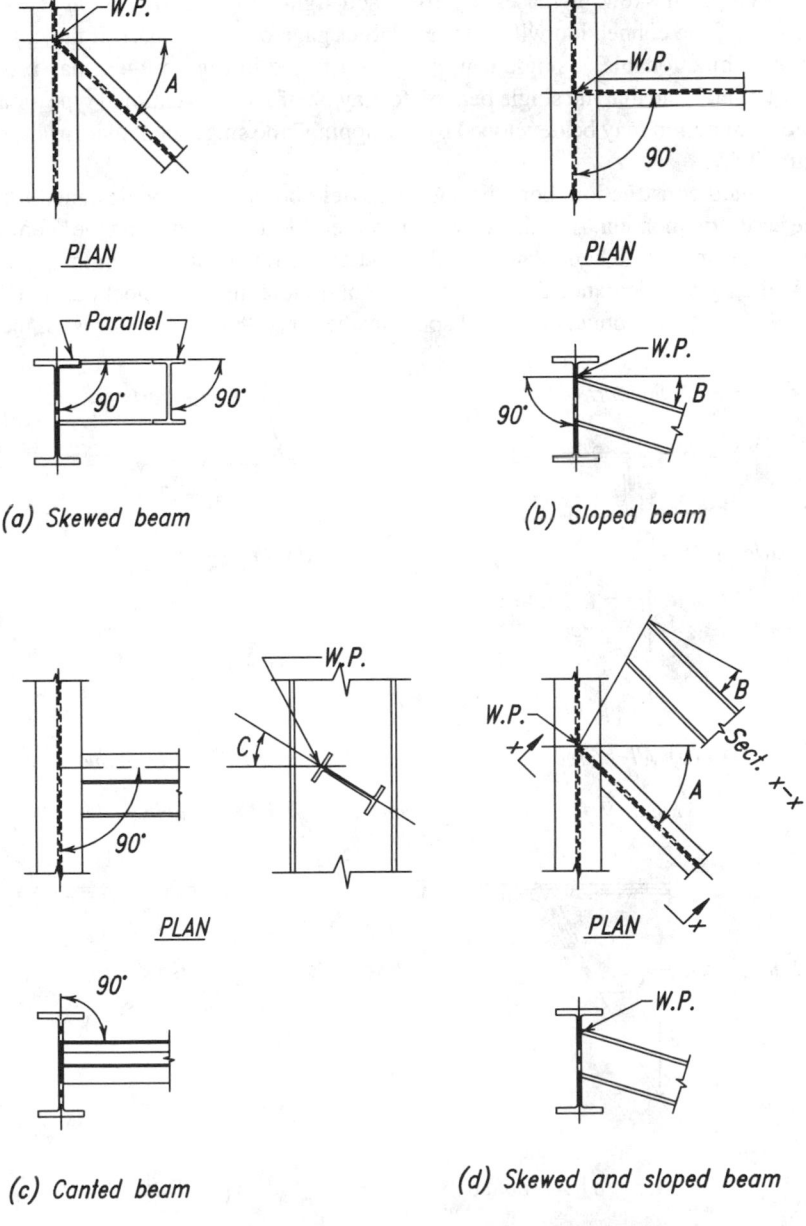

Figure 9-42. Non-rectangular connections.

Before bending, special attention should be given to the condition of plate edges transverse to the bend lines. Flame-cut edges of hardenable steels should be machined or softened by heat treatment. Nicks should be ground out and sharp corners should be rounded.

The strength of bent angles and bent plate connections may be calculated in the same manner as for square framed beams, making due allowances for eccentricity. The load is assumed to be applied at the point where the skewed beam center line intersects the face of the supporting member.

As the angle of skew increases, entering and tightening clearances on the acutely angled side of the connection will require a larger gage on the support. If the gage were to become objectionable, a single bent plate, illustrated in Figure 9-45, may provide a better solution. Note that the single bent plate may be of the conventional type, or a more compact connection may be developed by "wrapping" the single bent plate as illustrated in Figure 9-45c.

In all-bolted construction, both the shop and field bolts should be designed for shear and the eccentric moment. A C-shaped weld is preferable to avoid turning the beam during shop fabrication. Single bent plates should be checked for flexural strength.

Table 9-14 gives clearance dimensions for bent double-angle connections and double and single bent plate connections, and specifies beam set-backs and gages. Since these

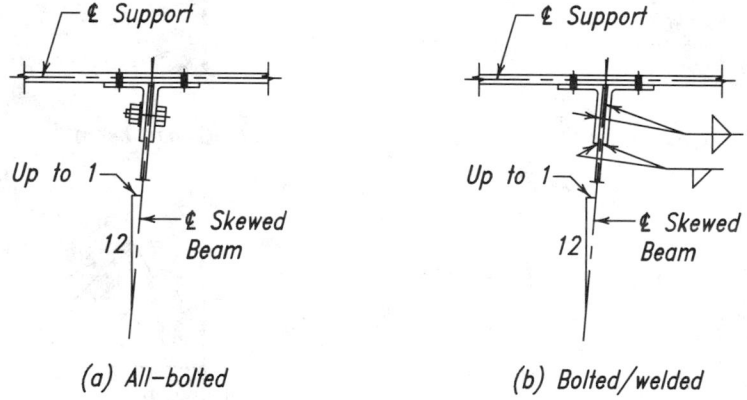

(a) All-bolted   (b) Bolted/welded

Figure 9-43. Skewed beam connection with bent double angles.

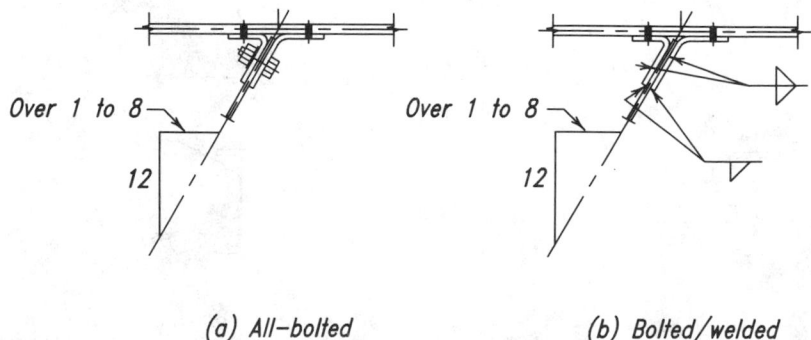

(a) All-bolted   (b) Bolted/welded

Figure 9-44. Skewed beam connection with double bent plates.

Table 9-13.
Minimum Radius for Cold Bending

ASTM Designation		Thickness, in.				
		Up to ¼	Over ¼ to ½	Over ½ to 1	Over 1 to 1½	Over 1½ to 2
A36		1½ t	1½ t	2t	3t	4t
A242		2t	3t	5t	—[a]	—[a]
A514[b]		2t	2t	2t	3t	3t
A529		2t	2t	—	—	—
A572[c]	Gr. 42	2t	2t	3t	4t	5t
	Gr. 50	2½t	2½t	4t	—[a]	—[a]
	Gr. 60	3½t	3½t	6t	—[a]	—
	Gr. 65	4t	4t	—[a]	—[a]	—
A588		2t	3t	5t	—[a]	—[a]
A852[b]		2t	2t	3t	3t	3t

[a] It is recommended that steel in this thickness range be bent hot. Hot bending, however, may result in a decrease in the as-rolled mechanical properties.
[b] The mechanical properties of ASTM A514 and A852 steels result from a quench-and-temper operation. Hot bending may adversely affect these mechanical properties. If necessary to hot-bend, fabricator should discuss procedure with the steel supplier.
[c] Thickness may be restricted because of columbium content. Consult supplier.

dimensions are based on the maximum material thicknesses and fastener sizes indicated, it is suggested that in cases where many duplicate connections with less than maximum material or fasteners are required, savings can be effected if these dimensions are developed from specific bevels, beam sizes, and fitting thicknesses.

Skewed single plate and skewed end plate connections, shown in Figures 9-46 and 9-47, provide a simple, direct connection with a minimum of fittings and multiple punching requirements. When fillet welded, these connections may be used for skews up to 30° (or a slope of $6\frac{5}{16}$ in 12) provided the root opening formed does not exceed $\frac{3}{16}$-in. as specified in AWS D1.1 paragraph 3.3.1. For skew angles greater than 30°, see AWS D1.1, Section 2.11.

The maximum beam web thickness which may be supported is a function of the maximum root opening and the angle of skew. If the thickness of the beam web were such that a larger root opening were encountered, the skewed single plate or the web connecting to the skewed end plate may be beveled, as shown in Figures 9-46b and 9-47b. Since no root opening occurs with the bevel, there is no limitation on the thickness of the beam web. However, beveling, especially of the beam web, requires careful finishing and is an expensive procedure which may outweigh its advantages.

The design of skewed end plate connections is similar to that discussed previously in "Shear End-Plate Connections" in this Part. However, when the gage of the bolts is not centered on the beam web, this eccentric loading should be considered. The design of skewed single-plate connections is similar to that discussed previously in "Single-Plate Connections" in this Part.

Table 9-14 specifies gages and the dimension A which is added to the fillet weld size to compensate for the root opening for skewed end-plate connections. This table is based conservatively on a gap of $\frac{1}{8}$-in. For beam webs beveled to the appropriate skew, $A = 0$ and the tabulated values do not apply. Table 9-14 also provides similar information for skewed single-plate connections. Additionally, this table provides clearances and dimensions for groove welded single-plate connections with backing bars for skews greater than 30°; refer to AWS D1.1 for prequalified welds for both types of joints.

When skewed stiffened seated connections are used, the stiffening element should be located so as to cross the skewed beam centerline well out on the seat. This can be accomplished by shifting the stiffener to the left or right of center to support beams which skew to the left or to the right, respectively. Alternatively, it may be possible to skew the stiffening element.

Example 9-26

Given: Refer to Figure 9-48. Design the skewed double-bent-plate connection for the W16×77 beam to W27×94 girder-web connection.

$R_u = 80$ kips

W16×77

$t_w = 0.455$ in., $d = 16.52$ in.
$F_y = 50$ ksi, $F_u = 65$ ksi

W27×94

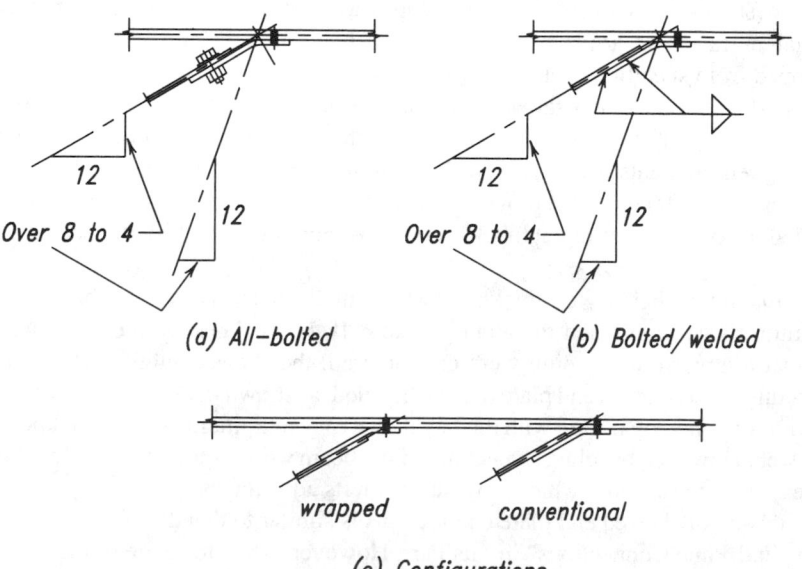

(a) All-bolted   (b) Bolted/welded

wrapped   conventional

(c) Configurations

Figure 9-45. Skewed beam connections with single bent plates.

$$t_w = 0.490$$
$$F_y = 50 \text{ ksi}, \ F_u = 65 \text{ ksi}$$

Use ⅞-in. diameter A325-N bolts in standard holes through the support. Use 70 ksi electrode welds to the supported beam. Assume plate material with $F_y = 36$ ksi and $F_u = 58$ ksi.

Solution:

From the scaled layout of Figure 9-48c, assuming the welds across the top and bottom of the plates will be 2½-in. long, the load is assumed to act at the intersection of the beam centerline and the support face. While the welds do not coincide on opposite faces of the beam web and the weld groups are offset, the locations of the weld groups will be averaged and considered identical.

Design welds

Assume plate length of 8½-in.

$$k = \frac{kl}{l}$$
$$= \frac{2\frac{1}{2}\text{-in.}}{8\frac{1}{2}\text{-in.}}$$
$$= 0.294$$

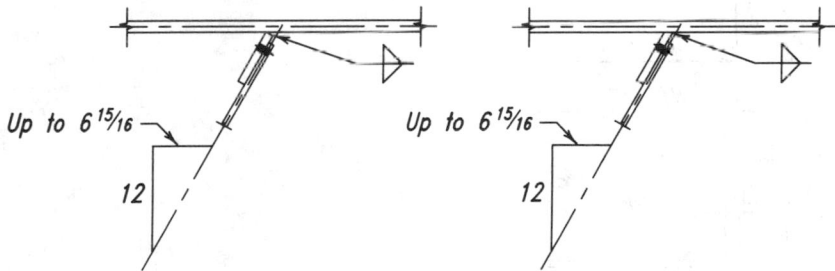

(a) Square edge (preferred)   (b) Beveled edge (alternative)

Figure 9-46. Skewed single-plate connections.

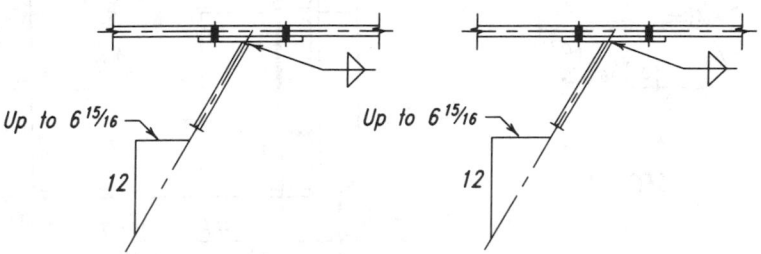

(a) Square edge (preferred)   (b) Beveled edge (alternative)

Figure 9-47. Skewed end-plate connections.

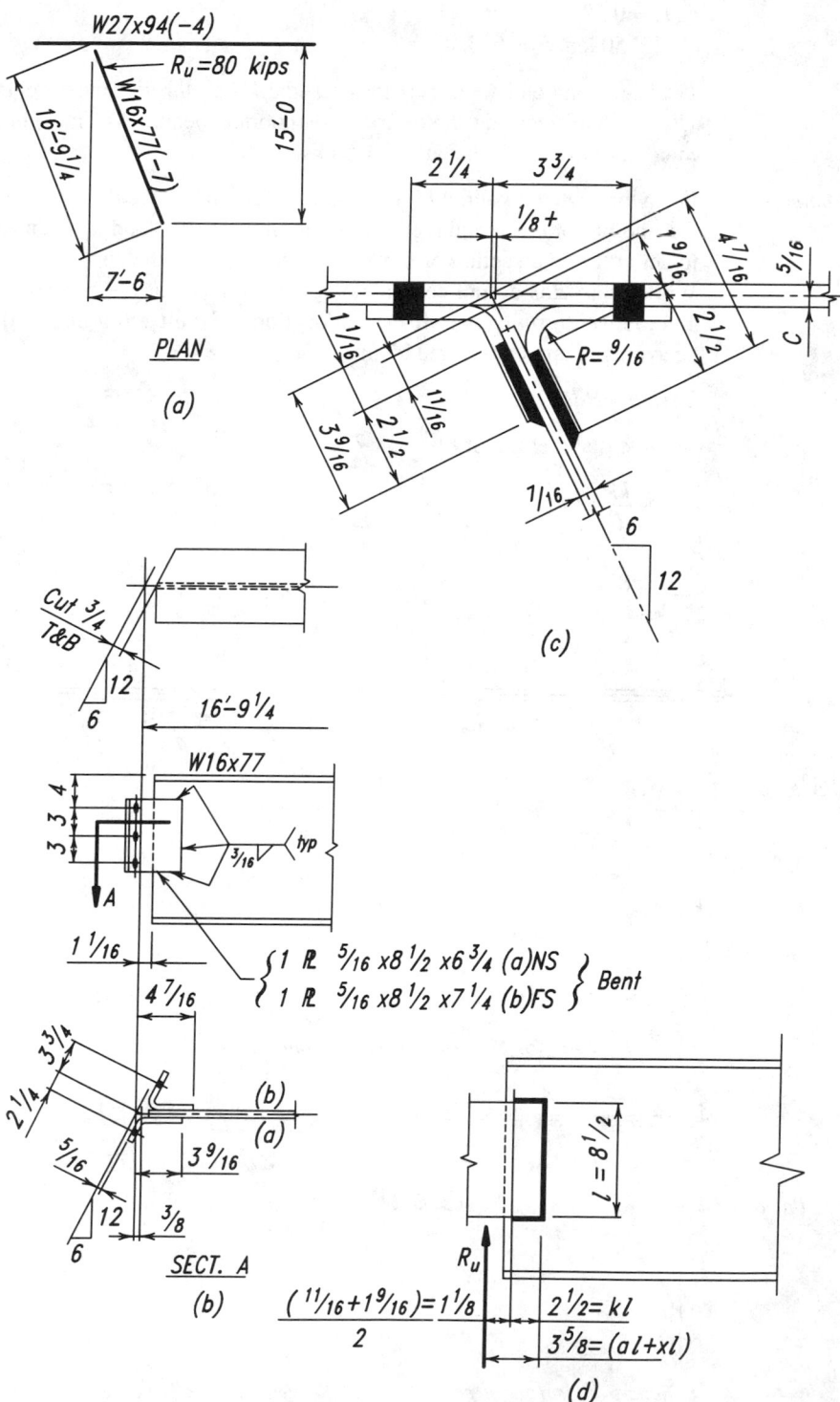

Figure 9-48. Illustration for Example 9-26.

From Table 8-42, with $\theta = 0°$ and $k = 0.294$

$x = 0.054$ by interpolation

Thus,

$$a = \frac{3\frac{5}{8}\text{-in.} - xl}{l}$$

$$= \frac{3\frac{5}{8}\text{-in.} - 0.054\,(8\frac{1}{2}\text{-in.})}{8\frac{1}{2}\text{-in.}}$$

$$= 0.372$$

Interpolation from Table 8-42 with $\theta = 0°$, $a = 0.372$, and $k = 0.294$,

$C = 1.84$

and the required weld size for two such welds is

$$D_{req} = \frac{R_u/2}{CC_1 l}$$

$$= \frac{40\text{ kips}}{(1.84)\,(1.0)\,(8\frac{1}{2}\text{-in.})}$$

$$= 2.56 \rightarrow 3 \text{ sixteenths}$$

Use $\frac{3}{16}$-in. fillet welds.

Check beam web thickness

$$t_{min} = \frac{5.16D}{F_y}$$

$$= \frac{5.16\,(3 \text{ sixteenths})}{50\text{ ksi}}$$

$$= 0.310\text{ in.} < 0.455\text{ in.} \quad \textbf{o.k.}$$

Design bolts

For an $8\frac{1}{2}$-in. plate length, use three rows of bolts.

$$\phi R_n = n \times \phi r_n$$

$$= 6 \text{ bolts} \times 21.6 \text{ kips/bolt}$$

$$= 130 \text{ kips} > 80 \text{ kips} \quad \textbf{o.k.}$$

Use six $\frac{7}{8}$-in. diameter A325-N bolts.

Check bearing on support

$$\phi R_n = n \times \phi(2.4 dt F_u)$$

$$= 6 \text{ bolts} \times 0.75(2.4 \times \frac{7}{8}\text{-in.} \times 0.490\text{ in.} \times 65\text{ ksi})$$

$$= 301 \text{ kips} > 80 \text{ kips} \quad \textbf{o.k.}$$

Design bent plates

Try PL$\frac{5}{16}$-in.

Check bearing on plates

$$\phi R_n = n \times \phi(2.4 dt F_u)$$
$$= 6 \text{ bolts} \times 0.75(2.4 \times \text{7/8-in.} \times 2 \times \text{5/16-in.} \times 58 \text{ ksi})$$
$$= 343 \text{ kips} > 80 \text{ kips} \quad \textbf{o.k.}$$

Check shear yielding of plates

$$\phi R_n = \phi(0.6 F_y)A_g$$
$$= 0.9(0.6 \times 36 \text{ ksi})(\text{8½-in.} \times 2 \times \text{5/16-in.})$$
$$= 103 \text{ kips} > 80 \text{ kips} \quad \textbf{o.k.}$$

Check shear rupture of plates

$$\phi R_n = \phi(0.6 F_u)A_n$$
$$= 0.75(0.6 \times 58 \text{ ksi})(\text{8½-in.} - 3 \times 1 \text{ in.})(2 \times \text{5/16-in.})$$
$$= 90.0 \text{ kips} > 80 \text{ kips} \quad \textbf{o.k.}$$

Check block shear rupture of the plates

From Tables 8-47 and 8-48, $0.6 F_u A_{nt} > F_u A_{nt}$. Thus,

$$\phi R_n = \phi[0.6 F_u A_{nv} + F_y A_{gt}]$$

From Tables 8-48a and 8-48b, with $n = 3$ and $L_{ev} = L_{eh} = 1\text{¼}$,

$$\phi R_n = (124 \text{ kips/in.} + 33.8 \text{ kips/in.})(2 \times \text{5/16-in.})$$
$$= 98.6 \text{ kips} > 80 \text{ kips} \quad \textbf{o.k.}$$

Sloped Connections

A beam is said to be sloped if its web is perpendicular to the web of the supporting member, but its flanges are not perpendicular to this face. The angle of slope B is shown in Figure 9-42b and represents the vertical angle to which the fittings must be set to the web of the sloped beam, or the amount that seat and top angles must be bent.

The design of sloped connections usually can be adapted directly from the rectangular connections covered earlier in this part, with consideration of the geometry of the connection to establish the location of fittings and fasteners. Note that sloped beams often require copes to clear supporting girders, as illustrated in Figure 9-49.

Figure 9-50 shows a sloped beam with double-angle connections, welded to the beam and bolted to the support. The design of this connection is essentially similar to that for rectangular double-angle connections. Alternatively, shear end-plate, tee, single-angle, single-plate, or seated connections could be used. Selection of a particular connection type may be influenced by fabrication economy, erectability, and/or by the types of connections used elsewhere in the structure.

Sloped seated beam connections may utilize either bent angles or plates, depending on the angle of slope. Dimensioning and entering and clearance requirements for sloped seated connections are generally similar to those for skewed connections. The bent seat and top plate shown in Figure 9-51 may be used for smaller bevels.

When the angle of slope is small, it is economical to place transverse holes in the beam web on lines perpendicular to the beam flange; this requires only one stroke of a multiple punch per line. Since non-standard hole arrangements, then, usually occur in the connecting materials (which are single punched), this requires that sufficient dimensions be provided for the connecting material to contain fasteners with adequate edges and gages, and at the same time fit the angle to the web without encroaching on the flange fillets of the beam. For the end connection of the beam, this was accomplished by using a 6-in. angle leg; a 4-in. or even a 5-in. leg would not have furnished sufficient edge distance at the extreme fastener.

As the angle of slope increases, however, bolts for the end connections cannot conveniently be lined up to permit simultaneous punching of all holes in a transverse row. In this case, the fabricator may choose to disregard beam gage lines and arrange the hole punching so that ordinary square framed connection material can be used throughout, as shown in Figure 9-52.

Canted Connections

A beam perpendicular to the face of a supporting member, but rotated so that its flanges are tilted with respect to those of the support, is said to be canted. The angle of cant C is shown in Figure 9-42c.

The design of canted connections usually can be adapted directly from the rectangular connections covered earlier in this part. In Figure 9-53, a double-angle connection is used. Alternatively, shear end-plate, seated, single angle, single-plate, and tee connections may also be used.

For channel B2, which is supported by a sloping member B1 (not shown), to match the hole pattern in supporting member B1, the holes in the connecting materials must be canted. As shown in Figure 9-54, the top flange of the channel and the connection angles d^R and d^L are cut to clear the flanges of beam B1. In this detail, with a 3 in 12 angle of cant, 4-in. legs were wide enough to contain the pattern of hole punching.

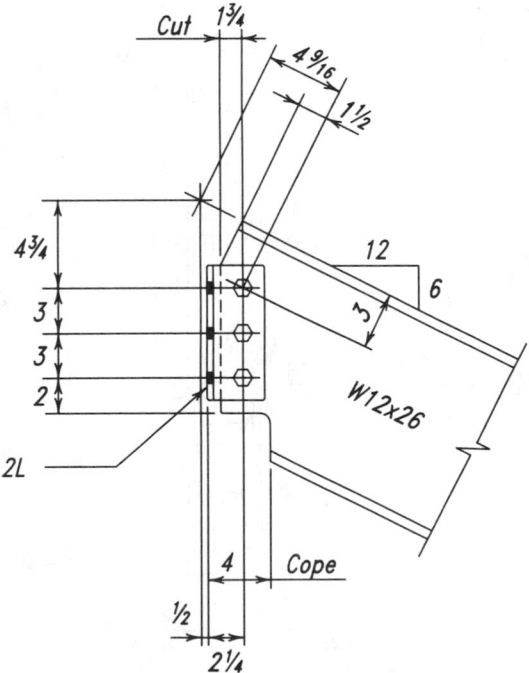

Figure 9-49. Sloped connection with cuts to clear supporting girder flange.

Since the multiple punching or drilling of column flanges requires strict adherence to column gage lines, punching is generally skewed in the fittings. When, for some reason, this is not possible, as in Figure 9-55, skewed reference lines are shown on the column to aid in matching connections.

When canted connecting materials are assembled on the beam, particular care must be used in determining the direction of skew for punching the connection angles. An error reversing this skew may permit matching of holes in both members, but the beam will be canted opposite to the intended direction.

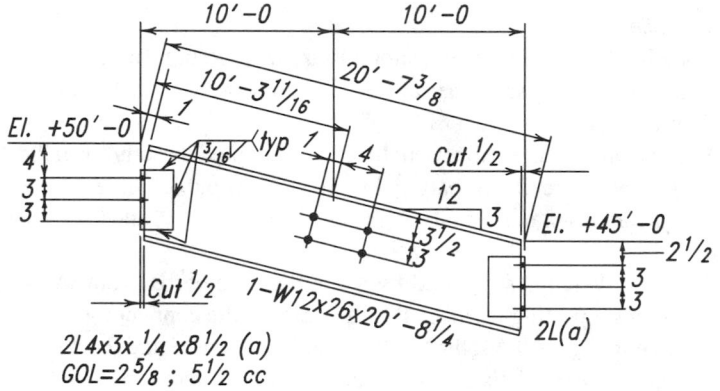

Figure 9-50. Sloped double-angle connection.

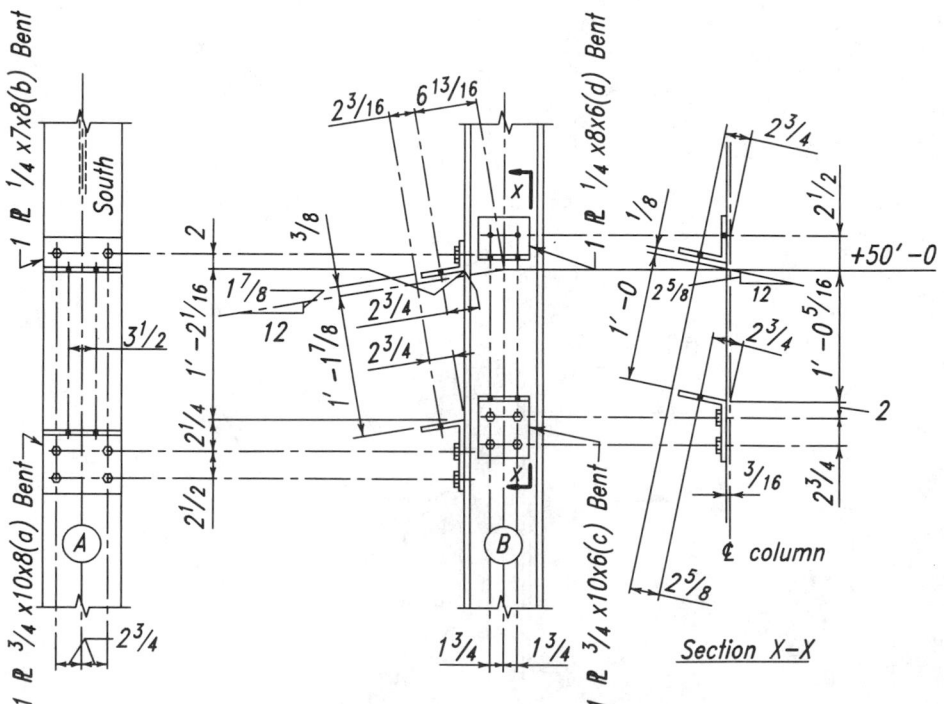

Figure 9-51. Sloped seated connection.

Note the connection angles in Figure 9-55 are shown shop welded to the beam. This was done to provide tightening clearance for ¾-in. high-strength field bolts in the opposite leg. Had the shop fasteners been bolts, it would have been necessary to stagger the field and shop fasteners and provide longer angles for the increased spacing.

Canted seated beams, shown in Figure 9-56, present few problems other than those in ordinary square-end seated beams. Sufficient width and length of angle leg must be provided to contain the gage line punching or drilling in the column face, as well as the off-center location of the holes matching the punching in the beam flange. The elevation of the top flange centerline and the bevel of the beam flange may be given for reference on the beam detail, although the bevel shown will not affect the fabrication.

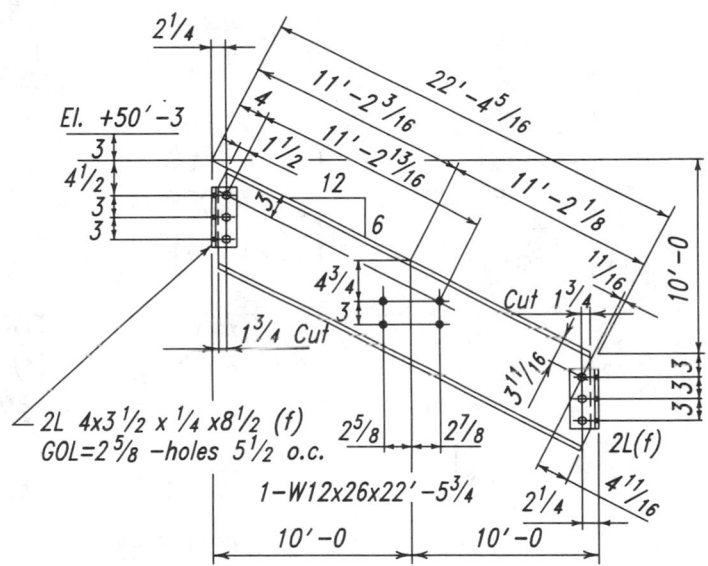

Figure 9-52. Sloped beam with rectangular connections.

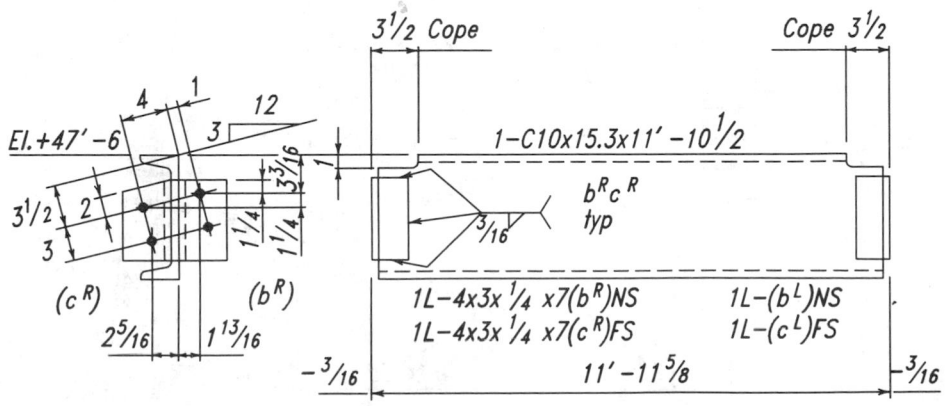

Figure 9-53. Canted double-angle connections.

Inclines in Two or More Directions (Hip and Valley Framing)
When a beam inclines in two or more directions with respect to the axis of its supporting member, it can be classified as a combination of those inclination directions. For example, the beam of Figure 9-42d is both skewed and sloped. Angle A shows the skew and angle B shows the slope. Note that, since the inclined beam is foreshortened in the elevation, the true angle B appears only in the auxiliary projection, Section X-X. The development of these details is quite complicated and graphical solutions to this compound angle work can be found in any textbook on descriptive geometry. Accurate dimensions may then be determined with basic trigonometry.

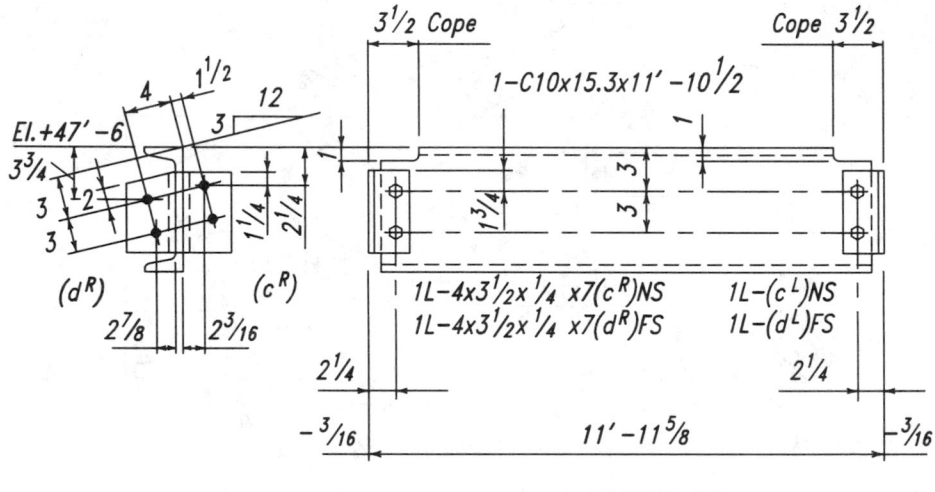

Figure 9-54. Canted connections to a sloping support.

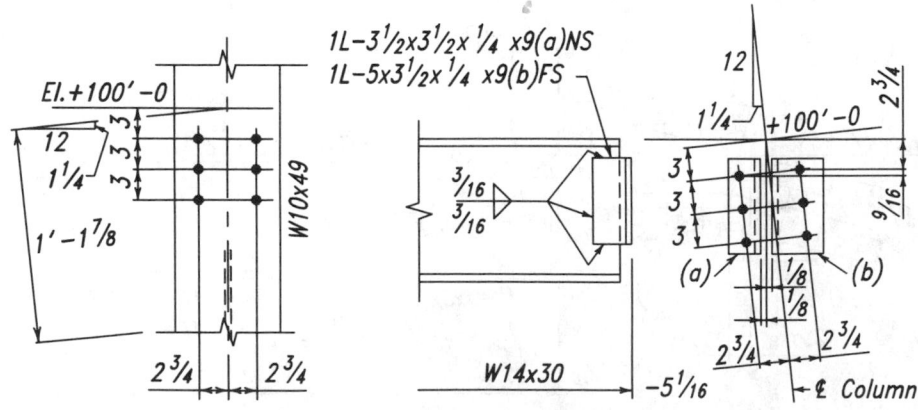

Figure 9-55. Canted connection to column flange.

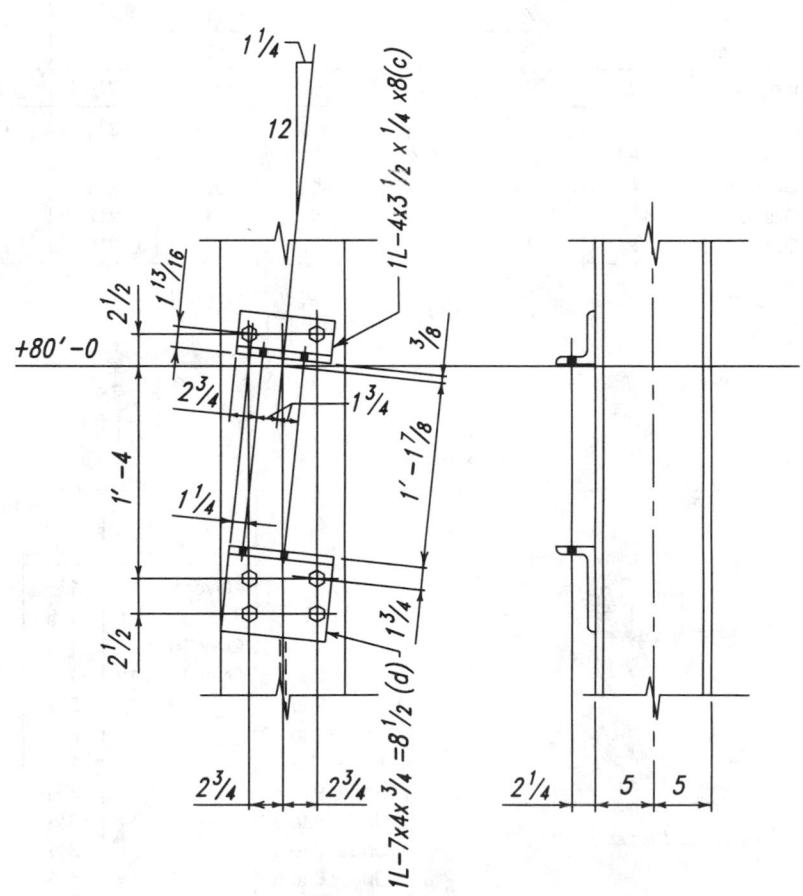

Figure 9-56. Canted seated connection.

Table 9-14.
Clearance Dimensions for Skewed Connections
All-Bolted

Values given are for webs up to $\frac{3}{4}$-in. thick, angles up to $\frac{5}{8}$-in. thick, and bent plates up to $\frac{1}{2}$-in. thick. Bolts are either $\frac{7}{8}$-in. diameter or 1 in. diameter, as noted. Values will be conservative for material thinner than the maximums listed, or for work with smaller bolts, and may be reduced to suit conditions by calculation or layout. For thicker material or larger bolts, check entering, driving, and tightening clearances and increase D and bolt gages as necessary. All dimensions are in inches. Enter bolts as shown.

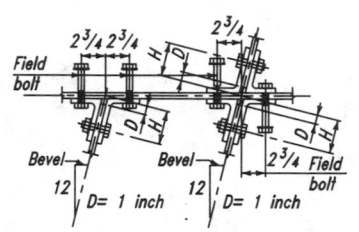

Bent angles

Values of H for Various Fastener Combinations

	Field Bolts	$\frac{7}{8}$	1
	Shop Bolts	$\frac{7}{8}$	1
Bevel	Up to 1	4*	$4\frac{1}{4}$*
	Over 1 to 2	$4\frac{1}{8}$	$4\frac{3}{8}$
	Over 2 to 3	$4\frac{3}{8}$	$4\frac{3}{4}$

*For back to back connections, stagger shop and field bolts or increase the $2\frac{3}{4}$-in. field bolt dimension to $3\frac{1}{4}$.

Values of H, H_1, H_2, and D for Various Bolt Combinations

Field Fastener		$\frac{7}{8}$			1			
Shop Fastener		$\frac{7}{8}$			1			
Dimension		H	H_1	H_2	H	H_1	H_2	D
Bevel	Over 3 to 4	$3\frac{3}{4}$	$3\frac{1}{4}$	$2\frac{1}{2}$	$4\frac{1}{4}$	$3\frac{1}{4}$	$2\frac{3}{4}$	$1\frac{1}{4}$
	Over 4 to 5	$3\frac{3}{4}$	$3\frac{1}{2}$	$2\frac{1}{4}$	$4\frac{1}{2}$	$3\frac{1}{2}$	$2\frac{1}{2}$	$1\frac{1}{4}$
	Over 5 to 6	4	$3\frac{3}{4}$	$2\frac{1}{4}$	$4\frac{3}{4}$	$3\frac{3}{4}$	$2\frac{1}{4}$	$1\frac{1}{2}$
	Over 6 to 7	$4\frac{1}{2}$	4	$2\frac{1}{4}$	5	4	$2\frac{1}{4}$	$1\frac{1}{2}$
	Over 7 to 8	$4\frac{3}{4}$	$4\frac{1}{4}$	$2\frac{1}{4}$	$5\frac{1}{4}$	$4\frac{1}{4}$	$2\frac{1}{4}$	$1\frac{1}{2}$

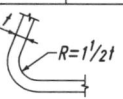

Double bent plates

Min. radius of cold bend for A 36 steel up to $\frac{1}{2}$ in. thick. For other bends see Table 9-13

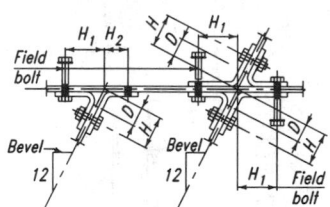

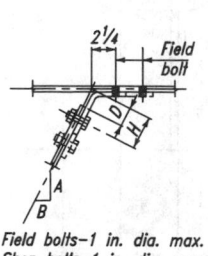

Field bolts-1 in. dia. max.
Shop bolts-1 in. dia. max

Single bent plates

			Shop Bolts	
	A	B	D	H
	12	Over 8 to 9	$1\frac{1}{2}$	3
	12	Over 9 to 10	$1\frac{5}{8}$	$3\frac{1}{8}$
	12	Over 10 to 11	$1\frac{3}{4}$	$3\frac{1}{4}$
	12	Over 11 to 12	$1\frac{7}{8}$	$3\frac{3}{8}$
	Under 12 to 11	12	$2\frac{1}{8}$	$3\frac{5}{8}$
	Under 11 to 10	12	$2\frac{1}{4}$	$3\frac{3}{4}$
	Under 10 to 9	12	$2\frac{1}{2}$	4
	Under 9 to 8	12	$2\frac{3}{4}$	$4\frac{1}{4}$
	Under 8 to 7	12	$3\frac{1}{4}$	$4\frac{3}{4}$
	Under 7 to 6	12	$3\frac{3}{4}$	$5\frac{1}{4}$
	Under 6 to 5	12	$4\frac{1}{2}$	6
	Under 5 to 4	12	$5\frac{5}{8}$	$7\frac{1}{8}$

Table 9-14 (cont.). Clearance Dimensions for Skewed Connections Bolted/Welded

Values given are for webs up to ¾-in. thick, angles up to ⅝-in. thick, and bent plates up to ½-in. thick, with bolts 1 in. diameter maximum. Values will be conservative for thinner material and for work with smaller bolts, and may be reduced to suit conditions by calculation or layout. For thicker material or larger bolts check entering and tightening clearances and increase beam set-back *D* and bolt gages as necessary. Enter bolts as shown. All dimensions are in inches.

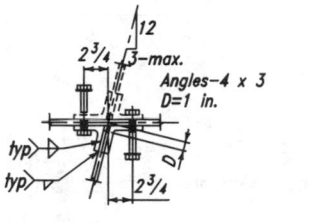

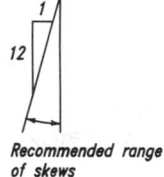

12
2¾
3–max.
Angles–4 x 3
D=1 in.
typ
typ
D
2¾
1
12
Recommended range of skews

Bent angles

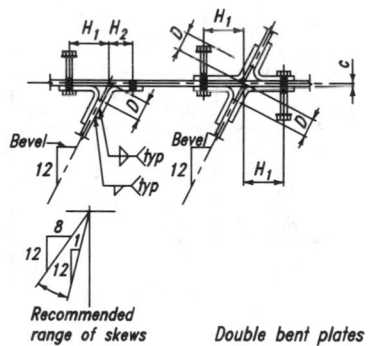

H1 H2
H1
D
c
D
D
Bevel
12
Bevel
12
typ
typ
H1
8
12
1
12
Recommended range of skews
R=1½t
Min. radius of cold bend for A 36 steel up to ½ in. thick. For other bends see Table 9–13

Double bent plates

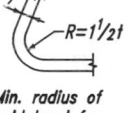

Bevel	D	H_1	H_2
Over 3 to 4	c + ⅝	3¼	2¾
Over 4 to 5	c + 11⁄16	3½	2½
Over 5 to 6	c + ¾	3¾	2¼
Over 6 to 7	c + 13⁄16	4	2¼
Over 7 to 8	c + ⅞	4¼	2¼

$$C = \frac{t_w}{2} + 1/16''$$

2¼
¾
D
4
12
12
8
Determine value of D by calculation or layout
Recommended range of skews

Single bent plates

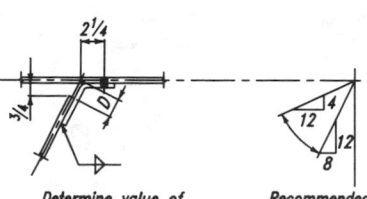

Table 9-14 (cont.).
Clearance Dimensions for Skewed Connections
Bolted/Welded

Values given are for material and bolt sizes noted below. See "Shear End-Plate Connections" in Part 9 for proportioning these connections. S indicates weld size required for strength, or a size suitable to the thickness of material. When the beam web is cut square, only that portion of the table above the heavy lines is applicable. Dimension A is added to the weld size to compensate for the root opening caused by the skew. When the beam web is beveled to the required skew, values of H_1 for the entire table are valid, and A = 0. In either case, where weld strength is critical, increase the weld size to obtain the required throat dimension. Enter bolts as shown. All dimensions are in inches.

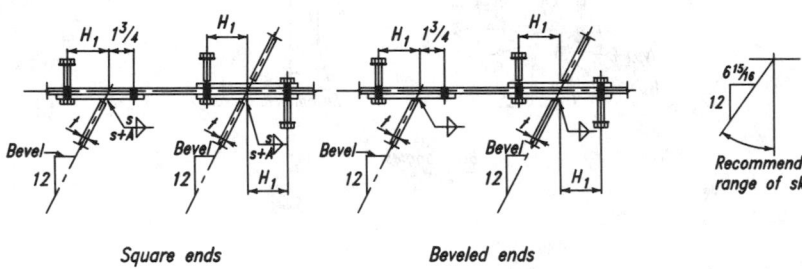

Square ends   Beveled ends

End plates

Recommended range of skews

	$t = \frac{1}{4}$		$t = \frac{5}{16}$		$t = \frac{3}{8}$		$t = \frac{7}{16}$		$t = \frac{1}{2}$		$t = \frac{5}{8}$		$t = \frac{3}{4}$	
Bevel	H_1	A	H_1	A	H_1	A	H_1	A	H_1	A	H_1	A	H_1	A
Up to $1\frac{5}{8}$	$1\frac{3}{4}$	0	$1\frac{3}{4}$	0	$1\frac{3}{4}$	$\frac{1}{16}$	$1\frac{3}{4}$	$\frac{1}{16}$	$1\frac{3}{4}$	$\frac{1}{16}$	$1\frac{7}{8}$	$\frac{1}{8}$	$1\frac{7}{8}$	$\frac{1}{8}$
Over $1\frac{5}{8}$ to $2\frac{1}{8}$	$1\frac{3}{4}$	0	$1\frac{3}{4}$	$\frac{1}{16}$	$1\frac{7}{8}$	$\frac{1}{16}$	$1\frac{7}{8}$	$\frac{1}{16}$	$1\frac{7}{8}$	$\frac{1}{8}$	2	$\frac{1}{8}$	2	$\frac{1}{8}$
Over $2\frac{1}{8}$ to $3\frac{1}{4}$	$1\frac{7}{8}$	$\frac{1}{16}$	$1\frac{7}{8}$	$\frac{1}{8}$	2	$\frac{1}{8}$	2	$\frac{1}{8}$	2	$\frac{1}{8}$	$2\frac{1}{8}$	0	$2\frac{1}{8}$	0
Over $3\frac{1}{4}$ to $4\frac{3}{8}$	$2\frac{1}{8}$	$\frac{1}{8}$	$2\frac{1}{8}$	$\frac{1}{8}$	$2\frac{1}{8}$	$\frac{1}{8}$	$2\frac{1}{8}$	0	$2\frac{1}{4}$	0	$2\frac{1}{4}$	0	$2\frac{3}{8}$	0
Over $4\frac{3}{8}$ to $5\frac{5}{8}$	$2\frac{1}{4}$	$\frac{1}{8}$	$2\frac{1}{4}$	$\frac{1}{8}$	$2\frac{3}{8}$	0	$2\frac{3}{8}$	0	$2\frac{3}{8}$	0	$2\frac{1}{2}$	0	$2\frac{1}{2}$	0
Over $5\frac{5}{8}$ to $6\frac{15}{16}$	$2\frac{1}{2}$	$\frac{1}{8}$	$2\frac{1}{2}$	0	$2\frac{1}{2}$	0	$2\frac{1}{2}$	0	$2\frac{5}{8}$	0	$2\frac{5}{8}$	0	$2\frac{3}{4}$	0

Bolts: $\frac{7}{8}$-in. diameter maximum
End Plate thickness: $\frac{3}{8}$-in. maximum
Supporting web thickness: $\frac{3}{4}$-in. maximum

Use of fillet welds is limited to connections with bevels of $6\frac{15}{16}$ in 12 and less.
For greater bevels consider use of double or single bent plates.

Table 9-14 (cont.).
Clearance Dimensions for Skewed Connections
Bolted/Welded

For Skews Up to 30 Degrees

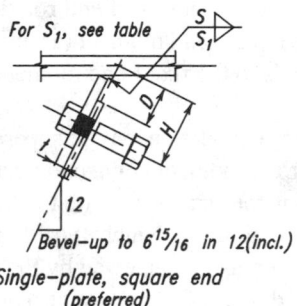

For S_1, see table

/12
Bevel–up to $6^{15}/_{16}$ in 12(incl.)

Single–plate, square end
(preferred)

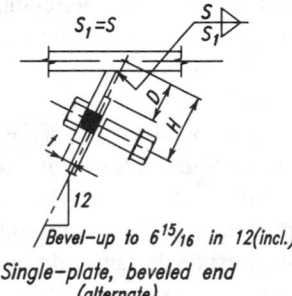

$S_1 = S$

/12
Bevel–up to $6^{15}/_{16}$ in 12(incl.)

Single–plate, beveled end
(alternate)

Values of S_1 for Single-Plate Skewed Connection

Plate Thickness, t, in.	$^1/_4$	$^5/_{16}$	$^3/_8$	$^7/_{16}$	$^1/_2$	$^5/_8$	$^3/_4$
Up to $1^5/_8$	S	S	$S + ^1/_{16}$	$S + ^1/_{16}$	$S + ^1/_{16}$	$S + ^1/_8$	$S + ^1/_8$
Over $1^5/_8$ to $2^1/_8$	S	$S + ^1/_{16}$	$S + ^1/_{16}$	$S + ^1/_{16}$	$S + ^1/_8$	$S + ^1/_8$	$S + ^1/_8$
Over $2^1/_8$ to $3^1/_4$	$S + ^1/_{16}$	$S + ^1/_8$	$S + ^1/_8$	$S + ^1/_8$	$S + ^1/_8$		
Over $3^1/_4$ to $4^3/_8$	$S + ^1/_8$	$S + ^1/_8$	$S + ^1/_8$				
Over $4^3/_8$ to $5^5/_8$	$S + ^1/_8$	$S + ^1/_8$					
Over $5^5/_8$ to $6^{15}/_{16}$	$S + ^1/_8$	For values not shown use alternate single-plate.					

S indiates weld size required for strength, or size suitable to thickness of material.
Where weld strength is critical, proportion size S_1 to obtain required throat dimension.

For Skews Over 30 to 45 Degrees	For Skews Over 45 to 70 Degrees
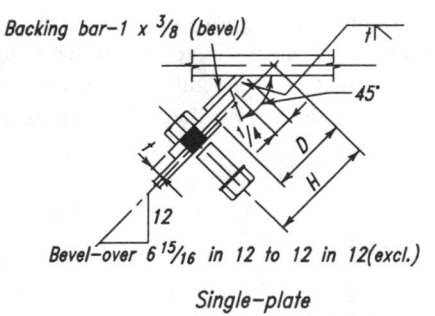 Backing bar–1 x $^3/_8$ (bevel) 45° /12 Bevel–over $6^{15}/_{16}$ in 12 to 12 in 12(excl.) Single–plate	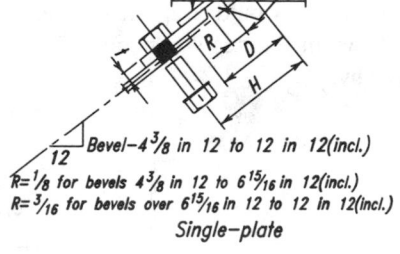 Backing bar–$1^1/_4$ x $^3/_8$ (bevel) /12 Bevel–$4^3/_8$ in 12 to 12 in 12(incl.) $R = ^1/_8$ for bevels $4^3/_8$ in 12 to $6^{15}/_{16}$ in 12(incl.) $R = ^3/_{16}$ for bevels over $6^{15}/_{16}$ in 12 to 12 in 12(incl.) Single–plate

Note:
Proportion dimensions D and H to provide field clearances with welds, or to permit bolt entry and
tightening. Enter bolts as shown.

PR MOMENT CONNECTIONS

The behavior of PR moment connections, as illustrated in Figure 9-57, is intermediate in degree between the flexibility of simple shear connections and the full rigidity of FR moment connections. PR moment connections are permitted upon evidence that the connections to be used are capable of furnishing, as a minimum, a predictable percentage of full end restraint.

A beam line represents the relationship between end moment and end rotation for a given beam. The maximum end rotation corresponds to zero end moment (a simple shear connection) whereas the zero end rotation corresponds to the fixed-end moment (an FR moment connection).

The moment-rotation curve of the given PR moment connection may be superimposed upon the beam line as illustrated in Figure 9-58. For PR moment connection curve A or B, the point of intersection of the connection moment-rotation curve with the beam line defines the beam end moment and the required strength for which the PR moment connection must be designed. In turn, the design of members connected by PR moment connections must then be predicated upon no greater degree of end restraint. Thus, when the moment-rotation curve is known, a dependable and known moment strength may be assumed. Since the exact location of this intersection point is largely dependent upon test results and experience with similar situations, thus, PR moment connections are only as good as the moment-rotation curves upon which they are based.

Modeling PR Moment Connections for Gravity Loads

The following simplified approach to PR moment connections for gravity loading is taken from Geschwindner (1991). For a discussion of PR moment connections for lateral loading, refer to Nethercot and Chen (1988)

Geschwindner (1991) models the full range of connection behavior, from the truly pinned to the fully restrained, as a rotational spring with a specified stiffness n; the moment in the spring will be given by:

$$M_{con} = n\theta_{con} \qquad (9\text{-}1)$$

If these connections are attached to the ends of a simply supported beam with a uniformly distributed load as shown in Figure 9-59, a classical indeterminate analysis may be performed to relate the moment in the spring to the load and to the spring and beam stiffnesses. Using the method of consistent deformations, the springs are first removed from the beam leaving a simply supported beam as shown in Figure 9-60a. Then the moments that would be applied by the springs are applied independently to the beam ends as shown in Figures 9-60b and 9-60c. The rotations at end A for these three cases are given by:

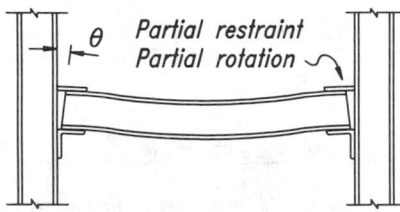

Figure 9-57. PR moment connection behavior.

$$\theta_a = \frac{WL^2}{24EI} \tag{9-2}$$

$$\theta_{aa} = \frac{-M_a L}{3EI} \tag{9-3}$$

$$\theta_{ab} = \frac{-M_b L}{6EI} \tag{9-4}$$

Superposition of these rotations yields the final rotation on the beam at end **a**. Thus,

$$\theta_{final} = \theta_a + \theta_{aa} + \theta_{ab} \tag{9-5}$$

Since the final beam rotation and the final spring rotation must be the same, substitutions of Equations 9-1 through 9-4 into Equation 9-5 yields

$$\frac{M_{con}}{n} = \frac{WL^2}{24EI} - \frac{M_a L}{3EI} - \frac{M_b L}{6EI} \tag{9-6}$$

Taking into account the symmetry of the structure and recognizing that the moment in the spring is the moment on the beam, $M_{con} = M_a = M_b$, Equation 9-6 may be rearranged to solve for this moment.

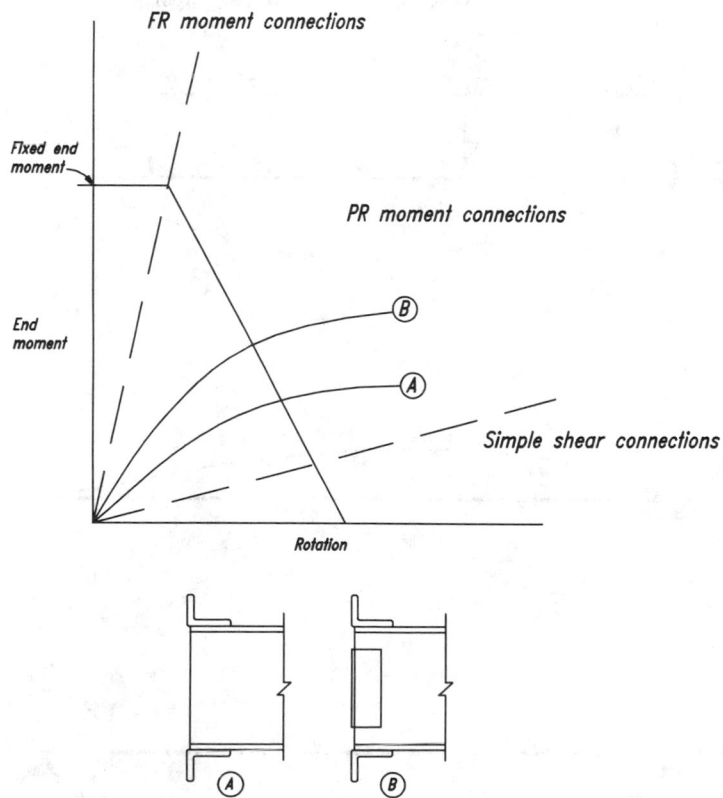

Figure 9-58. PR moment connection behavior.

$$M_{con} = \frac{\dfrac{WL^2}{24EI}}{\dfrac{1}{n} + \dfrac{L}{2EI}} \tag{9-7}$$

To simplify this expression, the ratio of the beam stiffness to spring stiffness is defined as

$$u = \frac{\left(\dfrac{EI}{L}\right)}{n} \tag{9-8}$$

or

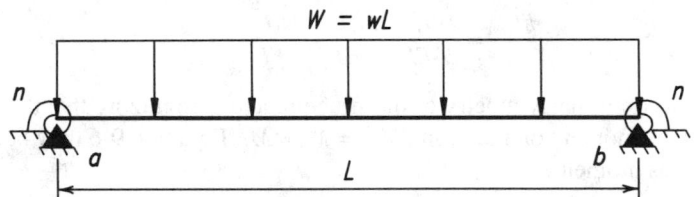

Figure 9-59. Beam and connection model.

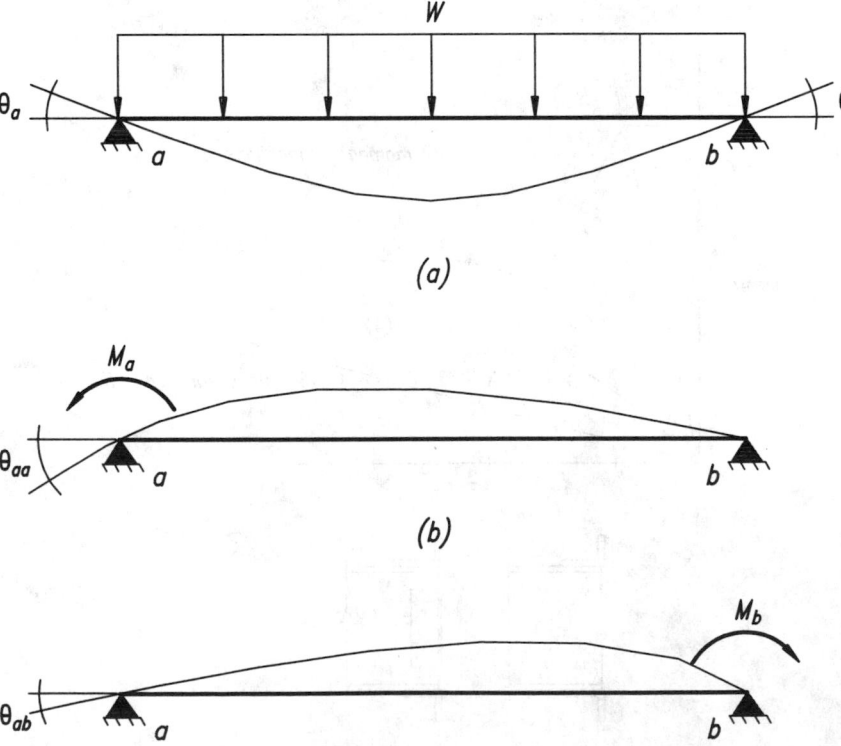

Figure 9-60. The cut-back structure.

$$n = \frac{EI}{uL} \tag{9-9}$$

and substitution of this new representation of the spring stiffness into Equation 9-7 and simplifying yields

$$M_{con} = \left(\frac{1}{(2u+1)}\right)\frac{WL}{12} \tag{9-10}$$

The beam moment diagram is shown in Figure 9-61. Superposition of the simply supported beam moment diagram on the beam with end moments yields a positive centerline moment of

$$M_{pos} = \frac{WL}{8} - \frac{\frac{WL}{12}}{(2u+1)} \tag{9-11}$$

or

$$M_{pos} = \left(\frac{6u+1}{4u+2}\right)\frac{WL}{12} \tag{9-12}$$

Both the connection and the centerline moments are written as a coefficient times the fixed end moment. If these coefficients are plotted as a function of the spring stiffness ratio, the full response of the beam can be represented as shown in Figure 9-62.

Deflections

The centerline deflection may now be determined using the method of conjugate beam. The beam and the corresponding conjugate beam are shown in Figure 9-63. The area of the M/EI diagram above the beam represents the influence of the load on the simply supported beam while that below the beam represents the influence of the negative end moments. The end rotation may be determined by taking moments of these areas about end B of the conjugate beam such that

$$R_a = \frac{M_s L}{3EI} - \frac{M_{con} L}{2EI} \tag{9-13}$$

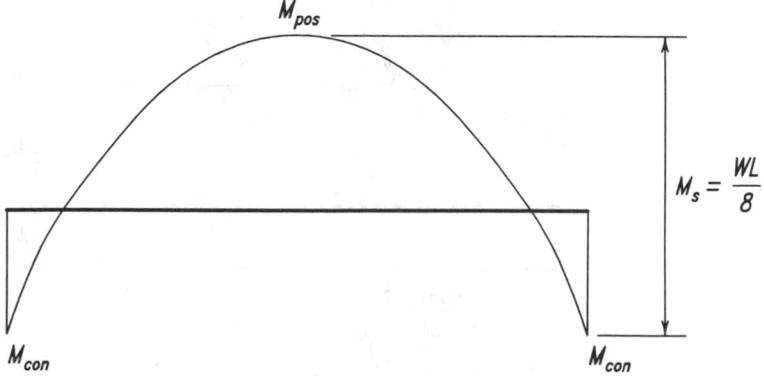

Figure 9-61. Bending moment diagram.

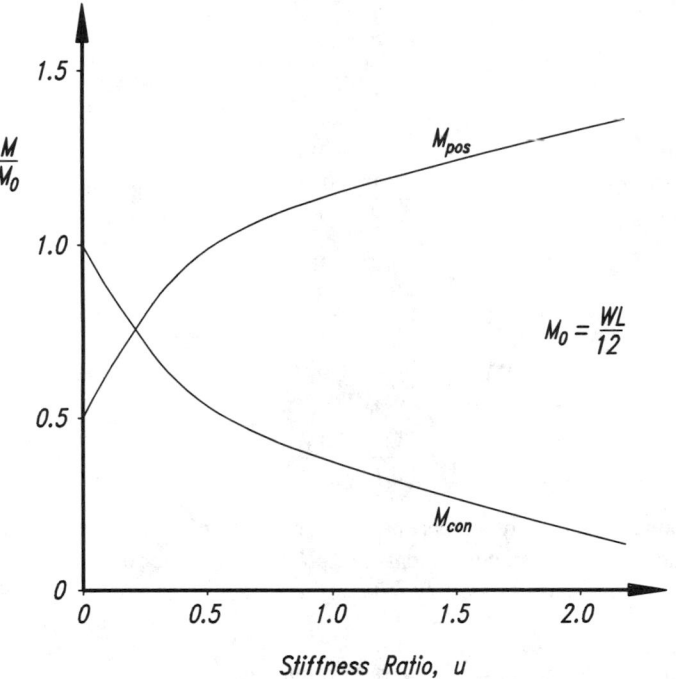

Figure 9-62. Bending moment coefficients vs. connection sitffness ratio.

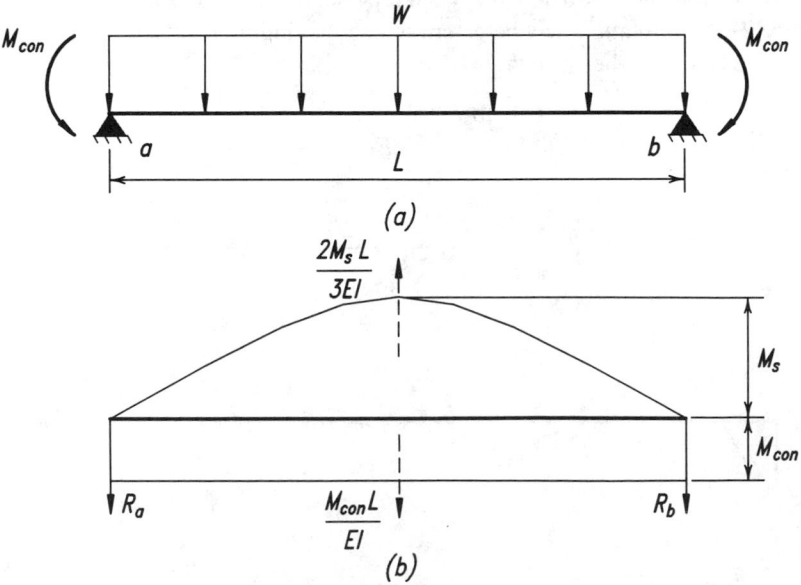

Figure 9-63. Conjugate beam.

The deflection at the centerline D may now be determined by taking moments about the conjugate beam centerline which yields, after simplification

$$D = \frac{5M_sL^2}{48EI} - \frac{M_{con}L^2}{8EI} \qquad (9\text{-}14)$$

The first term in this equation represents the centerline deflection of a uniformly loaded simply supported beam D_{simp}, while the second term represents the reduction in centerline deflection as a result of the end moments D_{-M}. The ratio of these terms will show the overall reduction in deflection due to the end restraint. If Equation 9-10 were substituted for the moment in the connection, the deflection ratio becomes

$$\frac{D_{-M}}{D_{simp}} = \frac{4}{5(2u+1)} \qquad (9\text{-}15)$$

The deflection ratio, given as a function of the spring stiffness ratio, is plotted in Figure 9-64. It can be seen that for the fixed-ended condition ($u = 0$) the deflection will be reduced by 80 percent of the simply supported beam deflection. For spring stiffness ratios greater than zero, the reduction in deflection will be correspondingly less.

The Beam Line
The relationship between moment and rotation on the end of a uniformly loaded prismatic beam, as shown in Figure 9-65, is the beam line (Blodgett, 1966). Note that the rotation is zero for a fixed-ended beam with the resulting fixed-end moment and the moment is zero for a simply supported beam with the resulting simply supported beam rotation. A straight line connects these two extreme conditions. Since the connection is represented

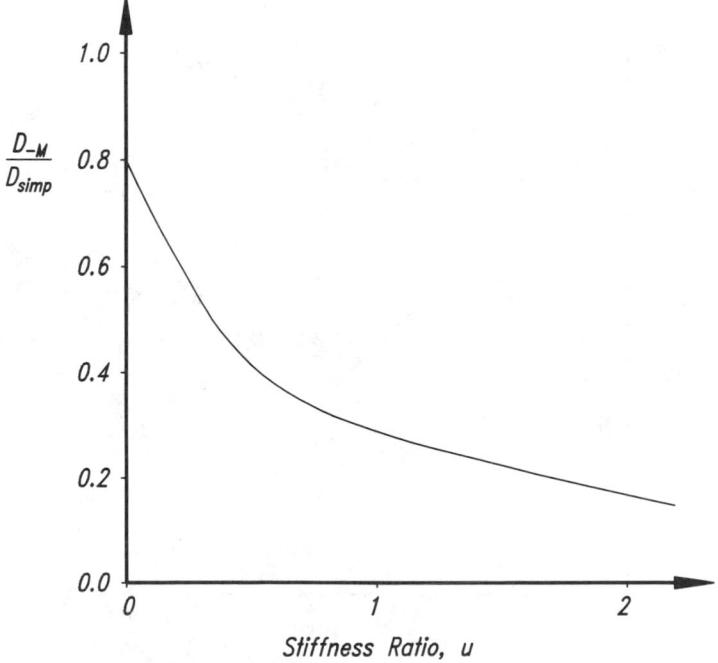

Figure 9-64. Deflection reduction vs. connection stiffness ratio.

by Equation 9-1, it too may be plotted on the graph of Figure 9-65 as a straight line with a slope of n. The intersection of these two lines represents the final equilibrium condition for the beam with the given PR moment connections. Thus for a connection with a known stiffness ratio u, the solution will again be given by Equation 9-12.

Elastic Design

Figure 9-66 combines the two views of the beam and connection interaction. The normal approach to design would have a connection capable of developing up to 20 percent of the fixed-end moment considered as a pinned connection and one capable of developing at least 90 percent of the fixed-end moment considered fixed (Blodgett, 1966). These two regions are shaded on both portions of Figure 9-66. They represent the area below a value of $u = 0.0555$ and above the value $u = 2.0$. Beam-connection combinations falling within the unshaded area should be treated so as to include the connection behavior. The LRFD Specification does not directly recommend these assumptions but rather suggests that any combination which is not fully pinned or fully rigid be treated in a way that reflects actual behavior.

In order to fully understand the impact that the use of flexible connections may have on beam design, it is important to consider further the results presented in Figure 9-66b. The maximum moment on the beam is indicated by the maximum coefficient. This will occur on the end of the beam for values of $u = 0$ to $u = 0.167$. For values of $u > 0.167$, the maximum moment will occur at the beam centerline. The most economical design from the standpoint of the beam would occur at the point where the end moment and the centerline moment would be the same, a connection with a value of $u = 0.167$. Unfortunately, any slight deviation from this value will result in a beam design moment larger

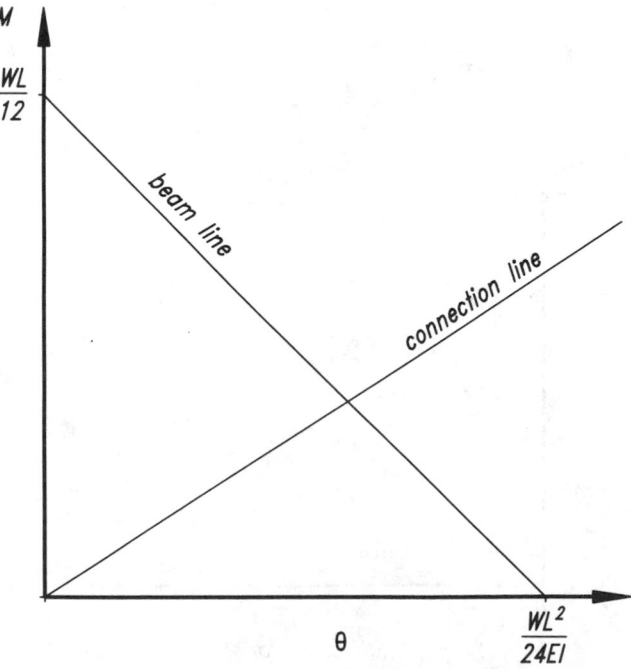

Figure 9-65. Moment-rotation diagram—the beam line.

than that anticipated. Thus, the beam would no longer be adequate to carry the design loads. Considering a beam designed for the fixed-end condition $u = 0$, it can be seen that a range of stiffness ratios up to $u = 0.5$ will still permit the beam to adequately carry the design moment, thus allowing for some inaccuracies in the determination of connection stiffness. If the beam is designed as a simply supported beam with $u = \infty$, any connection, regardless of its stiffness ratio will still result in an acceptable beam. For any connection with a stiffness ratio between these two extremes, there is always the potential that an inaccuracy in determining the connection stiffness could result in a beam moment larger than that for which it was designed.

Recent papers would seem to suggest that extreme care is not required in modeling connection stiffness (Gerstle and Ackroyd, 1989) or that the actual shape of the moment-rotation curve is not really critical (Deierlein et al., 1990). However, currently available connection models may actually predict a stiffness that varies from the actual stiffness by a factor of plus or minus 2 (Deierlein et al., 1990). Thus, from the above it would appear that connection stiffness, as measured by the stiffness ratio, may be quite important for a broad range of possible situations. In addition, if sufficient care is not exercised, the resulting design may be significantly inadequate.

In order to take advantage of connection strength and the incremental nature of beam sizes, it will be helpful to add the beam center line moment curve to Figure 9-65. This is shown in Figure 9-67 where, in addition to the centerline moment, an arbitrary beam flexural strength is shown. From the figure it is clear that as long as the negative moment is less than that given by point a, the beam flexural strength will not be exceeded in that region and as long as the positive moment is less than that given by point c, the beam will be adequate in that region. Thus, an acceptable connection will be one that yields an equilibrium condition between points a and b. This is where the flexibility of PR connection design can be most effectively implemented. Even though the exact connection curve might be somewhat elusive, a reasonable representation will be sufficient to provide an acceptable design condition.

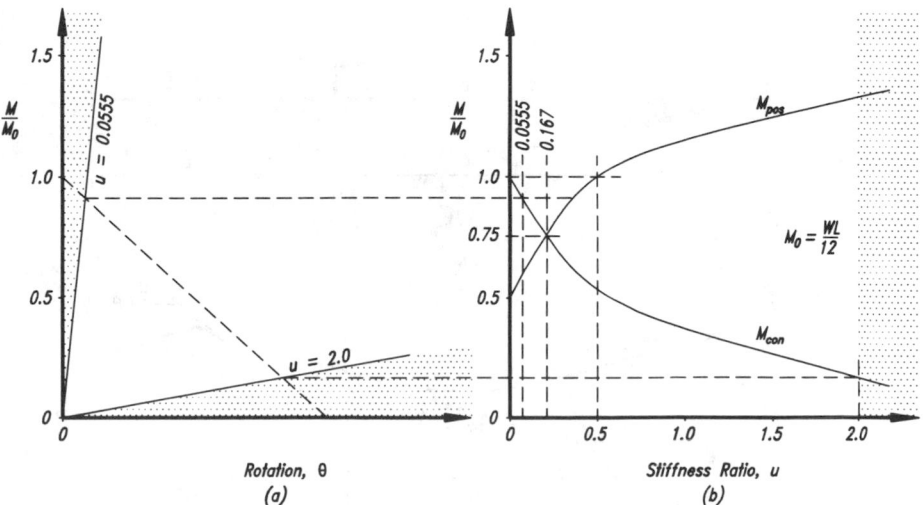

Figure 9-66. Combined views of moment-rotation-stiffness diagrams.

Non-Rigid Supports

The previously developed equations were based on the assumption that the connection was attached to a non-yielding support. Since in most real structures the beams are attached to columns or other flexible elements, it will be informative to investigate the situation presented in Figure 9-68. As with the single beam already considered, the beam

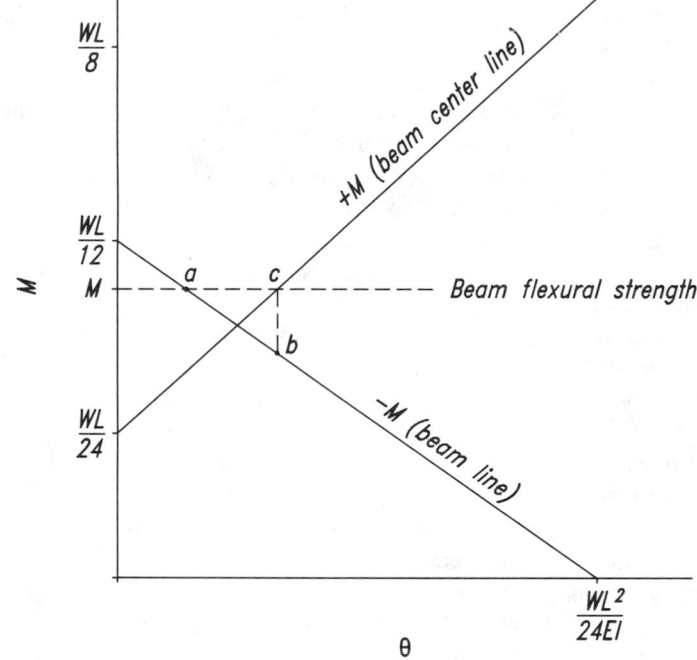

Figure 9-67. Moment-rotation diagram—negative and positive moment

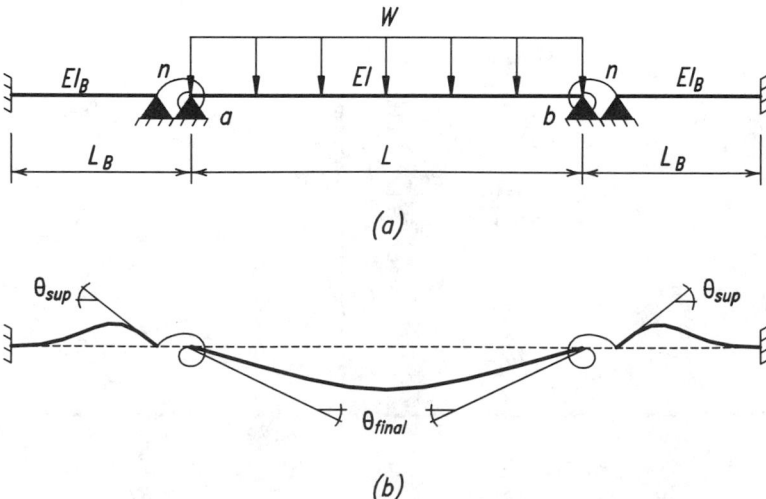

Figure 9-68. Semi-rigid connection with flexible supports.

of Figure 9-68 is symmetrical and loaded with a uniform load. The spring stiffness and stiffness ratio are defined as in Equation 9-1 and Equation 9-8. The support members are defined with the stiffness EI_B / L_B as shown in Figure 9-68a. In this situation, the connection rotation is no longer equal to the final beam rotation, but instead is equal to the final beam rotation less the support rotation as shown in Figure 9-68b. Thus, with the inclusion of the support rotation, Equation 9-6 becomes

$$\frac{M}{n} = \frac{WL^2}{24EI} - \frac{ML}{3EI} - \frac{ML}{6EL} - \frac{ML_B}{4EI_B} \qquad (9\text{-}16)$$

Simplifying Equation 9-16 and solving for the moment yields

$$M = \frac{\dfrac{WL^2}{24EI}}{\dfrac{1}{n} + \dfrac{L_B}{4EI_B} + \dfrac{L}{2EI}} \qquad (9\text{-}17)$$

Inspection of Equation 9-17 reveals that the first two terms in the denominator represent the spring and support respectively. If the support beam is infinitely rigid, the second term may be eliminated and Equation 9-17 becomes Equation 9-7. If, at the other extreme, the spring is made infinitely rigid, Equation 9-17 will yield the results for a three span

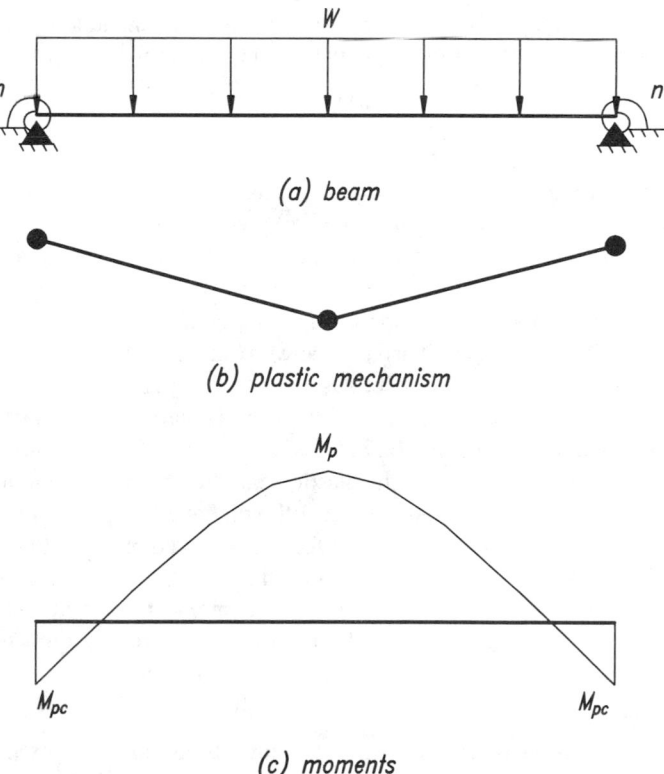

(a) beam

(b) plastic mechanism

(c) moments

Figure 9-69. Plastic analysis of beam with semi-rigid connections.

beam. If these two terms are combined and defined as an effective spring representing both the connection and the support, such that

$$\frac{1}{n_{eff}} = \frac{1}{n} + \frac{L_B}{4EI_B} \tag{9-18}$$

the moment on the end of the beam may be given by Equation 9-7 with n being replaced by n_{eff}. It then becomes clear that the range of responses available for the beam is the same as shown in Figure 9-66. In addition, regardless of the structure which may provide support, an effective spring can be defined which will dictate the beam response.

Plastic Analysis

A beam with PR moment connections may also be investigated through plastic analysis. The primary requirement is that the connection be capable of maintaining the plastic moment while undergoing significant rotation. If the plastic moment strength of the beam is defined as M_p and the plastic moment strength of the connection is defined as M_{pc}, the plastic mechanism and corresponding moment diagram are as shown in Figure 9-69. Equilibrium requires that the simply supported beam moment

$$M_s = M_p + M_{pc} \tag{9-19}$$

If the connection strength is taken as a certain portion of the beam strength such that

$$M_{pc} = aM_p \tag{9-20}$$

then for $a = 1.0$, the connection has the same strength as the beam, independent of rotation. Substituting Equation 9-20 into Equation 9-19 and rearranging,

$$M_p = \frac{M_s}{(1 + a)} \tag{9-21}$$

Equation 9-21 represents the plastic moment strength required for the beam to carry the applied load. A plot of Equation 9-21 is provided in Figure 9-70. Since the most economical beam design would result when the connection is capable of resisting the full plastic moment strength of the beam, $(a = 1.0)$, the design by plastic analysis would require only that the connection be capable of attaining that moment. Its actual moment-rotation characteristics (i.e., how it arrived there) would not be important.

Recognizing that PR moment connections will not always have a strength equal to that of the beam and that sufficient rotation must be assured in order for the plastic mechanism to develop, a plastic beam line can be developed. Figure 9-71 shows the plastic beam line. If the connection line intersects the plastic beam line between a and b, the negative moment will equal the positive moment and both will equal the plastic moment strength of the member, thus, a plastic mechanism forms. If the connection line intersects the plastic beam line between b and c, the beam centerline moment will be the plastic moment while the end moment will be that indicated by the intersection point and a mechanism will form. Plastic analysis for beams with PR moment connections requires that sufficient lateral support be provided to insure adequate member rotation.

Real Connections

The moment-rotation characteristics for real connections normally exhibit non-linear behavior. Two comprehensive collections of connection data have been reported (Goverdhan, 1984 and Kishi and Chen, 1986) which provide the designer with a starting

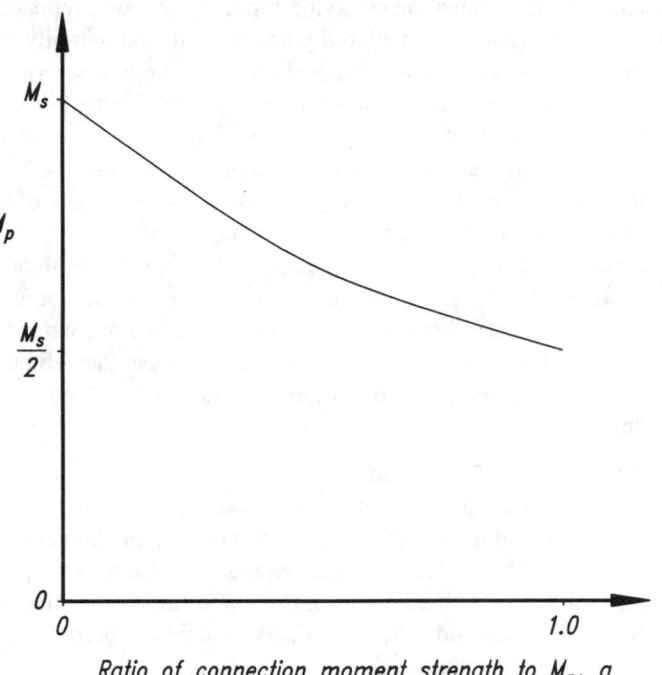

Figure 9-70. *Required plastic moment capacity of beam.*

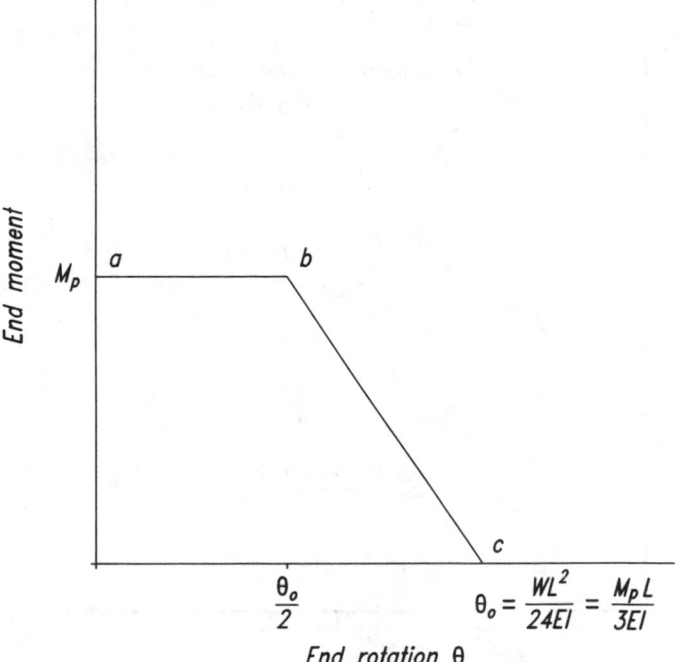

Figure 9-71. *Plastic beam line.*

point for considering true connection behavior. Figure 9-72 shows representative curves for connections which might be considered pinned, fixed, and partially restrained. It is obvious that the linear model used previously does not accurately describe the full range of behavior of these true connections. However, as shown in Figure 9-73, if the intersection of the beam and connection lines were known, an effective linear connection could be determined with a stiffness $1 / n_{eff}$ which would provide the same solution as the true connection curve. This again shows that, regardless of the complexity of the connection model, the beam will consistently respond as shown in Figure 9-66b.

In addition, for connections which behave linearly within the range of loading being considered, the linear spring model presented may prove quite useful. Historically, flange-plated connections have been treated as linear PR moment connections (Blodgett, 1966). The accuracy of this model will depend on the moment taken by the connection. Provided that the forces in the plates do not exceed the yield strength of the plates, the model is reasonable.

Flange-Plated PR Moment Connections

As illustrated in Figure 9-74, a flange-plated PR moment connection consists of a shear connection and top and bottom flange plates which connect the flanges of the supported beam to the supporting column. These flange plates are welded to the supporting column and may be bolted or welded to the flanges of the supporting beam. An unwelded length of 1½ times the flange-plate width b_A is normally assumed to permit the elongation of the plate necessary for PR behavior.

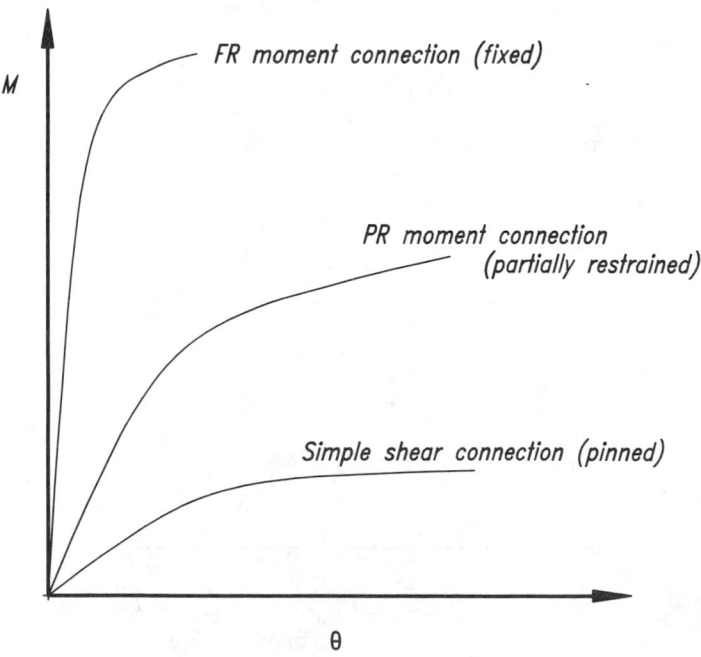

Figure 9-72. Connection moment-rotation curves.

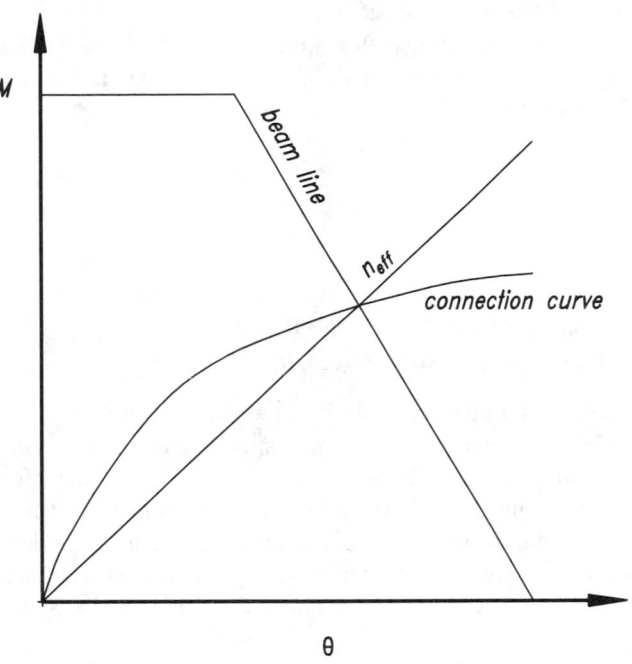

Figure 9-73. Beam line with true connection and effective stiffness.

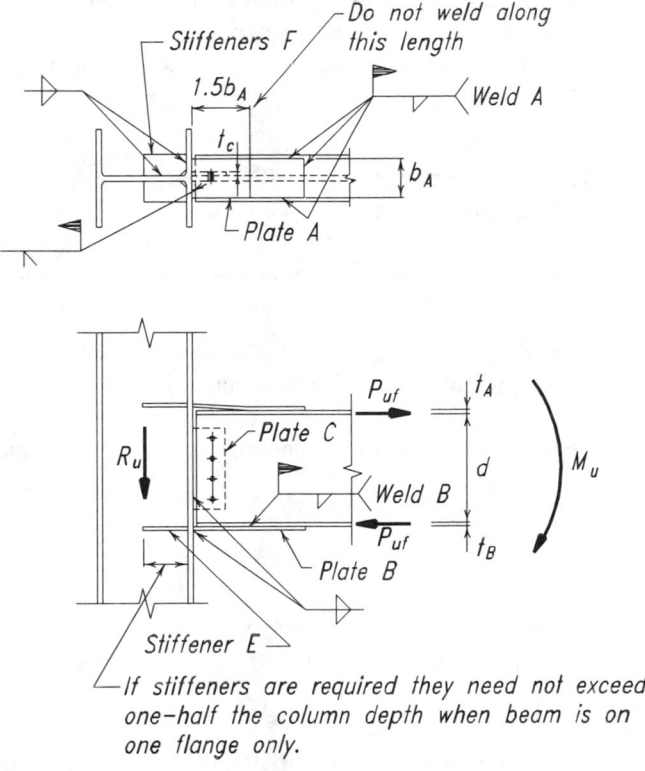

If stiffeners are required they need not exceed one-half the column depth when beam is on one flange only.

Figure 9-74. Flange-plated PR moment connections.

Force Transfer in PR Moment Connections

As with FR moment connections, the moment may be resolved into an effective tension-compression couple acting as axial forces at the beam flanges. The flange force P_{uf} may be calculated as:

$$P_{uf} = \frac{M_u}{d_m} \qquad (9\text{-}22)$$

where

P_{uf} = factored beam flange force, tensile or compressive, kips
M_u = beam end moment, kip-in.
d_m = moment arm between flange forces, in.

Shear is primarily transferred through the beam web connection. Axial forces, if present, are assumed to be distributed uniformly across the beam cross-sectional area, and are additive algebraically to the flange forces and vectorially to the shear force.

The supporting column must have sufficient strength and stiffness to develop the moment transferred to it through the PR moment connection. Additionally, the concentrated flange forces may require the selection of a larger column section or the stiffening of the existing column section; refer to "Column Stiffening" in Part 10.

Design Checks

The design strengths of the bolts and/or welds, connecting elements, and affected elements of connected members must be determined in accordance with the provisions of the LRFD Specification. The applicable limit states in each of the aforementioned design strengths are discussed in Part 8. In all cases, the design strength ϕR_n must exceed the required strength R_u.

Shop and Field Practices

The shop and field practices for flange-plated FR moment connections (see Part 10) are equally applicable to flange-plated PR moment connections.

Example 9-27

Given: Design a welded flange-plated PR moment connection for a 20-ft-long W18×50 beam to W14×109 column flange connection. The beam supports a 7.29 kip/ft uniform load. For structural members, $F_y = 50$ ksi and $F_u = 65$ ksi; for connecting materials, $F_y = 36$ ksi and $F_u = 58$ ksi.

$R_u = 73.0$ kips
$M_u = 225$ ft-kips

W18×50

$d = 17.99$ in.	$b_f = 7.495$ in.	$Z_x = 101$ in.^{3}
$t_w = 0.355$ in.	$t_f = 0.570$	$I_x = 800$ in.^{4}

W14×109

$d = 14.32$ in.	$b_f = 14.605$ in.	$k = 1\%_{16}$ in.
$t_w = 0.525$ in.	$t_f = 0.860$ in.	$T = 11\frac{1}{4}$ in.

Use 70 ksi electrodes and $\frac{7}{8}$-in. diameter A325-N bolts.

Solution: *Check beam design flexural strength*

$$Z_{req} = \frac{M_u \times 12 \text{ in. / ft}}{0.9 F_y}$$

$$= \frac{(225 \text{ ft-kips})(12 \text{ in ./ft})}{0.9 \ (50 \text{ ksi})}$$

$$= 60 \text{ in.}^3$$

$$Z_x = 101 \text{ in.}^3$$

Since $Z_x > Z_{req}$, the beam design flexural strength is **o.k.**

Design the single-plate web connection

Determine number of $\frac{7}{8}$-in. diameter A325-N bolts required for shear.

From Table 8-11

$$n_{min} = \frac{R_u}{\phi r_n}$$

$$= \frac{73.0 \text{ kips}}{21.6 \text{ kips / bolt}}$$

$$= 3.38 \rightarrow 4 \text{ bolts}$$

Try PL$\frac{3}{8}$

Determine number of $\frac{7}{8}$-in. diameter A325-N bolts required for material bearing, assuming $L_e = 1\frac{1}{2}$-in. and $s = 3$ in. The $\frac{3}{8}$-in. thick plate ($F_u = 58$ ksi) is more critical than the 0.355-in. thick beam web ($F_u = 65$ ksi). From Table 8-13,

$$n_{min} = \frac{R_u}{\phi r_n}$$

$$= \frac{73.0 \text{ kips}}{34.3 \text{ kips / bolt}}$$

$$= 2.13 \rightarrow 3 \text{ bolts}$$

Bolt shear is more critical. Try a four-bolt single-plate connection.

Check shear yielding of the plate

$$\phi R_n = 0.9 \ (0.6 F_y A_g)$$

$$= 0.9[0.6 \times 36 \text{ ksi } (12 \text{ in.} \times \tfrac{3}{8} \text{ in.)}]$$

$$= 87.5 \text{ kips} > 73.0 \text{ kips} \quad \textbf{o.k.}$$

Check shear rupture of the plate

$$\phi R_n = 0.75 \ (0.6 F_u A_n)$$

$$= 0.75[0.6 \times 58 \text{ ksi } (12 \text{ in.} - 4(\tfrac{7}{8}\text{-in.} + \tfrac{1}{8}\text{-in.}))\tfrac{3}{8}\text{-in.}]$$

$$= 78.3 \text{ kips} > 73.0 \text{ kips} \quad \textbf{o.k.}$$

Check block shear rupture of the plate

With L_{eh} = 1½-in. and L_{ev} = 1½-in., from Tables 8-47a and 8-48a, $0.6F_u A_{nv} > F_u A_{nt}$. Thus,

$$\phi R_n = \phi \, [0.6F_u A_{nv} + F_y A_{gt}]$$

From Tables 8-48a and 8-48b,

$$\phi R_n = (183 \text{ kips/in.} + 40.5 \text{ kips/in.})\tfrac{3}{8}\text{-in.}$$
$$= 83.8 \text{ kips} > 73.0 \text{ kips} \quad \textbf{o.k.}$$

Determine required weld size for fillet welds to supporting column flange:

$$D_{min} = \frac{R_u}{2 \times 1.392 l}$$
$$= \frac{73.0 \text{ kips}}{2 \times 1.392 \, (12 \text{ in.})}$$
$$= 2.19 \rightarrow 3 \text{ sixteenths}$$

From LRFD Specification Table J2.4, since the column flange thickness is over ¾-in., the minimum fillet weld size is $\tfrac{5}{16}$-in., use two $\tfrac{5}{16}$-in. fillet welds.

Design the tension flange plate and connection

Calculate the flange force P_{uf}

$$P_{uf} = \frac{M_u \times 12 \text{ in./ft}}{d}$$
$$= \frac{(225 \text{ ft-kips})(12 \text{ in /ft})}{17.99 \text{ in.}}$$
$$= 150 \text{ kips}$$

Determine tension flange-plate dimensions

From Figure 8-42, assume a shelf dimension of $\tfrac{5}{8}$-in. on both sides of the plate. The plate width, then, is 7.495 in. − 2($\tfrac{5}{8}$-in.) = 6.245. Try a ¾-in. × 6¼-in. flange plate.

Check tension yielding of the flange plate:

$$\phi R_n = \phi F_y A_g$$
$$= 0.9(36 \text{ ksi})(6\tfrac{1}{4}\text{-in.})(\tfrac{3}{4}\text{-in.})$$
$$= 152 \text{ kips} > 150 \text{ kips} \quad \textbf{o.k.}$$

Determine required weld size and length for fillet welds to beam flange.

Try a $\tfrac{5}{16}$-in. fillet weld. The minimum length of weld l_{min} is:

$$l_{min} = \frac{P_{uf}}{1.392D}$$
$$= \frac{150 \text{ kips}}{1.392 \, (5 \text{ sixteenths})}$$
$$= 21.6 \text{ in.}$$

Use 8 in. of weld along each side and 6¼-in. of weld along the end of the flange plate.

Select tension flange plate dimensions

To provide for an 8-in. weld length and an unwelded length of 1½ times the plate width, use PL¾-in.×6¼-in.×17½-in.

Determine required weld size for fillet welds to supporting column flange.

$$D_{min} = \frac{P_{uf}}{2 \times 1.392l}$$

$$= \frac{150 \text{ kips}}{2 \times 1.392 \, (6\frac{1}{4}\text{-in.})}$$

$$= 8.62 \rightarrow 9 \text{ sixteenths}$$

Use ⁹⁄₁₆-in. fillet welds.

Since these fillet welds are large, groove welds may be more economical.

Design the compression flange plate and connection

The compresssion flange plate should have approximately the same area as the tension flange plate (4.69 in.²). Assume a shelf dimension of ⅝-in. The plate width, then, is 7.495 in. + 2(⅝-in.) = 8.745 in. To approximately balance the flange-plate areas, try a ⅝-in.×8¾-in. compression flange plate.

Check design compressive strength of flange plate

Assuming $K = 0.65$ and $l = $ ¾-in. (½-in. setback plus ¼-in. tolerance).

$$\frac{Kl}{r} = \frac{0.65 \, (\frac{3}{4}\text{-in.})}{\sqrt{\dfrac{(8\frac{3}{4}\text{-in.})(\frac{5}{8}\text{-in.})^3 / 12}{(8\frac{3}{4}\text{-in.})(\frac{5}{8}\text{-in.})}}}$$

$$= 2.70$$

From LRFD Specification Table 3-36 with $\dfrac{Kl}{r} = 2.70$,

$$\phi_c F_{cr} = 30.59 \text{ ksi}$$

and the design compressive strength of the flange plate is

$$\phi R_n = \phi_c F_{cr} A$$
$$= (30.59 \text{ ksi})(8\frac{3}{4} \text{ in.} \times \frac{5}{8} \text{ in.})$$
$$= 167 \text{ kips} > 150 \text{ kips} \quad \textbf{o.k.}$$

Determine required weld size and length for fillet welds to beam flange

As before for the tension flange plate, with ⁵⁄₁₆-in. fillet welds, use 8 in. along each side and 6¼-in. along the end of the compression flange plate.

Select compression flange plate dimensions

Use PL⅝-in. × 8¾-in. × 8¾-in.

Determine required weld size for fillet welds to supporting column flange

$$D_{min} = \frac{P_{uf}}{2 \times 1.392l}$$

$$= \frac{150 \text{ kips}}{2 \times 1.392 \ (8¾\text{-in.})}$$

$$= 6.16 \rightarrow 7 \text{ sixteenths}$$

Use ⁷⁄₁₆-in. fillet welds.

Since these fillet welds are large, groove welds may be more economical.

Investigate connection stiffness

$$n = \frac{M_u}{\theta}$$

$$= \frac{t_{pl}d^2E}{3}$$

$$= \frac{(¾\text{-in.})(17.99 \text{ in.})^2(29,000 \text{ ksi})}{3}$$

$$= 2.35 \times 10^6 \ \frac{\text{in.-kips}}{\text{rad}} \quad \text{or} \quad 80.9E \ \frac{\text{in.}^3}{\text{rad}}$$

For the beam line,

$$FEM = \frac{wl^2}{12}$$

$$= \frac{(7.29 \text{ kips/ft})(20 \text{ ft})^2}{12}$$

$$= 243 \text{ ft-kips}$$

$$\theta_s = \frac{wl^3 \ (144 \text{ in.}^2/\text{ft}^2)}{24EI}$$

$$= \frac{(7.29 \text{ kips/ft})(20 \text{ ft})^3 \ (144 \text{ in.}^2/\text{ft}^2)}{24 \ (29,000 \text{ ksi})(800 \text{ in.}^4)}$$

$$= 0.0151 \text{ rad}$$

The beam line and connection line are plotted in Figure 9-75. The equilibrium condition may also be obtained from Equations 9-8 and 9-10 where

$$u = \frac{\left(\dfrac{EI}{L}\right)}{n}$$

$$= \frac{E \ (800 \text{ in.}^4)}{20 \text{ ft } (12 \text{ in./ft})(80.9E)}$$

= 0.0412

and

$$M_{conn} = \left(\frac{1}{(2u+1)}\right)\frac{wL^2}{12}$$

$$= \left(\frac{1}{2\times 0.0412 + 1}\right)(243 \text{ ft-kips})$$

$$= 225 \text{ ft-kips}$$

Since this is the moment for which the connection was designed, the stiffness and strength of the connection are consistent. It should also be noted that this is a very stiff connection and perhaps should only be considered when close to the full fixed-end moment is to be carried.

Comment: The column section should be checked for stiffening requirements. A check of the applicable limit states from LRFD Specification Section K1 (as described in Part 10) will show the W14×109 column in the above example is adequate without stiffening.

Flexible Wind Connections

Flexible wind connections are made with top and bottom angles and a simple shear connection. The flexible wind connection is designed in two stages. First, considering only the gravity loads, a simple shear connection is designed. Second, the lateral loads only are arbitrarily distributed to selected connections to form the wind frames and the

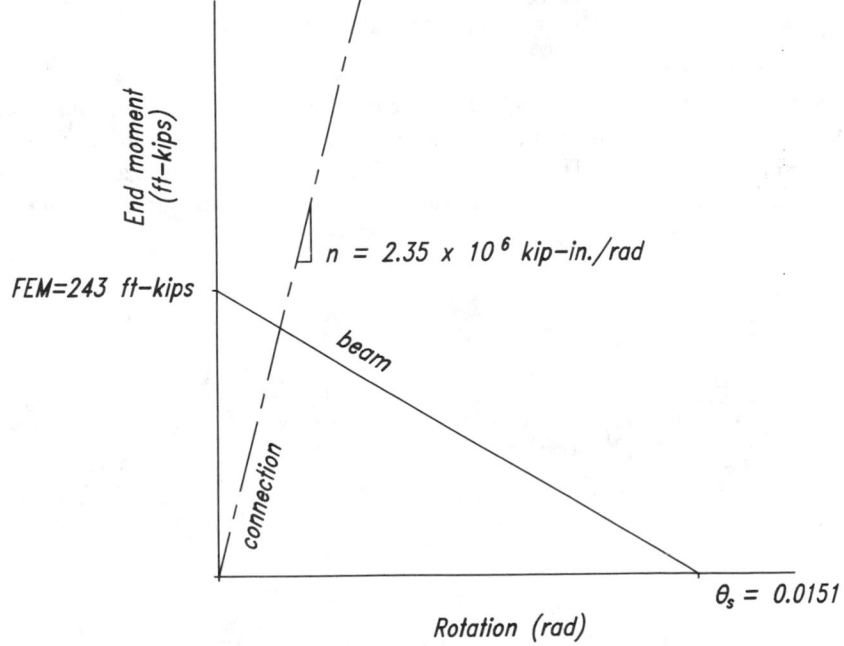

Figure 9-75. Moment-rotation diagram for Example 9-27.

resulting flexible wind connections are then designed as "fully restrained" for the calculated required strength.

While flexible wind connections (see Figure 9-76a) are not true PR moment connections, they do provide a simple, reliable, and economical alternative in the design of connections which must resist wind-induced moments. Flexible wind connections usually result in heavier beams, lighter columns, and reduced stiffening requirements. Additionally, there are several advantages to their use: (1) simplified analysis and calculations; (2) the beams and girders may be designed as simply connected members for gravity loads; and (3) the columns may be designed as axially loaded members with applied wind moments. Certain provisions, however, must be met when using this type of wind moment connection:

1. The wind frames must resist the wind moments throughout the entire structure from top to bottom.
2. The beams, columns, and their connections must resist the applied wind moments.
3. The girders must be capable of carrying the full gravity load as simply supported beams.
4. The connection material must have sufficient inelastic rotation capacity to prevent the welds and/or fasteners from failing due to combined gravity and wind loading.

The loading and unloading sequence which occurs in the flexible wind connections is described in detail by Disque (1964). The assumed distribution of this loading, the assumed angle deformation, and the locations of the points of inflection for use in calculating the bending moments are illustrated in Figure 9-76. Reasonably proportioned connections will result despite these apparently arbitrary assumptions which are required to overcome the complexities of an "exact" analysis. An in depth investigation of the analysis and design of flexibly connected wind frames is given in Ackroyd (1987). Ackroyd reports that the flexible wind frame approach is valid for frames less than 10 stories in height.

Design Checks
The design strengths of the bolts and/or welds, connecting elements, and affected elements of connected members must be determined in accordance with the provisions of the LRFD Specification. The applicable limit states in each of the aforementioned design strengths are discussed in Part 8. In all cases, the design strength ϕR_n must exceed the required strength R_u.

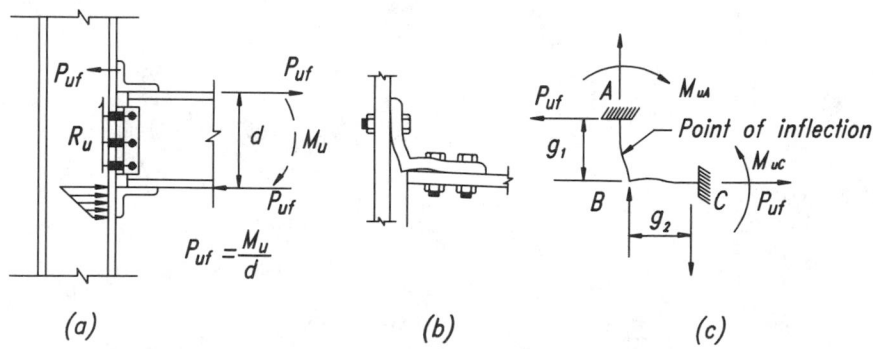

Figure 9-76. Flexible wind connections.

The tensile force is carried to the angle by the flange bolts, with the angle assumed to deform as in Figure 9-76. A point of inflection is assumed between the bolt gage line and the top face of the connection angle, for use in calculating the local bending moment and the corresponding required angle thickness. The effect of prying action must also be considered.

The strength of this type of connection is limited by the available angle thickness and the maximum number of fasteners which can be placed on a single gage line of the vertical leg of the connection angle at the tension flange. Figure 9-77 illustrates the column flange deformation and shows that only the fasteners closest to the column web are fully effective in transferring forces.

The column flange and web must be investigated by the designer for stiffening requirements at both the tension and compression flanges of the supported beam.

Example 9-28

Given: Refer to Figure 9-78. Design the flexible wind connection shown for the W16×36 beam to W14 column flange connection. From the simple beam gravity analysis with 3.4 kips/ft on a 20 ft span,

$$R_u = 34.0 \text{ kips}$$
$$M_{ug} = 170 \text{ ft-kips (at beam centerline)}$$

From the portal analysis shown in the sketch below the wind moment is

$$M_{uw} = 56.0 \text{ ft-kips (at connection)}$$

W16×36

$d = 15.86$ in.	$b_f = 6.985$ in.	$Z_x = 64.0$ in.^{3}
$t_w = 0.295$ in.	$t_f = 0.430$ in.	

Note that the W16×36 beam has been selected based upon a simple beam gravity analysis. Use ¾-in. diameter A325-N bolts. For structural members, assume $F_y = 50$ ksi and $F_u = 65$ ksi; for connecting materials, assume $F_y = 36$ ksi and $F_u = 58$ ksi.

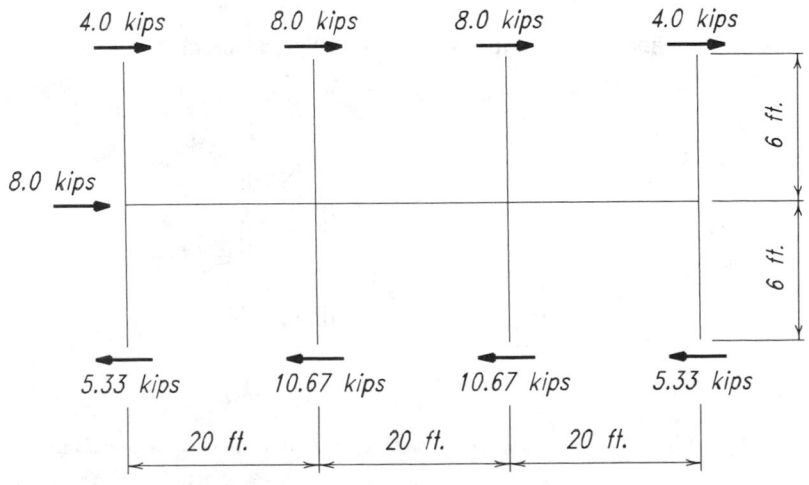

Solution: *Check beam design flexural strength (at connection)*

$$Z_{req} = \frac{M_{uw} \times 12 \text{ in./ft}}{0.9 F_y}$$

$$= \frac{(56.0 \text{ ft-kips})(12 \text{ in./ft})}{0.9 (50 \text{ ksi})}$$

$$= 14.9 \text{ in.}^3$$

Assuming two rows of ¾-in. diameter A325-N bolts in standard holes, from LRFD Specification Section B10:

$$A_{fg} = b_f \times t_f$$
$$= 6.985 \text{ in.} \times 0.430 \text{ in.}$$
$$= 3.00 \text{ in.}^2$$

$$A_{fn} = A_{fg} - 2 (d_b + \text{⅛-in.}) t_f$$
$$= 3.00 \text{ in.}^2 - 2 (\text{¾-in.} + \text{⅛-in.})(0.430 \text{ in.})$$
$$= 2.25 \text{ in.}^2$$

Since $0.75 F_u A_{fn}$ (= 110 kips) is less than $0.9 F_y A_{fg}$ (= 135 kips), the effective tension flange area A_{fe} is

$$A_{fe} = \frac{5 F_u}{6 F_y} A_{fn}$$

$$= \frac{5}{6} \left(\frac{65 \text{ ksi}}{50 \text{ ksi}} \right) 2.25 \text{ in.}^2$$

$$= 2.44 \text{ in.}^2$$

This is an 18.7 percent reduction from the gross flange area A_{fg} and the effective plastic section modulus Z_e is

$$Z_e \approx Z_x - 2 \left(0.187 A_{fg} \frac{d}{2} \right)$$

$$\approx 64.0 \text{ in.}^3 - 2 \left(0.187 \times 3.00 \text{ in.}^2 \times \frac{15.86 \text{ in.}}{2} \right)$$

$$\approx 55.1 \text{ in.}^3$$

Since $Z_e > Z_{req}$, the beam design flexural strength is **o.k.**

Design the double-angle web connection

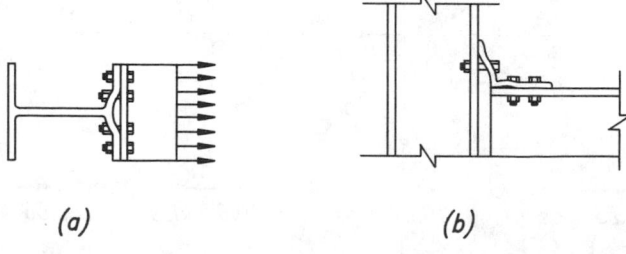

(a)   (b)

Fig. 9-77. Illustration of deformations in flexible wind connections

From Table 9-2, for ¾-in. diameter A325-N bolts and angle material with F_y = 36 ksi and F_u = 58 ksi, select three rows of bolts and ¼-in. angle thickness.

ϕR_n = 76.7 kips > 34.0 kips   **o.k.**

From Table 9-2, for three rows of bolts and an uncoped beam with F_y = 50 ksi and F_u = 65 ksi

ϕR_n = (263 kips/in.)(0.295 in.)
    = 77.6 kips > 34.0 kips   **o.k.**

Note: If the column section were given, it could also be checked using Table 9-2.

Design the tension flange angle and connection

Calculate the flange force P_{uf}

$$P_{uf} = \frac{M_{uw} \times 12 \text{ in./ft}}{d}$$

$$= \frac{(56.0 \text{ ft-kips})(12 \text{ in./ft})}{15.86 \text{ in.}}$$

$$= 42.4 \text{ kips}$$

Determine number of ¾-in. diameter A325-N bolts required for shear (bolts through beam flange)

From Table 8-11:

$$n_{\min} = \frac{P_{uf}}{\phi r_n}$$

$$= \frac{42.4 \text{ kips}}{15.9 \text{ kips/bolt}}$$

$$= 2.67 \rightarrow 4 \text{ bolts (even number required)}$$

Determine number of ¾-in. diamter A325-N bolts required for tension (bolts through column flange)

From Table 8-15:

$$n_{\min} = \frac{P_{uf}}{\phi r_n}$$

$$= \frac{42.4 \text{ kips}}{29.8 \text{ kips/bolt}}$$

$$= 1.42 \rightarrow 2 \text{ bolts}$$

Determine flange angle thickness for flexure

Try L6×4 8-in. long. The tributary load in bending is then 42.4 kips/8 in. = 5.3 kips / in. The preliminary angle thickness may now be selected from Table 11-1. Since this table is based upon a symmetrical connection, enter table with twice the tributary load or 10.6 kips/in. and $b = 1\frac{1}{2}$-in. For $F_y = 36$ ksi angle material, a $\frac{3}{4}$-in. thickness provides for 12.2 kips/in.

Try L6×4×$\frac{3}{4}$×8 in.

Check angle thickness for prying action assuming a 4-in. gage

$$r_{ut} = \frac{P_{uf}}{2 \text{ bolts}}$$

$$= \frac{42.4 \text{ kips}}{2 \text{ bolts}}$$

$$= 21.2 \text{ kips/bolt}$$

$b = 1\frac{1}{2}$-in.

$a = 4$ in. $- b - t$

$\quad = 4$ in. $- 1\frac{1}{2}$-in. $- \frac{3}{4}$-in.

$\quad = 1\frac{3}{4}$-in.

Since $a = 1\frac{3}{4}$-in. is less than 1.25b, use $a = 1\frac{3}{4}$-in. in calculations

$b' = b - d/2$

$\quad = 1\frac{1}{2}$-in. $- \dfrac{\frac{3}{4}\text{-in.}}{2}$

$\quad = 1\frac{1}{8}$-in.

$a' = a + d/2$

$\quad = 1\frac{3}{4}$-in. $+ \dfrac{\frac{3}{4}\text{-in.}}{2}$

$\quad = 2\frac{1}{8}$-in.

$\rho = \dfrac{b'}{a'}$

$\quad = \dfrac{1\frac{1}{8}\text{-in.}}{2\frac{1}{8}\text{-in.}}$

$\quad = 0.529$

$\beta = \dfrac{1}{\rho}\left(\dfrac{\phi r_n}{r_{ut}} - 1\right)$

$\quad = \dfrac{1}{0.529}\left(\dfrac{29.8 \text{ kips/bolt}}{21.2 \text{ kips/bolt}} - 1\right)$

$\quad = 0.767$

$\delta = 1 - \dfrac{d'}{p}$

$\quad = 1 - \dfrac{\frac{13}{16}\text{-in.}}{4 \text{ in.}}$

$\quad = 0.797$

Since $\beta < 1$, α' is equal to the lesser of 1.0 and

$$\frac{1}{\delta}\left(\frac{\beta}{1-\beta}\right) = \frac{1}{0.797}\left(\frac{0.767}{1-0.767}\right)$$

$$= 4.13$$

Thus, set $\alpha' = 1.0$ and

$$t_{req} = \sqrt{\frac{4.44 r_{ut} b'}{p F_y \, (1 + \delta \alpha')}}$$

$$= \sqrt{\frac{4.44 \, (21.2 \text{ kips} / \text{bolt})(1\frac{1}{8}\text{-in.})}{(4 \text{ in.})(36 \text{ ksi})[1 + (0.797)(1.0)]}}$$

$$= 0.640 \text{ in.} < \frac{3}{4}\text{-in.} \quad \textbf{o.k.}$$

Check tension yielding of the angle

$$\phi R_n = \phi F_y A_g$$
$$= 0.9(36 \text{ ksi})(8 \text{ in.} \times \frac{3}{4}\text{-in.})$$
$$= 194 \text{ kips} > 42.4 \text{ kips} \quad \textbf{o.k.}$$

Check tension rupture of the angle

$$\phi R_n = \phi F_u A_n$$
$$= 0.75(58 \text{ ksi})(8 \text{ in.} - 2 \times 0.875 \text{ in.})(\frac{3}{4}\text{-in.})$$
$$= 204 \text{ kips} > 42.4 \text{ kips} \quad \textbf{o.k.}$$

Check shear yielding of the angle

$$\phi R_n = \phi(0.6 F_y) A_g$$
$$= 0.9(0.6 \times 36 \text{ ksi})(8 \text{ in.} \times \frac{3}{4}\text{-in.})$$

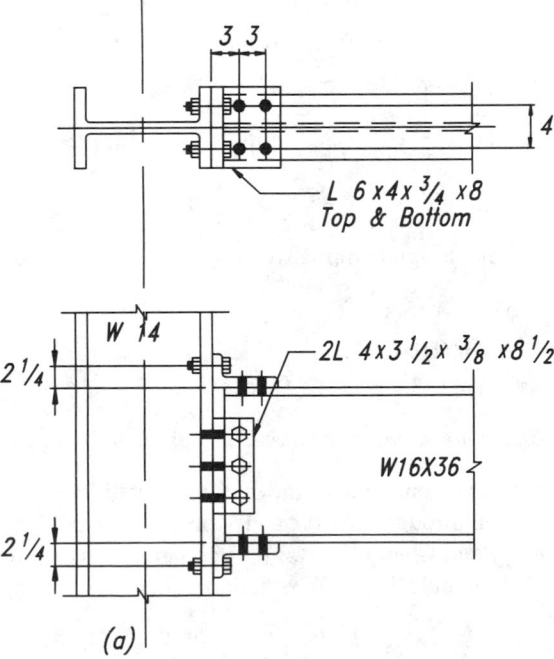

Figure 9-78. Illustration for Example 9-28.

= 116 kips > 42.4 kips   **o.k.**

Check shear rupture of the angle

$$\phi R_n = \phi(0.6F_u)A_n$$
$$= 0.75(0.6 \times 58 \text{ ksi})(8 \text{ in.} - 2 \times 0.875 \text{ in.})(\tfrac{3}{4}\text{-in.})$$
$$= 122 \text{ kips} > 42.4 \text{ kips} \quad \textbf{o.k.}$$

Check block shear rupture of the angle

From Tables 8-47 and 8-48, with L_{ev} = 2 in., L_{eh} = 2 in., and n = 2, $0.6F_u A_{nv} > F_u A_{nt}$. Thus, from Tables 8-48a and 8-48b,

$$\phi R_n = \phi \ (0.6F_u A_{nv} + F_y A_{gt})$$
$$= (96.0 \text{ kips/in.} + 54.0 \text{ kips/in.})(\tfrac{3}{4}\text{-in.})$$
$$= 113 \text{ kips} > 42.4 \text{ kips} \quad \textbf{o.k.}$$

Design the compression flange angle and connection

For symmetry, try L6×4×¾×8 in. with four ¾-in. diameter A325-N bolts through beam flange and two ¾-in. diamter A325-N bolts through column flange.

Check design compressive strength of angle assuming K = 0.65 and l = 3 in. (normal gage).

$$\frac{Kl}{r} = \frac{0.65 \ (3\text{in.})}{\sqrt{\dfrac{(8 \text{ in.})(\tfrac{3}{4}\text{-in.})^3 / 12}{(8 \text{ in.})(\tfrac{3}{4}\text{-in.})}}}$$

$$= 9.01$$

From LRFD Specification Table 3-36 with $\dfrac{KL}{r}$ = 9.01,

$$\phi_c F_{cr} = 30.47$$

and the design compressive strength ofthe angle is

$$\phi R_n = \phi_c F_{cr} A$$
$$= (30.47 \text{ ksi})(8 \text{ in.} \times \tfrac{3}{4}\text{-in.})$$
$$= 183 \text{ kips} > 42.4 \text{ kips} \quad \textbf{o.k.}$$

Check the moment-rotation characteristics of the resulting connection

For this connection, the moment-rotation characteristics may be viewed through the Frye and Morris (1975) polynomial as reported by Kishi and Chen (1986). The standardized moment-rotation curve for the top- and seat-angle with double-angle web connection is given by

$$\theta = 2.23 \times 10^{-5}KM + 1.85 \times 10^{-8}KM^3 + 3.19 \times 10^{-12}KM^5$$

where

$$K = \frac{\left(g_t - \dfrac{d_b}{2}\right)^{1.35}}{t_t^{1.13} d^{1.29} t_w^{0.415} l_t^{0.694}}$$

In the above equation,

- t_t = thickness of top angle, in.
- d = beam depth, in.
- t_w = web connection angle thickness, in.
- l_t = length of top angle, in.
- g_t = gage in vertical leg of top angle, in.
- d_b = bolt diameter, in.

Thus, for this connection,

$$K = \frac{\left(2\tfrac{1}{4}\text{-in.} - \dfrac{\tfrac{3}{4}\text{-in.}}{2}\right)^{1.35}}{(\tfrac{3}{4}\text{-in.})^{1.13}(15.86\text{ in.})^{1.29}(\tfrac{3}{8}\text{-in.})^{0.415}(8\text{ in.})^{0.694}}$$

$$= 0.0325$$

The standardized connection curve is given in Figure 9-79. The beam line is also shown in the figure with

$$KM_{FEM} = 0.0325(1{,}360\text{ in.-kips})$$
$$= 44.2\text{ in.-kips}$$

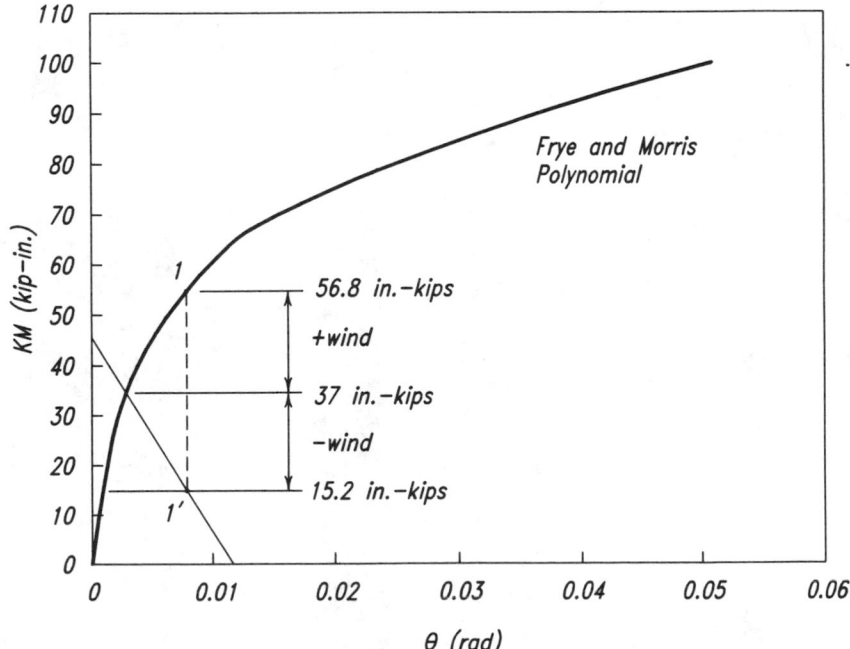

Figure 9-79. Standardized moment-rotation curve for Example 9-28.

and

$$\theta_{simple} = 0.0126 \text{ rad}$$

When the wind moment KM_{wind} is added and subtracted from the connection moment (see Disque, 1964) points 1 and 1' are reached, respectively. This final result shows that the connection has sufficient flexibility for this application since the final connction moment is

$$M_{u1} = \frac{56.8 \text{ in.-kips}}{0.0325 \ (12 \text{ in./ft})}$$

$$= 146 \text{ ft-kips} < \phi M_n = 173 \text{ ft-kips}$$

Comment: In all situations where flexible connections are used, the impact of connection rotation on drift of the wind frame must be checked. In addition, the column design must account for the reduced beam stiffness due to connection rotation.

REFERENCES

Ackroyd, M. H., 1987, "Simplified Frame Design of Type PR Construction," *Engineering Journal*, Vol. 24, No. 4, (4th Qtr.), pp. 141–146, AISC, Chicago, IL.

Astaneh, A. and M. N. Nader, 1989, "Design of Tee Framing Shear Connections," *Engineering Journal*, Vol. 26, No. 1, (1st Qtr.), pp. 9–20, AISC, Chicago, IL.

Astaneh, A., S. M. Call, and K. M. McMullin, 1989, "Design of Single-Plate Shear Connections," *Engineering Journal*, Vol. 26, No. 1, (1st Qtr.), pp. 21–32, AISC, Chicago, IL.

Blodgett, O. W., 1966, *Design of Welded Structures*, James F. Lincoln Arc Welding Foundation, Cleveland, OH.

Deierlein, G. G., S. H. Hseih, and Y. J. Shen, 1990, "Computer-Aided Design of Steel Structures with Flexible Connections," *Proceedings of the 1990 National Steel Construction Conference*, pp. 9.1–9.21, AISC, Chicago, IL.

Disque, R. O., 1964, "Wind Connections with Simple Framing," *Engineering Journal*, Vol. 1, No. 3, (July), pp. 101–103, AISC, Chicago, IL.

Frye, M. J. and Morris, G. A., 1975, "Analysis of Flexibly Connected Steel Frames," *Canadian Journal of Civil Engineering*, Vol. 2, pp. 280–291.

Garrett, J. H., Jr. and R. L. Brockenbrough, 1986, "Design Loads for Seated-Beam Connections in LRFD," *Engineering Journal*, Vol. 23, No. 2, (2nd Qtr.), pp. 84–88, AISC, Chicago, IL.

Gerstle, K. H., and M. H. Ackroyd, 1989, "Behavior and Design of Flexibly-Connected Building Frames," *Proceedings of the 1989 National Steel Construction Conference*, pp. 1.1–1.28, AISC, Chicago, IL.

Geschwindner, 1991, "A Simplified Look at Partially Restrained Connections," *Engineering Journal*, Vol. 28, No. 2, (2nd Qtr.), pp. 73–78, AISC, Chicago, IL.

Goverdhan, A. V., 1984, "A Collection of Experimental Moment Rotation Curves and Evaluation of Prediction Equations for Semi-Rigid Connections, Master of Science Thesis, Vanderbilt University, Nashville, TN.

Kishi, N. and W. F. Chen, 1986, "Data Base of Steel Beam-to-Column Connections," CE-STR-86-26, Purdue University, School of Engineering, West Lafayette, IN.

Kulak, G. L. and D. L. Green, 1990, "Design of Connectors in Web-Flange Beam or Girder Splices," *Engineering Journal*, Vol. 27, No. 2, (2nd Qtr.), pp. 41–48, AISC, Chicago, IL.

Nethercot, D. A. and W. F. Chen, 1988, "Effects of Connections on Columns," *Journal of Constructional Steel Research*, pp. 201–239, Elsevier Applied Science Publishers, Essex, England.

Palmer, F. J., 1990, "Tubular Connections," *Proceedings of the 1990 National Steel Construction Conference*, pp. 21.1–21.10, AISC, Chicago, IL.

Ricker, D. T., 1985, "Practical Tubular Connections," *Symposium on Hollow Sections in Building Construction* (ASCE Structures Congress, Chicago, IL), ASCE, New York, NY.

Roeder, C. W. and R. H. Dailey, 1989, "The Results of Experiments on Seated Beam Connections," *Engineering Journal*, Vol. 26, No. 3, (3rd Qtr.), pp. 90–95, AISC, Chicago, IL.

Salmon, C. G. and J. E. Johnson, 1990, *Steel Structures: Design and Behavior*, 3rd Edition, Harper Collins Publishers, New York, NY.

Sherman, D. R. and J. M. Ales, 1991, "The Design of Shear Tabs with Tubular Columns," *Proceedings of the 1991 National Steel Construction Conference*, pp.1.1–1.22, AISC, Chicago, IL.

Sherman, D. R. and S. M. Herlache, 1988, "Beam Connections to Rectangular Tube Columns," *Proceedings of the 1988 National Steel Construction Conference*, pp. 23.1–23.14, AISC, Chicago, IL.

Sputo, T. and D. S. Ellifritt, 1991, "Proposed Design Criteria for Stiffened Seated Connections to Column Webs," *1991 National Steel Construction Conference Proceedings*, pp. 8.1–8.26, AISC. Chicago, IL.

Thornton, W. A., 1992, "Eliminating the Guesswork in Connection Design," *Modern Steel Construction*, Vol. 32, No. 6, June, pp. 27–31, AISC, Chicago, IL.

Van Dalen, K. and J. MacIntyre, 1988, "The Rotational Behaviour of Clipped End-Plate Connections," *Canadian Journal of Civil Engineering*, Vol. 15, pp. 117–126, Canadian Steel Construction Council, Edmonton, Alberta, Canada.

PART 10

FULLY RESTRAINED (FR) MOMENT CONNECTIONS

OVERVIEW

Part 10 contains general information, design considerations, examples, and design aids for the design of fully restrained (FR) moment connections, column stiffening, moment splices, and special considerations in the aforementioned topics. It is based on the requirements of the 1993 LRFD Specification. Supplementary information may also be found in the Commentary on the LRFD Specification.

Following is a detailed list of the topics addressed.

FULLY RESTRAINED (FR) MOMENT CONNECTIONS

Fully restrained (FR) moment connections are also known as continuous or rigid-frame connections. As defined in LRFD Specification Section A2.2, FR moment connections possess sufficient rigidity to maintain the angles between intersecting members as illustrated in Figure 10-1. While connections considered to be fully restrained seldom provide for zero rotation between members, the small amount of flexibility present is usually neglected and the connection is idealized to prevent relative rotation. Connections A, B, and C in Figure 10-2 illustrate this.

Force Transfer in FR Moment Connections

LRFD Specification Section B9 states that end connections in FR construction shall be designed to carry the factored forces and moments, except that some inelastic but self-limiting deformation of a part of the connection is permitted. Huang, et al. (1973) showed that the moment may be resolved into an effective tension-compression couple acting as axial forces at the beam flanges. The flange force P_{uf} may be calculated as:

$$P_{uf} = \frac{M_u}{d_m}$$

where

P_{uf} = factored beam flange force, tensile or compressive, kips
M_u = beam end moment, kip-in.
d_m = moment arm between the flange forces, in. (varies for all FR connections and for stiffener design)

Furthermore, it was shown that shear is primarily transferred through the beam web shear connection. Since, by definition, the angle between the beam and column in an FR moment connection remains unchanged under loading, eccentricity may be neglected entirely in the shear connection. Additionally, it is permissible to use bolts in bearing in either standard or slotted holes perpendicular to the line of force. Axial forces, if present, are assumed to be distributed uniformly across the beam cross-sectional area, and are additive algebraically to the flange forces and vectorially to the shear force.

The supporting column must have sufficient strength and stiffness to develop the moment transferred to it through the FR moment connection. Additionally, the concentrated flange forces may require the selection of a larger column section or the stiffening of the existing column section; refer to "Column Stiffening" in this Part.

Temporary Support During Erection

Bolted construction provides a ready means to erect and temporarily connect members by use of the bolt holes. In contrast, FR moment connections in welded construction must

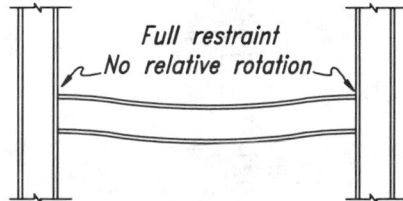

Fig. 10-1. Illustration of fully restrained (FR) moment connection.

be given special attention so that all pieces may be erected, fitted, and supported until the necessary welds are made, sometimes at a much later date. Temporary support can be provided in welded construction by furnishing holes for erection bolts, temporary seats, special lugs, or by other means.

Temporary erection aids should be carefully studied for their effect on the finished structure, particularly on members subjected to fatigue or tension loading. They should be permitted to remain in place whenever possible since they seldom are reusable and the cost to remove them can be significant. If left in place, erection aids should be located so as not to cause a stress concentration. If, however, erection aids are to be removed, care should be taken so that the base metal is not damaged.

Temporary supports should be sufficient to carry any loads imposed by the erection process, such as the dead weight of the member, additional construction equipment, or material storage. Additionally, they must be flexible enough to allow plumbing of the structure, particularly in tier buildings.

Welding Considerations for Fully Restrained Connections

Field welding should be arranged for down-hand or horizontal position welding and preference should be given to fillet welds over groove welds when possible. Additionally, the joint detail and welding procedure should be constructed to minimize distortion and the possibility of lamellar tearing.

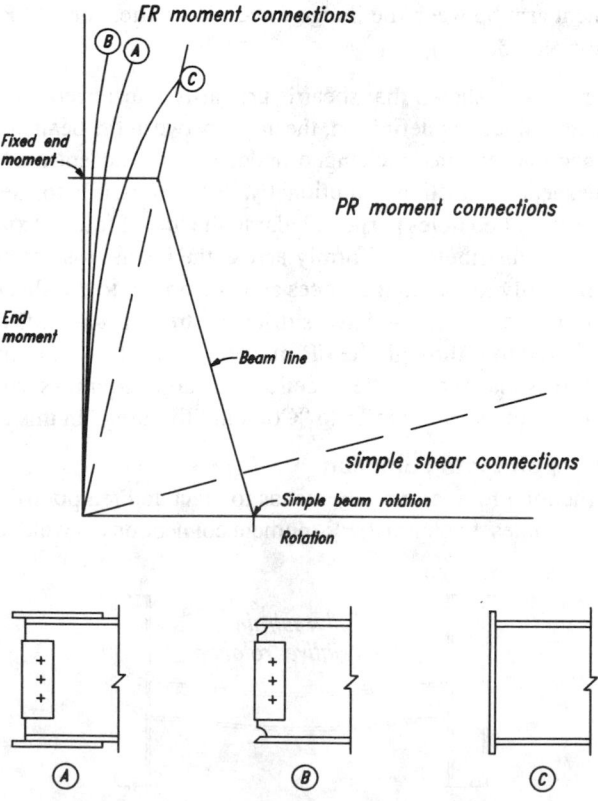

Fig. 10-2. FR moment connection behavior.

The typical complete-joint-penetration groove weld in a directly welded flange connection for a rolled beam can be expected to shrink about $\frac{1}{16}$-in. in the length dimension of the beam when it cools and contracts. Thicker welds, such as for welded plate-girder flanges will shrink even more—up to $\frac{1}{8}$-in. or $\frac{3}{16}$-in. This amount of shrinkage can cause erection problems in locating and plumbing the columns along lines of continuous beams. A method of calculating weld shrinkage may be found in Lincoln Electric Co. (1973)

Weld shrinkage can best be controlled by fabricating the beam longer than required by the amount of the anticipated weld shrinkage. Alternatively, the weld-joint opening could be increased; refer to AWS D1.1.

Unnecessarily thick stiffeners with complete-joint-penetration groove welds should be avoided since the accompanying weld shrinkage may contribute to lamellar tearing; refer to "Minimizing Weld Repairs—Lamellar Tearing" in Part 8.

Special Considerations for Seismic Loading
The effect of severe seismic loading on test specimens subjected to low-cycle fatigue tests is discussed in Krawinkler and Popov (1982). Slippage occured early in the inelastic cycles for slip-critical-bolted shear connections indicating the possible existence of bending and shearing forces in the beam flange close to the connecting weld. Thus, it is recommended that the shear connection be designed for a portion of the bending moment when deep rolled beams and plate girders are rigidly connected to a column flange support. Refer to AISC *Seismic Provisions for Structural Steel Buildings*.

Flange-Plated Connections
As illustrated in Figure 10-3, a flange-plated FR moment connection consists of a shear connection and top and bottom flange plates which connect the flanges of the supported beam to the supporting column. These flange plates are welded to the supporting column and may be bolted or welded to the flanges of the supported beam.

Design Checks
The design strengths of the bolts and/or welds and connected elements must be determined in accordance with the LRFD Specification; the applicable limit states are discussed in Part 8. The effect of eccentricity in the shear connection may be neglected. The strength of the supporting column (and thus the need for stiffening) must be checked; refer to "Column Stiffening" in this Part.

Shop and Field Practices
In a column flange connection, the flange plates are usually located with respect to the column web centerline. Because of the column-flange mill tolerance on out-of-squareness with the web, it is desirable to shop-fit long flange plates from the theoretical column-web centerline to assure good field fit-up with the beam. Misalignment on short connections, as illustrated in Figure 10-4, can be accommodated by providing oversized holes in the plates. Since mill tolerances in both the beam and the column may cause significant shop and/or field assembly problems, it may be desirable to ship the flange plates loose for field attachment to the column.

Example 10-1

Given: Design a bolted flange-plated FR moment connection for a W18×50 beam to W14×99 column-flange connection. For structural members, assume F_y = 50 ksi and F_u = 65 ksi; for connecting material, assume F_y = 36 ksi and F_u = 58 ksi. Use ⅞-in. diameter ASTM A325-N bolts and 70 ksi electrodes.

R_u = 45.0 kips

M_u = 250 ft-kips

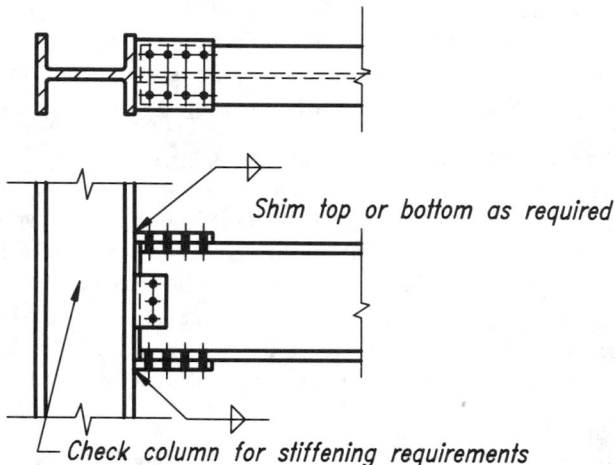

(a) Column flange support, bolted flange plates

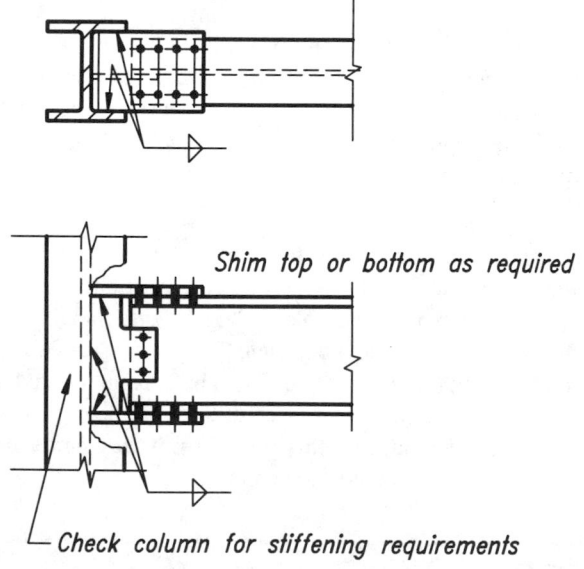

(b) Column web support, bolted flange plates

Fig. 10-3. Flange-plated FR moment connections.

W18×50

d = 17.99 in. b_f = 7.495 in. Z_x = 101 in.^{3}
t_w = 0.355 in. t_f = 0.570 in.

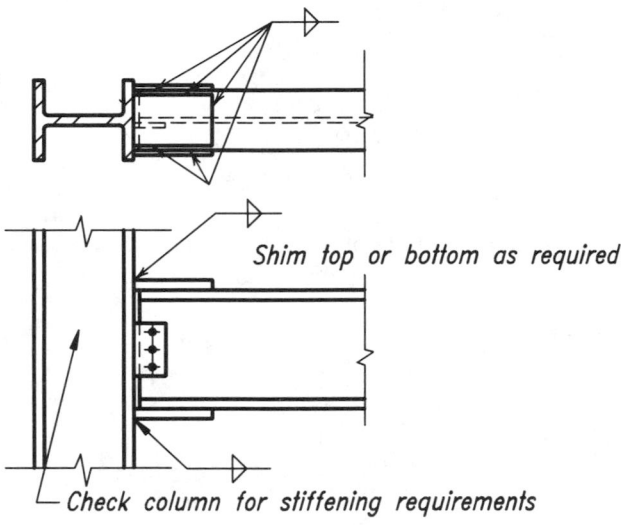

Shim top or bottom as required

Check column for stiffening requirements

(c) Column flange support, welded flange plates

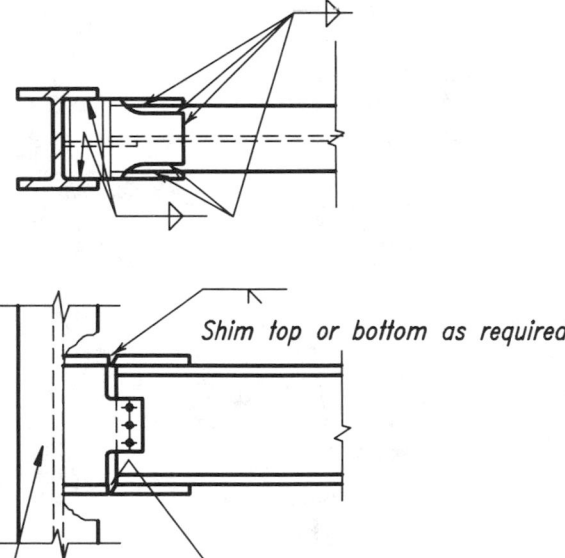

Shim top or bottom as required

Check column for stiffening requirements

(d) Column web support, welded flange plates

Fig. 10-3 (cont.). Flange-plated FR moment connections.

W14×99

d = 14.16 in. b_f = 14.565 in. k = 1⁷⁄₁₆-in.
t_w = 0.485 in. t_f = 0.780 in. T = 11¼-in.

Solution: *Check beam design flexural strength:*

$$Z_{req} = \frac{M_u \times 12 \text{ in.} / \text{ft}}{0.9F_y}$$

$$= \frac{(250 \text{ ft-kips}) (12 \text{ in.} / \text{ft})}{0.9(50 \text{ ksi})}$$

$$= 66.7 \text{ in.}^3$$

Assuming two rows of ⅞-in. diameter A325-N bolts in standard holes, from LRFD Specification Section B10:

$$A_{fg} = b_f \times t_f$$
$$= 7.495 \text{ in.} \times 0.570 \text{ in.}$$
$$= 4.27 \text{ in.}^2$$
$$A_{fn} = A_{fg} - 2(d_b + \text{⅛-in.})t_f$$
$$= 4.27 \text{ in.}^2 - 2(\text{⅞-in.} + \text{⅛-in.})(0.570 \text{ in.})$$
$$= 3.13 \text{ in.}^2$$

since $0.75F_uA_{fn}$ (= 153 kips) is less than $0.9F_yA_{fg}$ (= 192 kips), the effective tension flange area A_{fe} is:

$$A_{fe} = \frac{5 F_u}{6 F_y}A_{fn}$$

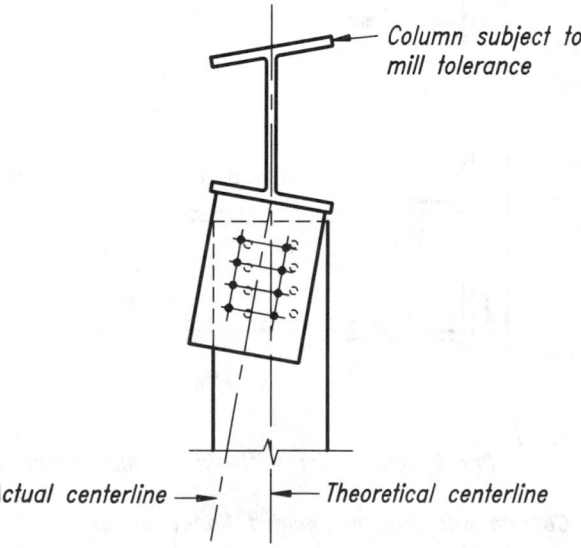

Fig. 10-4. Effect of mill tolerances on flange-plate alignment.

$$= \frac{5}{6} \left(\frac{65 \text{ ksi}}{50 \text{ ksi}} \right) 3.13 \text{ in.}^2$$

$$= 3.39 \text{ in.}^2$$

This is a 20.6 percent reduction from the gross flange area A_{fg} and the effective plastic section modulus Z_e is:

$$Z_e \approx Z_x - 2 \left(0.206 A_{fg} \frac{d}{2} \right)$$

$$\approx 101 \text{ in.}^3 - 2 \left(0.206 \times 4.27 \text{ in.}^2 \times \frac{17.99 \text{ in.}}{2} \right)$$

$$\approx 85.2 \text{ in.}^3$$

Since $Z_e > Z_{req}$, the beam design flexural strength is **o.k.**

Design the single-plate web connection.

Determine number of $\frac{7}{8}$-in. diameter A325-N bolts required for shear. From Table 8-11:

$$n_{min} = \frac{R_u}{\phi r_n}$$

$$= \frac{45.0 \text{ kips}}{21.6 \text{ kips / bolt}}$$

$$= 2.08 \rightarrow 3 \text{ bolts}$$

Try PL $\frac{5}{16} \times 9$

Determine number of $\frac{7}{8}$-in. diameter A325-N bolts required for material bearing, assuming $L_e = 1\frac{1}{2}$-in., and $s = 3$ in. The $\frac{5}{16}$-in. plate is more critical than the 0.355-in. thick beam web. From Table 8-13:

$$n_{min} = \frac{R_u}{\phi r_n}$$

$$= \frac{45.0 \text{ kips}}{28.6 \text{ kips / bolt}}$$

$$= 1.57 \rightarrow 2 \text{ bolts}$$

Bolt shear is more critical. Try a three-bolt single-plate connection.

Check shear yielding of the plate:

$$\phi R_n = 0.9(0.6 F_y A_g)$$
$$= 0.9[0.6 \times 36 \text{ ksi } (9 \text{ in.} \times \frac{5}{16}\text{-in.})]$$
$$= 54.7 \text{ kips} > 45.0 \text{ kips} \quad \textbf{o.k.}$$

Check shear rupture of the plate:

$$\phi R_n = 0.75 \, (\, 0.6 F_u A_n)$$

$$= 0.75 \, [0.6 \times 58 \text{ ksi } (9 \text{ in.} - 3 \, (\text{7/8-in.} + \text{1/8-in.})) \, \text{5/16-in.}]$$
$$= 48.9 \text{ kips} > 45.0 \text{ kips} \quad \textbf{o.k.}$$

Check block shear rupture of the plate:

With L_{eh} = 1½-in. and L_{ev} = 1½-in., from Tables 8-47a and 8-48a, $0.6F_uA_{nv} > F_uA_{nt}$. Thus,

$$\phi R_n = \phi[0.6F_uA_{nv} + F_yA_{gt}]$$

From Tables 8-48a and 8-48b,

$$\phi R_n = (131 \text{ kips / in.} + 40.5 \text{ kips / in.}) \text{5/16-in.}$$
$$= 53.6 \text{ kips} > 45.0 \text{ kips} \quad \textbf{o.k.}$$

Determine required weld size for fillet welds to supporting column flange:

$$D_{\min} = \frac{R_u}{2 \times 1.392l}$$

$$= \frac{45.0 \text{ kips}}{2 \times 1.392(9 \text{ in.})}$$

$$= 1.80 \rightarrow 2 \text{ sixteenths}$$

From LRFD Specification Table J2.4, since the column flange thickness is over ¾-in., the minimum fillet weld size is 5/16-in., use two 5/16-in. fillet welds.

Design the tension flange plate and connection.

Calculate the flange force P_{uf}:

$$P_{uf} = \frac{M_u \times 12 \text{ in. / ft}}{d}$$

$$= \frac{(250 \text{ ft-kips})(12 \text{ in. / ft})}{17.99 \text{ in.}}$$

$$= 167 \text{ kips}$$

Determine number of ⅞-in. diameter A325-N bolts required for shear. From Table 8-11:

$$n_{\min} = \frac{P_{uf}}{\phi r_n}$$

$$= \frac{167 \text{ kips}}{21.6 \text{ kips / bolt}}$$

$$= 7.73 \rightarrow 8 \text{ bolts}$$

Try PL ¾-in.×7in.

Determine number of $\frac{7}{8}$-in. diameter A325-N bolts required for material bearing on beam flange (more critical than flange plate), assuming $L_e \geq 1.5d$ and $s = 3$ in. From Table 8-13:

$$n_{min} = \frac{P_{uf}}{\phi r_n}$$

$$= \frac{167 \text{ kips}}{58.1 \text{ kips / bolt}}$$

$$= 2.9 \rightarrow 4 \text{ bolts (even number required)}$$

Bolt shear is more critical. Try two rows of four bolts on a 4-in. gage.

Check tension yielding of flange plate:

$$\phi R_n = \phi F_y A_g$$
$$= 0.9 \times 36 \text{ ksi} \times 7 \text{ in.} \times \frac{3}{4}\text{-in.}$$
$$= 170 \text{ kips} > 167 \text{ kips} \quad \textbf{o.k.}$$

Check tension rupture of flange plate:

$$\phi R_n = \phi F_u A_n$$
$$= 0.75 \times 58 \text{ ksi } [7 \text{ in.} - 2 \times (\tfrac{7}{8}\text{-in.} + \tfrac{1}{8}\text{-in.})] \tfrac{3}{4}\text{-in.}$$
$$= 163 \text{ kips} < 167 \text{ kips required} \quad \textbf{n.g.}$$

Try PL $\frac{3}{4}$-in.×$7\frac{1}{4}$-in.:

$$\phi R_n = 0.75 \times 58 \text{ ksi } [7\tfrac{1}{4}\text{-in.} - 2 \times (\tfrac{7}{8}\text{-in.} + \tfrac{1}{8}\text{-in.})] \tfrac{3}{4}\text{-in.}$$
$$= 171 \text{ kips} > 167 \text{ kips} \quad \textbf{o.k.}$$

Check block shear rupture of flange plate:

There are two cases for which block shear must be checked. The first case involves the tearout of the two blocks outside the two rows of bolt holes in the flange plate; for this case $L_{eh} = 1\frac{5}{8}$-in. and $L_{ev} = 1\frac{1}{2}$-in. The second case involves the tearout of the block between the two rows of holes in the flange plate. Tables 8-47 and 8-48 may be adapted for this calculation by considering the 4-in. width to be comprised of two 2-in. wide blocks where $L_{eh} = 2$ in. and $L_{ev} = 1\frac{1}{2}$-in. Thus, the former case is more critical. From Tables 8-47a and 8-48a, $0.6F_u A_{nv} > F_u A_{nt}$. Thus,

$$\phi R_n = \phi [0.6F_u A_{nv} + F_y A_{gt}]$$

From Tables 8-48a and 8-48b,

$$\phi R_n = 2 (183 \text{ kips/in.} + 43.9 \text{ kips/in.}) \tfrac{3}{4}\text{-in.}$$

$$= 340 \text{ kips} > 167 \text{ kips} \quad \textbf{o.k.}$$

Determine required weld size for fillet welds to supporting column flange:

$$D_{min} = \frac{P_{uf}}{2 \times 1.392(l)}$$

$$= \frac{167 \text{ kips}}{2 \times 1.392(7\frac{1}{4}\text{-in.})}$$

$$= 8.27 \rightarrow 9 \text{ sixteenths}$$

Use $\frac{9}{16}$-in. fillet weld.

Since these fillet welds are large, groove welds may be more economical.

Design the compression flange plate and connection.

Check design compressive strength of flange plate assuming $K = 0.65$ and $l = 2$ in. ($1\frac{1}{2}$-in. edge distance plus $\frac{1}{2}$-in. setback)

$$\frac{Kl}{r} = \frac{0.65 \ (2 \text{ in.})}{\sqrt{\frac{7\frac{1}{4}\text{-in.}) \ (\frac{3}{4}\text{-in.})^3 / 12}{(7\frac{1}{4}\text{-in.}) \ (\frac{3}{4}\text{-in.})}}}$$

$$= 6.00$$

From LRFD Specification Table 3-36 with $\frac{Kl}{r} = 6.00$,

$$\phi F_{cr} = 30.54 \text{ ksi}$$

and the design compressive strength of the flange plate is

$$\phi R_n = \phi_c F_{cr} A$$

$$= (30.54 \text{ ksi}) \ (7\frac{1}{4}\text{-in.} \times \frac{3}{4}\text{-in.})$$

$$= 167 \text{ kips}$$

Since the design strength equals the required strength, the flange plate is adequate.

The compression flange plate will be identical to the tension flange plate: a $\frac{3}{4}$-in.$\times 7\frac{1}{4}$-in. plate with eight bolts in two rows of four bolts on a 4-in. gage and $\frac{9}{16}$-in. fillet welds to the supporting column flange.

Check the column section for stiffening requirements; refer to Example 10-6.

Example 10-2

Given: Design a welded flange-plated FR moment connection for a W18×50 beam to W14×99 column flange connection. For structural members, $F_y = 50$ ksi; for connecting material $F_y = 36$ ksi. Use 70 ksi electrodes and ASTM A325-N bolts.

$R_u = 45.0$ kips
$M_u = 250$ kips

W18×50

$d = 17.99$ in.	$b_f = 7.495$ in.	$Z_x = 101$ in.^{3}
$t_w = 0.355$ in.	$t_f = 0.570$ in.	

W14×99

$d = 14.16$ in.	$b_f = 14.565$ in.	$k = 1\frac{7}{16}$-in.
$t_w = 0.485$ in.	$t_f = 0.780$ in.	$T = 11\frac{1}{4}$-in.

Solution: *Check beam design flexural strength:*

$$Z_{req} = \frac{M_u \times 12 \text{ in. / ft}}{0.9F_y}$$

$$= \frac{(250 \text{ ft-kips})(12 \text{ in. / ft})}{0.9(50 \text{ ksi})}$$

$$= 66.7 \text{ in.}^3$$
$$Z_x = 101 \text{ in.}^3$$

Since $Z_x > Z_{req}$, the beam design flexural strength is **o.k.**

Design the single-plate web connection.

From Example 10-1, a three-bolt, $\frac{5}{16}$-in. thick single plate with two $\frac{5}{16}$-in. fillet welds will be adequate.

Design the tension flange plate and connection.

Calculate the flange force P_{uf}.

$$P_{uf} = \frac{M_u \times 12 \text{ in. / ft}}{d}$$

$$= \frac{(250 \text{ ft-kips})(12 \text{ in. / ft})}{17.99 \text{ in.}}$$

$$= 167 \text{ kips}$$

Determine tension flange-plate dimensions.

From Figure 8-42, assume a shelf dimension of $\frac{5}{8}$-in. on both sides of the plate. The plate width, then, is 7.495 in. $- 2(\frac{5}{8}$-in.$) = 6.245$. Try a 1 in.×6¼-in. flange plate.

Check tension yielding of the flange plate:

$$\phi R_n = \phi F_y A_g$$
$$= 0.9 \times 36 \text{ ksi} \times 6\frac{1}{4}\text{-in.} \times 1 \text{ in.}$$
$$= 202.5 \text{ kips} \quad \textbf{o.k.}$$

Determine required weld size and length for fillet welds to beam flange. Try a $\frac{5}{16}$-in. fillet weld. The minimum length of weld l_{min} is:

$$l_{min} = \frac{P_{uf}}{2 \times 1.392(D)}$$

$$= \frac{167 \text{ kips}}{2 \times 1.392 \text{ (5 sixteenths)}}$$

$$= 12.0 \text{ in.}$$

Use 3 in. of weld along each side and 6¼-in. of weld along the end of the flange plate.

Determine required weld size for fillet welds to supporting column flange:

$$D_{min} = \frac{P_{uf}}{2 \times 1.392(l)}$$

$$= \frac{167 \text{ kips}}{2 \times 1.392(6\frac{1}{4}\text{-in.})}$$

$$= 9.60 \rightarrow 10 \text{ sixteenths}$$

Use ⅝-in. fillet welds.

Since these fillet welds are large, groove welds may be more economical.

Design the compression flange plate and connection:

The compression flange plate should have approximately the same area as the tension flange plate (6.25 in.²). Assume a shelf dimension of ⅝-in. The plate width, then, is 7.495 in. + 2(⅝-in.) = 8.745. To approximately balance the flange-plate areas, try a ¾-in.×8¾-in. compression flange plate.

Check design compressive strength of flange plate assuming K = 0.65 and l = 2 in. (1½-in. edge distance plus ½-in. setback).

$$\frac{Kl}{r} = \frac{0.65 \text{ (2 in.)}}{\sqrt{\dfrac{(8\frac{3}{4}\text{-in.}) (\frac{3}{4}\text{-in.})^3 / 12}{(8\frac{3}{4}\text{-in.}) (\frac{3}{4}\text{-in.})}}}$$

$$= 6.00$$

From LRFD Specification Table 3-36 with $\frac{Kl}{r} = 6.00$,

$$\phi_c F_{cr} = 30.54 \text{ ksi}$$

and the design compressive strength of the flange plate is
$$\phi R_n = \phi_c F_{cr} A$$
$$= (30.54 \text{ ksi})(8\frac{3}{4}\text{-in.} \times \frac{3}{4}\text{-in.})$$
$$= 200 \text{ kips} > 167 \text{ kips} \quad \textbf{o.k.}$$

Determine required weld size and length for fillet welds to beam flange.

As before for the tension flange plate, use a $\frac{5}{16}$-in. fillet weld and six inches of weld along each side of the beam flange.

Determine required weld size for fillet welds to supporting column flange.

As before for the tension flange plate, use $\frac{5}{8}$-in. fillet welds.

Check the column section for stiffening requirements; refer to Example 10-6.

Directly Welded Flange Connections

As illustrated in Figure 10-5, a directly welded flange FR moment connection consists of a shear connection and complete-joint-penetration groove welds which directly connect the top and bottom flanges of the supported beam to the supporting column. Note, in Figure 10-5b, the stiffener extends beyond the toe of the column flange to eliminate the effects of triaxial stresses.

The plastic moment of the supported beam ϕM_p can be developed with sufficient inelastic rotation and deformation capacity through such a connection. This apparent increase in beam strength above the prediction of elastic theory occurs because of strain hardening in the flanges. See Huang, et al. (1973), Krawinkler and Popov (1982), and Beedle, et al. (1973).

Design Checks

The design strengths of the bolts and/or welds and connected elements must be determined in accordance with the LRFD Specification; the applicable limit states are discussed in Part 8. The strength of the supporting column (and thus the need for stiffening) must be checked; refer to "Column Stiffening" in this Part.

Example 10-3

Given: Design a directly welded flange FR moment connection for a W18×50 beam to W14×99 column-flange connection. For structural members, assume $F_y = 50$ ksi and $F_u = 65$ ksi; for connecting material, assume $F_y = 36$ ksi and $F_u = 50$ ksi. Use 70 ksi electrodes and ASTM A325-N bolts.

$R_u = 45.0$ kips
$M_u = 250$ kips

W18×50

| $d = 17.99$ in. | $b_f = 7.495$ in. | $Z_x = 101$ in.^{3} |
| $t_w = 0.355$ in. | $t_f = 0.570$ in. | |

W14×99

| $d = 14.16$ in. | $b_f = 14.565$ in. | $k = 1\frac{7}{16}$-in. |
| $t_w = 0.485$ in. | $t_f = 0.780$ in. | $T = 11\frac{1}{4}$-in. |

Solution: *Check beam design flexural strength.*

From Example 10-2, the beam design flexural strength is   **o.k.**

Design the single-plate connection.

From Example 10-1, three $\frac{7}{8}$-in. diameter A325-N bolts, $\frac{5}{16}$-in. thick single plate with two $\frac{5}{16}$-in. fillet welds will be adequate.

A complete-joint-penetration groove weld will transfer the entire flange force in tension and compression.

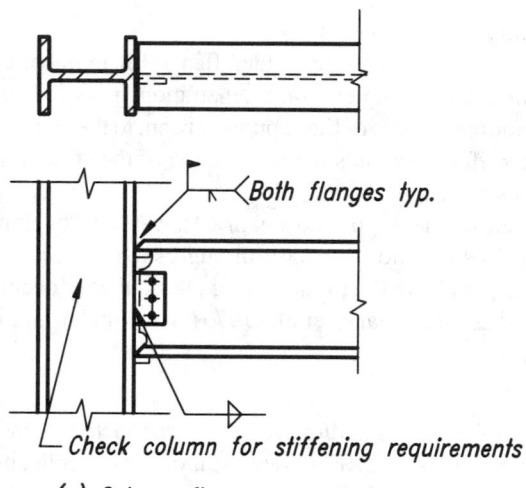

(a) Column flange support

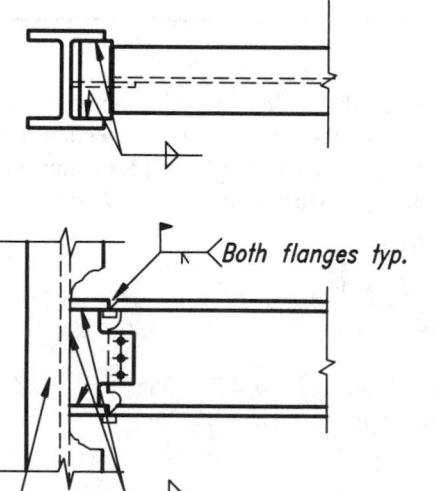

(b) Column web support

Fig. 10-5. Directly welded flange FR connections.

Check the column flange section for stiffening requirements; refer to Example 10-6.

Extended End-Plate Connections

Extended end-plate connections may be used only in statically loaded applications (buildings in seismic zone 1 and unimportant buildings in seismic zone 2 are considered statically loaded) because adequate research has not been conducted on their low-cycle fatigue strength. Wind, snow, and temperature loads are considered static loads.

As illustrated in Figure 10-6, an extended end-plate connection consists of a plate of length greater than the beam depth, perpendicular to the longitudinal axis of the supported beam. The end-plate is always welded to the web and flanges of the supported beam on each side and bolted to the supporting member with fully tensioned high-strength bolts.

As illustrated in Figure 10-7, extended end-plate connections are classified by the number of bolts at the tension flange and may be used with or without end-plate stiffeners. The four-bolt unstiffened extended end-plate connection of Figure 10-7a is generally limited by bolt strength to use with less than one-half of the available beam sections. The strength of this connection can be increased by increasing the number of bolts per row to four, as shown in Figure 10-7b. Note that the four-bolt-wide unstiffened case requires a wide supporting column flange. An alternative is the eight-bolt stiffened extended end-plate connection shown in Figure 10-7c.

Design assumptions and basic procedures for the four-bolt unstiffened and eight-bolt stiffened configurations follow. For the design procedure for four-bolt-wide unstiffened extended end-plate connections, or for a more detailed discussion of the aforementioned design procedures, refer to the AISC Design Guide *Extended End-Plate Moment Connections* (Murray, 1990).

Design Checks

The design strengths of the bolts and/or welds and connected elements must be determined in accordance with the LRFD Specification; the applicable limit states are discussed in Part 8. The strength of the supporting column (and thus the need for stiffening) must be checked; refer to "Column Stiffening" in this Part.

When fully-tensioned bearing bolts (N or X) are used, they must be designed using the shear-tension interaction equation of LRFD Specification Table J3.5. If bolts are to be slip-critical, all bolts may be designed for shear only and the shear-tension interaction equation may be ignored. From RCSC Specification Commentary Section C5, "Connections of the type...in which some of the bolts lose a part of their clamping force due to applied tension suffer no overall loss of frictional resistance. The bolt tension produced by the moment is coupled with a compensating compressive force on the other side of the axis in bending." Thus, the net clamping force is maintained in the connection.

Shop and Field Practices

This type of connection requires extra care in shop fabrication and field erection. The fit-up of extended end-plate connections is sensitive to the column flanges and may be affected by column flange-to-web squareness, beam camber, or squareness of the beam end. The beam is frequently fabricated short to accommodate the column overrun tolerances with shims furnished to fill any gaps which might result.

Design Assumptions

Several assumptions have been made in the design procedures which follow for four-bolt unstiffened and eight-bolt stiffened extended end-plate connections. These assumptions are as follows:

1. Fully-tensioned ASTM A325 or A490 high-strength bolts in diameters not greater than 1½-in. must be used, except that ASTM A490 bolts should not be used in the eight-bolt stiffened configuration.

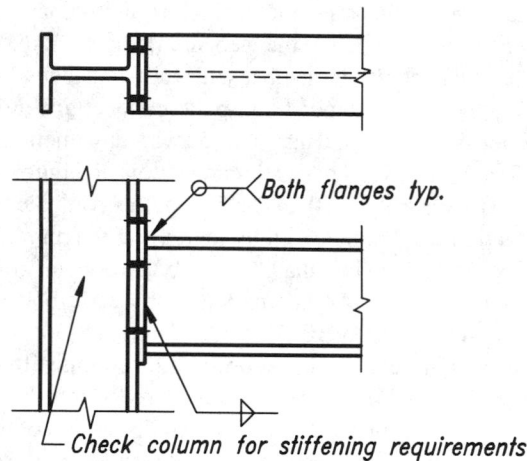

Both flanges typ.

Check column for stiffening requirements

(a) Column flange support

Accessibility may limit weld length

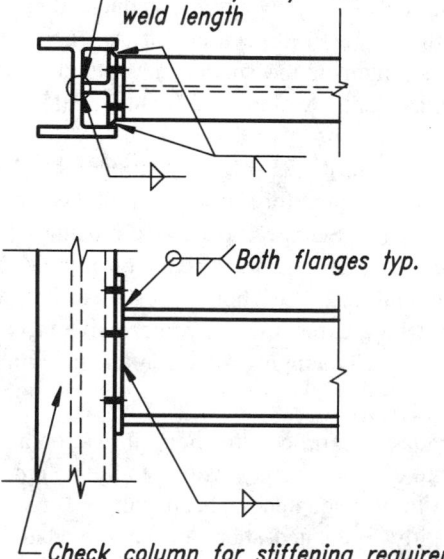

Both flanges typ.

Check column for stiffening requirements

(b) Column web support

Fig. 10-6. Extended end-plate FR connections.

2. End-plate material should preferably be ASTM A36.
3. Only static loading is permitted (wind, snow, and temperature loads are considered static loads).
4. The recommended minimum distance from the face of the beam flange to the nearest bolt centerline is the bolt diameter d_b plus ½-in. Note that, although the smallest possible distance will generally result in the most economical connection, many fabricators prefer to use a standard dimension, usually two inches, which is adequate for all bolt diameters.
5. The end-plate width which is effective in resisting the applied moment is not greater than the beam flange width b_f plus 1 in.
6. The gage of the tension bolts (horizontal distance between vertical bolt lines) should not exceed the beam tension flange width.
7. When the applied moment is less than the design flexural strength of the beam, the bolts and end plate may be designed for the applied moment only. However, beam-web-to-end-plate welds in the vicinity of the tension bolts should be designed to develop 60 percent of the minimum specified yield strength of the beam web. This is recommended even if the full design flexural strength of the beam is not required for frame strength.
8. Only the web-to-end-plate weld between the mid-depth of the beam and the inside face of the beam compression flange or the weld between the inner row of tension bolts plus $2d_b$ and the inside face of the beam compression flange, whichever is smaller, is considered effective in resisting the beam end shear.

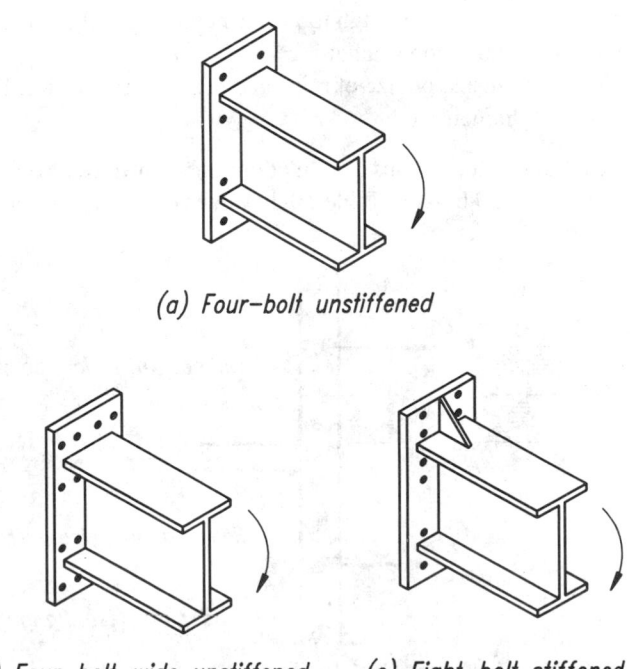

(a) Four-bolt unstiffened

(b) Four-bolt-wide unstiffened   (c) Eight-bolt stiffened

Fig. 10-7. Configurations of extended end-plate FR connection.

Four-Bolt Unstiffened Extended End-Plate Design

The following design procedure is based on Krishnamurthy (1978), Hendrick and Murray (1984), and Curtis and Murray (1989). In Krishnamurthy's design procedure, prying action forces are considered to be negligible and the tensile flange force is distributed equally among the four tension bolts. Possible local yielding of the tension flange and tensile area of the web is neglected.

The required end-plate thickness is determined using the tee-stub analogy, as illustrated in Figure 10-8, with the effective critical moment in the end plate given by

$$M_{eu} = \frac{\alpha_m P_{uf} p_e}{4}$$

where

P_{uf} = factored beam flange force, kips

$\alpha_m = C_a C_b (A_f / A_w)^{1/3} (p_e / d_b)^{1/4}$

C_a = constant from Table 10-1

$C_b = (b_f / b_p)^{1/2}$

b_f = beam flange width, in.

b_p = effective end-plate width, in., not to exceed $b_f + 1$ in.

A_f = area of beam tension flange, in.^{2}

A_w = area of beam web, clear of flanges, in.^{2}

p_e = effective pitch, in.

     = $p_f - (d_b / 4) - w_t$

p_f = distance from centerline of bolt to nearer surface of the tension flange, in. Generally, $d_b + \frac{1}{2}$-in. is enough to provide entering and tightening clearance; two inches is a common standard.

w_t = fillet weld throat size or size of reinforcement for groove weld, in.

d_b = nominal bolt diameter, in.

Values of C_a are tabulated for various combinations of beam and end-plate material grades and ASTM A325 or A490 bolts in Table 10-1. Values of A_f / A_w for the W-shapes listed

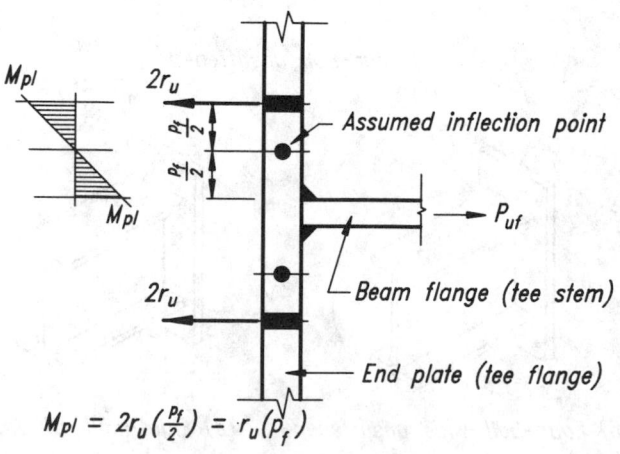

Fig. 10-8. Tee-stub analogy for end-plate moment.

Table 10-1.
Values of C_a for Extended End-Plate Design

ASTM Bolt Desig.	Beam F_y, ksi	End-Plate F_y, ksi	C_a
A325	36	36	1.36
		50	1.23
	50	36	1.45
		50	1.31
A490	36	36	1.38
		50	1.25
	50	36	1.48
		50	1.33

Notes:

$$C_a = 1.2\left[1.29\left(\frac{F_{avg}}{F_{bu}}\right)^{2/5}\left(\frac{F_{bt}}{F_b}\right)^{1/2}\right]$$

$F_{bu} = 93.0$ ksi for A325 bolts; 115 ksi for A490 bolts.
$F_{bt} = 44.0$ ksi for A325 bolts; 54.0 ksi for A490 bolts.

$$F_{avg} = \frac{F_{y\,beam} + F_{y\,end\,plate}}{2}$$

$F_b = 0.75 F_{y\,end\,plate}$

in Part 1 are found in Table 10-2. The required end-plate thickness $t_{p_{req}}$ is then determined as

$$t_{p_{req}} = \sqrt{\frac{4M_{eu}}{\phi F_y b_p}}$$

where F_y is the specified minimum yield stress of the end-plate material, ksi, and $\phi = 0.90$.

The strength of the column should then be investigated for stiffening requirements; refer to "Column Stiffening" in this Part. Note that, since column web stiffeners add considerable fabrication expense and may interfere with weak-axis framing, it is often advantageous to eliminate the need for stiffening. The designer should therefore consider increasing the column size to a section with adequate strength. Alternatively, if the column were inadequate due to local flange bending strength, increasing the tension-bolt pitch p_f or switching to an eight-bolt stiffened extended end-plate configuration may increase the length of column flange effective in flange bending and thereby eliminate the need for stiffening.

Example 10-4

Given: Design a four-bolt unstiffened extended end-plate FR moment connection for a W18×50 beam to W14×99 column-flange connection. For structural members, assume $F_y = 50$ ksi and $F_u = 65$ ksi; for connecting material, assume $F_y = 36$ ksi and $F_u = 58$ ksi. Use ASTM A325-SC bolts (Class A surfaces) and 70 ksi electrodes.

Table 10-2.
Values of A_f / A_w for Extended End-Plate Design

Section	A_f/A_w	Section	A_f/A_w	Section	A_f/A_w	Section	A_f/A_w
W44×335	0.684	W33×354	0.925	W21×166	1.140	W12×87	1.748
×290	0.711	×318	0.926	×147	1.011	×79	1.732
×262	0.700	×291	0.913	×132	1.002	×72	1.720
×230	0.669	×263	0.909	×122	1.003	×65	1.706
		×241	0.853	×111	0.994	×58	1.631
W40×593	0.824	×221	0.829	×101	0.995	×53	1.527
×503	0.806	×201	0.807	× 93	0.683	×50	1.281
×431	0.782	×169	0.667	× 83	0.686	×45	1.266
×372	0.777	×152	0.612	× 73	0.683	×40	1.281
×321	0.771	×141	0.583	× 68	0.667	×35	0.992
×297	0.768	×130	0.541	× 62	0.641	×30	0.963
×277	0.822	×118	0.492	× 57	0.532	×26	0.936
×249	0.816			× 50	0.465	×22	0.575
×215	0.809	W30×235	0.961	× 44	0.423	×19	0.520
×199	0.706	×211	0.905			×16	0.419
×174	0.550	×191	0.887	W18×143	1.204	×14	0.390
		×173	0.861	×130	1.186		
W40×466	0.611	×148	0.672	×119	1.082	W10×60	1.842
×392	0.600	×132	0.606	×106	1.059	×54	1.882
×331	0.582	×124	0.590	× 97	1.076	×49	1.859
×278	0.581	×116	0.558	× 86	1.056	×45	1.603
×264	0.588	×108	0.516	× 76	1.048	×39	1.516
×235	0.617	× 99	0.476	× 71	0.741	×33	1.348
×211	0.610			× 65	0.751	×30	1.045
×183	0.607	W27×217	1.003	× 60	0.751	×26	1.033
×167	0.510	×194	0.986	× 55	0.722	×22	0.913
×149	0.426	×178	0.909	× 50	0.714	×19	0.672
		×161	0.902	× 46	0.604	×17	0.583
W36×848	0.976	×146	0.885	× 40	0.595	×15	0.497
×798	0.971	×129	0.710	× 35	0.504	×12	0.463
×650	0.946	×114	0.646				
×527	0.932	×102	0.635	W16×100	1.170	W8×35	1.796
×439	0.912	× 94	0.597	× 89	1.152	×31	1.711
×393	0.909	× 84	0.545	× 77	1.146	×28	1.495
×359	0.899			× 67	1.149	×24	1.487
×328	0.903	W24×176	1.021	× 57	0.789	×21	1.127
×300	0.887	×162	0.994	× 50	0.781	×18	1.007
×280	0.882	×146	0.959	× 45	0.768	×15	0.690
×260	0.850	×131	0.904	× 40	0.772	×13	0.593
×245	0.835	×117	0.877	× 36	0.679	×10	0.635
×230	0.818	×104	0.848	× 31	0.589		
×256	0.648	×103	0.711	× 26	0.506	W6×25	1.580
×232	0.644	× 94	0.683			×20	1.545
×210	0.588	× 84	0.655	W14×120	1.855	×15	1.238
×194	0.587	× 76	0.616	×109	1.899	×16	1.148
×182	0.579	× 68	0.560	× 99	1.859	×12	0.890
×170	0.573	× 62	0.428	× 90	1.860	× 9	0.911
×160	0.554	× 55	0.397	× 82	1.348		
×150	0.530			× 74	1.394	W5×19	1.867
×135	0.463			× 68	1.382	×16	1.748
				× 61	1.364		
				× 53	1.141	W4×13	1.442
				× 48	1.115		
				× 43	1.103		
				× 38	0.861		
				× 34	0.824		
				× 30	0.734		
				× 26	0.633		
				× 22	0.557		

R_u = 45.0 kips
M_u = 250 ft-kips

W18×50

d = 17.99 in.	b_f = 7.495 in.	Z_x = 101 in.[3]
t_w = 0.355 in.	t_f = 0.570 in.	A_f / A_w = 0.714

W14×99

d = 14.16 in.	b_f = 14.565 in.	k = $1\frac{7}{16}$-in.
k_1 = $\frac{7}{8}$-in.	t_w = 0.485 in.	t_f = 0.780 in.
T = $11\frac{1}{4}$-in.		

Solution: *Check beam design flexural strength.*

From Example 10-2, the beam design flexural strength is   **o.k.**

Design the bolts (a minimum of four bolts is required at the tension flange; a minimum of two bolts is required at the compression flange).

Calculate the flange force P_{uf}.

$$P_{uf} = \frac{M_u \times 12 \text{ in.} / \text{ft}}{(d - t_f)}$$

$$= \frac{(250 \text{ ft-kips})(12 \text{ in.} / \text{ft})}{17.99 \text{ in.} - 0.570 \text{ in.}}$$

$$= 172 \text{ kips}$$

Determine number of 1-in. diameter A325-SC bolts required for tension (Note that fully tensioned bearing-type bolts would also be acceptable). From Table 8-15

$$n_{\min} = \frac{P_{uf}}{\phi r_n}$$

$$= \frac{172 \text{ kips}}{53.0 \text{ kips} / \text{bolt}}$$

$$= 3.25 \rightarrow 4 \text{ bolts}$$

Determine number of 1-in. diameter A325-SC bolts required for slip resistance. From Table 8-17

$$n_{\min} = \frac{R_u}{\phi r_n}$$

$$= \frac{45 \text{ kips}}{19.0 \text{ kips} / \text{bolt}}$$

$$= 2.37 \rightarrow 3 \text{ bolts}$$

Minimum of four bolts at tension flange and two bolts at compression flange controls. Try six 1-in. diameter A325-SC bolts (N for bolt shear check).

Check bolt shear:

From Table 8-11 for six 1 in. diameter A325-N bolts:

ϕR_n = 6 × 28.3 kips / bolt
        = 170 kips > 45.0 kips   **o.k.**

Try ¾-in. thick end plate.

Check material bearing.

Assuming for the end plate $L_e \geq 1.5d$ and $s \geq 3d$, the thickness of the end plate is more critical than the column flange. From Table 8-13, with the conservative assumption that only the bolts at the compression flange are in bearing,

ϕR_n = 2 bolts × 104 kips / bolt / in. × ¾-in.
        = 156 kips > 45.0 kips   **o.k.**

Design the end plate and its connection to beam.

Calculate the effective end-plate width b_p.

Try an end plate with L_e = 1½-in., g = 5½-in., and $p_f = d_b + ½$-in. = 1½-in.

$b_p = 2L_e + g$
     = 2(1½-in.) + 5½-in.
     = 8½-in.

Since $b_p \approx b_f + 1$ in., the full width of the end plate may be considered effective.

Determine the required end-plate thickness.

$$M_{eu} = \alpha_m P_{uf} \frac{p_e}{4}$$

$$\alpha_m = C_a C_b \left(\frac{A_f}{A_w}\right)^{1/3} \left(\frac{p_e}{d_b}\right)^{1/4}$$

C_a = 1.45 from Table 10-1.

$$C_b = \sqrt{\frac{b_f}{b_p}} = \sqrt{\frac{7.495}{8½}} = 0.939$$

$$p_e = p_f - \frac{d_b}{4} - w_t \text{ (assuming ½-in. fillet weld)}$$

$$= 1½\text{-in.} - \frac{1 \text{ in.}}{4} - ½\text{-in.}$$

$$= 0.75 \text{ in.}$$

$$\alpha_m = 1.45 \times 0.939 \times (0.714)^{\frac{1}{3}} \left(\frac{0.75 \text{ in.}}{1 \text{ in.}} \right)^{\frac{1}{4}}$$

$$= 1.13$$

$$M_{eu} = 1.13 \times 172 \text{ kips} \times \frac{0.75 \text{ in.}}{4}$$

$$= 36.4 \text{ in.-kips}$$

$$t_{p \ min} = \sqrt{\frac{4 M_{eu}}{\phi F_y b_p}}$$

$$= \sqrt{\frac{4(36.4 \text{ in.-kips})}{0.9 \times 36 \text{ ksi} \times 8\frac{1}{2}\text{-in.}}}$$

$$= 0.727 \rightarrow \frac{3}{4}\text{-in.}$$

Try a ¾-in.×8½-in. end plate.

Check shear yielding of the end plate.

From LRFD Specification Section J5.3:

$$\phi R_n = 2 \times \phi(0.60 F_y A_g)$$
$$= 2 \times 0.9(0.6 \times 36 \text{ ksi} \times 8\frac{1}{2}\text{-in.} \times \frac{3}{4}\text{-in.})$$
$$= 248 \text{ kips} > 172 \text{ kips} \quad \textbf{o.k.}$$

Determine required fillet weld for beam-web-to-end-plate connection.

From LRFD Specification Table J2.4, the minimum size is ⁵⁄₁₆-in. Determine size required to develop web flexural strength near tension bolts:

$$D_{min} = \frac{0.9 F_y t_w}{2 \times 1.392}$$

$$= \frac{0.9 \times 36 \text{ ksi} \times 0.355 \text{ in.}}{2 \times 1.392}$$

$$= 4.13 \rightarrow 5 \text{ sixteenths}$$

Use ⁵⁄₁₆-in. fillet weld on both sides of the beam web from the inside face of the beam flange to the centerline of the inside bolt holes plus two bolt diameters.

Determine size required for the factored shear R_u. R_u is resisted by weld between the mid-depth of the beam and the inside face of the compression flange or between the inner row of tension bolts plus two bolt diameters, whichever is smaller. By inspection the former governs for this example.

$$l = \frac{d}{2} - t_f$$

$$= \frac{17.99 \text{ in.}}{2} - 0.570 \text{ in.}$$

$$= 8.43 \text{ in.}$$

$$D_{min} = \frac{R_u}{2 \times 1.392l}$$

$$= \frac{45.0 \text{ kips}}{2 \times 1.392(8.43 \text{ in.})}$$

$$= 1.92 \rightarrow 5 \text{ sixteenths (minimum size)}$$

Use $\frac{5}{16}$-in. fillet weld on both sides of the beam web below the tension-bolt region.

Determine required fillet weld size for beam flange to end-plate connection.

$$l = 2(b_f + t_f) - t_w$$

$$= 2(7.495 \text{ in.} + 0.570 \text{ in.}) - 0.355 \text{ in.}$$

$$= 15.8 \text{ in.}$$

$$D_{min} = \frac{P_{uf}}{1.392l}$$

$$D_{min} = \frac{172 \text{ kips}}{1.392 \times 15.8 \text{ in.}}$$

$$= 7.82 \rightarrow 8 \text{ sixteenths}$$

Use $\frac{1}{2}$-in. fillet welds at beam tension flange. Welds at compression flange may be $\frac{5}{16}$-in. fillet welds (minimum size from LRFD Specification Table J2.4)

Check the column section for stiffening requirements; refer to Example 10-6.

Eight-Bolt Stiffened Extended End-Plate Design

The following design procedure is based on Murray and Kukreti (1988), Hendrick and Murray (1984), and Curtis and Murray (1989). Murray and Kukreti (1988) present two methods for determining the required end-plate thickness and bolt diameter; both methods are limited to the use of ASTM A36 end-plate material with ASTM A325 bolts and include the effects of prying action.

The first method was developed from a regression analysis of finite-element-analysis data including second-order geometric effects and inelastic plate and bolt material properties. The resulting equations are elaborate and beyond the scope of this Manual; refer to the AISC Design Guide *Extended End-Plate Moment Connections* (Murray, 1990) and Murray and Kukreti (1988).

The second method offers a simplified approach which was developed with the first method by generating end-plate thicknesses and bolt diameters for all W-shapes listed in Part 1 assuming ASTM A36 steel and beam sections at various moment levels. The number of bolts effective in resisting the tensile flange force was then determined for each connection; a conservative lower bound of six effective bolts was established. Next, it was assumed that the plate thickness could be established from tee-stub analogy bending, as illustrated in Figure 10-8, where

$$M_{eu} = 2r_u \left(\frac{p_{eff}}{2} \right) = r_u \, p_{eff}$$

where r_u is the force per bolt based on six effective bolts, kips

From the generated designs, it was determined the effective pitch p_{eff} is

$$p_{eff} = \frac{p_f}{4.17} \sqrt{g^2 + p_f^2}$$

The required end-plate thickness is then determined from

$$t_{p_{req}} = \sqrt{\frac{4M_{eu}}{\phi F_y \, b_p}}$$

where $\phi = 0.9$

In addition to the design assumptions listed previously, the following limitations must be met for the eight-bolt stiffened configuration:

1. The supported beam must be a hot-rolled W-shape listed in Part 1.
2. The vertical pitch p_f from the face of the beam tension flange to the centerline of the first row of bolts must not exceed 2½-in. The recommended minimum pitch is d_b plus ½-in.; entering and tightening clearance may require a larger pitch.
3. The vertical spacing between bolt rows p_b must not exceed $3d_b$.
4. The horizontal gage g must be between 5½-in. and 7½-in.
5. Bolt diameter d_b must be not less than ¾-in. nor greater than 1½-in.

The strength of the column should then be investigated for stiffening requirements. The recommendations of Hendrick and Murray (1984) can be used to check column web strengths in local yielding, buckling, and panel zone shear; refer to "Column Stiffening" in this Part.

Unless the column flange is 1.5 to 2 times thicker than the end plate, transverse stiffening is required. If effective-flange-length effects are neglected, the behavior of the column flange is identical to that of the end plate. Therefore, the column flange must be at least as thick as the end plate and the transverse stiffeners must be at least as thick as the beam flange. Additionally, the weld connecting the transverse stiffener to the flange must be sufficient to develop the strength of the full thickness of the stiffener plate.

A column flange which is 1.5 to 2 times thicker than the end plate may not require transverse stiffening. From Curtis and Murray (1989), an unstiffened flange may be evaluated according to the flange bending equation presented for extended end-plate connections in "Column Stiffening" in this Part with $b_s = 3.5 \, p_b + c$. Because this reference considered only ASTM A36 steel, it is recommended that column material with greater yield strength be checked as if ASTM A36 material were used.

Note that, since column web stiffeners add considerable fabrication expense and may interfere with weak-axis framing, it is often advantageous to eliminate the need for stiffening. The designer should therefore consider increasing the column size to a section with adequate strength.

Example 10-5

Given: Design an eight-bolt stiffened extended end-plate FR moment connection for a W33×118 beam to W14×311 column-flange connection. For structural members, assume $F_y = 50$ ksi and $F_u = 65$ ksi; for connecting material, assume $F_y = 36$ ksi and $F_u = 58$ ksi. Use ASTM A325-SC bolts (Class A surfaces) and 70 ksi electrodes.

R_u = 135 kips
M_u = 1,050 ft-kips

W33×118

$d = 32.86$ in.	$b_f = 11.48$ in.	$Z_x = 415$ in.^{3}
$t_w = 0.550$ in.	$t_f = 0.740$ in.	

W14×311

$d = 17.12$ in.	$b_f = 16.230$ in.	$k = 2^{15}/_{16}$-in.
$k_1 = 1^5/_{16}$-in.	$t_w = 1.410$ in.	$t_f = 2.260$ in.
$T = 11^1/_4$-in.		

Solution: *Check beam design flexural strength:*

$$Z_{req} = \frac{M_u \times 12 \text{ in.} / \text{ft}}{0.9 F_y}$$

$$= \frac{(1,050 \text{ ft-kips})(12 \text{ in.} / \text{ft})}{0.9(50 \text{ ksi})}$$

$$= 280 \text{ in.}^3$$
$$Z_x = 415 \text{ in.}^3$$

Since $Z_x > Z_{req}$, the beam design flexural strength is   **o.k.**

Design the bolts (a minimum of eight bolts is required at the tension flange; a minimum of two bolts is required at the compression flange).

Calculate the flange force P_{uf}:

$$P_{uf} = \frac{M_u \times 12 \text{ in.} / \text{ft}}{(d - t_p)}$$

$$= \frac{(1,050 \text{ ft-kips})(12 \text{ in.} / \text{ft})}{(32.86 \text{ in.} - 0.740 \text{ in.})}$$

= 392 kips

Try eight $1\frac{1}{8}$-in. diameter A325-SC bolts (six effective).

$$\phi r_{n_{req}} = \frac{P_{uf}}{6 \text{ bolts}}$$

$$= \frac{392 \text{ kips}}{6 \text{ bolts}}$$

$$= 65.3 \text{ kips/bolt} < 67.1 \text{ kips/bolt} \quad \textbf{o.k. for tension}$$

Check slip resistance with eight bolts at tension flange and two bolts at compression flange.

From Table 8-17:

ϕR_n = 10 bolts × 20.9 kips/bolt
  = 209 kips > 135 kips **o.k.**

Try eight $1\frac{1}{8}$-in. diameter A325-SC bolts (N for bolt shear check).

Check bolt shear.

From Table 8-11 for ten $1\frac{1}{8}$-in. diameter A325-N bolts:

ϕR_n = 10 bolts × 35.8 kips / bolt
  = 358 kips > 135 kips **o.k.**

Try PL $1\frac{1}{4}$.

Check material bearing.

From Table 8-13, the design bearing strength of one bolt is

ϕr_n = 147 kips/bolt

Since this exceeds the design shear strength of the bolts, bearing is not critical.

Design the end plate and its connection to the beam.

Calculate the effective end-plate width b_p.

Try an end plate with $L_e = 1\frac{3}{4}$-in., g = 6 in., $p_f = d_b + \frac{1}{2}$-in. = $1\frac{5}{8}$-in., $p_b = 3d_b$, and stiffener thickness $t_s = \frac{5}{8}$-in. ($t_{w_{beam}}$ = 0.550 in.). Note that all of the specified limitations for this simplified method have been met.

$$b_p = 2L_e + g = 2(1\frac{3}{4}\text{-in.}) + 6 \text{ in.} = 9.5 \text{ in.}$$

This dimension is less than the flange width of the beam b_f. Thus, use a plate with $b_p = b_f + 1$ in. ≈ $12\frac{1}{2}$-in. This allows for runoff.

Determine the required end-plate thickness:

$$p_{eff} = \frac{p_f}{4.17} \sqrt{g^2 + p_f^2}$$

$$= \frac{1\frac{5}{8}\text{-in.}}{4.17}\sqrt{(6 \text{ in.})^2 + (1\frac{5}{8}\text{-in.})^2}$$

$$= 2.42 \text{ in.}$$

$$M_{eu} = r_u \, p_{eff}$$

$$= \frac{392 \text{ kips}}{6 \text{ bolts effective}} \times 2.42 \text{ in.}$$

$$= 158 \text{ in.-kips}$$

$$t_{p_{req}} = \sqrt{\frac{4M_{eu}}{\phi F_y b_p}}$$

$$= \sqrt{\frac{4 \times 158 \text{ in.-kips}}{0.9 \times 36 \text{ ksi} \times 12\frac{1}{2}\text{-in.}}}$$

$$= 1.25 \rightarrow 1\frac{1}{4}\text{-in.}$$

Try PL 1¼-in.×12½-in.

Check shear yielding of the end plate.

From LRFD Specification Section J5.3.

$$\phi R_n = 2 \times \phi(0.60 F_y A_g)$$
$$= 2 \times 0.9(0.60 \times 36 \text{ ksi} \times 12\frac{1}{2}\text{-in.} \times 1\frac{1}{4}\text{-in.})$$
$$= 607.5 \text{ kips} > 392 \text{ kips} \quad \textbf{o.k.}$$

Determine required fillet weld size for beam web to end-plate connection.

From LRFD Specification Table J2.4, the minimum size is ⁵⁄₁₆-in.

Determine size required to develop web flexural strength near tension bolts:

$$D_{\min} = \frac{0.9 F_y t_w}{2 \times 1.392}$$

$$= \frac{0.9 \times 50 \text{ ksi} \times 0.550 \text{ in.}}{2 \times 1.392}$$

$$= 8.9 \rightarrow 9 \text{ sixteenths}$$

Use ⁹⁄₁₆-in. fillet welds on both sides of the beam web from the inside face of the beam flange to the centerline of the inside bolt holes plus two bolt diameters.

Determine size required for the factored shear R_u.

R_u is resisted by weld between the mid-depth of the beam and the inside face of the compression flange or between the inner row of tension bolts plus two bolt diameters, whichever is smaller. By inspection, the former governs for this example.

$$l = \frac{d}{2} - t_f$$

$$= \frac{32.86 \text{ in.}}{2} - 0.740 \text{ in.}$$

$$= 15.7 \text{ in.}$$

$$D_{min} = \frac{R_u}{2 \times 1.392l}$$

$$= \frac{135 \text{ kips}}{2 \times 1.392(15.7 \text{ in.})}$$

$$= 3.09 \rightarrow 5 \text{ sixteenths (minimum size)}$$

Use ⁵⁄₁₆-in. fillet welds on both sides of the beam web below the tension-bolt region.

Determine required weld for beam flange to end-plate connection.

By inspection, fillet welds at the tension flange will be impractical. Use a complete-joint-penetration groove weld at the tension flange. Welds at the compression flange may be ⁵⁄₁₆-in. fillet welds (minimum size from LRFD Specification Table J2.4).

Check the column section for stiffening requirements; refer to Example 10-7.

COLUMN STIFFENING AT FR AND PR MOMENT CONNECTIONS

As illustrated in Figure 10-9, FR and PR moment connections produce double concentrated forces, one tensile and one compressive, forming a couple on the same side of the supporting column. From LRFD Specification Section K1, the following limit states determine if the column section is adequate to carry these concentrated forces.

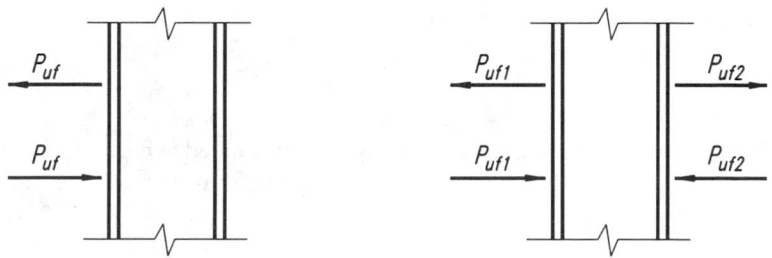

(a) Double concentrated forces   *(b) A pair of double concentrated forces*

Figure 10-9. Illustration of FR and PR moment connection flange force terminology.

At the location of the tensile component of the double concentrated force, the limit states of local flange bending and local web yielding must be checked. At the location of the compressive component of the double concentrated force, the limit states of local web yielding and web crippling must be checked. If FR and/or PR moment connections are made to both flanges of a column, the limit state of compression buckling of the web must be checked at the location of the compressive components of the pair of double concentrated forces. Finally, the limit state of panel zone web shear must be checked.

Following are discussions of: (1) economical considerations; (2) the aforementioned limit states and their applicability at intermediate column locations and column end locations with flange-plated, directly welded flange, and extended end-plate FR moment connections; and, (3) design of transverse stiffeners, doubler plates, and diagonal stiffeners.

Economical Considerations

If the design strength of the investigated column is inadequate, the designer has two options. First, the designer should consider selecting a heavier column section which will eliminate the need for stiffening. Although this will increase the material cost of the column, it may well be that this heavier section will provide a more economical solution due to the reduction in labor cost associated with the elimination of stiffening (Ricker, 1992 and Thornton, 1992). Alternatively, the designer may stiffen the original column section with transverse stiffeners and/or doubler plate(s) or diagonal stiffeners as provided in LRFD Specification Section K1.

Local Flange Bending

This requirement applies only to the tensile component of the double concentrated force created by the FR or PR moment connection. If the required strength P_{uf} exceeds the design strength ϕR_n, a pair of transverse stiffeners, one on each side of the column web, must be provided and must extend at least one-half the depth of the column web.

Intermediate Column Locations, Flange-Plated and Directly Welded Flange Connections
The tensile concentrated force causes bending distortions to occur in the column flange, as shown in Figure 10-10. Such deformation causes a concentration of stress in the area which is stiffened by the column web and creates a zone of possible fracture in the connecting weld. From LRFD Specification Section K1.2, the design strength of the column flange is ϕR_n, where $\phi = 0.90$ and

$$R_n = 6.25 t_f^2 F_{yf}$$

The design local flange bending strength is tabulated as P_{fb} for W and HP shapes in the Properties section of the Column Tables in Part 3 where,

$$\phi R_n = P_{fb}$$

Intermediate Column Locations, Extended End-Plate Connections
In bolted FR moment connections, flange bending must be limited to prevent yielding of the column flange in the tension region. The design strength of the column flange is ϕR_n, where $\phi = 0.90$ and

$$R_n = \left(\frac{b_s}{\alpha_m p_e} \right) t_f^2 F_{yf}$$

In the above equation,

b_s = 2.5 $(2p_f + t_{fb})$, in., for a four-bolt unstiffened extended end plate

   = $2p_f + t_{fb} + 3.5p_b$, in., for an eight-bolt stiffened extended end plate

p_b = vertical pitch of bolt group above and bolt group below tension flange, in.

$$\alpha_m = 1.36 \left(\frac{p_e}{d_b}\right)^{1/4} \text{ for a four-bolt unstiffened extended end plate}$$

$$= 1.13 \left(\frac{p_e}{d_b}\right)^{1/4} \text{ for an eight-bolt stiffened extended end plate}$$

$$p_e = \frac{g}{2} - \frac{d_b}{4} - k_1$$

Note that this equation was developed from research which considered only ASTM A36 steel. If columns with higher material yield strengths are used, it is recommended that F_{yf} be taken conservatively as 36 ksi in the calculation of the design strength of the column in local flange bending (Curtis and Murray, 1989).

Column-End Locations, Flange-Plated, Directly Welded Flange, and Extended End-Plate Connections

From LRFD Specification Commentary Section K1.2, the effective column flange length for local flange bending is $12t_f$ (Graham et al., 1959). Thus, it is assumed that yield lines form in the flange at $6t_f$ in each direction from the point of the applied concentrated force. To develop the fixed edge consistent with the assumptions of this model, an additional $4t_f$ (resulting in a total of $10t_f$) is required for the full flange bending strength given by LRFD Specification Equation K1-1. Thus, if the distance from the column end to the top

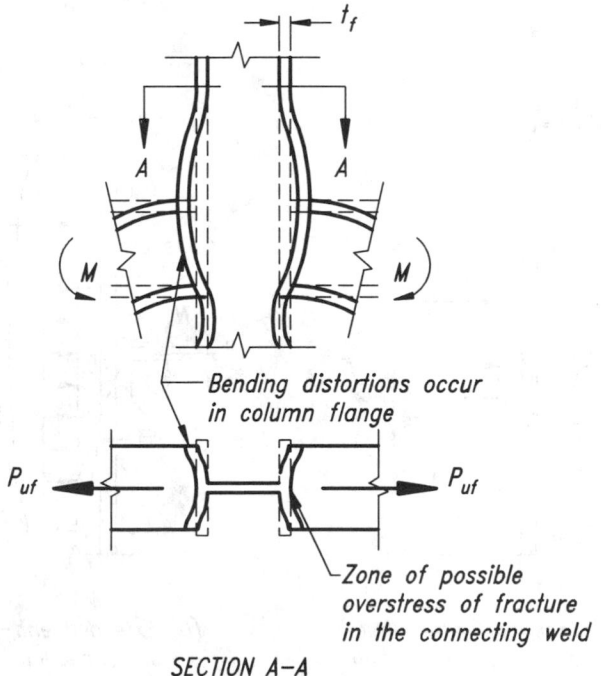

SECTION A-A

Fig. 10-10. Illustration of local flange bending.

of the connected beam tension flange or flange plate is less than $10t_f$, LRFD Specification Section K1.2 states that the flange bending strength at this column-end location must be reduced by 50 percent from the strength at an intermediate column location.

Local Web Yielding

This requirement applies to both the tensile and compressive components of the double concentrated force created by the FR or PR moment connection. If the required strength P_{uf} exceeds the design strength ϕR_n, either a pair of transverse stiffeners, one on each side of the column web, or a doubler plate must be provided and must extend at least one-half the depth of the column web.

Intermediate Column Locations, Flange-Plated and Directly Welded Flange Connections
From LRFD Specification Section K1.3, the design strength of the column web is ϕR_n, where $\phi = 1.0$ and

$$R_n = (5k + N)F_{yw}t_w$$

The derivation of this equation is illustrated in Figure 10-11a.

The design local flange bending strength is tabulated as P_{wo} and P_{wi} for W and HP shapes in the Properties section of the Column Tables in Part 3, where

$$P_{wi} = \phi F_{yw}t_w$$
$$P_{wo} = \phi 5 F_{yw}t_w k$$
$$\phi R_n = P_{wi} t_b + P_{wo}$$

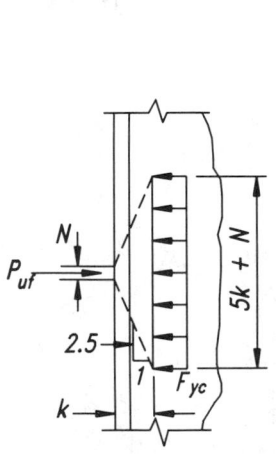

(a) Flange–plated or directly
welded flange connection

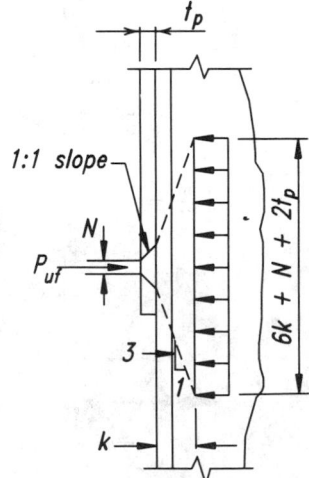

(b) Extended end–plate
connection

Fig. 10-11. Derivation of local web yielding.

Intermediate Column Locations, Extended End-Plate Connections
With minor modification of LRFD Specification Equation K1-2 to account for the effects of the end-plate thickness and fillet weld leg size or groove weld reinforcement leg size, the design strength of the column web is ϕR_n, where $\phi = 1.0$ and

$$R_n = (6k + N + 2t_p)F_{yw}t_w$$

where

   N = thickness of the beam flange delivering the concentrated force plus $2w$, in.
   t_p = end-plate thickness, in.
   w = leg size of fillet weld or groove weld reinforcement, in.

The derivation of this equation is illustrated in Figure 10-11b.

Column-End Locations, Flange-Plated and Directly Welded Flange Connections
From LRFD Specification Section K1.3, when the concentrated tensile or compressive force to be resisted is applied at a distance from the column end which is less than or equal to the depth of the column, the design strength of the column web is ϕR_n, where $\phi = 1.0$ and

$$R_n = (2.5k + N)F_{yw}t_w$$

Column-End Locations, Extended End-Plate Connections
With minor modification of LRFD Specification Equation K1-3 to account for the effects of the end-plate thickness and fillet weld leg size or groove weld reinforcement leg size, the design strength of the column web is ϕR_n, where $\phi = 1.0$ and

$$R_n = (3k + N + t_p)F_{yw}t_w$$

where

   N = thickness of the beam flange delivering the concentrated force plus $2w$, in.
   t_p = end-plate thickness, in.
   w = leg size of fillet weld or groove weld reinforcement, in.

Web Crippling
This requirement applies only to the compressive component of the double concentrated force created by the fully restrained connection. From LRFD Specification Commentary Section K1.4, for the rolled shapes listed in Part 1 with F_y not greater than 50 ksi, the web crippling limit state will never control the design in an FR or PR moment connection except to a W12×50 or W10×33 column; note that the less than 3 percent overstress for these two column shapes is considered negligible. Therefore, the limit state of web crippling is not included in the discussion of column stiffening.

Compression Buckling of the Web
This requirement applies only to the compressive components of a pair of double concentrated forces (see Figure 10-9b) created by two FR or PR moment connections as illustrated in Figure 10-12. If the required strength P_{uf} exceeds the design strength ϕR_n, either a single transverse stiffener, a pair of transverse stiffeners, one on each side of the column web, or a doubler plate must be provided and must extend the full depth of the column web.

Intermediate Column Locations, Flange-Plated, Directly Welded Flange, and Extended End-Plate Connections

From LRFD Specification Section K1.6, the design strength of the column web is ϕR_n, where $\phi = 0.9$ and

$$R_n = \frac{4,100 t_w^3 \sqrt{F_{yw}}}{d_c}$$

In the above equation, d_c is the column-web depth clear of fillets, in.

The design compression buckling strength of the web is tabulated as P_{wb} for W and HP shapes in the Properties section of the Column Tables in Part 3 where,

$$\phi R_n = P_{wb}$$

Column End Locations, Flange-Plated, Directly Welded Flange,
and Extended End-Plate Connections

In the absence of applicable research, if the distance from the column end to the location of the pair of compressive forces is less than one-half the depth of the column, LRFD Specification Section K1.6 states that the compression buckling strength of the unreinforced web at this column-end location is reduced by 50 percent from the strength at an intermediate column location. From LRFD Specification Section K1.9, when stiffeners are required, the length of the column web effective in resisting the pair of compressive forces applied at an intermediate column location is $25 t_w$ or $12.5 t_w$ on either side of the location of the compressive forces.

Panel Zone Web Shear

This requirement applies to the web of the column within the boundary of the column flanges and the tensile and compressive concentrated forces imposed by the FR or PR moment connection(s). If the required strength P_{uf} exceeds the design strength ϕR_v, either a doubler plate or a pair of diagonal stiffeners, one on each side of the column web, must be provided.

From LRFD Specification Section K1.7, the design strength ϕR_v may be determined from LRFD Specification Equation K1-9, K1-10, K1-11, or K1-12, depending upon whether the effect of plastic panel zone deformation on frame stability is or is not considered and upon the axial force in the column; refer to LRFD Specification Section K1.7. These equations are applicable at intermediate column locations and column-end locations for flange-plated, directly welded flange, and extended end-plate connections.

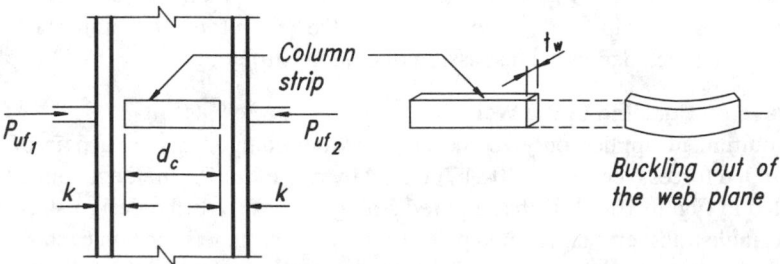

Fig. 10-12. Exaggerated illustration of compression buckling of the web.

Transverse Stiffener Design

At locations of FR and PR moment connections, transverse stiffening may be used to stiffen a column flange which is inadequate in local bending, or a web which is inadequate in local yielding or compression buckling. Transverse stiffeners, when required, should be designed to provide the strength required in excess of the design strength of the column web or flange. The designer should be aware of the increased fabrication costs incurred by the addition of transverse stiffeners to a column. It frequently is less costly to select a member with a thicker flange and/or web or higher yield strength than it is to add the transverse stiffening.

Concentric Transverse Stiffeners

A concentric transverse stiffener is one which coincides with the axis of the flange which delivers the concentrated force. The factored force delivered to the stiffener $R_{u\,st}$ is

$$R_{u\,st} = P_{uf} - \phi R_{n\,min}$$

where

P_{uf} = factored beam flange force (required strength), kips

$\phi R_{n\,min}$ = the lesser of the design strengths in flange bending and web yielding at the location of the tensile concentrated force, or the lesser of the design strengths in web yielding and compression buckling of the web (if applicable) at the location of the compressive concentrated force, kips

If $R_{u\,st}$ is negative, transverse stiffeners are not required. If $R_{u\,st}$ is positive, A_{st} the area of transverse stiffeners required for strength may be calculated as

$$A_{st} = \frac{R_{u\,st}}{\phi F_{y\,st}}$$

where $F_{y\,st}$ is the yield strength of the stiffener material and $\phi = 0.90$. Note that stiffeners are generally made of material with $F_y = 36$ ksi.

Additionally, when stiffeners are required, LRFD Specification Section K1.9 establishes minimum width and thickness dimensions. The minimum width of each stiffener is a function of the width of the beam flange or flange plate connected to the column flange. As illustrated in Figure 10-13, this minimum stiffener width $b_{s\,min}$ may be calculated from the following relationship, where t_w is the thickness of the column web.

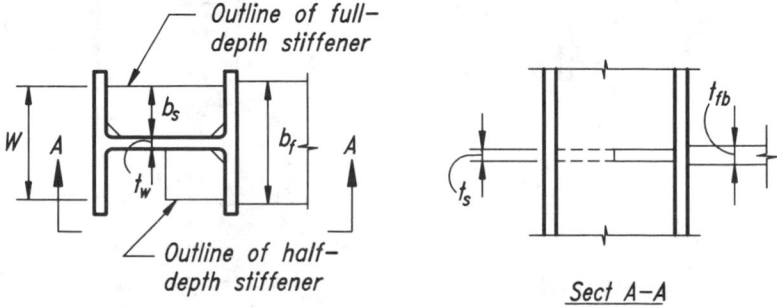

Fig. 10-13. Minimum dimensions for transverse stiffeners.

$$b_{s\,min} = \frac{W - t_w}{2}$$

where

$$W_{min} = \frac{2b_f}{3}$$

Note, for a flange-plated connection, b_f should be taken as the flange-plate width. The minimum stiffener thickness $t_{s\,min}$ is

$$t_{s\,min} = \frac{t_{fb}}{2} \geq \frac{b_s \sqrt{F_y}}{95}$$

where t_{fb} is the flange thickness of the beam.

Full-depth and partial-depth transverse stiffeners are illustrated in Figure 10-14a and 10-14b, respectively. In order to resist tensile concentrated forces, the stiffener must be welded directly to the flange upon which the tensile concentrated force is imposed to develop the strength of the welded portion of the stiffener. While fillet welds are preferable, complete-joint-penetration groove welds may be required when the force in the stiffener is large.

When the concentrated force is always compressive, one end of a full-depth stiffener is sometimes finished for bearing with the other end welded. At partial-depth stiffeners for compressive concentrated forces, some fabricators prefer to finish the end in contact for bearing.

If concentrated forces from opposed FR or PR moment connections are equal, as in the case of balanced moments, they may be theoretically transferred entirely through the stiffeners with no attachment to the column web, except as required for the web limit state of compression buckling and/or to prevent the stiffener from buckling as a column. More often, the moments are not balanced and the differential axial forces must be transferred to the column web. In this case, appropriate weld sizes are required.

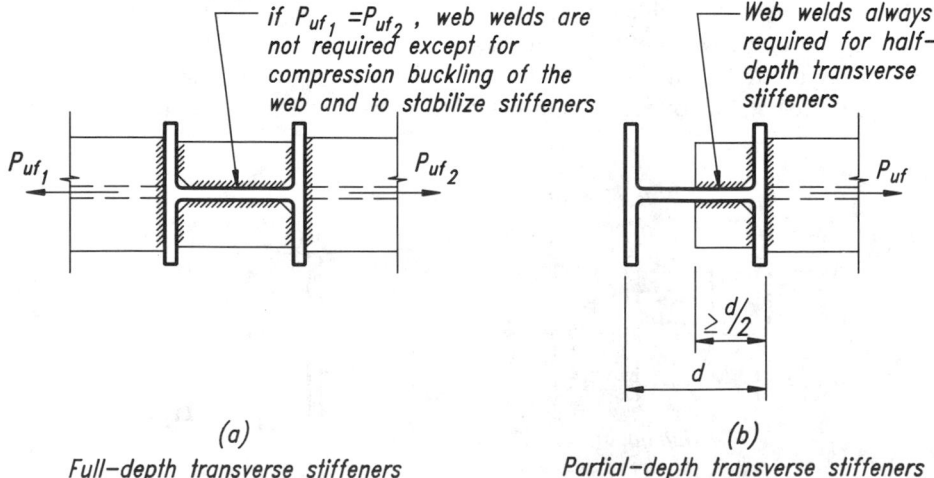

(a)   (b)
Full-depth transverse stiffeners   Partial-depth transverse stiffeners

Fig. 10-14. Full-depth and partial-depth transverse stiffeners.

It is obvious from Figure 10-14b that a web weld is always required for a partial-depth stiffener. Note that it may be desirable to extend the partial-depth stiffener beyond one-half the column-web depth in order to reduce the weld size. Fillet welds are preferable and complete- or partial-joint-penetration groove welds are seldom required for connection between the stiffener and the column web.

Example 10-6

Given: Refer to Examples 10-1, 10-2, 10-3, and 10-4. The FR moment connections developed in these examples deliver double concentrated forces, one tensile and one compressive, to the flange of the W14×99 column. Determine:

A. if the column is adequate for the flange forces delivered by the flange-plated connections of Examples 10-1 and 10-2 where $P_{uf} = 167$ kips.

B. if the column is adequate for the flange forces delivered by the directly welded flange connections of Example 10-3 where $P_{uf} = 172$ kips.

C. if the column is adequate for the flange forces delivered by the four-bolt unstiffened extended end-plate connection of Example 10-4 where $P_{uf} = 172$ kips.

D. the column size required in the above cases to eliminate the need for transverse stiffening.

E. the transverse stiffeners required in the above cases with the W14×99 column.

F. if transverse stiffening would be required if there were an identical W18×50 beam and connection opposite and adjacent to the existing one.

W18×50

$d = 17.99$ in.	$b_f = 7.495$ in.	$Z_x = 101$ in.^{3}
$t_w = 0.355$ in.	$t_f = 0.570$ in.	

W14×99

$d = 14.16$ in.	$b_f = 14.565$ in.	$k = 1\frac{7}{16}$-in.
$k_1 = \frac{7}{8}$-in.	$t_w = 0.485$ in.	$t_f = 0.780$ in.
$T = 11\frac{1}{4}$-in.		

Solution A: *Determine the design strength of the column in local flange bending:*

$$\phi R_n = \phi[6.25 t_f^2 F_{yf}]$$
$$= 0.90[6.25(0.780 \text{ in.})^2(50 \text{ ksi})]$$
$$= 172 \text{ kips} > 167 \text{ kips} \quad \textbf{o.k.}$$

Determine the design strength of the column in local web yielding:

$$\phi R_n = \phi[(5k + N)F_{yw}t_w]$$
$$= 1.0[(5 \times 1\tfrac{7}{16}\text{-in.} + 0.570 \text{ in.})(50 \text{ ksi})(0.485 \text{ in.})]$$
$$= 188 \text{ kips} > 167 \text{ kips} \quad \textbf{o.k.}$$

For the flange-plated FR connections of Examples 10-1 and 10-2, transverse stiffening is not required at either the tensile or compressive component of the double concentrated force.

Solution B: From Solution A, the design strengths in local flange bending and local web yielding are $\phi R_n = 172$ kips and $\phi R_n = 188$ kips, respectively. Thus at the tensile and compressive components of the double concentrated force, the design strength is adequate with respect to the required strength of 172 kips.

Solution C: *Determine the design strength of the column in local flange bending assuming $F_{yf} = 36$ ksi:*

$$\phi R_n = \phi \left[\frac{b_s}{p_e \alpha_m} \right] t_f^2 F_{fy}$$

$$b_s \quad = 2.5 \, (2_{pf} + t_{fb})$$
$$= 2.5 \, (2 \times 1\tfrac{1}{2}\text{-in.} + 0.570 \text{ in.})$$
$$= 8.93 \text{ in.}$$

$$p_e \quad = \frac{g}{2} - \frac{d_b}{4} - k_1$$

$$= \frac{5\tfrac{1}{2}\text{-in.}}{2} - \frac{1 \text{ in.}}{4} - \tfrac{7}{8}\text{-in.}$$

$$= 1.63 \text{ in.}$$

$$\alpha_m \quad = 1.36 \left(\frac{p_e}{d_b} \right)^{\tfrac{1}{4}}$$

$$= 1.36 \left(\frac{1.63 \text{ in.}}{1 \text{ in.}} \right)^{\tfrac{1}{4}}$$

$$= 1.54$$

$$\phi R_n = 0.90 \left[\frac{(8.93 \text{ in.})}{(1.63 \text{ in.})(1.54)} \right] (0.780 \text{ in.})^2 \, (36 \text{ ksi})$$

$$= 70.1 \text{ kips} < 172 \text{ kips} \quad \textbf{n.g.}$$

Determine the design strength of the column in local web yielding:

$$\phi R_n = \phi(6k + N + 2t_p)F_{yw}t_w$$
$$= 1.0[(6 \times 1\tfrac{7}{16}\text{-in.} + 0.570 \text{ in.} + 2 \times \tfrac{3}{4}\text{-in.})(50 \text{ ksi})(0.485 \text{ in.})]$$
$$= 259 \text{ kips} > 172 \text{ kips} \quad \textbf{o.k.}$$

The W14×99 is not adequate for the tensile component of the double concentrated force imposed by the four-bolt unstiffened extended end-plate connection of the W18×50 beam. Transverse stiffeners will be required; refer to Solutions D and E which follow. At the compressive component of the double concentrated force, transverse stiffening is not required.

Solution D: For the flange-plated and directly welded flange connections of Solutions A and B, transverse stiffening is not required and the W14×99 column is adequate.

For the extended end-plate connection of Solution C, the local flange bending strength of the W14×99 column is not adequate. The required flange thickness may be calculated as:

$$t_{f\,req} = \sqrt{\frac{P_{uf}\,p_e\alpha_m}{\phi F_{yf}\,b_s}}$$

where from Solution C,

b_s = 8.93 in.
p_e = 1.63 in.
α_m = 1.54

Thus,

$$t_{f\,req} = \sqrt{\frac{172\ \text{kips}(1.63\ \text{in.})(1.54)}{0.90(36\ \text{ksi})(8.93\ \text{in.})}}$$

$$= 1.22\ \text{in.}$$

and the lightest W14 which satisfies this flange thickness requirement is a W14×176. The cost of the additional 77 pounds per foot of column must be compared with the cost of adding stiffeners; see Solution E for the stiffening design.

Solution E: The transverse stiffening must be sized for the difference between the required strength P_{uf} and the least design strength ϕR_n. Thus, the force in the two stiffeners $R_{u\,st}$ will be:

$$R_{u\,st} = P_{uf} - \phi R_{n\,min}$$
$$= 172\ \text{kips} - 70.1\ \text{kips}$$
$$= 102\ \text{kips}$$

and the required area of stiffeners

$$A_{st} = \frac{R_{ust}}{\phi F_{yst}}$$

$$= \frac{102\ \text{kips}}{0.9 \times 36\ \text{ksi}}$$

$$= 3.15 \text{ in.}^2$$

The minimum stiffener size, from LRFD Specification Section K1.9 is:

$$b_{s\,min} = \frac{W_{min} - t_w}{2}$$

$$W_{min} = \frac{2b_f}{3} = \frac{2(7.495 \text{ in.})}{3}$$

$$= 5.00 \text{ in.}$$

$$b_{s\,min} = \frac{5.00 \text{ in.} - 0.485 \text{ in.}}{2}$$

$$= 2.26 \text{ in.}$$

The minimum stiffener thickness from LRFD Specification Section K1.9 is:

$$t_{s\,min} = \frac{t_{fb}}{2} \leq \frac{b_{s\,min}\sqrt{F_y}}{95}$$

$$= \frac{0.570 \text{ in.}}{2}$$

$$= 0.285 \text{ in.}$$

$$\frac{b_{s\,min}\sqrt{F_y}}{95} = \frac{2.26 \text{ in.}\sqrt{36 \text{ ksi}}}{95}$$

$$= 0.14 \text{ in.} < 0.285 \text{ in.} \quad \textbf{does not control}$$

and the minimum stiffener length is

$$l_{min} = \frac{d}{2} - k$$

$$= \frac{14.16 \text{ in.}}{2} - 1\frac{7}{16}\text{-in.}$$

$$= 5.64 \text{ in.}$$

Try two ½-in.×4½-in. stiffeners with ¾-in. corner clips.

$$A_{st} = 2 \times \frac{1}{2}\text{-in. } (4\frac{1}{2}\text{-in.} - \frac{3}{4}\text{-in.})$$

$$= 3.75 \text{ in.}^2 > 3.15 \text{ in.}^2 \quad \textbf{o.k.}$$

Determine required stiffener-to-column-flange weld (weld must be sized to develop the strength of the welded portion of the stiffener):

$$D_{min} = \frac{0.9F_y t_s}{2 \times 1.392 \times 1.5}$$

(Note: 1.5 in denominator per LRFD Specification Appendix J2.4)

$$= \frac{0.9 \times 36 \text{ ksi} \times \frac{1}{2}\text{-in.}}{2 \times 1.392 \times 1.5}$$

$$= 3.88 \rightarrow 4 \text{ sixteenths}$$

Use $\frac{1}{4}$-in. fillet welds on both sides of each stiffener.

Determine required stiffener to column web weld:

From LRFD Specification Table J2.4, the minimum weld size is $\frac{3}{16}$-in. Try $\frac{3}{16}$-in. fillet welds on both sides of each stiffener. The minimum length of the stiffeners is then:

$$l_{min} = \frac{R_{u\,st}}{4 \times 1.392D} + \text{clip}$$

$$= \frac{102 \text{ kips}}{4 \times 1.392(3 \text{ sixteenths})} + \frac{3}{4}\text{-in.}$$

$$= 6.86 \text{ in.}$$

Use $l = 7$ in. with $\frac{3}{16}$-in. fillet welds both sides.

Solution F: If W18×50 beams were rigidly connected at both flanges of the W14×99 column, the compression buckling strength of the web would have to be checked in addition to the design checks in Solutions A, B, and C.

Determine the design compression buckling strength of the column web:

$$\phi R_n = \phi \frac{4,100t_w^3 \sqrt{F_{yw}}}{d_c}$$

$$= 0.9 \left[\frac{4,100 \times (0.485 \text{ in.})^3 \times \sqrt{50 \text{ ksi}}}{11\frac{1}{4}\text{-in.}} \right]$$

$$= 265 \text{ kips} \quad \textbf{o.k.}$$

The W14×99 would not require transverse stiffening for compression buckling of the web.

Example 10-7

Given: Refer to Example 10-5. Determine if transverse stiffening of the W14×311 is required.

$P_{uf} = 392$ kips

W33×118

$d = 32.86$ in.	$b_f = 11.48$ in.	$Z_x = 415$ in.^{3}
$t_w = 0.550$ in.	$t_f = 0.740$ in.	

W14×311

$d = 17.12$ in. $b_f = 16.230$ in. $k = 2^{15}/_{16}$-in.

$k_1 = 1^5/_{16}$-in. $t_w = 1.410$ in. $t_f = 2.260$ in.

$T = 11^1/_4$-in.

Solution: Determine the design strength of the column in local flange bending conservatively assuming $F_{yf} = 36$ ksi:

$$\phi R_n = \phi \left[\frac{b_s}{p_e \alpha_m} \right] t_f^2 F_{yf}$$

b_s $= 3.5 p_b + 2 p_f + t_{fb}$

  $= 3.5(3^3/_8\text{-in.}) + 2(1^5/_8\text{-in.}) + 0.740$ in.

  $= 15.8$ in.

p_e $= \dfrac{g}{2} - \dfrac{d_b}{4} - k_1$

  $= \dfrac{6 \text{ in.}}{2} - \dfrac{1^1/_8\text{-in.}}{4} - 1^5/_{16}\text{-in.}$

  $= 1.41$ in.

α_m $= 1.13 \left(\dfrac{p_e}{d_b} \right)^{1/4}$

  $= 1.13 \left(\dfrac{1.41 \text{ in.}}{1^1/_8\text{-in.}} \right)^{1/4}$

  $= 1.20$

ϕR_n $= 0.9 \left[\dfrac{15.8 \text{ in.}}{(1.41 \text{ in.})(1.20)} \right] (2.26 \text{ in.})^2 (36 \text{ ksi})$

  $= 1{,}550$ kips > 392 kips   **o.k.**

Determine the design strength of the column in local web yielding:

ϕR_n $= \phi(6k + N + 2t_p) F_{yw} t_w$

  $= 1.0[(6 \times 2^{15}/_{16}\text{-in.} + 0.740 \text{ in.} + 2 \times 1^1/_4\text{-in.})](50 \text{ ksi})(1.410 \text{ in.})$

  $= 1{,}470$ kips > 392 kips   **o.k.**

TheW14×311 is adequate without transverse stiffening.

Eccentric Transverse Stiffeners

Frequently, beams of differing depths are connected with FR or PR moment connections to opposite flanges of a column at the same location. Since, in general, it is advantageous to use as few stiffeners as possible, the two partial-depth stiffeners in Figure 10-15a could be replaced with one full-depth eccentric stiffener as shown in Figure 10-15b.

In full-scale tests, Graham, et. al. (1959) showed that stiffeners with 2-in. eccentricity e provided 65 percent of the strength of identical concentric stiffeners and rapidly declined in effectiveness at greater spacing. It was thus recommended that "for design purposes it would probably be advisable to neglect the resistance of stiffeners having eccentricities greater than two inches." Given this, the required stiffener area, width, and thickness may be established by the same criteria as for concentric transverse stiffeners.

Alternatively, the sloped full-depth transverse stiffener as shown in Figure 10-15c may provide a more economical alternative. The design of this transverse stiffener is similar to that for diagonal stiffeners, refer to "Column Stiffening—Diagonal Stiffener Design" in this Part.

Concurrent Strong-Axis and Weak-Axis FR Connections
When transverse stiffeners are required for FR or PR moment connections made to both the flange and the web of a column at the same location, adequate clearance must be provided to install the stiffeners. A detail such as that in Figure 10-16 may provide an economical solution; it is recommended that the vertical spacing of transverse stiffeners located on the same side of a column web be no less than three inches to ensure adequate clearance for welding. Note that the bottom plate for the weak-axis connection also serves as an eccentric transverse stiffener for the strong-axis connection of the left beam; refer to "Eccentric Transverse Stiffeners" above.

Doubler Plate Design
At locations of FR or PR moment connections, a doubler plate or pair of doubler plates may be used to stiffen a column web which is inadequate in local yielding, compression buckling, or panel zone shear. The designer should be aware of the increased fabrication costs incurred by the addition of doubler plates to a column. It frequently is less costly to select a member with a thicker web or higher yield strength than it is to add the doubler plate.

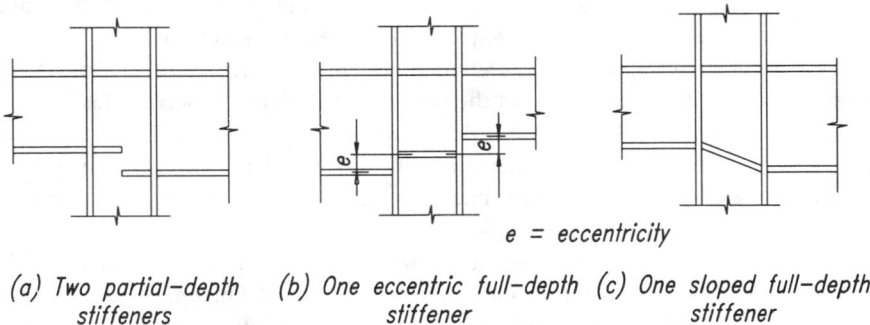

e = eccentricity

(a) Two partial–depth   (b) One eccentric full–depth (c) One sloped full–depth
stiffeners   stiffener   stiffener

Fig. 10-15. Eccentric and sloped transverse stiffeners.

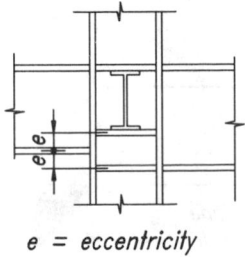

e = eccentricity

Fig. 10-16. Stiffeners for concurrent strong-axis and weak-axis connections.

For Local Web Yielding or Compression Buckling of the Web
From LRFD Specification Section K1.10, when required for local web yielding or compression buckling of the web, the thickness and extent of the doubler plate shall provide the additional material necessary to equal or exceed the required strength. Additionally, the doubler plate shall be welded to develop the proportion of the total force transmitted to the doubler plate.

For Panel Zone Web Shear
When a doubler plate is required for panel zone web shear, the required thickness $t_{p\ req}$ is

$$t_{p\ req} = t_e - t_{wc}$$

where

> t_e = total required effective thickness, in.
> t_{wc} = actual column web thickness, in.

For the doubler plate to be effective in shear, it must be effectively welded to the column flange. In Section A in Figure 10-18, the doubler plate is stopped short of the flange fillet and the edge is beveled in preparation for a complete-joint-penetration groove weld. Partial-joint-penetration groove welds could be used instead as long as the weld effectively bridges the reduced section as shown in Section A—Thin Plate of Figure 10-18. Alternatively, if the plate is thick enough, it can be beveled to clear the column fillet radius and then be fillet welded as shown in Section A—Thick Plate of Figure 10-18. Note that the effective thickness of a beveled doubler plate may have to be reduced. As illustrated in Figure 10-17, the cross section of the doubler plate at the toe of the fillet weld is reduced by the beveled edge. Thus, the required thickness of the doubler plate $t_{p\ req}$ must be adjusted so that the total required effective thickness is present.

While a doubler plate appears to be a simple solution, it requires a great deal of welding and can cause significant distortion of the column flanges if the doubler plate is thick. Thus, although thicker doubler plates allow a greater shear strength in the weld with respect to the base metal, if a doubler plate thicker than the column web or ¾-in. is required, the use of two thinner plates, one on either side of the column web, should be considered.

Thin doubler plates may be subject to local buckling; refer to LRFD Specification Section F2.2. Additionally, to reduce the risk of buckling the doubler plate due to the heat of welding, doubler plates less than ¼-in. thick are not normally used. However, welds connecting such doubler plates may be sized for the required thickness instead of the actual thickness.

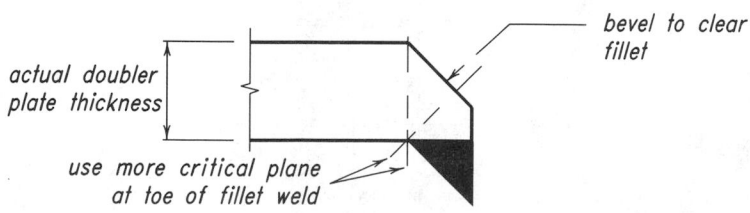

Fig. 10-17. Effective doubler plate thickness.

At Locations of Weak-Axis Connections

In many cases, some provision must be made for the attachment of a weak-axis FR or PR moment connection to the web of the column through the doubler plate. The shear from the end reaction of the supported beam must be added algebraically to the vertical shear in the doubler plate to determine the required thickness and weld size. If the beam also is subjected to axial tension, localized bending would be a major consideration in sizing the doubler plate. In either case, eliminating the need for a doubler plate through the selection of a column section with a thicker web may be the most reasonable alternative.

Example 10-8

Given: Refer to Examples 10-1, 10-2, 10-3, and 10-4. Assuming the effect of panel zone deformation on frame stability is not considered in the analysis and $P_u / P_y = 0.7$ (thus, $P_u > 0.4 P_y$ per LRFD Specification Section K1.7), determine:

A. if the column web is adequate for the web shear induced by the flange-plated connection of Examples 10-1 and 10-2.

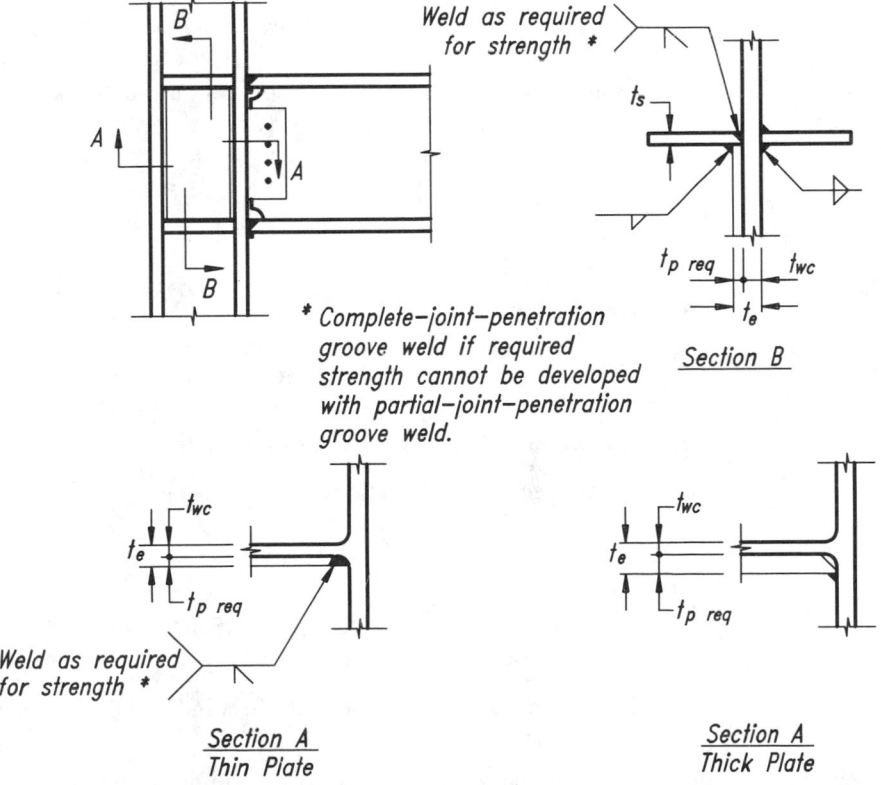

*Weld as required for strength**

* Complete-joint-penetration groove weld if required strength cannot be developed with partial-joint-penetration groove weld.

Section B

Section A
Thin Plate

Section A
Thick Plate

Fig. 10-18. Doubler plate welding.

B. if the column web is adequate for the web shear induced by the directly welded flange and four-bolt unstiffened extended end-plate connections of Examples 10-3 and 10-4.

C. the column size required to eliminate the need for the doubler plate

D. the doubler plate required in the above cases with the W14×99 column.

Neglect the effect of story shear for the purposes of this example.

W18×50

$d = 17.99$ in. $b_f = 7.495$ in. $Z_x = 101$ in.^{3}
$t_w = 0.355$ in. $t_f = 0.570$ in.

W14×99

$d = 14.16$ $b_f = 14.565$ in. $k = 1\frac{7}{16}$-in.
$k_1 = \frac{7}{8}$-in. $t_w = 0.485$ in. $t_f = 0.780$ in.
$T = 11\frac{1}{4}$-in.

Solution A: From LRFD Specification Commentary Section K1.7, the panel zone web shear force ΣF_u is:

$$\Sigma F_u = \frac{M_{u_1}}{d_{m_1}} + \frac{M_{u_2}}{d_{m_2}} - V_u$$

Since Example 10-1 has an FR moment connection to only one side of the column and the effect of story shear is to be conservatively neglected, this equation may be reduced to:

$$\Sigma F_u = \frac{M_{u_1}}{d_{m_1}}$$

From Example 10-1

$\Sigma F_u = P_{uf}$
     $= 167$ kips

Determine the design shear strength of the column web panel zone.

From LRFD Specification Section K1.7:

$$\phi R_v = \phi \left[0.60 F_y d_c t_w \left(1.4 - \frac{P_u}{P_y} \right) \right]$$

     $= 0.90[0.60 \times 50 \text{ ksi} \times 14.16 \text{ in.} \times 0.485 \text{ in.} (1.4 - 0.7)]$
     $= 130$ kips < 167 kips   **n.g.**

The W14×99 is not adequate for the web shear induced by the flange-plated connections of Examples 10-1 and 10-2.

Solution B: In a manner similar to that developed in Solution A, the panel zone web shear force ΣF_u from Example 10-3 is

$$\Sigma F_u = P_{uf}$$
$$= 172 \text{ kips}$$

Determine the design strength of the column web panel zone.

As developed in Solution A:

$$\phi R_v = 130 \text{ kips} < 172 \text{ kips} \quad \textbf{n.g.}$$

The W14×99 is not adequate for the web shear induced by the directly welded flange and four-bolt unstiffened extended end-plate connections of Examples 10-3 and 10-4.

Solution C: For the connections of Solutions A and B, the required thickness t_{req} is:

$$t_{req} = \frac{\Sigma F_u}{\phi \left[0.60 F_y d_c \left(1.4 - \frac{P_u}{P_y} \right) \right]}$$

$$= \frac{P_{uf}}{0.90 \left[0.60 F_y d_c \left(1.4 - \frac{P_u}{P_y} \right) \right]}$$

For convenience, P_{uf} will be taken as 172 kips, the larger value from Examples 10-1, 10-2, 10-3, and 10-4.

$$t_{req} = \frac{172 \text{ kips}}{0.90[0.60 \times 50 \text{ ksi} \times 14.16 \text{ in.}(1.4 - 0.7)]}$$

$$= 0.643 \text{ in.}$$

The lightest W14 which satisfies this web thickness requirement is a W14×132. The cost of the additional of 33 pounds per foot of column must be compared with the cost of adding the doubler plate; see Solution D for a design of the doubler plate for the W14×99 column.

Solution D: The thickness of doubler plate required for the W14×99 column is:

$$t_{p\,req} = t_e - t_{wc}$$
$$= 0.643 \text{ in.} - 0.485 \text{ in.}$$
$$= 0.158 \text{ in.}$$

Try ¼-in.×11¼-in.×18 in. doubler plate with a ³⁄₁₆-in. groove weld.

Check doubler plate buckling.

From LRFD Specification Appendix F2.2, the full design shear strength of the doubler plate may be used if

$$\frac{h}{t_w} \leq 187 \sqrt{\frac{k_v}{F_y}}$$

where

$$\frac{h}{t_w} = \frac{11\frac{1}{4}\text{-in.}}{\frac{1}{4}\text{-in.}}$$

$$= 45.0$$

$$k_v = 5 + \frac{5}{(a/h)^2}$$

$$= 5 + \frac{5}{(18 \text{ in.}/11\frac{1}{4}\text{-in.})^2}$$

$$= 6.95$$

$$187\sqrt{\frac{k_v}{F_y}} = 187\sqrt{\frac{6.95}{36 \text{ ksi}}}$$

$$= 82.2$$

Since $\dfrac{h}{t_w} < 187\sqrt{\dfrac{k_v}{F_y}}$ doubler plate is **o.k.**

Use $\frac{1}{4}$-in.×$11\frac{1}{4}$-in.×18-in. doubler plate with a $\frac{3}{16}$-in. groove weld.

Note that, for the four-bolt unstiffened extended end-plate connection, the doubler-plate size will have to be adjusted for the transverse stiffener required at the tension flange as determined in Example 10-4.

Diagonal Stiffeners

At locations of FR or PR moment connections, a pair of diagonal stiffeners may be used as an alternative to doubler plates to stiffen a column web which is inadequate in panel zone shear. The designer should be aware of the increased fabrication costs incurred by the addition of diagonal stiffeners to a column. It frequently is less costly to select a member with a thicker web or higher yield strength than it is to add the diagonal stiffening.

Diagonal stiffeners are sized for the strength required in excess of the design strength of the web. The full force in the stiffener must be developed at each end, as for any truss diagonal, by use of either fillet or groove welds. The diagonal stiffeners will prevent column web buckling with only a nominal attachment to the web.

From Figure 10-19, the combined horizontal and vertical shear forces may be resolved as a diagonal compressive stress in the column web. Thus, a diagonal stiffener may be used to "truss" the column as a compression strut with node points at interior panel corners A and C.

For static equilibrium, the panel zone shear ΣF_u must be resisted by the column web and the horizontal component of the diagonal stiffener resistance. Thus,

$$\Sigma F_u = \phi R_v + (P_{uf} \times \cos\theta)$$

Where, for a connection to one side of a column,

$$\Sigma F_u = \frac{M_u}{d_m} - V_{us}$$

and the force in the diagonal stiffener T_s is

$$T_s = \phi_c P_n = \phi_c A_s F_{cr}$$

Assuming $d_m = 0.9d$ and substituting terms,

$$\frac{M_u}{0.9d} - V_{us} = \phi R_v + (\phi_c F_{cr} A_s \times \cos\theta)$$

Solving for the required stiffener area,

$$A_{s\,req} = \frac{1}{\cos\theta} \left(\frac{M_u}{(0.9d) \times \phi_c F_{cr}} - \frac{V_{us}}{\phi_c F_{cr}} - \frac{\phi R_v}{\phi_c F_{cr}} \right)$$

where

A_s = the required diagonal stiffener area, in.[2]

M_u = $M_{uL} + M_{uG}$, the sum of the factored moments due to lateral load and gravity load on the leeward side of the connection, kip-in.

$\phi_c F_{cr}$ = the design compressive strength as given LRFD Specification Section E2, kips

ϕR_v = the design shear strength as given in LRFD Specification Section K1.7, kips

V_{us} = the factored story shear due to the lateral load, kips

Letting $\phi F_{cr} = 0.85 F_y$ (assumes for stiffener $\dfrac{Kl}{r} = 0$) and $\phi R_v = 0.90(0.60 F_y\, d_c t_w)$,

$$A_{s\,req} = \frac{1}{\cos\theta} \left(\frac{1.31 M_u}{d_b F_y} - \frac{V_{us}}{0.85 F_y} - 0.64 t_w d_c \right)$$

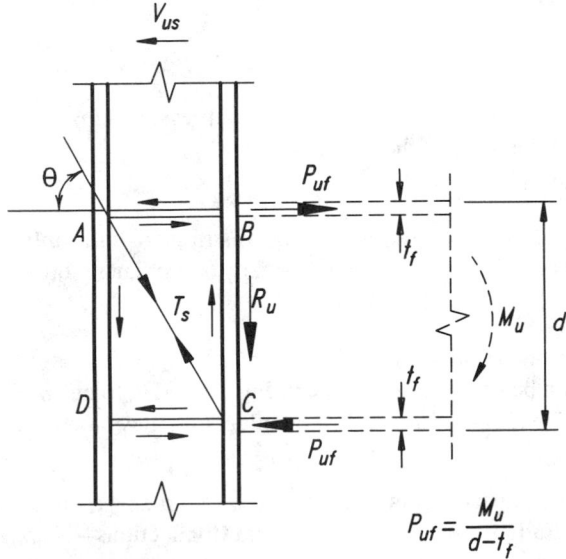

Fig. 10-19. Force diagram for diagonal stiffeners.

MOMENT SPLICES

Beams and girders sometimes are spliced in locations where both shear and moment must be transferred across the splice. Some design specifications require that the strength of the splice be fully equivalent to the strength, in shear and flexure, of the uncut section. Alternatively, other specifications allow the splice to be designed for an arbitrarily established minimum percentage of the strength of the uncut section. However, since the maximum shear and maximum moment seldom occur at the same location, these requirements may be overly conservative.

LRFD Specification Section J7 requires that the full strength of the smaller section being spliced be developed in groove-welded butt splices. Other types of beam or girder splices must develop the strength required by the actual forces at the point of the splice.

Location of Moment Splices

A careful analysis is particularly important in continuous structures where a splice may be located at or near the point of contraflexure—the point of zero moment. Since this inflection point can and does migrate under service loading, actual forces and moments may differ significantly from those assumed. Further, since loading application and frequency can change in the lifetime of the structure, it is prudent for the designer to specify some minimum strength requirement at the splice. Hart and Milek (1965) propose that splices in fixed-ended beams be located at the one-sixth point of the span and be adequate to resist a moment equal to one-sixth of the flexural strength of the member.

Force Transfer in Moment Splices

Force transfer in moment splices may be assumed to occur in a manner similar to that developed for FR moment connections. That is, the shear R_u is primarily transferred through the beam web connection and the moment may be resolved into an effective tension-compression couple where the force at each flange is P_{uf} where:

$$P_{uf} = \frac{M_u}{d_m}$$

where

P_{uf} = factored beam flange force, tensile or compressive, kips
M_u = moment in the beam at the splice, kip-in.
d_m = moment arm, in.

Axial forces, if present, are assumed to be distributed uniformly across the beam cross-sectional area, and are additive algebraically to the flange forces and vectorially to the shear force.

Flange-Plated Moment Splices

Moment splices can be designed as shown in Figure 10-20, to utilize flange plates and a web connection. The flange plates and web connection may be bolted or welded.

Design Checks

The splice and spliced beams should be checked in a manner similar to that described previously under "Fully Restrained (FR) Moment Connections—Flange-Plated Connections," except that the web connection should be designed as illustrated previously in "Shear Splices" in Part 9.

Shop and Field Practices

Figure 10-20 is a composite detail illustrating two types of splices, bolted and welded. The left side of the splice in Figure 10-20 illustrates the detail of a bolted flange-plated moment splice. For this case, the flange plates are normally made approximately the same width as the beam flange as shown in Section A-A.

Alternatively, the right side of the splice in Figure 10-20 illustrates the detail of a welded splice. As shown in Section B-B, the top plate is narrower and the bottom plate is wider than the beam flange, permitting the deposition of weld metal in the downhand or horizontal position without inverting the beam. While this is a benefit in shop fabrication (the beam does not have to be turned over), it is of extreme importance in the field where the weld can be made in the horizontal instead of the overhead position since the beam cannot be turned over. This detail also provides tolerance for field alignment, since the joint gap can be opened or closed. When splices are field welded, some means for temporary support must be provided. Refer to "Fully Restrained (FR) Moment Connections—Temporary Support During Erection".

If the beam or girder flange is thick and the flange forces are large, it may be desirable to place additional plates on the insides of the flanges. In a bolted splice (Section A-A), the bolts are then loaded in double shear and a more compact joint may result. Note that these additional plates must have sufficient area to develop their share of the double-shear bolt load.

In a welded splice (Section B-B), these additional plates must have sufficient area to match the strength of the welds which connect them. Additionally, these plates must be set away from the beam web a distance sufficient to permit deposition of weld metal as shown in Figure 10-21a. This distance is a function of the beam depth and flange width,

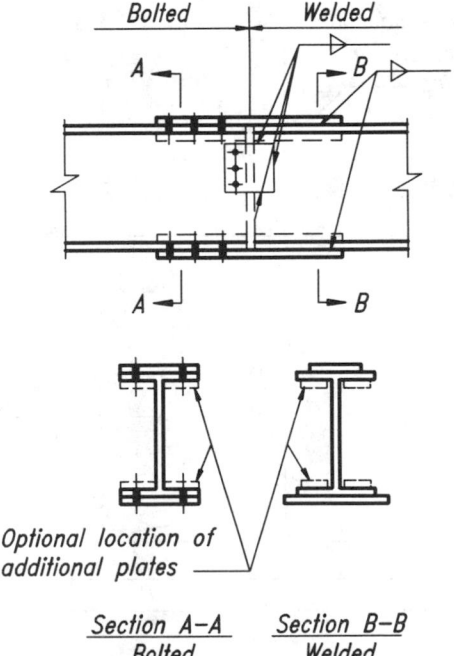

Optional location of additional plates

Section A–A Section B–B
Bolted   Welded

Fig. 10-20. Bolted and welded flange-plated moment splices.

as well as the welding equipment to be used; a distance of 2 to 2½-in. or more may be required for this access. One alternative is to bevel the bottom edge of the plate to clear the beam fillet and place the plate tight to the beam web with a fillet weld as illustrated in Figure 10-21a. The effects of this bevel on the area of the plate must be considered in determining the required plate width and thickness. Another alternative would be to use unbeveled inclined plates as shown in Figure 10-21b.

Directly Welded Flange Moment Splices

Moment splices can be designed, as shown in Figure 10-22, to utilize a complete-joint-penetration groove weld connecting the flanges of the members being spliced. The web connection may then be bolted or welded.

Design Checks

The splice and spliced beams should be checked in a manner similar to that described previously under "Fully Restrained (FR) Moment Connections—Directly Welded Flange Connections," except that the web connection should be designed as illustrated previously in "Shear Splices" in Part 9.

Shop and Field Practices

When the flange thickness or width varies across the splice and the calculated stress is greater than one-third of the specified tensile stress F_u, Figure 10-23 shows the detail required at the tension butt joint by AWS D1.1 Section 8.10. A transition slope of not less than 1 in 2½ must be provided by "chamfering the thicker part, tapering the wider part, sloping the weld metal, or by any combination of these." When the calculated stress does

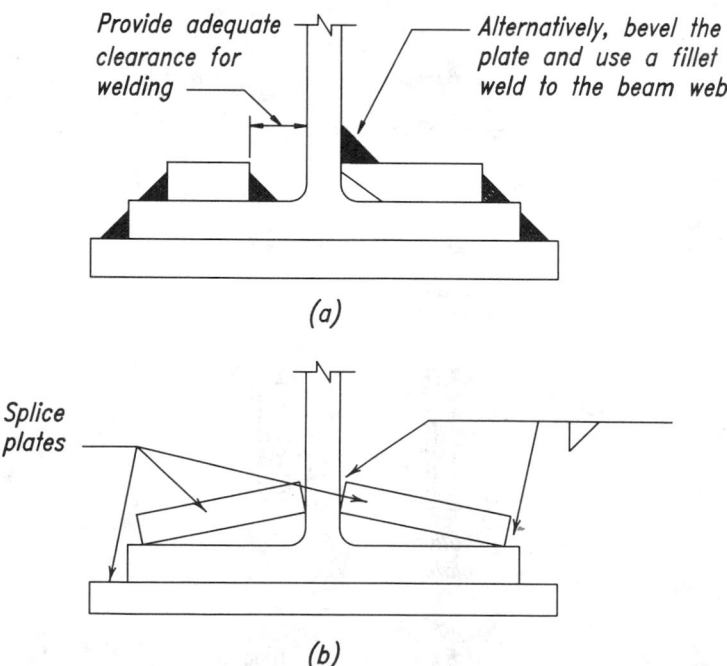

Fig. 10-21. Welding clearances required for flange-plated moment splices.

not exceed one-third of the specified tensile stress F_u, no transition is required in statically loaded structures. Compression butt joints do not require transitional tapering.

Although rare in occurence, some spliced members must be level on top. Where the depths of these spliced members differ, consideration should be given to the use of a flange plate of uniform thickness for the full length of the shallower member. This avoids the fabrication problems created by an inverted transition.

In Figure 10-23, the web depth is kept constant (this is always the case with rolled shapes of the same nominal depth). This avoids an offset cut with a transition in the web for a built-up girder. Eccentricity resulting from differing flange thicknesses is usually ignored in the design. The web plates normally are aligned to their center lines and the 1 in 2½ slope is chamfered into the plate or the weld is sloped, depending upon the relative thicknesses.

The groove (butt) welded splice preparation shown in Figure 10-22 may be used for either shop or field welding. Alternatively, for shop welding where the beam may be turned over, the joint preparation of the bottom flange could be inverted.

In splices subjected to dynamic or fatigue loading, the backing bar should be removed and the weld should be ground flush when it is normal to the applied stress (AISC, 1977). The access holes should be free of notches and should provide a smooth transition at the juncture of the web and flange.

Extended End-Plate Moment Splices
Moment splices can be designed as shown in Figure 10-24, to utilize four-bolt unstiffened extended end-plates connecting the members being spliced. If the end-plate and the bolts are designed properly, it is possible to load this type of connection to reach the full plastic moment capacity of the beam, ϕM_p.

Design Checks
The splice and spliced beams should be checked in a manner similar to that described previously under "Fully Restrained (FR) Moment Connections—Extended End-Plate Connections."

Shop and Field Practices
The comments for "Extended End-Plate Connections" are equally applicable to extended end-plate moment splices.

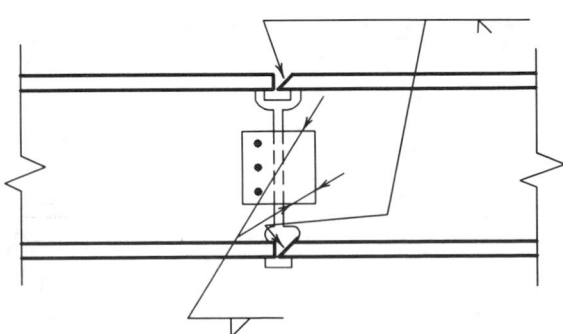

Fig. 10-22. Directly welded flange moment splice.

SPECIAL CONSIDERATIONS

FR Moment Connections to Column-Web Supports

It is frequently required that FR moment connections be made to column web supports. While the mechanics of analysis and design do not differ from FR moment connection to column flange supports, the details of the connection design as well as the ductility considerations required are significantly different.

Recommended Details

When an FR moment connection is made to a column web, it is normal practice to stop the beam short and locate all bolts outside of the column flanges as illustrated in Figures

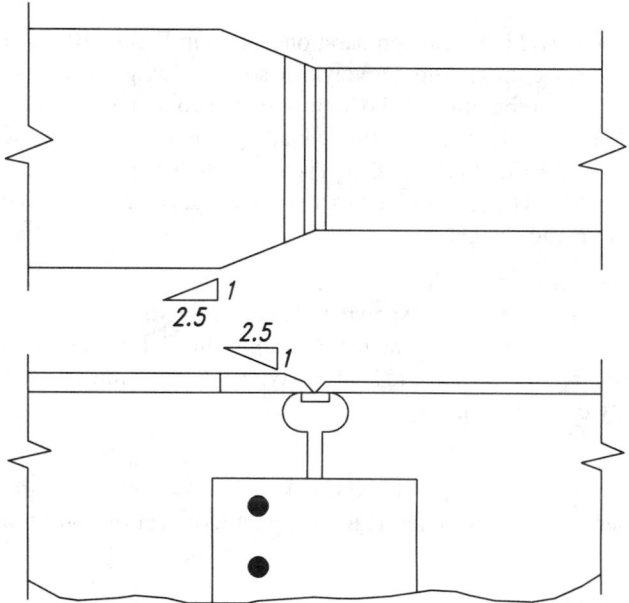

Fig. 10-23. Transition detail at tension flange for directly welded flange moment splices.

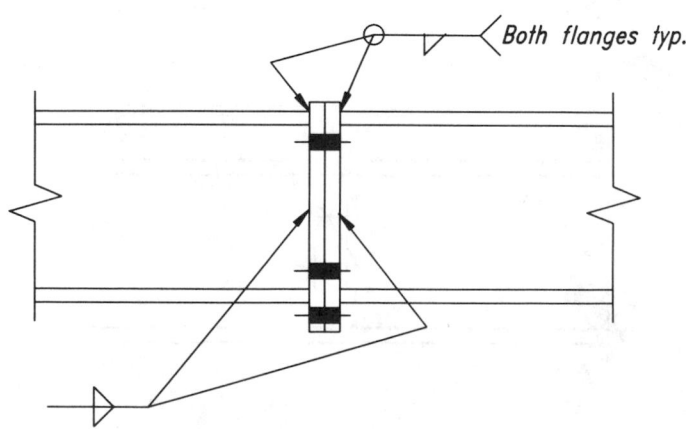

Fig. 10-24. Extended end-plate moment splice.

10-3b and 10-5b. This simplifies the erection of the beam and permits the use of an impact wrench to tighten all bolts. It is also preferable to locate welds outside the column flanges to provide adequate clearance.

Ductility Considerations

Driscoll and Beedle (1982) discuss the testing and failure of two FR moment connections to column-web supports: a directly welded flange connection and a bolted flange-plated connections, shown respectively in Figures 10-25a and 10-25b. Although the connections in these tests were proportioned to be "critical," they were expected to provide inelastic rotations at full plastic load. Failure occurred unexpectedly, however, on the first cycle of loading; brittle fracture occurred in the tension connection plate at the load corresponding to the plastic moment before significant inelastic rotation had occurred.

Examination and testing after the unexpected failure revealed that the welds were of proper size and quality and that the plate had normal strength and ductility. The following is quoted, with minor editorial changes relative to figure numbers, directly from Driscoll and Beedle (1982).

"Calculations indicate that the failures occurred due to high strain concentrations. These concentrations are: (1) at the junction of the connection plate and the column flange tip and (2) at the edge of the butt weld joining the beam flange and the connection plate.

"Figure 10-26 illustrates the distribution of longitudinal stress across the width of the connection plate and the concentration of stress in the plate at the column flange tips. It also illustrates the uniform longitudinal stress distribution in the connection plate at some distance away from the connection. The stress distribution shown represents schematically the values measured during the load tests and those obtained from finite element analysis. (σ_o is a nominal stress in the elastic range.) The results of the analyses are valid up to the loading that causes the combined stress to equal the yield point. Furthermore, the analyses indicate that localized yielding could begin when the applied uniform stress is less than one-third of the

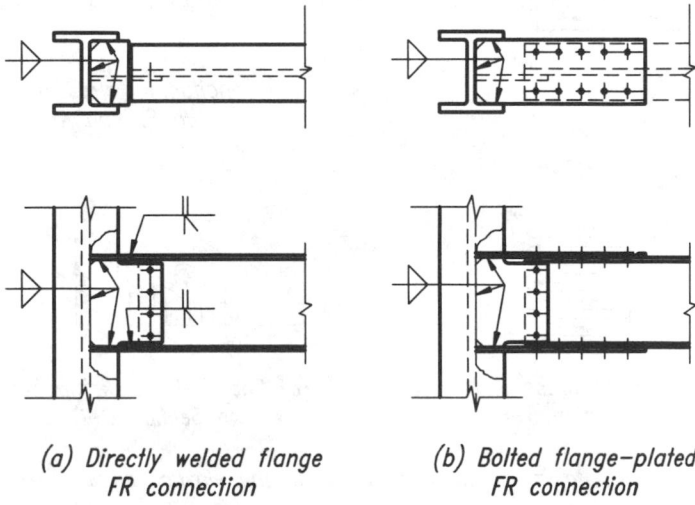

(a) Directly welded flange   (b) Bolted flange-plated
    FR connection   FR connection

Fig. 10-25. Test specimens used by Driscoll and Beedle (1982).

yield point. Another contribution of the non-uniformity is the fact that there is no back-up stiffener. This means that the welds to the web near its center are not fully effective.

"The longitudinal stresses in the moment connection plate introduce strains in the transverse and the through-thickness directions (the Poisson effect). Because of the attachment of the connection plate to the column flanges, restraint is introduced; this causes tensile stresses in the transverse and the through-thickness directions. Thus, referring to Figure 10-26, tri-axial tensile stresses are present along Section A-A, and they are at their maximum values at the intersections of Sections A-A and C-C. In such a situation, and when the magnitudes of the stresses are sufficiently high, materials that are otherwise ductile may fail by premature brittle fracture."

The results of nine simulated weak-axis FR moment connection tests performed by Driscoll, et. al. (1983) are summarized in Figure 10-27. In these tests, the beam flange was simulated by a plate measuring either 1 in.×10 in. or 1⅛-in.×9 in. The fracture strength exceeds the yield strength in every case, and sufficient ductility is provided in all cases except for that of Specimen D. Also, if the rolling direction in the first five specimens (A, B, C, D, and E) were parallel to the loading direction, which would more closely approximate an actual beam flange, the ductility ratios for these would be higher. The connections with extended connection plates (i.e., projection of three inches), with extensions either rectangular or tapered, appeared equally suitable for the static loads of the tests.

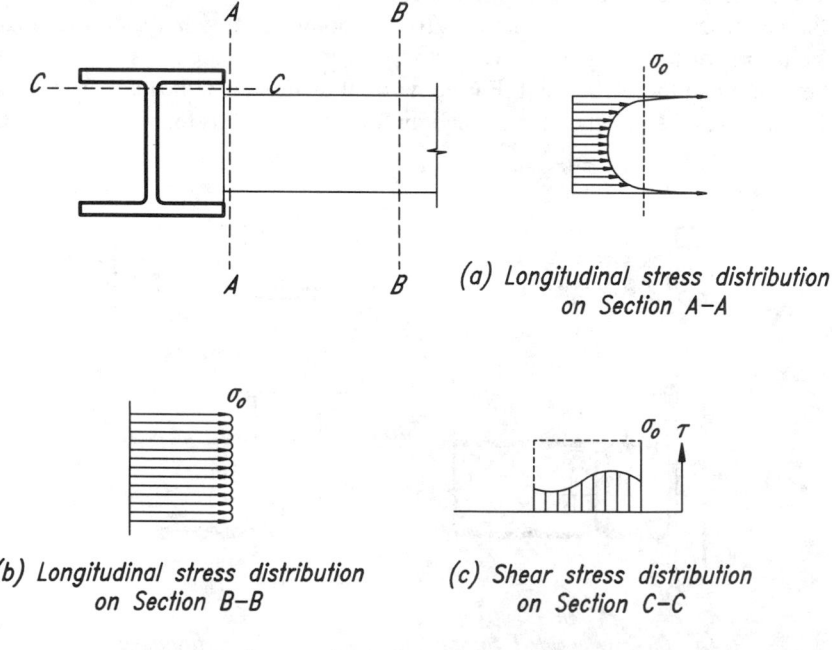

(a) Longitudinal stress distribution on Section A–A

(b) Longitudinal stress distribution on Section B–B

(c) Shear stress distribution on Section C–C

σ_o = the nominal stress in the elastic range

Fig. 10-26. Stress distributions in test specimens used by Driscoll and Beedle (1982).

Based on the tests, Driscoll, et. al. (1983) report that those specimens with extended connection plates have better toughness and ductility and are preferred in design for seismic loads, even though the other connection types (except D) may be deemed adequate to meet the requirements of many design situations.

In accordance with the preceeding discussion, the following suggestions are made regarding the design of this type of connection:

1. For directly welded (butt) flange-to-plate connections, the connection plate should be thicker than the beam flange. This greater area accounts for shear lag and also provides for misalignment tolerances.

   AWS D1.1, Section 3.3.3 restricts the misalignment of abutting parts such as this to 10 percent of the thickness, with ⅛-in. maximum for a part restrained against

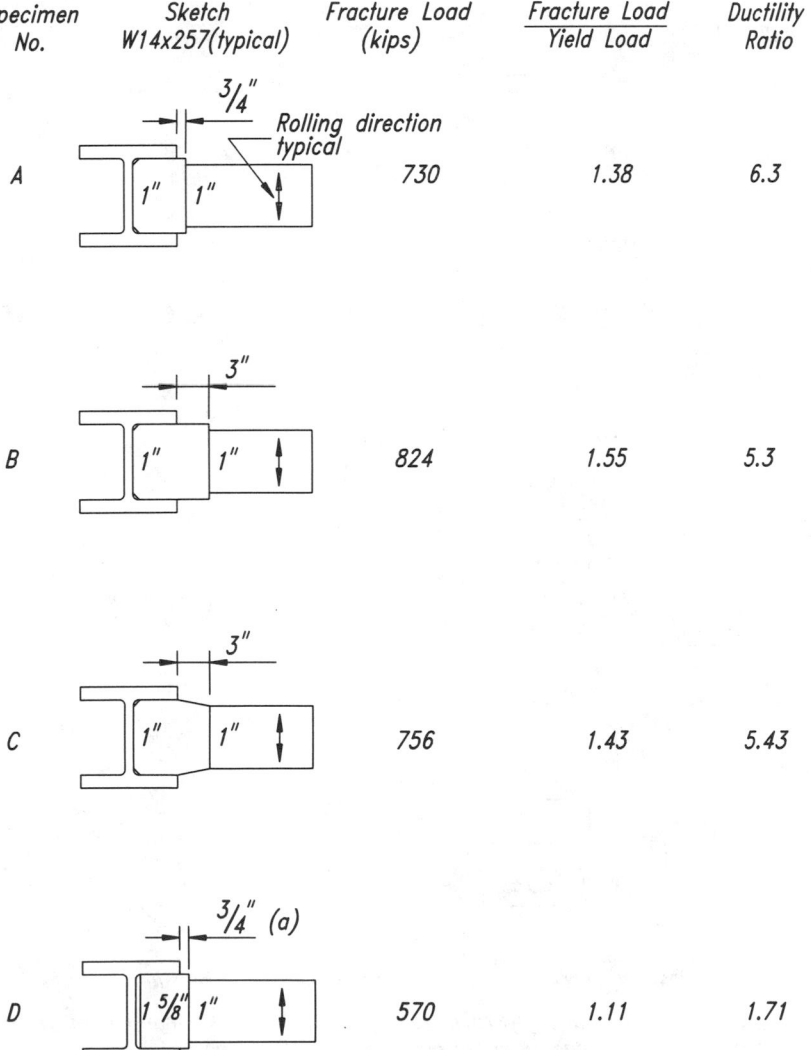

Specimen No.	Sketch W14x257(typical)	Fracture Load (kips)	Fracture Load / Yield Load	Ductility Ratio
A		730	1.38	6.3
B		824	1.55	5.3
C		756	1.43	5.43
D		570	1.11	1.71

Fig. 10-27a. Results of weak-axis FR connection ductility tests performed by Driscoll, et al.

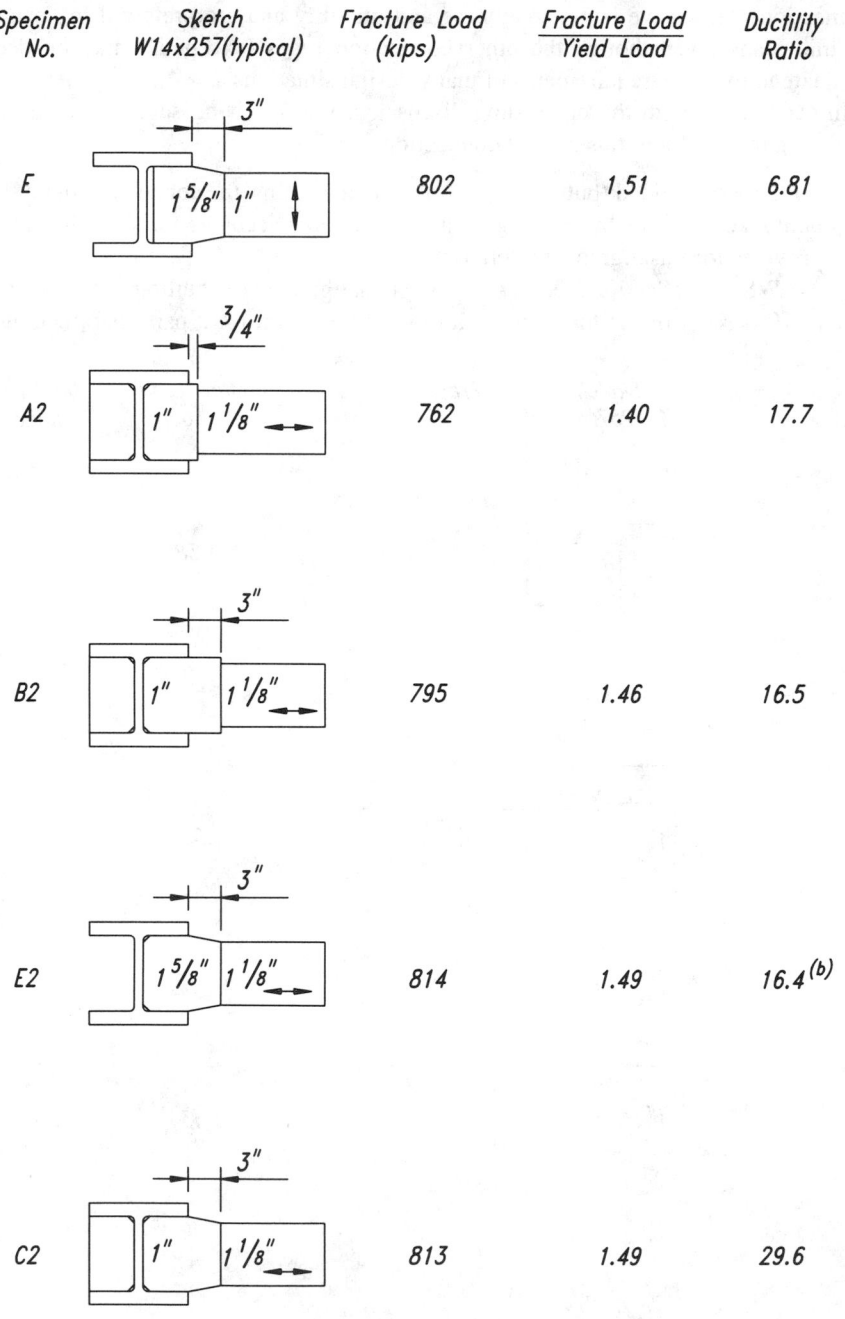

Specimen No.	Sketch W14x257(typical)	Fracture Load (kips)	Fracture Load / Yield Load	Ductility Ratio
E		802	1.51	6.81
A2		762	1.40	17.7
B2		795	1.46	16.5
E2		814	1.49	16.4 [b]
C2		813	1.49	29.6

Notes: (a) ³/₄" dimension is estimated–no dimension given.

(b) Ductility ratio estimated. Actual value not known due to malfuntion in deflection gage.

Fig. 10-27b. Results of weak-axis FR connection ductility tests performed by Driscoll, et al.

bending due to eccentricity of alignment. Considering the various tolerances in mill rolling (±⅛-in. for W-shapes), fabrication, and erection, it is prudent design to call for the stiffener thickness to be increased to accommodate these tolerances and avoid the subsequent problems encountered at erection. An increase of ⅛-in. to ¼-in. generally is used.

Frequently, this connection plate also serves as the stiffener for a strong axis FR or PR moment connection. The welds which attach the plate/stiffener to the column flange may then be subjected to combined tensile and shearing or compression and shearing forces. Vector analysis is commonly used to determine weld size and stress.

It is good practice to use fillet welds whenever possible. Welds should not be made in the column fillet area for strength.

2. The connection plate should extend at least ¾-in. beyond the column flange to avoid intersecting welds and to provide for strain elongation of the plate. The extension should also provide adequate room for runout bars when required.
3. Tapering an extended connection plate is only necessary when the connection plate is not welded to the column web (Specimen E, Figure 10-27). Tapering is not necessary if the flange force is always compressive (e.g., at the bottom flange of a cantilevered beam).
4. To provide for increased ductility under seismic loading, a tapered connection plate should extend three inches. Alternatively, a backup stiffener and an untapered connection plate with 3-in. extension could be used.

Normal and acceptable quality of workmanship for connections involving gravity and wind loading in building construction would tolerate the following:

1. Runoff bars and backing bars may be left in place for Groups 4 and 5 beams (subject to tensile stress only) where they are welded to columns or used as tension members in a truss.
2. Welds need not be ground, except as required for nondestructive testing.
3. Connection plates that are made thicker or wider for control of tolerances, tensile stress, and shear lag need not be ground or cut to a transition thickness or width to match the beam flange to which they connect.
4. Connection plate edges may be sheared or plasma or gas cut.
5. Intersections and transitions may be made without fillets or radii.
6. Burned edges may have reasonable roughness and notches within AWS tolerances.

If a structure is subjected to loads other than gravity and wind loads, such as seismic, dynamic, or fatigue loading, more stringent control of the quality of fabrication and erection with regard to stress risers, notches, transition geometry, welding, and testing may be necessary; refer to AISC's *Seismic Provisions for Structural Steel Buildings* in Part 6.

FR Moment Connections Across Girder Supports

Frequently, beam-to-girder-web connections must be made continuous across a girder-web support as with continuous beams and with cantilevered beams at wall, roof-canopy, or building lines. While the same principles of force transfer discussed previously for FR moment connections may be applied, the designer must carefully investigate the relative stiffness of the assembled members being subjected to moment or torsion and provide the fabricator and erector with reliable camber ordinates.

Additionally, the design should still provide some means for final field adjustment to accommodate the accumulated tolerances of mill production, fabrication, and erection; it is very desirable that the details of field connections provide for some adjustment during erection. Figure 10-28 illustrates several details that have been used in this type of connection and the designer may select the desirable components of one or more of the sketches to suit a particular application. Therefore, these components are discussed here as a top flange, bottom flange, and web connection.

Top Flange Connection

As shown in Figure 10-28a, the top flange connection may be directly welded to the top flange of the supporting girder. Figures 10-28b and 10-28c illustrate an independent splice plate that ties the two beams together by use of a longitudinal fillet weld or bolts. This tie plate does not require attachment to the girder flange, although it is sometimes so connected to control noise if the connection is subjected to vibration.

Bottom Flange Connection

When the bottom flanges deliver a compressive force only, the flange forces are frequently developed by directly welding these flanges to the girder web as illustrated in Figure 10-28a. Figure 10-28b illustrates the use of an angle or channel extending beyond the beam flange to provide for a horizontal fillet weld; Figure 10-28c is similar, but uses bolts instead of welds to develop the flange force.

Web Connection

While a single-plate connection is shown in Figure 10-28a and unstiffened seated connections are shown in Figures 10-28b and 10-28c, any of the shear connections in

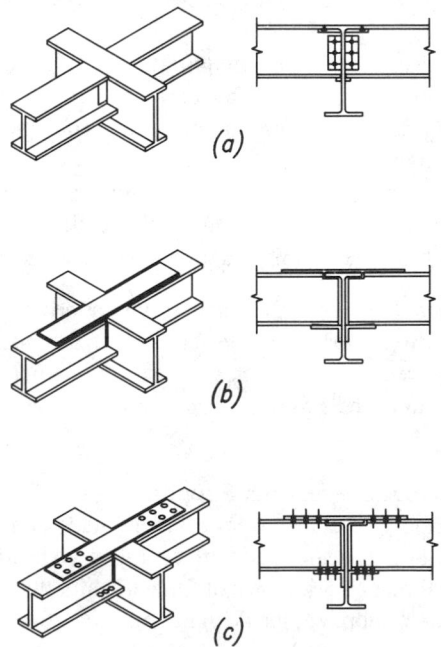

Fig. 10-28. Typical FR connections across girder web supports.

Part 9 may be used. Note that the effect of eccentricity in the shear connection may be neglected.

Knee or Corner Connections

Knee or corner connections, illustrated in Figure 10-29, are used frequently in single-story structures that are designed using FR construction. The knee connection must transfer the fixed-end moment from the beam into the column as well as the shear at the top of the column into the beam. The bending moment and axial forces are assumed to be carried by the flanges and the shear is assumed to be carried by the web.

This type of connection must be designed as part of the main member design and is beyond the scope of this volume. Additionally, the shape of the knee may be established as part of the architectural aesthetics or for structural considerations. For more information, refer to Blodgett (1966), Beedle, et al. (1964), and Salmon and Johnson (1980).

Non-Rectangular FR Moment Connections

Although FR moment connections are not often specified where skews and slopes are pronounced, framing requirements sometimes dictate their use. When required, the flange-plated, directly welded flange, and extended end-plate FR moment connections discussed previously for rectangular framing may be adapted to non-rectangular applications.

When flange-plated and directly welded flange connections are used, the web connection, usually a single-plate connection, may be designed as illustrated previously in "Non-Rectangular Simple Shear Connections" in Part 9. In general, the comments in that section apply equally to non-rectangular FR moment connections.

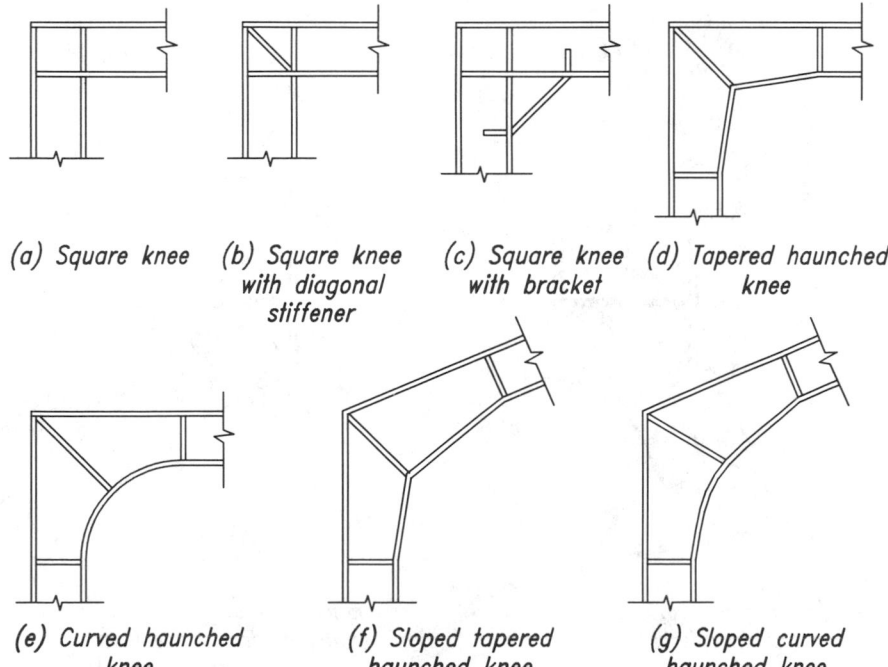

(a) Square knee   (b) Square knee   (c) Square knee   (d) Tapered haunched
                      with diagonal   with bracket   knee
                       stiffener

(e) Curved haunched   (f) Sloped tapered   (g) Sloped curved
        knee   haunched knee   haunched knee

Fig. 10-29. Knee or corner connections.

Skewed Connections
Large angles of skew can produce very awkward connections, particularly when the connection is to the column web where the projecting column flange interferes with the supported beam flange. The designer should consider altering the structural geometry if possible; in Figure 10-30, a slight relocation of the work point simplifies the connection. Alternatively, rotation of the supporting column orientation may permit more normal framing. Other skewed FR moment connections are illustrated in Figure 10-31.

Sloped Connections
Sloped FR moment connections are illustrated in Figure 10-32.

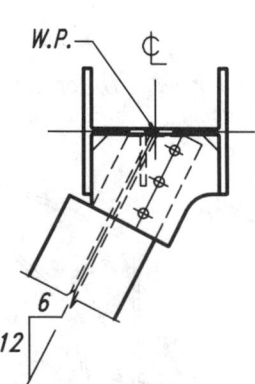

(a) Original working point at column centerline results in an awkward connection

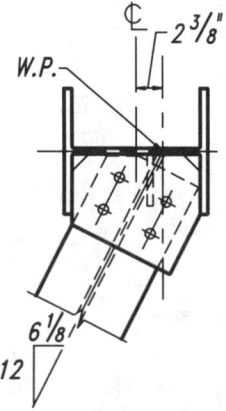

(b) Relocation of working point simplifies the connection

Fig. 10-30. Simplifying skewed FR connection details.

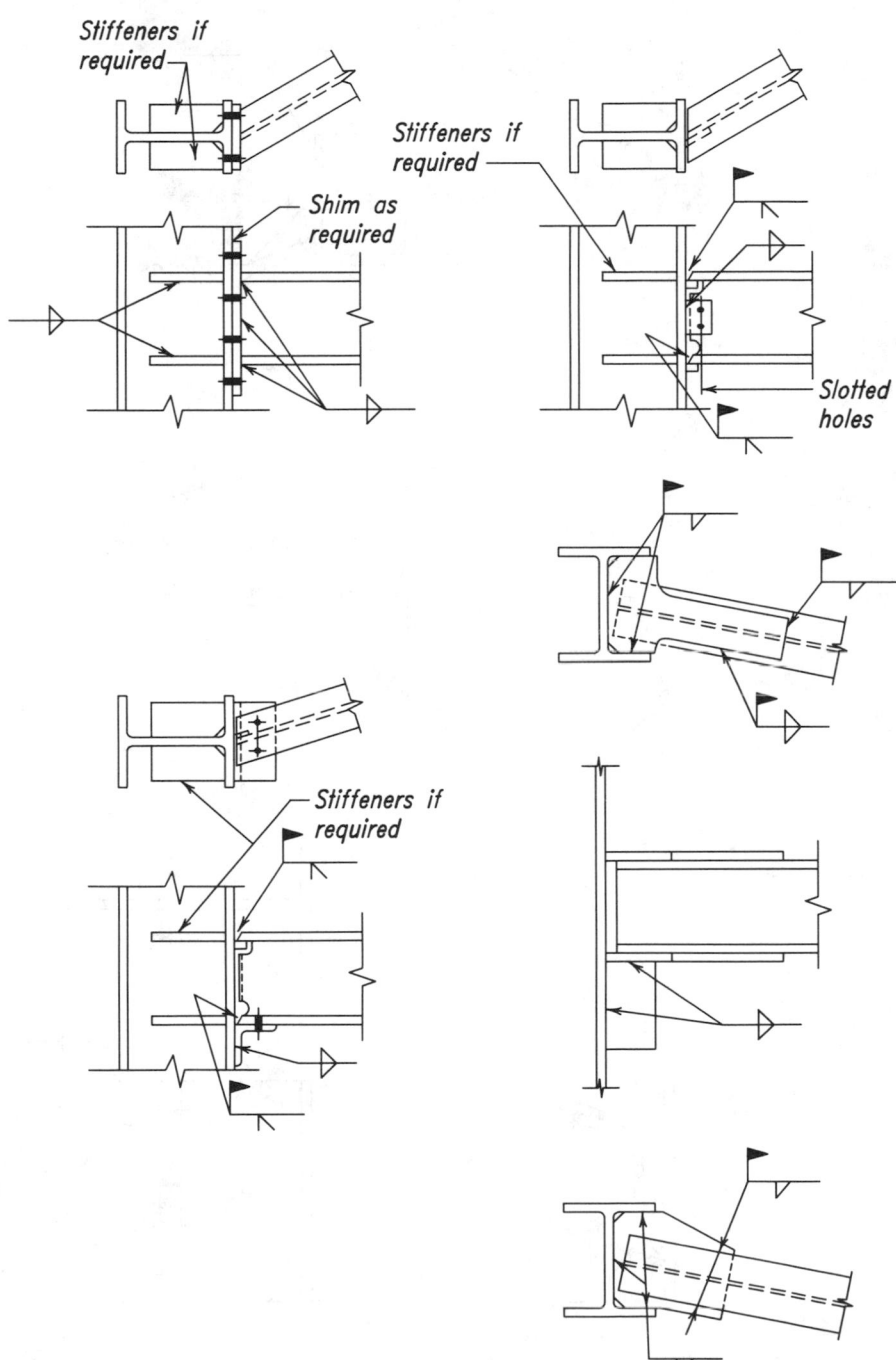

Figure 10-31. Skewed FR moment connections.

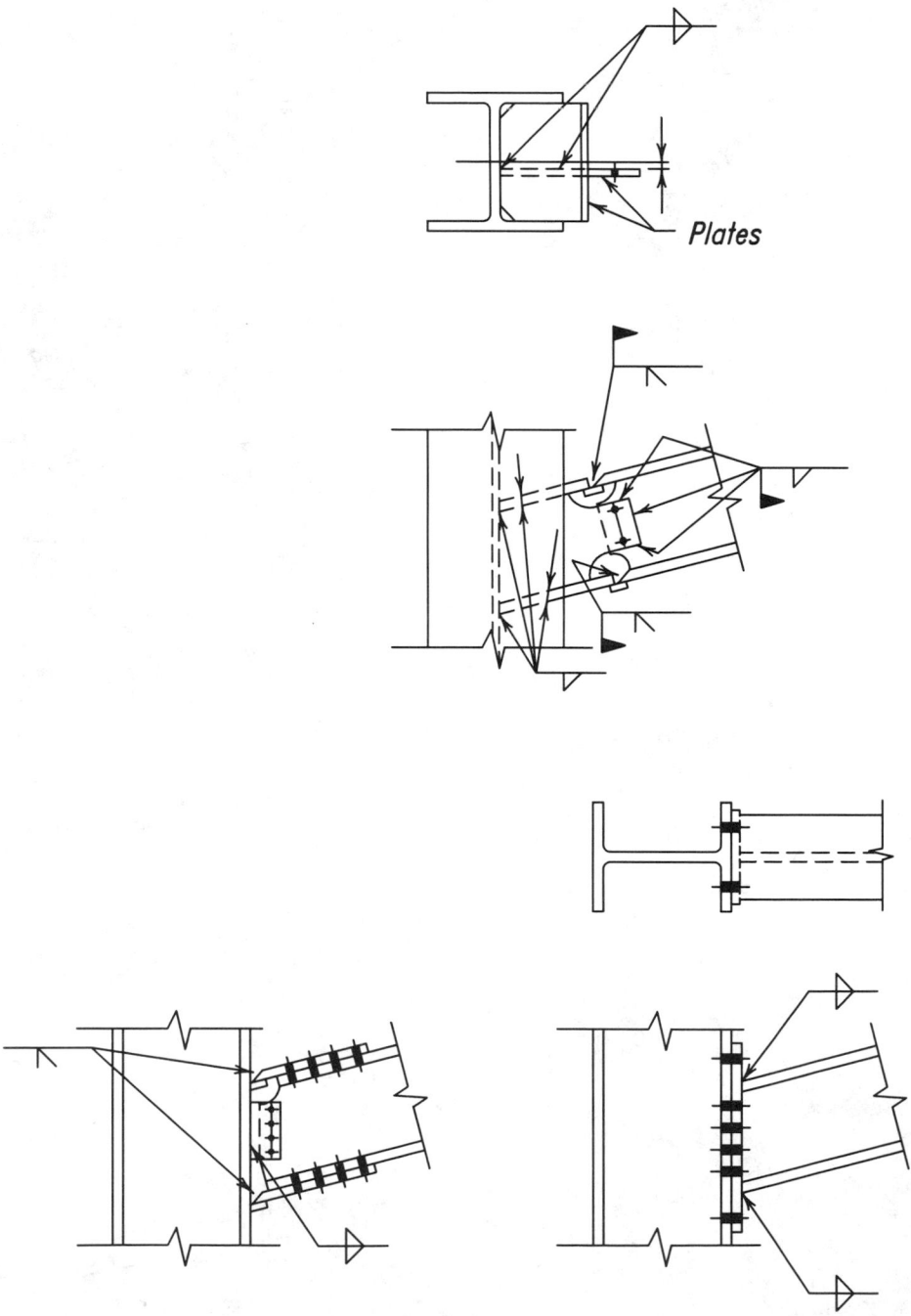

Figure 10-32. Sloped FR moment connections.

REFERENCES

American Institute of Steel Construction, 1977, *Bridge Fatigue Guide Design and Details*, AISC, Chicago, IL.

Beedle, et al., 1964, *Structural Steel Design*, The Ronald Press Co., New York, NY.

Beedle, L. S., L. W. Lu, and E. Ozer, 1973, "Recent Developments in Steel Building Design," *Engineering Journal*, Vol. 10, No. 4, (4th Qtr.), pp. 98–111, AISC, Chicago, IL.

Blodgett, O. W., 1966, *Design of Welded Structures*, James F. Lincoln Arc Welding Foundation, Cleveland, OH.

Curtis, L. E. and T. M. Murray, 1989, "Column Flange Strength at Moment End-Plate Connections," *Engineering Journal*, Vol. 26, No. 2, (2nd Qtr), pp. 41–50, AISC, Chicago, IL.

Driscoll, G. C. and L. S. Beedle, 1982, "Suggestions for Avoiding Beam-to-Column Web Connection Failures," *Engineering Journal*, Vol. 19, No. 1, (1st Qtr.), pp. 16–19, AISC, Chicago, IL.

Driscoll, G. C., A. Pourbohloul, and X. Wang, 1983, "Fracture of Moment Connections— Tests on Simulated Beam-to-Column Web Moment Connection Details," *Fritz Engineering Laboratory Report No. 469.7*, Lehigh University, Bethlehem, PA.

Graham, J. D., A. N. Sherbourne, R. N. Khabbaz, and C. D. Jensen, 1959, "Welded Interior Beam-to-Column Connections," *Report for AISC*, AISC, Chicago, IL.

Hart, W. H. and W. A. Milek, 1965, "Splices in Plastically Designed Continuous Structures," *Engineering Journal*, Vol. 2, No. 2, (April), pp. 33–37, AISC, Chicago, IL.

Hendrick, R. A. and T. M. Murray, 1984, "Column Web Compression Strength at End-Plate Connections," *Engineering Journal*, Vol. 21, No. 3, (3rd Qtr.), pp. 161–169, AISC, Chicago, IL.

Huang, J. S., W. F. Chen, and L. S. Beedle, 1973, "Behavior and Design of Steel Beam-to-Column Moment Connections," *Bulletin 188*, October, Welding Research Council, New York, NY.

Krawinkler, H. and E. P. Popov, 1982, "Seismic Behavior of Moment Connections and Joints," *Journal of the Structural Division*, Vol. 108, No. ST2, (February), pp. 373–391, ASCE, New York, NY.

Krishnamurthy, N., 1978, "A Fresh Look at Bolted End-Plate Behavior and Design," *Engineering Journal*, Vol. 15, No. 2, (2nd Qtr.), pp. 39–49, AISC, Chicago, IL.

Lincoln Electric Company, 1973, *The Procedure Handbook of Arc Welding*, Lincoln Electric Company, Cleveland, OH.

Murray, T. M., 1990, *Extended End-Plate Moment Connections*, AISC, Chicago, IL.

Murray, T. M. and A. Kukreti, 1988, "Design of Eight-Bolt Stiffened Moment End-Plates," *Engineering Journal*, Vol. 25, No. 2, (2nd Qtr.), pp. 45–52, AISC, Chicago, IL.

Ricker, D. T., 1992, "Value Engineering and Steel Economy," *Modern Steel Construction*, Volume 32, No. 2, (February), pp. 22–26, AISC, Chicago, IL.

Salmon, C. G. and J. E. Johnson, 1980, *Steel Structures—Design and Behavior*, 2nd Edition, Harper & Row, New York, NY.

Thornton, W. A., 1992, "Designing for Cost Efficient Fabrication," *Modern Steel Construction*, Vol. 25, No. 2, (February), pp. 12–20, AISC, Chicago, IL.

PART 11

CONNECTIONS FOR TENSION AND COMPRESSION

T/C

OVERVIEW

Part 11 contains general information, design considerations, examples, and design aids for the design of hanger connections, diagonal bracing connections, beam bearing plates, column base plates and splices, and truss connections. It is based on the provisions of the 1993 LRFD Specification. Supplementary information may also be found in the Commentary on the LRFD Specification.

Following is a detailed list of the topics addressed.

HANGER CONNECTIONS

Hanger connections, illustrated in Figure 11-1 are usually composed of a plate, tee, angle, or pair of angles which transfers the tensile force from the tension member to the support.

Design Checks

The design strengths of the bolts and/or welds and connected elements must be determined in accordance with the provisions of the LRFD Specification. The applicable limit states in each of the aforementioned design strengths are discussed in Part 8. Additionally, hanger connections produce tensile single concentrated forces acting on the support; the limit states of local flange bending and local web yielding must be checked. In all cases, the design strength ϕR_n must exceed the required strength R_u.

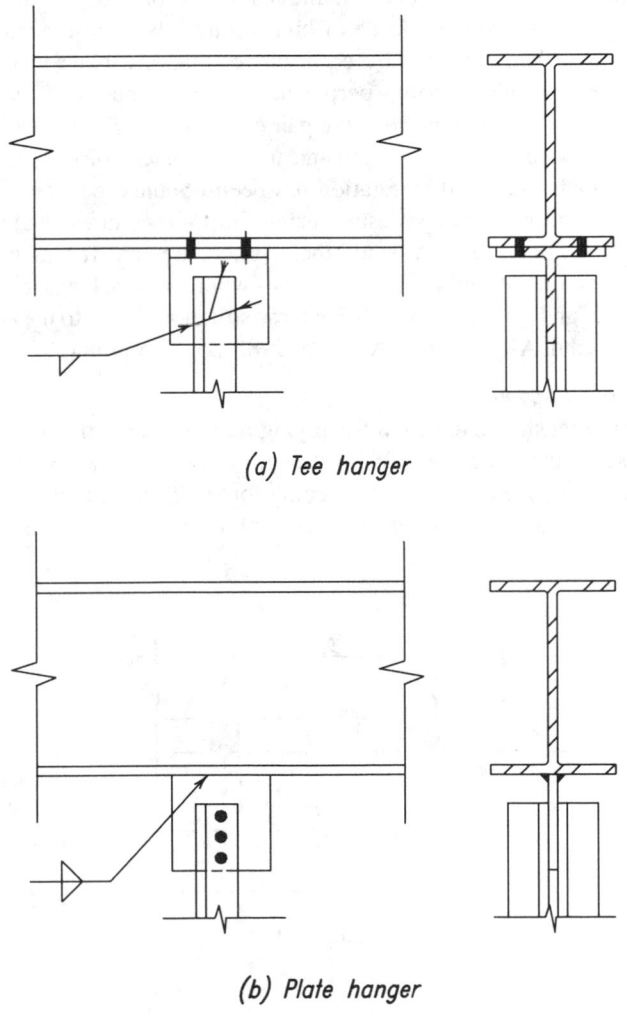

(a) Tee hanger

(b) Plate hanger

Fig. 11-1. Typical hanger connections.

Prying Action

Prying action is a phenomenon associated with bolted construction and tensile loads only where either the connected fitting or the support deforms and thereby increases the tensile force in the bolt. LRFD Specification Section J3.6 states that any tension resulting from prying action must be considered in determining the required strength of bolts. However, prying action is primarily a function of the connected elements. Furthermore, while the connected elements must have adequate flexural strength, it is their stiffness which is the key to satisfactory performance.

Consider the tee used in a hanger connection in Figure 11-2. To ensure adequate flange stiffness, dimension b should be made as small as the bolt entering and tightening clearances will permit; see Tables 8-4 and 8-5. The actual distribution of stresses resulting from prying action is extremely complex. Since dimension b is only slightly larger than the thickness of the fitting, the classical moment diagram as shown on Figure 11-2 does not truly represent all the restraining forces at the bolt line. Consequently, this model overestimates the actual prying force. In addition, local deformation of the fitting, known as "quilting", under the clamping force of high-strength bolts also accounts for a less critical prying force than indicated by earlier investigations. Note that the maximum tributary length p per pair of bolts (perpendicular to the plane of the page) should preferably not exceed the gage between the pair of bolts g.

The following procedures for designing and analyzing a tension connection for prying action are recommended. Good correlation has been obtained between estimated connection strength and observed test results using these procedures (Kulak, Fisher, and Struik, 1987). Note, however, that since these procedures are formulated in terms of factored loads, they are not applicable to situations where service loads must be used (i.e., fatigue, deflection, and drift limitations). For these situations, refer to the allowable stress procedures outlined in AISC (1989), Astaneh (1985), or Thornton (1985).

Designing for Prying Action

When designing a tension connection for prying action, select the number and size of bolts required such that the design tensile strength of one bolt ϕr_n exceeds the factored tensile force per bolt r_{ut} (exclusive of tightening force). Then use Table 11-1 to make a preliminary selection of a trial fitting for steels with F_y equal to 36 ksi or 50 ksi.

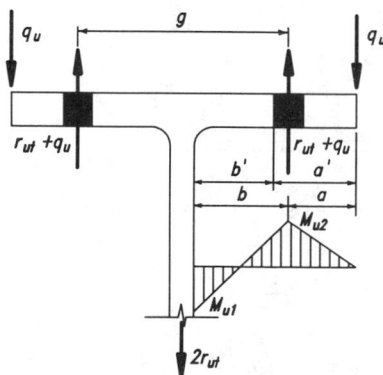

Fig. 11-2. Variables in prying action.

In this table, it is assumed that equal critical moments exist at the face of the tee stem (M_{u1}) and at the bolt line (M_{u2}). From LRFD Specification Section F1.1, the design flexural yielding strength of the tee flange is $\phi_b M_n$, where $\phi_b = 0.90$ and

$$M_n = M_p = F_y Z_x$$

In the above equation, the plastic section modulus Z_x per unit length of the tee flange is

$$Z_x = \frac{t^2}{4}$$

where t is the thickness of the angle or tee flange, in. Thus, for a unit length of the tee flange

$$\phi_b M_n = \frac{0.90 F_y t^2}{4}$$

and the factored tensile force on the fitting $2r_{ut}$ must be such that

$$2r_{ut} \leq \frac{0.9 F_y t^2}{b}$$

where b is the distance from bolt centerline to face of the angle leg or tee stem, in. For $F_y = 36$ksi, the above equation may be simplified as

$$2r_{ut} \leq \frac{32.4 t^2}{b}$$

and for $F_y = 50$ ksi, the above equation may be simplified as

$$2r_{ut} \leq \frac{45.0 t^2}{b}$$

With the preliminary fitting selected from Table 11-1, its strength must be investigated. Given the above relationship, b, and the flange thickness t of the selected trial section, calculate b', a', and ρ as

$$a' = \left(a + \frac{d}{2} \right)$$

$$b' = \left(b - \frac{d}{2} \right)$$

$$\rho = \frac{b'}{a'}$$

In the above equations, a is the distance from the bolt centerline to the edge of the fitting; for calculation purposes, a should not be taken to be greater than $1.25b$.

Next, calculate β as follows:

$$\beta = \frac{1}{\rho} \left(\frac{\phi r_n}{r_{ut}} - 1 \right)$$

if $\beta \geq 1$, set $\alpha' = 1.0$

F_y = 36 ksi

Table 11-1.
Preliminary Hanger Connection Selection Table
Design tensile strength, kips per linear in.,
limited by flexural yielding of the flange

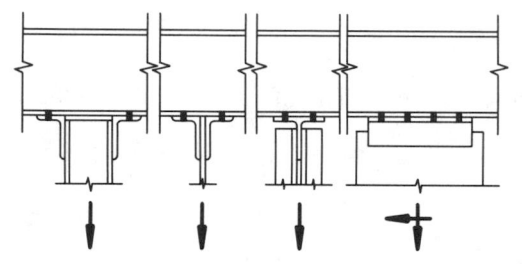
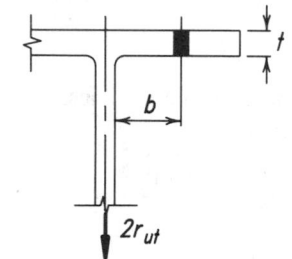

					b, in.					
t, in.	1	$1\frac{1}{4}$	$1\frac{1}{2}$	$1\frac{3}{4}$	2	$2\frac{1}{4}$	$2\frac{1}{2}$	$2\frac{3}{4}$	3	$3\frac{1}{4}$
$\frac{5}{16}$	3.16	2.53	2.11	1.81	1.58	1.41	1.27	1.15	1.05	0.974
$\frac{3}{8}$	4.56	3.65	3.04	2.60	2.28	2.03	1.82	1.66	1.52	1.40
$\frac{7}{16}$	6.20	4.96	4.13	3.54	3.10	2.76	2.48	2.26	2.07	1.91
$\frac{1}{2}$	8.10	6.48	5.40	4.63	4.05	3.60	3.24	2.95	2.70	2.49
$\frac{9}{16}$	10.3	8.20	6.83	5.86	5.13	4.56	4.10	3.73	3.42	3.15
$\frac{5}{8}$	12.7	10.1	8.44	7.23	6.33	5.63	5.06	4.60	4.22	3.89
$\frac{11}{16}$	15.3	12.3	10.2	8.75	7.66	6.81	6.13	5.57	5.10	4.71
$\frac{3}{4}$	18.2	14.6	12.2	10.4	9.11	8.10	7.29	6.63	6.08	5.61
$\frac{13}{16}$	21.4	17.1	14.3	12.2	10.7	9.51	8.56	7.78	7.13	6.58
$\frac{7}{8}$	24.8	19.8	16.5	14.2	12.4	11.0	9.92	9.02	8.27	7.63
$\frac{15}{16}$	28.5	22.8	19.0	16.3	14.2	12.7	11.4	10.4	9.49	8.76
1	32.4	25.9	21.6	18.5	16.2	14.4	13.0	11.8	10.8	9.97
$1\frac{1}{16}$	36.6	29.3	24.4	20.9	18.3	16.3	14.6	13.3	12.2	11.3
$1\frac{1}{8}$	41.0	32.8	27.3	23.4	20.5	18.2	16.4	14.9	13.7	12.6
$1\frac{3}{16}$	45.7	36.6	30.5	26.1	22.8	20.3	18.3	16.6	15.2	14.1
$1\frac{1}{4}$	50.6	40.5	33.8	28.9	25.3	22.5	20.3	18.4	16.9	15.6

$F_y = 50$ ksi

Table 11-1 (cont.).
Preliminary Hanger Connection Selection Table
Design tensile strength, kips per linear in.,
limited by flexural yielding of the flange

t, in.	b, in.									
	1	$1\frac{1}{4}$	$1\frac{1}{2}$	$1\frac{3}{4}$	2	$2\frac{1}{4}$	$2\frac{1}{2}$	$2\frac{3}{4}$	3	$3\frac{1}{4}$
$\frac{5}{16}$	4.39	3.52	2.93	2.51	2.20	1.95	1.76	1.60	1.46	1.35
$\frac{3}{8}$	6.33	5.06	4.22	3.62	3.16	2.81	2.53	2.30	2.11	1.95
$\frac{7}{16}$	8.61	6.89	5.74	4.92	4.31	3.83	3.45	3.13	2.87	2.65
$\frac{1}{2}$	11.3	9.00	7.50	6.43	5.63	5.00	4.50	4.09	3.75	3.46
$\frac{9}{16}$	14.2	11.4	9.49	8.14	7.12	6.33	5.70	5.18	4.75	4.38
$\frac{5}{8}$	17.6	14.2	11.7	10.0	8.79	7.81	7.03	6.39	5.86	5.41
$\frac{11}{16}$	21.3	17.0	14.2	12.2	10.6	9.45	8.51	7.73	7.09	6.54
$\frac{3}{4}$	25.3	20.3	16.9	14.5	12.7	11.3	10.1	9.20	8.44	7.79
$\frac{13}{16}$	29.7	23.8	19.8	17.0	14.9	13.2	11.9	10.8	9.90	9.14
$\frac{7}{8}$	34.5	27.6	23.0	19.7	17.2	15.3	13.8	12.5	11.5	10.6
$\frac{15}{16}$	39.6	31.6	26.4	22.6	19.8	17.6	15.8	14.4	13.2	12.2
1	45.0	36.0	30.0	25.7	22.5	20.0	18.0	16.4	15.0	13.8
$1\frac{1}{16}$	50.8	40.6	33.9	29.0	25.4	22.6	20.3	18.5	16.9	15.6
$1\frac{1}{8}$	57.0	45.6	38.0	32.5	28.5	25.3	22.8	20.7	19.0	17.5
$1\frac{3}{16}$	63.5	50.8	42.3	36.3	31.7	28.2	25.4	23.1	21.2	19.5
$1\frac{1}{4}$	70.3	56.3	46.9	40.2	35.2	31.3	28.1	25.6	23.4	21.6

if $\beta < 1$, set $\alpha' =$ the lesser of 1.0 and

$$\frac{1}{\delta}\left(\frac{\beta}{1-\beta}\right)$$

where δ, the ratio of the net area at the bolt line to the gross area at the face of the stem or angle leg, is

$$\delta = 1 - \frac{d'}{p}$$

The required flange thickness t_{req} may then be calculated as:

$$t_{req} = \sqrt{\frac{4.44 r_{ut} b'}{p F_y (1 + \delta\alpha')}}$$

and

$d' =$ width of bolt hole parallel to the tee stem or angle leg, in.

$p =$ length of flange, parallel to the tee stem or angle leg, tributary to each bolt, in. Note that p should preferably not exceed the gage between bolts illustrated in Figure 11-2.

If $t_{req} \leq t$, the preliminary fitting is satisfactory. Otherwise, a section with a thicker flange, or a change in geometry (i.e., b and p) is required.

The factored prying force per bolt q_u may be calculated from α as follows:

$$\alpha = \frac{1}{\delta}\left[\frac{r_{ut}}{\phi r_n}\left(\frac{t_c}{t}\right)^2 - 1\right] \geq 0$$

$$q_u = \phi r_n\left[\delta\alpha\rho\left(\frac{t}{t_c}\right)^2\right]$$

and the factored force per bolt including prying action is $r_{ut} + q_u$. In the above equations, t_c, the flange or angle thickness required to develop the design strength of the bolt ϕr_n with no prying action, is calculated as:

$$t_c = \sqrt{\frac{4.44 \phi r_n b'}{p F_y}}$$

Designing to Minimize Prying Action

In applications where the prying force q_u must be reduced to an insignificant amount, set $\alpha' = 0$ and calculate t_{req} as:

$$t_{req} = \sqrt{\frac{4.44 r_{ut} b'}{p F_y}}$$

Analyzing a Connection for Prying Action
The foregoing procedure is somewhat simplified when analyzing a connection for prying action. As before, check that $r_{ut} \leq \phi r_n$. Then calculate α' as:

$$\alpha' = \frac{1}{\delta(1+\rho)} \left[\left(\frac{t_c}{t}\right)^2 - 1 \right]$$

If $\alpha' < 0$, r_{ut} must be such that

$$r_{ut} \leq \phi r_n$$

If $0 \leq \alpha' \leq 1$, r_{ut} must be such that

$$r_{ut} \leq \phi r_n \left(\frac{t}{t_c}\right)^2 (1 + \delta\alpha')$$

If $\alpha' > 1$, r_{ut} must be such that

$$r_{ut} \leq \phi r_n \left(\frac{t}{t_c}\right)^2 (1 + \delta)$$

If desired, the factored prying force per bolt q_u may be determined as before.

EXAMPLE 11-1

Given: Refer to Figure 11-3. Design a WT tension-hanger connection for a 2L3×3×5⁄16 tension member to W24×94 beam connection. For the beam and WT, $F_y = 50$ ksi and $F_u = 65$ ksi; for the double angles, $F_y = 36$ ksi

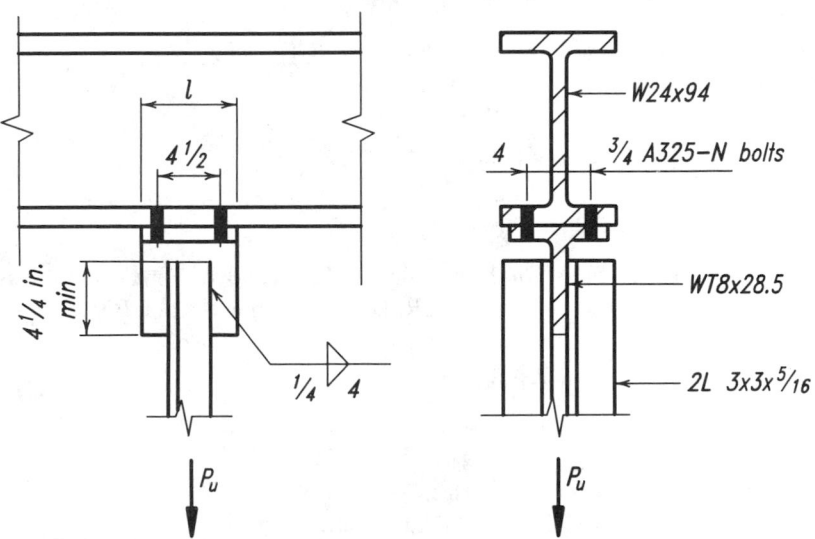

Fig. 11-3. Illustration for Example 11-1.

and $F_u = 58$ ksi. Use ¾-in. diameter ASTM A325-N bolts and 70 ksi electrodes.

$P_u = 80$ kips

W24×94

> $d = 24.31$ in. $b_f = 9.065$ in.
> $t_w = 0.515$ in. $t_f = 0.875$ in.

2L3×3×⁵⁄₁₆

> $A = 3.55$ in.²

Solution: *Check tension yielding of angles*

$$\phi R_n = \phi F_y A_g$$
$$= 0.9(36 \text{ ksi})(3.55 \text{ in.}^2)$$
$$= 115 \text{ kips} > 80 \text{ kips} \quad \textbf{o.k.}$$

Check tension rupture of angles (Design welds to find length of connection and U)

Try ¼-in. fillet welds

$$L_{min} = \frac{P_u}{1.392D}$$
$$= \frac{80 \text{ kips}}{1.392 \text{ (4 sixteenths)}}$$
$$= 14.4 \text{ in.}$$

Use four 4-in. welds (16 in. total), one at each toe and heel of each angle.

Calculate effective net area

From LRFD Specification Section B3

$$U = 1 - \frac{\overline{x}}{L} \le 0.9$$
$$= 1 - \frac{0.865 \text{ in.}}{4 \text{ in.}}$$
$$= 0.784$$

(Note: in lieu of the calculation shown above, U may be taken as 0.75 since $1.5w > 1 \ge w$ per LRFD Specification Section B3.2d)

$$A_e = UA_n$$
$$= 0.784(3.55 \text{ in.}^2)$$
$$= 2.78 \text{ in.}^2$$
$$\phi R_n = \phi F_u A_e$$
$$= 0.75(58 \text{ ksi})(2.78 \text{ in.}^2)$$
$$= 121 \text{ kips} > 80 \text{ kips} \quad \textbf{o.k.}$$

Select preliminary **WT** *using beam gage g = 4 in.*

With four ¾-in. diameter A325-N bolts,

$$r_{ut} = \frac{P_u}{n} = \frac{80 \text{ kips}}{4 \text{ bolts}} = 20 \text{ kips/bolt}$$

Since for ¾-in. diameter A325N bolts $\phi r_n = 29.8$ kips (> 20 kips), the bolts are **o.k.**

With four bolts, the maximum effective length is $2g = 8$ in. Thus, there are 4 in. of tee length tributary to each pair of bolts and

$$\frac{2 \text{ bolts } (20 \text{ kips/bolt})}{4 \text{ in.}} = 10.0 \text{ kips/in.}$$

The minimum depth of WT that can be used is equal to the sum of the weld length plus the weld size plus the k-dimension for the selected section. From Table 11-1 with an assumed $b = 4$ in./2 = 2 in., $t_o \approx \text{}^{11}\!/_{16}$-in., and $d_{min} = 4$ in. + ¼-in. + $k \approx 6$ in., appropriate selections include:

WT6×39.5   WT8×28.5
WT7×34   WT9×30

Try **WT8×28.5**; $b_f = 7.12$ in., $t_f = 0.715$ in., $t_w = 0.430$ in.

Check prying action with WT8×28.5×0′–8

$$b = \frac{g - t_w}{2}$$

$$= \frac{4 \text{ in.} - 0.430 \text{ in.}}{2}$$

$$= 1.79 \text{ in.} > 1\text{¼-in. entering and tightening clearance,} \quad \textbf{o.k.}$$

$$a = \frac{b_f - g}{2}$$

$$= \frac{7.12 \text{ in.} - 4 \text{ in.}}{2}$$

$$= 1.56 \text{ in.}$$

Since $a = 1.56$ in. is less than $1.25b = 2.24$ in., use $a = 1.56$ in.

$$b' = b - d/2$$

$$= 1.79 \text{ in.} - \frac{\text{¾-in.}}{2}$$

$$= 1.42 \text{ in.}$$

$$a' = a + \frac{d}{2}$$

$$= 1.56 \text{ in.} + \frac{\text{¾-in.}}{2}$$

$$= 1.94 \text{ in.}$$

$$\rho = \frac{b'}{a'}$$

$$= \frac{1.42 \text{ in.}}{1.94 \text{ in.}}$$

$$= 0.732$$

$$\beta = \frac{1}{\rho}\left(\frac{\phi r_n}{r_{ut}} - 1\right)$$

$$= \frac{1}{0.732}\left(\frac{29.8 \text{ kips / bolt}}{20 \text{ kips / bolt}} - 1\right)$$

$$= 0.669$$

Since $\beta < 1.0$,

$$\delta = 1 - \frac{d'}{p}$$

$$= 1 - \frac{^{13}\!/_{16}\text{-in.}}{4 \text{ in.}}$$

$$= 0.797$$

$$\alpha' = \frac{1}{\delta}\left(\frac{\beta}{1-\beta}\right) \le 1.0$$

$$= \frac{1}{0.797}\left(\frac{0.669}{1-0.669}\right)$$

$$= 2.54 \rightarrow 1.0$$

$$t_{req} = \sqrt{\frac{4.44\, r_{ut}\, b'}{p F_y\, (1 + \delta\alpha')}}$$

$$= \sqrt{\frac{4.44(20 \text{ kips / bolt})(1.42 \text{ in.})}{(4 \text{ in.})(50 \text{ ksi})[1 + (0.797)(1.0)]}}$$

$$= 0.592 \text{ in.} < t_f = 0.715 \text{ in.} \quad \textbf{o.k.}$$

Check design tensile strength of bolts.

(Note this calculation is optional; the required thickness t_{req}, calculated above, will keep the total bolt tensile force $r_{ut} + q_u$ less than the design strength ϕr_n. It is included for information only.)

Calculate q_u

$$t_c = \sqrt{\frac{4.44(\phi r_n) b'}{p F_y}}$$

$$= \sqrt{\frac{4.44(29.8 \text{ kips / bolt})(1.42 \text{ in.})}{4 \text{ in.} \times 50 \text{ ksi}}}$$

$$= 0.969 \text{ in.}$$

$$\alpha = \frac{1}{\delta}\left[\frac{r_{ut}}{\phi r_n}\left(\frac{t_c}{t}\right)^2 - 1\right] \ge 0$$

$$= \frac{1}{0.797} \left[\frac{20 \text{ kips / bolt}}{29.8 \text{ kips / bolt}} \left(\frac{0.969 \text{ in.}}{0.715 \text{ in.}} \right)^2 - 1 \right]$$
$$= 0.292$$

$$q_u = \phi r_n \left[\delta \alpha \rho \left(\frac{t}{t_c} \right)^2 \right]$$

$$= 29.8 \text{ kips/bolt} \left[0.797(0.292)(0.732) \left(\frac{0.715 \text{ in.}}{0.969 \text{ in.}} \right)^2 \right]$$
$$= 2.76 \text{ kips/bolt}$$

Total tension on bolt

$$r_{ut} + q_u = 20 \text{ kips/bolt} + 2.76 \text{ kips/bolt}$$
$$= 22.8 \text{ kips/bolt} < 29.8 \text{ kips/bolt} \quad \textbf{o.k.}$$

Check the **WT** hanger as follows:

Check tension yielding of the tee stem on the Whitmore section (see sketch below)

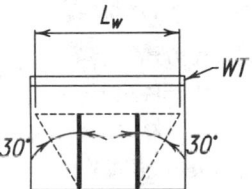

The effective width of the tee stem (which cannot exceed the actual width of 8 in.) is

$$L_w = 3 \text{ in.} + 2(4 \text{ in.} \times \tan 30°) \le 8 \text{ in.}$$
$$= 7.62 \text{ in.}$$

and the design strength is

$$\phi R_n = \phi F_y A_{g \text{ eff}}$$
$$= 0.9(50 \text{ ksi})(7.62 \text{ in.} \times 0.430 \text{ in.})$$
$$= 147 \text{ kips} > 80 \text{ kips} \quad \textbf{o.k.}$$

Check shear yielding of the base metal along the toe and heel of each weld line.

$$\phi R_n = \phi (0.6 F_y) A_g$$
$$= 0.9(0.6 \times 50 \text{ ksi})(4 \times 4 \text{ in.} \times 0.430 \text{ in.})$$
$$= 186 \text{ kips} > 80 \text{ kips} \quad \textbf{o.k.}$$

Check shear rupture of the base metal along the toe and heel of each weld line.

$$\phi R_n = \phi (0.6 F_u) A_n$$
$$= 0.75(0.6 \times 65 \text{ ksi})(4 \times 4 \text{ in.} \times 0.430 \text{ in.})$$
$$= 201 \text{ kips} > 80 \text{ kips} \quad \textbf{o.k.}$$

Check shear rupture of the flanges.

$\phi R_n = \phi [0.6F_u A_n]$
$= 0.75[0.6(65 \text{ ksi})(2 \times 8 \text{ in.} - 4 \times 0.875 \text{ in.})(0.715 \text{ in.})]$
$= 261 \text{ kips} > 80 \text{ kips}$ **o.k.**

Check shear yielding of the flanges.

$\phi R_n = \phi[0.6F_y A_g]$
$= 0.9[0.6 (50 \text{ ksi})(2 \times 8 \text{ in.} \times 0.715 \text{ in.})]$
$= 309 \text{ kips} > 80 \text{ kips}$ **o.k.**

Check block shear rupture of the tee stem.

From LRFD Specification Section J4.3

$0.6F_u A_{nv} = 0.6(65 \text{ ksi})(2 \times 4 \text{ in.} \times 0.430 \text{ in.})$
$= 134 \text{ kips}$
$F_u A_{nt} = (65 \text{ ksi})(3 \text{ in.} \times 0.430 \text{ in.})$
$= 83.9 \text{ kips}$

Since $0.6F_u A_{nv} > F_u A_{nt}$,

$\phi R_n = \phi [0.6F_u A_{nv} + F_y A_{gt}]$
$= 0.75[134 \text{ kips} + (50 \text{ ksi})(3 \text{ in.} \times 0.430 \text{ in.})]$
$= 149 \text{ kips} > 80 \text{ kips}$ **o.k.**

Comments: Alternatively, a **WT** tension hanger could be selected with a flange thickness which would reduce the effect of prying action to an insignificant amount, i.e., $q_u \approx 0$. Using $b' = 1.42$ as an assumption,

$$t_{req} = \sqrt{\frac{4.44r_{ut}b'}{pF_y}}$$

$$= \sqrt{\frac{4.44(20 \text{ kips}/\text{bolt})(1.42 \text{ in.})}{4 \text{ in.}/\text{bolt} (50 \text{ ksi})}}$$

$$= 0.794 \text{ in.}$$

A **WT**9×35.5 with $t_f = 0.810$ in., $t_w = 0.495$ in. (> 0.430 in.), and $b_f = 7.635$ in. is adequate.

DIAGONAL BRACING CONNECTIONS

If the members in the unbraced frame of Figure 11-4a were connected with simple shear connections, the lateral force H_u acting from the left would cause the building to "rack" or deflect laterally as shown by the dashed lines. In fact, the frame would be unstable under gravity loading. In lieu of a frame with moment connections, frame stability and resistance to lateral loads can be provided by diagonal bracing members. Whereas moment connections resist lateral loads through flexure in the beams and columns which comprise the frame, diagonal bracing members create a vertical truss which transfers the lateral loads through the members of the truss as axial forces. Although a diagonally braced frame is, in general, more efficient than a frame with moment connections, the use of diagonal bracing may be precluded by interference with architectural features such as corridors, windows, and doors.

Diagonal bracing may be concentric or eccentric. Eccentrically braced frames are commonly used in seismic regions; their design is beyond the scope of this book; refer to Ishler (1992), Popov, et al. (1989) and Lindsay and Goverdahn (1989). The following discussion is limited to concentric diagonal bracing.

The concentric diagonal brace shown in Figure 11-4b will provide for stability and lateral forces acting from the left; the diagonal brace is in tension (+) and induces only axial forces in the other members of the frame. Since the lateral forces may be incident from either the right or the left, two diagonal braces would be used, as shown in Figure 11-4c.

As the stiffnesses of the diagonal bracing members increase, lateral forces will divide (not necessarily equally) between the two diagonal braces with one in tension and the other in compression. It is normal practice to neglect the strength of the diagonal in compression and design each diagonal for the tension which results from the lateral loads; this is called tension-only bracing.

Figure 11-5a shows the vertical arrangement of X-bracing in a single bay of a multistory building. Figure 11-5b shows a common type of K-bracing. Figures 11-5c and 11-5d show bracing which is composed of members subjected to both tension and compression; as shown, this bracing occupies a single bay of a multistory building. Figure 11-5e is similar except the bracing occupies two adjacent bays of a multistory building. Other arrangements, such as the one shown in Figure 11-5f, are also possible.

When possible, diagonal bracing should be located in a bay or bays at the mid-section of a building. In buildings with expansion joints, diagonal bracing should be located in a bay or bays at the mid-section between expansion joints. Furthermore, this bracing

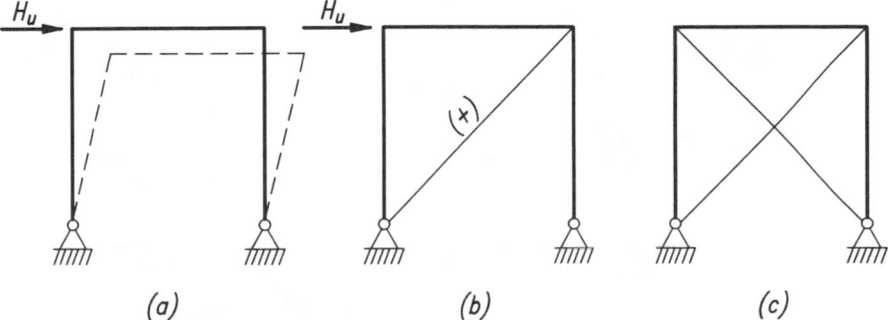

Fig. 11-4. Lateral forces and diagonal bracing.

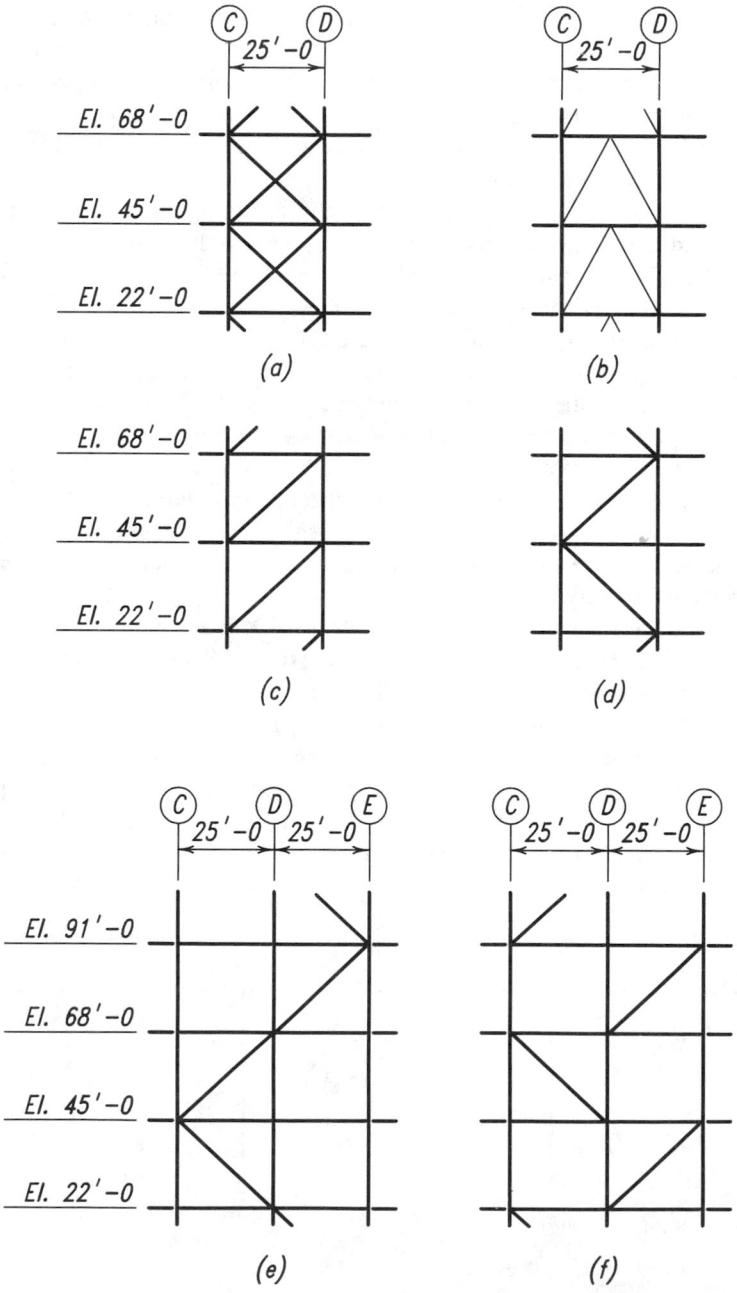

Fig. 11-5. *Diagonal bracing in multistory buildings.*

should preferably be located in the corresponding bay or bays in other frames across the width of the building; symmetrically located diagonal bracing minimizes torsional effects of lateral load on the overall structure.

Diagonal Bracing Members
Diagonal bracing members may be rods, single angles or channels, double angles or channels, tees, W shapes, or tubes as required by the lateral loads.

Slender diagonal bracing members are relatively flexible and, thus, vibration and sag may be considerations. In slender tension-only bracing, these problems can be minimized with "draw" or pretension created by shortening the fabricated length of the diagonal brace from the theoretical length L between member working points. In general, the following deductions will be sufficient: no deduction for $L \leq 10$ ft; deduct $\frac{1}{16}$-in. for 10 ft $< L \leq 20$ ft; deduct $\frac{1}{8}$-in. for $20 < L \leq 35$ ft; and, deduct $\frac{3}{16}$-in. for $L > 35$ ft. This principle is not applicable to diagonal bracing members other than light angles since it is difficult to stretch heavier members; vibration and sag are not usually design considerations in heavier diagonal bracing members. In any diagonal bracing member, however, it is permissible to deduct an additional $\frac{1}{32}$-in. when necessary to avoid dimensioning to thirty-seconds of an inch.

When double-angle diagonal bracing members are separated, as at "sandwiched" end connections to gussets, intermittent connections must be provided if the unsupported length of the diagonal brace exceeds the limits specified in LRFD Specification Section D2 for tension members or LRFD Specification Section E4 for compression members; note that a minimum of two stitch-fillers is required. These may be made with either bolted or welded stitch-fillers. Many fabricators prefer ring or rectangular bolted stitch-fillers when the angles require other punching, as at the end connections. In welded construction, a stitch-filler with protruding ends, as shown in Figure 11-6a is preferred because it is easy to fit and weld. The short stitch-filler shown in Figure 11-6b is used if a smooth appearance is desired.

When a full-length filler is provided, as in corrosive environments, the maximum spacing of stitch bolts should be as specified in LRFD Specification Section J3.5. Alternatively, the edges of the filler may be seal welded.

Force Transfer in Diagonal Bracing Connections
There has been some controversy as to which of several available analysis methods provides the best means for the safe and economical design and analysis of diagonal

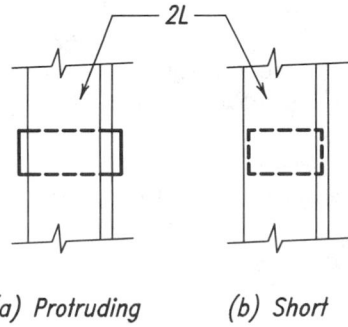

(a) Protruding   (b) Short

Fig. 11-6. Welded stitch fillers.

bracing connections. To resolve this situation, starting in 1981, AISC sponsored extensive computer studies of this connection by Richard (1986). Associated with Richard's work, full scale tests were performed by Bjorhovde and Chakrabarti (1985), Gross and Cheok (1988), and Gross (1990). Also, AISC and ASCE formed a task group to recommend a design method for this connection. In 1990, this task group recommended three methods for further study; refer to Appendix A of Thornton (1991).

Using the results of the aforementioned full scale tests, Thornton (1991) showed that these three methods yield safe designs, and that of the three methods, the Uniform Force Method (see Model 3 of Thornton, 1991) best predicts both the design strength and critical limit state of the connection. Furthermore, Thornton (1992) showed that the Uniform Force Method yields the most economical design through comparison of actual designs by the different methods and through consideration of the efficiency of force transmission. For the above reasons, and also because it is the most versatile method, the Uniform Force Method has been adopted for use in this book.

The Uniform Force Method—The essence of the Uniform Force Method is to select the geometry of the connection so that moments do not exist on the three connection interfaces; i.e., gusset-to-beam, gusset-to-column, and beam-to-column. In the absence of moment, these connections may then be designed for shear and/or tension only, hence the origin of the name Uniform Force Method.

With the working point chosen at the intersection of the centerlines of the beam, column, and diagonal brace as shown in Figure 11-7a, four geometric parameters e_b, e_c, α, and β can be identified, where

e_b = one-half the depth of the beam, in.

e_c = one-half the depth of the column, in. Note that, for a column web support, $e_c \approx 0$.

α = distance from the face of the column flange or web to the centroid of the gusset-to-beam connection, in.

β = distance from the face of the beam flange to the centroid of the gusset-to-column connection, in.

For the force distribution shown in the free-body diagrams of Figures 11-7b, 11-7c, and 11-7d to remain free of moments on the connection interfaces, the following expression must be satisfied.

$$\alpha - \beta\tan\theta = e_b\tan\theta - e_c \qquad (11\text{-}1)$$

Since the variables on the right of the equal sign (e_b, e_c, and θ) are all defined by the members being connected and the geometry of the structure, the designer may select values of α and β for which the equation is true, thereby locating the centroids of the gusset-to-beam and gusset-to-column connections.

Once α and β have been determined, the factored axial and shear forces for which these connections must be designed can be determined from the following equations.

$$V_{uc} = \frac{\beta}{r}P_u \qquad H_{uc} = \frac{e_c}{r}P_u$$

$$H_{ub} = \frac{\alpha}{r}P_u \qquad V_{ub} = \frac{e_b}{r}P_u$$

where

$$r = \sqrt{(\alpha + e_c)^2 + (\beta + e_b)^2}$$

The gusset-to-beam connection must be designed for the factored shear force H_{ub} and the factored axial force V_{ub}, the gusset-to-column connection must be designed for the

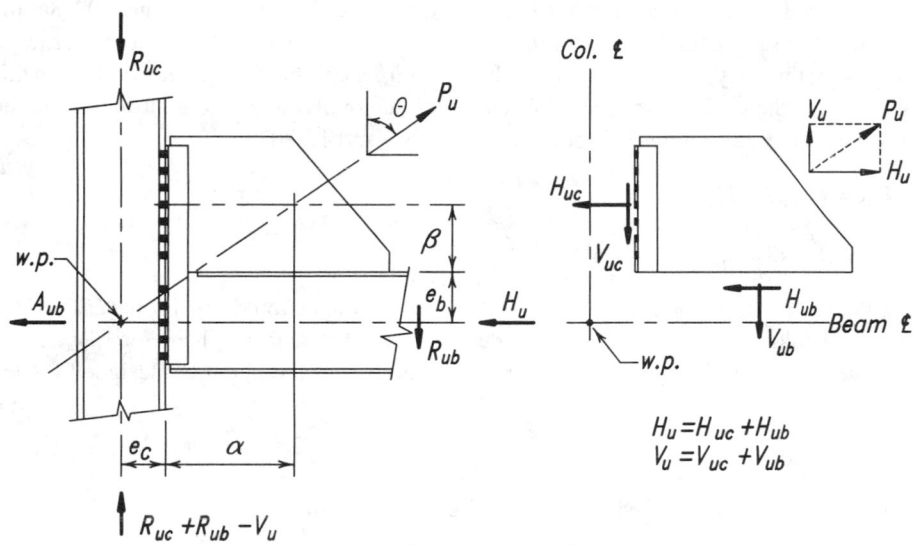

$$H_u = H_{uc} + H_{ub}$$
$$V_u = V_{uc} + V_{ub}$$

(a) Diagonal bracing connection and external forces

(b) Gusset free-body diagram

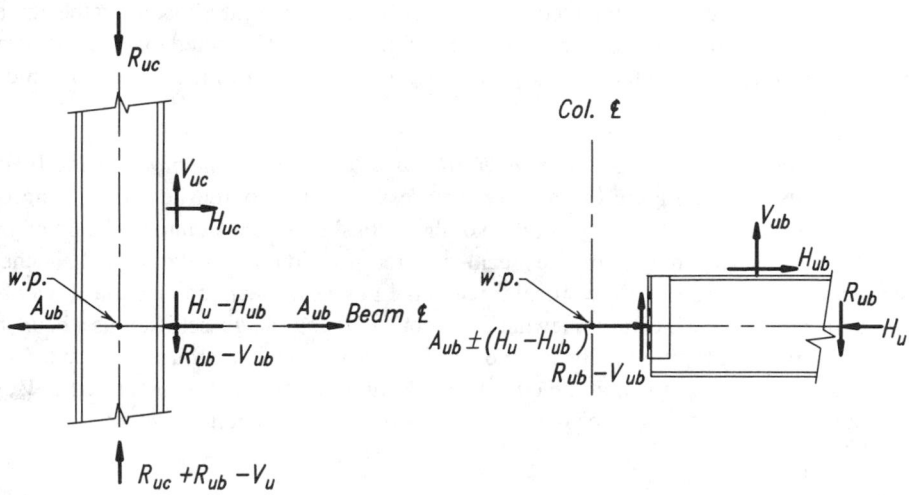

(c) Column free-body diagram

(d) Beam free-body diagram

Fig. 11-7. Force transfer, Uniform Force Method.

factored shear force V_{uc} and the factored axial force H_{uc}, and the beam-to-column connection must be designed for the factored shear $R_u - V_{ub}$ and the factored axial force $A_{ub} \pm (H_u - H_{ub})$, where R_u is the factored end reaction of the beam and A_{ub} is the factored axial force in the beam (see Figure 11-7). Note that, while P_u is shown as a tensile force, it may also be a compressive force; were this the case the signs of the resulting gusset forces would change.

Special Case 1, Modified Working Point Location—As illustrated in Figure 11-8a, the working point in Special Case 1 of the Uniform Force Method is chosen at the corner of the gusset; this may be done to simplify layout or for a column web connection. With this assumption, the terms in the gusset force equations involving e_b and e_c drop out and the interface forces, as shown in Figures 11-8b, 11-8c, and 11-8d, are:

$$H_{ub} = P_u \sin\theta = H_u \qquad V_{ub} = 0$$

$$V_{uc} = P_u \cos\theta = V_u \qquad H_{uc} = 0$$

The gusset-to-beam connection must be designed for the factored shear force H_{ub} and the gusset-to-column connection must be designed for the factored shear force V_{uc}. Note, however, that the change in working point requires that the beam be designed for the factored moment M_{ub}, where

$$M_{ub} = H_{ub}e_b$$

and the column must be designed for the factored moment M_{uc} where

$$M_{uc} = \frac{V_{uc}e_c}{2}$$

An example demonstrating this eccentric special case is presented in AISC (1984). This eccentric case was endorsed by the AISC/ASCE task group (Thornton, 1991) as a reduction of the three recommended methods when the work point is located at the gusset corner. While calculations are somewhat simplified, it should be noted that resolution of the factored force P_u into the shears V_{uc} and H_{ub} may not result in the most economical connection.

Special Case 2, Minimizing Shear in the Beam-to-Column Connection—If the brace force, as illustrated in Figure 11-9a, were compressive instead of tensile and the factored beam reaction R_{ub} were high, the addition of the extra shear force V_{ub} into the beam might exceed the design strength of the beam and require doubler plates or a haunched connection. Alternatively, the vertical force in the gusset-to-beam connection V_{ub} can be limited in a manner which is somewhat analogous to using the gusset itself as a haunch.

As illustrated in Figure 11-9b, assume that V_{ub} is reduced by an arbitrary amount ΔV_{ub}. By statics, the vertical force at the gusset-to-column interface will be increased to $V_{uc} + \Delta V_{ub}$, and a moment M_{ub} will result on the gusset-to-beam connection, where

$$M_{ub} = (\Delta V_{ub})\alpha$$

If ΔV_{ub} is taken equal to V_{ub}, none of the vertical component of the brace force is transmitted to the beam; the resulting procedure is that presented by AISC (1984) for concentric gravity axes, extended to connections to column flanges. This method was also recommended by the AISC/ASCE task group as the "Engineering for Steel Construction" method.

Design by this method may be uneconomical. It is very punishing to the gusset and beam because of the moment M_{ub} induced on the gusset-to-beam connection. This moment will require a larger connection and a thicker gusset. Additionally, the limit state

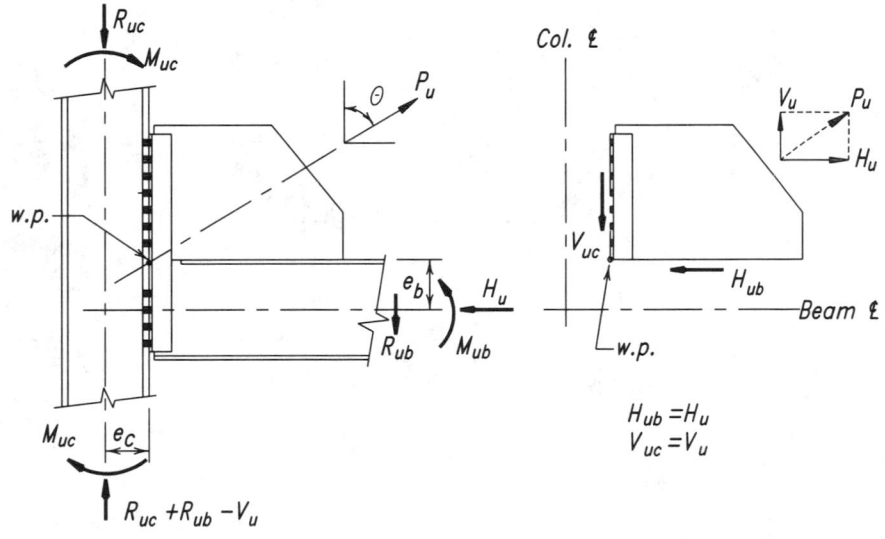

(a) Diagonal bracing connection (b) Gusset free-body diagram

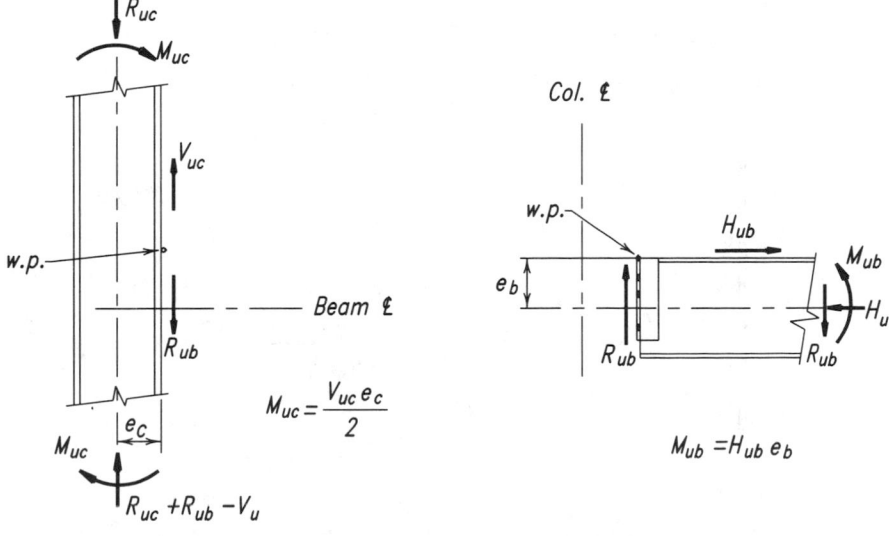

(c) Column free-body diagram (d) Beam free-body diagram

Fig. 11-8. Force transfer, Special Case 1.

of local web yielding may limit the strength of the beam. This special case interrupts the natural flow of forces assumed in the Uniform Force Method and thus is best used when the beam-to-column interface is already highly loaded, independently of the brace, by a high shear R_u in the beam-to-column connection.

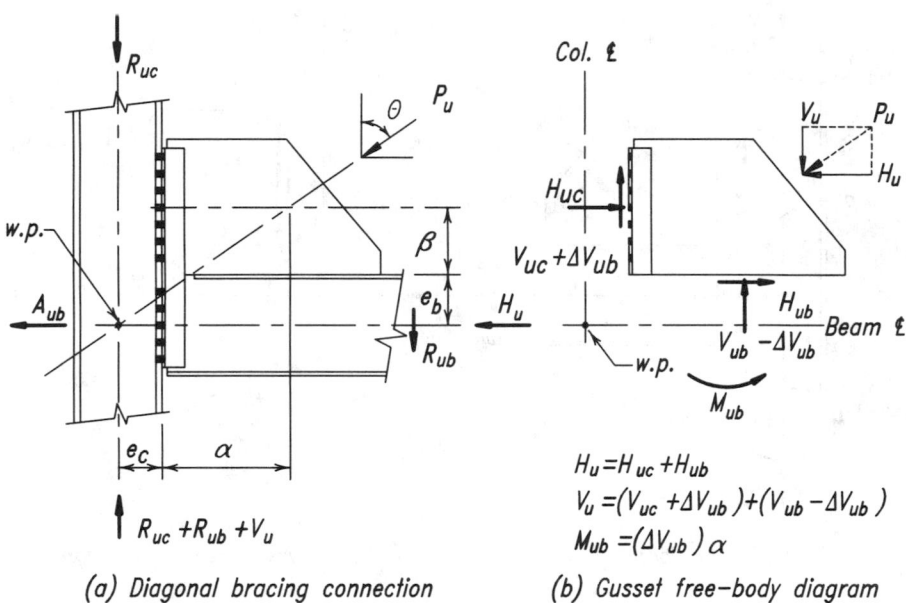

$$H_u = H_{uc} + H_{ub}$$
$$V_u = (V_{uc} + \Delta V_{ub}) + (V_{ub} - \Delta V_{ub})$$
$$M_{ub} = (\Delta V_{ub})\,\alpha$$

(a) Diagonal bracing connection   (b) Gusset free-body diagram

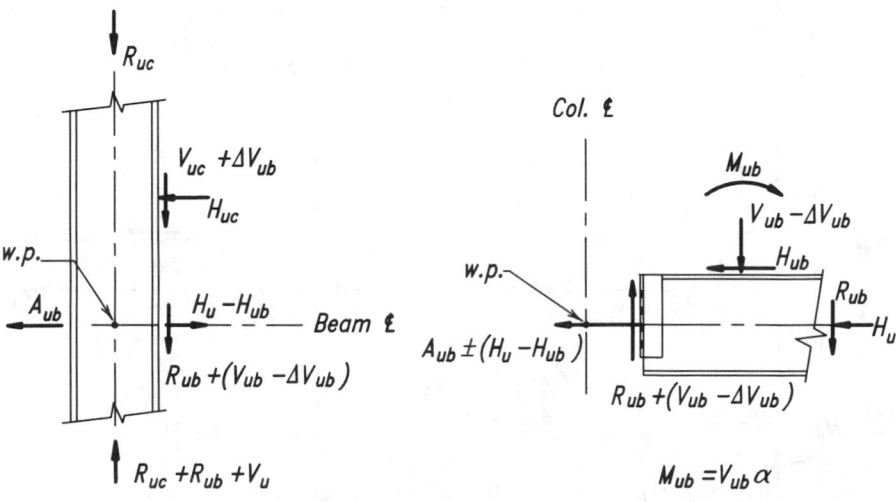

$$M_{ub} = V_{ub}\,\alpha$$

(c) Column free-body diagram   (d) Beam free-body diagram

Fig. 11-9. Force transfer, Special Case 2.

Special Case 3, No Gusset-to-Column Web Connection—When the connection is to a column web and the brace is shallow (as for large θ) or the beam is deep, it may be more economical to eliminate the gusset-to-column connection entirely and connect the gusset to the beam only. The Uniform Force Method can be applied to this situation by setting β and e_c equal to zero as illustrated in Figure 11-10. Since there is to be no gusset-to-column connection, V_{uc} and H_{uc} also equal zero. Thus, $V_{ub} = V_u$ and $H_{ub} = H_u$.

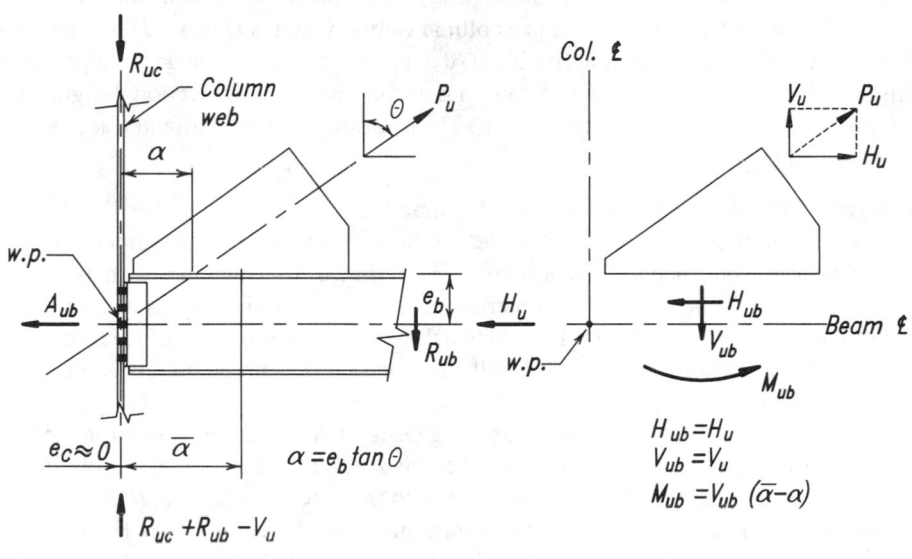

(a) Diagonal bracing connection

(b) Gusset free-body diagram

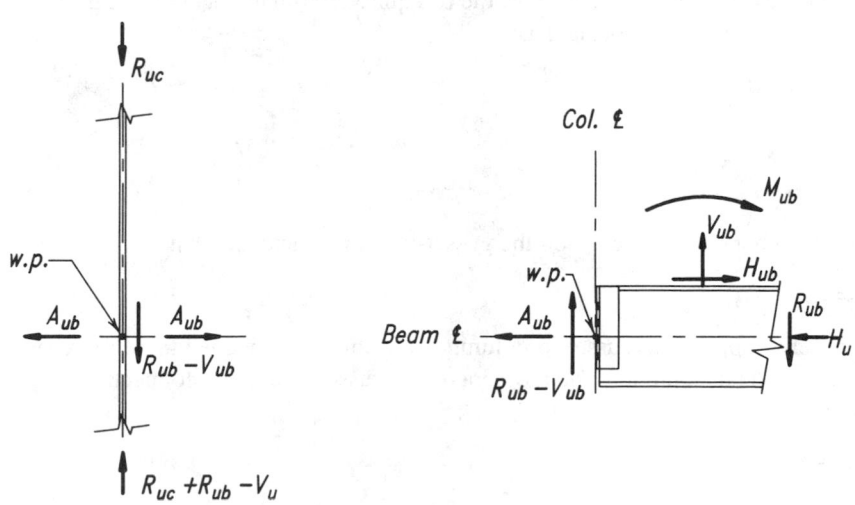

(c) Column free-body diagram

(d) Beam free-body diagram

Fig. 11-10. Force transfer, Special Case 3.

If $\bar{\alpha} = \alpha = e_b\tan\theta$, there is no moment on the gusset-to-beam interface and the gusset-to-beam connection can be designed for the factored shear force H_{ub} and the factored axial force V_{ub}. If $\bar{\alpha} \neq \alpha = e_b\tan\theta$, the gusset-to-beam interface must be designed for the moment M_{ub} in addition to H_{ub} and V_{ub}, where

$$M_{ub} = V_{ub}(\alpha - \bar{\alpha})$$

The beam-to-column connection must be designed for the factored shear force $R_u + V_{ub}$.

Note that, since the connection is to a column web, e_c is zero and hence H_c is also zero. For a connection to a column flange, if the gusset-to-column-flange connection is eliminated, the beam-to-column connection must be a moment connection designed for the moment $V_u\,e_c$ in addition to the shear V_u. Thus, uniform forces on all interfaces are no longer possible.

Analysis of Existing Diagonal Bracing Connections

A combination of α and β which provides for no moments on the three interfaces can usually be achieved when a connection is being designed. However, when analyzing an existing connection or when other constraints exist on gusset dimensions, the values of α and β may not satisfy Equation 11-1. When this happens, uniform interface forces will not satisfy equilibrium and moments will exist on one or both gusset edges or at the beam-to-column interface.

To illustrate this point, consider an existing design where the actual centroids of the gusset-to-beam and gusset-to-column connections are at $\bar{\alpha}$ and $\bar{\beta}$, respectively. If the connection at one edge of the gusset is more rigid than the other, it is logical to assume that the more rigid edge takes all of the moment necessary for equilibrium. For instance, the gusset of Figure 11-7 is shown welded to the beam and bolted with double angles to the column. For this configuration, the gusset-to-beam connection will be much more rigid than the gusset-to-column connection.

Take α and β as the ideal centroids of the gusset-to-beam and gusset-to-column connections, respectively. Setting $\beta = \bar{\beta}$, the α required for no moment on the gusset-to-beam connection may be calculated as:

$$\alpha = K + \bar{\beta}\tan\theta$$

where

$$K = e_b\tan\theta - e_c$$

If $\alpha \neq \bar{\alpha}$, a moment M_{ub} will exist on the gusset-to-beam connection where,

$$M_{ub} = V_{ub}(\alpha - \bar{\alpha})$$

Conversely, suppose the gusset-to-column connection were judged to be more rigid. Setting $\alpha = \bar{\alpha}$, the β required for no moment on the gusset-to-column connection may be calculated as:

$$\beta = \frac{\bar{\alpha} - K}{\tan\theta}$$

If $\beta \neq \bar{\beta}$, a moment M_{uc} will exist on the gusset-to-column connection where,

$$M_{uc} = H_{uc}(\beta - \bar{\beta})$$

If both connections were equally rigid and no obvious allocation of moment could be made, the moment could be distributed based on minimized eccentricities $\alpha - \overline{\alpha}$ and $\beta - \overline{\beta}$ by minimizing the objective function ϕ, where

$$\phi = \left(\frac{\alpha - \overline{\alpha}}{\overline{\alpha}}\right)^2 + \left(\frac{\beta - \overline{\beta}}{\overline{\beta}}\right)^2 - \lambda(\alpha - \beta\tan\theta - K)$$

In the above equation, λ is a Lagrange multiplier.
The values of α and β which minimize ϕ are:

$$\alpha = \frac{K'\tan\theta + K\left(\dfrac{\overline{\alpha}}{\overline{\beta}}\right)^2}{D}$$

and

$$\beta = \frac{(K' - K\tan\theta)}{D}$$

where

$$K' = \overline{\alpha}\left(\tan\theta + \frac{\overline{\alpha}}{\overline{\beta}}\right)$$

$$D = \tan^2\theta + \left(\frac{\overline{\alpha}}{\overline{\beta}}\right)^2$$

Design Checks

The design strengths of the bolts and/or welds and connected elements must be determined in accordance with the provisions of the LRFD Specification. The applicable limit states in each of the aforementioned design strengths are discussed in Part 8. In all cases, the design strength ϕR_n must exceed the required strength R_u. Note that when the gusset is directly welded to the beam or column, the connection must be designed for the larger of the peak stress and 1.4 times the average stress; this 40 percent increase is recommended to provide ductility to allow adequate force redistribution in the weld group. Additionally, the gusset must be checked on the Whitmore Section for yielding and for column buckling under compressive brace forces.

EXAMPLE 11-2

Given: Refer to Figure 11-11. Design the diagonal bracing connection between the W12×87 brace and the W18×106 beam and the W14×605 column. Use $\frac{7}{8}$-in. diameter A325-N bolts in standard holes and 70 ksi electrodes. For structural members, assume $F_y = 50$ ksi and $F_u = 65$ ksi; for connecting material, assume $F_y = 36$ ksi and $F_u = 58$ ksi.

W12×87

$d = 12.53$ in. $b_f = 12.125$ in. $A = 25.6$ in.^{2}

$t_w = 0.515$ in. $t_f = 0.810$ in.

W18×106

$d = 18.73$ in. $b_f = 11.200$ in. $k = 1\frac{5}{8}$-in.
$t_w = 0.590$ in. $t_f = 0.940$ in.

W14×605

$d = 20.92$ in. $b_f = 17.415$ in.
$t_w = 2.595$ in. $t_f = 4.160$ in.

Solution: *Brace-to-gusset connection*

Distribute brace force in proportion to web and flange areas.

Force in flange

$$P_{uf} = \frac{P_u (b_f t_f)}{A}$$

$$= \frac{675 \text{ kips} (12.125 \text{ in.} \times 0.810 \text{ in.})}{25.6 \text{ in.}^2}$$

$$= 259 \text{ kips}$$

Force in web

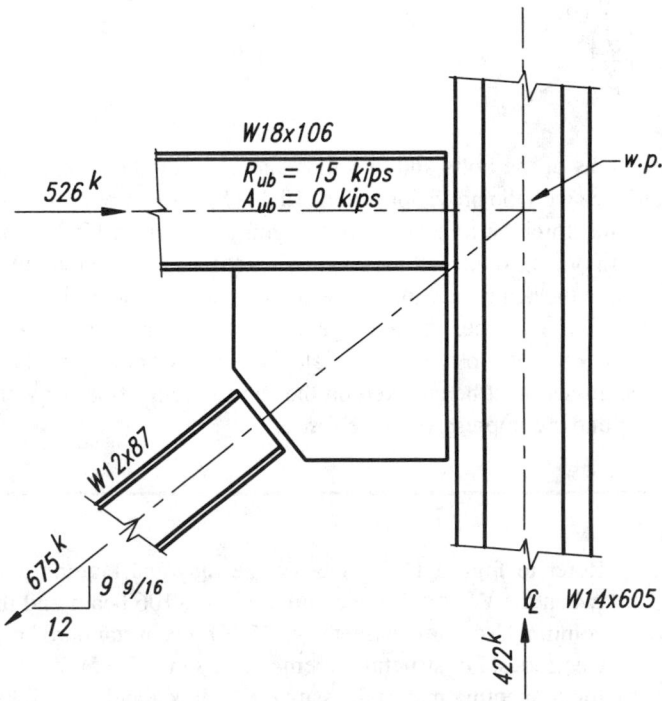

(a) members and forces

Figure 11-11a. Illustration for Example 11-2.

$$P_{uw} = P_u - 2P_{uf}$$
$$= 675 \text{ kips} - 2(259 \text{ kips})$$
$$= 157 \text{ kips}$$

Design brace-flange-to-gusset connection.

Determine number of $\frac{7}{8}$-in. diameter A325-N bolts required on the brace side (single shear) for shear.

$$n_{\min} = \frac{P_{uf}}{\phi r_n}$$
$$= \frac{259 \text{ kips}}{21.6 \text{ kips}/\text{bolt}}$$
$$= 11.99 \rightarrow 12 \text{ bolts}$$

On the gusset side, since these bolts are in double shear, half as many bolts will be required. Try six rows of two bolts each through the flange, six bolts through the gusset, and 2L4×4×¾ angles (A = 10.9 in.^{2}, $\bar{x}$ = 1.27 in.).

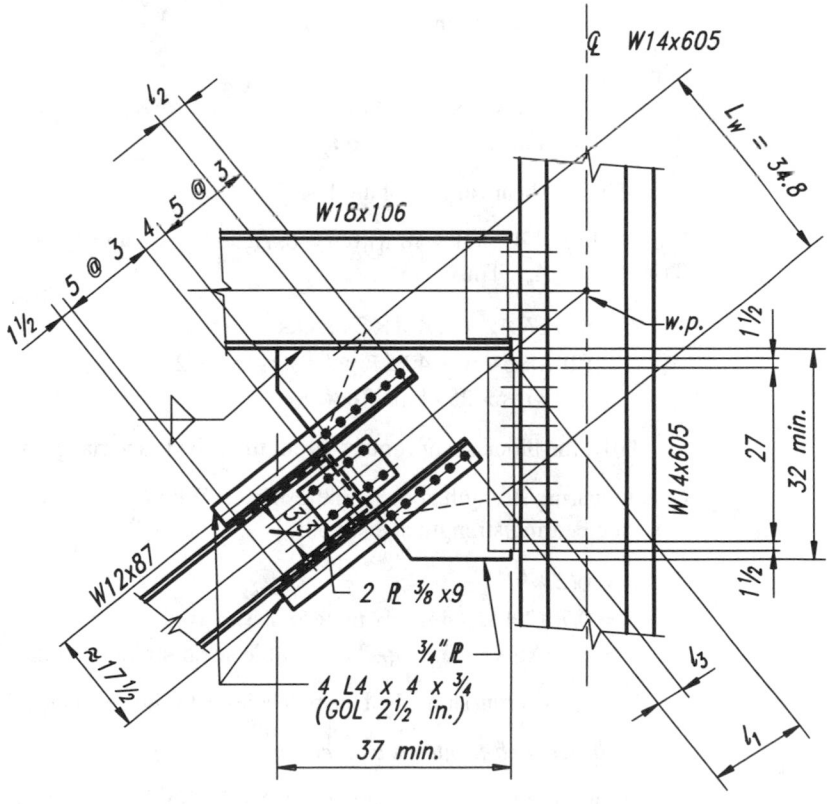

(b) connection

Figure 11-11b. Illustration for Example 11-2.

Check tension yielding of the angles

$$\phi R_n = \phi F_y A_g$$
$$= 0.90(36 \text{ ksi})(10.9 \text{ in.}^2)$$
$$= 353 \text{ kips} > 259 \text{ kips} \quad \textbf{o.k.}$$

Check tension rupture of the angles.

Taking A_e as the lesser of UA_n and $0.85A_g$, from LRFD Specification Sections B3.2 and J5.2, respectively

$$U = 1 - \frac{\bar{x}}{l} \le 0.9$$
$$= 1 - \frac{1.27 \text{ in.}}{15 \text{ in.}}$$
$$= 0.92 \rightarrow 0.9$$
$$UA_n = 0.9(10.9 \text{ in.}^2 - 2 \times 0.75 \text{ in.} \times 1 \text{ in.})$$
$$= 8.46 \text{ in.}^2$$
$$0.85A_g = 0.85(10.9 \text{ in.}^2)$$
$$= 9.27 \text{ in.}^2$$

Thus $A_e = UA_n = 8.46 \text{ in.}^2$

$$\phi R_n = \phi F_u A_e$$
$$= 0.75(58 \text{ ksi})8.46 \text{ in.}^2$$
$$= 368 \text{ kips} > 259 \text{ kips} \quad \textbf{o.k.}$$

Check block shear rupture of angles.

From Tables 8-47 and 8-48 with $n = 6$, $L_{ev} = 1\frac{1}{2}$-in., and $L_{eh} = 1\frac{1}{2}$-in., $0.6F_u A_{nv} > F_u A_{nt}$. Thus,

$$\phi R_n = \phi[0.6F_u A_{nv} + F_y A_{gt}] \times 2 \text{ blocks}$$
$$= (287 \text{ kips/in.} + 40.5 \text{ kips/in.})(\frac{3}{4}\text{-in.}) \times 2$$
$$= 491 \text{ kips} > 259 \text{ kips} \quad \textbf{o.k.}$$

Similarly, the block shear rupture strength of the brace flange is   **o.k.**

Check bearing strength at bolts in the angles. With $L_e = 1\frac{1}{2}$-in. and $s = 3$ in., the bearing strength at each bolt is

$$\phi r_n = \phi(2.4dtF_u)$$
$$= 0.75(2.4 \times \frac{7}{8}\text{-in.} \times \frac{3}{4}\text{-in.} \times 58 \text{ ksi})$$
$$= 68.5 \text{ kips} > 43.3 \text{ kips/bolt double shear strength} \quad \textbf{o.k.}$$

Similarly, the bearing strength of the bolt holes in the brace flange is   **o.k.**

Design brace-web-to-gusset connection

Determine number of $\frac{7}{8}$-in. diameter A325-N bolts required on the brace side (double shear) for shear.

$$n_{\min} = \frac{P_{uw}}{\phi r_n}$$

$$= \frac{157 \text{ kips}}{43.3 \text{ kips/ bolt}}$$

$$= 3.63 \rightarrow 4 \text{ bolts}$$

On the gusset side, the same number of bolts are required. Try two rows of two bolts and 2PL⅜×9.

Check tension yielding of the plates.

$\phi R_n = 0.90 F_y A_g$
$\quad = 0.90(36 \text{ ksi})(2 \times \text{⅜-in.} \times 9 \text{ in.})$
$\quad = 219 \text{ kips} > 157 \text{ kips}$ **o.k.**

Check tension rupture of the plates (LRFD Specification Section J5.2).

Taking A_e as the lesser of A_n and $0.85 A_g$,

$A_n = 2 \times \text{⅜-in.} \times 9 \text{ in.} - 4 \times \text{⅜-in.} \times 1 \text{ in.}$
$\quad = 5.25 \text{ in.}^2$
$0.85 A_g = 0.85(2 \times \text{⅜-in.} \times 9 \text{ in.})$
$\quad = 5.74 \text{ in.}^2$

Thus, $A_e = A_n = 5.25 \text{ in.}^2$

$\phi R_n = 0.75 F_u A_e$
$\quad = 0.75(58 \text{ ksi})5.25 \text{ in.}^2$
$\quad = 228 \text{ kips} > 157 \text{ kips}$ **o.k.**

Check block shear rupture of the plates (outer blocks) from Tables 8-47 and 8-48 with $n = 2$, $L_{ev} = 1\frac{1}{2}$, and $L_{eh} = 1\frac{1}{2}$, $0.6 F_u A_{nv} > F_u A_{nt}$. Thus,

$\phi R_n = \phi[0.6 F_u A_{nv} + F_y A_{gt}] \times 2 \text{ blocks} \times 2 \text{ plates}$
$\quad = (78.0 \text{ kips/in.} + 40.5 \text{ kips/in.})(\text{⅜-in.}) \times 4$
$\quad = 178 \text{ kips} > 157 \text{ kips}$ **o.k.**

Similarly, the block shear rupture strength of the interior blocks of the brace-web plates and the brace web are   **o.k.**

Check bearing strength of bolt holes in the plates. As before, with $L_e = 1\frac{1}{2}$-in. and $s = 3$ in., the bearing strength at each bolt hole is 68.5 kips which exceeds the double shear strength of the bolt and is   **o.k.**

Check tension yielding of the brace.

$\phi R_n = \phi F_y A_g$
$\quad = 0.90(50 \text{ ksi})(25.6 \text{ in.}^2)$
$\quad = 1,150 \text{ kips} > 675 \text{ kips}$ **o.k.**

Check tension rupture of the brace.

Taking A_e as A_n,

$A_n = 25.6 \text{ in.}^2 - (4 \times 0.810 \text{ in.} + 2 \times 0.515 \text{ in.})(1 \text{ in.})$
$\quad = 21.3 \text{ in.}^2$

Thus, $A_e = A_n = 21.3$

$$\phi R_n = \phi F_u A_n$$
$$= 0.75(65 \text{ ksi})21.3 \text{ in.}^2$$
$$= 1{,}040 \text{ kips} > 675 \text{ kips} \quad \textbf{o.k.}$$

Design gusset

From edge distance, spacing, and clearance requirements, try PL $\frac{3}{4}$-in. Check bearing strength at bolt holes. With $L_e = 1\frac{1}{2}$-in. and $s = 3$ in., the bearing strength at each bolt hole is

$$\phi r_n = \phi(2.4 dt F_u)$$
$$= 0.75(2.4 \times \tfrac{7}{8}\text{-in.} \times \tfrac{3}{4}\text{-in.} \times 58 \text{ ksi})$$
$$= 68.5 \text{ kips} > 43.3 \text{ kips/bolt double shear strength} \quad \textbf{o.k.}$$

Check block shear rupture for force transmitted through web.

From Tables 8-47 and 8-48 with $n = 2$, $L_{ev} = 1\frac{1}{2}$-in., $L_{eh} = 3$ in., $F_u A_{nt} > 0.6 F_u A_{nv}$. Thus

$$\phi R_n = \phi[0.6 F_y A_{gv} + F_u A_{nt}] \times 2 \text{ blocks}$$
$$= (81 \text{ kips/in.} + 109 \text{ kips/in.})(\tfrac{3}{4}\text{-in.}) \times 2 \text{ blocks}$$
$$= 285 \text{ kips} > 157 \text{ kips} \quad \textbf{o.k.}$$

Check block shear rupture for total brace force.

With $A_{gv} = 24.8$ in.^{2}, $A_{gt} = 13.2$ in.^{2}, $A_{nv} = 15.8$ in.^{2}, and $A_{nt} = 12.4$ in.^{2}, $F_u A_{nt} > 0.6 F_u A_{nv}$. Thus

$$\phi R_n = \phi[0.6 F_y A_{gv} + F_u A_{nt}]$$
$$= 0.75[0.6(36 \text{ ksi})(24.8 \text{ in.}^2) + (58 \text{ ksi})(12.4 \text{ in.}^2)]$$
$$= 941 \text{ kips} > 675 \text{ kips} \quad \textbf{o.k.}$$

Check tension yielding on the Whitmore section of the gusset. The Whitmore section, as illustrated with dashed lines in Figure 11-11b, is 34.8 in. long; 30.9 in. occurs in the gusset and 3.90 in. occurs in the beam web. Thus

$$\phi R_n = \phi F_y A_w$$
$$= 0.90[(36 \text{ ksi})(30.9 \times \tfrac{3}{4}\text{-in.}) + (50 \text{ ksi})(3.90 \text{ in.} \times 0.590 \text{ in.})]$$
$$= 854 \text{ kips} > 675 \text{ kips} \quad \textbf{o.k.}$$

The beam web thickness is used, conservatively ignoring the larger thickness in the beam-flange and flange-to-web-fillet area.

Note that, were this a compressive force, gusset buckling would have to be checked; refer to the comments at the end of this example.

Distribution of brace force to beam and column

From the members and frame geometry

$$e_b = \frac{d_b}{2} = \frac{18.73 \text{ in.}}{2} = 9.37 \text{ in.}$$

$$e_c = \frac{d_c}{2} = \frac{20.92 \text{ in.}}{2} = 10.5 \text{ in.}$$

$$\tan\theta = \frac{12}{9\%_{16}} = 1.25$$

and

$$e_b\tan\theta - e_c = 9.37 \text{ in.}(1.25) - 10.5 \text{ in.}$$
$$= 1.21 \text{ in.}$$

Try gusset PL¾-in. × 42 in. horizontally × 33 in. vertically (Several intermediate gusset dimensions were inadequate). With connection centroids at the midpoint of the gusset edges

$$\alpha = \frac{42 \text{ in.}}{2} + \tfrac{1}{2} \text{ in.}$$
$$= 21.5 \text{ in.}$$

where ½-in. is allowed for the setback between the gusset and the column, and

$$\overline{\beta} = \frac{33\text{in.}}{2}$$
$$= 16.5 \text{ in.}$$

Choosing $\beta = \overline{\beta}$, the $\overline{\alpha}$ required for uniform forces is

$$\overline{\alpha} = e_b\tan\theta - e_c + \beta \tan\theta$$
$$= 1.21 \text{ in.} + (16.5 \text{ in.})(1.25)$$
$$= 21.8 \text{ in.}$$

The resulting eccentricity is $\alpha - \overline{\alpha}$, where

$$\alpha - \overline{\alpha} = 21.5 \text{ in.} - 21.8 \text{ in.}$$
$$= -0.3 \text{ in.}$$

This slight eccentricity is negligible. Use $\alpha = 21.8$ in. and $\beta = 16.5$ in.

Calculate gusset interface forces

$$r = \sqrt{(\alpha + e_c)^2 + (\beta + e_b)^2}$$
$$= \sqrt{(21.8 \text{ in.} + 10.5 \text{ in.})^2 + (16.5 \text{ in.} + 9.37 \text{ in.})^2}$$
$$= 41.4 \text{ in.}$$

On the gusset-to-column connection

$$H_{uc} = \frac{e_c}{r} P_u$$
$$= \frac{10.5 \text{ in.}}{41.4 \text{ in.}} (675 \text{ kips})$$
$$= 171 \text{ kips}$$

$$V_{uc} = \frac{\beta}{r} P_u$$

$$= \frac{16.5 \text{ in.}}{41.4 \text{ in.}} (675 \text{ kips})$$

$$= 269 \text{ kips}$$

On the gusset-to-beam connection

$$H_{ub} = \frac{\alpha}{r} P_u$$

$$= \frac{21.8 \text{ in.}}{41.4 \text{ in.}} (675 \text{ kips})$$

$$= 355 \text{ kips}$$

$$V_{ub} = \frac{e_b}{r} P_u$$

$$= \frac{9.37 \text{ in.}}{41.4 \text{ in.}} (675 \text{ kips})$$

$$= 153 \text{ kips}$$

Design gusset-to-column connection

Try 2L4×4×⅝×2′-6 welded to the gusset and bolted with 10 rows of ⅞-in. diameter A325-N bolts in standard holes to the column flange.

Calculate tensile force per bolt r_{ut}.

$$r_{ut} = \frac{H_{uc}}{n}$$

$$= \frac{171 \text{ kips}}{20 \text{ bolts}}$$

$$= 8.55 \text{ kips/bolt}$$

Check design strength of bolts for tension-shear interaction.

$$r_{uv} = \frac{V_{uc}}{n}$$

$$= \frac{269 \text{ kips}}{20 \text{ bolts}}$$

$$= 13.5 \text{ kips/bolt} < 21.6 \text{ kips/bolt} \quad \textbf{o.k.}$$

$$F_t = 117 \text{ ksi} - 1.9 f_v \le 90 \text{ ksi}$$

$$= 117 \text{ ksi} - 1.9 \left(\frac{13.5 \text{ kips / bolt}}{\frac{\pi}{4} (⅞\text{-in.})^2} \right)$$

$$= 74.3 \text{ ksi}$$

$$\phi r_n = \phi F_t A_b$$

$$= 0.75(74.3 \text{ ksi}) \left[\frac{\pi}{4}(⅞\text{-in.})^2 \right]$$

$$= 33.5 \text{ kips} > 8.55 \text{ kips/bolt} \quad \textbf{o.k.}$$

Check bearing strength at bolt holes.

With $L_e = 1\frac{1}{2}$-in. and $s = 3$ in., the bearing strength per bolt is

$\phi r_n = \phi(2.4dtF_u)$
$= 0.75\ (\ 2.4 \times \frac{7}{8}$-in. $\times \frac{5}{8}$-in. $\times 58$ ksi$)$
$= 57.1$ kips/bolt

Since this exceeds the single-shear strength of the bolts, bearing strength is   **o.k.**

Check prying action.

$b = g - t$
$= 2\frac{1}{2}$-in. $- \frac{5}{8}$-in.
$= 1.875$ in. $> 1\frac{1}{4}$-in. entering and tightening clearance,   **o.k.**

$a = 4$ in. $- g$
$= 4$ in. $- 2\frac{1}{2}$-in.
$= 1.5$ in.

Since $a = 1.5$ in. is less than $1.25b = 2.34$ in., use $a = 1.5$ in.

$b' = b - d/2$
$= 1.875$ in. $- \dfrac{\frac{7}{8}\text{-in.}}{2}$
$= 1.44$ in.

$a' = a + d/2$
$= 1.5$ in. $+ \dfrac{\frac{7}{8}\text{-in.}}{2}$
$= 1.94$ in.

$\rho = \dfrac{b'}{a'}$
$= \dfrac{1.44 \text{ in.}}{1.94 \text{ in.}}$
$= 0.742$

$\beta = \dfrac{1}{\rho}\left(\dfrac{\phi r_n}{r_{ut}} - 1\right)$
$= \dfrac{1}{0.742}\left(\dfrac{33.5 \text{ kips / bolt}}{8.55 \text{ kips / bolt}} - 1\right)$
$= 3.93$

Since $\beta \geq 1$, set $\alpha' = 1.0$

$\delta = 1 - \dfrac{d'}{p}$
$= 1 - \dfrac{\frac{15}{16}\text{-in.}}{3 \text{ in.}}$
$= 0.688$

$$t_{req} = \sqrt{\frac{4.44 r_{ut} b'}{p F_y (1 + \delta \alpha')}}$$

$$= \sqrt{\frac{4.44(8.55 \text{ kips / bolt})(1.44 \text{ in.})}{(3 \text{ in.})(36 \text{ ksi})[1 + (0.688)(1)]}}$$

$$= 0.548 \text{ in.}$$

Since $t = \frac{5}{8}$-in. > 0.548 in., angles are   **o.k.**

Design welds

Try fillet welds around perimeter (3 sides) of both angles.

$$P_{uc} = \sqrt{H_{uc}^2 + V_{uc}^2}$$
$$= \sqrt{(171 \text{ kips})^2 + (269 \text{ kips})^2}$$
$$= 319 \text{ kips}$$

$$\theta = \tan^{-1}\left(\frac{H_{uc}}{V_{uc}}\right)$$
$$= \tan^{-1}\left(\frac{171 \text{ kips}}{269 \text{ kips}}\right)$$
$$= 32.4°$$

From Table 8-42 with $\theta = 30°$,

l = 30 in.
kl = 3½-in.
k = 0.117

By interpolation

x = 0.011
xl = 0.011(30 in.)
   = 0.33 in.
al = 4 in. $- xl$
   = 4 in. $-$ 0.33 in.
   = 3.67 in.
a = 0.122

By interpolation

$C = 1.95$

and

$$D_{req} = \frac{P_{uc}}{C C_1 l}$$

$$= \frac{319 \text{ kips}}{1.95 \times 1.0 \times (2 \text{ welds} \times 30 \text{ in.})}$$

$$= 2.73 \rightarrow 3 \text{ sixteenths required for strength}$$

From LRFD Specification Table J2.4, minimum weld size is $\frac{1}{4}$-in. Use $\frac{1}{4}$-in. fillet welds.

Check gusset thickness (against weld size required for strength)

For two fillet welds

$$t_{min} = \frac{5.16D}{F_y}$$
$$= \frac{5.16(2.73 \text{ sixteenths})}{36 \text{ ksi}}$$
$$= 0.391 \text{ in.} < \frac{3}{4}\text{-in.} \quad \textbf{o.k.}$$

Check strength of angles.

Shear yielding (due to V_{uc})

$$\phi R_n = \phi(0.60F_y A_g)$$
$$= 0.90[0.60(36 \text{ ksi})(2 \times 30 \text{ in.} \times \frac{5}{8}\text{-in.})]$$
$$= 729 \text{ kips} > 269 \text{ kips} \quad \textbf{o.k.}$$

Similarly, shear yielding of the angles due to H_{uc} is not critical.

Shear rupture

$$\phi R_n = \phi(0.60F_u A_{nv})$$
$$= 0.75[0.60(58 \text{ ksi})(2 \times \frac{5}{8}\text{-in.} \times 30 \text{ in.} - 20 \times \frac{5}{8}\text{-in.} \times 1 \text{ in.})]$$
$$= 653 \text{ kips} > 269 \text{ kips} \quad \textbf{o.k.}$$

Block shear rupture

From Tables 8-47 and 8-48, with $n = 10$, $L_{ev} = 1\frac{1}{2}$-in., and $L_{eh} = 1\frac{1}{2}$-in., $0.6F_u A_{nv} > F_u A_{nt}$. Thus

$$\phi R_n = \phi[0.6F_u A_{nv} + F_y A_{gt}] \times 2 \text{ blocks}$$
$$= (496 \text{ kips/in.} + 40.5 \text{ kips/in.})(\frac{5}{8}\text{-in.}) \times 2 \text{ blocks}$$
$$= 671 \text{ kips} > 269 \text{ kips} \quad \textbf{o.k.}$$

Check column flange.

By inspection, the 4.16-in. thick column flange has adequate flexural strength, stiffness, and bearing strength.

Design gusset-to-beam connection

$$P_{ub} = \sqrt{H_{ub}^2 + V_{ub}^2}$$
$$= \sqrt{(355 \text{ kips})^2 + (153 \text{ kips})^2}$$
$$= 387 \text{ kips}$$

From Richard (1986) it is recommended that the design factored force be increased by 40 percent to ensure adequate force redistribution in the weld group and the validity of the Uniform Force Method. Thus,

$$D_{req} = \frac{1.4P_{ub}}{1.392l}$$

$$= \frac{1.4(387 \text{ kips})}{1.392(2 \times 42 \text{ in.})}$$

$$= 4.63 \rightarrow 5 \text{ sixteenths}$$

(Note that, if a moment existed on this interface, the connection would be designed for the larger of the peak stress and 1.4 times the average stress.)

This is equal to the minimum weld size from LRFD Specification Table J2.4.

Check gusset thickness (against weld size required for strength)

For two fillet welds

$$t_{min} = \frac{5.16D}{F_y}$$

$$= \frac{5.16(4.64 \text{ sixteenths})}{36 \text{ ksi}}$$

$$= 0.664 \text{ in.} < \sqrt[3]{4}\text{-in.} \quad \textbf{o.k.}$$

Check local web yielding of the beam.

$$\phi R_n = \phi(N + 2.5k)F_{yw}t_w$$

$$= 1.0 \, [2.5 \, (1\tfrac{5}{8}\text{-in.}) + 42 \text{ in.}](50 \text{ ksi})(0.590 \text{ in.})$$

$$= 1{,}360 \text{ kips} > 153 \text{ kips} \quad \textbf{o.k.}$$

Design beam-to-column connection

Since the brace may be in tension or compression, the required strength of the beam-to-column connection is as follows. The required shear strength is

$$R_{ub} \pm V_{ub} = 15 \text{ kips} \pm 153 \text{ kips}$$

$$= 168 \text{ kips}$$

and the required axial strength is

$$A_{ub} \pm (H_u - H_{ub}) = 0 \text{ kips} \pm (171 \text{ kips})$$

$$= 171 \text{ kips}$$

Try 2L8×6×1×1'–2½ (Leg gage = 2¾-in.) welded to the beam web and bolted with five rows of ⅞-in. diameter A325-N bolts in standard holes to the column flange.

Calculate tensile force per bolt r_{ut}.

$$r_{ut} = \frac{171 \text{ kips}}{10 \text{ bolts}}$$

$$= 17.1 \text{ kips/bolt}$$

Check design strength of bolts for tension-shear interaction.

$$r_{uv} = \frac{168 \text{ kips}}{10 \text{ bolts}}$$

$$= 16.8 \text{ kips/bolt} < 21.6 \text{ kips/bolt} \quad \textbf{o.k.}$$

$$F_t = 117 \text{ ksi} - 1.9 f_v \leq 90 \text{ ksi}$$

$$= 117 \text{ ksi} - 1.9 \left(\frac{16.8 \text{ kips/bolt}}{\frac{\pi}{4} (\text{⅞-in.}^2)} \right)$$

$$= 63.9 \text{ ksi}$$

$$\phi r_n = \phi F_t A_b$$

$$= 0.75 \ (63.9 \text{ ksi}) \left[\frac{\pi}{4} (\text{⅞-in.})^2 \right]$$

$$= 28.8 \text{ kips/bolt} > 17.1 \text{ kips/bolt} \quad \textbf{o.k.}$$

Check bearing strength at bolt holes.

With $L_e = 1\frac{1}{4}$-in. $(<1.5d = 1.31$ in.$)$ and $s = 3$ in., the bearing strength of the top bolt from LRFD Specification Section J3.10 is

$$\phi r_n = \phi (L_e t F_u) \leq \phi (2.4 d t F_u)$$

$$= 0.75 \ (1\frac{1}{4}\text{-in.})(1 \text{ in.})(58 \text{ ksi}) \leq 91.4 \text{ kips/bolt}$$

$$= 54.4 \text{ kips/bolt}$$

and the bearing strength of each remaining bolt is

$$\phi r_n = \phi (s - d/2) t F_u \leq \phi (2.4 d t F_u)$$

$$= 0.75 \left(3 \text{ in.} - \frac{\text{⅞-in.}}{2} \right) (1 \text{ in.})(58 \text{ ksi}) \leq 91.4 \text{ kips/bolt}$$

$$= 111 \text{ kips/bolt}$$

Since the strength of each bolt exceeds the single shear strength of the bolts, bearing strength is **o.k.**

Check prying action

$$b = g - t$$

$$= 2\frac{3}{4}\text{-in.} - 1 \text{ in.}$$

$$= 1\frac{3}{4}\text{-in.} > 1\frac{1}{4}\text{-in. entering and tightening clearance, } \textbf{o.k.}$$

$$a = 6 \text{ in.} - g$$

$$= 6 \text{ in.} - 2\frac{3}{4}\text{-in.}$$

$$= 3\frac{1}{4}\text{-in.}$$

Since $a = 3\frac{1}{4}$-in. exceeds $1.25b = 2.19$ in., use $a = 2.19$ in. for calculation purposes.

$$b' = b - d/2$$

$$= 1\frac{3}{4}\text{-in.} - \frac{\text{⅞-in.}}{2}$$

$$= 1.31 \text{ in.}$$

$$a' = a + d/2$$

$$= 2.19 \text{ in.} + \frac{\text{⅞-in.}}{2}$$

$$= 2.63 \text{ in.}$$

$$\rho = \frac{b'}{a'}$$

$$= \frac{1.31 \text{ in.}}{2.63 \text{ in.}}$$

$$= 0.498$$

$$\beta = \frac{1}{\rho}\left(\frac{\phi r_n}{r_{ut}} - 1\right)$$

$$= \frac{1}{0.498}\left(\frac{28.8 \text{ kips / bolt}}{17.1 \text{ kips / bolt}} - 1\right)$$

$$= 1.37$$

Since $\beta \geq 1$, set $\alpha' = 1.0$

$$p = \frac{14\frac{1}{2}\text{-in.}}{5 \text{ bolts}}$$

$$= 2.90 \text{ in./bolt}$$

$$\delta = 1 - \frac{d'}{p}$$

$$= 1 - \frac{{}^{15}\!/_{16}\text{-in.}}{2.90 \text{ in.}}$$

$$= 0.677$$

$$t_{req} = \sqrt{\frac{4.44 \, r_{ut} \, b'}{p F_y \, (1 + \delta\alpha')}}$$

$$= \sqrt{\frac{4.44 \, (17.1 \text{ kips / bolt})(1.31 \text{ in.})}{(2.90 \text{ in.}) (36 \text{ ksi}) \, [1 + (0.677)\,(1.0)]}}$$

$$= 0.754 \text{ in.}$$

Since $t = 1$ in. > 0.754 in., angles are   **o.k.**

Design welds

Try fillet welds around perimeter (three sides) of both angles.

$$P_u = \sqrt{(171 \text{ kips})^2 + (168 \text{ kips})^2}$$

$$= 240 \text{ kips}$$

$$\theta = \tan^{-1}\left(\frac{171 \text{ kips}}{168 \text{ kips}}\right)$$

$$= 45.5°$$

From Table 8-42 with $\theta = 45°$

$l = 14\frac{1}{2}$-in.

$kl = 7\frac{1}{2}$-in.

$k = 0.517$

By interpolation

$x = 0.132$

$$xl = 0.132 \, (14\tfrac{1}{2}\text{-in.})$$
$$= 1.91 \text{ in.}$$
$$al = 8 \text{ in.} - xl$$
$$= 8 \text{ in.} -1.91 \text{ in.}$$
$$= 6.09 \text{ in.}$$
$$a = 0.420$$

By interpolation

$$C = 2.66$$

and

$$D_{req} = \frac{P_u}{CC_1 l}$$
$$= \frac{240 \text{ kips}}{2.66 \times 1.0 \times (2 \text{ welds} \times 14\tfrac{1}{2}\text{-in.})}$$
$$= 3.11 \rightarrow 4 \text{ sixteenths required for strength}$$

From LRFD Specification Table J2.4, minimum weld size is $\tfrac{5}{16}$-in. Use $\tfrac{5}{16}$-in. fillet welds.

Check beam web thickness (against weld size required for strength)

For two fillet welds,

$$t_{min} = \frac{5.16D}{F_y}$$
$$= \frac{5.16(3.11 \text{ sixteenths})}{50 \text{ ksi}}$$
$$= 0.321 \text{ in.} < 0.590 \text{ in.} \quad \textbf{o.k.}$$

Check the strength of angles

Shear yielding

$$\phi R_n = \phi(0.60F_y A_g)$$
$$= 0.9[0.60(36 \text{ ksi})(2 \times 14\tfrac{1}{2}\text{-in.} \times 1 \text{ in.})]$$
$$= 564 \text{ kips} > 168 \text{ kips} \quad \textbf{o.k.}$$

Similarly, shear yielding of the angles due to H_{uc} is not critical.

Shear rupture

$$\phi R_n = \phi(0.60F_u A_{nv})$$
$$= 0.75[0.60(58 \text{ ksi})(2 \times 1 \text{ in.} \times 14\tfrac{1}{2}\text{-in.} - 10 \times 1 \text{ in.} \times 1 \text{ in.})]$$
$$= 496 \text{ kips} > 168 \text{ kips} \quad \textbf{o.k.}$$

Block shear rupture

With $n = 5$, $L_{ev} = 1\tfrac{1}{4}$-in., $L_{eh} = 3\tfrac{1}{4}$-in., $0.6F_u A_{nv} > F_u A_{nt}$. Thus

$$\phi R_n = \phi[0.6F_u A_{nv} + F_y A_{gt}] \times 2 \text{ blocks}$$

$$= 0.75[0.6(58 \text{ ksi})(13\frac{1}{4}\text{-in.} - 5 \times 1 \text{ in.})(1 \text{ in.})$$
$$+ (36 \text{ ksi})(3\frac{1}{4}\text{-in.})(1 \text{ in.})](2)$$
$$= 606 \text{ kips} > 168 \text{ kips} \quad \textbf{o.k.}$$

Check column flange.

By inspection, the 4.16-in. thick column flange has adequate flexural strength, stiffeners, and bearing strength.

Comments: Were the brace in compression, the buckling strength of the gusset would have to be checked, where

$$\phi R_n = \phi_c F_{cr} A_w$$

In the above equation $\phi_c F_{cr}$ may be determined from $\frac{kl_1}{r}$ with LRFD Specification Table C-36, where l_1 is the perpendicular distance from the Whitmore section to the interior corner of the gusset. Alternatively, the average value of

$$\frac{l_1 + l_2 + l_3}{3}$$

may be substituted (AISC, 1984), where these quantities are illustrated in Figure 11-11. Note that, for this example, l_2 is negative since part of the Whitmore section is in the beam web.

The effective length factor K has been established as 0.5 by full scale tests on bracing connections (Gross, 1990). It assumes that the gusset is supported on both edges as is the case in Figure 11-11. In cases where the gusset is supported on one edge only, such as that illustrated in Figure 11-12d (and possibly Figure 11-12a) the brace can more readily move out-of-plane and a sidesway mode of buckling can occur in the gusset. For this case, K should be taken as 1.2.

EXAMPLE 11-3

Given: Refer to Figure 11-12. Each of the four designs shown for the diagonal bracing connection between the W14×68 brace, W24×55 beam, and W14×211 column web have been developed using the Uniform Force Method (the General Case, and Special Cases 1, 2, and 3) for the load case of 1.2D + 1.3W. Refer the AISC (1992) for the unfactored loads and complete designs. For the given values of α and β, determine the interface forces on the gusset-to-column and gusset-to-beam connections for

A. General Case of Figure 11-12a.

B. Special Case 1 of Figure 11-12b.

C. Special Case 2 of Figure 11-12c.

D. Special Case 3 of Figure 11-12d.

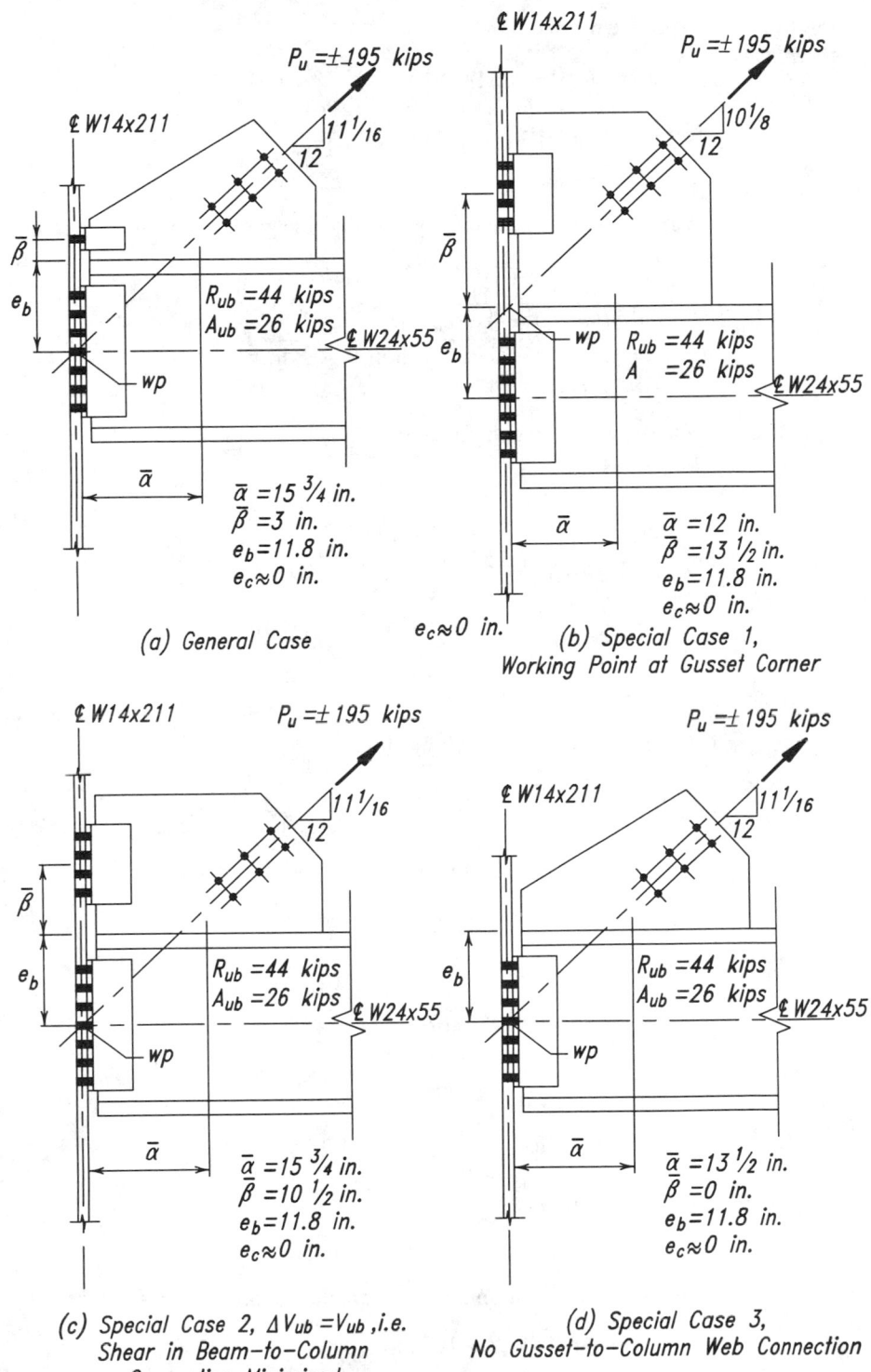

(a) General Case

(b) Special Case 1,
Working Point at Gusset Corner

(c) Special Case 2, $\Delta V_{ub} = V_{ub}$, i.e.
Shear in Beam–to–Column
Connection Minimized

(d) Special Case 3,
No Gusset–to–Column Web Connection

Fig. 11-12. Uniform force method.

Solution A: Assume $\beta = \bar{\beta} = 3$ in.
(General Case)

$$\alpha = e_b \tan\theta - e_c + \beta\tan\theta$$

$$= 11.8 \text{ in.} \left(\frac{12}{11\frac{1}{16}}\right) - 0 + 3 \text{ in.} \left(\frac{12}{11\frac{1}{16}}\right)$$

$$= 16.1 \text{ in.}$$

Since $\alpha \neq \bar{\alpha}$, an eccentricity exists on the gusset-to-beam connection.

Calculate the interface forces:

$$r = \sqrt{(\alpha + e_c)^2 + (\beta + e_b)^2}$$

$$= \sqrt{(16.1 \text{ in.} + 0 \text{ in.})^2 + (3 \text{ in.} + 11.8 \text{ in.})^2}$$

$$= 21.9 \text{ in.}$$

On the gusset-to-column connection

$$V_{uc} = \frac{\beta}{r} P_u$$

$$= \frac{3 \text{ in.}}{21.9 \text{ in.}} (195 \text{ kips})$$

$$= 26.7 \text{ kips}$$

$$H_{uc} = \frac{e_c}{r} P_u$$

$$= 0 \text{ kips}$$

On the gusset-to-beam connection

$$H_{ub} = \frac{\alpha}{r} P_u$$

$$= \frac{16.1 \text{ in.}}{21.9 \text{ in.}} (195 \text{ kips})$$

$$= 143 \text{ kips}$$

$$V_{ub} = \frac{e_b}{r} P_u$$

$$= \frac{11.8 \text{ in.}}{21.9 \text{ in.}} (195 \text{ kips})$$

$$= 105 \text{ kips}$$

$$M_{ub} = V_{ub}(\alpha - \bar{\alpha})$$

$$= \frac{105 \text{ kips } (15\frac{3}{4}\text{-in.} - 16.1 \text{ in.})}{12 \text{ in.} / \text{ft}}$$

$$= -3.06 \text{ kip-ft}$$

In this case, this small moment is negligible.

On the beam-to-column connection, the factored shear is

$$R_{ub} + V_{ub} = 44 \text{ kips} + 105 \text{ kips}$$
$$= 149 \text{ kips}$$

and the factored axial force is

$$A_{ub} \pm H_{uc} = 26 \text{ kips} \pm 0 \text{ kips}$$
$$= 26 \text{ kips}$$

For a discussion of the sign to use between A_{ub} and H_{uc}, refer to AISC (1992).

Solution B:
(Special Case 1) In this case, the centroidal positions of the gusset-edge connections are irrelevant; $\bar{\alpha}$ and $\bar{\beta}$ are given to define the geometry of the connection, but are not needed to determine the gusset edge forces.

The angle of the brace from the vertical is

$$\theta = \tan^{-1}\left(\frac{12}{10\frac{1}{8}}\right)$$
$$= 49.8°$$

The horizontal component of the brace force is

$$H_u = P_u \sin\theta$$
$$= 195 \text{ kips} \times \sin(49.8°)$$
$$= 149 \text{ kips}$$

and the vertical component of the brace force is

$$V_u = P_u \cos\theta$$
$$= 195 \text{ kips} \times \sin(49.8°)$$
$$= 126 \text{ kips}$$

On the gusset-to-column connection

$$V_{uc} = V_u = 126 \text{ kips}$$
$$H_{uc} = 0 \text{ kips}$$

On the gusset-to-beam connection

$$V_{ub} = 0 \text{ kips}$$
$$H_{ub} = H_u = 149 \text{ kips}$$

On the beam-to-column connection

$$R_{ub} = 44 \text{ kips (shear)}$$
$$A_{ub} = 26 \text{ kips (axial transfer force)}$$

In addition to the forces on the connection interfaces, the beam is subjected to a moment M_{ub} (see Figure 11-8d), where

$$M_{ub} = H_{ub}e_b$$
$$= \frac{149 \text{ kips} \times 11.8 \text{ in.}}{12 \text{ in. / ft}}$$
$$= 147 \text{ kips-ft}$$

This moment, as well as the beam axial load $H_u = 149$ kips and the moment and shear in the beam associated with the end reaction R_{ub}, must be considered in the design of the beam.

Solution C:
(Special Case 2)

Assume $\beta = \bar{\beta} = 10\frac{1}{2}$-in.

$$\alpha = e_b\tan\theta - e_c + \beta\tan\theta$$

$$= 11.8 \text{ in.} \left(\frac{12}{11\frac{1}{16}}\right) - 0 + 10\frac{1}{2}\text{-in.} \left(\frac{12}{11\frac{1}{16}}\right)$$

$$= 24.2 \text{ in.}$$

Calculate the interface forces for the general case before applying Special Case 2.

$$r = \sqrt{(\alpha + e_c)^2 + (\beta + e_b)^2}$$

$$= \sqrt{(24.2 \text{ in.} + 0 \text{ in.})^2 + (10\frac{1}{2}\text{-in.} + 11.8 \text{ in.})^2}$$

$$= 32.9 \text{ in.}$$

On the gusset-to-beam connection

$$H_{ub} = \frac{\alpha}{r} P_u$$

$$= \frac{24.2 \text{ in.}}{32.9 \text{ in.}} (195 \text{ kips})$$

$$= 143 \text{ kips}$$

$$V_{ub} = \frac{e_B}{r} P_u$$

$$= \frac{11.8 \text{ in.}}{32.9 \text{ in.}} (195 \text{ kips})$$

$$= 69.9 \text{ kips}$$

On the gusset-to-column connection

$$H_{uc} = \frac{e_c}{r} P_u$$

$$= 0 \text{ kips}$$

$$V_{uc} = \frac{\beta}{r} P_u$$

$$= \frac{10.5 \text{ in.}}{32.9 \text{ in.}} (195 \text{ kips})$$

$$= 62.2 \text{ kips}$$

On the beam-to-column connection, the factored shear is

$$R_{ub} + V_{ub} = 44.0 \text{ kips} + 66.9 \text{ kips}$$

$$= 111 \text{ kips}$$

and the factored axial force is

$$A_{ub} \pm H_{uc} = 26.0 \text{ kips} \pm 0 \text{ kips}$$

$$= 26.0 \text{ kips}$$

Next, applying Special Case 2 with $\Delta V_{ub} = V_{ub} = 69.9$ kips, calculate the interface forces.

On the gusset-to-beam connection (where V_{ub} is replaced by $V_{ub} - \Delta V_{ub}$)

$$H_{ub} = 143 \text{ kips (unchanged)}$$

$$V_{ub} = 69.9 \text{ kips} - 69.9 \text{ kips}$$
$$= 0 \text{ kips}$$

$$M_{ub} = (\Delta V_{ub})\alpha$$
$$= \frac{(69.9 \text{ kips}) (24.2 \text{ in.})}{12 \text{ in. / ft}}$$
$$= 141 \text{ kips-ft}$$

On the gusset-to-column connection (where V_{uc} is replaced by $V_{uc} + \Delta V_{ub}$)

$$H_{uc} = 0 \text{ kips (unchanged)}$$

$$V_{uc} = 62.2 \text{ kips} + 69.9 \text{ kips}$$
$$= 132 \text{ kips}$$

On the beam-to-column connection, the factored shear is

$$R_{ub} + \Delta V_{ub} - \Delta V_{ub} = 44 \text{ kips} + 69.9 \text{ kips} - 69.9 \text{ kips}$$
$$= 44 \text{ kips}$$

and the factored axial force is

$$A_{ub} \pm H_{uc} = 26 \text{ kips} \pm 0 \text{ kips}$$
$$= 26 \text{ kips}$$

Solution D:
(Special Case 3)

Assume $\beta = \overline{\beta} = 0$ in.

$$\alpha = e_h \tan\theta$$

$$= 11.8 \text{ in.} \left(\frac{12}{11\frac{1}{16}}\right)$$

$$= 12.8 \text{ in.}$$

Since $\alpha \neq \overline{\alpha}$, an eccentricity exists on the gusset-to-beam connection.

Calculate the interface forces.

$$r = \sqrt{\alpha^2 + e_b^2}$$
$$= \sqrt{(12.8 \text{ in.})^2 + (11.8 \text{ in.})^2}$$
$$= 17.4 \text{ in.}$$

On the gusset-to-beam connection

$$H_{ub} = \frac{\alpha}{r} P_u$$

$$= \frac{12.8 \text{ in.}}{17.4 \text{ in.}} (195 \text{ kips})$$

$$= 143 \text{ kips}$$

$$V_{ub} = \frac{e_b}{r} P_u$$

$$= \frac{11.8 \text{ in.}}{17.4 \text{ in.}} (195 \text{ kips})$$

$$= 132 \text{ kips}$$

$$M_{ub} = V_{ub}(\alpha - \overline{\alpha})$$

$$= \frac{132 \text{ kips } (12.8 \text{ in.} - 13\frac{1}{2}\text{-in.})}{12 \text{ in. / ft}}$$

$$= -7.70 \text{ kip-ft}$$

In this case, this small moment is negligible.

On the beam-to-column connection the factored shear is

$$R_{ub} + V_{ub} = 44 \text{ kips} + 132 \text{ kips}$$
$$= 176 \text{ kips}$$

and the factored axial force is

$$A_{ub} \pm H_{uc} = 26 \text{ kips} \pm 0 \text{ kips}$$
$$= 26 \text{ kips}$$

Comments: From the foregoing results, designs by Special Case 3 and the General Case of the Uniform Force Method provide more economical designs. Additionally, note that designs by Special Case 1 and Special Case 2 result in moments on the beam and/or column which must be considered.

BEAM-BEARING PLATES

When required, a beam-bearing plate is provided to distribute the beam end reaction over an area of the concrete or masonry support which is sufficient to keep the average pressure on the suppport within the limits of its design strength.

Design Checks

A beam-bearing plate produces a compressive single concentrated force at the beam end; the limit states of the web design strength in local yielding and crippling must be checked. The design compressive strength of the concrete or masonry must be checked. The limit state of flexural yielding must be checked to determine the design strength of the beam-bearing plate. In all cases, the design strength ϕR_n must exceed the required strength R_u.

Local Web Yielding

From LRFD Specification Section K1.3, the local yielding design strength of the beam web at the member end is ϕR_n, where $\phi = 1.0$ and:

$$R_n = (2.5k + N)F_{yw}t_w$$

The length of bearing N required for a beam end reaction R_u, may be calculated from constants ϕR_1 and ϕR_2 in the factored uniform load tables in Part 4 as

$$N_{\min} = \frac{R_u - \phi R_1}{\phi R_2}$$

where

$$\phi R_1 = \phi(2.5kF_{yw}t_w)$$
$$\phi R_2 = \phi F_{yw}t_w$$

Web Crippling

From LRFD Specification Section K1.4, the crippling design strength of the beam web at the member end is ϕR_n, where $\phi = 0.75$ and, when $N/d \leq 0.2$:

$$R_n = 68t_w^2 \left[1 + 3 \left(\frac{N}{d} \right) \left(\frac{t_w}{t_f} \right)^{1.5} \right] \sqrt{\frac{F_{yw}t_f}{t_w}}$$

The length of bearing N required for a beam end reaction R_u, may be calculated from constants ϕR_3 and ϕR_4 in the factored uniform load tables in Part 4 as

$$N_{req} = \frac{R_u - \phi R_3}{\phi R_4}$$

where

$$\phi R_3 = \phi \left(68t_w^2 \sqrt{\frac{F_{yw}t_f}{t_w}} \right)$$

$$\phi R_4 = \phi \left[68t_w^2 \left[3 \left(\frac{N}{d} \right) \left(\frac{t_w}{t_f} \right)^{1.5} \right] \sqrt{\frac{F_{yw}t_f}{t_w}} \right]$$

When $N/d > 0.2$,

$$R_n = 68t_w^2 \left[1 + \left(\frac{4N}{d} - 0.2 \right) \left(\frac{t_w}{t_f} \right)^{1.5} \right] \sqrt{\frac{F_{yw}t_f}{t_w}}$$

The length of bearing N required for a beam end reaction R_u may be calculated from constants ϕR_5 and ϕR_6 in the factored uniform load tables in Part 4 as

$$N_{req} = \frac{R_u - \phi R_5}{\phi R_6}$$

where

$$\phi R_5 = \phi \left[68t_w^2 \left[1 - 0.2 \left(\frac{t_w}{t_f} \right)^{1.5} \right] \sqrt{\frac{F_{yw}t_f}{t_w}} \right]$$

$$\phi R_6 = \phi \left[68t_w^2 \left(\frac{4}{d} \right) \left(\frac{t_w}{t_f} \right)^{1.5} \sqrt{\frac{F_{yw}t_f}{t_w}} \right]$$

Concrete Compressive Strength

The bearing plate is assumed to distribute the beam end reaction uniformly to the area of the concrete under the bearing plate. In the absence of other code specifications, the required bearing-plate area A_1 may then be determined from LRFD Specification Section J9 such that $R_u \leq \phi_c P_p$. On the full area of a concrete support

$$A_1 = \frac{R_u}{\phi_c(0.85f_c')}$$

and on less than the full area of a concrete support,

$$A_1 = \frac{1}{A_2}\left(\frac{R_u}{\phi_c(0.85f_c')}\right)^2$$

where

A_2 = maximum area of the portion of the supporting surface that is geometrically similar to and concentric with the loaded area, in.^{2}

f_c' = compressive strength of concrete, ksi

ϕ_c = 0.60

The length of bearing N may be established by available wall thickness, clearance requirements, or by the minimum requirements based on local web yielding or web crippling. The required bearing-plate width may be determined as

$$B_{req} = \frac{A_1}{N}$$

The selected dimensions B and N should preferably be in full inches.

Required Bearing-Plate Thickness

As illustrated in Figure 11-13, the beam end reaction R_u is assumed to be uniformly distributed from the beam to the bearing plate over an area equal to $N \times 2k$. Based on cantilevered bending of the bearing plate under the uniformly distributed load, the minimum bearing-plate thickness is

$$t = \sqrt{\frac{2.22R_u n^2}{A_1 F_y}}$$

where

$n = (B/2) - k$, in.

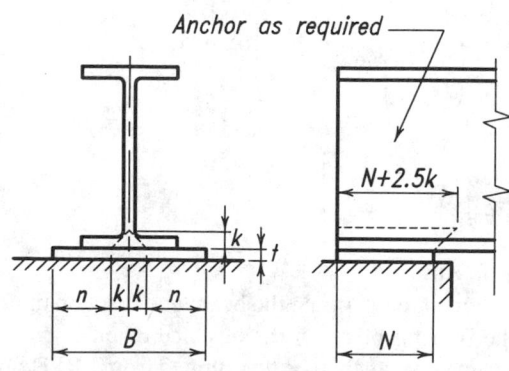

Fig. 11-13. Beam bearing-plate variables.

F_y = yield strength of the bearing plate, ksi.

In the rare case where a bearing plate is not required, the beam end reaction R_u is assumed to be uniformly distributed from the beam to the concrete over an area equal to $N \times b_f$. Additionally, when this is the case, the cantilever distance n used to determine the minimum bearing-plate thickness is taken as

$$n = (b_f/2) - k_1, \text{ in.}$$

EXAMPLE 11-6.

Given:

AW18×50 beam with a factored end reaction of 85 kips is supported by a 10-in. thick concrete wall. If the beam has F_y = 50 ksi, the concrete has f_c' = 3 ksi, and the bearing plate has F_y = 36 ksi, determine:

A. if a bearing plate is required if the beam is supported by the full wall thickness,

B. the bearing plate required if N = 10 in. (the full wall thickness),

C. the bearing plate required if N = 6 in. and the bearing plate is centered on the thickness of the wall.

W18×50

d = 17.99 in.	b_f = 7.495 in.	k = 1¼-in.
t_w = 0.355 in.	t_f = 0.570 in.	k_1 = ¹³⁄₁₆-in.

Solution A:

N = 10 in.

Check local web yielding

From the factored uniform load tables in Part 4,

$$N_{req} = \frac{R_u - \phi R_1}{\phi R_2}$$

$$= \frac{85 \text{ kips} - 55.5 \text{ kips}}{17.8 \text{ kips / in.}}$$

$$= 1.66 \text{ in.} < 10 \text{ in.} \quad \textbf{o.k.}$$

Check web crippling

$$N/d = \frac{10 \text{ in.}}{17.99 \text{ in.}}$$

$$= 0.556$$

Since $\dfrac{N}{d} > 0.2$, from the factored uniform load tables in Part 4,

$$N_{req} = \frac{R_u - \phi R_5}{\phi R_6}$$

$$= \frac{85 \text{ kips} - 51.9 \text{ kips}}{6.29 \text{ kips / in.}}$$

$$= 5.26 \text{ in.} < 10 \text{ in.} \quad \textbf{o.k.}$$

Check bearing strength of concrete

$$\phi_c P_p = \phi_c(0.85f_c')A_1$$
$$= 0.60\ (0.85 \times 3\ \text{ksi})(7.495\ \text{in.} \times 10\ \text{in.})$$
$$= 115\ \text{kips} > 85\ \text{kips}\quad \textbf{o.k.}$$

Check beam flange thickness

$$n = \frac{b_f}{2} - k_1$$
$$= \frac{7.495}{2} - {}^{13}\!/_{16}\text{-in.}$$
$$= 2.94\ \text{in.}$$

$$t_{req} = \sqrt{\frac{2.22R_u n^2}{A_1 F_y}}$$

$$= \sqrt{\frac{2.22(85\ \text{kips})(2.94\ \text{in.})^2}{(7.495\ \text{in.} \times 10\ \text{in.})(50\ \text{ksi})}}$$

$$= 0.660\ \text{in.} > 0.570\ \text{in.}\quad \textbf{n.g.}$$

A bearing plate is required.

Solution B: $N = 10$ in.

From Solution A, local web yielding and web crippling are not critical.

Calculate required bearing-plate width.

$$A_{1\ req} = \frac{R_u}{\phi_c(0.85f_c')}$$
$$= \frac{85\ \text{kips}}{0.60(0.85 \times 3\ \text{ksi})}$$
$$= 55.6\ \text{in.}^2$$

$$B_{req} = \frac{A_{1\ req}}{N}$$
$$= \frac{55.6\ \text{in.}^2}{10\ \text{in.}}$$
$$= 5.56\ \text{in.}$$

Use $B = 8$ in. (least whole-inch dimension which exceeds b_f)

Calculate required bearing-plate thickness.

$$n = \frac{B}{2} - k$$
$$= \frac{8\ \text{in.}}{2} - 1.25\ \text{in.}$$
$$= 2.75\ \text{in.}$$

$$t_{min} = \sqrt{\frac{2.22 R_u n^2}{A_1 F_y}}$$

$$= \sqrt{\frac{2.22(85 \text{ kips})(2.75 \text{ in.})^2}{(10 \text{ in.} \times 8 \text{ in.})(36 \text{ ksi})}}$$

$$= 0.704 \text{ in.}$$

Use PL$\frac{3}{4}$×10×0'-8

Solution C: $N = 6$ in.

From Solution A, local web yielding and web crippling are not critical.

Try $B = 8$ in.

$$A_1 = B \times N$$
$$= (8 \text{ in.})(6 \text{ in.})$$
$$= 48 \text{ in.}^2$$

Given these dimensions and $N_1 = 10$ in. (the full wall thickness), the dimension which makes the support area geometrically similar to the bearing plate is

$$B_1 = B\left(\frac{N_1}{N}\right)$$
$$= \frac{8 \text{ in.}(10 \text{ in.})}{6 \text{ in.}}$$
$$= 13.3 \text{ in.}$$

and

$$A_2 = B_1 \times N_1$$
$$= 13.3 \text{ in. } (10 \text{ in.})$$
$$= 133 \text{ in.}^2$$

Check $\sqrt{A_2 / A_1} = 1.66 \leq 2$ **o.k.**

$$A_{1 \, req} = \frac{1}{A_2}\left(\frac{R_u}{\phi_c(0.85 f_c')}\right)^2$$

$$= \frac{1}{133 \text{ in.}^2}\left(\frac{85 \text{ kips}}{0.6(0.85 \times 3 \text{ ksi})}\right)^2$$

$$= 23.2 \text{ in.}^2 < 48 \text{ in.}^2 \quad \textbf{o.k.}$$

Calculate required bearing-plate thickness

$$n = \frac{B}{2} - k$$

$$= \frac{8 \text{ in.}}{2} - 1\frac{1}{4} \text{ in.}$$

$$= 2.75 \text{ in.}$$

$$t_{min} = \sqrt{\frac{2.22 R_u n^2}{A_1 F_y}}$$

$$= \sqrt{\frac{2.22(85 \text{ kips})(2.75 \text{ in.})^2}{(6 \text{ in.} \times 8 \text{ in.})(36 \text{ ksi})}}$$

$$= 0.909 \text{ in.}$$

Use PL1×6×0'-8.

COLUMN BASE PLATES

Column base plates distribute the forces at the base of the column to an area of foundation large enough to prevent crushing the concrete. Base plate thicknesses should be specified in multiples of ⅛-in. up to 1¼-in. and in multiples of ¼-in. thereafter.

Typical base plates, illustrated in Figure 11-14, are often attached to the bottoms of columns in the shop. For anchor rod diameters not greater than 1¼-in., angles bolted or welded to the column as shown in Figure 11-15a are generally adequate to transfer uplift forces resulting from axial loads and moments. When greater resistance is required, stiffeners may be used with horizontal plates or angles as illustrated in Figure 11-15b. These stiffeners are not usually considered to be part of the column area in bearing on the base-plate. The angles preferably should be set back from the column end about ⅛-in. Stiffeners preferably should be set back about one inch from the base plate to eliminate a pocket that might prevent drainage and, thus, protect the column and column base plate from corrosion.

For extremely heavy loads in major structures, or where subsoil conditions are poor, a grillage as shown in Figure 11-16 may be required. This grillage consists of one or more layers of closely spaced beams (usually S shapes because of the thicker webs) encased in the concrete foundation.

The criteria for fit-up of column splices are also applicable to column base plates. For anchor rod design, refer to Part 8.

Finishing Requirements

The following base-plate finishing requirements are from LRFD Specification Section M2.8. Base plates not greater than two inches thick need not be milled if satisfactory contact in bearing is present. Base plates greater than two inches thick, but not greater than four inches thick must be either straightened by pressing or milled to obtain satisfactory contact in bearing, at the option of the fabricator. Base plates greater than four inches thick must be finished if the bearing area does not meet flatness tolerances. Note that finishing of base plates is not required in the following cases: (1) bottom surfaces of base plates when grout is used to ensure full contact on foundations; and, (2) top surfaces of base plates when complete joint-penetration groove welds are provided between the column and the base plate.

When base plates must be finished, the plate must be ordered thicker than the specified finished dimension to allow for the material removed in finishing. Table 11-2 provides finishing allowances for carbon steel base plates based on the width, thickness, and whether one or two sides are to be finished. These allowances are derived from the Standard Mill Practice flatness tolerances in Part 1. Allowances for alloy steel base plates should be adjusted for the Standard Mill Practice flatness tolerances specified in Part 1.

Holes for Anchor Rods and Grouting

Holes in base plates for anchor rods may be punched, drilled, or flame cut. Depending upon the hole diameter and base-plate thickness, machine capacity may limit the fabricator's ability to punch holes in base plates. Furthermore, many fabricators are limited by a 1½-in. diameter maximum drill size. Thus, flame-cut holes should be permitted for any plate thickness when the hole diameter is larger than one inch. Note that the walls of flame-cut holes will have a slight taper and should be inspected to assure

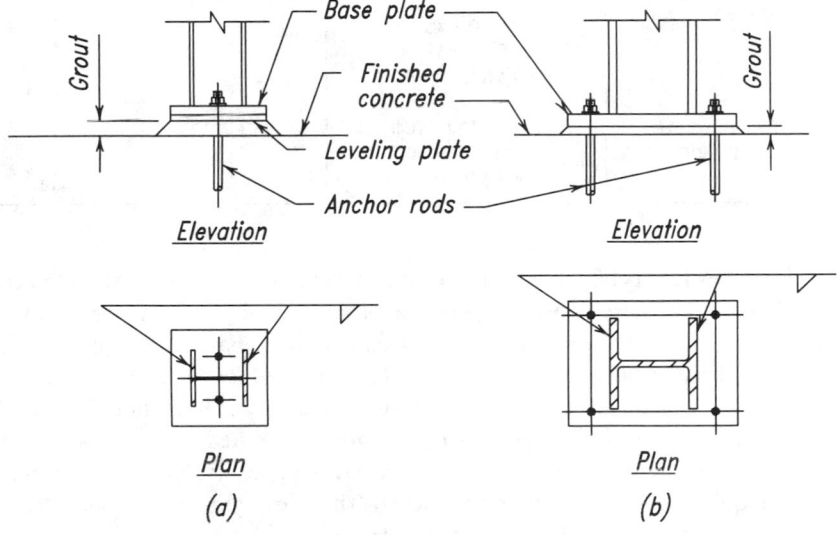

Fig. 11-14. Typical column base plates.

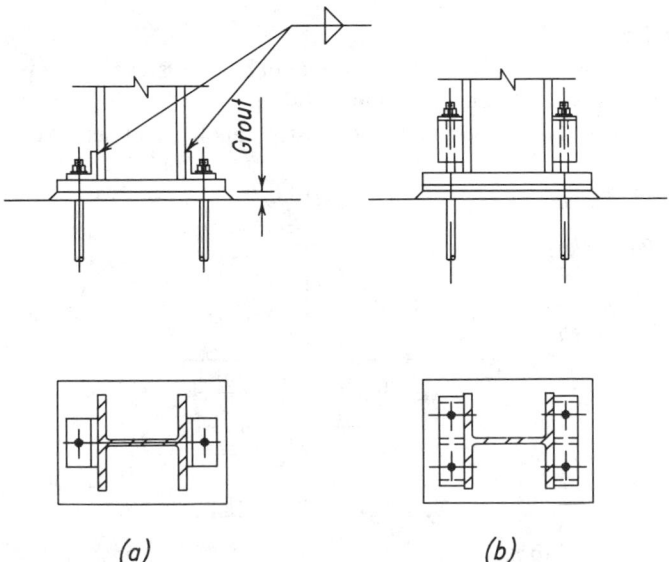

Fig. 11-15. Base plates for uplift.

Table 11-2. Finish Allowances			
Size	Thickness (in.)	Add to Fin. One Side (in.)	Add to Fin. Two Sides (in.)
Maximum dimension 24 in. or less	1¼ or less	¹⁄₁₆	⅛
	over 1¼ to 2, incl.	⅛	¼
Maximum dimension over 24 in.	1¼ or less	⅛	¼
	over 1¼ to 2, incl.	³⁄₁₆	⅜
56 in. wide or less	over 2 to 7½, incl.	¼	⅜
	over 7½ to 10, incl.	½	⅝
	over 10 to 15, incl.	¾	⅞
Over 56 in. wide to 72 in. wide	over 2 to 6, incl.	¼	⅜
	over 6 to 10, incl.	½	⅝
	over 10 to 15, incl.	¾	⅞

proper clearances for anchor rods. Table 11-3 gives recommended hole sizes to accommodate anchor rods. These hole sizes permit a reasonable tolerance for misalignment in setting the bolts and more precision in the adjustment of the base plate or column to the correct centerlines. An adequate washer should be provided for each anchor rod.

When base plates with large areas are used, at least one grout hole should be provided near the center of the plate through which grout may be poured; this will provide for a more even distribution of the grout and also prevent air pockets. Note that a grout hole may not be required when the grout is dry-packed. The size of grout holes usually requires that they be flame cut. Grout holes do not require the same accuracy for size and location as anchor-rod holes. The area of holes for grouting and anchor rods is not usually deducted when determining the required base-plate area.

Leveling Methods

Light Base Plates—For light base plates, a smooth bearing area may be provided with a steel leveling plate as illustrated in Figure 11-14a. Since leveling plates need only be approximately ¼-in. thick, they are more easily handled and set level to the proper

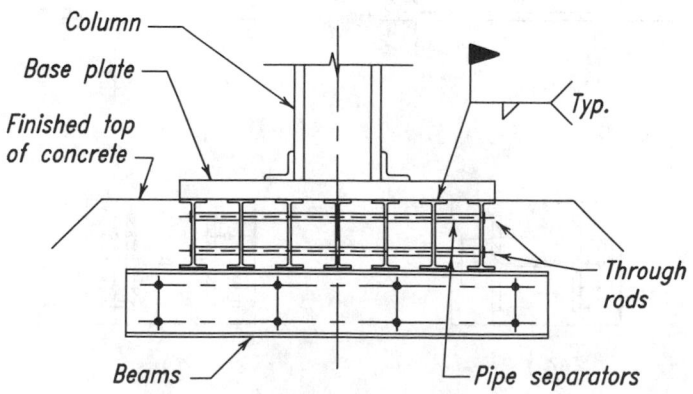

Fig. 11-16. Grillage foundation.

Table 11-3.
Sizes for Anchor-Rod Holes in Base Plates

Anchor Rod Diameter	Hole Diameter	Anchor Rod Diameter	Hole Diameter
$3/4$	$1 5/16$	$1 1/2$	$2 5/16$
$7/8$	$1 9/16$	$1 3/4$	$2 3/4$
1	$1 13/16$	2	$3 1/4$
$1 1/4$	$2 1/16$	$2 1/2$	$3 3/4$

elevation prior to erection of the columns. Leveling plates should meet the Standard Mill Practice flatness tolerances specified in Part 1. The leveling plate may be larger than the base plate to accommodate tolerances of anchor-rod placement. The leveling plate can serve as a setting template for the anchor rods. Alternatively, leveling nuts could be supplied on the anchor rods to level the base plate as illustrated in Figure 11-17. However, to ensure stability during erection, leveling nuts should not be used with less than four anchor rods.

Leveling plates and loose base plates that are small enough to be set manually are placed by the foundation contractor. Larger base plates that must be lifted by a derrick or crane are usually set by the steel erector.

Heavy Base Plates—For heavy base plates, three-point leveling bolts, illustrated in Figure 11-18, are commonly used. These threaded attachments may consist of a nut or an angle and nut welded to the base plate. Leveling bolts must be of sufficient length to compensate for the space provided for grouting. Rounding the point of the leveling bolt will prevent it from "walking" or moving laterally as it is turned. Additionally, a small steel pad under the point reduces friction and prevents damage to the concrete.

Leveling bolts or nuts should not be used to support the column during erection. If grouting is delayed until after steel erection, the base plate must be shimmed to properly distribute loads to the foundation without overstressing either the base plate or the concrete. This difficulty of supporting columns while leveling and grouting their bases makes it advisable that footings be finished to near the proper elevation (Ricker, 1989). The top of the rough footing should be set approximately one inch below the bottom of the base plate to provide for adjustment. Alternatively, an angle frame as illustrated in Figure 11-19 could be constructed to the proper elevation and filled with grout prior to erection.

Heavy base plates should be provided with some means of handling at the erection site. Lifting holes may be provided in the vertical legs of the connection angles which are shop-attached to the base plate.

Design of Axially Loaded Base Plates

Three distinct methods for base-plate analysis and design, the cantilever method for large base plates, the Murray Stockwell method (Murray, 1983) for small, lightly loaded base plates, and a yield-line-theory method (Thornton, 1990a) based on Fling (1970), have been combined by Thornton (1990b) into a single method which treats all base-plate configurations. Base plates subjected to moment and base plates subjected to tensile loads are treated by DeWolf and Ricker (1990).

Design Checks—The design compressive strength of the concrete must be checked. The limit state of flexural yielding must be checked to determine the design strength of the column base plate. In all cases, the design strength ϕR_n must exceed the required strength R_u.

Concrete Compressive Strength—The base plate is assumed to distribute R_u, the axial force in the column, uniformly to the area of the concrete under the base plate. In the absence of other code specifications, the required base-plate area A_1 may then be determined from LRFD Specification Section J9 such that $R_u \leq \phi_c P_p$. Thus, on the full area of a concrete support

$$A_1 = \frac{R_u}{\phi_c(0.85f_c')}$$

and on less than the full area of a concrete support,

$$A_1 = \frac{1}{A_2}\left(\frac{R_u}{\phi_c(0.85f_c')}\right)^2$$

where

    A_2 = maximum area of the portion of the supporting surface that is geometrically similar to and concentric with the loaded area, in.^{2}

    f_c' = compressive strength of concrete, ksi

    ϕ_c = 0.60

The base-plate dimensions B and N may then be established such that

$$B \times N \geq A_1$$

The selected dimensions B and N should be in full inches.

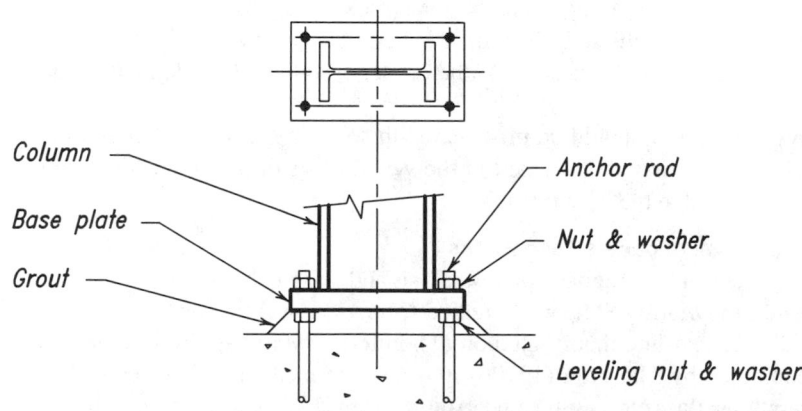

Fig. 11-17. Leveling nuts.

Required Base-Plate Thickness—The required base-plate thickness may be calculated as

$$t_{req} = l \sqrt{\frac{2P_u}{0.9F_y BN}}$$

In the above equation, l is the larger of m, n, and $\lambda n'$ where

$$m = \frac{(N - 0.95d)}{2}$$

$$n = \frac{(B - 0.8b_f)}{2}$$

$$n' = \frac{\sqrt{db_f}}{4}$$

and

$$\lambda = \frac{2\sqrt{X}}{1 + \sqrt{1 - X}} \leq 1$$

In the above equation,

$$X = \left(\frac{4db_f}{(d + b_f)^2}\right) \frac{P_u}{\phi_c P_p}$$

Note that, since both the term in parentheses and the ratio of P_u to $\phi_c P_p$ are always less than or equal to one, the value of X will always be less than or equal to one. From LRFD Specification Section J9

$$\phi_c = 0.6$$

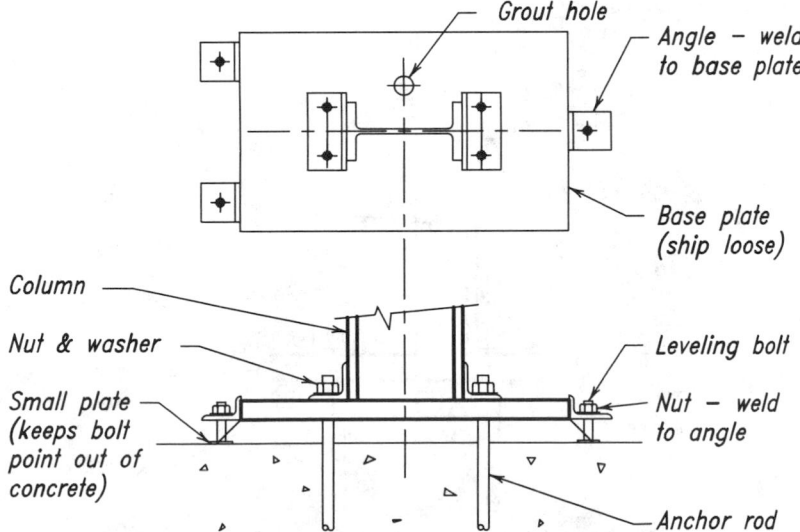

Fig. 11-18. Three-point leveling.

and, on the full area of a concrete support,

$$P_p = 0.85f_c'A_1$$

On less than the full area of the concrete support

$$P_p = 0.85f_c'A_1\sqrt{\frac{A_2}{A_1}}$$

Note that $\sqrt{\dfrac{A_2}{A_1}}$ must be less than or equal to two.

The physical variables in the above equations are illustrated in Figure 11-20.

EXAMPLE 11-7.

Given: A W12×170 column with a factored axial load of 1,100 kips bears on a concrete pedestal. If the column has F_y = 50 ksi, the concrete has f_c' = 3 ksi, and the base-plate has F_y = 36 ksi, determine:

       A. the base-plate and pedestal dimensions required if the base-plate is to cover the full pedestal area,

       B. the base-plate dimensions required for a 30 in.×30 in. concrete pedestal.

       W12×170

       d = 14.03 in. b_f = 12.570 in.
       t_w = 0.960 in. t_f = 1.560 in.

Solution A: Calculate required base-plate area.

$$A_{1\,req} = \frac{P_u}{\phi_c(0.85f_c')}$$
$$= \frac{1,100 \text{ kips}}{0.6(0.85 \times 3 \text{ ksi})}$$
$$= 719 \text{ in.}^2$$

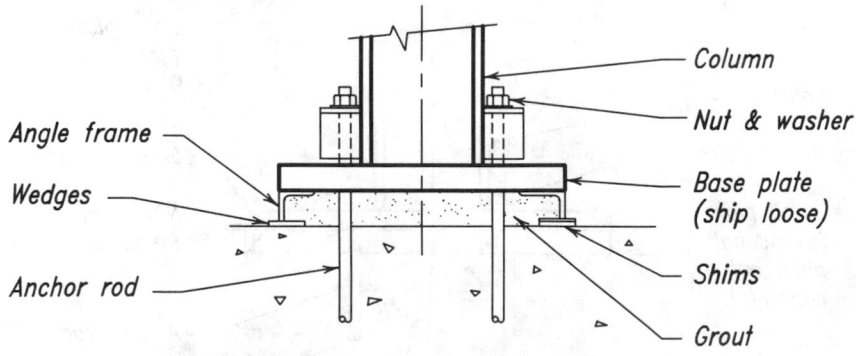

Fig. 11-19. Angle-framed leveling.

Optimize base-plate dimensions.

$$\Delta = \frac{0.95d - 0.8b_f}{2}$$
$$= \frac{0.95(14.03 \text{ in.}) - 0.8(12.570 \text{ in.})}{2}$$
$$= 1.64 \text{ in.}$$
$$N \approx \sqrt{A_1} + \Delta$$
$$\approx \sqrt{719 \text{ in.}^2} + 1.64 \text{ in.}$$
$$\approx 28.5 \text{ in.}$$

Try $N = 28$ in. and $B = 26$ in. (pedestal dimensions same)

Calculate required base-plate thickness

$$m = \frac{N - 0.95d}{2}$$
$$= \frac{28 \text{ in.} - 0.95(14.03 \text{ in.})}{2}$$
$$= 7.34 \text{ in.}$$
$$n = \frac{B - 0.8b_f}{2}$$
$$= \frac{26 \text{ in.} - 0.8(12.570 \text{ in.})}{2}$$
$$= 7.97 \text{ in.}$$

$$\phi_c P_p = 0.6 \, (0.85 f_c' A_1)$$
$$= 0.6 \, (0.85 \times 3 \text{ ksi} \times 26 \text{ in.} \times 28 \text{ in.})$$
$$= 1{,}110 \text{ kips}$$

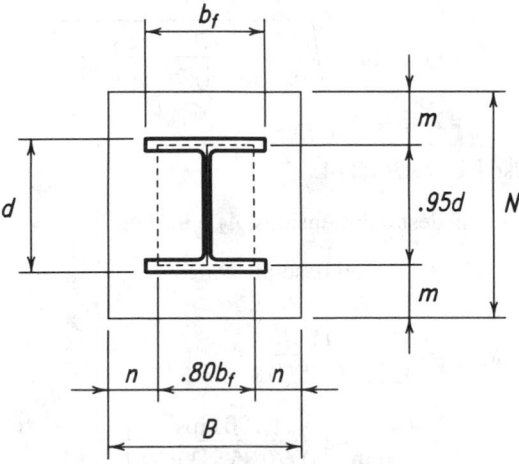

Fig. 11-20. Column base-plate design variables.

$$X = \left(\frac{4db_f}{(d+b_f)^2}\right)\left(\frac{P_u}{\phi_c P_p}\right)$$

$$= \left(\frac{4(14.03 \text{ in.})(12.570 \text{ in.})}{(14.03 \text{ in.} + 12.570 \text{ in.})^2}\right)\left(\frac{1{,}100 \text{ kips}}{1{,}110 \text{ kips}}\right)$$

$$= 0.988$$

$$\lambda = \frac{2\sqrt{X}}{1 + \sqrt{1-X}} \le 1$$

$$= \frac{2\sqrt{0.988}}{1 + \sqrt{1-0.988}}$$

$$= 1.79 \rightarrow 1$$

$$\lambda n' = \frac{\lambda\sqrt{db_f}}{4}$$

$$= \frac{(1)\sqrt{14.03 \text{ in.} \times 12.570 \text{ in.}}}{4}$$

$$= 3.32 \text{ in.}$$

From this

$$l = \max(m, n, \lambda n')$$
$$= \max (7.34 \text{ in.}, 7.97 \text{ in.}, 3.32 \text{ in.})$$
$$= 7.97 \text{ in.}$$

and

$$t_{req} = l\sqrt{\frac{2P_u}{0.9F_y BN}}$$

$$= (7.97 \text{ in.})\sqrt{\frac{2(1{,}100 \text{ kips})}{0.9(36 \text{ ksi})(26 \text{ in.})(28 \text{ in.})}}$$

$$= 2.43 \text{ in.}$$

Use PL2½×26×2'-4.

Solution B: From pedestal dimensions, $A_2 = 900 \text{ in.}^2$

Calculate required base-plate area.

$$A_{1\,req} = \frac{1}{A_2}\left(\frac{P_u}{\phi_c(0.85f_c')}\right)^2$$

$$= \frac{1}{900 \text{ in.}^2}\left(\frac{1{,}100 \text{ kips}}{0.6(0.85 \times 3 \text{ ksi})}\right)^2$$

$$= 575 \text{ in.}^2$$

Optimize base-plate dimensions.

From Solution A, $\Delta = 1.64$ in.

$$N \approx \sqrt{A_1} + \Delta$$
$$\approx \sqrt{575 \text{ in.}^2} + 1.64 \text{ in.}$$
$$\approx 25.6 \text{ in.}^2$$

Try $N = 25$ in. and $B = 23$ in.

Calculate required base-plate thickness.

$$m = \frac{N - 0.95d}{2}$$
$$= \frac{25 \text{ in.} - 0.95(14.03 \text{ in.})}{2}$$
$$= 5.84 \text{ in.}$$

$$n = \frac{B - 0.8b_f}{2}$$
$$= \frac{23 \text{ in.} - 0.8(12.570 \text{ in.})}{2}$$
$$= 6.47 \text{ in.}$$

$$\phi_c P_p = \phi_c \left(0.85 f_c' A_1 \sqrt{\frac{A_2}{A_1}} \right)$$
$$= 0.6 \left(0.85 \times 3 \text{ ksi} \times 575 \text{ in.}^2 \sqrt{\frac{900 \text{ in.}^2}{575 \text{ in.}^2}} \right)$$
$$= 1{,}101 \text{ kips}$$

$$X = \left(\frac{4db_f}{(d + b_f)^2} \right) \left(\frac{P_u}{\phi_c P_p} \right)$$
$$= \left(\frac{4(14.03 \text{ in.})(12.570 \text{ in.})}{(14.03 \text{ in.} + 12.570 \text{ in.})^2} \right) \left(\frac{1{,}100 \text{ kips}}{1{,}101 \text{ kips}} \right)$$
$$= 0.996$$

$$\lambda = \frac{2\sqrt{X}}{1 + \sqrt{1 - X}} \leq 1$$
$$= \frac{2\sqrt{0.996}}{1 + \sqrt{1 - 0.996}}$$
$$= 1.88 \rightarrow 1$$

$$\lambda n' = \frac{\lambda \sqrt{db_f}}{4}$$
$$= \frac{(1)\sqrt{14.03 \text{ in.} \times 12.570 \text{ in.}}}{4}$$

$$= 3.32 \text{ in.}$$

From this

$$l = \max(m, n, \lambda n')$$
$$= \max(5.84 \text{ in.}, 6.47 \text{ in.}, 3.32 \text{ in.})$$
$$= 6.47 \text{ in.}$$

and

$$t_{req} = l\sqrt{\frac{2P_u}{0.9F_y BN}}$$

$$= (6.47 \text{ in.})\sqrt{\frac{2(1,100 \text{ kips})}{0.9(36 \text{ ksi})(23 \text{ in.})(25 \text{ in.})}}$$

$$= 2.22$$

Use $PL2\frac{1}{2}\times23\times2'\text{-}1$.

COLUMN SPLICES

When the height of a building exceeds the available length of column sections, or when it is economically advantageous to change the column size at a given floor level, it becomes necessary to splice two columns together. When required, column splices should preferably be located about four feet above the finished floor to accommodate the attachment of safety cables which may be required at floor edges or openings.

Fit-Up of Column Splices

From LRFD Specification Section M2.6, the ends of columns in a column splice which depend upon contact bearing for the transfer of axial forces must be finished to a common plane by milling, sawing, or other suitable means. In theory, if this were done and the pieces were erected truly plumb, there would be full-contact bearing across the entire surface; this is true in most cases. However, LRFD Specification Section M4.4 recognizes that a perfect fit on the entire available surface will not exist in all cases.

A $\frac{1}{16}$-in. gap is permissible with no requirements for repair or shimming. During erection, at the time of tightening the bolts or depositing the welds, columns will usually be subjected to loads which are significantly less than the design loads. Full scale tests (Popov and Steven, 1977) which progressed to column failure have demonstrated that subsequent loading to the design loads does not result in distress in the bolts or welds of the splice.

If the gap exceeds $\frac{1}{16}$-in., but is less than $\frac{1}{4}$-in., non-tapered steel shims are required if sufficient contact area does not exist. Mild steel shims are acceptable regardless of the steel grade of the column or bearing material. If required, these shims must be contained, usually with a tack weld, so that they cannot be worked out of the joint.

There is no provision in the LRFD Specification for gaps larger than $\frac{1}{4}$-in. When such a gap exists, an engineering evaluation should be made of this condition based upon the type of loading transfered by the column splice. Tightly driven tapered shims may be required or the required strength may be developed through flange and web splice plates. Alternatively, the gap may be ground or gouged to a suitable profile and filled with weld metal.

Lifting Devices

As illustrated in Figure 11-21, lifting devices are typically used to facilitate the handling and erection of columns. When flange-plated or web-plated column splices are used for W-shape columns, it is convenient to place lifting holes in these flange plates as illustrated in Figure 11-21a. When butt-plated column splices are used, additional temporary plates with lifting holes may be required as illustrated in Figure 11-21b. W-shape column splices which do not utilize web-plated or butt-plated column splices, i.e., groove welded column splices, may be provided with a lifting hole in the column web as illustrated in Figure 11-21c. While a hole in the column web reduces the cross-sectional area of the column, this reduction will seldom be critical since the column is sized for the loads at the floor below and the splice is located above the floor. Alternatively, auxilliary plates with lifting holes may be connected to the column so that they do not interfere with the welding. Typical column splices for tubes and box-columns are illustrated in Figure 11-21d. Holes in lifting devices may be drilled, reamed, or flame cut with a mechanically guided torch. In the latter case, the bearing surface of the hole in the direction of the lift must be smooth.

The lifting device and its attachment to the column must be of sufficient strength to support the weight of the column as it is brought from the horizontal position (as delivered) to the vertical position (as erected); the lifting device and its attachment to the column must be adequate for the tensile forces, shear forces, and moments induced during handling and erection.

A suitable shackle and pin are connected to the lifting device while the column is on the ground. The size and type of shackle and pin to be used in erection is usually established by the steel erector and this information must be transmitted to the fabricator prior to detailing. Except for excessively heavy lifting pieces, it is customary to select a single pin and pinhole diameter to accommodate the majority of structural steel members, whether they are columns or other heavy structural steel members. The pin is attached to the lifting hook and a lanyard trails to the ground or floor level. After the column is erected and connected, the pin is removed from the device by means of the lanyard, eliminating the need for an ironworker to climb the column. The shackle pin, as assembled with the column, must be free and clear, so that it may be withdrawn laterally after the column has been landed and stabilized.

The safety of the structure, equipment, and personnel is of utmost importance during the erection period. It is recommended that all welds that are used on the lifting devices and stability devices be inspected very carefully, both in the shop and later in the field, for any damage that may have occurred in handling and shipping. Groove welds frequently are inspected with ultrasonic methods (UT) and fillet welds are inspected with magnetic particle (MT) or liquid dye penetrant (DPT) methods.

Column Alignment and Stability During Erection

Column splices should provide for safety and stability during erection when the columns might be subjected to wind, construction, and/or accidental loading prior to the placing of the floor system. The nominal flange-plated, web-plated, and butt-plated column splices developed here consider this type of loading.

In other splices, column alignment and stability during erection are achieved by the addition of temporary lugs for field bolting as illustrated in Figure 11-22. The material thickness, weld size, and bolt diameter required are a function of the loading. A conservative resisting moment arm is normally taken as the distance from the compressive toe or flange face to the gage line of the temporary lug. The overturning moment should be checked about both axes of the column. The recommended minimum plate or angle thickness is ½-in.; the recommended minimum weld size is 5⁄16-in.; additionally, high-strength bolts are normally used for stability devices.

Temporary lugs are not normally used as lifting devices. Unless required to be removed in the contract documents, these temporary lugs may remain.

Column alignment is provided with centerpunch marks which are useful in centering the columns in two directions.

Force Transfer in Column Splices

As illustrated in Figure 11-23, for the W-shapes most frequently used as columns, the distance between the inner faces of the flanges is constant throughout any given nominal depth; as the nominal weight per foot increases for each nominal depth, the flange and web thicknesses increase. From LRFD Specification Section J8, the design bearing strength of the contact area of a milled surface is

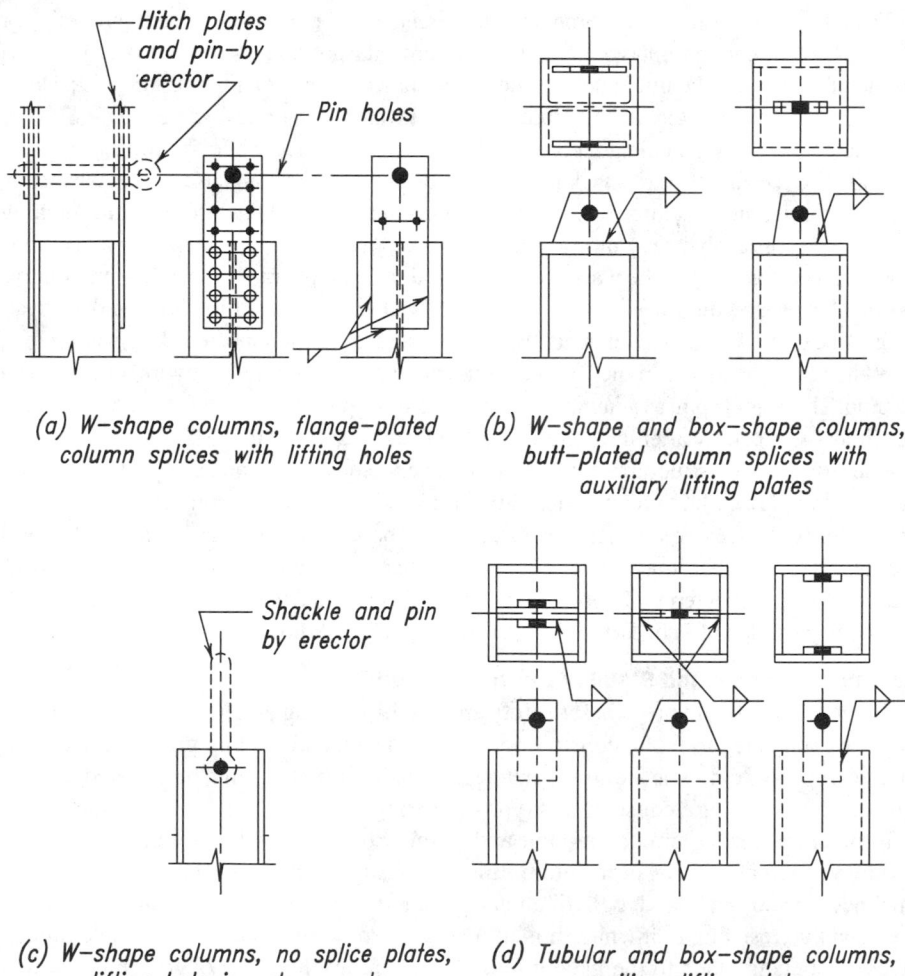

(a) W–shape columns, flange–plated column splices with lifting holes

(b) W–shape and box–shape columns, butt–plated column splices with auxiliary lifting plates

(c) W–shape columns, no splice plates, lifting hole in column web

(d) Tubular and box–shape columns, auxiliary lifting plates

Fig. 11-21. Lifting devices for columns.

$$\phi R_n = 0.75(1.8 F_y A_{pb})$$

This bearing strength is much greater than the axial strength of the column and will seldom prove critical in the member design. In column splices transferring only axial forces, then, complete axial force transfer may be achieved through bearing on finished surfaces; bolts or welds are required by LRFD Specification Section J1.4 to be sufficient to hold all parts securely in place.

In addition to axial forces, from LRFD Specification Section J1.4, column splices must be proportioned to resist tension developed by the factored loads specified by load combination A4-6 which is $0.9D \pm (1.3W$ or $1.0E)$. Note that it is not permissible to use forces due to live load to offset the tensile forces from wind or seismic loads.

For dead and wind loads, the required strength is $0.9D - 1.3W$, where D is the compressive force due to the dead load and W is the tensile force due to wind load. If $0.9D \geq 1.3W$, the splice is not subjected to tension and a nominal splice may be selected from those in Tables 11-4. When $0.9D < 1.3W$, the splice will be subjected to tension and the nominal splices from Table 11-4 are acceptable if the design tensile strength of the splice $\phi_t P_n$ is greater than or equal to the required strength. Otherwise, a splice must be designed with sufficient area and attachment.

When shear from lateral loads is divided among several columns, the force on any single column is relatively small and can usually be resisted by friction on the contact bearing surfaces and/or by the flange plates, web plates, or butt plates. If the required shear strength exceeds the design shear strength of the column splice selected from Tables 11-4, a column splice must be designed with sufficient area and attachment.

Flange-Plated Column Splices

Tables 11-4 give typical flange-plated column splice details for W-shape columns. These details are not splice standards, but rather, typical column splices in accordance with LRFD Specification provisions and typical erection requirements. Other splice designs may also be developed. It is assumed in all cases that the lower shaft will be the heavier, although not necessarily the deeper, section.

Full-contact bearing is always achieved when lighter sections are centered over heavier sections of the same nominal depth. If the upper column is not centered on the lower column, or if columns of different nominal depths must bear on each other, some areas of the upper column will not be in contact with the lower column. These areas are hatched in Figure 11-24.

When additional bearing area is not required, unfinished fillers may be used. These fillers are intended for "pack-out" of thickness and are usually set back ¼-in. or more from the finished column end. Since no force is transferred by these fillers, only nominal attachment to the column is required.

When additional bearing area is required, fillers finished to bear on the larger column may be provided. Such fillers are proportioned to carry bearing loads at the bearing strength calculated from LRFD Specification Section J8 and must be connected to the column to transfer this calculated force.

Although flange plates are shown shop assembled to the lower column, it is equally acceptable to invert this arrangement and place them on the upper column. This will usually require fills of increased thickness to maintain erection clearances.

In Tables 11-4, Cases I and II are for all-bolted flange-plated column splices for W-shape columns. Bolts in column splices are usually the same size and type as for other bolts on the column. Bolt spacing, end distance, and edge distances resulting from the billed plate sizes permit the use of ¾-in. and ⅞-in. bolts in the splice details shown. Larger diameter bolts may require an increase in edge or end distances. Refer to LRFD Specification

Chapter J. The use of high-strength bolts in bearing-type connections is assumed in all field and shop splices. However, when slotted or oversized holes are utilized, or in splices employing under-developed fillers over ¼-in. thick, slip-critical connections may be required; refer to LRFD Specification Section J6. For ease of erection, field clearances for lap splices fastened by bolts range from ⅛-in. to 3/16-in. under each plate.

Cases IV and V are for all-welded flange-plated column splices for W-shape columns. Splice welds are assumed to be made with E70XX electrodes and are proportioned as

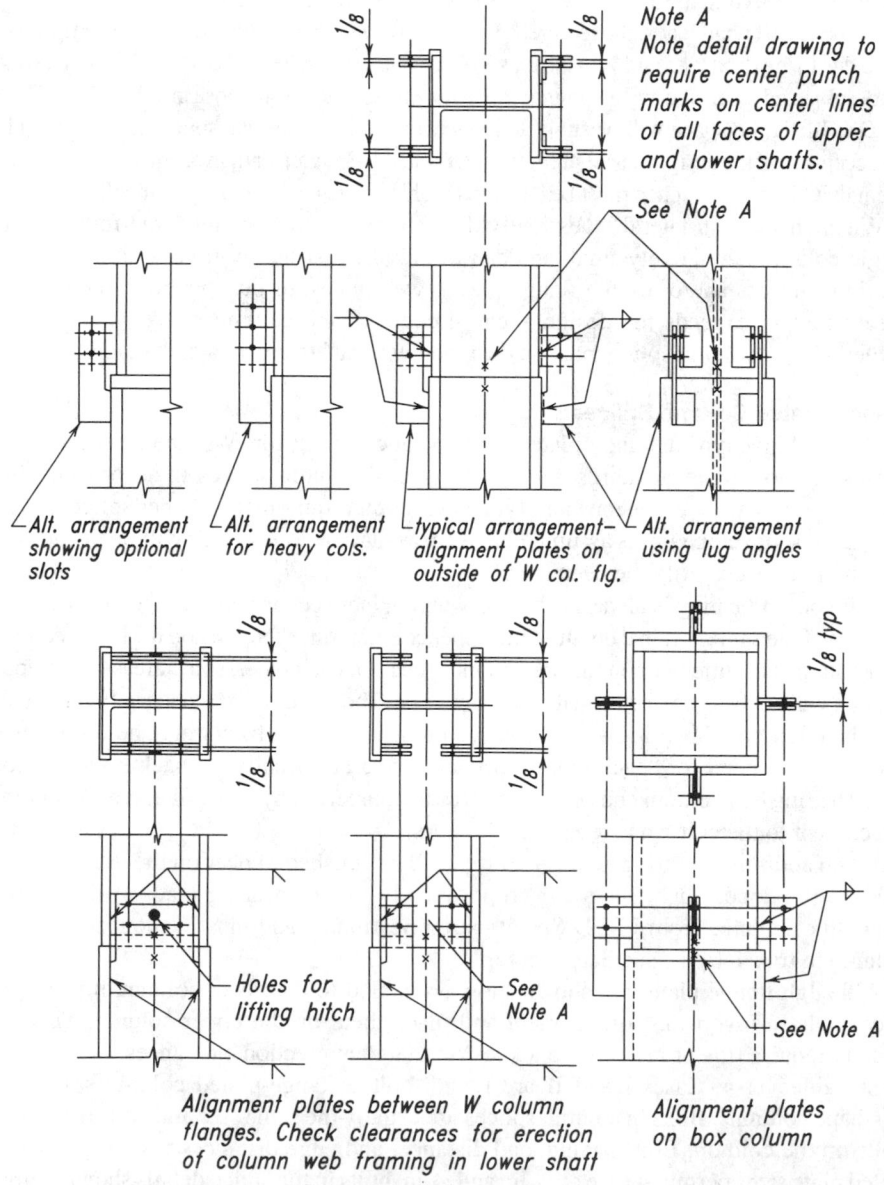

Fig. 11-22. Column stability and alignment devices.

required by the LRFD Specification provisions. The SAW, GMAW, and FCAW equivalents to E70XX electrodes may be substituted if desired. Field clearance for welded splices are limited to $\frac{1}{16}$-in. to control the expense of building up welds to close openings. Note that the fillet weld lengths Y as compared to the lengths $L/2$, provide 2-in. unwelded distance below and above the column shaft finish line. This provides a degree of flexibility in the splice plates to assist the erector.

Cases VI and VII are for combination bolted and welded column splices. Since the design strength of the welds will, in most cases, exceed the strength of the bolts, the weld and splice lengths shown may be reduced, if desired, to balance the strength of the fasteners to the upper or lower column, provided that the design strength of the splice is still greater than the required strength of the splice, including erection loading.

Directly Welded Flange Column Splices

Tables 11-4 also include typical directly welded flange column splice details for W-shape and tubular or box-shaped columns. These details are not splice standards, but rather, typical column splices in accordance with LRFD Specification provisions and typical erection requirements. Other splice designs may also be developed. It is assumed in all cases that the lower shaft will be the heavier, although not necessarily the deeper, section.

Case VIII is for W-shape columns spliced with either partial-joint-penetration or complete-joint-penetration groove welds. Case X is for tubular or box-shaped columns spliced with partial-joint-penetration or complete-joint-penetration groove welds.

Butt-Plated Column Splices

Tables 11-4 further include typical butt-plated column splice details for W-shape and tubular or box-shaped columns. These details are not splice standards, but rather, present typical column splices in accordance with LRFD Specification provisions and typical erection requirements. Other splice designs may also be developed. It is assumed in all cases that the lower shaft will be the heavier, although not necessarily the deeper, section.

Butt plates are used frequently on welded splices where the upper and lower columns are of different nominal depths, but may not be economical for bolted splices since fillers cannot be eliminated. Typical butt plates are $1\frac{1}{2}$-in. thick for a W8 over W10 splice, and 2-in. thick for other W-shape combinations such as W10 over W12 and W12 over W14. Butt plates which are subjected to substantial bending stresses, such as required on boxed columns, will require a more careful review and analysis. One method of extensive experience is to assume forces are transferred through the butt plate on a 45° angle and check the thickness obtained for shear and bearing strength. Finishing requirements for butt plates are specified in LRFD Specification Section M2.8.

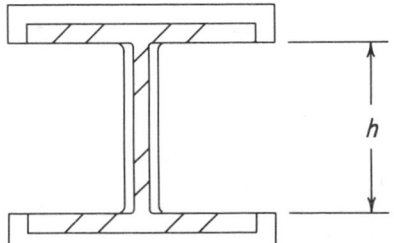

Column Size	h (in.)
W8×24–67	7.13
W10×33–112	8.86
W12×40–336	10.91
W14×43–730	12.60

Fig. 11-23. Distance between flanges for typical W-shape columns.

Case III is a combination flange-plated and butt-plated column splice for W-shape columns. Case IX is for welded butt-plated column splices for W-shape columns. Case XI is for welded butt-plated column splices for tubular or box-shaped columns. Case XII is for welded butt-plated column splices between W-shape and tubular or box-shaped columns.

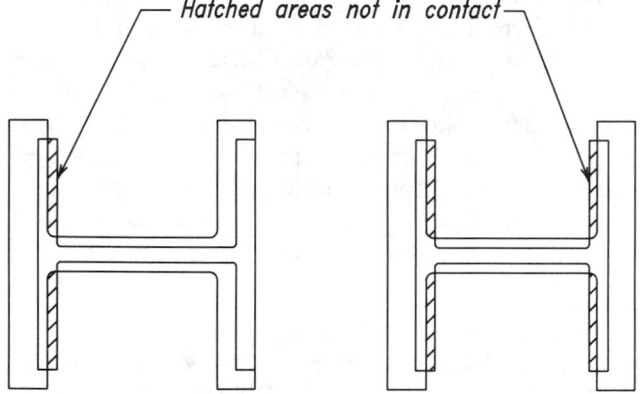

Fig. 11-24. Columns not centered or of different nominal depth.

Table 11-4.
Typical Column Splices

Case I:
All-bolted flange-plated column splices between columns
with depth d_u and d_l nominally the same.

Column Size	Gage g_u or g_l	Flange Plates			
		Type	Width	Thk.	Length
W14×455 to 730	13½	1	16	¾	1'-6½
257 to 426	11½	1	14	⅝	1'-6½
145 to 233	11½	1	14	½	1'-6½
90 to 132	11½	2	14	⅜	1'-0½
43 to 82	5½	2	8	⅜	1'-0½
W12×120 to 336	5½	2	12	⅝	1'-0½
40 to 106	5½	2	8	⅜	1'-0½
W10×33 to 112	5½	2	8	⅜	1'-0½
W8×31 to 67	5½	2	8	⅜	1'-0½
24 & 28	3½	2	6	⅜	1'-0½

Gages shown may be modified if necessary to accommodate fittings
elsewhere on the column.

Case I-A:
$d_l = (d_u + \frac{1}{4}\text{-in.})$
to $(d_u + \frac{5}{8}\text{-in.})$

Flange plates: Select g_u for upper column; select g_l and
flange plate dimensions for lower columns (see table
above).
Fillers: None.
Shims: Furnish sufficient strip shims 2½×1⅛ to provide
0 to $\frac{1}{16}$-in. clearance each side.

Case I-B:
$d_l = (d_u - \frac{1}{4}\text{-in.})$
to $(d_u + \frac{1}{8}\text{-in.})$

Flange plates: Same as Case I-A.
Fillers (shop bolted under flange plates): Select thickness
as ⅛-in. for $d_l = d_u$ and
$d_l = (d_u + \frac{1}{8}\text{-in.})$ or as ¼-in. for
$d_l = (d_u - \frac{1}{8}\text{-in.})$ and $d_l = (d_u - \frac{1}{4}\text{-in.})$
Select width to match flange plate and length as 0'-9
for Type 1 or 0'-6 for Type 2.
Shims: Same as Case I-A.

Case I-C:
$d_l = (d_u + \frac{3}{4}\text{-in.})$
and over.

Flange plates: Same as Case I-A.
Fillers (shop bolted to upper column): Select thickness as
$(d_l - d_u)/2$ minus ⅛-in., whichever results in ⅛-in.
multiples of filler thickness. Select width to match flange
plate, but not greater than upper column flange width.
Select length as 1'-0 for Type 1 or 0'-9 for Type 2.
Shims: Same as Case I-A.

For lifting devices, see Figure 11-21.

Table 11-4 (cont.).
Typical Column Splices

Case I:
All-bolted flange-plated column splices between columns
with depth d_u and d_l nominally the same.

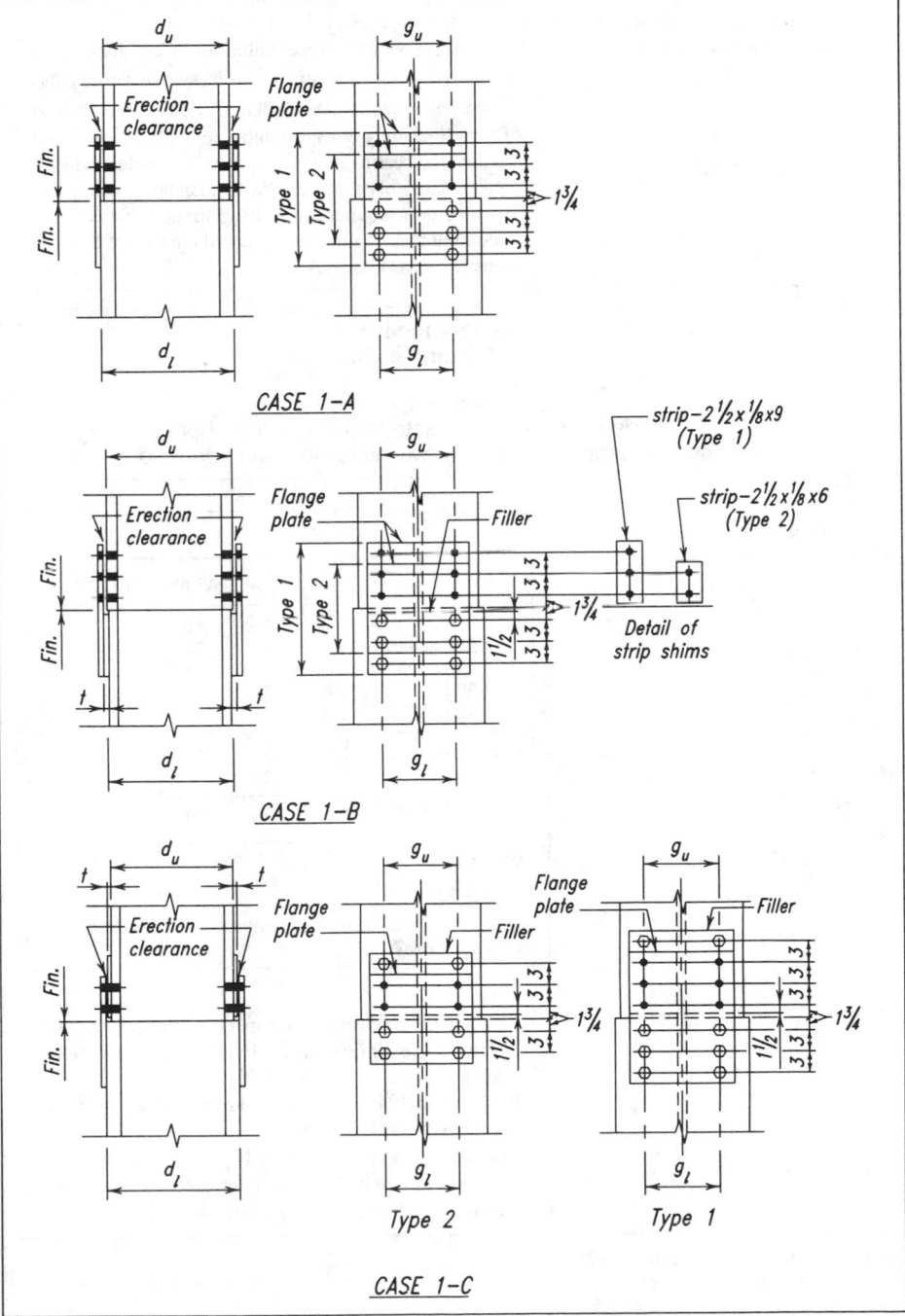

CASE 1-A

strip-$2\frac{1}{2} \times \frac{1}{8} \times 9$ (Type 1)

strip-$2\frac{1}{2} \times \frac{1}{8} \times 6$ (Type 2)

Detail of strip shims

CASE 1-B

Type 2   Type 1

CASE 1-C

Table 11-4 (cont.).
Typical Column Splices

Case II:
All-bolted flange-plated column splices between columns with depth d_u nominally two inches less than depth d_l.

Fillers on upper column developed for bearing on lower column.	Flange plates: Same as Case I-A. Fillers (shop bolted to upper column): Select thickness as $(d_l - d_u) / 2$ minus $\frac{1}{8}$-in. or $\frac{3}{16}$-in., whichever results in $\frac{1}{8}$-in. multiples of filler thickness. Select bolts through fillers (including bolts through flange plates) on each side to develop bearing strength of the filler. Select width to match flange plate, but not greater than upper column flange width unless required for bearing strength. Select length as required to accommodate required number of bolts. Shims: Same as Case I-A.

Table 11-4 (cont.).
Typical Column splices

Case III:
All-bolted flange-plated and butt-plated column splices between columns with depth d_u nominally two inches less than depth d_l.

Fillers on upper column developed for bearing on lower column.

Column Size	Gage g_u or g_l	Flange Plates			
		Type	Width	Thk.	Length
W14×455 to 730	$13\frac{1}{2}$	1	16	$\frac{3}{4}$	1'-8$\frac{1}{2}$
257 to 426	$11\frac{1}{2}$	1	14	$\frac{5}{8}$	1'-8$\frac{1}{2}$
145 to 233	$11\frac{1}{2}$	1	14	$\frac{1}{2}$	1'-8$\frac{1}{2}$
90 to 132	$11\frac{1}{2}$	2	14	$\frac{3}{8}$	1'-2$\frac{1}{2}$
43 to 82	$5\frac{1}{2}$	2	8	$\frac{3}{8}$	1'-2$\frac{1}{2}$
W12×120 to 336	$5\frac{1}{2}$	2	12	$\frac{5}{8}$	1'-2$\frac{1}{2}$
40 to 106	$5\frac{1}{2}$	2	8	$\frac{3}{8}$	1'-2$\frac{1}{2}$
W10×33 to 112	$5\frac{1}{2}$	2	8	$\frac{3}{8}$	1'-2
W8×31 to 67	$5\frac{1}{2}$				
24 & 28	$3\frac{1}{2}$				

Gages shown may be modified if necessary to accommodate fittings elsewhere on the column.

Flange plates: Select g_u for upper column, select g_l and flange plate dimensions for lower column (see table above).

Fillers (shop bolted to upper column): Same as Case I-C.

Shims: Same as Case I-A.

Butt plate: Select thickness as $1\frac{1}{2}$-in. for W8 upper column or two inches for others. Select width the same as upper column and length as $d_l - \frac{1}{4}$-in.

For lifting devices, see Figure 11-21.

Table 11-4 (cont.).
Typical Column Splices

Case II:
All-bolted flange-plated column splices between columns with depth d_u nominally two inches less than depth d_l.

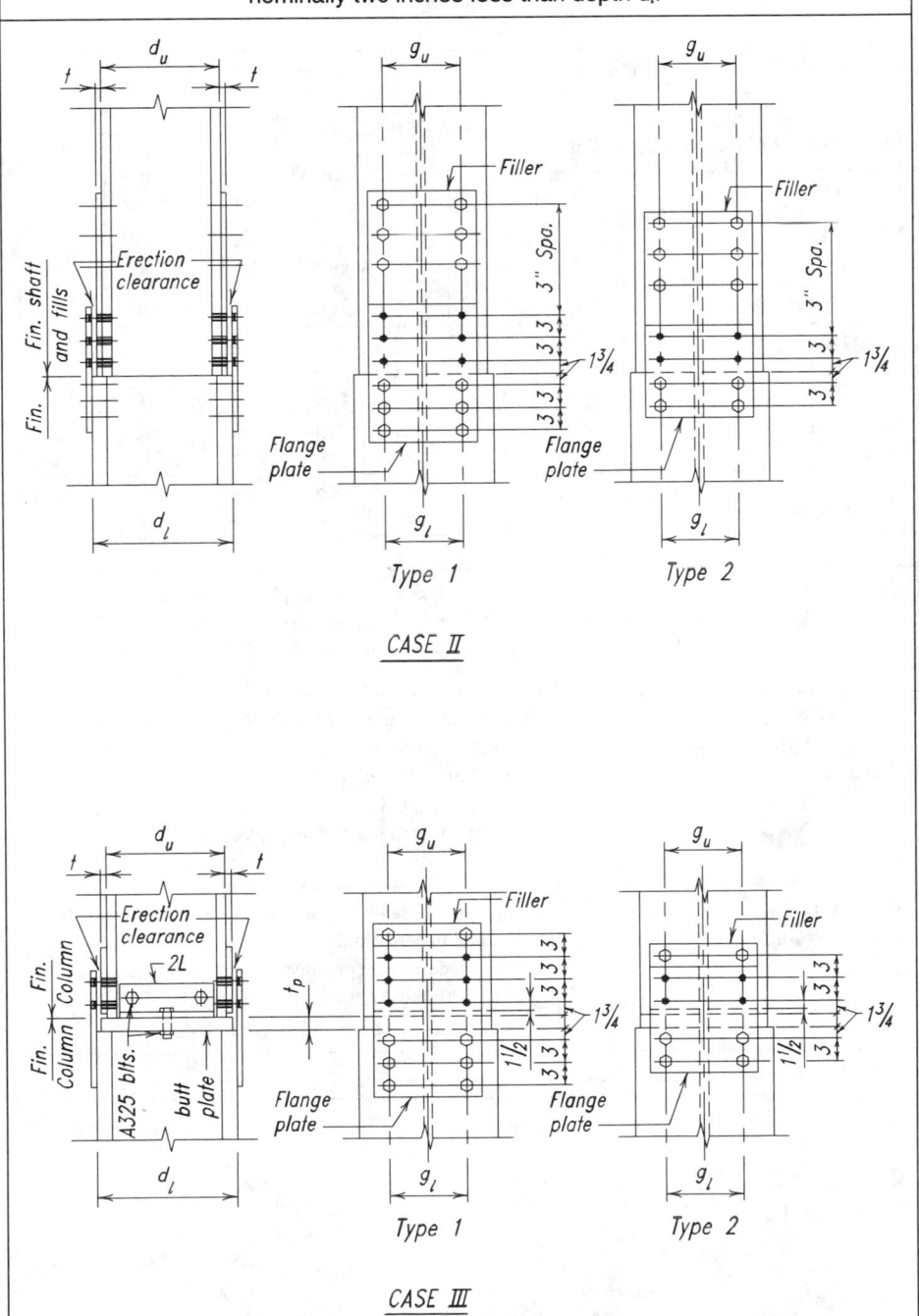

Table 11-4 (cont.).
Typical Column Splices

Case IV:
All-welded flange-plated column splices between columns with
depths d_u and d_l nominally the same.

Column Size	Flange Plate			Welds				Minimum Space for Welding	
	Width	Thk.	Length L	Size A	Length			M	N
					X	Y			
W14×455 & over	14	$\frac{5}{8}$	1'-6	$\frac{1}{2}$	5	7		$\frac{13}{16}$	$\frac{11}{16}$
311 to 426	12	$\frac{5}{8}$	1'-4	$\frac{1}{2}$	4	6		$\frac{13}{16}$	$\frac{11}{16}$
211 to 283	12	$\frac{1}{2}$	1'-4	$\frac{3}{8}$	4	6		$\frac{11}{16}$	$\frac{9}{16}$
90 to 193	12	$\frac{3}{8}$	1'-4	$\frac{5}{16}$	4	6		$\frac{5}{8}$	$\frac{1}{2}$
61 to 82	8	$\frac{3}{8}$	1'-4	$\frac{5}{16}$	3	6		$\frac{5}{8}$	$\frac{1}{2}$
43 to 53	6	$\frac{5}{16}$	1'-2	$\frac{1}{4}$	2	5		$\frac{9}{16}$	$\frac{7}{16}$
W12×120 to 336	8	$\frac{1}{2}$	1'-4	$\frac{3}{8}$	3	6		$\frac{11}{16}$	$\frac{9}{16}$
53 to 106	8	$\frac{3}{8}$	1'-4	$\frac{5}{16}$	3	6		$\frac{5}{8}$	$\frac{1}{2}$
40 to 50	6	$\frac{5}{16}$	1'-2	$\frac{1}{4}$	2	5		$\frac{9}{16}$	$\frac{7}{16}$
W10×49 to 112	8	$\frac{3}{8}$	1'-4	$\frac{5}{16}$	3	6		$\frac{5}{8}$	$\frac{1}{2}$
33 to 45	6	$\frac{5}{16}$	1'-2	$\frac{1}{4}$	2	5		$\frac{9}{16}$	$\frac{7}{16}$
W8×31 to 67	6	$\frac{3}{8}$	1'-2	$\frac{5}{16}$	2	5		$\frac{5}{8}$	$\frac{1}{2}$
24 & 28	5	$\frac{5}{16}$	1'-0	$\frac{1}{4}$	2	4		$\frac{9}{16}$	$\frac{7}{16}$

Case IV-A: $d_l = (d_u + \frac{1}{8})$	Flange plates: Select flange-plate width and length and weld lengths for upper (lighter) column; select flange-plate thickness and weld size for lower (heavier) column. Fillers: None.
Case IV-B: $d_l = (d_u - \frac{1}{4}\text{-in.})$ to d_u	Flange plates: Same as Case IV-A, except use weld size $A + t$ on lower column. Fillers (undeveloped on lower column, shop welded under flange plates): Select thickness t as $(d_l - d_u)/2 + \frac{1}{16}$-in. Select width to match flange plate and length as $L/2 - 2$ in.
Case IV-C: $d_l = (d_u + \frac{1}{4}\text{-in.})$ to $(d_u + \frac{1}{2}\text{-in.})$	Flange plates: Same as Case IV-A, except use weld size $A + t$ on upper column. Fillers (undeveloped on upper column, shipped loose): Select thickness t as $(d_l - d_u)/2 - \frac{1}{16}$-in. Select width to match flange plate and length as $L/2 - 2$ in.

For lifting devices, see Figure 11-21.

Table 11-4 (cont.).
Typical Column Splices

Case IV:
All-welded flange-plated column splices between columns with depths d_u and d_l nominally the same.

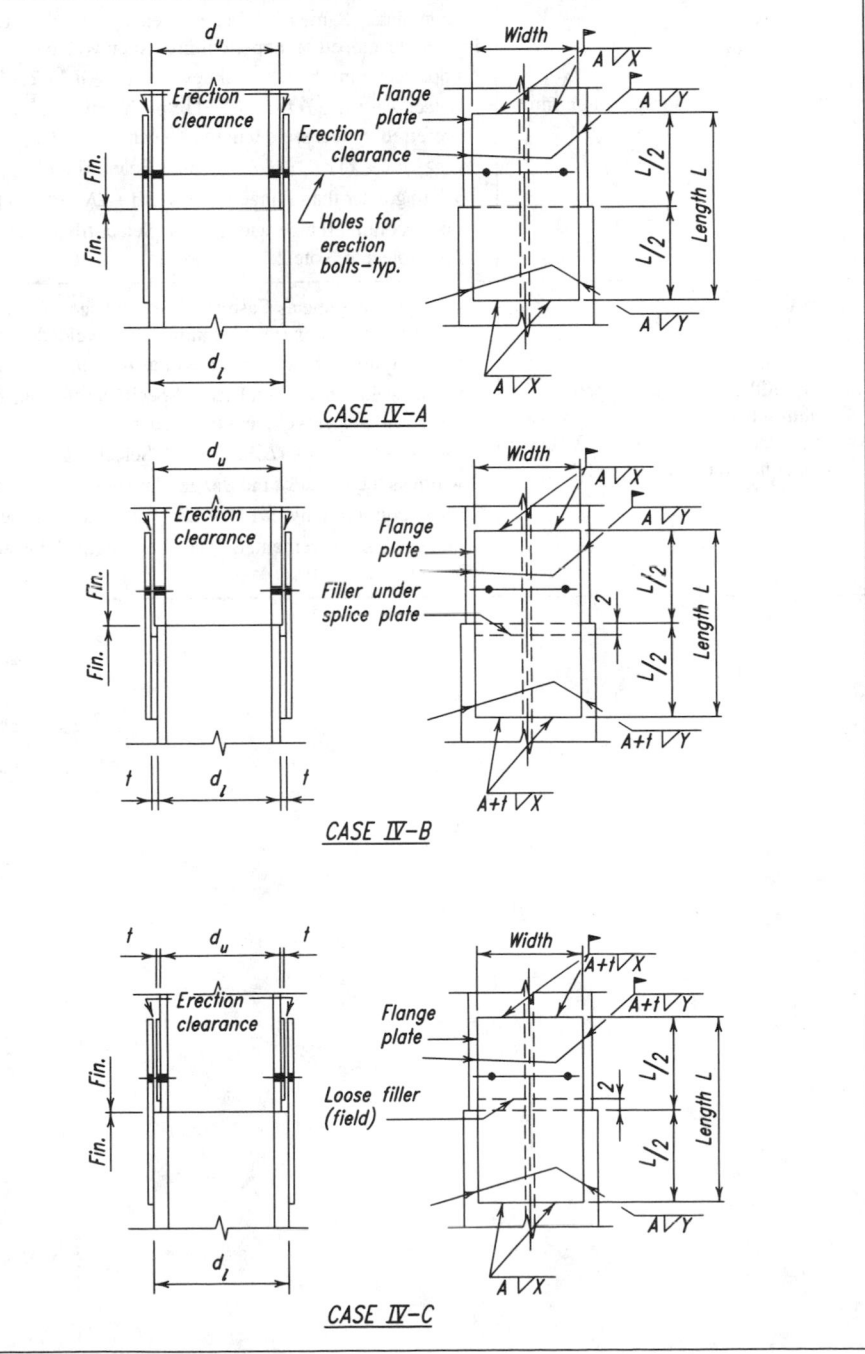

Table 11-4 (cont.). Typical Column Splices
Case IV: All-welded flange-plated column splices between columns with depths d_u and d_l nominally the same

Case IV-D: $d_l = (d_u + \frac{5}{8}\text{-in.})$ and over Filler width less than upper column flange width.	Flange plates: Same as Case IV-A, except see Note 1. Fillers (developed on upper column, shop welded to upper column): Select thickness t as $(d_l - d_u) / 2 - \frac{1}{16}$-in. Select weld size B from LRFD Specification; $\leq\frac{5}{16}$-in. preferred. Select weld length L_B such that $L_B \geq A(X + Y) / B \geq (L / 2 + 1 \text{ in.})$. Select filler width greater than flange plate width + $2N$ but less than upper column flange width – $2M$. Select filler length as L_B, subject to Note 2.
Case IV-E: $d_l = (d_u + \frac{5}{8}\text{-in.})$ and over Filler width greater than upper column flange width. Use this case only when M or N in Case IV-D are inadequate for welds B and A.	Flange plates: Same as Case IV-A, except see Note 1. Fillers (developed on upper column, shop welded to upper column): Select thickness t as $(d_l - d_u) / 2 - \frac{1}{16}$-in. Select weld size B from LRFD Specification; $\leq\frac{5}{16}$-in. preferred. Select weld length L_B such that $L_B \geq A(X + Y) / B \geq (L / 2 + 1 \text{ in.})$. Select filler width as the larger of the flange plate width + $2N$ and the upper column flange width + $2M$, rounded to the next higher $\frac{1}{4}$-in. increment. Select filler length as L_B subject to Note 2.

Table 11-4 (cont.).
Typical Column Splices

Case IV:
All-welded flange-plated column splices between columns with
depths d_u and d_l nominally the same

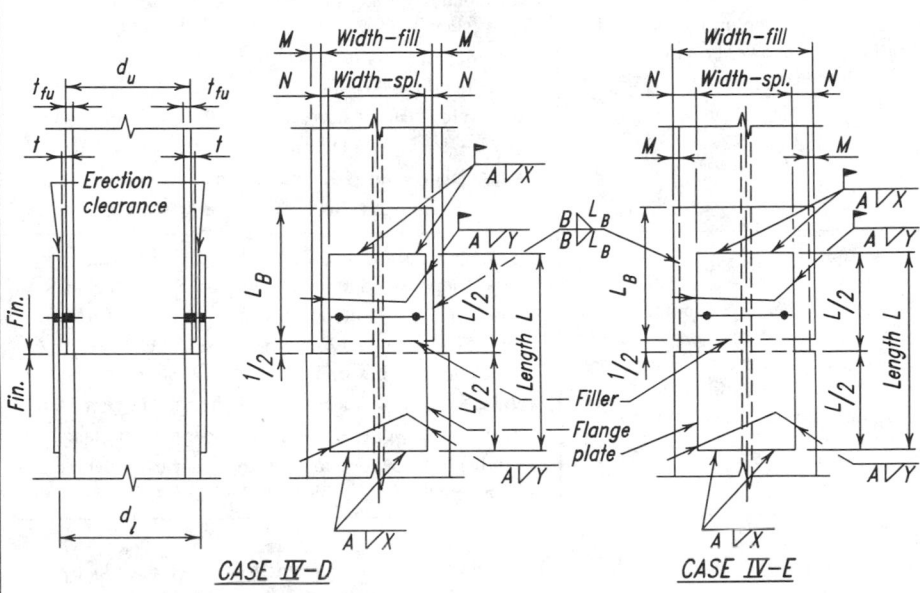

CASE IV–D   CASE IV–E

Note 1:

Where welds fasten flange plates to developed fillers, or developed fillers to column flanges (Cases IV-E and V-B), use the table to the right to check minimum fill thickness for balanced fill and weld shear strength.

Assume that an E70XX weld with $A = \frac{1}{2}$, $X = 4$, and $Y = 6$ is to be used

Weld A E70XX	Minimum Fill Thickness for Balanced Weld and Plate Shear F_y	
	36	50
$\frac{1}{4}$	0.26	0.19
$\frac{5}{16}$	0.32	0.23
$\frac{3}{8}$	0.38	0.28
$\frac{7}{16}$	0.45	0.33
$\frac{1}{2}$	0.51	0.37

at full strength on an A36 fill $\frac{1}{4}$-in.

thick. Since this table shows that the minimum fill thickness to develop this $\frac{1}{2}$-in. weld is 0.51 in., the $\frac{1}{4}$-in. fill will be overstressed. A balanced condition is obtained by multiplying the length $(X + Y)$ by the ratio of the minimum to the actual thickness of fill, thus:

$$(4 + 6) \times \frac{0.51}{0.25} = 20.4$$

use $(X + Y) = 20\frac{1}{2}$-in.

Placing this additional increment of $(X + Y)$ can be done by making weld lengths X continuous across the end of the splice plate and by increasing lengths Y (and therefore the plate length) if required.

Note 2:

If fill length, based on L_B, is excessive, place weld of size B across one or both ends of fill and reduce L_B accordingly, but not to less than $(L / 2 + 1)$. Omit return welds in Cases IV-E and V-B.

Table 11-4 (cont.).
Typical Column Splices

Case V:
All-welded flange-plated column splices between columns with
depth d_u nominally two inches less than depth d_l.

Case V-A: Fillers on upper column developed for bearing on lower column. Filler width less than upper column flange width.	Flange plates: Same as Case IV-A, except see Note 1. Fillers (shop welded to upper column): Select thickness as $(d_l - d_u) / 2 - \frac{1}{16}$-in. Select weld size B from LRFD Specification; $\leq\frac{5}{16}$-in. preferred. Select weld length L_B to develop bearing strength of the filler but not less than $(L/2 + 1\frac{1}{2}$-in.$)$. Select filler width greater than the flange plate width + $2N$ but less than the upper column flange width $- 2M$. See Case IV for M and N.
Case V-B: Same as Case V-A except filler width is greater than upper column flange width. Use this case only when M or N in Case V-A are inadequate for weld A, or when additional filler bearing area is required.	Flange plates: Same as Case IV-A, except see Note 1. Fillers (shop welded to upper column): Select thickness as $(d_l - d_u) / 2 - \frac{1}{16}$-in. Select weld size B from LRFD Specification; $\leq\frac{5}{16}$-in. preferred. Select weld length L_B to develop bearing strength of the filler but not less than $(L/2 + 1\frac{1}{2}$-in.$)$. Select filler width as the larger of the flange plate width + $2N$ and the upper column flange width + $2M$, rounded to the next higher $\frac{1}{4}$-in. increment. Filler length as L_B, subject to Note 3.

Note 3:
If fill length, based on L_B, is excessive, place weld of size B across end of fill and reduce L_B by
one-half of such additional weld length, but not to less than $(L/2 + 1\frac{1}{2})$. Omit return welds in
Case V-B.

Table 11-4 (cont.).
Typical Column Splices

Case V:
All-welded flange-plated column splices between columns with depth d_u nominally two inches less than depth d_l.

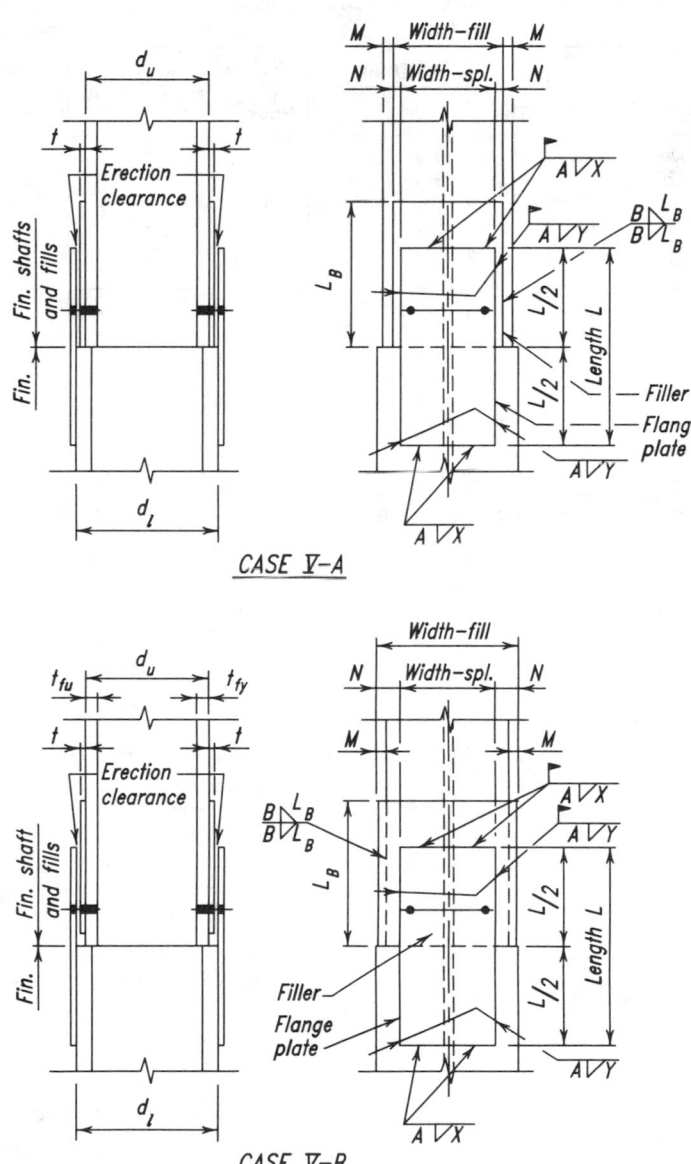

CASE V-A

CASE V-B

Table 11-4 (cont.).
Typical Column Splices

Case VI:
Combination bolted and welded column splices between columns with depths d_n and d_l nominally the same.

Column Size	Flange Plate					Bolts			Welds		
	Width	Thk.	Length			No. of Rows	Gage g	Size A		Length	
			L_U	L_L						X	Y
W14×455 & over	14	$\frac{5}{8}$	$9\frac{1}{4}$	9		3	$11\frac{1}{2}$	$\frac{1}{2}$		5	7
311 to 426	12	$\frac{5}{8}$	$9\frac{1}{4}$	8		3	$9\frac{1}{2}$	$\frac{1}{2}$		4	6
211 to 283	12	$\frac{1}{2}$	$9\frac{1}{4}$	8		3	$9\frac{1}{2}$	$\frac{3}{8}$		4	6
90 to 193	12	$\frac{3}{8}$	$6\frac{1}{4}$	8		2	$9\frac{1}{2}$	$\frac{5}{16}$		4	6
61 to 82	8	$\frac{3}{8}$	$6\frac{1}{4}$	8		2	$5\frac{1}{2}$	$\frac{5}{16}$		3	6
43 to 53	6	$\frac{5}{16}$	$6\frac{1}{4}$	7		2	$3\frac{1}{2}$	$\frac{1}{4}$		2	5
W12×120 to 336	8	$\frac{1}{2}$	$6\frac{1}{4}$	8		2	$5\frac{1}{2}$	$\frac{3}{8}$		4	6
53 to 106	8	$\frac{3}{8}$	$6\frac{1}{4}$	8		2	$5\frac{1}{2}$	$\frac{5}{16}$		3	6
40 to 50	6	$\frac{5}{16}$	$6\frac{1}{4}$	7		2	$3\frac{1}{2}$	$\frac{1}{4}$		2	5
W10×49 to 112	8	$\frac{3}{8}$	$6\frac{1}{4}$	8		2	$5\frac{1}{2}$	$\frac{5}{16}$		3	6
33 to 45	6	$\frac{5}{16}$	$6\frac{1}{4}$	7		2	$3\frac{1}{2}$	$\frac{1}{4}$		2	5
W8×31 to 67	6	$\frac{3}{8}$	$6\frac{1}{4}$	7		2	$3\frac{1}{2}$	$\frac{5}{16}$		2	5
24 & 28	5	$\frac{5}{16}$	$6\frac{1}{4}$	6		2	$3\frac{1}{2}$	$\frac{1}{4}$		2	4

Gages shown may be modified if necessary to accommodate fittings elsewhere on the columns.

Case VI-A:

$d_l = (d_u + \frac{1}{4}\text{-in.})$
to $(d_u + \frac{5}{8}\text{-in.})$

Flange plates: Select flange plate width, bolts, and length L_U for upper column; select flange plate thickness, weld size A, weld lengths X and Y, and length L_L for lower column. Total flange plate length is $L_U + L_L$ (see table above).
Fillers: None.
Shims: Furnish sufficient strip shims $2\frac{1}{2} \times \frac{1}{8}$ to obtain 0 to $\frac{1}{16}$-in. clearance on each side.

Case VI-B:

$d_l = (d_u - \frac{1}{4}\text{-in.})$
to $(d_u + \frac{1}{8}\text{-in.})$

Flange plates: Same as Case VI-A, except use weld size $A + t$ on lower column.
Fillers (shop welded to lower column under flange plate): Select thickness t as $\frac{1}{8}$-in. for for $d_l = d_u$ and $d_l = (d_u + \frac{1}{8}\text{-in.})$ or as $\frac{3}{16}$-in. for $d_l = (d_u - \frac{1}{8}\text{-in.})$ and $d_l = (d_u - \frac{1}{4}\text{-in.})$. Select width to match flange plate and length as $L_L - 2$ in.
Shims: Same as Case VI-A.

Case VI-C:

$d_l = (d_u + \frac{3}{4}\text{-in.})$
and over

Flange plates: Same as Case VI-A.
Fillers (shop welded to upper column): Select thickness t as $(d_l - d_u) / 2$ minus $\frac{1}{8}$-in. or $\frac{3}{16}$-in., whichever results in $\frac{1}{8}$-in. multiples of fill thickness. Select weld size B as minimum size from LRFD Specification Section J2. Select weld length as $L_U - \frac{1}{4}$-in. Select filler width as flange plate width and filler length as $L_U - \frac{1}{4}$-in.
Shims: Same as Case VI-A.

Table 11-4 (cont.).
Typical Column Splices

Case VI:
Combination bolted and welded column splices between columns with depths d_n and d_l nominally the same.

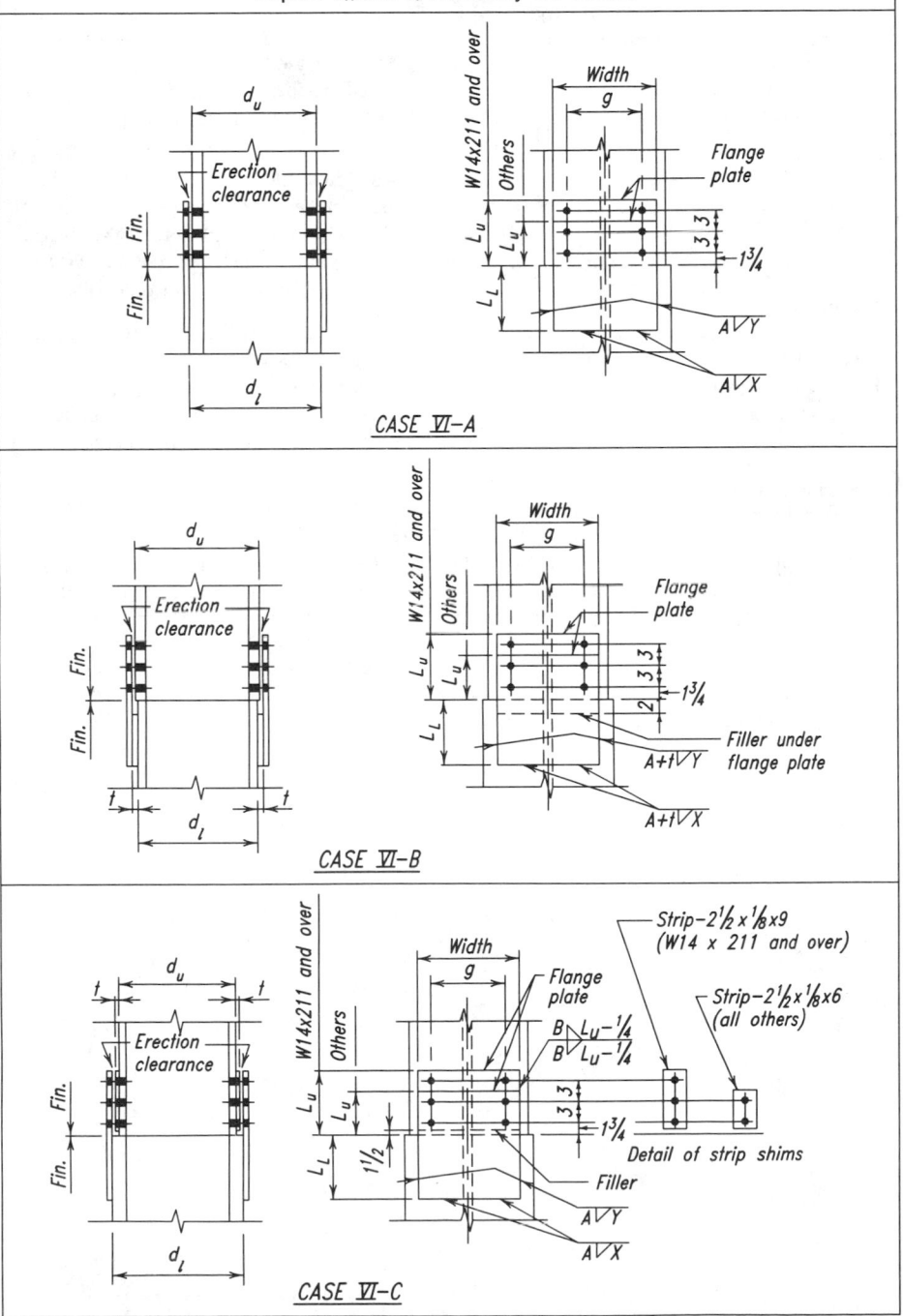

Table 11-4 (cont.).
Typical Column Splices

Case VII:
Combination bolted and welded flange-plated column splices between
columns with depth d_u nominally two inches less than depth d_l
Fillers developed for bearing.

Case VII-A: Fillers of width less than upper column flange width.	Flange plates: Same as Case VI-A. Fillers (shop welded to upper column): Select filler thickness t as $(d_l - d_u) / 2$ minus $\frac{1}{8}$-in. or $\frac{3}{16}$-in., whichever results in $\frac{1}{8}$-in. multiples of filler thickness. Select weld size B from LRFD Specification; $\leq \frac{5}{16}$-in. preferred. Select weld length L_B to develop bearing strength of filler. Select filler width not less than flange plate width but not greater than upper column flange width $-2M$ (see Case IV). Select filler length as L_B, subject to Note 4.
Case VII-B: Filler of width greater than upper column flange width. Use Case VII-B only when fillers must be widened to provide additional bearing area.	Flange plates: Same as Case VI-A. Fillers (shop welded to upper columns): Same as Case VII-A except select filler width as upper column flange width $+ 2M$ (see Case IV) rounded to the next larger $\frac{1}{2}$-in. increment.

Note 4:
If fill length based on L_B is excessive, place weld of size B across end of fill and reduce L_B by one-half of such additional weld length, but not less than L_U. Omit return welds, Case VII-B.

Table 11-4 (cont.).
Typical Column Splices

Case VII:
Combination bolted and welded flange-plated column splices between columns with depth d_u nominally two inches less than depth d_l
Fillers developed for bearing.

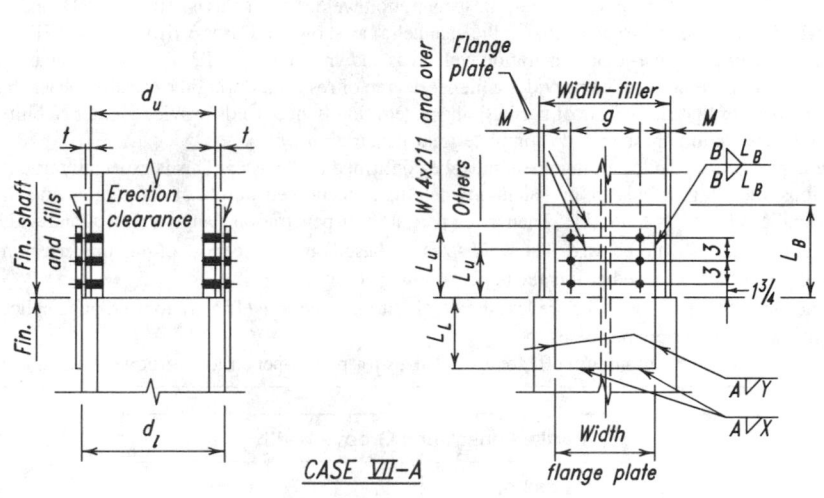

CASE VII-A

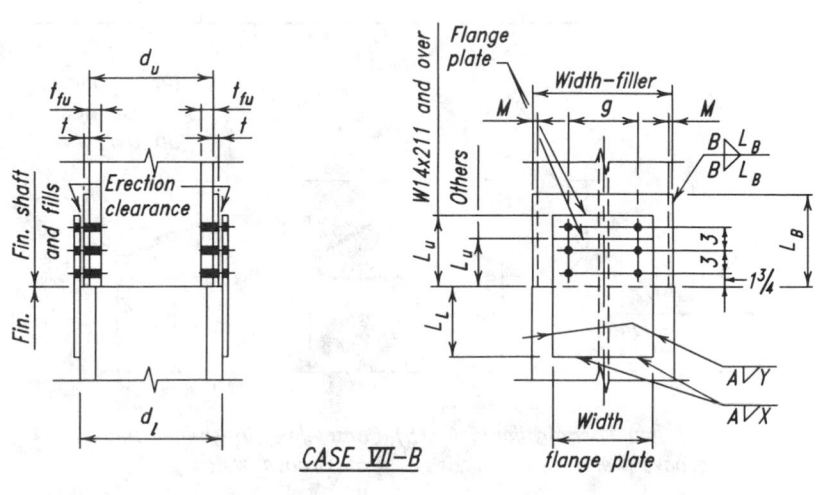

CASE VII-B

Table 11-4 (cont.).
Typical Column Splices

Case VIII:
Directly welded flange column splices between columns with
depths d_u and d_l nominally the same.

These types of splices exhibit versatility. The flanges may be partial-joint-penetration welded as in
Cases VIIIA and VIIIB, or complete-joint-penetration welded as in Cases VIIIC, VIIID, and
VIIIE. The webs may be spliced using the channel(s) as shown in Cases VIIIA, VIIIB, VIIIC,
and VIIID, or complete-joint-penetration welded as shown in Case VIIIE. The use of a channel
or channels at the web splice provides a higher degree of restraint during the erection phase than
does a plate or plates. The use of partial-joint-penetration flange welds provide greater stability
during the erection phase than do complete-joint-penetration welds.
The adequacy of any splice arrangement must be confirmed by the user. This is especially true in
regions where high winds are prevalent or when the concentrated weight of the fabricated column
is significantly off its centerline. Then using partial-joint-penetration flange welds, a land width of
$1/4$-in. or greater should be used. The weld sizes are based on the thickness of the thinner column
flange, regardless of whether it is the upper or lower column.
When column flange thicknesses are less than $1/2$-in. it may be more efficient to use flange splice
plates as shown in previous cases.
See the table below for minimum effective weld sizes for partial-penetration groove welds.

Partial Penetration Groove Width	
[a]**Thickness of Column Material** T_u	**Minimum Effective Welds Size** E
[b]Over $1/2$ to $3/4$, incl.	$1/4$
Over $3/4$ to $1\,1/2$, incl.	$5/16$
Over $1\,1/2$ to $2\,1/4$, incl.	$3/8$
Over $2\,1/4$ to 6, incl.	$1/2$
Over 6	$5/8$

[a]Thickness of thicker part joined.
[b]For less than $1/2$, use splice plates.

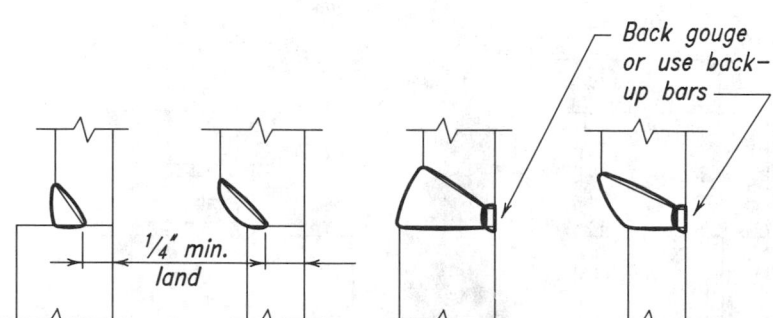

(a) Partial–joint–penetration   (b) Complete–joint–penetration
          groove welds   groove welds

Table 11-4 (cont.).
Typical Column Splices

Directly welded flange column splices between columns with
depths d_u and d_l nominally the same.

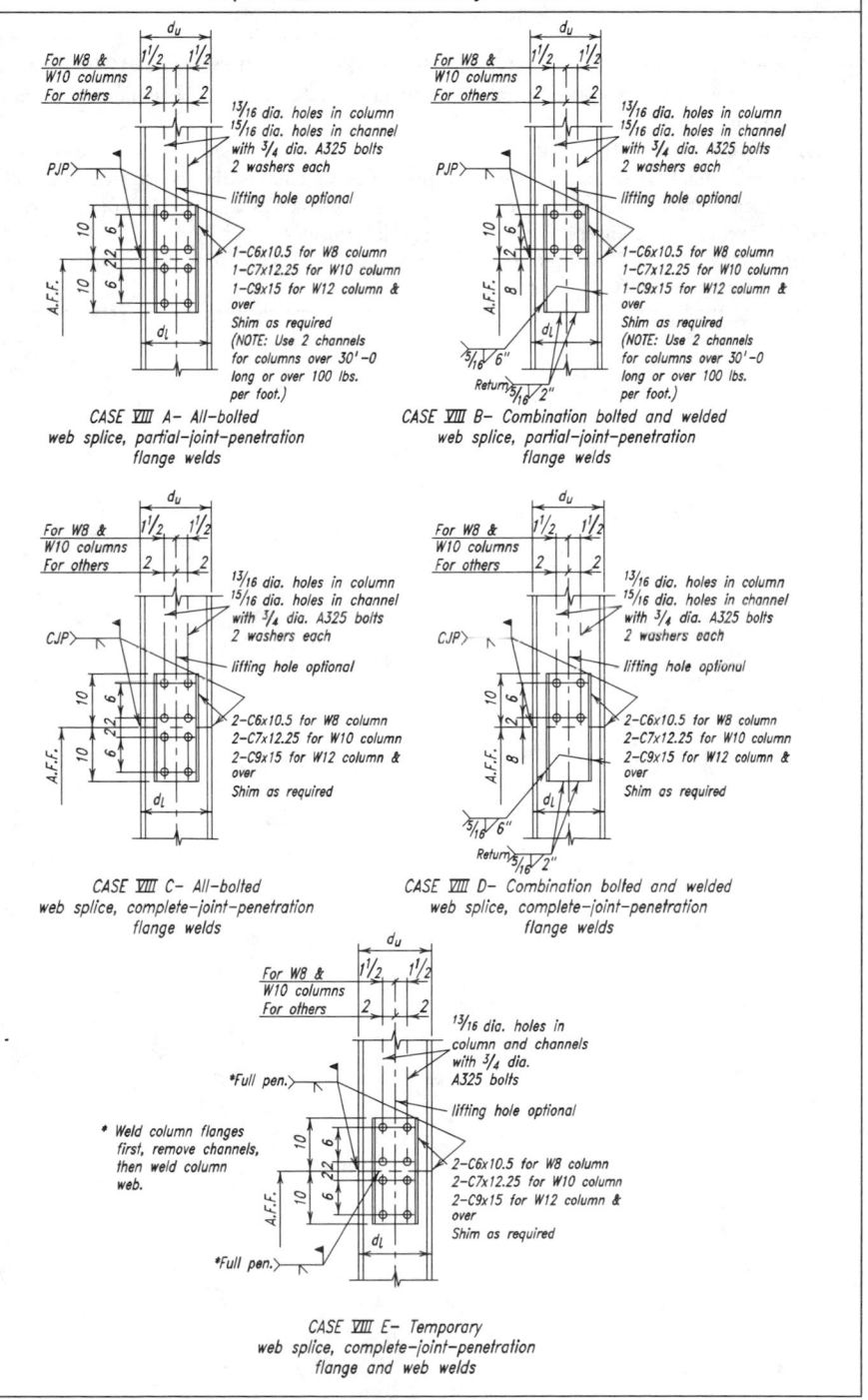

CASE VIII A- All-bolted
web splice, partial-joint-penetration
flange welds

CASE VIII B- Combination bolted and welded
web splice, partial-joint-penetration
flange welds

CASE VIII C- All-bolted
web splice, complete-joint-penetration
flange welds

CASE VIII D- Combination bolted and welded
web splice, complete-joint-penetration
flange welds

CASE VIII E- Temporary
web splice, complete-joint-penetration
flange and web welds

Table 11-4 (cont.).
Typical Column Splices

Case IX:
Butt-plated column splices between columns with
depth d_u nominally 2 in. less than depth d_l.

Butt plate: Select a butt plate thickness of $1\frac{1}{2}$-in. for W8 over W10 columns and 2 in. for all other
combinations. Select butt plate width and length not less than w_l and d_l assuming the lower is the
larger column shaft.

Weld: Select weld to upper column based on the thicker of t_{fu} and t_p. Select weld to lower column
based on the thicker of t_{fl} and t_p. The edge preparation required by the groove weld is usually
performed on the column shafts. However, special cases such as when the butt plate must
be field welded to the lower column require special consideration.

Erection: clip angles, such as those shown in the sketch below, help to locate and stabilize the upper
column during the erection phase.

Table 11-4 (cont.).
Typical Column Splices

Case IX:
Butt-plated column splices between columns with depth d_u nominally 2 in. less than depth d_l.

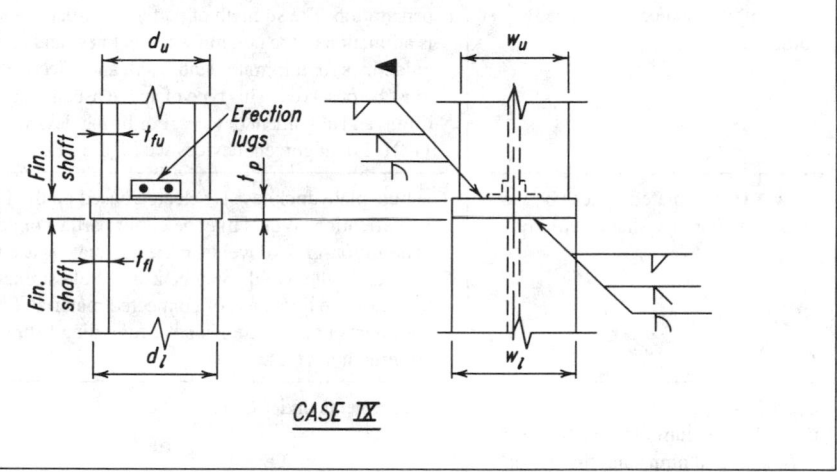

CASE IX

Table 11-4 (cont.).
Typical Column Splices

Cases X, XI, XII
Special column splices

Case X: Directly welded splice between tubular and/or box-shaped columns.	Welds may be either partial-joint- or complete-joint-penetration. The strength of partial-joint-penetration welds is a function of the column wall thickness and appropriate guidelines for minimum land width and effective weld size must be observed. This type of splice usually requires lifting and alignment devices. For lifting devices see Figure 11-21. For alignment devices see Figure 11-22.
Case XI: Butt-plated splices between tubular and/or box-shaped columns.	The butt-plate thickness is selected based on the LRFD Specification. Welds may be either partial- or complete-penetration-groove welds, or, if adequate space is provided, fillet welds may be used. Weld strength is based on the thickness of connected material. See comments under Case X above regarding lifting and alignment devices.
Case XII: Butt-plated column splices between W-shape columns and tubular or box-shaped columns.	See comments under Case XI above.

Table 11-4 (cont.).
Typical Column Splices

Cases X, XI, XII
Special column splices

CASE X

$t_u \leq t_l$

CASE XI

CASE XII

TRUSS CONNECTIONS

Members in Trusses

For light loads, trusses are commonly composed of tees for the top and bottom chords with single-angle or double-angle web members. In welded construction, the single-angle and double-angle web members may, in many cases, be welded to the stem of the tee, thus, eliminating the need for gussets. When single-angle web members are used, all web members should be placed on the same side of the chord; staggering the web members causes a torque on the chord, as illustrated in Figure 11-25.

Double-angle truss members are designed to act as a single composite unit. When unequal-leg angles are used, long legs are normally assembled back to back. A simple notation for this is LLBB (long legs back-to-back) and SLBB (short legs back-to-back). Alternatively, the notation might be graphical in nature as $\rfloor\llcorner$ and $\underline{\quad}\rfloor\llcorner\underline{\quad}$.

For large loads, W-shapes may be used with the web vertical and gussets welded to the flange for the truss connections. Web members may be single angles or double angles, although W-shapes are sometimes used for both chord and web members as shown in Figure 11-26. Heavy sections in trusses must meet the design and fabrication restrictions and special requirements of LRFD Specification Sections A3.1c, J1.5, J1.6, J2, and M2.2. With member orientation as shown for the field-welded truss joint in Figure 11-26a, connections usually are made by groove welding flanges to flanges and fillet welding webs directly or indirectly by the use of gussets. Fit-up of joints in this type of construction are very sensitive to dimensional variations in the rolled shapes; fabricators sometimes prefer to use built-up shapes in these cases.

The web connection plate in Figure 11-26a is a typical detail. While the diagonal member could theoretically be cut so that the diagonal web would be extended into the web of the chord for a direct connection, such a detail is difficult to fabricate. Additionally, welding access becomes very limited; note the obvious difficulty of welding the gusset or diagonal directly to the chord web. As illustrated, this weld is usually omitted.

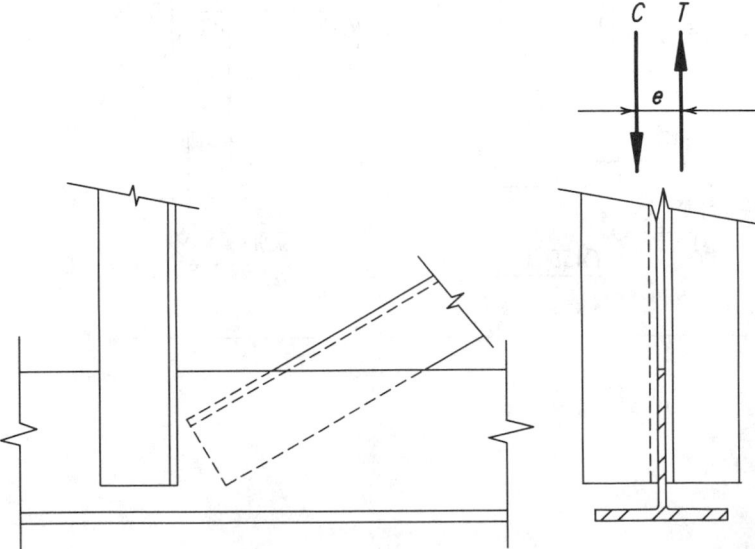

Fig. 11-25. Staggered truss-web members result in a torque on the truss chord.

When stiffeners and doubler plates are required for concentrated flange forces, the designer should consider selecting a heavier section which would eliminate the need for stiffening. Although this will increase the material cost of the member, the heavier section will likely provide a more economical solution due to the reduction in labor cost associated with the elimination of stiffening (Ricker, 1992 and Thornton, 1992).

Minimum Connection Strength

From LRFD Specification Section J1.7, truss connections must be designed for a minimum factored load of 10 kips. Additionally, when trusses are shop assembled or field assembled on the ground for subsequent erection, consideration should be given to loads induced during handling, shipping, and erection. A common requirement for these cases is that the connection be designed for a minimum of 50 percent of the member strength or a lesser amount as determined by the engineer.

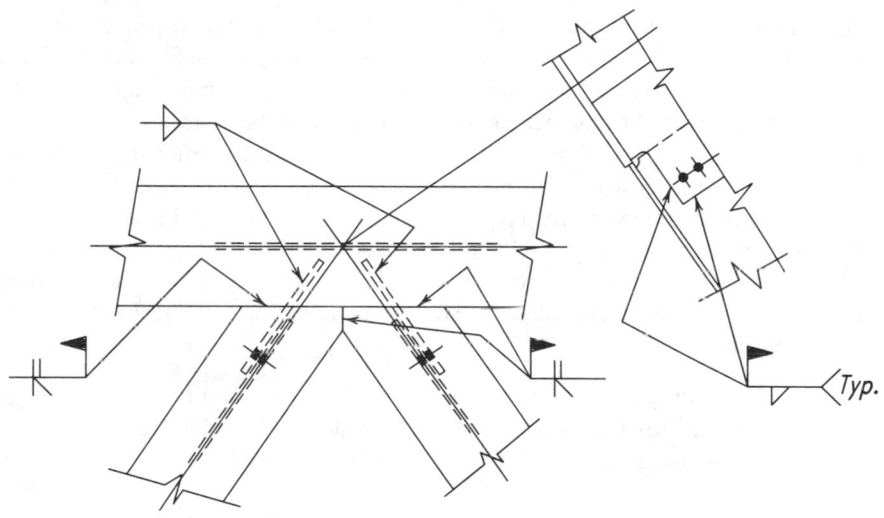

(a) Shop and field welding

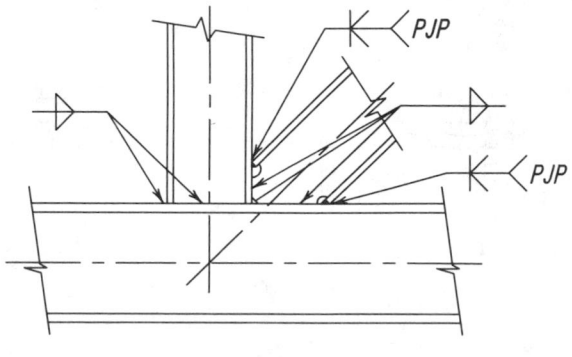

(b) Shop welding

Fig. 11-26. Truss-panel-point connections for W-shape members.

Panel-Point Connections

A panel-point connection connects diagonal and/or vertical web members to the chord member of a truss. These web members deliver axial forces, tensile or compressive, to the truss chord. In bolted construction, a gusset is usually required because of bolt spacing and edge distance requirements. In welded construction, it is sometimes possible to eliminate the need for a gusset.

Design Checks—The design strengths of the bolts and/or welds, connecting elements, and affected elements of the connected members must be determined in accordance with the provisions of the LRFD Specification. The applicable limit states in each of the aforementioned design strengths are discussed in Part 8. In all cases, the design strength ϕR_n must exceed the required strength R_u.

In the panel-point connection of Figure 11-27, the neutral axes of the vertical and diagonal truss members intersect on the neutral axis of the truss chord. As a result, the forces in all members of the truss are axial. It is common practice, however, to modify working lines slightly from the gravity axes to establish repetitive panels and avoid fractional dimensions less than $\frac{1}{8}$-in. or to accommodate a larger panel-point connection or a connection for bottom-chord lateral bracing, a purlin, or a sway-frame. This eccentricity and the resulting moment must be considered in the design of the truss chord.

In contrast, for the design of the truss web members, LRFD Specification Section J1.8 states that the center of gravity of the end connection of a statically loaded truss member need not coincide with the gravity axis of the connected member. This is because tests have shown that there is no appreciable difference in the static design strength between balanced and unbalanced connections of this nature. Accordingly, the truss web members and their end connections may be designed for the axial load, neglecting the effect of this minor eccentricity.

Shop and Field Practices—In bolted construction, it is convenient to use standard gage lines of the angles as truss working lines; where wider angles with two gage lines are used, the gage line nearest the heel of the angle is the one which is substituted for the gravity axis.

To provide for stiffness in the finished truss, the web members of the truss are extended to near the edge of the fillet of the tee (*k*-distance). If welded, the required welds are then applied along the heel and toe of each angle, beginning at their ends rather than at the edge of the tee stem.

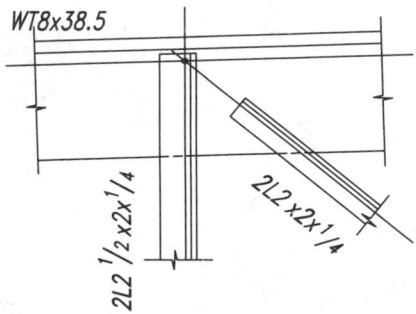

Fig. 11-27. Panel-point truss connection.

EXAMPLE 11-8

Given: Refer to Figure 11-28. Determine the requirements for the following cases:

A. joint L_1

B. joint U_1

Assume 70 ksi electrodes. For the WT truss chord, assume $F_y = 50$ ksi and $F_u = 65$ ksi. For angle and splice material, assume $F_y = 36$ ksi and $F_u = 58$ ksi.

Solution A: *Check shear yield of the tee stem (on Section A-A)*

$$\phi R_n = \phi(0.6F_y A_w)$$
$$= 0.90(0.6 \times 50 \text{ ksi} \times 8.215 \text{ in.} \times 0.430 \text{ in.})$$
$$= 95.4 \text{ kips} < 104 \text{ kips} \quad \textbf{n.g.}$$

Additional shear area must be provided.

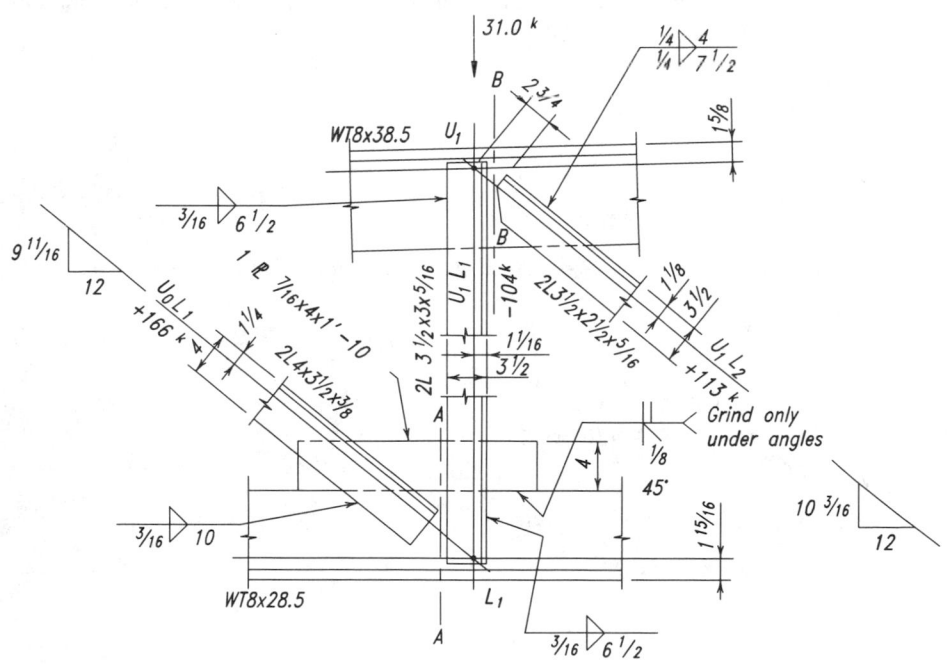

Joint Detail for U_1 & L_1

Fig. 11-28.

Try PL $\frac{7}{16}$-in. $\times$ 4 in. complete-joint-penetration groove welded to the stem of the WT.

$$\phi R_n = 95.4 \text{ kips} + 0.9(0.6 \times 36 \text{ ksi} \times 4 \text{ in.} \times \tfrac{7}{16}\text{-in.})$$
$$= 129 \text{ kips} > 104 \text{ kips} \quad \textbf{o.k.}$$

Design welds for member U_1L_1

The minimum weld size from LRFD Specification Table J2.4 is $\frac{3}{16}$-in. The maximum weld size cannot exceed $\frac{5}{16}$-in. Calculate the minimum length of $\frac{3}{16}$-in. fillet weld:

$$L_{\min} = \frac{R_n}{1.392D}$$
$$= \frac{104 \text{ kips}}{1.392 \text{ (3 sixteenths)}}$$
$$= 24.9 \text{ in.}$$

Use $6\frac{1}{2}$-in. of $\frac{3}{16}$-in. weld at the heel and toe of both angles for a total of 26 inches.

Design welds for member U_0L_1

The minimum weld size from LRFD Specification Table J2.4 is $\frac{3}{16}$-in. The maximum weld size cannot exceed $\frac{1}{4}$-in. Calculate the minimum length of $\frac{3}{16}$-in. fillet weld:

$$L_{\min} = \frac{R_u}{1.392D}$$
$$= \frac{166 \text{ kips}}{1.392 \text{ (3 sixteenths)}}$$
$$= 39.8 \text{ in.}$$

Use 10 in. of $\frac{3}{16}$-in. weld at the heel and toe of both angles for a total of 40 inches.

Check tension yielding of angles ($U_0\,L_1$)

$$\phi R_n = \phi F_y A_g$$
$$= 0.9(36 \text{ ksi})(2 \times 2.67 \text{ in.}^2)$$
$$= 173 \text{ kips} > 166 \text{ kips} \quad \textbf{o.k.}$$

Check tension rupture of angles ($U_0\,L_1$)

From LRFD Specification Section B3,

$$U = 1 - \frac{\overline{x}}{L} \le 0.9$$
$$= 1 - \frac{1.21 \text{ in.}}{10 \text{ in.}}$$
$$= 0.879$$

However, from LRFD Specification Section J5.2, $A_n \le 0.85A_g$. Thus, from LRFD Specification Section J5.2,

$$\phi R_n = \phi F_u A_n$$
$$= 0.75(58 \text{ ksi})(0.85 \times 2 \times 2.67 \text{ in.}^2)$$
$$= 197 \text{ kips} > 166 \text{ kips} \quad \textbf{o.k.}$$

Solution B: *Check shear yielding of the tee stem (on Section B-B)*

$$\phi R_n = \phi(0.6 F_y A_w)$$
$$= 0.90(0.6 \times 50 \text{ ksi} \times 8.26 \text{ in.} \times 0.455)$$
$$= 101 \text{ kips} > 73.1 \text{ kips} \quad \textbf{o.k.}$$

Design welds for member $U_1 L_1$

As calculated previously in Solution A, use 6½-in. of ³⁄₁₆-in. weld at the heel and toe of both angles for a total of 26 inches.

Design welds for member $U_1 L_2$

The minimum weld size from LRFD Specification Table J2.4 is ³⁄₁₆-in. The maximum weld size cannot exceed ¼-in. Calculate the minimum length of ¼-in. fillet weld:

$$L_{min} = \frac{R_n}{1.392 D}$$
$$= \frac{113 \text{ kips}}{1.392 \text{ (4 sixteenths)}}$$
$$= 20.3 \text{ in.}$$

Use 7½-in. of fillet weld at the heel and four inches of fillet weld at the toe of each angle for a total of 23 inches.

Check tension yielding of angles ($U_1 L_2$)

$$\phi R_n = \phi F_y A_g$$
$$= 0.9(36 \text{ ksi})(2 \times 1.78 \text{ in.}^2)$$
$$= 115 \text{ kips} > 113 \text{ kips} \quad \textbf{o.k.}$$

Check tension rupture of angles ($U_1 L_2$)

From LRFD Specification Section B3,

$$U = 1 - \frac{\bar{x}}{L} \le 0.9$$
$$= 1 - \frac{1.14 \text{ in.}}{(4 \text{ in.} + 7\tfrac{1}{2}\text{-in.}) / 2}$$
$$= 0.802$$

Thus,

$$A_e = U A_g$$
$$= 0.802(2 \times 1.78 \text{ in.}^2)$$
$$= 2.86 \text{ in.}^2$$

From LRFD Specification Section J5.2,

$$\phi R_n = \phi F_u A_e$$
$$= 0.75(58 \text{ ksi})(2.86 \text{ in.}^2)$$
$$= 124 \text{ kips} > 113 \text{ kips} \quad \textbf{o.k.}$$

Check block shear rupture

Because of the cut end of the angle, the block shear rupture model presented in Part 8 does not directly apply. Conservatively, the block shear rupture strength will be based on the shear rupture strength of the WT stem along the length of the welds. Thus, the design strength is:

$$\phi R_n = \phi(0.6 F_u A_w)$$
$$= 0.90[0.6 \times 50 \text{ ksi} \times (7.5 \text{ in.} + 4 \text{ in.}) \times 0.455 \text{ in.}]$$
$$= 141 \text{ kips} > 113 \text{ kips} \quad \textbf{o.k.}$$

Support Connections

A truss support connection connects the ends of trusses to supporting columns.

Design Checks—The design strengths of the bolts and/or welds, connecting elements, and affected elements of the connected members must be determined in accordance with the provisions of the LRFD Specification. The applicable limit states in each of the aforementioned design strengths are discussed in Part 8. Additionally, truss support connections produce tensile or compressive single concentrated forces at the beam end; the limit states of the flange design strength in local bending and the limit states of the web design strength in local yielding, crippling, and compression buckling may have to be checked. In all cases, the design strength ϕR_n must exceed the required strength R_u.

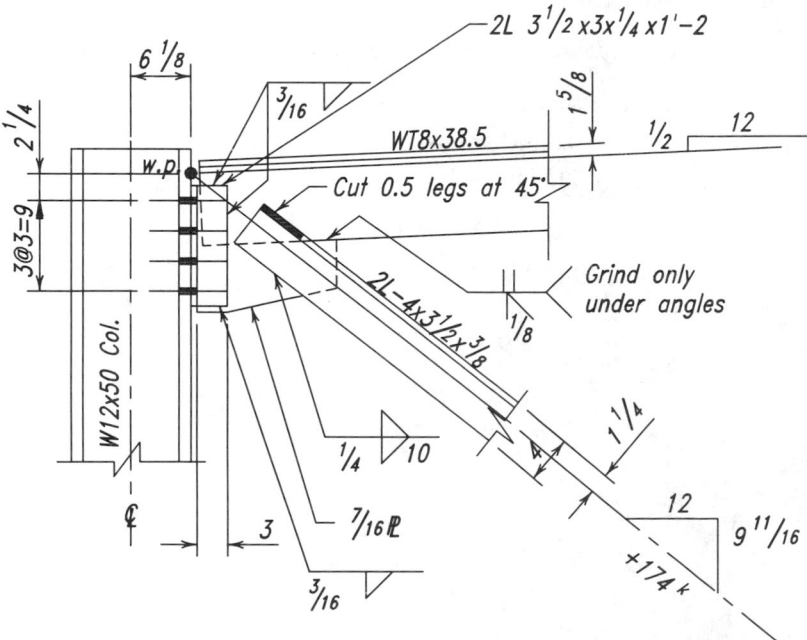

Fig. 11-29. Truss-support connection, working point on column face.

At the end of a truss supported by a column, all member axes may not intersect at a common point. When this is the case, an eccentricity results. Typically, it is the neutral axis of the column which does not meet at the working point.

If trusses with similar reactions line up on opposite sides of the column, consideration of eccentricity would not be required since any moment would be transfered through the column and into the other truss. However, if there is little or no load on the opposite side of the column, the resulting eccentricity must be considered.

In Figure 11-29, the truss chord and diagonal intersect at a common working point on the face of the column flange. In this detail, there is no eccentricity in the gusset, gusset-to-column connection, truss chord, or diagonal. However, the column must be designed for the moment due to the eccentricity of the truss reaction from the neutral axis of the column.

For the truss support connection illustrated in Figure 11-30, this eccentricity results in a moment. Assuming the connection between the members is adequate, joint rotation is resisted by the combined flexural strength of the column, the truss top chord, and the truss diagonal. However, the distribution of moment between these members will be proportional to the stiffness of the members. Thus, when the stiffness of the column is much greater than the stiffness of the other elements of the truss suport connection, it is good practice to design the column and gusset-to-column connection for the full eccentricity.

Due to its importance, the truss support connection is frequently shown in detail on the design drawing.

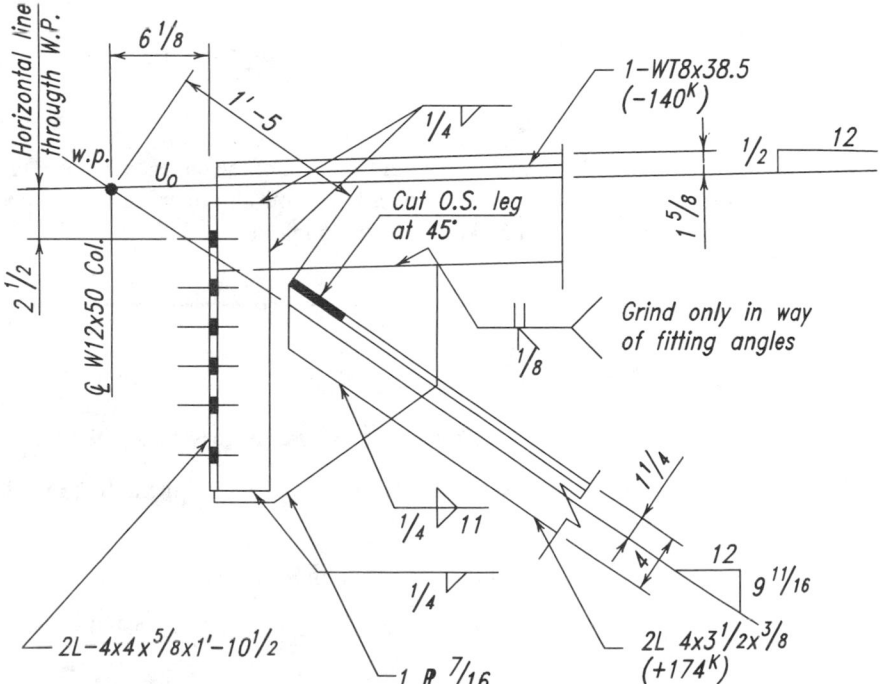

Fig. 11-30. Truss-support connection, working point at column centerline.

Shop and Field Practices—When a truss is erected in place and loaded, truss members in tension will lengthen and truss members in compression will shorten. At the support connection, this may cause the tension chord of a "square-ended" truss to encroach on its connection to the supporting column. When the connection is shop-attached to the truss, erection clearance must be provided with shims to fill out whatever space remains after the truss is erected and loaded. In field erected connections, however, provision must be made for the necessary adjustment in the connection.

When the tension chord delivers no calculated force to the connection, adjustment can usually be provided with slotted holes. For short spans with relatively light loads, the comparatively small deflections can be absorbed by the normal hole clearances provided for bolted construction. Slightly greater misalignment can be corrected in the field by reaming the holes. If appreciable deflection is expected, the connection may be welded or bolt holes may be field-drilled; this is an expensive operation which should be avoided if at all possible.

An approximation of the elongation which may be expected can be determined from the relationship between stress σ and strain ε, where E, the modulus of elasticity, is

$$E = \frac{\sigma}{\varepsilon}$$

With $\sigma = P/A$, $\varepsilon = \Delta/l$, and $E = 29{,}000$ ksi, Δ, the elongation in in. will be:

$$\Delta = \frac{Pl}{29{,}000A}$$

In the above equation,

   P = unfactored axial force, kips
   A = gross area of the truss chord, in.^{2}
   l = length, in.

The total change in length of the truss chord is $\Sigma\Delta_i$, the sum of the changes in the lengths of the individual panel segments of the truss chord. The misalignment at each support connection of the tension chord is one-half the total elongation.

EXAMPLE 11-9

Given: Refer to Figure 11-31. Determine:

                            A. the connection requirements between the gusset and column,

                            B. the required gusset size and the weld requirements for member U_0L_1 at the gusset.

Solution A: *Design bolts connecting angles to column (shear and tension)*

                            From Table 8-11, the number of $\frac{7}{8}$-in. diameter A325-N bolts required for shear only is

$$n_{\min} = \frac{R_u}{\phi r_n}$$

$$= \frac{111 \text{ kips}}{21.6 \text{ kips / bolt}}$$

$$= 5.14 \rightarrow 6 \text{ bolts}$$

Assuming an angle thickness of ⅝-in., bearing is not critical.

For a trial calculation, the number of bolts was increased to 12 in pairs at 3-in. spacing; the flexural strength of the angles was found to be insufficient. Subsequently, the spacing was revised to 4½-in. between the two rows of bolts at the top and bottom of the connection as illustrated in Figure 11-31.

The eccentric moment at the faying surface is

$$M_u = R_u e$$
$$= (111 \text{ kips})(6.10 \text{ in.})$$
$$= 677 \text{ kip-in.}$$

For the bolt group of Figure 11-31, the moment of inertia and section modulus are as follows.

$$I = A_b \, (\Sigma d^2)$$
$$= 0.6013 \text{ in.}^2 \, [(4 \times (1.5 \text{ in.})^2) + (4 \times (4.5 \text{ in.})^2) + (4 \times (9 \text{ in.})^2)]$$
$$= 249 \text{ in.}^4$$
$$S = \frac{I}{C}$$
$$= \frac{249 \text{ in.}^4}{9 \text{ in.}}$$
$$= 27.7 \text{ in.}^3$$

and the maximum tensile force per bolt is

$$r_{ut} = \left(\frac{M_u}{S}\right) A$$

$$= \left(\frac{677 \text{ kips-in.}}{27.7 \text{ in.}^3}\right)(0.6013 \text{ in.}^2)$$

$$= 14.7 \text{ kips}$$

Check design tensile strength of bolts

From LRFD Specification Section J3.7,

$$F_t = 117 \text{ ksi} - 1.9 f_v \leq 90 \text{ ksi}$$

$$= 117 \text{ ksi} - 1.9 \left(\frac{111 \text{ kips}}{12 \times 0.6013 \text{ in.}^2}\right)$$

$$= 87.8 \text{ ksi}$$

$$\phi r_n = \phi F_t A_b$$
$$= 0.75(87.8 \text{ ksi })(0.6013 \text{ in.}^2)$$
$$= 39.6 \text{ kips} > 14.7 \text{ kips} \quad \textbf{o.k.}$$

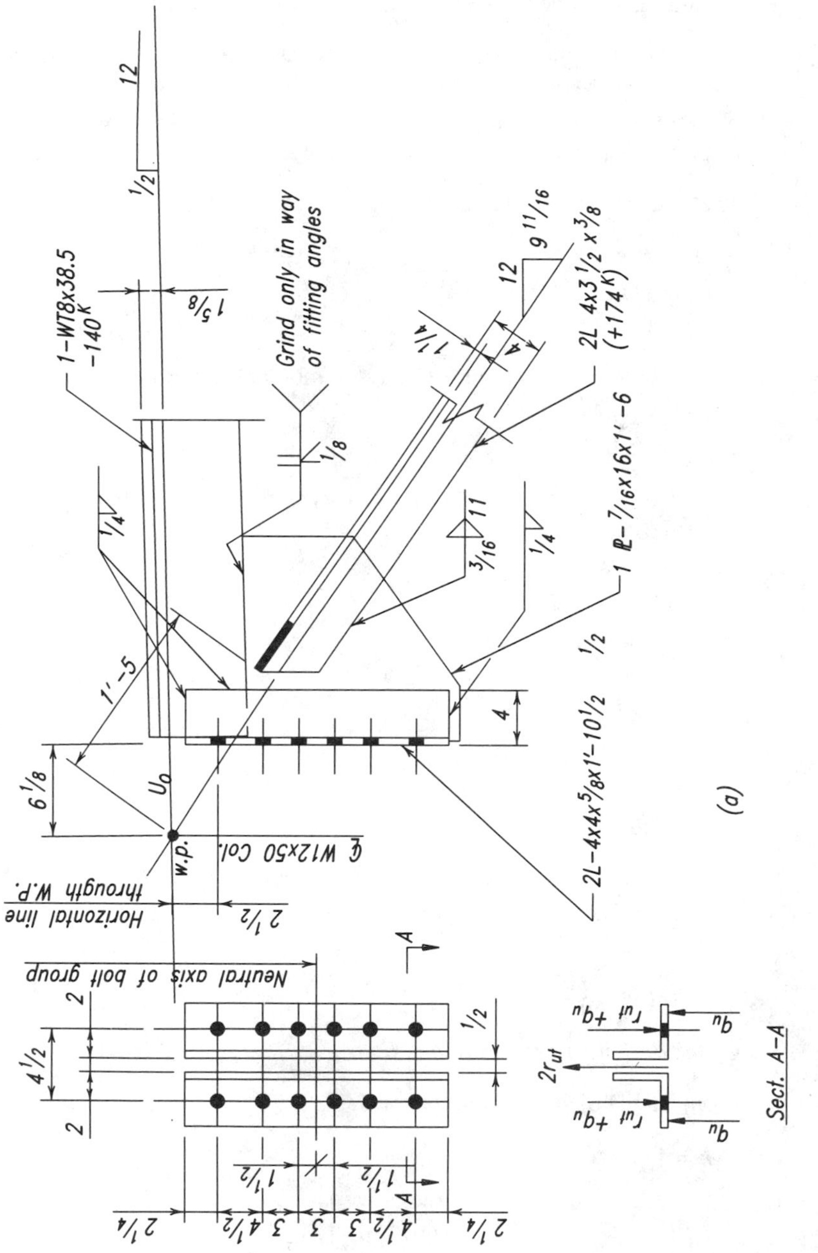

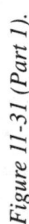

Figure 11-31 (Part 1).

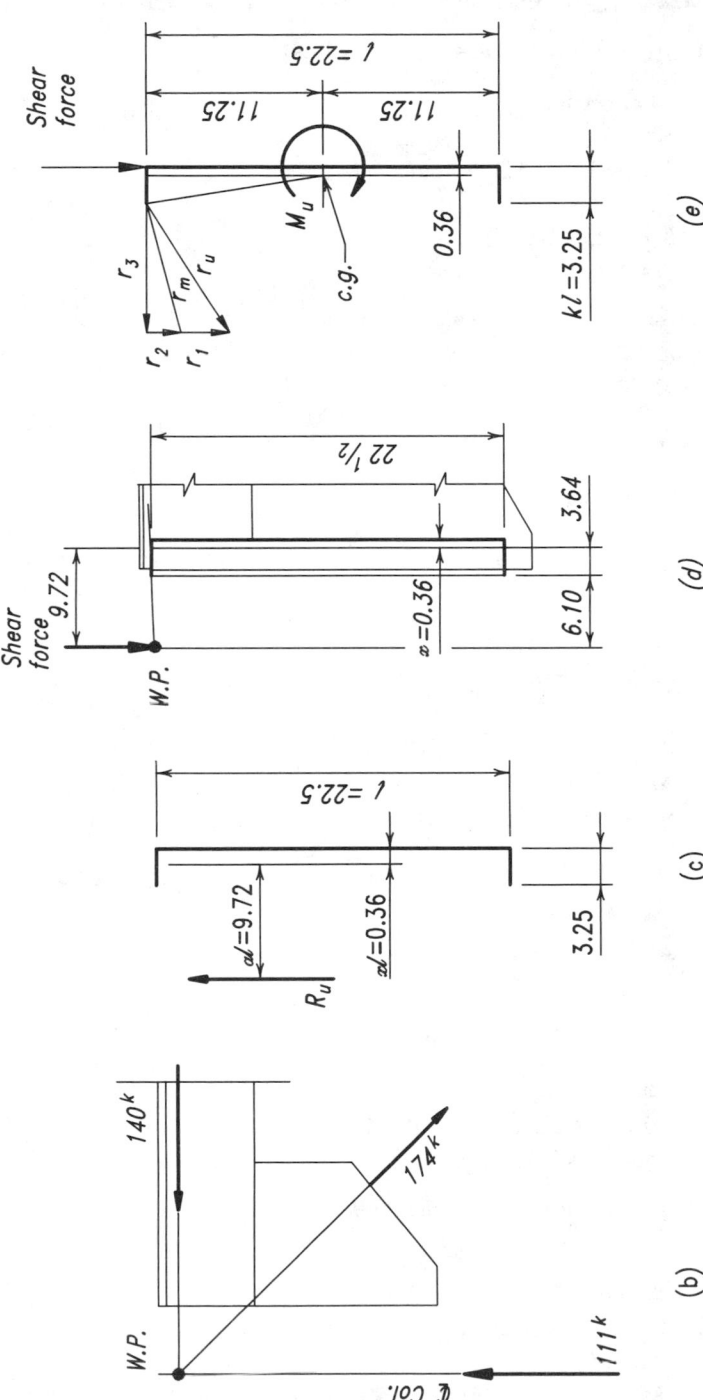

Figure 11-31 (Part 2).

Design angles (note angle thickness will be chosen to preclude prying action, i.e., $q_u = 0$, since bolt group has already been designed)

With $r_{ut} = 14.7$ kips and $p = 4\frac{1}{2}$-in., try 2L4×4×⅝

$\quad b = 2$ in. $- $ ⅝-in.
$\quad\quad = 1.38$ in.
$\quad a = 4$ in. $- 2$ in. $\leq 1.25b$ (for calculation purposes)
$\quad\quad = 2$ in.
$\quad 1.25b = 1.25(1.38$ in.$)$
$\quad\quad\quad = 1.73$ in.

Thus, use $a = 1.73$ in. for calculation purposes.

$$b' = b - \frac{d}{2}$$

$$= 1.38 \text{ in.} - \frac{⅞\text{-in.}}{2}$$

$$= 0.943 \text{ in.}$$

$$t_{req} = \sqrt{\frac{4.44 r_{ut} \, b'}{pF_y}}$$

$$= \sqrt{\frac{4.44 \, (14.7 \text{ kips})(0.943 \text{ in.})}{(4\frac{1}{2}\text{-in.})(36 \text{ ksi})}}$$

$$= 0.616 \text{ in.} < 0.625 \text{ in.} \quad \textbf{o.k.}$$

(Note: Alternatively, a lesser required angle thickness may be determined by designing the connection for pryng action, i.e., $q_u > 0$.

Check shear yielding of the angles.

$\quad \phi R_n = \phi(0.6F_y) A_g$
$\quad\quad = 0.9(0.6 \times 36 \text{ ksi})(2 \times 22.5 \text{ in.} \times ⅝\text{-in.})$
$\quad\quad = 547 \text{ kips} > 111 \text{ kips} \quad \textbf{o.k.}$

Check shear rupture of the angles.

$\quad \phi R_n = \phi(0.6F_u) A_n$
$\quad\quad = 0.75(0.6 \times 58 \text{ ksi})[2 \times (22.5 \text{ in.} - 6 \times 1 \text{ in.}) \times ⅝\text{-in.}]$
$\quad\quad = 538 \text{ kips} > 111 \text{ kips} \quad \textbf{o.k.}$

Check block shear rupture of the angles.

$\quad A_{gv} = 2 \times (22\frac{1}{2}\text{-in.} - 2\frac{1}{4}\text{-in.})(⅝\text{-in.})$
$\quad\quad = 25.3 \text{ in.}^2$
$\quad A_{gt} = 2 \times (2 \text{ in.} \times ⅝\text{-in.})$
$\quad\quad = 2.50 \text{ in.}^2$
$\quad A_{nv} = 25.3 \text{ in.}^2 - 2 \times [5.5(1 \text{ in.})(⅝\text{-in.})]$
$\quad\quad = 18.4 \text{ in.}^2$
$\quad A_{nt} = 2.5 \text{ in.}^2 - 2 \times [0.5(1 \text{ in.})(⅝\text{-in.})]$

$$= 1.88 \text{ in.}^2$$

Since $0.6F_u A_{nv} > F_u A_{nt}$,

$$\phi R_n = \phi[0.6F_u A_{nv} + F_y A_{gt}]$$
$$= 0.75[0.6(58 \text{ ksi})(18.4 \text{ in.}^2) + (36 \text{ ksi})(2.50 \text{ in.}^2)]$$
$$= 548 \text{ kips} > 111 \text{ kips} \quad \textbf{o.k.}$$

Use 2L4×4×⅝

Design angle-to-gusset connection

From LRFD Specification Table J2.4, the minimum weld size is ¼-in.

From Table 8-42 with $\theta = 0°$

$$kl = 3.25 \text{ in.}$$
$$l = 22.5 \text{ in.}$$
$$k = \frac{3.25 \text{ in.}}{22.5 \text{ in.}}$$
$$= 0.144$$

by interpolation, $x = 0.017$ and

$$al + xl = 10.1 \text{ in.}$$
$$a = \frac{10.1 \text{ in.} - 0.017 \, (22.5 \text{ in.})}{22.5 \text{ in.}}$$
$$= 0.432$$

By interpolation, $C = 1.33$ and

$$D_{req} = \frac{R_u}{2 \times CC_1 l}$$
$$= \frac{111 \text{ kips}}{2 \times (1.33)(1.0)(22.5 \text{ in.})}$$
$$= 1.86 \rightarrow 2 \text{ sixteenths}$$

Use ¼-in. fillet welds.

Solution B: *Design gusset*

The gusset thickness must match that of the tee stem; approximately ⁷⁄₁₆-in.

Check tension yielding of the gusset on the Whitmore section.

$$L_w = 4 \text{ in.} + 2 \times (11 \text{ in.} \times \tan 30°)$$
$$= 16.7 \text{ in.}$$
$$\phi R_n = \phi F_y A_g$$
$$= 0.9(36 \text{ ksi})(16.7 \text{ in.} \times ⁷⁄₁₆\text{-in.})$$
$$= 237 \text{ kips} > 174 \text{ kips} \quad \textbf{o.k.}$$

Check block shear rupture of the gusset.

From LRFD Specification Section J4.3,

$$0.6F_u A_{nv} = 0.6(58 \text{ ksi})(2 \times 11 \text{ in.} \times \tfrac{7}{16}\text{-in.})$$
$$= 335 \text{ kips}$$
$$F_u A_{nt} = (58 \text{ ksi})(4 \text{ in.} \times \tfrac{7}{16}\text{-in.})$$
$$= 102 \text{ kips}$$

Since $0.6F_u A_{nv} > F_u A_{nt}$,

$$\phi R_n = \phi[0.6F_u A_{nv} + F_y A_{gt}]$$
$$= 0.75[335 \text{ kips} + (36 \text{ ksi})(4 \text{ in.} \times \tfrac{7}{16}\text{-in.})]$$
$$= 299 \text{ kips} > 174 \text{ kips} \quad \textbf{o.k.}$$

The gusset width must be such that the groove weld connecting it to the stem of the tee can transfer the 140 kip force between the gusset and the top chord (note that the slight slope of the top chord has been ignored). The required length is

$$L_{req} = \frac{R_u}{\phi\,(0.6F_u)\,t}$$

$$= \frac{140 \text{ kips}}{0.75\,(0.6 \times 58 \text{ ksi})\,(\tfrac{7}{16}\text{-in.})}$$

$$= 12.3 \text{ in.}$$

Use $L = 16$ in. to allow for weld runout and offset between the gusset and tee stem at the end of the chord.

The gusset length depends upon the connection angles. From a scaled layout, the gusset must extend 1′-6 below the tee stem.

Use PL$\tfrac{7}{16}$-in.×16 in.×1′-6. Note that fabricators may prefer to use ½-in. plate from stock instead of ordering $\tfrac{7}{16}$-in. plate. Were this the case, the weld joining the angles to the tee stem could be increased slightly to accommodate the resulting small gap.

Design weld connecting diagonal to gusset

From LRFD Specification Table J2.4, the minimum weld size is $\tfrac{3}{16}$-in.

Try $\tfrac{3}{16}$-in. fillet weld.

$$L_{req} = \frac{R_u}{2 \times D \times 1.392}$$

$$= \frac{174 \text{ kips}}{2\,(3 \text{ sixteenths})(1.392)}$$

$$= 20.8 \text{ in.}$$

Use 11 in. at the heel and 11 in. at the toe.

Check gusset thickness

For two $\tfrac{3}{16}$-in. fillet welds

$$t_{min} = \frac{5.16D}{F_y}$$

$$= \frac{5.16(3 \text{ sixteenths})}{36 \, \text{ksi}}$$

$$= 0.430 \text{ in.} < \frac{7}{16}\text{-in.} \quad \textbf{o.k.}$$

Truss Chord Splices

Truss chord splices are expensive to fabricate and should be avoided whenever possible. In general, chord splices in ordinary building trusses are confined to cases where: (1) the finished truss is too large to be shipped in one piece; (2) the truss chord exceeds the available material length; (3) the reduction in member size of the chord justifies the added cost of a splice; or, (4) a sharp change in direction occurs in working line of the chord and bending does not provide a satisfactory alternative.

REFERENCES

American Institute of Steel Construction, Inc., 1984, *Engineering for Steel Construction*, pp. 7.55–7.62, AISC, Chicago, IL.

American Institute of Steel Construction, Inc., 1989, *Manual of Steel Construction—Allowable Stress Design and Plastic Design*, AISC, Chicago, IL.

American Institute of Steel Construction, Inc., 1992, *Manual of Steel Construction, Volume II—Connections*, ASD 9th Ed./LRFD 1st Ed., AISC, Chicago, IL.

Astaneh, A., 1985, "Procedure for Design Analysis of Hanger-Type Connections," *Engineering Journal*, Vol. 22, No. 2, (2nd Qtr.), pp. 63–66, AISC, Chicago, IL.

Bjorhovde, R. and S. K. Chakrabarti, 1985, "Tests of Full-Size Gusset Plate Connections," *Journal of Structural Engineering*, Vol. 111, No. 3, (March), pp. 667–684, ASCE, New York, NY.

DeWolf, J. T. and D. T. Ricker, 1990, *Column Base Plates*, AISC, Chicago, IL.

Fling, R. S., 1970, "Design of Steel Bearing Plates," *Engineering Journal*, Vol. 7, No. 2, (April), pp. 37–39, AISC, Chicago, IL.

Gross, J. L. and G. Cheok, 1988, *Experimental Study of Gusseted Connections for Laterally Braced Steel Buildings*, National Institute of Standards and Technology Report NISTIR 88-3849, NIST, Gaithersburg, MD.

Gross, J. L., 1990, "Experimental Study of Gusseted Connections," *Engineering Journal*, Vol. 27, No. 3, (3rd Qtr.), pp. 89–97, AISC, Chicago, IL.

Ishler, M., 1992, "Seismic Design Practice for Eccentrically Braced Frames," *Steel TIPS*, Structural Steel Education Council, Moraga, CA.

Kulak, G. L., J. W. Fisher, and J. H. A. Struik, 1987, *Guide to Design Criteria for Bolted and Riveted Joints*, 2nd Edition, pp. 274–286, John Wiley & Sons, New York, NY.

Lindsay, S. D. and A. V. Goverdahn, 1989, "Eccentrically Braced Frames: Suggested Design Procedures for Wind and Low Seismic Forces," *National Steel Construction Conference Proceedings*, pp. 17.1–17.25, AISC, Chicago, IL.

Murray, T. M., 1983, "Design of Lightly Loaded Column Base Plates," *Engineering Journal*, Vol. 20, No. 4, (4th Qtr.), pp. 143–152, AISC, Chicago, IL.

Popov, E. P., M.D. Englehardt, and J. M. Ricles, 1989, "Eccentrically Braced Frames: U.S. Practice," *Engineering Journal*, Vol. 26, No. 2, (2nd Qtr.), pp. 66–80, AISC, Chicago, IL.

Popov, E. P. and R. M. Stephen, 1977, "Capacity of Columns with Splice Imperfections," *Engineering Journal*, Vol. 14, No. 1, (1st Qtr.), pp. 16–23, AISC Chicago, IL.

Richard, R. M., 1986, "Analysis of Large Bracing Connection Designs for Heavy Construction," *National Steel Construction Conference Proceedings*, pp. 31.1–31.24, AISC, Chicago, IL.

Ricker, D. T., 1989, "Some Practical Aspects of Column Base Selection," *Engineering Journal*, Vol. 26, No. 3, (3rd Qtr.), AISC, Chicago, IL.

Ricker, D. T., 1992, "Value Engineering and Steel Economy," *Modern Steel Construction*, Vol. 32, No. 2, (February), AISC, Chicago, IL.

Thornton, W. A., 1985, "Prying Action—A General Treatment," *Engineering Journal*, Vol. 22, No. 2, (2nd Qtr.), pp. 67–75, AISC, Chicago, IL.

Thornton, W. A., 1990a, "Design of Small Base Plates for Wide-Flange Columns," *Engineering Journal*, Vol. 27, No. 3, (3rd Qtr.), pp. 108–110, AISC, Chicago, IL.

Thornton, W. A., 1990b, "Design of Small Base Plates for Wide-Flange Columns—A Concatenation of Methods," *Engineering Journal*, Vol. 27, No. 4, (4th Qtr.), pp. 173–174, AISC, Chicago, IL.

Thornton, W. A., 1991, "On the Analysis and Design of Bracing Connections," *National Steel Construction Conference Proceedings*, pp. 26.1–26.33, AISC, Chicago, IL.

Thornton, W. A., 1992, "Designing for Cost Efficient Fabrication and Construction," *Constructional Steel Design—An International Guide*, (Chapter 7), pp. 845–854, Elsevier, London, UK.

PART 12

OTHER CONNECTIONS AND TOPICS

OTHER

OVERVIEW

Part 12 contains general information, design considerations, and examples for the following detailed list of topics. It is based on the provisions of the 1993 LRFD Specification. Supplementary information may also be found in the Commentary on the LRFD Specification.

BRACKET PLATES

A bracket plate, illustrated in Figure 12-1, acts as a cantilevered beam. The design strengths of the bolts and/or welds and connected elements must be determined in accordance with the LRFD Specification; the applicable limit states are discussed in Part 8. Additionally, the design must consider flexural yielding and rupture at the section of maximum moment as well as local buckling and flexural yielding on the free edge of the bracket.

In lieu of a more detailed analysis, Salmon and Johnson (1990) recommend that the design strength of a triangular-shaped bracket plate in flexural yielding on the free edge be determined as:

$$\phi P_n = 0.85 F_y zbt$$

where

$$z = 1.39 - 2.2\left(\frac{b}{a}\right) + 1.27\left(\frac{b}{a}\right)^2 - 0.25\left(\frac{b}{a}\right)^3$$

b = width of bracket plate as shown in Figure 12-1, in.
a = depth of bracket plate as shown in Figure 12-1, in.
t = thickness of bracket plate, in.

For flexural yielding, the design strength of the bracket plate is ϕM_n where $\phi = 0.90$, and

$$M_n = F_y S_x$$

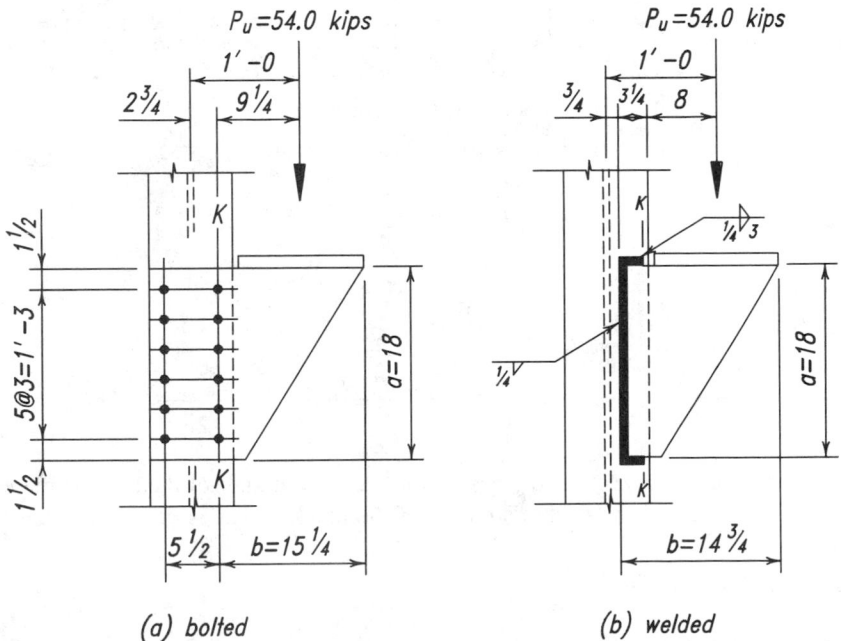

Fig. 12-1. Bracket plate.

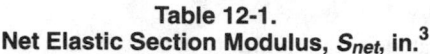

Table 12-1.
Net Elastic Section Modulus, S_{net}, in.³

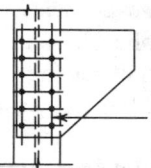

Net elastic section modulus taken along this line

# Bolts in One Vertical Row, n	Bracket Plate Depth, in.	Nominal Bolt Diameter, in.														
		¾					⅞					1				
		Bracket Plate Thickness, in.														
		¼	⅜	½	⅝	¾	⅜	½	⅝	¾	⅞	½	⅝	¾	⅞	1
2	6	1.17	1.76	2.34	2.93	3.52	1.69	2.25	2.81	3.38	3.94	2.16	2.70	3.23	3.77	4.31
3	9	2.50	3.75	5.00	6.25	7.50	3.56	4.75	5.94	7.13	8.31	4.50	5.63	6.75	7.88	9.00
4	12	4.36	6.54	8.72	10.9	13.1	6.19	8.25	10.3	12.4	14.4	7.78	9.7	11.7	13.6	15.6
5	15	6.75	10.1	13.5	16.9	20.3	9.6	12.8	15.9	19.1	22.3	12.0	15.0	18.0	21.0	24.0
6	18	9.67	14.5	19.3	24.2	29.0	13.7	18.3	22.8	27.4	31.9	17.2	21.4	25.7	30.0	34.3
7	21	13.1	19.7	26.3	32.8	39.4	18.6	24.8	30.9	37.1	43.3	23.3	29.1	34.9	40.7	46.5
8	24	17.1	25.7	34.2	42.8	51.3	24.2	32.3	40.3	48.4	56.4	30.3	37.9	45.4	53.0	60.6
9	27	21.6	32.4	43.3	54.1	64.9	30.6	40.8	50.9	61.1	71.3	38.3	47.8	57.4	66.9	76.5
10	30	26.7	40.0	53.3	66.7	80.0	37.7	50.3	62.8	75.4	87.9	47.2	58.9	70.7	82.5	94.3
12	36	38.4	57.5	76.7	95.9	115	54.2	72.3	90.3	108	126	67.8	84.7	102	119	136
14	42	52.2	78.3	104	130	157	73.7	98.3	123	147	172	92.2	115	138	161	184
16	48	68.1	102	136	170	204	96.2	128	160	192	224	120	150	180	210	241
18	54	86.2	129	172	215	259	122	162	203	243	284	152	190	228	266	304
20	60	106	160	213	266	319	150	200	250	300	350	188	235	282	329	376
22	66	129	193	257	322	386	182	242	303	363	424	227	284	341	398	454
24	72	153	230	306	383	459	216	288	360	432	504	270	338	405	473	541
26	78	180	270	359	449	539	254	338	423	507	592	317	396	476	555	634
28	84	208	313	417	521	625	294	392	490	588	686	368	460	552	644	736
30	90	239	359	478	598	718	338	450	563	675	788	422	528	633	739	844
32	96	272	408	544	680	816	384	512	640	768	896	480	600	720	840	961
34	102	307	461	614	768	922	434	578	723	867	1010	542	678	813	949	1080
36	108	344	517	689	861	1030	486	648	810	972	1130	608	760	912	1060	1220

Notes
Diameter of holes assumed ⅛-in. larger than nominal bolt diameter.
Bolts spaced 3 in. vertically with 1½-in. edge distance at top and bottom.
Interpolate for intermediate plate thicknesses. Otherwise, use general equation

$$S_{net} = \frac{t}{6}\left[d^2 - \frac{s^2 n\,(n^2 - 1)\,(d_b + 0.125\ \text{in.})}{d} \right]$$

where

t = bracket plate thickness, in.
d = bracket plate depth, in.
s = bolt spacing, in.
n = number of bolts in one vertical row
d_b = nominal bolt diameter, in.

In the above equation, S_x is the gross elastic section modulus of the bracket plate.

For flexural rupture, the design strength of the bracket plate is ϕM_n where $\phi = 0.75$, and

$$M_n = F_u S_{net}$$

In the above equation, S_{net} is the net elastic section modulus of the bracket plate. Table 12-1 gives values of the net elastic section modulus S_{net} for various hole diameters and numbers of fasteners spaced three inches on center, the usual spacing for these connections.

Furthermore, local buckling will not occur (Salmon and Johnson, 1990) if, for $0.5 < b/a < 1.0$,

$$\frac{b}{t} \leq \frac{250}{\sqrt{F_y}}$$

nor if, for $1.0 \leq b/a \leq 2.0$,

$$\frac{b}{t} \leq \frac{250}{\sqrt{F_y}}\left(\frac{b}{a}\right)$$

Two assumptions must be satisfied if this simplified approach is to be used. First, the centroid of the applied load must be approximately $0.6b$ from the line of support (line K in Figure 12-1a). Second, the lateral movement of the outstanding portion of the bracket plate must be prevented.

Example 12-1

Given: Refer to Figure 12-1a. Design a bracket plate to support the factored force indicated. Use $\frac{7}{8}$-in. diameter A325-N bolts. For the plate, assume $F_y = 36$ ksi and $F_u = 58$ ksi.

Solution: *Select bolt group.*

For $\frac{7}{8}$-in. diameter A325-N bolts in single shear, $\phi r_n = 21.6$ kips (Table 8-11). Thus,

$$\begin{aligned}C_{\min} &= \frac{R_u}{\phi r_n} \\ &= \frac{54.0 \text{ kips}}{21.6 \text{ kips}} \\ &= 2.50\end{aligned}$$

From Table 8-20 with $\theta = 0°$, a $5\frac{1}{2}$-in. gage with $s = 3$ in., $e_x = 12$ in., and $n = 6$ provides

$C = 4.53 > 2.50$ **o.k.**

Try PL $\frac{3}{8}$-in.×18 in.

Check bolt bearing.

With $l_v = 1\frac{1}{2}$-in. and $s = 3$ in., $\phi r_n = 91.4$ kips/bolt (Table 8-13)

Since this is greater than the single-shear strength of one bolt, bolt bearing is not critical.

Check flexure in the bracket plate.

On line K, the required strength M_u is

$$M_u = P_u e_b$$
$$= 54.0 \text{ kips } (12 \text{ in.} - 2\frac{3}{4}\text{-in.})$$
$$= 500 \text{ in.-kips}$$

For flexural yielding on line K,

$$\phi M_n = \phi F_y S_x$$
$$= 0.9 \ (36 \text{ ksi}) \frac{(\frac{3}{8}\text{-in.}) \ (18 \text{ in.})^2}{6}$$
$$= 656 \text{ in.-kips} > 500 \text{ in.-kips} \quad \textbf{o.k.}$$

For flexural rupture on line K, with $S_{net} = 13.7$ in.^{3} from Table 12-1,

$$\phi M_n = \phi F_u S_{net}$$
$$= 0.75(58 \text{ ksi})(13.7 \text{ in.}^3)$$
$$= 596 \text{ in.-kips} > 500 \text{ in.-kips} \quad \textbf{o.k.}$$

For flexural yielding on the free edge of the triangular plate,

$$z = 1.39 - 2.2 \left(\frac{b}{a}\right) + 1.27 \left(\frac{b}{a}\right)^2 - 0.25 \left(\frac{b}{a}\right)^3$$
$$= 1.39 - 2.2 \left(\frac{15\frac{1}{4}\text{-in.}}{18 \text{ in.}}\right) + 1.27 \left(\frac{15\frac{1}{4}\text{-in.}}{18 \text{ in.}}\right)^2 - 0.25 \left(\frac{15\frac{1}{4}\text{-in.}}{18 \text{ in.}}\right)^3$$
$$= 0.286$$

$$\phi P_n = 0.85 F_y z b t$$
$$= 0.85(36 \text{ ksi})(0.286)(15\frac{1}{4}\text{-in.})(\frac{3}{8}\text{-in.})$$
$$= 50.0 \text{ kips} < 54.0 \text{ kips} \quad \textbf{n.g.}$$

Try PL $\frac{1}{2}$-in.×18 in.

$$\phi P_n = 0.85(36 \text{ ksi})(0.286)(15\frac{1}{4}\text{-in})(\frac{1}{2}\text{-in.})$$
$$= 66.7 \text{ kips} > 54.0 \text{ kips} \quad \textbf{o.k.}$$

Check local buckling of the bracket plate.

$$\frac{b}{a} = \frac{15\frac{1}{4}\text{-in.}}{18 \text{ in.}} = 0.847$$

Since $0.5 \le \dfrac{b}{a} < 1.0$

$$t_{min} = b \left(\frac{\sqrt{F_y}}{250}\right)$$
$$= 15\frac{1}{4}\text{-in.} \left(\frac{\sqrt{36 \text{ ksi}}}{250}\right)$$
$$= 0.366 \text{ in.} < \frac{1}{2}\text{-in.} \quad \textbf{o.k.}$$

Check shear yielding of the bracket plate.

$$\phi R_n = \phi(0.6 F_y) A_g$$
$$= 0.9(0.6 \times 36 \text{ ksi})(18 \text{ in.} \times \frac{1}{2}\text{-in.})$$
$$= 175 \text{ kips} > 54.0 \text{ kips} \quad \textbf{o.k.}$$

Check shear rupture of the bracket plate.

$$\phi R_n = \phi(0.6F_u)A_n$$
$$= 0.75(0.6 \times 58 \text{ ksi})[18 - (6 \times 1 \text{ in.})](\tfrac{1}{2}\text{-in.})$$
$$= 157 \text{ kips} > 54.0 \text{ kips} \quad \textbf{o.k.}$$

Check block shear rupture of the bracket plate (shear plane on line K, tension plane across bottom two bolts).

$$A_{gv} = (18 \text{ in.} - 1\tfrac{1}{2}\text{-in.})(\tfrac{1}{2}\text{-in.})$$
$$= 8.25 \text{ in.}^2$$
$$A_{nv} = 8.25 \text{ in.}^2 - 5.5(1 \text{ in.})(\tfrac{1}{2}\text{-in.})$$
$$= 5.50 \text{ in.}^2$$
$$A_{gt} = (5\tfrac{1}{2}\text{-in.} + 1\tfrac{1}{2}\text{-in.})(\tfrac{1}{2}\text{-in.})$$
$$= 3.50 \text{ in.}^2$$
$$A_{nt} = 3.50 \text{ in.}^2 - 1.5(1 \text{ in.})(\tfrac{1}{2}\text{-in.})$$
$$= 2.75 \text{ in.}^2$$

Since $0.6F_u A_{nv} > F_u A_{nt}$,

$$\phi R_n = \phi[0.6F_u A_{nv} + F_y A_{gt}]$$
$$= 0.75[0.6(58 \text{ ksi})(5.50 \text{ in.}^2) + (36 \text{ ksi})(3.50 \text{ in.}^2)]$$
$$= 238 \text{ kips} > 54.0 \text{ kips} \quad \textbf{o.k.}$$

Example 12-2

Given:

Refer to Figure 12-1b. Design a bracket plate to support the factored force indicated. Use 70 ksi electrodes. For the plate, assume $F_y = 36$ ksi and $F_u = 58$ ksi.

Solution:

Select weld group assuming PL ½-in.×18 in.

Try "C"-shaped weld with $kl = 3$ in. and $l = 18$ in.

Interpolating from Table 8-42 with $\theta = 0°$,

$x = 0.0221$

and

$al + xl = 11\tfrac{1}{4}\text{-in.}$

$a(18 \text{ in.}) + (0.0221 \times 18 \text{ in.}) = 11\tfrac{1}{4}\text{-in.}$

$a = 0.603$

Interpolating from Table 8-42 with $\theta = 0°$, $k = 0.167$, and $a = 0.647$,

$C = 1.09$

Thus, the weld size required for strength is

$$D_{req} = \frac{P_u}{CC_1 l}$$

$$= \frac{54.0 \text{ kips}}{(1.09)\,(1.0)\,(18 \text{ in.})}$$

$$= 2.75 \rightarrow 3 \text{ sixteenths}$$

Minimum weld size from LRFD Specification Table J2.4 is $\frac{3}{16}$-in. Therefore, use $\frac{3}{16}$-in. fillet weld.

Check flexure on the bracket plate.

Conservatively taking the moment in the plate equal to the moment on the weld group,

$$M_u = P_u(al)$$
$$= 54.0 \text{ kips } (10.85 \text{ in.})$$
$$= 586 \text{ in.-kips}$$

For flexural yielding of the plate,

$$\phi M_n = \phi F_y S_x$$
$$= 0.9\,(36 \text{ ksi})\,\frac{(\frac{1}{2}\text{-in.})\,(18 \text{ in.})^2}{6}$$
$$= 875 \text{ in.-kips} > 586 \text{ in.-kips} \quad \textbf{o.k.}$$

For yielding on the free edge of the triangular plate,

$$z = 1.39 - 2.2\left(\frac{b}{a}\right) + 1.27\left(\frac{b}{a}\right)^2 - 0.25\left(\frac{b}{a}\right)^3$$

$$= 1.39 - 2.2\left(\frac{10.85 \text{ in.}}{18 \text{ in.}}\right) + 1.27\left(\frac{10.85 \text{ in.}}{18 \text{ in.}}\right)^2 - 0.25\left(\frac{10.85 \text{ in.}}{18 \text{ in.}}\right)^3$$

$$= 0.381$$

$$\phi P_n = 0.85 F_y zbt$$
$$= 0.85\,(36 \text{ ksi})\,(0.381)\,(10.85 \text{ in.})\,(\frac{1}{2}\text{-in.})$$
$$= 63.2 \text{ kips} > 54.0 \text{ kips} \quad \textbf{o.k.}$$

Check local buckling of the bracket plate.

$$\frac{b}{a} = \frac{10.85 \text{ in.}}{18 \text{ in.}} = 0.603$$

Since $0.5 \le \dfrac{b}{a} < 1.0$

$$t_{min} = b\left(\frac{\sqrt{F_y}}{250}\right)$$

$$= 10.85 \text{ in.}\left(\frac{\sqrt{36 \text{ ksi}}}{250}\right)$$

$$= 0.260 \text{ in.} < \frac{1}{2}\text{-in.} \quad \textbf{o.k.}$$

Check shear yielding of the bracket plate.

$$\phi R_n = \phi(0.6 F_y)A_g$$

$$= 0.9 \ (0.6 \times 36 \text{ ksi}) \ (18 \text{ in.} \times \tfrac{1}{2} \text{-in.})$$
$$= 175 \text{ kips} > 54.0 \text{ kips} \quad \textbf{o.k.}$$

BEAM-WEB PENETRATIONS

Beam-web penetrations, illustrated in Figure 12-2, may be used to accommodate the passage of ductwork and/or other utilities. This integration of structural and other building systems minimizes story height, reducing cost and maximizing the number of stories that can be built when height limitations exist. Beam-web penetrations are usually rectangular, although circular openings are sometimes used; the latter are analyzed as an equivalent rectangular opening.

Depending upon the size and location of the beam-web penetration, stiffeners may or may not required based on an ultimate strength evaluation of flexure, shear, and the interaction between them on the reduced section. High local stress concentrations at the corners of beam-web penetrations preclude the usefulness of elastic analysis.

Stiffening requirements can be minimized by selecting a favorable location for the beam-web penetrations. In general, the most significant effect of an opening is a reduction in shear strength. A beam-web penetration, then, should be located in a region of low shear when possible; avoid locations near beam-support reactions where shear is high. Since an opening also reduces the flexural strength, regions of high moment should also be avoided.

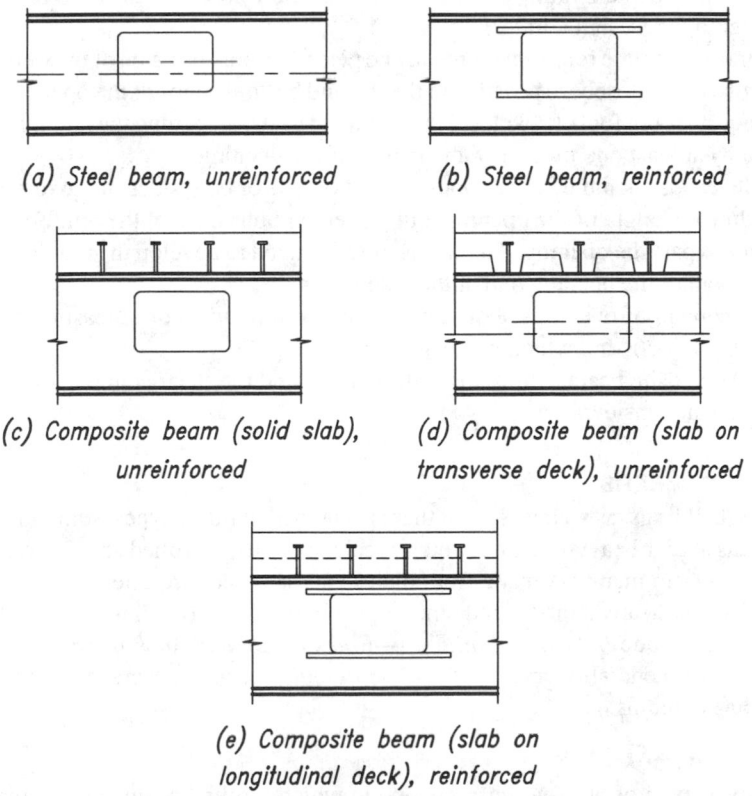

(a) Steel beam, unreinforced   (b) Steel beam, reinforced

(c) Composite beam (solid slab),   (d) Composite beam (slab on
       unreinforced   transverse deck), unreinforced

(e) Composite beam (slab on
longitudinal deck), reinforced

Fig. 12-2. Beam-web penetrations.

When stiffening is required, various reinforcing schemes have been used, including horizontal stiffeners, vertical stiffeners, and stiffeners around the entire periphery of the opening. However, horizontal stiffeners above and below the opening, as illustrated in Figures 12-2b and 12-2e, can effectively and economically provide the needed reinforcement. More elaborate schemes are unnecessary.

A summary of past research in beam-web penetrations in both steel and composite beams and the resulting design procedures is available in the AISC Design Guide *Steel and Composite Beams with Web Openings* (Darwin, 1990). These procedures are also available electronically as WEBOPEN, AISC's computer program. Although the complete design of a beam-web penetration is beyond the scope of this Manual, some general guidelines for proportioning and detailing taken from the aforementioned Design Guide are presented below. Refer to the Design Guide for more specific information.

1. Steel yield strength is limited to 65 ksi (specified minimum value), and sections must meet the compact-section requirements of LRFD Specification Section B5.1.
2. Opening depth cannot exceed 70 percent of the member depth.
3. Multiple openings in the same member should be spaced far enough apart so that they will not interact; otherwise the beam must be treated as a castellated beam.
4. The edge of an opening should be no closer to a support than the member depth d.
5. Concentrated loads should not be placed above beam-web penetrations; the load may be placed a distance d or $d/2$ from the edge of an opening, depending on the depth-to-thickness ratio of the web and the width-to-thickness ratio of the flange. Refer to the Design Guide.
6. In most cases, the reinforcement may be placed on only one side of the web. It should be placed as closely as possible to the top and bottom edges of the opening, but with adequate room for fillet welds. It must extend past the opening the required distance, and by at least one-quarter of the length of the opening.
7. Fillet welds should be continuous, placed on one or both sides of the reinforcement within the length of the opening, but placed on both sides of the reinforcement that extends past the opening. The welds must be sized to develop the required strength, both within the opening and at the extensions.
8. The corners of openings should have a minimum radius of at least twice the web thickness, or ⅝-in., whichever is greater.
9. For composite beams, the slab reinforcement and the shear connector locations are important considerations.

BUILT-UP MEMBERS

Industrial buildings, as well as some other specialized building types, sometimes require clear spans and/or heavy loadings which preclude the use of rolled shapes. When this is the case, built-up members made from plates and/or shapes are often used. A complete reference on built-up members and other topics in industrial buildings is available in the AISC Design Guide *Industrial Buildings—Roofs to Column Anchorage* (Fisher, 1993). Following is a general overview of built-up members: girders, crane-runway girders, trusses, and columns.

Built-Up Girders

The simplest type of built-up girder is one in which a rolled beam is reinforced by the addition of coverplates to its flanges, as shown in Figure 12-3a. Deeper girders, built up entirely from plates, are shown in Figures 12-3b and 12-3c. The girder of Figure 12-3b is comprised of a web plate and two flange plates. Each flange usually is made of a single thickness of plate, but plates of varying thickness may be spliced end-to-end with groove welds to provide greater strength in areas of high moment. The box girder in Figure 12-3c is comprised of two web plates and two flange plates. Box girders are particularly useful where lateral stability and torsional resistance are required.

Crane-Runway Girders

In addition to vertical wheel loads, overhead cranes in buildings impose substantial lateral and longitudinal forces on their supports. To provide for the necessary strength, crane-runway girders, as illustrated Figure 12-4a, are typically built-up using a rolled beam with a channel attached horizontally to its top flange; the channel provides lateral bending strength. Bolts or welds connecting the channel to the beam must be of sufficient strength to ensure that these two components act together in resisting both the vertical and horizontal forces. The use of intermittent fillet welds could be investigated as an alternative, but some codes require a continuous weld.

When lateral crane loads exceed the strength of a channel, the top flange of the girder may be connected to a separate longitudinal member which functions as a horizontal girder, as illustrated in Figure 12-4b. The web of this girder may be solid or composed of lacing bars or angles to form a lattice girder; a solid web will also serve as an inspection or access walkway. Note that the horizontal plate is interrupted at the column center and there is no direct connection between the tops of the abutting crane girders, avoiding continuity.

Heavy crane loads sometimes require built-up girders (illustrated in Figure 12-4b) for greater strength. In built-up crane-runway girders, when the web plate and flange plate are in tight contact before welding, wheel loads will be transferred through bearing directly into the web. This may be accomplished by edge planing the web or by trimming the web with a mechanically guided torch; either process will provide the smooth straight edge necessary for continuous tight contact. If tight contact does not exist, the flange-to-web welds must be designed to transfer concentrated loads from the crane wheels. Additionally, transverse bending of the top flange under heavy crane loads has led to fatigue failures in fillet-welded joints. Consequently, many designers require a complete-joint-penetration groove weld for the joint of top plate-to-web for an active crane, particularly for heavy cranes.

Crane-runway girders supporting heavy loads should be designed as simple spans; continuous construction is not desirable. Longitudinal forces usually are distributed

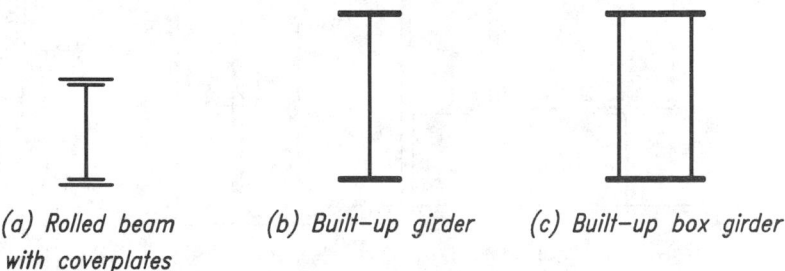

(a) Rolled beam   (b) Built-up girder   (c) Built-up box girder
with coverplates

Fig. 12-3. Built-up girders.

through lateral bracing located in the plane of the crane columns. Note that knee braces under crane-runway girders are usually avoided unless the resulting intermediate support (and potential for resulting continuity) are considered in the design.

While the design of crane-runway girders is beyond the scope of this volume, following are some general recommendations and sources of further information. The design of crane-runway girders must be in accordance with the LRFD Specification, but may additionally be controlled by the *Guide for the Design and Construction of Mill Buildings—Technical Report No. 13* (AISE, 1969), AWS (1991), or the *Standard Specifications*

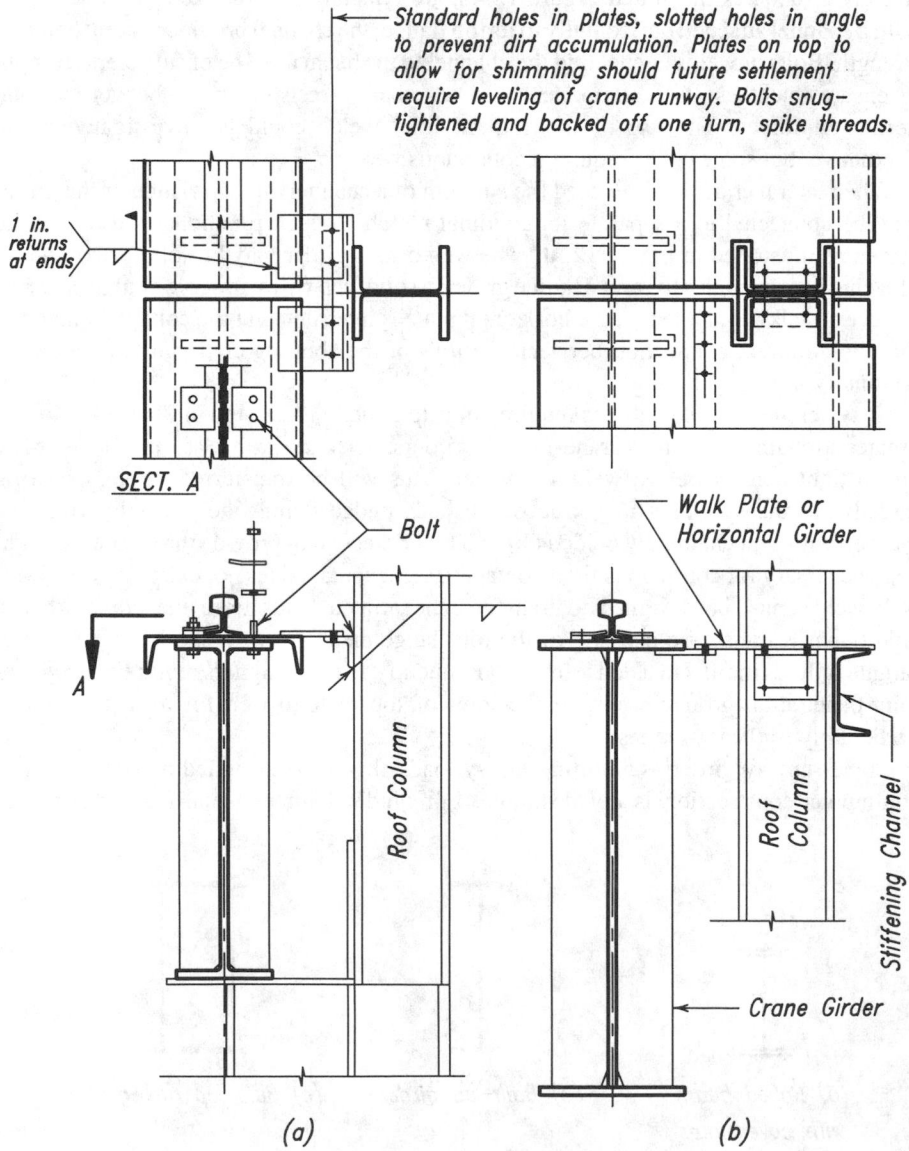

Fig. 12-4. Typical crane-runway girders.

for Highway Bridges (AASHTO, 1992). Additionally, fatigue must be considered in the design of crane-runway girders. Refer to the AISC publication *Bridge Fatigue Guide Design and Details* (Fisher, 1977); while this book was developed specifically as a guide for highway and railroad bridge design, it is equally applicable to crane girders. Additionally, refer to Fisher (1993) and Ricker (1982) for practical considerations in crane runway girders.

The crane rail may be fastened to the crane runway girder with bolted clamp plates, J-bolts, or other proprietary devices. The crane rail should not be welded directly to the crane-runway girder. The floating-type rail clamp, which is specified frequently, permits both longitudinal and lateral movement to accommodate thermal and alignment adjustments. Typical details, including proprietary items, for each particular case must be supplied by the designer.

A typical end connection, as shown in the plan view of Figure 12-4a, is designed to allow for the necessary end rotation of the crane-runway girder. Short-slotted holes in the angles with snug-tightened bolts backed off one turn (threads spiked) are used to alleviate fatigue cracking in the connection. Other details can be used (Ricker, 1982).

Bearing stiffeners should be used where required and must be finished to bear or welded sufficiently to transmit the reaction. Stiffeners should be used in pairs on each side of the web and should be welded to the top flange to prevent flange rotation. Intermediate stiffeners should be cut to clear the bottom flange by four to six times the web thickness, but not less than two inches. The stiffener corner at the juncture of the flange and web should be clipped to avoid intersecting welds and a reduction in fatigue strength. This clip should be four to six times the web thickness, with a 2-in. minimum. Stiffeners frequently are required to be punched for brackets (usually supplied by others) to support the electrical conductors from which the crane draws its power. The designer should obtain this information from the electrical contractor and show it on the design drawings.

Local stresses in the upper web plate, due to large concentrated wheel loads, can be critical to the life of a girder. These local stresses are normal to the flexural stresses and are compressive. The web plate adjacent to the weld may be subject to a residual tensile stress that is at or near the yield strength of the material. Since each passage of a crane wheel can reduce the residual tensile stress, the result is a cyclical loading in the tensile range in this area which must be considered in the design.

The crane runway must be kept almost exactly straight for safe operation and minimum wear on the crane wheels and rails. Crane stops are always provided at both ends of each line rail. These shock absorbing devices are used to stop the crane and to provide a means of realigning a crane which has become skewed slightly on its runway; the resulting forces must be considered in the design.

If crane stops are attached to the girders and the crane rail ends near the face of the stop, the total length of the rail is shortened by several inches. Since crane rails are not usually shop-fabricated items, this must be accounted for when ordering the rails and splice plates. For light-duty cranes, where the stops are clamped to the rails, the rails are ordered for the full length of the runway. Medium- and heavy-duty crane rails are usually ordered with "tight joints." Refer also to the discussion of crane rails in Part 1.

Rails are usually ordered as two runs of the total length necessary. Frequently, it is stipulated that not more than one rail in each run be less than the standard length of either 33 ft or 39 ft. It is good practice to stagger the rail joints on opposite sides of the runway by ordering one odd-length piece for each line of rail, to be placed at opposite ends of the runway.

Trusses

Because of their greater depth, trusses usually provide a greater stiffness and, therefore, reduced deflection when compared by weight with rolled or built-up girders of equal strength. Six general types of trusses frequently used in building frames are shown in Figure 12-5.

The Pratt truss of Figure 12-5a and the Warren truss of Figure 12-5c (and modifications of these types shown in Figures 12-5b and 12-5d) are commonly used as the principal supporting members in floor and roof framing. Note that the Pratt and Warren trusses shown have a top chord which is not quite parallel with the bottom chord. Such an arrangement is used to provide a slope for drainage on "flat" roofs. Most of the connections for the roof beams or purlins supported by these trusses can be identical. This would not be the case if the top chord were truly level and the elevation of the purlins had to be varied. When used in floor framing, the Pratt and Warren trusses are designed with parallel chords.

The Fink truss of Figure 12-5e and the Scissors truss of Figure 12-5h, (and modifications of these types shown in Figures 12-5f, 12-5g, and 12-5i) serve a similar function in symmetrical roofs having a pronounced pitch.

As discussed previously in Part 11, truss chord and web members are placed with the work lines intersecting at common panel points, resulting in purely axially loaded members when loads are applied only at the panel points of the trusses. Two exceptions follow.

The Bowstring truss of Figure 12-5j is used to support a curved roof. The deviation of the arched top chord from a straight line between panel points will produce a moment. The Vierendeel truss of Figure 12-5k is used to provide free passage through deep floor trusses or an orderly and orthogonal arrangement for exposed steelwork. In the absence of diagonal members, the members in a Vierendeel truss are subjected to both axial forces and flexure; the truss-member connections must be adequate to transfer the required moment and the design details of all joint connections must be shown by the design engineer.

In any truss, when vertical loads are imposed at locations other than at panel points, bending moments result. For example, the forces not at panel points in Figure 12-5c create moment in the top chord members which must be considered in addition to the direct compressive force in the top chord. Ordinarily, if loads must be supported at these intermediate points, vertical struts are usually placed under them in the truss to transmit the load directly to the joint or panel point in the bottom chord, as indicated in Figure 12-5d. Similarly, if loads are applied away from panel points to the bottom chord, additional vertical tension members are added as indicated by the dashed lines in Figure 12-5d.

Built-Up Columns

Built-up columns of the types shown in Figure 12-6 support the transverse beams, girders or trusses, crane-runway girders, and other structural members in an industrial building. When the construction is relatively light, single-shaft columns, either one-piece or stepped, may be used as shown in Figures 12-6a and 12-6b. The use of a bracket support for a crane-runway girder such as shown in Figure 12-6a usually is limited to about a 75-kip factored reaction.

For heavy mill buildings with heavy bridge cranes, the double- and triple-shaft columns shown in Figures 12-6c and 12-6d, respectively, provide a means of separately

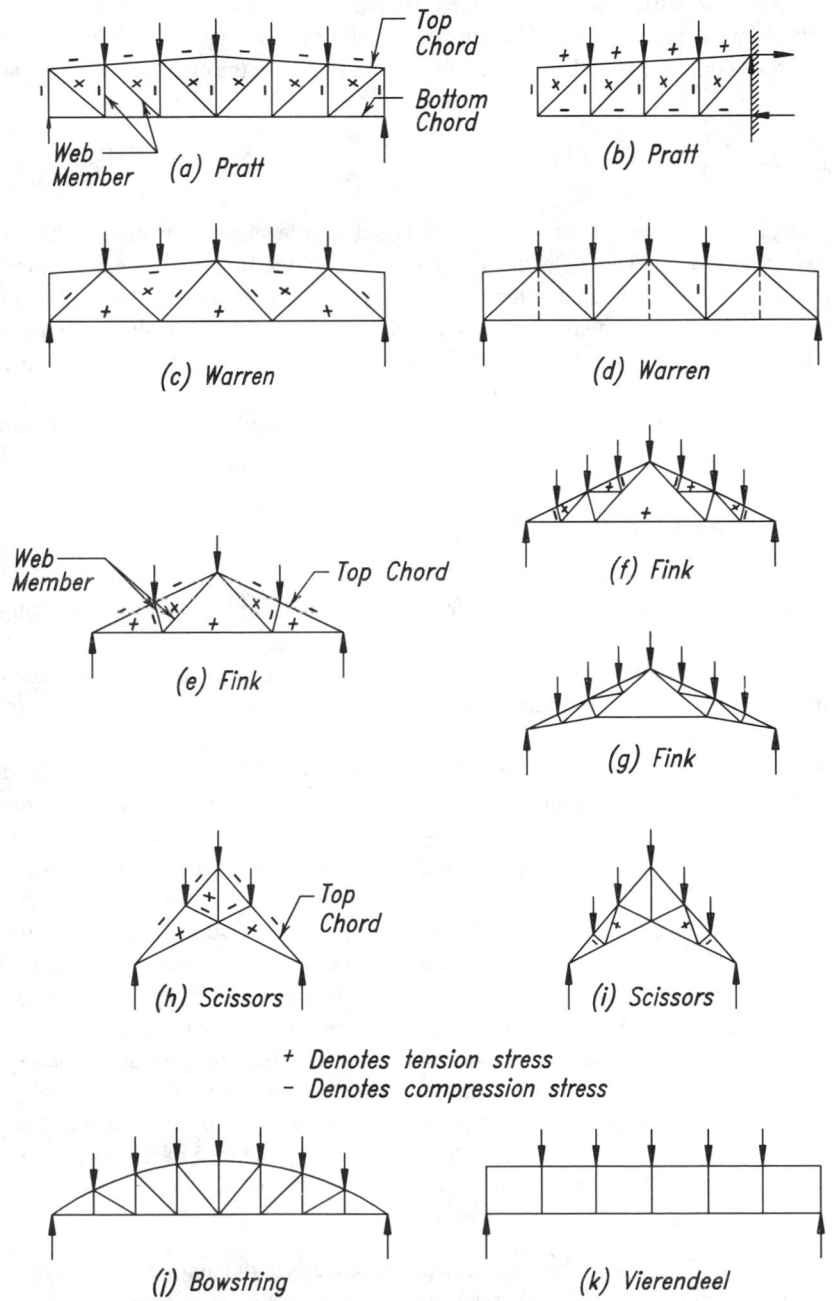

+ Denotes tension stress
- Denotes compression stress

Fig. 12-5. Typical trusses.

supporting the crane girder reactions and the roof girder or truss loads. The multiple shaft columns consist of one or two crane columns tied to the building columns by diaphragms. Refer to the AISC Design Guide *Industrial Buildings—Roofs to Column Anchorage* (Fisher, 1993).

EFFECT OF CAMBER ON END CONNECTIONS

Note that when a cambered beam bearing on a wall or other support is loaded, expansion of the unrestrained end must be considered. In Figure 12-7a, the end will move a distance Δ, where

$$\Delta = \frac{4Cd}{L}$$

If instead the cambered beam is supported on a simple shear connection at both ends, the top and bottom flange will each move a distance of one-half Δ since end rotation will occur approximately about the neutral axis. The designer should be aware of the magnitude of these movements and make provisions to accommodate them. Figure 12-7a considers the geometry of a girder in the horizontal position, and Figure 12-7b illustrates the condition when the girder is not level.

In general for building design, connections are fabricated square with the cambered beam end.

PURLIN AND GIRT CONNECTIONS

Girts

Girts, usually channels or angles, transfer wind forces from the siding to the columns. Intermediate wind columns are sometimes provided to reduce the unsupported length of girts. In general, channel girts should be placed with the toes down, to avoid collecting dirt and debris. Openings for doors and sash, however, will sometimes require that channel girts toe up.

Since the gravity load of the siding and girts is carried to the eave struts through a system of sag rods, each girt should be designed as a beam resisting the wind load incident upon its tributary area. As a common rule of thumb, girts supporting typical metal siding should be proportioned such that their depth in the direction of the wind load is not less than $\frac{1}{60}$ of the span; girts supporting steel sash should be proportioned such that their depth is not less than $\frac{1}{48}$ of the span. Wind columns for girts supporting metal sheeting are frequently proportioned for a ratio of depth-to-length of $\frac{1}{32}$. More stringent requirements may be required in areas of high wind or buildings with blast-pressure exposures.

Figure 12-8 shows a typical girt-to-column connection in which a clip angle is bolted or welded to the column flange and positioned to avoid coping the girt. Since the sag rods transfer the gravity forces, this is a nominal connection and two bolts are normally used at each girt end. When girt alignment is critical, e.g., at sash or wall panels, it is good practice to provide a clearance of $\frac{1}{4}$-in. to $\frac{1}{2}$-in. between the face of the column and the back of the girt with slotted holes for adjustment.

Purlins

Purlins, usually W shapes or channels, transfer roof loads to the major structural elements supporting the roof; the type and spacing of purlins is a design consideration which depends upon the incident roof loads as well as the limiting lengths of sheeting to be used.

When channels are used on a sloping roof, the channel should toe upward to permit the erector to walk on the member. Additionally, a channel that is toed upward will be more nearly loaded through its shear center by the gravity roof loads. When corrosion is a

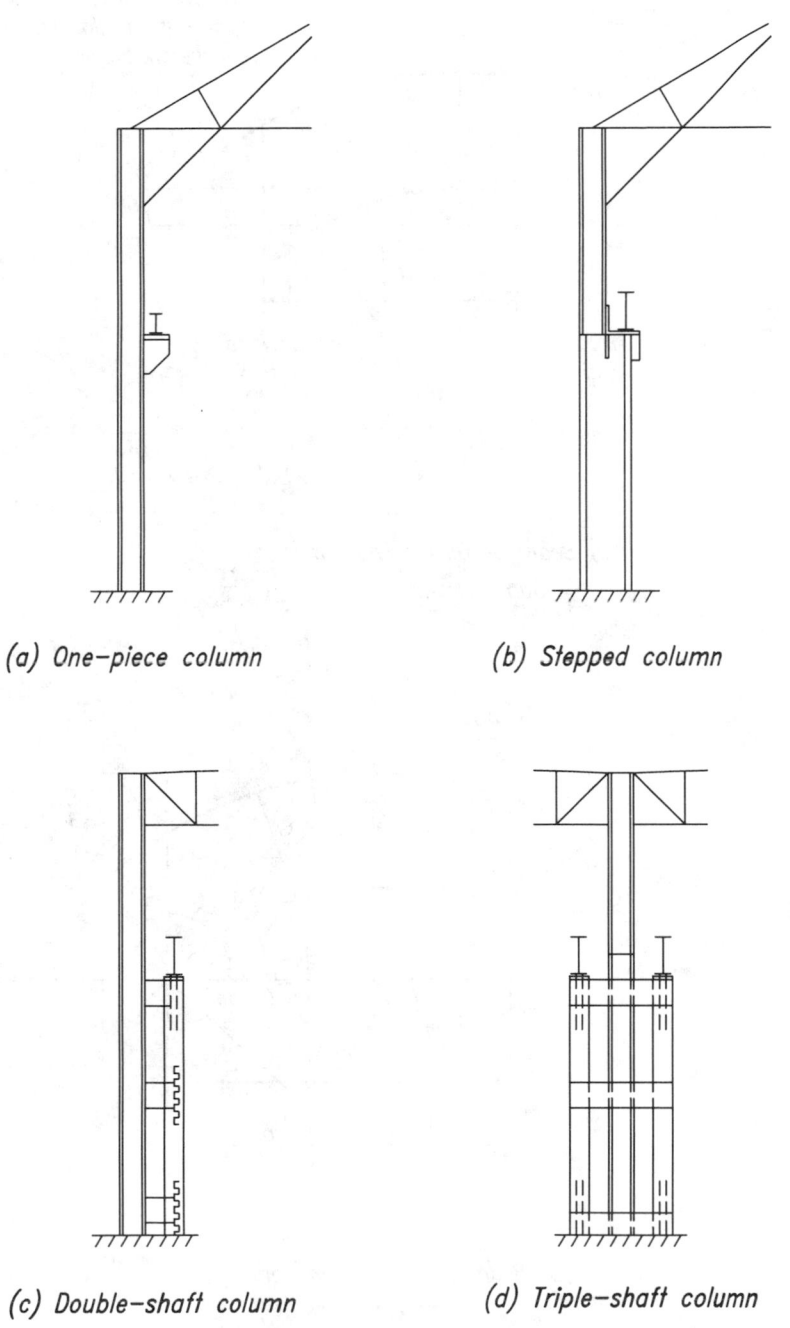

(a) One-piece column   *(b) Stepped column*

(c) Double-shaft column   *(d) Triple-shaft column*

Fig. 12-6. Built-up columns.

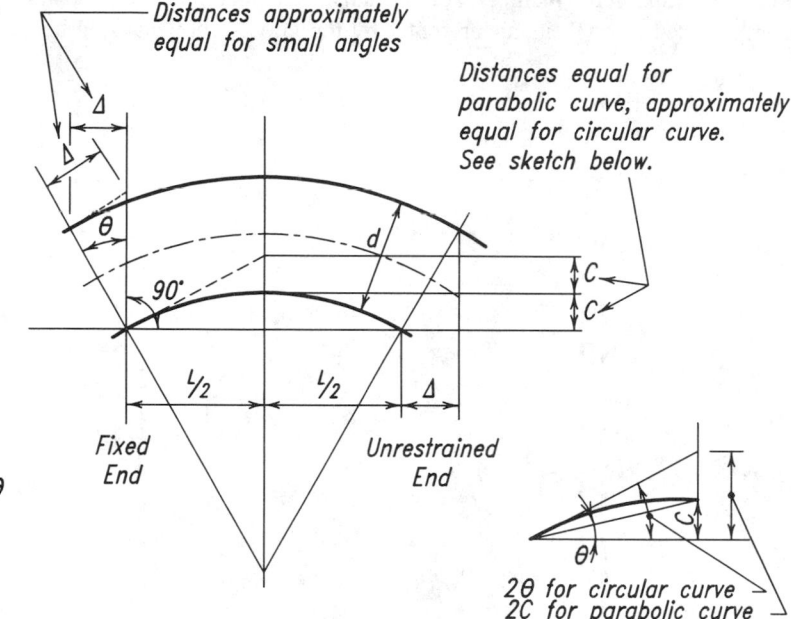

$$tan\theta=\frac{2C}{L/2}$$

$$\Delta =d\ tan\theta$$

$$\Delta =\frac{4Cd}{L}$$

(a) Beam or Girder Ends at
Same Elevation

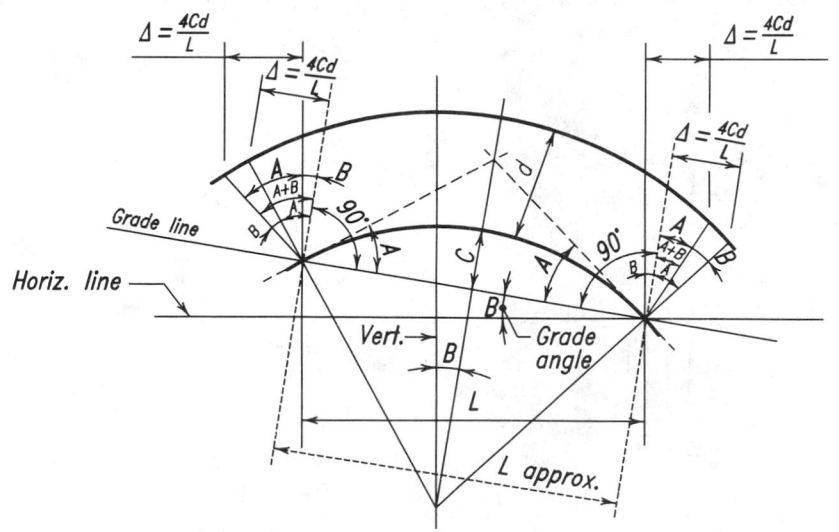

(b) Beam or Girder Ends at
Different Elevations

Fig. 12-7. Camber for beams and girders.

consideration, however, the channel should toe down, despite the inconvenience and additional expense of erection.

Roof purlins supporting metal sheeting are frequently proportioned for a ratio of depth-to-length of $\frac{1}{32}$. Other materials, unusual loadings, or deflection requirements must be investigated by the designer.

When channels are used, the ridge purlin is placed as close to the peak of the truss as possible in order to shorten the connection to the purlin on the opposite side of the centerline (see Figure 12-9). This also serves to decrease the overhang of the roof sheeting where it extends beyond the purlin to the ridge.

Sag Rods

Sag rods are usually furnished to transmit the gravity load of girts to a supporting member. Additionally, sag rods are used to control the deflection of and stiffen girts and purlins. Typical sag rods are $\frac{5}{8}$-in. or $\frac{3}{4}$-in. in diameter with lines spaced approximately six to eight feet apart.

To be effective, the force in the sag rods must be carried across the roof ridge and must be balanced by a corresponding force on the opposite side of the ridge. Several ridge-purlin connections are illustrated in Figure 12-9. Ridge purlins also are fastened together at other points along their length to increase their transverse stiffness, and thus permit them to be more effective if also used as struts.

Sag rods efficiently distribute gravity forces in the girts or purlins. Sag rods are useful for buildings of moderate width and height and near symmetrical dimension. The gravity loads of the siding are carried up sequentially; each sag rod carries the force of the previous rod plus the additional gravity load of the girt between them. This may continue

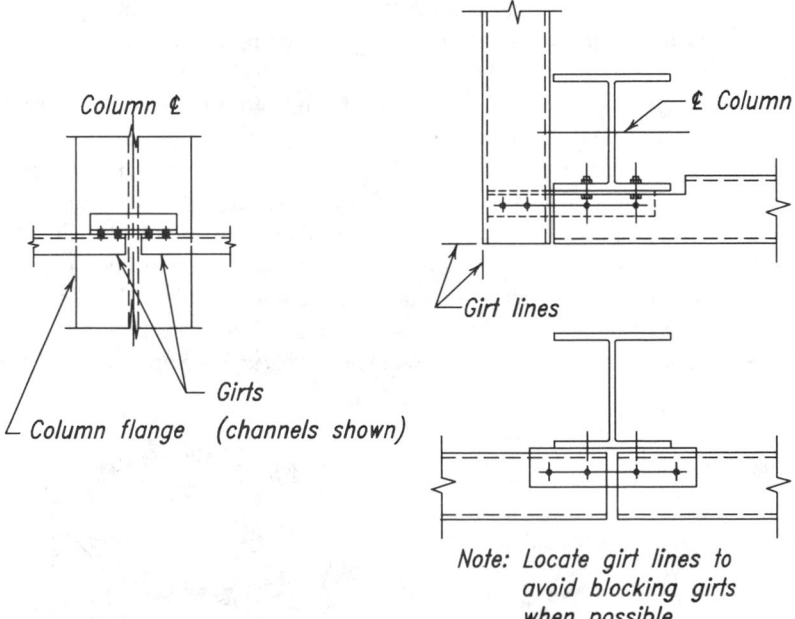

Fig. 12-8 Typical girt end connection.

up to the eave strut and through the roof slope to the ridge where the horizontal components of the gravity forces from either side of the building then offset for both the siding and the roofing. The sag rods are installed in each space and a suitable diaphragm is used at the ridge strut for load transfer. To avoid double punching the purlins, the sag-rod size should not exceed the size of the field bolts.

Sag rods are not usually required to be bent, since the slope gradient is usually quite small. Sag rods are usually connected with one nut on each end. If oversized holes are present, a washer should be used.

For the single diagonal sag rod shown in Figure 12-10a, four to six 8 in. nominal channel girts can generally be carried on a bay size of 20 to 25 feet before excessive twist occurs. Alternatively, girts may be stabilized with blocking or by other means. Otherwise, double diagonal sag rods should be used as illustrated in Figure 12-10b.

Negative wind pressure will cause compression in the interior flange of girts and purlins. When the exterior flange is laterally supported by the siding, sag rods can be used to provide lateral support to the inside flange of girts and purlins at intermediate positions if hole pattern A, illustrated in Figure 12-11, is used. Hole pattern B does not provide this same control. For single diagonal sag rods, a nut must be placed on the sag rod on both sides of the girt or purlin. Double diagonal sag rods are a tension-only system and therefore do not require double nuts.

LATERAL BRACING OF STRUCTURAL MEMBERS

In general, concrete slabs and concrete slabs on metal deck provide adequate lateral bracing to the compression flange of a beam. However, the question remains, how to design bracing for the compression flange of a beam not restrained by a slab or for a column or other compression member when it is required.

To provide adequate lateral restraint, the brace must possess both sufficient strength and stiffness. An approximate and conservative procedure of long practice is to design the brace to resist a force of two percent of the factored compressive force in the restrained member.

Several more rigorous empirical approaches have been proposed: (1) cross bracing, which depends on the axial stiffness of the brace to prevent relative lateral movement of two points on the braced member; (2) single-point or discreet bracing, which depends upon the flexural stiffness of transversely framed bracing members; (3) continuous bracing, provided by light-gage metal decking or other material; and, (4) leaning column bracing, wherein two or more compression members are linked at one or more points along their length requiring buckling to occur simultaneously in all linked members. These analyses are beyond the scope of this volume. The first three approaches are detailed in Salmon and Johnson (1990); Geschwindner (1993) summarizes and compares

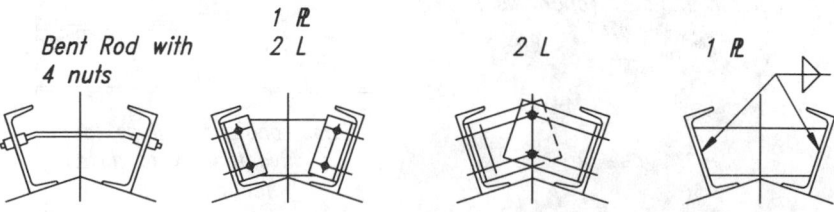

Fig. 12-9 Typical connections between ridge purlins.

three approaches to leaning-column bracing taken by Yura (1971), Lim and McNamara (1972), and LeMessurier (1977).

WALL ANCHORS
Figure 12-12 illustrates two common types of wall anchors. For the design of concrete embedments, refer ACI 349.

SHELF ANGLES
Figure 12-13 illustrates typical shelf angle configurations which provide for adjustment. Slotted holes may be used to provide for horizontal and/or vertical adjustment. Alternatively, shims may be used to provide vertical adjustment. Alignment tolerances are specified in AISC *Code of Standard Practice* Section 7.11.3.3.

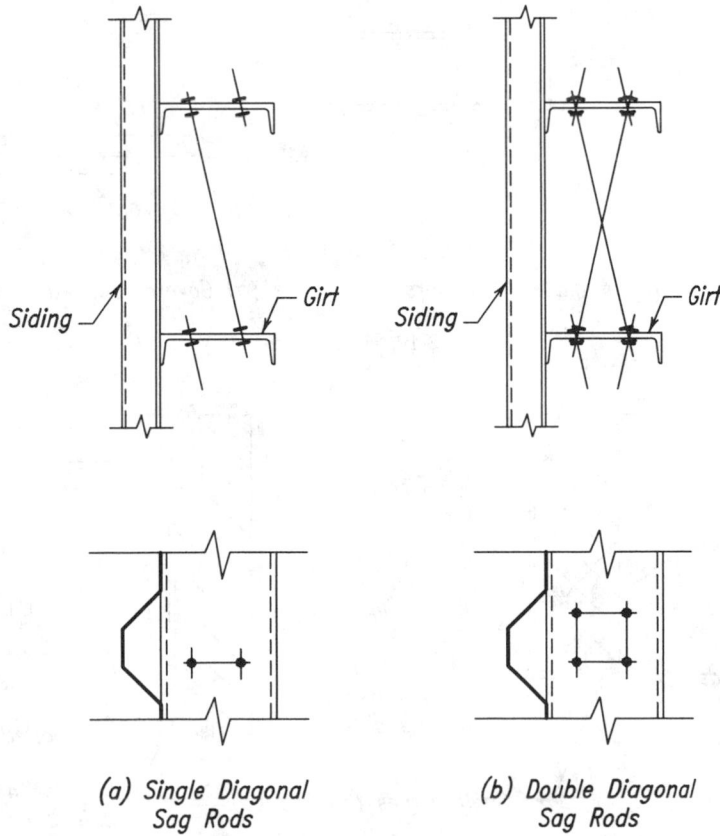

(a) Single Diagonal Sag Rods   *(b) Double Diagonal Sag Rods*

Fig. 12-10. Sag rods.

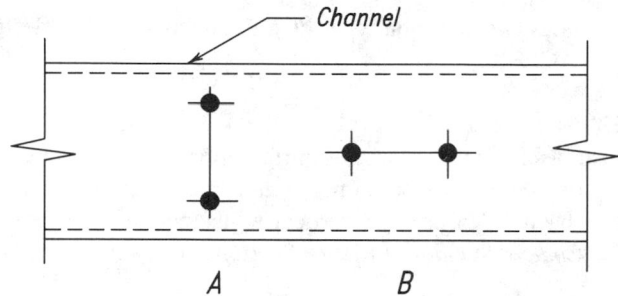

Fig. 12-11. Hole patterns for sag rods in girts and purlins.

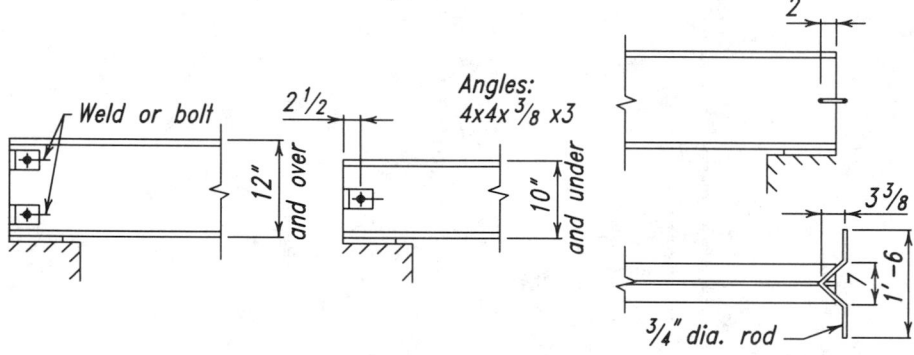

(a) Angle wall anchors   (b) Government wall anchor

Fig. 12-12. Wall anchors.

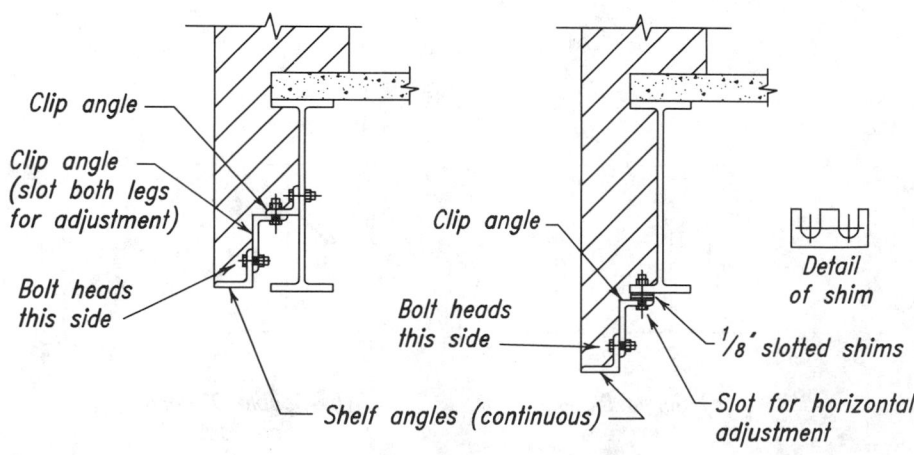

Fig. 12-13. Shelf angles.

REFERENCES

American Association of State Highway and Transportation Officials, 1992, *Standard Specifications for Highway Bridges,* AASHTO, Washington, DC.

American Welding Society, 1991, *Welding Handbook*, AWS, Miami, FL.

Association of Iron and Steel Engineers, 1969, *Guide for the Design and Construction of Mill Buildings—Technical Report No. 13*, AISE, Pittsburgh, PA.

Darwin, D., 1990, *Steel and Composite Beams with Web Openings*, AISC, Chicago, IL.

Fisher, J. M., 1993, *Industrial Buildings—Roofs to Column Anchorage*, AISC, Chicago, IL.

Fisher, J. W., 1977, *Bridge Fatigue Guide Design and Details*, AISC, Chicago, IL.

Geschwindner, L. F., 1993, "The Leaning Column in ASD and LRFD," *Proceedings of the 1993 National Steel Construction Conference*, pp. 19.1–19.17, AISC, Chicago, IL.

LeMessurier, W. J., 1977, "A Practical Method of Second Order Analysis," *Engineering Journal*, Vol. 14, No. 2, (2nd Qtr.), pp. 49–67, AISC, Chicago, IL.

Lim, L. C. and R. J. McNamara, 1972, "Stability of Novel Building System," *Structural Design of Tall Steel Buildings, Volume II-16, Proceedings of the ASCE-IABSE International Conference on the Planning and Design of Tall Buildings*, pp. 499–524, ASCE, New York, NY.

Ricker, D. T., 1982, "Tips for Avoiding Crane Runway Problems," *Engineering Journal*, Vol. 19, No. 4, (4th Qtr.), pp. 181–205, AISC, Chicago, IL.

Salmon, C. G. and J. E. Johnson, 1990, *Steel Structures—Design and Behavior*, Third Edition, Harper & Row, New York, NY.

Yura, J. A., 1971, "The Effective Length of Columns in Unbraced Frames," *Engineering Journal*, Vol. 8, No. 2, (2nd Qtr.), pp. 37–42, AISC, Chicago, IL.

PART 13

CONSTRUCTION INDUSTRY ORGANIZATIONS

PHONE

OVERVIEW

Part 13 lists private construction industry, government and government related, and foreign organizations which are potential sources of technical information for those engaged in steel design, detailing, fabrication, erection, project management, and building operation.

Following is a summary of the organizations listed. Statements which appear in the text of this Part were provided in whole or part by the respective organizations.

PRIVATE AND NON-GOVERNMENT-RELATED ORGANIZATIONS

Aluminum Association (AA)
900 19th Street, N.W., Washington, DC 20006
(202) 862-5100
(202) 862-5164 (fax)

The Aluminum Association (AA) is the trade association for domestic producers of primary and secondary aluminum and semi-fabricated aluminum products. Member companies operate 300 plants in 40 states.

American Concrete Institute (ACI)
P.O. Box 9094, Farmington Hills, MI 48333-9094
(248) 848-3700
(248) 848-3701 (fax)

The American Concrete Institute (ACI) is a non-profit organization which represents the public agency, engineer, architect, owner, contractor, educator, or other specialist interested in the design, construction, or maintenance of concrete structures.

American Galvanizers Association (AGA)
12200 East Iliff Avenue, #204, Aurora, CO 80014
(303) 750-2900
(303) 750-2909 (fax)

The American Galvanizers Association (AGA) promotes corrosion prevention through the use of post-fabrication hot-dip galvanizing. The AGA produces over 50 different publications, videos, and slide programs discussing various aspects of galvanizing for long-term corrosion prevention. These materials are provided at no charge to specifiers. Other complimentary services include educational seminars and the 1-800-HOT-SPEC line for answering questions about galvanizing and its applications. The AGA represents galvanizing companies in the United States, Canada, Mexico, and 18 other countries.

American Institute for Hollow Structural Sections (AIHSS)
now merged with Steel Tube Institute of North America
929 McLaughlin Run Road, Suite 8, Pittsburgh, PA 15017
(412) 221-8926
(412) 221-9119 (fax)

The American Institute for Hollow Structural Sections (AIHSS) is a non-profit technical organization committed to advancing and improving the use of structural steel tubing and pipe in buildings, bridges, and special structures. AIHSS encourages knowledgeable decisions concerning hollow structural sections in construction applications through the development and publication of engineering data and design aids, seminars, research and development, and specifications and standards activities. Among its publications are *HSS/Structural Steel Tubing—Dimensions and Section Properties. HSS—Column Load Tables*, and *HSS—Beam Load Tables*.

American Institute of Architects (AIA)
1735 New York Avenue, N.W., Washington, DC 20006
(202) 626-7300
(202) 626-7587 (fax)

Since 1857, The American Institute of Architects has represented the professional interests of America's architects. The AIA works to meet the needs and interests of the nation's architects and the public they serve by developing public awareness in the value of architecture and the importance of good design. In partnership with The American Architectural Foundation, the AIA strives for a national design literacy in the belief that a well-trained, creative profession and an informed public are prerequisites for a community's quality of life.

American Institute of Mining, Metallurgical, and Petroleum Engineers (AIME)
345 East 47th Street, 14th Floor, New York, NY 10017
(212) 705-7695
(212) 371-9622 (fax)

Constituent societies of AIME include the Iron and Steel Society (see separate entry), the Society of Petroleum Engineers, the Society of Mining Engineers, and the Minerals, Metals, and Materials Society.

American Institute of Steel Construction (AISC)
One East Wacker Drive, Suite 3100, Chicago, IL 60601-2001
(312) 670-2400
(312) 670-5403 (fax)

The American Institute of Steel Construction (AISC) is a non-profit trade association representing and serving the fabricated structural steel industry as well as engineers practicing structural steel design in the United States. For over 70 years, its purpose has been to advance the technology and competitiveness of steel construction through standardization, research and development, education, technical assistance, and quality control. AISC's programs include: the development of specifications and technical publications, research, technical and management seminars, engineering fellowships, and programs for quality control, productivity, and safety. AISC represents the combined experience, judgment, and strength of the steel fabricating industry and the structural engineering design profession.

American Institute of Timber Construction (AITC)
7012 South Revere Parkway #140, Englewood, CO 80112
(303) 792-9559
(303) 792-9559 (fax)

The American Institute of Timber Construction (AITC) is the oldest national technical trade association of the structural glued-laminated (glulam) timber industry. AITC was formed in 1952 to further the development, production, and promotion of laminated timber systems through the application of sound engineering practices and research. AITC has established design and product standards and developed industry quality control and inspection procedures that help assure economical, efficient, and reliable performance in structural applications.

American Iron and Steel Institute (AISI)
1101 17th Street, N.W., Suite 1300, Washington, DC 20036-4700
(202) 452-7100
(202) 463-6573 (fax)

The American Iron and Steel Institute (AISI) is a non-profit association of companies and individuals in the Western Hemisphere engaged in the iron and steel industry. The Construction Marketing Committee promotes the use of steel buildings, bridges, pipe/tank, and construction products through research, education, and promotion programs. The Committee on Construction Codes and Standards oversees efforts to achieve competitive provisions in applicable building codes and standards. AISI publishes the *Specification for the Design of Cold-Formed Steel Structural Members.*

American National Standards Institute (ANSI)
11 West 42nd Street, New York, NY 10036
(212) 642-4900
(212) 398-0023 (fax)

The American National Standards Institute (ANSI) is a private non-profit membership organization that coordinates the United States voluntary standards system, bringing together interests from the private and public sectors to develop voluntary standards for a wide array of United States industries. ANSI is the official United States member body to the world's leading standards bodies: the International Organization for Standardization (ISO) and the International Electrotechnical Commission (IEC), via the United States National Committee (USNC).

American Nuclear Society (ANS)
555 N. Kensington Avenue, LaGrange Park, IL 60526
(708) 352-6611
(708) 352-0499 (fax)

American Petroleum Institute (API)
1220 L Street, N.W., Washington, DC 20005
(202) 682-8000
(202) 682-8232 (fax)

The American Petroleum Institute (API), founded in 1919, is a non-profit corporation that represents the domestic petroleum industry. Its membership consists of a broad cross section of the petroleum and allied industries, including such functional segments as exploration, production, transportation, refining, and marketing.

American Railway Engineering Association (AREA)
50 F Street, N.W., Suite 7702, Washington, DC 20001
(202) 639-2190
(202) 639-2183 (fax)

The American Railway Engineering Association (AREA) is a professional organization concerned with engineering and maintenance work on railways in North America. It covers the track and bridge aspects of railroading, as well as roadbed, electrification, scales, and the mechanics of track maintenance machinery. AREA's twenty-two technical committees determine the content of the *Manual for Railway Engineering.* This standard reference in its field is revised annually to reflect the latest field-proven procedures and designs for railway engineering.

American Society for Metals International (ASMI)
9639 Kinsman Road, Materials Park, OH 44073
(216) 338-5151
(216) 338-4634 (fax)

American Society for Nondestructive Testing (ASNT)
P.O. Box 28518, 1711 Arlingate Lane, Columbus, OH 43228-0518
(614) 274-6003
(614) 274-6899 (fax)

American Society for Testing and Materials (ASTM)
100 Barr Harbor Dr., West Conshohocken, PA 19428
(610) 832-9500
(610) 832-9555 (fax)

Organized in 1898, ASTM has grown into one of the world's largest voluntary, full-consensus standards development organizations. From the work of 132 technical standards-writing committees, ASTM publishes standard testing methods, specifications, practices, guides, classifications, and terminology for materials, products, systems, and services. Related scientific and technical information is also published in various books and journals. ASTM's activities encompass metals, paints, plastics, textiles, petroleum, construction, energy, the environment, consumer products, medical services and devices, electronics, and many other areas. Technical research and testing is performed voluntarily by 34,000 members worldwide. Almost 9,000 standards are published each year in the 69 volumes of the Annual Book of ASTM Standards. These standards and related information are widely used and accepted throughout the world.

American Society of Civil Engineers (ASCE)
1801 Alexander Bell Dr., Reston, VA 20191-4400
(800) 548-ASCE or (703) 295-6000
(703) 295-6222 (fax)

The mission of the American Society of Civil Engineers is to advance professional knowledge and improve the practice of civil engineering in service to humanity by: improving the quality of life worldwide; developing and promoting standards of excellence; providing life-long education for civil engineers; serving members' needs, to meet the challenges at the frontiers of developing technology and societal change. The building load standard ASCE-7 is one of several that ASCE produces.

American Society of Mechanical Engineers (ASME)
345 East 47th Street, New York, NY 10017-2392
(212) 705-7722 or (800) THE-ASME
(212) 705-7674 (fax)

The American Society of Mechanical Engineers (ASME) is a non-profit educational and technical organization. Founded in 1880, ASME serves its members, industry, and government by encouraging the development of new technologies and finding solutions to the problems of an increasingly global technological society. Its programs include publishing, technical conferences and exhibits, engineering education, government relations, and public education, as well as the development of codes and standards.

American Water Works Association (AWWA)
6666 West Quincy Avenue, Denver, CO 80235-3098
(303) 794-7711
(303) 794-7310 (fax)
(303)794-8915 (fax)

The American Water Works Association (AWWA) is composed of over 54,000 professionals and 4,000 companies in the water supply field. AWWA is dedicated to the promotion of public health and welfare by assuring drinking water of unquestionable quality and sufficient quantity. As a leader for the public drinking water profession, AWWA is an effective instrument of education and change, setting standards, and advancing technology, science, and governmental policies relative to the management, collection, storage, treatment, and distribution of public water supplies.

American Welding Institute (AWI)
10628 Dutchtown Road, Knoxville, TN 37932
(615) 675-2150
(615) 675-6081 (fax)

The American Welding Institute (AWI) is a member owned non-profit organization. AWI promotes quality improvement, along with productivity, as top priorities for the United States welding industry. The mission of AWI is to put America's best ideas about welding to productive use in American industry. AWI provides services to the welding industry including welding engineering, equipment evaluation, mechanical testing, customized software, onsite trouble-shooting, metallurgical analysis, specialized training, and failure analysis.

American Welding Society (AWS)
550 N.W. LeJeune Road, Miami, FL 33126
(305) 443-9353 or (800) 443-9353
(305) 443-7559 (fax)

The American Welding Society (AWS) provides services to its members and the industry that advance the science, technology, and applications of welding and materials joining throughout the world. In its leadership role, AWS is recognized as the authority on joining technology and the source for coordinating matters pertaining to codes, standards, materials, education, certification, and research. Services include the AWS International Welding Exposition, publishing the *Welding Journal*, developing and publishing consensus standards, and offering a broad range of educational and welding certification programs.

Association of American Railroads (AAR)
50 F Street N.W., Washington, DC 20001
(202) 639-2429
(202) 639-2806 (fax)

Association of Iron and Steel Engineers (AISE)
Three Gateway Center, Suite 1900, Pittsburgh, PA 15222-1097
(412) 281-6323
(412) 281-4657 (fax)

The Association of Iron and Steel Engineers (AISE) is a technical society serving the steel industry worldwide through the collection and dissemination of technical information relating to the production of iron and steel. This is accomplished through a monthly technical journal, national conventions, local and regional meetings, technical publications, equipment specifications, a biennial industrial trade show, and technical committees which represent both user and supplier. Founded in 1907, AISE has developed into a multi-disciplined organization with over 10,000 members covering all phases of steel industry operations.

Building Officials and Code Administrators International (BOCA)
4051 West Flossmoor Road, Country Club Hills, IL 60478-5795
(708) 799-2300
(708) 799-4981 (fax)

Building Officials and Code Administrators (BOCA) International, Inc., is a not-for-profit organization which publishes the *National Building Code*. Founded in 1915, BOCA International is the original professional association of construction code officials. The organization was specifically established to provide a forum for the exchange of knowledge and ideas concerning building safety and construction regulation. BOCA came into being because its founders had a desire for excellence and professionalism in code enforcement. Today, BOCA offers a wide variety of membership services to promote code professionalism. The organization maintains ongoing model code development activity, conducts regular training and education programs, offers a wide variety of model construction codes and code-related publications, provides code interpretation assistance to members, and provides various other code-related services in the public interest.

Concrete Reinforcing Steel Institute (CRSI)
933 North Plum Grove Road, Schaumburg, IL 60173-4758
(847) 517-1200
(847) 517-1206 (fax)

The Concrete Reinforcing Steel Institute represents reinforcing steel producers and fabricators, epoxy coating applicators and powder manufacturers, and suppliers of other products used in concrete construction and fabricating equipment manufacturing. Technical activities are conducted by the CRSI Engineering Practice Committee and subcommittees on bar supports, placing reinforcing bars, concrete joist construction, detailing reinforced concrete, epoxy coating, and splicing reinforcing steel.

Construction Specifications Institute (CSI)
601 Madison Street, Alexandria, VA 22314-1791
(703) 684-0300
(703) 684-0465 (fax)

The Construction Specifications Institute (CSI), founded in 1948, is a not-for-profit organization dedicated to the advancement of construction technology through communication, education, research, and service. CSI serves the interest of architects, engineers, specifiers, contractors, product manufacturers, and others in the construction industry.

Corrugated Steel Pipe Institute (CSPI)
652 Bishop Street N., Unit 2A, Cambridge, Ontario, Canada, N3H 4V6
(519) 650-8080
(519) 650-8081 (fax)

The Corrugated Steel Pipe Institute (CSPI) was formed in 1961 to promote wider use of corrugated steel pipe and corrugated structural plate structures for drainage and other uses across Canada. CSPI provides product information, recommends standards and specifications, and recommends practices in the design, selection, application, and installation of corrugated steel pipe. CSPI provides liaison with the Canadian Standards Association, the National Corrugated Steel Pipe Association, and the American Iron and Steel Institute.

Crane Manufacturers Association of America (CMAA)
8720 Red Oak Boulevard, #201, Charlotte, NC 28217
(704) 676-1190
(704) 676-1199 (fax)

Electronic Industries Association (EIA)
2500 Wilson Boulevard, Arlington, VA 22201
(703) 907-7500
(703) 907-7501 (fax)

For more than 68 years, the Electronic Industries Association (EIA) has been the national trade organization representing the United States electronics manufacturers. Committed to the competitiveness of the American producer, EIA represents the entire spectrum of companies involved in the manufacture of electronic components, parts, systems, and equipment for communications, industrial, government, and consumer-end uses.

Engineering Foundation
345 East 47th Street, Suite 303, New York, NY 10017
(212) 705-7943
(212) 705-7441 (fax)

Factory Mutual Engineering and Research Company
1151 Boston-Providence Turnpike, Norwood, MA 02062
(781) 762-4300
(781) 762-9375 (fax)

Gypsum Association
810 First Street NE, #510, Washington, DC 20002
(202) 289-5440
(202) 289-3707 (fax)

Industrial Fasteners Institute (IFI)
1717 East Ninth Street, Suite 1105, Cleveland, OH 44114
(216) 241-1482
(216) 241-5901 (fax)

The Industrial Fasteners Institute (IFI) is an association of North American manufacturers of bolts, nuts, screws, rivets, and special formed parts. IFI members combine their technical knowledge to advance the technology and application engineering of fasteners and formed parts through planned programs of research and education. IFI and its members work closely with leading national and international technical organizations in developing standards and other technical practices. IFI is comprised of 90 fastener manufacturers and 35 suppliers of goods and services commonly used in the manufacture of fasteners.

Institute of the Ironworking Industry (III)
1750 New York Avenue N.W., Suite 400, Washington, DC 20006
(202) 783-3998
(202) 393-1507 (fax)

The Institute of the Ironworking Industry (III) is a non-profit labor-management trade association representing over 8,500 erection firms and 150,000 ironworkers. A board of directors equally apportioned from management and the Ironworkers International Union (AFL-CIO) sets policy to develop ways of eliminating problems which reduce the competitiveness and inhibit the economic development of the erection industry in the United States and Canada. Cooperation with other associations related to steel construction is encouraged to enhance safety, productivity, and the quality of the delivered product.

International Conference of Building Officials (ICBO)
5360 Workman Mill Road, Whittier, CA 90601-2258
(562) 699-0541
(562) 692-6031 (fax)

The International Conference of Building Officials is dedicated worldwide to public safety in the built environment through the development, maintenance, and promotion of uniform codes and standards, enhancement of professionalism in code administration, and the facilitation of the acceptance of innovative building products and systems. The Conference works toward these objectives through the publication of the *Uniform Building Code* and its associated family of codes and standards and through the offering of high quality training, technical assistance, and certification examinations based on these documents.

Iron and Steel Society (ISS)
410 Commonwealth Drive, Warrendale, PA 15086
(412) 776-1535
(412) 776-0430 (fax)

The Iron and Steel Society (ISS) is a constituent society of the American Institute of Mining, Metallurgical, and Petroleum Engineers (AIME). ISS members are active in the field of iron and steel processing and technology. ISS provides a medium of communication and cooperation among those interested in any phase of ferrous metallurgy and materials science and technology.

James F. Lincoln Arc Welding Foundation (JFLF)
22801 St. Claire, P.O. Box 17035, Cleveland, OH 44117-1199
(216) 481-4300
(216) 486-1751 (fax)

The James F. Lincoln Arc Welding Foundation, incorporated as a non-profit entity in 1936, is the only organization in the United States specifically dedicated to educating the public about the art and science of arc welding. The Lincoln Foundation recognizes technical achievement with substantial monetary awards and publishes educational materials for dissemination to the public. International Assistant Secretaries now carry out Lincoln Foundation programs in Argentina, Australia, Canada, Croatia, Hungary, Japan, New Zealand, the People's Republic of China, Russia, Southern Africa, and the United Kingdom.

Material Handling Industry (MHI)
8720 Red Oak Boulevard, Suite 201, Charlotte, NC 28217
(704) 522-8644
(704) 522-7826 (fax)

Materials Properties Council
345 E. 47th Street, New York, NY 10017
(212) 705-7693
(212) 371-9622 (fax)

Metal Building Manufacturers Association (MBMA)
1300 Sumner Avenue, Cleveland, OH 44115-2851
(216) 241-7333
(216) 241-0105 (fax)

The Metal Building Manufacturers Association (MBMA) was formed in 1956 with the goal of developing sound design criteria for verifying the performance of structures under various loads. MBMA has promoted the benefits of metal building systems to building code officials, architects, and engineers. MBMA has 27 member manufacturing firms that employ 10,000 persons and operate 57 manufacturing facilities in 24 states and three foreign countries.

National Association of Architectural Metals Manufacturers (NAAMM)
8 South Michigan Avenue, Suite 1000, Chicago, IL 60603
(312) 782-4951
(312) 580-0165 (fax)

The National Association of Architectural Metal Manufacturers (NAAMM) is the Chicago-based trade association representing manufacturers of metal products. NAAMM develops, maintains, and publishes technical information on products from members in its five divisions: Architectural Metals Products Division (metal stairs, railing systems, and miscellaneous and ornamental products), Flagpole Division, Hollow Metal Manufacturers Association Division (hollow metal doors and frames), Metal Bar Grating Division, and Metal Lath/Steel Framing Association Division.

National Association of Corrosion Engineers (NACE)
11440 S. Creek Drive, Houston, TX 77084-4906
(281) 492-0535
(281) 228-6300 (fax)

NACE develops and distributes high-quality technology to prevent and control degradation of materials in engineered systems. NACE promotes: (1) the application of all materials, e.g., metals, polymers, concrete, ceramics, natural materials, composites, and electronic materials; (2) the integration of all degradation phenomena, e.g., corrosion, wear, and fracture; and, (3) the integration of corrosion science and engineering into the design process. NACE is a professional association with more than 16,000 members across many industries. Programs include professional recognition and certification, education, training, seminars, committee work weeks, and an annual conference. NACE also publishes two monthly journals, standards, books, and computer software.

National Concrete Masonry Association (NCMA)
2302 Horse Pen Road, Herndon, VA 20171-3499
(703) 713-1900
(703) 713-1910 (fax)

National Corrugated Steel Pipe Association (NCSPA)
1255 23rd Street, N.W., Washington, DC 20037
(202) 452-1700
(202) 833-3636 (fax)

The National Corrugated Steel Pipe Association (NCSPA) was founded in 1956 to promote sound public policy relating to the use of corrugated steel drainage structures in private and public construction. The association collects and distributes technical information, assists in the formulation of specifications and designs, and conducts seminars to increase the awareness of the product. Among publications are *Design Data Sheets, Drainage Technology Bulletins*, two installation manuals, and two cost analyses of pipe materials.

National Erectors Association (NEA)
1501 Lee Highway, Suite 202, Arlington, VA 22209
(703) 524-3336
(703) 524-3364 (fax)

The National Erectors Association (NEA) is a national trade association of union contractors dedicated to providing its members with the highest level of labor relations and safety services, the promotion of positive labor-management programs in construction, and the advancement of a dynamic union construction industry. Membership includes steel erectors, industrial maintenance contractors, specialty contractors, general contractors, and construction managers. Active standing committees include its nationally-known Labor Committee and Safety & Health Committee.

National Fire Protection Association (NFPA)
1 Batterymarch Park, P.O. Box 9101, Quincy, MA 02269-9101
(617) 770-3000
(617) 770-0700 (fax)

The National Fire Protection Association (NFPA), an international non-profit organization, is recognized as the premier institution dedicated exclusively to protecting lives and property from fire and related hazards. NFPA publishes over 270 nationally recognized codes and standards, as well as numerous fire service training and educational programs. More than 62,500 members work voluntarily to further NFPA's mission.

National Fire Sprinkler Association (NFSA)
4 Robin Hill Corporate Park, Route 22, P.O. Box 1000, Patterson, NY 12563
(914) 878-4200
(914) 878-4215 (fax)

National Institute of Steel Detailing (NISD)
P.O. Box 121484, Arlington, TX 76012
(817) 860-9890
(817) 860-9891 (fax)

The National Institute of Steel Detailing (NISD) was formed in 1969 to create a better understanding and bond between individuals engaged in the detailing profession. NISD strives to eliminate practices which are injurious, to promote the efficiency of their work, and to uphold the proper standards for the steel detailer in relations with other members of the construction industry. The institute is a non-profit association of regional chapters, firms, and individuals in the United States who serve the fabricated structural and miscellaneous steel industry.

National Society of Architectural Engineers (NSAE)
700 S.W. Jackson, Suite 702, Topeka, KS 66603-3758
(913) 232-5707
(785) 357-6629 (fax)

Nickel Development Institute (NiDI)
214 King Street W, #510, Toronto, Ontario, Canada M5H 3S6
(416) 591-7999
(416) 591-7987 (fax)

The Nickel Development Institute (NiDI) provides technical service to nickel consumers and others concerned with nickel/nickel alloys and their uses. NiDI's information services are available to designers, specifiers, and educators as well as nickel users. Inquiries are welcomed from architects, engineers, specification writers, and others responsible for selection of materials for manufacturing and construction. NiDI looks forward to cooperating with colleges and universities by furnishing relevant information and materials for engineering, materials science, and industrial design education.

Portland Cement Association (PCA)
5420 Old Orchard Road, Skokie, IL 60077-1083
(847) 966-6200
(847) 966-9781 (fax)

Post-Tensioning Institute (PTI)
1717 West Northern Avenue, Suite 114, Phoenix, AZ 85021
(602) 870-7540
(602) 870-7541 (fax)

The Post-Tensioning Institute, a not-for-profit organization, provides research, technical development, marketing, and promotional activities for companies engaged in post-tensioned prestressed construction. Its publications are a major communications system for disseminating information on p/t design and construction technology. In addition, PTI publishes a quarterly newsletter dealing with developments in the p/t industry. Members include p/t materials fabricators, manufacturers of prestressing materials, companies supplying miscellaneous materials, services, and equipment used in p/t construction, and more than 700 professional engineers, architects, and contractors.

Prestressed Concrete Institute (PCI)
175 W. Jackson Boulevard, Suite 1859, Chicago, IL 60604
(312) 786-0300
(312) 786-0353 (fax)

Southern Building Code Congress International (SBCCI)
900 Montclair Road, Birmingham, AL 35213-1206
(205) 591-1853
(205) 592-7001 (fax)

The Southern Building Code Congress International, Inc. (SBCCI) was established in 1940 as a membership organization dedicated to promulgating and maintaining a comprehensive set of model building codes and to providing support services to users of the code. It continues that tradition today with the *Standard Codes*™ which cover every aspect of commercial and residential construction. The SBCCI also provides technical and educational services to assist code enforcement professionals and others in providing the most efficient, effective, and skilled service to the building industry.

Steel Deck Institute (SDI)
P.O. Box 25, Fox River Grove, IL 60021-0025
(847) 462-1930

Since 1939, the Steel Deck Institute (SDI) has provided uniform industry standards for the engineering, design, manufacture, and field usage of steel decks. The SDI is concerned with cold-formed steel products, with various configurations distinctive to individual manufacturers, used to support finished roofing materials, or to serve as a permanent form and/or positive reinforcement for concrete floor slabs. Members of SDI are manufacturers of steel floor and roof decks. Associate members are manufacturers of fasteners, coatings, and other related components.

Steel Joist Institute (SJI)
3127 10th Ave., North Ext., Myrtle Beach, SC 29577-6760
(803) 626-1995
(803) 626-5565 (fax)

The Steel Joist Institute (SJI) is a not-for-profit organization. Besides setting standards for the steel joist industry, SJI works closely with major building code bodies throughout the country helping to develop code regulations regarding steel joists and joist girders. SJI also invests thousands of dollars in ongoing research related to steel joists and joist girders, and offers a complete library of publications and other training and research aids.

Steel Service Center Institute (SSCI)
Society Center, Suite 2400, 127 Public Square, Cleveland, OH 44114-1216
(216) 694-3630
(216) 694-3940 (fax)

The Steel Service Center Institute (SSCI) was established in 1907 to enhance the financial return of member companies by providing information, education, governmental representation, networking opportunities, and a forum to enhance the quality of products and services in meeting customer, supplier, and employee expectations. Steel service centers purchase basic steel products, add value to them through services such as inventory management, pre-production processing, just-in-time delivery, electronic data interchange, and barcoding, and subsequently sell production-ready metal pieces and parts to manufacturers. Producing mills are Associate Members. International members are welcome.

Steel Structures Painting Council (SSPC)
40 24th Street, Pittsburgh, PA 15222
(412) 281-2331
(412) 281-9992 (fax)

Steel Tank Institute (STI)
570 Oakwood Road, Lake Zurich, IL 60047
(847) 438-8265
(847) 438-8766 (fax)

The Steel Tank Institute (STI) is a trade association and standards-setting body representing steel tank fabricators and affiliated corporations. STI develops technical standards for fabrication, corrosion control, installation, and secondary containment of underground and aboveground storage tanks. STI members manufacture single- and double-wall steel UST's with sti-P3 or ACT-100® corrosion protection systems, new Permatank™ double-wall UST's and F911™ and F921™ secondarily contained aboveground tanks.

Steel Tube Institute of North America (STI)
8500 Station Street, Suite 270, Mentor, OH 44060
(216) 974-6990
(216) 974-6994 (fax)

The Steel Tube Institute of North America (STI), founded in 1930, promotes the responsible growth, prosperity, and competitiveness of the steel tubing industry. STI collects and disseminates information on manufacturing techniques and data and analysis on growth areas, market trends, and product applications. STI provides information to customers on tubular products. Active members are producers of mechanical, pressure, and structural tubing. Associates are suppliers of raw materials and equipment to the tubular products industry.

Structural Stability Research Council (SSRC)
Fritz Engineering Laboratory, 13 East Packer Avenue, Lehigh University,
  Bethlehem, PA 18015
(610) 758-3522
(610) 758-6405 (fax)

The Structural Stability Research Council (SSRC), founded in 1944, offers guidance, through its 16 task groups and 8 task reporters, to specification writers and practicing engineers by developing both simplified and refined calculation procedures for the solution of stability problems, and assessing the limitations of these procedures. SSRC holds regular annual meetings to report on research activities and to indicate where deficiencies exist in our present understanding of structural behavior. The membership of the SSRC is made up of representatives from organizations, consulting firms, and individuals.

Underwriters Laboratories Inc. (UL)
333 Pfingsten Road, Northbrook, IL 60062-2096
(847) 272-8800
(847) 272-8129 (fax)

Underwriters Laboratories Inc. (UL), an independent, not-for-profit, safety testing and certification organization, evaluates products, materials, and systems in the interest of public safety. Founded in 1894, UL is neither a commercial enterprise nor a government agency, but a member of the private sector whose primary objective is to help manufacturers bring safer products to U.S. and global markets. More than 6 billion UL Marks are placed on products annually by more than 40,000 manufacturers. A UL Listing Mark on a product means samples of the product have been tested to nationally recognized safety standards and have been found to be reasonably free from fire, electric shock, and related safety hazards.

Welding Research Council (WRC)
345 E. 47th Street, New York, NY 10017
(212) 705-7956
(212) 371-9622 (fax)

FEDERAL AND STATE GOVERNMENT AND RELATED AGENCIES

Army Corps of Engineers
Office of the Chief of Engineers, Hdqr., U.S. Army, 20 Massachusetts Avenue,
  Washington, DC 20314-1000
(202) 761-0660
(202) 761-1373 (fax)

American Association of State Highway and Transportation Officials (AASHTO)
444 N. Capitol Street, N.W., Suite 249, Washington, DC 20001
(202) 624-5800
(202) 624-5806 (fax)

Bureau of Labor Statistics
Department of Labor, 200 Constitution Avenue, N.W., Washington, DC 20210
(202) 606-7828
(202) 606-7891 (fax)

Department of Housing and Urban Development (HUD)
451 Seventh Street, S.W., Washington, DC 20410
(202) 708-1422
(202) 619-8365 (fax)

Environmental Protection Agency (EPA)
401 M Street S.W., Washington, DC 20460
(202) 260-4111

Federal Construction Council (FCC)
c/o National Academy of Sciences, 2101 Constitution Avenue N.W.,
  Washington, DC 20418
(202) 334-3378

Federal Highway Administration (FHA)
Department of Transportation, 400 Seventh Street, S.W., Washington, DC 20590
(202) 366-0630

Federal Railroad Administration
Department of Transportation, 400 Seventh Street, S.W., Washington, DC 20590
(202) 632-3124

General Services Administration (GSA)
General Services Building, 18th & F Streets, N.W., Washington, DC 20405
(202) 708-5082

National Institute of Building Sciences (NIBS)
1090 Vermont Avenue, #700, Washington, DC 20005
(202) 289-7800
(202) 289-1092 (fax)

National Institute of Standards and Technology (NIST)
Department of Commerce, Building 101, Room A903, Gaithersburg, MD 20899
(301) 975-3058
(301) 926-1630 (fax)

National Science Foundation (NSF)
4201 Wilson Boulevard, Arlington, VA 22230
(703) 306-1234
(703) 306-0202 (fax)

National Technical Information Service (NTIS)
NTIS Operations Center, 5285 Port Royal Road, Springfield, VA 22161
(703) 605-6040
(703) 321-8547 (fax)

Occupational Safety and Health Administration (OSHA)
Department of Labor, 820 First Street, N.E., Washington, DC 20002
(202) 523-1452

United States Information Agency
301 Fourth Street, S.W., Washington, DC 20547
(202) 619-4700
(202) 619-4510 (fax)

United States Government Printing Office
Superintendent of Documents, Washington, DC 20402
(202) 512-0000
(202) 512-2250 (fax)

FOREIGN ORGANIZATIONS

Australian Institute of Steel Construction (AISC)
Level 13, 99 Mount Street, North Sydney, Australia NSW 2060
PO Box 6366, North Sydney, Australia NSW 2059
011-61-2/99296666
011-61-2/99555406 (fax)

British Constructional Steelwork Association (BCSA)
4 Whitehall Court
London, SW1A 2ES, United Kingdom
011-44171-839-8566
011-44171-976-1634 (fax)

Canadian Institute of Steel Construction (CISC)
201 Consumers Road, Suite 300, Willowdale, Ontario, Canada M2J 4G8
(416) 491-4552
(416) 491-6461 (fax)

The Canadian Institute of Steel Construction (CISC), a national association, represents the structural steel, steel platework, and open-web steel joist industries by promoting good design, safety, and efficient and economical use of steel as a means of expanding markets for its Fabricator, Mill, Honorary, and Associate Members. Services encompass steel design information, technical publications, such as the *Handbook of Steel Construction*, computer programs, continuing education courses, marketing, and industry-government relations. CISC manages the Steel Structures Education Foundation and the Canadian Steel Construction Council.

European Convention for Constructional Steelwork (ECCS)
Avenue des Ombrages, 32/36 boite 20, B1200, Brussels, Belgium
011-322-762-0429
011-322-762-0935 (fax)

Japanese Society of Steel Construction (JSSC)
848 Shin Tokyo Building, 3-3-1 Marunouchi Chiyoda-Ku, J-Tokyo 100
011-81-3/32120875
011-81-3/32120878 (fax)

Mexican Institute of Steel Construction (MISC)
Antonio Zubieta No. 10, Los Reyes, Tlalnepantla, Mexico 54090
011-525-565-6800
011-525-390-1416 (fax)

South African Institute of Steel Construction (SAISC)
7th Floor, Metal Industries House, 42 Anderson Street, Johannesburg, South Africa 2001
PO Box 1338, Johannesburg, South Africa 2000
011-27-22-838-1665
011-27-11-834-4301 (fax)

GENERAL NOMENCLATURE

A	Cross-sectional area, in.^{2}
A	Horizontal distance from end panel point to mid-span of a truss, ft.
A	Minimum side dimension for square or rectangular beveled washer, in.
A_B	Loaded area of concrete, in.^{2}
A_{BM}	Cross-sectional area of base metal for a welded joint, in.^{2}
A_b	Nominal body area of a fastener, in.^{2}
A_b	Nominal bolt area, in.^{2}
A_c	Area of concrete in a composite column, in.^{2}
A_c	Area of concrete slab within effective width, in.^{2}
A_{cp}	Projected surface area of concrete cone surrounding headed anchor rods, in.^{2}
A_e	Effective net area, in.^{2}
A_f	Area of flange, in.^{2}
A_{fe}	Effective tension flange area, in.^{2}
A_{fg}	Gross area of flange, in.^{2}
A_{fn}	Net area of flange, in.^{2}
A_g	Gross area, in.^{2}
A_{gt}	Gross area subject to tension, in.^{2}
A_{gv}	Gross area subject to shear, in.^{2}
A_n	Net area, in.^{2}
A_{nt}	Net area subject to tension, in.^{2}
A_{nv}	Net area subject to shear, in.^{2}
A_{pb}	Projected bearing area, in.^{2}
A_r	Area of reinforcing bars, in.^{2}
A_s	Area of steel cross section, in.^{2}
A_{sc}	Cross-sectional area of stud shear connector, in.^{2}
A_{sf}	Shear area on the failure path, in.^{2}
A_{st}	Cross-sectional area of stiffener or pair of stiffeners, in.^{2}
A_v	Seismic coefficient representing the effective peak velocity-related acceleration
A_w	Area of web clear of flanges, in.^{2}
A_w	Effective area of weld, in.^{2}
A_0	Initial amplitude of a floor system due to a heel-drop excitation, in.
A_1	Area of steel bearing concentrically on a concrete support, in.^{2}
A_2	Total cross-sectional area of a concrete support, in.^{2}
B	Factor for bending stress in tees and double angles, defined by LRFD Specification Equation F1-16
B	Factor for bending stress in web-tapered members, defined by LRFD Specification Equations A-F3-8 through A-F3-11, in.
B	Horizontal distance from mid-span of a truss to a given panel point, ft.
B	Base plate width, in.
B_1, B_2	Factors used in determining M_u for combined bending and axial forces when elastic, first order analysis is employed

BF	A factor that can be used to calculate the flexural strength for unbraced length L_b between L_p and L_r, defined in Part 4
C	Required mid-span camber, in.
C	Width across points of square or hex bolt head or nut, or maximum diameter of countersunk bolt head, in.
C	Coefficient for eccentrically loaded bolt and weld groups
C_{PG}	Plate girder coefficient
C_{Tot}	Sum of compressive forces in a composite beam, kips
C_a, C_b	Coefficients used in extended end-plate connection design
C_b	Bending coefficient dependent upon moment gradient
C_c	Beam reaction coefficient (Part 5)
C_{con}	Effective concrete flange force for a composite beam, kips
C_m	Coefficient applied to bending term in interaction formula for prismatic members and dependent upon column curvature caused by applied moments
C_m'	Coefficient applied to bending term in interaction formula for tapered members and dependent upon axial stress at the small end of the member
C_p	Ponding flexibility coefficient for primary member in a flat roof
C_s	Ponding flexibility coefficient for secondary member in a flat roof
C_s	Seismic response factor related to the fundamental period of the building
C_{stl}	Compressive force in steel in a composite beam, kips
C_v	Ratio of "critical" web stress, according to linear buckling theory, to the shear yield stress of web material
C_w	Warping constant, in.[6]
C_1	Loading constant used in deflection calculations (Part 4)
C_1	Clearance for tightening, in. (see Tables 8-4 and 8-5)
C_1	Electrode coefficient for relative strength of electrodes where, for E70 electrodes, $C_1 = 1.00$ (see Table 8-37)
C_2	Clearance for entering, in. (see Tables 8-4 and 8-5)
C_3	Clearance for fillet based on one standard hardened washer, in. (see Tables 8-4 and 8-5)
CG	Center of gravity
D	Outside diameter of circular hollow section, in.
D	Dead load, due to the weight of the structural elements and permanent features on the structure
D	Factor used in LRFD Specification Equation A-G4-2, dependent on the type of transverse stiffeners used in a plate girder
D	Offset from the base line at a panel point of a truss, in.
D	Damping in percent of critical
D	Slip probability factor for bolts
D	Number of sixteenths-of-an-inch in the weld size
DLF	Dynamic load factor
E	Modulus of elasticity of steel (29,000 ksi)
E	Earthquake load
E	Minimum edge distance for clipped washer, in.

E	Minimum effective throat thickness for partial-joint-penetration groove weld, in.
E_c	Modulus of elasticity of concrete, ksi
E_m	Modified modulus of elasticity for the design of composite columns, ksi
EBF	Eccentrically braced frame (Seismic Specification)
ENA	Elastic neutral axis
F	Width across flats of bolt head, in.
F	Clearance for tightening staggered bolts, in. (see Tables 8-4 and 8-5)
F_{BM}	Nominal strength of the base material to be welded, ksi
F_{EXX}	Classification strength of weld metal, ksi
F_L	Smaller of $(F_{yf} - F_r)$ or F_{yw}, ksi
$F_{b\gamma}$	Flexural stress for tapered members defined by LRFD Specification Equations A-F4-4 and A-F4-5, ksi
F_{cr}	Critical stress, ksi
$F_{crft}, F_{cry}, F_{crz}$	Flexural-torsional buckling stresses for double-angle and tee-shaped compression members, ksi
F_e	Elastic buckling stress, ksi
F_{ex}	Elastic flexural buckling stress about the major axis, ksi
F_{ey}	Elastic flexural buckling stress about the minor axis, ksi
F_{ez}	Elastic torsional buckling stress, ksi
F_{my}	Modified yield stress for the design of composite columns, ksi
F_n	Nominal shear rupture strength, ksi
F_n, F_{nt}	Nominal strength of bolt, ksi
F_p	Nominal bearing stress on fastener, ksi
F_r	Compressive residual stress in flange, ksi
$F_{s\gamma}$	Stress for tapered members defined by LRFD Specification Equation A-F3-6, ksi
F_t	Nominal tensile strength of bolt from LRFD Specification Table J3.2, ksi
F_u	Specified minimum tensile strength of the type of steel being used, ksi
F_v	Nominal shear strength of bolt from LRFD Specification Table J3.2, ksi
F_w	Nominal strength of the weld electrode material, ksi
$F_{w\gamma}$	Stress for tapered members defined by LRFD Specification Equation A-F3-7, ksi
F_y	Specified minimum yield stress of the type of steel being used, ksi. As used in the LRFD Specification, "yield stress" denotes either the specified minimum yield point (for steels that have a yield point) or specified yield strength (for steels that do not have a yield point)
F_y'''	The theoretical maximum yield stress (ksi) based on the web depth-thickness ratio (h / t_w) above which the web of a column is considered a "slender element" (See LRFD Specification Table B5.1)

$$= \left[\frac{253}{h/t_w} \right]^2$$

Note: In the tables, — indicates $F_y''' > 65$ ksi.

F_{yb}	F_y of a beam, ksi
F_{yc}	F_y of a column, ksi

F_{yf}	Specified minimum yield stress of the flange, ksi
F_{yr}	Specified minimum yield stress of the longitudinal reinforcing bars, ksi
$F_{y\,st}$	Specified minimum yield stress of the stiffener material, ksi
F_{yw}	Specified minimum yield stress of the web, ksi
G	Shear modulus of elasticity of steel (11,200 ksi)
G	Ratio of the total column stiffness framing into a joint to that of the stiffening members framing into the same joint
H	Horizontal force, kips
H	Flexural constant in LRFD Specification Equation E3-1
H	Average story height
H	Height of bolt head or nut, in.
H	Theoretical thread height, in. (see Table 8-7)
H_s	Length of stud connector after welding, in.
H_1	Height of bolt head, in. (see Tables 8-4 and 8-5)
H_2	Maximum bolt shank extension based on one standard hardened washer, in. (see Tables 8-4 and 8-5)
I	Moment of inertia, in.^{4}
I_{LB}	Lower bound moment of inertia for composite section, in.^{4}
I_c	Moment of inertia of column section about axis perpendicular to plane of buckling, in.^{4}
I_d	Moment of inertia of the steel deck supported on secondary members, in.^{4}
I_g	Moment of inertia of girder about axis perpendicular to plane of buckling, in.^{4}
I_p	Moment of inertia of primary member in flat roof framing, in.^{4}
I_p	Polar moment of inertia of bolt and weld groups ($= I_x + I_y$), in.^{4} per in.^{2}
I_s	Moment of inertia of secondary member in flat roof framing, in.^{4}
I_{st}	Moment of inertia of a transverse stiffener, in.^{4}
I_t	Transformed moment of inertia of the composite section, in.^{4}
I_x	Moment of inertia of bolt and weld groups about X-axis, in.^{4} per in.^{2}
I_y	Moment of inertia of bolt and weld groups about Y-axis, in.^{4} per in.^{2}
I_{yc}	Moment of inertia of compression flange about y axis or if reverse curvature bending, moment of inertia of smaller flange, in.^{4}
IC	Instantaneous center of rotation
ID	Nominal inside diameter of flat circular washer, in.
J	Torsional constant for a section, in.^{4}
K	Effective length factor for a prismatic member
K	Coefficient for estimating the natural frequency of a beam (Part 4)
K	Minimum root diameter of threaded fastener, in. (see Table 8-7)
K_{area}	An idealized area representing the contribution of the fillet to the steel beam area, as defined in the composite beam model of Part 5, in.^{2}
K_{dep}	Fillet depth, $(k - t_f)$, in.
K_i'	Modified effective length factor of a column
K_z	Effective length factor for torsional buckling
K_γ	Effective length factor for a tapered member
L	Unbraced length of member measured between the centers of gravity of the bracing members, in. or ft, as indicated

L	Span length, ft
L	Length of connection in the direction of loading, in.
L	Story height, in. or ft, as indicated
L	Live load due to occupancy and moveable equipment
L	Edge distance or center-to-center distance for holes, in.
L'	Total live load
L_b	Laterally unbraced length; length between points which are either braced against lateral displacement of the compression flange or braced against twist of the cross section, in. or ft, as indicated
L_c	Length of channel shear connector, in.
L_c	Unsupported length of a column section, ft
L_e	Edge distance, in.
L_{eh}	Horizontal edge distance, in.
L_{ev}	Vertical edge distance, in.
L_g	Unsupported length of a girder or other restraining member, ft
L_h	Hook length for hooked anchor rods, in.
L_m	Limiting laterally unbraced length for full plastic flexural strength, in. or ft, as indicated
L_m'	Limiting laterally unbraced length for the maximum design flexural strength for noncompact shapes, in. or ft, as indicated
L_p	Column spacing in direction of girder, ft
L_p	Limiting laterally unbraced length for full plastic flexural strength, uniform moment case ($C_b = 1.0$), in. or ft, as indicated
L_p'	Limiting laterally unbraced length for the maximum design flexural strength for noncompact shapes, uniform moment case ($C_b = 1.0$), in. or ft, as indicated
L_{pd}	Limiting laterally unbraced length for plastic analysis, in. or ft, as indicated
L_r	Limiting laterally unbraced length for inelastic lateral-torsional buckling, in. or ft, as indicated
L_r	Roof live load
L_s	Column spacing perpendicular to direction of girder, ft
M	Beam bending moment, kip-in. or kip-ft, as indicated
M_A	Absolute value of moment at quarter point of the unbraced beam segment, kip-in.
M_B	Absolute value of moment at centerline of the unbraced beam segment, kip-in.
M_C	Absolute value of moment at three-quarter point of the unbraced beam segment, kip-in.
M_{LL}	Beam moment due to live load, kip-in. or kip-ft, as indicated
M_T	Applied torsional moment, kip-in.
M_{cr}	Elastic buckling moment, kip-in. or kip-ft, as indicated
M_{eu}	Required flexural strength for extended end-plate connections, kip-in.
M_{lt}	Required flexural strength in member due to lateral frame translation, kip-in.
M_{max}	Maximum bending moment, kip-in. or kip-ft, as indicated

$M_{\max}$	Absolute value of maximum moment in the unbraced beam segment, kip-in.
M_n	Nominal flexural strength, kip-in. or kip-ft, as indicated
M_n'	Maximum design flexural strength for noncompact shapes, when $L_b \leq L_m'$, kip-in. or kip-ft, as indicated
M_{nt}	Required flexural strength in member assuming there is no lateral translation of the frame, kip-in.
M_{nx}', M_{ny}'	Flexural strength defined in LRFD Specification Equations A-H3-7 and A-H3-8, for use in the alternate interaction equations for combined bending and axial force, kip-in. or kip-ft, as indicated
M_p	Plastic bending moment, kip-in. or kip-ft, as indicated
M_p'	Moment defined in LRFD Specification Equations A-H3-5 and A-H3-6, for use in the alternate interaction formulas for combined bending and axial force, kip-in. or kip-ft, as indicated
M_{pa}	Plastic bending moment modified by axial load ratio, kip-in.
M_r	Limiting buckling moment, M_{cr}, when $\lambda = \lambda_r$ and $C_b = 1.0$, kip-in. or kip-ft, as indicated
M_u	Required flexural strength, kip-in. or kip-ft, as indicated
M_y	Initial yield bending moment, kip-in.
M_{ob}	Elastic lateral-torsional buckling moment, kip-in. or kip-ft, as indicated
M_1	Smaller moment at end of unbraced length of beam or beam-column, kip-in.
M_2	Larger moment at end of unbraced length of beam or beam-column, kip-in.
N	Length of bearing, in.
N	Ratio of the factored gravity load supported by all columns in a story to that supported by the columns in the rigid frame
N	Length of base plate, in.
N_b	Number of bolts in a joint
N_{eff}	Number of beams effective in resisting floor vibration (Part 4)
N_r	Number of stud connectors in one rib at a beam intersection, not to exceed 3 in calculations
N_s	Number of slip planes
OD	Nominal outside diameter of flat circular washer, in.
P	Concentrated load, kips
P	Bolt stagger, in.
P	Thread pitch, in. (see Table 8-7)
P_D	Unfactored dead load, kips
P_E	Unfactored earthquake load, kips
P_L	Unfactored live load, kips
P_S	Unfactored snow load, kips
P_{bf}	Applied factored beam flange force in moment connections, kips
P_e, P_{e1}, P_{e2}	Euler buckling strengths, kips
P_{fb}	Resistance to local flange bending per LRFD Specification Equation K1-1 (used to check need for column web stiffeners), kips
P_n	Nominal axial strength (tension or compression), kips
P_p	Bearing load on concrete, kips
P_u	Factored concentrated beam load, kips

P_u	Required axial strength (tension or compression), kips
$P_u e$	Induced moment due to eccentricity e in an eccentrically loaded bolt or weld group, kip-in.
P_{uf}	Factored beam flange force, tensile or compressive, kips
P_{wb}	Resistance to compression buckling of the web per LRFD Specification Equation K1-8 (used to check need for column web stiffening), kips
P_{wi}	A factor consisting of terms from the second portion of LRFD Specification Equation K1-2 (used in a column web stiffener check for local web yielding), kips/in.
P_{wo}	A factor consisting of the first portion of LRFD Specification Equation K1-2 (used in a column web stiffener check for local web yielding), kips
P_y	Yield strength, kips
PNA	Plastic neutral axis
Q	Full reduction factor for slender compression elements
Q_a	Reduction factor for slender stiffened compression elements
Q_f	Statical moment for a point in the flange directly above the vertical edge of the web, in.^{3}
Q_i	Load effects
Q_n	Nominal strength of one stud shear connector, kips
Q_s	Reduction factor for slender unstiffened compression elements
Q_w	Statical moment at mid-depth of the section, in.^{3}
R	Nominal load due to initial rainwater or ice exclusive of the ponding contribution
R	Nominal reaction, kips
R	Earthquake response modification coefficient
R_{PG}	Plate girder bending strength reduction factor
R_e	Hybrid girder factor
R_n	Nominal resistance or strength, kips
R_s	Nominal slip resistance of a bolt, kips
R_u	Required strength determined from factored loads; must be less than or equal to design strength ϕR_n
$R_{u\,st}$	Required strength for transverse stiffener (factored force delivered to stiffener), kips
R_v	Web shear strength, kips
R_1	An expression consisting of the first portion of LRFD Specification Equation K1-3, kips
R_2	An expression consisting of terms from the second portion of LRFD Specification Equation K1-3, kips/in.
R_3	An expression consisting of the first portion of LRFD Specification Equation K1-5a, kips
R_4	An expression consisting of terms from the second portion of LRFD Specification Equation K1-5a, kips/in.
R_5	An expression consisting of terms from LRFD Specification Equation K1-5b, kips
R_6	An expression consisting of terms from LRFD Specification Equation K1-5b, kips/in.

S	Elastic section modulus, in.^{3}
S	Spacing, in. or ft, as indicated
S	Snow load
S	Groove depth for partial-joint-penetration groove welds, in.
S'	Additional elastic section modulus corresponding to $\frac{1}{16}$-in. increase in web thickness for built-up wide flange sections, in.^{3}
S_c	Elastic section modulus to the tip of the angle in compression, in.^{3}
S_{eff}	Effective section modulus about major axis, in.^{3}
S_{net}	Net elastic section modulus, in.^{3}
S_w	Warping statical moment at a point on the cross section, in.^{4}
S_x	Elastic section modulus about major axis, in.^{3}
S_x'	Elastic section modulus of larger end of tapered member about its major axis, in.^{3}
S_{xt}, S_{xc}	Elastic section modulus referred to tension and compression flanges, respectively, in.^{3}
SRF	Stiffness reduction factors (Table 3-1), for use with the alignment charts (Figure 3-1) in the determination of effective length factors K for columns
T	Distance between web toes of fillets at top and at bottom of web, in. $= d - 2k$
T	Tension force due to service loads, kips
T	Thickness of flat circular washer or mean thickness of square or rectangular beveled washer, in.
T	Unfactored tensile force on slip-critical connections designed at service loads, kips
T_b, T_m	Minimum bolt tension for fully tensioned bolts from LRFD Specification Table J3.1, kips
T_{stl}	Tensile force in steel in a composite beam, kips
T_{Tot}	Sum of tensile forces in a composite beam, kips
T_u	Factored tensile force, kips
U	Reduction coefficient, used in calculating effective net area
V	Shear force, kips
V_b	Shear force component, kips
V_h	Total horizontal force transferred by the shear connections, kips
V_n	Nominal shear strength, kips
V_u	Required shear strength, kips
W	Wind load
W	Uniformly distributed load, kips
W	Weight, lbs or kips, as indicated
W	Width across flats of nut, in.
W_c	Uniform load constant for beams, kip-ft
W_{no}	Normalized warping function at a point at the flange edge, in.^{2}
W_u	Total factored uniformly distributed load, kips
X_1	Beam buckling factor defined by LRFD Specification Equation F1-8
X_2	Beam buckling factor defined by LRFD Specification Equation F1-9
Y_{ENA}	Distance from bottom of steel beam to elastic neutral axis, in.
Y_{con}	Distance from top of steel beam to top of concrete, in.

$Y1$	Distance from top of steel beam to the plastic neutral axis, in.
$Y2$	Distance from top of steel beam to the concrete flange force in a composite beam, in.
Z	Plastic section modulus, in.^{3}
Z'	Additional plastic section modulus corresponding to $\frac{1}{16}$-inch increase in web thickness for built-up wide flange section, in.^{3}
Z_e	Effective plastic section modulus, in.^{3}
a	Clear distance between transverse stiffeners, in.
a	Distance between connectors in a built-up member, in.
a	Effective concrete flange thickness of a composite beam, in.
a	Shortest distance from edge of pinhole to edge of member measured parallel to direction of force, in.
a	Coefficient for eccentrically loaded weld group
a	Distance from bolt centerline to edge of fitting subjected to prying action, but not greater than $1.25b$, in.
a	Depth of bracket plate, in.
a_r	Ratio of web area to compression flange area
b	Compression element width, in.
b	Effective concrete flange width in a composite beam, in.
b	Width of composite column section, in.
b	Minimum shelf dimension for deposition of fillet weld, in.
b	Width of bracket plate, in.
b	Distance from bolt centerline to face of fitting subjected to prying action, in.
b_e	Reduced effective width for slender compression elements, in.
b_{eff}	Effective edge distance, in.
b_f	Flange width of rolled beam or plate girder, in.
b_s	Width of transverse stiffener, in.
b_s	Width of extended end-plate, in.
c	Distance from the neutral axis to the extreme fiber of the cross section, in.
c	Cope length, in.
c_1, c_2, c_3	Numerical coefficients used in the calculation of the modified yield stress and modulus of elasticity for composite columns
d	Nominal fastener diameter, in.
d	Overall depth of member, in.
d	Pin diameter, in.
d	Roller diameter, in.
d_L	Depth at larger end of unbraced tapered segment, in.
d_b	Nominal bolt diameter, in.
d_c	Column depth, in.
d_c	Cope depth, in.
d_{ct}	Top-flange cope depth, in.
d_{cb}	Bottom-flange cope depth, in.
d_h	Hole diameter, in.
d_m	Moment arm between resultant tensile and compressive forces due to a moment or eccentric force, in.

d_z	Overall panel-zone depth, in.
d_0	Depth at smaller end of unbraced tapered segment, in.
e	Eccentricity, in.
e	Base of natural logarithms = 2.71828...
e	Link length in eccentrically braced frame (EBF), in.
e_o	Horizontal distance from the outer edge of a channel web to its shear center, in.
f	Computed compressive stress in the stiffened element, ksi
f	Natural frequency, hz
f	Plate buckling model adjustment factor for beams coped at top flange only
f_b	Maximum bending stress, ksi
f_{b1}	Smallest computed bending stress at one end of a tapered segment, ksi
f_{b2}	Largest computed bending stress at one end of a tapered segment, ksi
f_c'	Specified compressive strength of concrete, ksi
f_d	Adjustment factor for beams coped at both flanges
f_{un}	Required normal stress, ksi
f_{uv}	Required shear stress, ksi
f_v	Computed shear stress, ksi
f_o	Stress due to $1.2D + 1.2R$, ksi
g	Transverse center-to-center spacing (gage) between fastener gage lines, in.
g	Acceleration due to gravity = 32.2 ft/sec^2 = 386 in./sec^2
h	Clear distance between flanges less the fillet or corner radius for rolled shapes; and for built-up sections, the distance between adjacent lines of fasteners or the clear distance between flanges when welds are used, in.
h	Depth of composite column section, in.
h_c	Twice the distance from the centroid to the following: the inside face of the compression flange less the fillet or corner radius, for rolled shapes; the nearest line of fasteners at the compression flange or the inside face of the compression flange when welds are used, for built-up sections, in.
h_r	Nominal rib height, in.
h_s	Factor used in LRFD Specification Equation A-F3-6 for web-tapered members
h_w	Factor used in LRFD Specification Equation A-F3-7 for web-tapered members
h_o	Remaining web depth of coped beam, in.
j	Factor defined by LRFD Specification Equations A-F2-4 for minimum moment of inertia for a transverse stiffener
k	Distance from outer face of flange to web toe of fillet, in.
k	Slenderness parameter
k	Plate buckling coefficient for beams coped at top flange only
k_s	Bolt slip coefficient
k_v	Web plate buckling coefficient
k_1	Distance from web center line to flange toe of fillet, in.
l	Unbraced length of member, in.
l	Span length, in.
l	Length of bearing, in.

l	Length of connection in the direction of loading, in.
l	Length of weld, in.
l	Characteristic length of weld group (see Tables 8-38 through 8-45), in.
l_o	Distance from center of gravity (CG) to instantaneous center of rotation (IC) of bolt or weld group, in.
m	Ratio of web to flange yield stress or critical stress in hybrid beams
m	Coefficient for converting bending to an approximate equivalent axial load in beam-columns (Part 3)
m	Cantilever dimension for base plate (see Part 11), in.
n	Number of shear connectors between point of maximum positive moment and the point of zero moment to each side
n	Number of bolts in a vertical row
n	Number of threads per inch on threaded fasteners
n	Cantilever dimension for base plate (see Part 11), in.
n'	Number of bolts above the neutral axis (in tension)
p	Length of supporting flange parallel to stem or leg of hanger tributary to each bolt in determinimg prying action, in.
p_e	Effective span used to compute M_{eu} for extended end-plate connections, in.
p_f	Distance from centerline of bolt to nearer surface of tension flange in extended end-plate connections, in.
q_u	Additional tension per bolt resulting from prying action produced by deformation of the connected parts, kips/bolt
r	Governing radius of gyration, in.
r_T	Radius of gyration of compression flange plus one third of the compression portion of the web taken about an axis in the plane of the web, in.
r_{To}	Radius of gyration, r_T, for the smaller end of a tapered member, in.
r_i	Minimum radius of gyration of individual component in a built-up member, in.
r_{ib}	Radius of gyration of individual component relative to centroidal axis parallel to member axis of buckling, in.
r_m	Radius of gyration of steel shape, pipe, or tubing in composite columns. For steel shapes it may not be less than 0.3 times the overall thickness of the composite section, in.
r_n	Nominal strength per bolt from LRFD Specification
r_x, r_y	Radius of gyration about x and y axes respectively, in.
r_{ut}	Required tensile strength per bolt or per inch of weld (factored force per bolt or per inch of weld due to a tensile force), kips/bolt
r_{ut}	Required shear strength per bolt or per inch of weld (factored force per bolt or per inch of weld due to a shear force), kips/bolt
r_{yc}	Radius of gyration about y axis referred to compression flange, or if reverse curvature bending, referred to smaller flange, in.
$\bar{r}_o$	Polar radius of gyration about the shear center, in.
r_{ox}, r_{oy}	Radius of gyration about x and y axes at the smaller end of a tapered member respectively, in.
s	Longitudinal center-to-center spacing (pitch) of any two consecutive holes, in.
s	Bolt spacing, in.

t	Thickness, in.
t	Change in temperature, degrees Fahrenheit or Celsius, as indicated
t_b	Thickness of beam flange or connection plate delivering concentrated force, in.
t_c	Flange or angle thickness required to develop design tensile strength of bolts with no prying action, in.
t_e	Total required effective thickness of column web with doubler plate, in.
t_f	Flange thickness, in.
t_f	Flange thickness of channel shear connector, in.
t_p	Thickness of base plate, in.
t_p	Panel zone thickness including doubler plates, in.
$t_{p\ req}$	Required doubler plate thickness, in.
t_s	Extended end-plate thickness, in.
t_w	Web thickness, in.
t_w	Web thickness of channel shear connector, in.
t_{wb}	Beam web thickness, in.
t_{wc}	Column web thickness, in.
t_z	Panel zone thickness, in.
u	Factor for approximate design of beam-columns (Part 3)
w	Uniformly distributed load per unit of length, kips/in.
w	Fillet weld size, in.
w	Plate width; distance between welds, in.
w	Subscript relating symbol to strong principal axis of angle
w	Unit weight of concrete, lbs/ft^3
w_r	Average width of concrete rib or haunch, in.
w_z	Panel zone width, in.
x	Subscript relating symbol to strong axis bending
x	Horizontal distance, in.
$\bar{x}$	Horizontal distance from the outer edge of a channel web to its centroid, in.
$\bar{x}$	Connection eccentricity, in.
x_p	Horizontal distance from the designated edge of member to its plastic neutral axis, in.
x_o	Horizontal distance, in.
x_o, y_o	Coordinates of the shear center with respect to the centroid, in.
y	Moment arm between centroid of tensile forces and compressive forces, in.
y	Subscript relating symbol to weak axis bending
y_p	Vertical distance from the designated edge of member to its plastic neutral axis, in.
y_1, y_2	Vertical distance from designated edge of member to center of gravity, in.
z	Distance from the smaller end of tapered member used in LRFD Specification Equation A-F3-1 for the variation in depth, in.
z	Subscript relating symbol to weak principal axis of angle
z	Coefficient for buckling of triangular-shaped bracket plate
Δ	Deflection, in.
Δ_{LL}	Live load deflection, in.

Δ_{oh}	Translation deflection of the story under consideration, in.
α	Separation ratio for built-up compression members, LRFD Specification Equation E4
α	Fraction of member force transferred across a particular net section
α	Ratio of moment at bolt line to moment at stem line for determining prying action in hanger connections
α	Ideal distance from face of column flange or web to centroid of gusset-to-beam connection for bracing connections and uniform force method, in.
$\bar{\alpha}$	Actual distance from face of column flange or web to centroid of gusset-to-beam connection for bracing connections and uniform force method, in.
α_m	Coefficient for calculating M_{eu} for extended end-plate connections
β	Ideal distance from face of beam flange to centroid of gusset-to-column connection for bracing connections and uniform force method, in.
$\bar{\beta}$	Actual distance from face of beam flange to centroid of gusset-to-column connection for bracing connections and uniform force method, in.
β_w	Special section property for unequal-leg angles (Single Angle Specification)
γ	Depth tapering ratio
γ	Subscript relating symbol to tapered members
γ_i	Load factor
δ	Deflection, in.
δ	Ratio of net area at bolt line to gross area at face of stem or angle leg used to determine prying action for hanger connections
ϵ	Coefficient of linear expansion, with units as indicated
ζ	Exponent for alternate beam-column interaction equation
η	Exponent for alternate beam-column interaction equation
λ	Slenderness parameter
λ_c	Column slenderness parameter
λ_e	Equivalent slenderness parameter
λ_{eff}	Effective slenderness ratio defined by LRFD Specification Equation A-F3-2
λ_p	Limiting slenderness parameter for compact element
λ_r	Limiting slenderness parameter for noncompact element
μ	Coefficient of friction; mean slip coefficient for bolts
ρ	Ratio of P_u to V_u of a link in an eccentrically braced frame (EBF)
ϕ	Resistance factor
ϕ_b	Resistance factor for flexure
ϕ_c	Resistance factor for compression
ϕ_c	Resistance factor for axially loaded composite columns
ϕ_r	Resistance factor for compression, used in web crippling equations
ϕ_{sf}	Resistance factor for shear on the failure path
ϕ_t	Resistance factor for tension
ϕ_v	Resistance factor for shear
ϕ_w	Resistance factor for welds
ϕF_{bc}	Design buckling stress for coped beams, ksi

ϕR_n Design strength from LRFD Specification; must equal or exceed required strength R_u

ϕr_n Design strength per bolt or per inch of weld from LRFD Specification; must equal or exceed required strength per bolt or per inch of weld r_u

kip   1,000 pounds

ksi   Stress, kips/in.^{2}

INDEX